Materials, Junctions, and Devices

SOLID-STATE devices are small but versatile units that can perform an amazing variety of control functions in electronic equipment. Like other electron devices, they have the ability to control almost instantly the movement of charges of electricity. They are used as rectifiers, detectors, amplifiers, oscillators, electronic switches, mixers, and modulators.

In addition, solid-state devices have many important advantages over other types of electron devices. They are very small and light in weight (some are less than an inch long and weigh just a fraction of an ounce). They have no filaments or heaters, and therefore require no heating power or warm-up time. They consume very little power. They are solid in construction, extremely rugged, free from microphonics, and can be made impervious to many severe environmental conditions. The circuits required for their operation are usually simple.

SEMICONDUCTOR MATERIALS

Unlike other electron devices, which depend for their functioning on the flow of electric charges through a vacuum or a gas, solid-state devices make use of the flow of current in a solid. In general, all materials may be classified in three major categories—conductors, semiconductors, and insulators—depending upon their ability to conduct an electric current. As the name indicates, a semiconductor material has poorer conductivity than a conductor, but better conductivity than an insulator. The material most often used in semiconductor devices is silicon.

Resistivity

The ability of a material to conduct current (conductivity) is directly proportional to the number of free (loosely held) electrons in the material. Good conductors, such as silver, copper, and aluminum, have large numbers of free electrons, their resistivities are of the order of a few millionths of an ohm-centimeter. Insulators such as glass, rubber, and mica, which have very few loosely held electrons, have resistivities of several million ohm-centimeters.

Semiconductor materials lie in the range between these two extremes, as shown in Fig. 1. Pure silicon has a resistivity, in the order of 60,000 ohm-centimeters. As used in semiconductor devices, however, semiconductor materials contain carefully

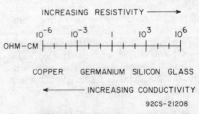

Fig. 1—Resistivity of typical conductor, semiconductor, and insulator.

controlled amounts of certain impurities which reduce their resistivity to about 2 ohm-centimeters at room temperature (this resistivity decreases rapidly as temperature rises).

Impurities

Carefully prepared semiconductor materials have a crystal structure. In this type of structure, which is called a lattice, the outer or valence electrons of individual atoms are tightly bound to the electrons of adjacent atoms in electron-pair bonds, as shown in Fig. 2. Because such a

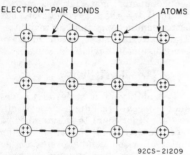

Fig. 2—Crystal lattice structure.

structure has no loosely held electrons, semiconductor materials are poor conductors under normal conditions. In order to separate the electron-pair bonds and provide free electrons for electrical conduction, it would be necessary to apply high temperatures or strong electric fields.

Another way to alter the lattice structure and thereby obtain free electrons, however, is to add small amounts of other elements having a different atomic structure. By the addition of almost infinitesimal amounts of such other elements called "impurities", the basic electrical properties of pure semiconductor materials can be modified and controlled. The ratio of impurity to the semiconductor material is usually extremely small, in the order of one part in ten million.

When the impurity elements are added to the semiconductor material,

impurity atoms take the place of semiconductor atoms in the lattice structure. If the impurity atoms added have the same number of valence electrons as the atoms of the original semiconductor material, they fit neatly into the lattice, forming the required number of electron-pair bonds with semiconductor atoms. In this case, the electrical properties of the material are essentially unchanged.

When the impurity atom has one more valence electron than the semiconductor atom, however, this extra electron cannot form an electron-pair bond becaue no adjacent valence electron is available. The excess electron is then held very loosely by the atom, as shown in Fig. 3, and

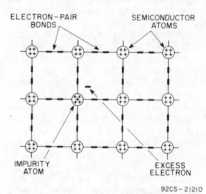

Fig. 3—Lattice structure of n-type material.

requires only slight excitation to break away. Consequently, the presence of such excess electrons makes the material a better conductor, i.e., its resistance to current flow is reduced.

Impurity elements which are added to silicon crystals to provide excess electrons include arsenic and antimony. When these elements are introduced, the resulting material is called **n-type** because the excess free electrons have a negative charge. (It should be noted, however, that the negative charge of the electrons is balanced by an equivalent positive charge in the center of the impurity atoms. Therefore, the net electrical charge of the semiconductor material is not changed.)

A different effect is produced when an impurity atom having one less valence electron than the semiconductor atom is substituted in the lattice structure. Although all the valence electrons of the impurity atom form electron-pair bonds with electrons of neighboring semiconductor atoms, one of the bonds in the lattice structure cannot be completed because the impurity atom lacks the final valence electron. As a result, a vacancy or "hole" exists in the lattice, as shown in Fig. 4. An electron from an adjacent electron-pair bond may then absorb enough energy to break its bond and move through the lattice to fill the hole. As in the

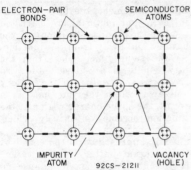

Fig. 4—*Lattice structure of p-type material.*

case of excess electrons, the presence of "holes" encourages the flow of electrons in the semiconductor material; consequently, the conductivity is increased and the resistivity is reduced.

The vacancy or hole in the crystal structure is considered to have a positive electrical charge because it represents the absence of an electron. (Again, however, the net charge of the crystal is unchanged.) Semiconductor material which contains these "holes" or positive charges is called **p-type** material. P-type materials are formed by the addition of aluminum, gallium, or indium.

Although the difference in the chemical composition of n-type and p-type materials is slight, the differences in the electrical characteristics of the two types are substantial, and are very important in the operation of solid-state devices.

P-N JUNCTIONS

When n-type and p-type materials are joined together, as shown in Fig. 5, an unusual but very important phenomenon occurs at the interface

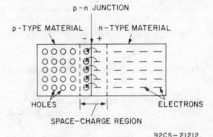

Fig. 5—*Interaction of holes and electrons at p-n junction.*

where the two materials meet (called the **p-n junction**). An interaction takes place between the two types of material at the junction as a result of the holes in one material and the excess electrons in the other.

When a p-n junction is formed, some of the free electrons from the n-type material diffuse across the junction and recombine with holes in the lattice structure of the p-type material; similarly, some of the holes in the p-type material diffuse across the junction and recombine with free electrons in the lattice structure of the n-type material. This interaction or diffusion is brought into equilibrium by a small space-charge region (sometimes called the **transition region or depletion layer**). The p-type material thus acquires a slight negative charge and the n-type material acquires a slight positive charge.

Thermal energy causes charge carriers (electrons and holes) to diffuse from one side of the p-n junction to the other side; this flow of charge carriers is called **diffusion current**. As a result of the diffusion process, however, a potential gradient builds up across the space-charge region. This potential gradient can be represented, as shown in Fig. 6, by an imaginary battery connected across the p-n junction. (The battery symbol

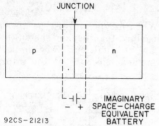

Fig. 6—Potential gradient across space-charge region.

is used merely to illustrate internal effects; the potential it represents is not directly measurable.) The potential gradient causes a flow of charge carriers, referred to as **drift current,** in the opposite direction to the diffusion current. Under equilibrium conditions, the diffusion current is exactly balanced by the drift current so that the net current across the p-n junction is zero. In other words, when no external current or voltage is applied to the p-n junction, the potential gradient forms an **energy barrier** that prevents further diffusion of charge carriers across the junction. In effect, electrons from the n-type material that tend to diffuse across the junction are repelled by the slight negative charge

induced in the p-type material by the potential gradient, and holes from the p-type material are repelled by the slight positive charge induced in the n-type material. The potential gradient (or energy barrier, as it is sometimes called), therefore, prevents total interaction between the two types of materials, and thus preserves the differences in their characteristics.

Current Flow

When an external battery is connected across a p-n junction, the amount of current flow is determined by the polarity of the applied voltage and its effect on the space-charge region. In Fig. 7(a), the positive terminal of the battery is connected to the n-type material and the negative terminal to the p-type material. In this arrangement, the free electrons in the n-type material are attracted toward the posiitve terminal of the battery and away from the junction. At the same time, holes from the p-type material are attracted toward the negative terminal of the battery and away from the junction. As a result, the space-charge region at the junction becomes effectively wider, and the potential gradient increases until it approaches the potential of the external battery. Current flow is then extremely small because no voltage difference (electric field) exists across either the p-type or the n-type region. Under these conditions, the p-n junction is said to be **reverse-biased.**

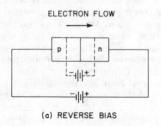

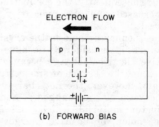

(a) REVERSE BIAS   (b) FORWARD BIAS

92CS-21214

Fig. 7—Electron current flow in biased p-n junctions.

In Fig. 7(b), the positive terminal of the external battery is connected to the p-type material and the negative terminal to the n-type material. In this arrangment, electrons in the p-type material near the positive terminal of the battery break their electron-pair bonds and enter the battery, creating new holes. At the same time, electrons from the negative terminal of the battery enter the n-type material and diffuse toward the junction. As a result, the space-charge region becomes effectively narrower, and the energy barrier decreases to an insignificant value. Excess electrons from the n-type material can then penetrate the space-charge region, flow across the junction, and move by way of the holes in the p-type material toward the positive terminal of the battery. This electron flow continues as long as the external voltage is applied. Under these conditions, the junction is said to be **forward-biased.**

The generalized voltage-current characteristic for a p-n junction in Fig. 8 shows both the reverse-bias and forward-bias regions. In the forward-bias region, current rises rapidly as the voltage is increased and is quite high. Current in the reverse-bias region is usually much lower. Excessive voltage (bias) in either direction should be avoided in normal applications because excessive currents and the resulting high temperatures may permanently damage the solid-state device.

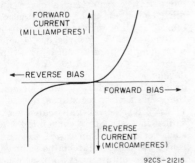

Fig. 8—Voltage-current characteristic for a p-n junction.

Diodes

The simplest type of solid-state devce is the diode, which is represented by the symbol shown in Fig. 9. Structurally, the diode is basically a p-n junction similar to those shown in Fig. 7. The n-type material which serves as the negative electrode is referred to as the **cathode,** and the p-type material which serves as the positive electrode is referred to as the **anode.** The arrow symbol used for the anode represents the direction of "conventional current flow";

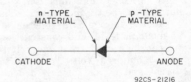

Fig. 9—Schematic symbol for a solid-state diode.

electron current flows in a direction opposite to the arrow.

Because the junction diode conducts current more easily in one direction than in the other, it is an effective rectifying device. If an ac signal is applied, as shown in Fig. 10, electron current flows freely during the positive half cycle, but little or no current flows during the negative half cycle.

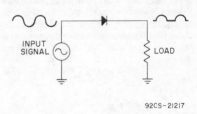

Fig. 10—Simple diode rectifying circuit.

Silicon Rectifiers

One of the most widely used types of solid-state diode is the silicon rectifier. These devices are available in a wide range of current capabilities, ranging from tenths of an ampere to several hundred amperes or more, and are capable of operation at voltages as high as 1000 volts or more. Parallel and series arrangements of silicon rectifiers permit even further extension of current and voltage limits.

Zener diodes are silicon rectifiers in which the reverse current remains small until the breakdown voltage is reached and then increases rapidly with little further increase in voltage. The schematic symbol for a zener diode is shown in Fig. 11; a typical zener characteristic curve is shown in Fig. 12 in comparison with that of a rectifying diode. The breakdown voltage is a function of the diode material and construction, and can be varied from one volt to several hundred volts for various current and power ratings, depending on the junction area and the method of cooling. Zener diodes are useful as stabilizing devices and as reference voltage sources.

Current stability in a transistor can be achieved by use of a **compensating diode.** Because the forward characteristic of a compensating diode is similar to the transfer characteristics of a transistor, the diode can maintain transistor bias voltages within ±0.015 volt of a desired value despite supply-voltage variations up to 40 per cent and simultaneously compensate for a wide range of ambient-temperature variations.

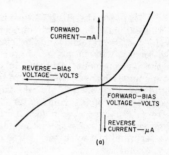

(a)

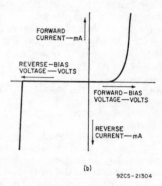

(b)

92CS-21304

Fig. 12—Typical characteristic curves for (a) a rectifying diode and (b) a zener diode.

N-P-N AND P-N-P STRUCTURES

Fig. 7 shows that a p-n junction biased in the reverse direction is equivalent to a high-resistance element (low current for a given applied voltage), while a junction biased in the forward direction is equivalent to a low-resistance element (high current for a given applied voltage). Because the power developed by a given current is greater in a high-resistance element than in a low-resistance element ($P = I^2R$), power gain can be obtained in a structure containing two such resistance elements if the current flow is not materially reduced. A device containing two p-n junctions biased in opposite directions can operate in this fashion.

92CS-21303

Fig. 11—Schematic symbol for a zener diode.

Bipolar Transistors

All bipolar transistors consist of three layers of semiconductor material (usually silicon) referred to as **emitter**, **base**, and **collector**. The resultant structure forms two back-to-back p-n junctions. The input (emitter-base) junction serves as the source, or injector, of current carriers; the output (base-collector) junction collects the injected current carriers. During normal operation, the emitter-base p-n junction is forward-biased, and the collector-base p-n junction is reverse-biased.

As explained in the section on **Silicon Rectifiers**, a p-n junction biased in the reverse direction is equivalent to a high-resistance element, while a junction biased in the forward direction is equivalent to a low-resistance element. The electric field across the forward-biased junction overcomes the energy barrier at the junction and causes holes to be injected into the n-type region and electrons to be injected into the p-type region. Because of the large number of free electrons in the n-type region and of holes in the p-type region, the injected holes and electrons are referred to as minority-charge carriers. A forward-biased p-n junction, therefore, is a minority-carrier injector, and the number of minority carriers injected is dependent upon the magnitude of the forward-bias voltage.

Charge-Carrier Flow—When a symmetrical p-n junction is forward-biased, the lifetime of the injected minority carrier is very short. Because of the many free electrons in the n-type region, a hole injected into this region is not likely to penetrate very far before it meets an electron and is annihilated (i.e., neutralized), as shown in Fig. 13. Similarly, any electron injected into the p-type region is usually quickly neutralized by one of the numerous holes in this region. In a symmetrical p-n junction, therefore, injected minority carriers cannot penetrate

very far or last very long before they are annihilated.

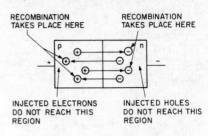

Fig. 13—Diagram showing that recombination limits travel of injected carriers in a symmetrical forward-biased p-n junction.

Fig. 14(a) shows a nonsymmetrical p-n junction in which the n-type region is made very thin and the p-type region is much more heavily doped. When this junction is forward-biased, an injected hole is much less likely to be annihilated

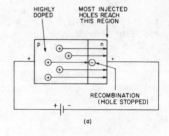

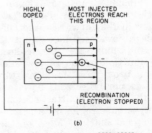

Fig. 14—Diagrams showing that (a) hole injection is improved by unequal doping and a thin n-type section and (b) electron injection is improved by unequal doping and a thin p-type section.

by an electron before it crosses to the end of the thin n-type region. Moreover, because of the heavy doping of the p-type region, more holes are injected into the n-type region than there are free electrons in this thin region. Consequently, even though some injected holes are annihilated by free electrons, most of them are able to survive and penetrate the full width of the n-type region, as shown in Fig. 14(a). Similarly, in a forward-biased p-n junction in which the p-type region is very thin and the n-type region is much more heavily doped, an injected electron is unlikely to meet (and be neutralized by) a hole before it penetrates to the end of the thin p-type region, as shown in Fig. 14(b).

In bipolar transistors, a thin lightly doped semiconductor layer (base region) is sandwiched between two wider (emitter and collector) semiconductor layers that are much more heavily doped with an opposite type of impurity from the dopant used in the thin base layer. The two nonsymmetrical back-to-back p-n junctions that result may form either a p-n-p or an n-p-n transistor. Fig. 15 shows the layer structure and the corresponding schematic symbol for each type of transistor.

N-P-N Types—Fig. 16 shows the basic biasing arrangements for an n-p-n bipolar transistor. External batteries bias the emitter-base (n-p) junction in the forward direction to provide a low-resistance input section, and bias the base-collector (p-n) junction in the reverse direction to provide a high-resistance output section. Electrons flow easily from the n-type emitter region to the p-type base region as a result of the forward biasing. Most of these electrons diffuse through the thin p-type region, however, and are attracted by the positive potential of the external bias supply across the base-collector (p-n) junction. In practical devices, approximately 95

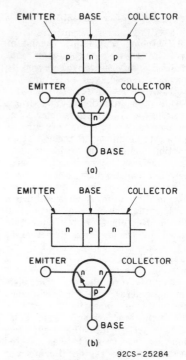

Fig. 15—Transistor nomenclature and symbols: (a) p-n-p type; (b) n-p-n type.

to 99.5 per cent of the injected electrons reach the n-type collector region. This high percentage of current penetration makes possible power gain in the high-resistance output circuit and is the basis for the amplification capability of a transistor.

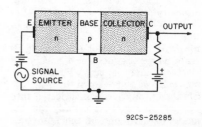

Fig. 16—Basic biasing and input-signal connections for an n-p-n transistor.

P-N-P Types—The operation of a p-n-p transistor is essentially identical to that of an n-p-n transistor except that hte polarities of the bias voltages are reversed and the main current carriers are holes instead of electrons. Fig. 17 shows the basic biasing and input-signal connections for a p-n-p transistor.

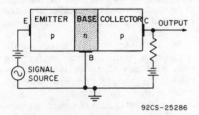

92CS-25286

Fig. 17—Basic biasing and input-signal connections for a p-n-p transistor.

Diacs

A diac is a two-electrode, three-layer bidirectional avalanche diode that can be switched from the off state to the on state for either polarity of applied voltage. Fig. 18 shows the junction diagram and schematic symbol for a diac; Fig. 19 shows the voltage-current characteristic.

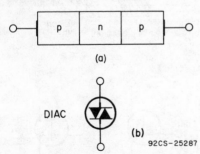

(a)

DIAC

(b)

92CS-25287

Fig. 18—(a) Junction diagram and (b) schematic symbol for a diac.

This three-layer trigger diode is similar in construction to a bipolar transistor, but differs from it in that the doping concentrations at the two junctions are approximately the same and there is no

contact made to the base layer. The equal doping levels result in a symmetrical bidirectional switching characteristic, as shown in Fig. 19. When an increasing positive or negative voltage is applied across the terminals of the diac, a minimum (leakage) current $I_{(BO)}$ flows through the device until the voltage reaches the breakover point $V_{(BO)}$. The reverse-biased junction then undergoes avalanche breakdown and, beyond this point, the device exhibits a negative-resistance characteristic, i.e., current through the device increases substantially with decreasing voltage.

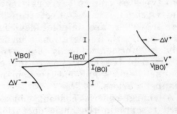

92CS-25288

Fig. 19—Voltage-current characteristic for a diac.

Diacs are primarily used as triggering devices in thyristor phase-control circuits used for light dimming, universal motor-speed control, heat control, and similar applications. Fig. 20 shows the general circuit diagram for a diac/triac phase-control circuit.

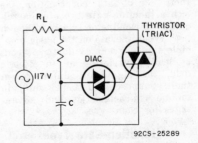

92CS-25289

Fig. 20—General circuit diagram for a diac/triac phase-control circuit.

FIELD-EFFECT TRANSISTORS

The field-effect transistor is another type of solid-state device that is becoming increasingly popular in electronic circuits. These transistors derive their name from the fact that current flow in them is controlled by variation of an **electric field** established by application of a voltage to a control electrode, referred to as the gate. In contrast, current flow in bipolar transistors is controlled by variation of the current injected into the base terminal. Moreover, the performance of bipolar transistors depends on the interaction of two types of charge carriers (holes and electrons). Field-effect transistors, however, are unipolar devices; as a result, their operation is basically a function of only one type of charge carrier, holes in p-channel devices and electrons in n-channel devices.

A charge-control concept can be used to explain the basic operation of field-effect transistors. A charge on the gate (control electrode) induces an equal, but opposite, charge in a semiconductor layer, referred to as the channel, located directly beneath the gate. The charge induced in the channel controls the conduction of current through the channel and, therefore, between the source and drain terminals which are connected to opposite ends of the channel.

Discrete-device field-effect transistors are classified, on the basis of their conrol-gate construction, as either **junction-gate** types or **metal-oxide-semiconductor** types. Although both types operate on the basic principle that current conducton is controlled by variation of an electric field, the significant difference in their gate construction results in unique characteristics and advantages for each type.

Junction-Gate Types

Junction-gate field-effect transistors, which are commonly referred to

as JFET's, may be either n-channel or p-channel devices. Fig. 21 shows the structure of an p-channel junction-gate field-effect transistor, together with the schematic symbols for both n-channel and p-channel versions of these devices. The structure for a p-channel device is identical to that of an n-channel device with the exception that n- and p-type semiconductor materials are replaced by p- and n-type materials, respectively.

In both types of junction-gate devices, a thin channel under the gate provides a conductive path between the source and drain terminals with zero gate-bias voltage. A p-n junction is formed at the interface of the gate and the source-to-drain layer. When this junction is reverse-biased, current conduction in the channel between the source and drain terminals is controlled by the magnitude of reverse-bias voltage, which if sufficient

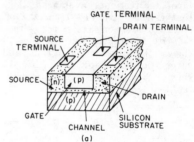

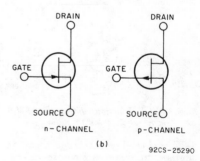

Fig. 21—Junction-gate field-effect transistor (JFET): (a) side-view cross section of an n-channel device; (b) schematic symbols for n- and p-channel devices.

can virtually cut off the flow of current through the channel. If the junction becomes forward-biased, the input resistance (i.e., resistance between the gate and the source-to-drain layer) decreases sharply, and an appreciable amount of gate current flows. Under such conditions, the gate loading reduces the amplitude of the input signal, and a significant reduction in power gain results. This characteristic is a major disadvantage of junction-gate field-effect transistors. Another undesirable feature of these devices is that the leakage currents across the reverse-biased p-n junction can vary markedly with changes in ambient temperature. This latter factor tends to complicate circuit design considerations. Nonetheless, the junction-gate field-effect transistor is a very useful device in many small-signal-amplifier and chopper applications.

Metal-Oxide-Semiconductor Types

Figs. 22 and 23 shows the structures and schematic symbols for both

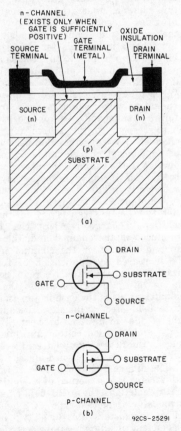

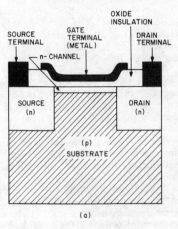

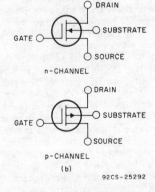

92CS-25291

92CS-25292

Fig. 22—Enhancement-type metal-oxide-semiconductor field-effect transistor (MOS/FET): (a) side-view cross section of an n-channel device; (b) schematic symbols of n- and p-channel devices.

Fig. 23—Depletion-type metal-oxide-semiconductor field-effect transistor (MOS/FET): (a) side-view cross section of an n-channel device; (b) schematic symbols for n- and p-channel devices.

enhancement and depletion types of metal-oxide-semiconductor field-effect transistors (MOS/FET's). In these devices, the metallic gate is electrically insulated from the semiconductor surface by a thin layer of silicon dioxide. These devices, which are commonly referred to as MOS field-effect transistors or, more simply, as MOS transistors, derive their name from the tri-layer construction of metal, oxide, and semiconductor material. Insulation of the gate from the remainder of the transistor structure results in an exceedingly high input resistance (i.e., in the order of 10^{14} ohms). It should be realized that the metal gate and the semiconductor channel form a capacitor in which the oxide layer serves as the dielectric insulator.

The marked differences in the construction of enhancement and depletion types of MOS field-effect transistors, as is apparent from a comparison of Figs. 22(a) and 23(a), results in significant differences in the characteristics of these devices and, therefore, in the applications in which they are normally employed.

As indicated by the interruptions in the channel line of the schematic symbols shown in Fig. 22(b), **enhancement-type MOS field-effect transistors** are characterized by the fact that they have a "normally open" channel so that no useful channel conductivity exists for either zero or reverse gate bias. Consequently, this type of device is ideal for use in digital and switching applications. The gate of the enhancement type of MOS field-effect transistor must be forward-biased with respect to the source to produce the active charge carriers in the channel required for conduction. When sufficient forward-bias (positive) voltage is applied to the gate of an n-channel device, the region under the gate changes from p-type to n-type and provides a conduction path between the n-type source and drain regions. Similarly, in p-channel devices, application of sufficient nega-

tive gate voltage draws holes into the region below the gate so that this channel region changes from n-type to p-type to provide a source-to-drain conduction path.

The technology for enhancement-type MOS field-effect transistors is making its greatest impact in the fabrication of integrated circuits for digital applications, particularly in large-scale-integration. (LSI) circuits.

Depletion-type MOS field-effect transistors are characterized by the fact that, with zero gate bias, the thin channel under the gate region provides a conductive path between the source and drain terminals. In the schematic symbols for these devices, shown in Fig. 23(b), the channel line is drawn continuous to indicate this "normally on" condition. When the gate is reverse-biased (negative with respect to the source for n-channel devices, or positive with respect to the source for p-channel devices), the channel can be depleted of charge carriers; conduction in the channel, therefore, can be cut off if the gate potential is sufficiently high.

A unique characteristic of depletion-type MOS transistors is that addtional charge carriers can be produced in the channel and, therefore, condnction in the channel can be increased by application of forward bias to the gate. No reduction in power gain occurs under these conditions, as is the case in junction-gate field-effect transistors, because the oxide insulation between the gate and the source-to-drain layer blocks the flow of gate current even when the gate is forward-biased.

The diagram shown in Fig. 23(a) illustrates the structure of a **single-gate depletion-type MOS field-effect transistor**. Depletion-type MOS field-effect transistors that have two independent insulated gate electrodes are also available. These devices offer unique advantages and represent the most important category of MOS field-effect transistors.

Fig. 24(a) shows a cross-sectional diagram of an n-chanel depletion-type **dual-gate MOS field-effect transstor**. The transistor includes three terminating (n-diffused) regions connected by two conductive channels, each of which is controlled by its own independent gate terminal. For convenience of explanation, the transistor is shown divided into two units. Unit No. 1 consists of the source, gate No. 1, channel No. 1, and the central n-region which functios as drain No. 1. These elements act as a conventional single-gate depletion-type MOS field-effect transistor for which unit No. 2 functions as a load resistor. Unit No. 2 consists of the central n-region, which functions as source No. 2, gate No. 2, channel No. 2, and the drain. This unit may also be used as an independent single-gate transistor for which unit No. 1 acts as a source resistor. Fig. 24(b) shows the schematic symbol for an n-channel dual-gate MOS field-effect transistor.

Equivalent-circuit representations of the two units in a dual-gate MOS transistor are shown in Fig. 25. Current can be cut off if either gate is sufficiently reverse-biased with respect to the source. When one gate

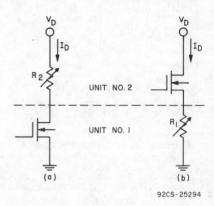

92CS-25294

Fig. 25—Equivalent-circuit representation of the two units in a dual-gate MOS field-effect transistor.

is biased to cutoff, a change in the voltage on the other gate is equivalent to a change in the value of a resistor in series with a cut-off transistor.

The dual-gate MOS field-effect transistor provides exceptional versatility for circuit applications. The independent pair of gates makes this device attractive for use in rf amplifiers, gain-controlled amplifiers, mixers, and demodulators. In a gain-controlled amplifier, the signal is applied to gate No. 1, and the gain-control voltage is applied to gate No. 2. This arrangement is recommended because the forward transconductance obtained with gate No. 1 is higher than that obtained with gate No. 2. Moreover, unit No. 2 is very effective for isolation of the drain and gate No. 1. This unit provides sufficient isolation so that the dual-gate devices can be operated at frequencies into the uhf range without the need for neutralization.

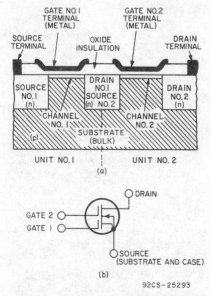

92CS-25293

Fig. 24—Dual-gate n-channel depletion-type metal-oxide-semiconductor field-effect transistor (MOS/FET): (a) side-view cross section; (b) schematic symbol.

P-N-P-N STRUCTURES (THYRISTORS)

When alternate layers of p-type and n-type semiconductor materials are arranged in a series array, various types of thyristors can be produced. The term **thyristor** is the generic name for solid-state devices that have electrical characteristics similar to those of thyratron tubes. Three popular types of thyristors are the reverse-blocking triode thyristor called the silicon controlled rectifier (SCR), the bidirectional triode thyristor, called the triac, and a four-terminal thyristor called the bilateral switch.

Silicon Controlled Rectifiers

Just as a transistor may be considered as basically a solid-state diode with a third semiconductor layer added to form two back-to-back diode junctions, the SCR may be considered as a transistor with an additional semiconductor region. Simple models of the "lower order" devices may be analyzed, therefore, to show the effect of the additional semiconductor regions on the operation of the devices.

An SCR is basically a four-layer p-n-p-n unidirectional device designed to provide bistable switching when operated in the forward-bias mode. The device has three electrodes, referred to as the **cathode**, the **anode**, and the **gate**. The gate is the control electrode for the device. For forward-bias operation, the anode potential must be positive with respect to the cathode. During normal operation, the SCR is turned on by application of a positive voltage to the gate electrode. The SCR then remains on, even though the gate voltage is removed or made negative, until the anode-to-cathode voltage is reduced to a value below that required to sustain regeneration, or forward current. Faster turn-off can be achieved by a reversal of the forward-current flow.

As shown in Fig. 26, the basic p-n-p-n SCR structure is analogous to a pair of complementary n-p-n and p-n-p bipolar transistors. Fig. 26(a) shows the schematic symbols for an SCR and equivalent connection of the complementary pair of transistors, and Fig. 26(b) shows the equivalent relationship of the p-n-p-n SCR structure and the interconnected transistor structures.

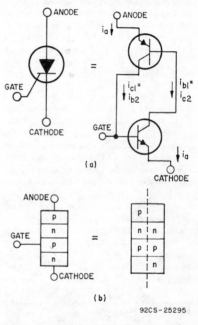

92CS-25295

Fig. 26—Two-transistor analogy of an SCR: (a) schematic symbols of an SCR and the equivalent two-transistor model; (b) structure of an SCR and of the equivalent two-transistor model.

The n-p-n and p-n-p transistors in the equivalent model are interconnected so that regenerative action occurs when a proper gating signal is applied to the base of the n-p-n transistor.

When the two-transistor model is connected in a circuit to simulate normal SCR operation, the emitter of the p-n-p transistor is returned to the positive terminal of a dc supply through a limiting resistor R_2, and the emitter of the n-p-n transistor Q_2 is returned to the negative terminal of the dc supply to provide a complete electrical path, as shown in Fig. 27. When the model

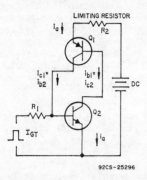

Fig. 27—Two-transistor model connected to show a complete electrical path.

is in the off state, the initial value of principal-current flow is zero. If a positive pulse is then applied to the base of the n-p-n transistor, the transistor turns on and forces the collector (which is also the base of the n-p-n transistor) to a low potential; as a result, a current I_a begins to flow. Because the p-n-p transistor Q_1 is then in the active state, its collector current flows into the base of the n-p-n transistor ($I_{c1} = I_{b2}$) and sets up the conditions for regeneration. If the external gate drive is removed, the model remains in the on state as a result of the division of currents associated with the two transistors, provided that sufficient principal current (I_a) is available.

Theoretically, the model shown in Fig. 27 remains in the on state until the principal current flow is reduced to zero. Actually, turn-off occurs at some value of current greater than zero. This effect can be explained by observation of the division of currents as the value of the limiting resistor is gradually increased. As the principal current is gradually reduced to the zero current level, the division of currents within the model can no longer sus-

tain the required regeneration, and the model reverts to the blocking state.

The two-transistor model illustrates three features of thyristors: (1) a gate trigger current is required to initiate regeneration, (2) a minimum principal current (referred to as "latching current") must be available to sustain regeneration, and (3) reduction of principal-current flow results in turn-off at some level of current flow (referred to as "holding current") that is slightly greater than zero.

Fig. 28 shows the effects of a resistive termination at the base of the n-p-n transistor on the latching and holding currents. The collector current through the p-n-p transistor must be increased to supply both the base current for the n-p-n transistor and the shunt current through the terminating resistor. Because the principal-current flow must be increased to supply this increased collector current, latching and holding current requirements also increase. The use of the two-transistor model provides a more concise meaning to the mechanics of thyristors. In thyristor fabrication, it is generally good practice to use a low-beta p-n-p unit and to include internal resistance termination for the base of the n-p-n unit. Termination of the n-p-n provides immunity from

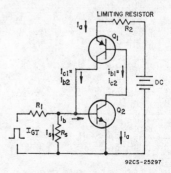

Fig. 28—Two-transistor model of SCR with resistive termination of the n-p-n transistor base.

"false" (non-gated) turn-on, and
the use of the low-beta p-n-p units
permits a wider base region to be
used to support the high voltage
encountered in thyristor applications.

Triacs

A triac is a bidirectional device
designed to provide bilateral switching characteristics for either polarity of applied voltage. The three
electrodes of this device are referred to as **main terminal 1,
main terminal 2,** and the **gate.**
The gate is specially designed so
that either positive or negative gate
voltage can trigger the triac into
conduction for either polarity of
the voltage across the main terminals. As with the SCR, however,
once the triac is turned on, the gate
has no further control. The device
remains in the on state until the
current through the main terminals
is reduced below the value required
to sustain conduction. Unlike the
SCR, however, the triac cannot be
turned off by a reversal of the polarity of the voltage across the main
terminals. A reversal of this voltage merely causes current to flow in
the opposite direction. The triac,
therefore, exhibits the forward-
blocking/forward-conducting voltage-current characteristic of the
SCR structure for either direction
of voltage applied to the main terminals. Fig. 29 shows the junction diagram and schematic symbol for a
triac.

Functionally, a triac may be considered as two parallel SCR
(p-n-p-n) structures oriented in
opposite directions, as shown in Fig.
30. The same approach used to explain gating, latching, and holding
currents in the SCR can be extended
to include the two-SCR model of a
triac.

In triacs, the gate-trigger-pulse
polarity is usually measured with
respect to main terminal No. 1,
which is comparable to the cathode

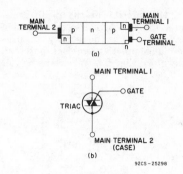

Fig. 29—Junction diagram (a) and schematic symbol (b) for a triac.

terminal of an SCR. The triac can
be triggered by a gate-trigger pulse
which is either positive or negative
with respect to main terminal No. 1
when main terminal No. 2 is either
positive or negative with respect to
main terminal No. 1. The triac,
therefore, can be triggered in any
of four operating modes, as sum-
marized in Table I. The quadrant
designations refer to the operating
quadrant on the principal voltage-
current characteristics (either I or
III), and the polarity symbol represents the gate-to-main-terminal-
No. 1 voltage. Fig. 31 shows the

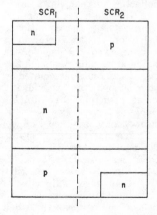

92CS-25299

Fig. 30—Diagram of a triac structure
which shows that this device is basically
two SCR's structures in an inverse parallel
arrangement.

Table I—Triac Triggering Modes

Gate-to-Main-Terminal-No. 1 Voltage	Main-Terminal-No. 2-to-Main-Terminal-No. 1 Voltage	Operating Quadrant
Positive	Positive	I(+)
Negative	Positive	I(−)
Positive	Negative	III(+)
Negative	Negative	III(−)

flow of current in a triac for each of the four triggering modes.

The gate-trigger requirements of the triac are different in each operating mode. The I(+) mode (gate positive with respect to main terminal No. 1 and main terminal No. 2 positive with respect to main terminal No. 1), which is comparable to equivalent SCR operation, is usually the most sensitive. The smallest gate current is required to trigger the triac in this mode. The other three operating modes require slightly higher gate-trigger currents. For RCA triacs, the maximum trigger-current rating in the published data is the largest value of gate current that is required to trigger the selected device in any operating mode.

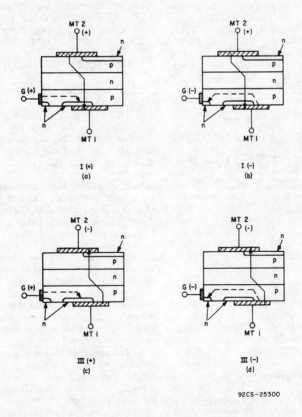

Fig. 31—Current flow in the four triggering modes of a triac: (a) Mode I(+); (b) Mode I(−); (c) Mode III(+); (d) Mode III(−).

Bilateral Switch

A bilateral switch is a four-layer, p-n-p-n device in which all layers are accessible as shown in the juncton diagram in Fig. 32. Fig. 33 shows the forward-bias anode-to-cathode characteristics of the device. The switch can be turned on by application of a forward bias voltage, or by increasing the anode current through the application of a current to either of the gates. The latter method permits the switch to be turned on even when the junction voltages are well below breakdown. Once the bilateral switch is turned on, it stays on until the anode current decreases below a value called the holding current. To turn the switch off, the anode current may be reduced below the holding value by reverse-biasing the anode, by diverting the current by means of a shunt current path, or by including the switch in an under-damped tuned circuit.

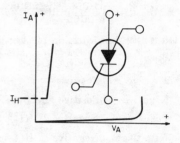

92CS-25302

Fig. 33—Forward-bias anode-to-cathode characteristics of a bilateral switch.

Bilateral switch applications include voltage-level detectors, bistable memory elements, binary counters, shift registers, time delay, pulse, and tone generators, relay drivers, and indicator lamp drivers.

INTEGRATED CIRCUITS

The distinguishing feature of any integrated circuit (IC) is that all (or nearly all) the components (active and passive) required to perform a particular electronic function are combined and interconnected on a common substrate. Viewed macroscopically, the constituent elements of an IC lose their identities as discrete components, and the device assumes the appearance of a "microminiaturized" function block. In comparison to its discrete-component counterpart, an integrated circuit offers equipment designers an essentially complete solid-state circuit in a package not significantly larger than that of a conventional discrete transistor. In addition to reducing the size and weight of electronic equipment, use of an integrated circuit can provide enhanced performance and has established new plateaus of reliability, at reduced costs. An integrated circuit, therefore, may be defined as "a combination of interconnected circuit elements inseparably associated on or within a continuous substrate" (e.g., a silicon "chip").

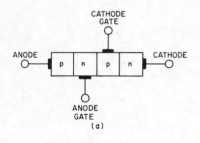

(a)

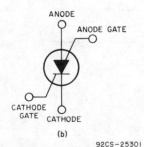

(b)

92CS-25301

Fig. 32—(a) Junction diagram and (b) schematic symbol for a bilateral switch.

Basic Principle of Integration

An integrated circuit is an electron device capable of performing an electronic function normally accomplished by interconnection of several individual electronic components such as transistors, resistors, capacitors, and the like. For the purpose of illustration, it can be assumed that the very elementary transistorized amplifier shown in Fig. 34 is to be redesigned to use a monolithic integrated circuit instead of individual (discrete) electronic components. This simple amplifier employs one discrete n-p-n transistor and two discrete resistors (R_i and R_L), or a total of three discrete electronic components, as shown in Fig. 35. Integrated-circuit technology permits the integration of these components on a single (monolithic) small "clip" of silicon, housed in a single small package, as shown in Fig. 36. Furthermore, the three components are already electrically interconnected by means of metallic wiring on the silicon clip. Elementary descriptions of the techniques by which this integration is accomplished are presented in a later section of this manual.

The simple illustration just described is intended to convey the principle of integration, but it does not describe the magnitudes of circuitry which may be encompassed by an integrated circuit. The monolithic integrated circuit capitalizes on the economies inherent in "batch" fabrication of electronic elements on the surface of a "single stone" (monolith). For example, contemporary monolithic integrated cir-

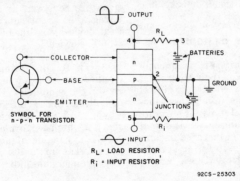

Fig. 34—Schematic diagram of elementary transistor amplifier.

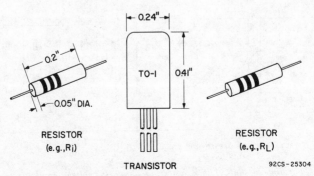

Fig. 35—Examples of "discrete" electronic components.

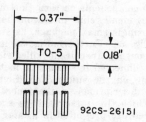

Fig. 36—Typical package for monolithic
integrated circuits.

cuits are fabricated on a single chip
of silicon with dimensions in the
order of 0.1 inch square and 0.01
inch thick. A chip of these dimensions may contain a "population"
in the order of 1000 interconnected
electronic elements. As the technology advances, the dimensions of
a practical chip can be expected to
increase, with corresponding increases in "population", density, and
circuit complexity.

Illustrative Examples

Integrated circuits are frequently
classified in terms of their functional
end-use for either linear or digital
circuit applications. Linear (or analog) types are a family of circuits
that operates on an electrical signal
to change its shape, increase its amplitude, or modify it for a specific
end-function. Digital types are a
family of circuits that operates effectively as "on-off" switches, in accordance with the absence or presence of a signal.

With the advent of more complex
integrated circuits, however, even
this distinction is becoming out-
moded. For example, the process
of decoding composite stereo audio
signals from an FM receiver into
"left" and "right" audio signals
has usually been considered as a
task for analog circuitry. Nevertheless, a complex integrated-circuit
stereo multiplex decoder (Type
CA3090AQ) employs three complex
digital flip-flops to actuate the analog circuits with fewer external
components. Such intermixtures of
linear and digital circuits on a monolithic chip will become more common.

Two illustrative examples of integrated circuits and their applications are reviewed in the following
paragraphs as introductions to practical devices for linear- and digital-
circuit service.

Linear IC In Voltage Regulator
Application—The circuit in Fig. 37
shows the simplicity with which a
monolithic linear integrated circuit
(Type CA3085) can be used to design a high-performance voltage
regulator capable of delivering output current up to 100 milliamperes.
The number of individual compo-

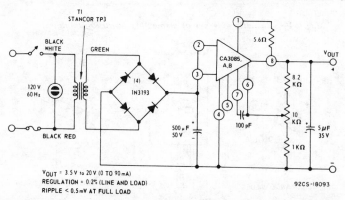

Fig. 37—Application of the CA3085 Series monolithic IC voltage regulator in a typical
power supply.

nents required to construct this regulator circuit is minimal in comparison to the requirements for a regulator of similar characteristics built with discrete components. The high-valued capacitors are used for ripple-filtering; the output voltage required is selected by the 10-Kilohm potentiometer.

The actual circuit schematic of the elements comprising the CA3085 integrated circuit, shown in Fig. 38, reveals that this integrated circuit contains two zener diodes, five diodes, eleven transistors, and five resistors. Circuitry to the left of transistor Q_5 provides a stable reference voltage of 1.6 volts (typical) to the base of Q_5 despite variations of the unregulated input voltage over the range from 7.5 to 40 volts. Transistors Q_5 and Q_6 comprise the basic differential amplifier that is used as a voltage-error amplifier to compare the stable reference voltage applied at the base of Q_5 with a sample of the regulator output voltage applied at terminal 6. This voltage-error amplifier stage controls the Darlington-pair transistors Q_{13} and Q_{14} that perform the basic series-pass regulation function between the unregulated input voltage at terminal 5 and the regulated output voltage at terminal 1. Transistor Q_{15} is used to provide current-

limiting as protection for the integrated circuit and/or to limit the load current.

All the circuitry in Fig. 38 is integrated on the single silicon chip (0.05-inch square) shown in Fig. 39. Metallic wiring is used on the surface of the chip to interconnect the various components and terminate at the edge of the chip as metallic terminal connections (called bonding pads) for 1.5-mil-diameter aluminum wires which are used to connect the integrated-circuit chip to its case terminals, as shown in Fig. 40. The assembly is completed by welding a metal-cap to the package-stem, providing a hermetic package with the chip sealed in an atmosphere of dry nitrogen. A completed TO-5 style integrated-circuit package is shown in Fig. 41(a).

Although the linear circuit shown in Figs. 38 and 39 is not the most complex being produced today, it is typical of linear integrated circuits being supplied in high-volume quantities at prices which are rapidly forcing the abandonment of linear circuit designs that use discrete small-signal transistors. The degree of integration achievable in linear monolithic integrated circuits is already sufficiently comprehensive to permit high-volume production of

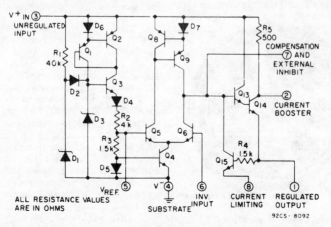

Fig. 38—Schematic diagram of CA3085 Series monolithic IC voltage regulator.

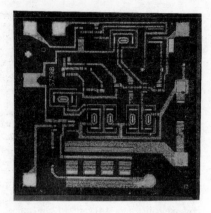

92CS-2213 4

Fig. 39—IC "chip" ($\cong$ 0:05" square) for CA3085 Series voltage regulator.

(a)

H-1528

8-LEAD TO-5

(b)

H-1517

14-LEAD DUAL-IN-LINE

single packages that contain the electronics for the entire "pix-if" portion of a TV receiver or the decoding function for an AM stereo receiver. Bipolar transistor technology has been responsible for the spectacular development of linear integrated circuits. Starting in 1973, however, the first significant linear-IC products predicated on MOS/FET technology made their appearance on the market.

(c)

H1383R1

14-LEAD FLAT-PACK

Fig. 41—Package styles commonly used with monolithic ICs: (a) TO-5, (b) Dual-In-Line (c) Flat-pack.

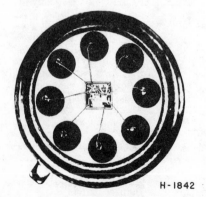

H-1842

Fig. 40—IC "chip" mounted in TO-5 style package assembly.

Digital IC in Clock-Circuit Application—The basic "logic elements" of digital circuit functions include "gate" circuits (e.g., OR, NOR, AND, and NAND), flip-flops, memory cells, inverters, and transmission gates. They are used to implement "logic operations" in computers, calculators, control systems, digital-type meters, and the like. "Logic element" circuits can be designed with discrete transistors, but the advent of IC's has brought the digital-circuit designer an abundance of specific "logic-element building-block" integrated-circuit devices that offer enhanced reliability, lower cost, greater compactness, simplicity for the user, and higher performance characteristics.

The designers of integrated circuits have made these "logic element" functions available in a number of distinctly different "logic families", e.g., resistor-transistor logic (RTL), diode transistor logic (DTL), transistor-transistor logic (TTL), emitter-coupled logic (ECL), P-MOS, COS/MOS, and others. These "logic families" are differentiated from a user standpoint in that they offer a choice of characteristics such as operating speed, power consumption, supply-voltage, and noise immunity. Integrated-circuit technology has also enabled the assemblage of "logic element" conglomerates (e.g., gates, flip-flops, and other basic logic elements) on a single monolithic chip capable of performing a more comprehensive function (e.g., registers and adders). Moderately complex conglomerates can generically be defined as "MSI" (Medium-Scale Integration), while the most complex conglomerates of logic elements and functions can be categorized as "LSI" (Large-Scale Integration). For example, the very complex circuity needed in a pocket-size calculator can be implemented with a few "LSI"-type IC chips.

The evolution of "LSI"-type digital integrated circuits has also revolutionized the design and manufacture of electronic timepieces, permitting the use of very complex circuitry to harness the time-period accuracy of crystal oscillators and the decoding of the information necessary to present a digital read-out of the time. Although the description of a digital read-out system is beyond the scope of this introductory treatment, a description of the crystal-oscillator frequency "count-down" circuitry is instructive in illustrating the contributions to design and manufacturing simplicity that accrue from use of a digital integrated circuit of LSI complexity to accomplish the "count-down" function.

Fig. 42 shows a 21-stage integrated circuit digital counter used to count-down the 2.097152 MHz crystal-oscillator frequency to an output pulse rate of one pulse per second. The "push-pull" output pulses provided by this counter can drive miniature synchronous motors, (or stepping motors) as prime movers in timepiece mechanisms. The choice of a 2-MHz frequency standard is a compromise between crystal size, cost, and frequency stability. The first "inverter" stage is used as a micropower crystal-oscillator, followed by four "inverter" stages that provide gain, pulse-shaping, and push-pull pulse signals (Φ, $\bar{\Phi}$) at 2.097152 MHz to drive the first flip-flop (F/F1) divider stage. Each of the succeeding 21-digital counter stages performs a "divide-by-two" function and ultimately produce one output pulse per second to flip-flops F/F22 and F/F23, which shape the output pulses to 1/32-second duration and drive the output "inverters" (I_6 and I_7) to obtain approximately 5 milliamperes of output drive current.

The circuit in Fig. 42 is an illustration of contemporary large-scale integration (LSI). It contains a total of 490 MOS field-effect transistors on a single chip about 0.08-inch square. The integrated circuit is supplied in dual-in-line or flat-pack packages of the generic type shown in Fig. 41(b) and (c) respectively. In dynamic operation with a 2 MHz crystal, the total circuit only consumes about 5-milliwatts of power at a supply voltage (V_{DD}) of 10 volts. The circuit continues to operate even though the supply-voltage (V_{DD}) varies over the range from 3 to 15 volts. A 16.5-volt zener-diode string is integrated on the chip to provide "voltage-clipping" protection against voltage transients encountered in automotive applications.

IC Packages

The simplicity of the packages shown in Fig. 41 permits the equip-

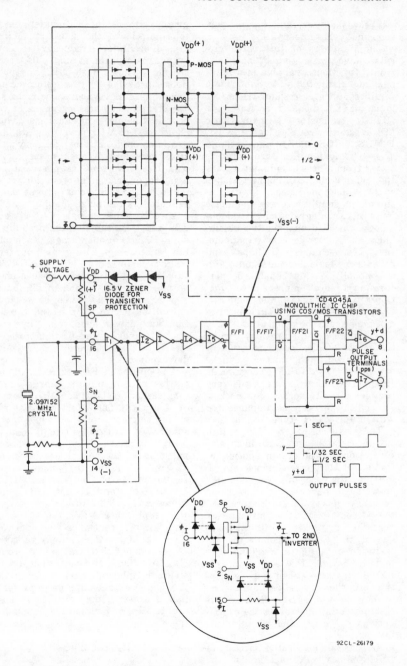

Fig. 42—21-Stage Counter IC (Digital "LSI") used to provide one pulse per second outputs for clock and other timing applications.

ment builder to procure, handle, and install complex blocks of circuit function with comparative ease. The TO-5 style packages are commonly used with circuits requiring 8, 10 or 12-leads. Dual-in-line packages are currently the most popular style of package. They are easily installed in sockets or soldered into printed-circuit boards. The 14-lead arrangements are currently the most popular, although 8-lead and 16-lead packages are also in high volume use. Most of the dual-in-line pack-

ages currently produced are plastic (nonhermetic) types, but large volumes of dual-in-line metal-and-ceramic (hermetic) type packages are also being manufactured. Flat-pack packages are used in applications in which space is at a premium, for example as is usually the case in electronic equipment for the military. Most flat-packs are hermetic packages. Additonal information on integrated-circuit packages is given in the section **Guide to RCA Solid-State Devices.**

Basic Rating Factors

RATINGS are established for solid-state devices to help circuit and equipment designers use the performance and service capabilities of each type to maximum advantage. They define the limiting conditions within which a device must be maintained to assure satisfactory and reliable operation in equipment applications. A designer must thoroughly understand the constraints imposed by the device ratings if he is to achieve effective, economical, and reliable equipment designs. Reliability and performance considerations dictate that he select devices for which no ratings will be exceeded by any operating conditions of his application, including equipment malfunction. He should also realize, however, that selection of devices that have overly conservative ratings may significantly add to the cost of his equipment.

BASIS FOR DEVICE RATINGS

Three systems of ratings (the absolute maximum system, the design center system, and the design maximum system) are currently in use in the electronics industry. The ratings given in the technical data for solid-state devices are based on the **absolute maximum system**. A definition for this system of ratings has been formulated by the **Joint Electron Devices Engineering Council (JEDEC)** and standardized by the National Electrical Manufacturers Association (NEMA) and the Electronic Industries Association (EIA), as follows:

"**Absolute-Maximum ratings** are limiting values of operating and environmental conditions applicable to any electron device of a specified type as defined by its published data, and should not be exceeded under the worst probable conditions.

"The device manufacturer chooses these values to provide acceptable serviceability of the device, taking no responsibility for equipment variations, environmental variations, and the effects of changes in operating conditions due to variations in device characteristics.

"The equipment manufacturer should design so that initially and throughout life no absolute-maximum value for the intended service is exceeded with any device under the worst probable operating conditions with respect to supply-voltage variation, equipment component variation, equipment control adjustment, load variation, signal variation, environmental conditions, and variations in device characteristics."

The rating values specified in the technical data for RCA solid-state devices are determined on the basis of extensive operating and life tests and comparison measurements of critical device parameters. These tests and measurements define the limiting capabilities of a specific device type in relation to the rating factors being considered. The test and measurement conditions simulate, as closely as possible, the worst-case conditions that the device is likely to encounter in actual equipment applications.

Rating tests are expensive, time-consuming, and often destructive. Obviously, therefore, all individual solid-state devices of a given type designation cannot be subjected to these tests. The validity of the ratings is assured, however, by use of stringent processing and fabrication controls and extensive quality checks at each stage in the manufacturing process to assure product uniformity among all devices of a specific type designation and by testing of a statistically significant number of samples.

Ratings are given for those stress factors that careful study and experience indicate may lead to severe degradation in performance characteristics or eventual failure of a device unless they are constrained within certain limits. Table II lists the critical rating factors used to specify the safe operating capabilities of different types of solid-state devices. These ratings are also applicable to the active devices included in monolithic integrated circuits and power hybrid circuits.

VOLTAGE RATINGS

A number of voltage ratings are provided for solid-state devices. These ratings are established with respect to a specified electrode (e.g., collector-to-emitter voltage or collector-to-base voltage for transistors) and indicate the maximum potential, for both steady-state and transient operation, that can be safely applied across the two specified electrodes before damage to the crystal occurs. These ratings are specified for particular conditions (e.g., with the third electrode open, or with a specific bias voltage or external resistance, for transistors and thyristors).

Excessive voltage potentials produce high leakage (or reverse) currents in solid-state devices. In silicon rectifiers, the high reverse currents that result from excessive reverse-bias voltages can lead to crystal breakdown and the consequent destruction of the devices. Similar junction breakdown can occur because of the high leakage currents that result from excessive collector-to-emitter or collector-to-base voltages in transistors or excessive off-state voltages in thyristors. Leakage currents flow in solid-state devices at all voltage levels, and device operation is significantly affected by the magnitude of these currents, even at voltages significantly below the breakdown value. For example, in transistors, the collector leakage currents critically affect biasing levels, and consequently the gain and stability of the over-all circuit. In thyristors, high leakage current levels can cause unwanted switching of device conduction states. In addition to their dependence on voltage, leakage currents also vary with temperature. In the technical data on solid-state devices, therefore, these currents are usually specified for particular voltage and temperature conditions.

FORWARD-CURRENT RATINGS

If the current in a solid-state device becomes sufficiently large, the semiconductor pellet could be melted by the excessive junction temperatures that result. Maximum current ratings, however, are not usually based on the current-carrying capacity of the semiconductor pellet. Such ratings are usually based on the degradation of specific device performance characteristics that result when the current density exceeds a critical value or on the fusing current of an internal connecting wire.

For devices in which the fusing current of internal connecting wires is not the limiting factor, different

Table II—Ratings and Limiting Characteristics for Solid State Devices

GENERAL

Quantity	Symbol
Ambient temperature	T_A
Case temperature	T_C
Junction temperature	T_J
Storage temperature	T_{stg}
Thermal Resistance	Θ
Junction to ambient	Θ_{J-A}
Junction to case	Θ_{J-C}
Case-to-ambient	Θ_{C-A}
Case-to-heat sink	Θ_{C-S}
Transient thermal impedance	$\Theta_{(t)}$
Junction-to-ambient	$\Theta_{J-A(t)}$
Junction-to-case	$\Theta_{J-C(t)}$
Delay time	t_d
Rise time	t_r
Fall time	t_f

SILICON RECTIFIERS

Quantity	Symbol
Forward current:	
Total rms value	$I_{F(RMS)}$
DC value, no alternating component	I_F
DC value, with alternating component	$I_{F(AV)}$
Instantaneous total	i_F
Maximum (peak) total value	I_{FM}
Repetitive peak	I_{FRM}
Surge (non-repetitive)	I_{FSM}
Forward voltage:	
Total rms value	$V_{F(RMS)}$
DC value, no alternating component	V_F
DC value, with alternating component	$V_{F(AV)}$
Instantaneous total value	v_F
Maximum (peak) value	V_{FM}
Reverse current:	
Total rms value	$I_{R(RMS)}$
DC value, no alternating component	I_R
DC value, with alternating component	$I_{R(AV)}$
Instantaneous total value	i_{RM}
Reverse recovery time	t_{rr}
Reverse voltage:	
Total rms value	$V_{R(RMS)}$
DC value, no alternating component	V_R
DV value, with alternating component	$V_{R(AV)}$
Instananeous total value	v_R
Maximum (peak) total value	V_{RM}
Working peak	V_{RWM}
Repetitive peak	V_{RRM}
Non-repetitive peak	V_{RSM}
Reverse breakdown voltage:	
DC value, no alternating component	$V_{(BR)R}$
Instantaneous total value	$v_{(BR)R}$
Forward Power Loss:	
DC value, no alternating component	P_F
DC value, with alternating component	$P_{F(AV)}$
Instantaneous total value	p_F
Maximum (peak) total value	P_{FM}

SILICON RECTIFIERS (Cont.)

Quantity	Symbol
Reverse power loss:	
DC value, no alternating component	P_R
DC value, with alternating component	$P_{R(AV)}$
Instantaneous total value	p_R
Maximum (peak) total value	P_{RM}

THYRISTORS AND DIACS

Quantity	Symbol
On-state current:	
Total rms value	$I_{T(RMS)}$
DC value, no alternating component	I_T
DC value, with alternating component	$I_{T(AV)}$
Instantaneous total value	i_T
Maximum (peak) total value	I_{TM}
Surge (non-repetitive)	I_{TSM}
Overload	$I_{T(OV)}$
Breakover current:	
DC value, no alternating component	$I_{(BO)}$
Instantaneous total value	$i_{(BO)}$
Off-state current:	
Total rms value	$I_{D(RMS)}$
DC value, no alternating component	I_D
DC value, with alternating component	$I_{D(AV)}$
Instantaneous total value	i_D
Maximum (peak) total value	I_{DM}
Repetitive peak	I_{DRM}
Reverse current:	
Total rms value	$I_{R(RMS)}$
DC value, no alternating component	I_R
DC value, with alternating component	$I_{R(AV)}$
Instantaneous total value	i_R
Maximum (peak) total value	I_{RM}
Repetitive peak	I_{RRM}
Reverse breakdown current:	
DC value, no alternating component	$I_{(BR)R}$
Instantaneous total	$i_{(BR)R}$
On-state voltage:	
Total rms value	$V_{T(RMS)}$
DC value, no alternating component	V_T
DC value, with alternating component	$V_{T(AV)}$
Instantaneous total value	v_T
Maximum (peak) total value	V_{TM}
Breakover voltage:	
DC value, no alternating component	$V_{(BO)}$
Instantaneous total value	$v_{(BO)}$
Off-state voltage:	
Total rms value	$V_{D(RMS)}$
DC value, no alternating component	V_D
DC value, with alternating component	$V_{D(AV)}$
Instantaneous total value	v_D
Maximum (peak) total value	V_{DM}
Working peak	V_{DWM}
Repetitive peak	V_{DRM}
Repetitive peak, with gate open	V_{DROM}
Non-repetitive peak	V_{DSM}
Non-repetitive peak with gate open	V_{DSOM}

Table II—Ratings and Limiting Characteristics for Solid-State Devices (cont'd)

Quantity	Symbol
THYRISTORS AND DIACS (Cont.)	
Reverse voltage:	
Total rms value	$V_{R(RMS)}$
DC value, no alternating component	V_R
DC value, with alternating component	$V_{R(AV)}$
Instantaneous total value	v_R
Maximum (peak) total value	V_{RM}
Working peak	V_{RWM}
Repetitive peak, with specified gate-to-cathode resistance	V_{RRM}
Repetitive peak, with gate open	V_{RROM}
Non-repetitive peak, with specified gate-to-cathode resistance	V_{RSM}
Non-repetitive peak, with gate open	V_{RSOM}
Reverse breakdown voltage:	
DC value, no alternating component	$V_{(BR)R}$
Instantaneous total	$v_{(BR)R}$
Holding current:	
DC value, no alternating component	I_H
Instantaneous total value	i_H
Latching current:	
DC value, no alternating component	I_L
Instantaneous total value	i_L
Gate current:	
DC value, no alternating component	I_G
DC value, with alternating component	$I_{G(AV)}$
Maximum (peak) total value	I_{GM}
Gate trigger current:	
DC value, no alternating component	I_{GT}
Maximum (peak) total value	I_{GTM}
Gate non-trigger current:	
DC value, no alternating component	I_{GD}
Maximum (peak) total value	I_{GDM}
Gate voltage:	
DC value, no alternating component	V_{GT}
Maximum (peak) total value	V_{GTM}
Gate trigger voltage:	
DC value, no alternating component	V_{GT}
Instantaneous total value	v_{GT}
Maximum (peak) total value	V_{GTM}
Gate non-trigger voltage:	
DC value, no alternating component	V_{GD}
Instantaneous total value	v_{GD}
Maximum (peak) total value	V_{GDM}
Gate power dissipation:	
DC value, no alternating component	P_G
DC value, with alternating component	$P_{G(AV)}$
Instantaneous total value	p_G
Maximum (peak) total value	P_{GM}
POWER TRANSISTORS	
Base current:	
DC value, no alternating component	I_B

Quantity	Symbol
POWER TRANSISTORS (Cont.)	
RMS value of alternating component	I_b
Instantaneous total value	i_B
Collector current:	
DC value, no alternating component	I_C
RMS value of alternating component	I_c
Instantaneous total value	i_C
Emitter current:	
DC value, no alternating component	I_E
RMS value of alternating component	I_e
Instantaneous total value	i_E
Collector-to-base cutoff current*	
dc value with emitter open	I_{CBO}
Collector-to-emitter cutoff current,* dc value:	
With base open	I_{CEO}
With specified resistance between base and emitter	I_{CER}
With base shorted to emitter	I_{CES}
With specified voltage between base and emitter	I_{CEV}
With specified circuit between base and emitter	I_{CEX}
Emitter-to-base cutoff current,* dc value with collector open	I_{EBO}
Power (common-emitter connection):	
DC input to base	P_{BE}
Instantaneous total input to base	p_{BE}
Large-signal output	P_{OE}
Total nonreactive input to all terminals	P_T
Instantaneous total nonreactive input to all terminals	p_T
Emitter-to-base open-circuit dc voltage (floating potential)	$V_{EB(fl)}$
Collector-to-base dc voltage, with emitter open	V_{CBO}
Collector-to-emitter dc voltage:	
With base open	V_{CEO}
With specified resistance between base and emitter	V_{CER}
With base shorted to emitter	V_{CES}
With specified voltage between base and emitter	V_{CEV}
With specified circuit between base and emitter	V_{CEX}
Emitter-to-base dc voltage, with collector open	V_{EBO}
Second Breakdown	
Forward-bias energy level	$I_{s/b}$
Reverse-bias energy level	$E_{s/b}$
Thermal-cycling capability (number of cycles)	N

* Cutoff current is also referred to as reverse current or leakage current.

parameters are used as the basis for the current ratings of the various types of devices. In silicon rectifiers and thyristors, the maximum on-state current rating is determined on the basis of the maximum permissible forward power dissipation. In power transistors, the current gain is significantly decreased at high current densities. The maximum forward-current rating, therefore, is established on the basis of an arbitrary minimum acceptable gain value.

POWER-DISSIPATION RATINGS

Power is dissipated in the semiconductor material of a solid-state device in the form of heat, which if excessive can cause irreversible changes in the crystal structure or melting of the pellet. This dissipation is equal to the difference between the input power applied to the device and the power delivered to the load circuit. Because of the sensitivity of semiconductor materials to variations in thermal con-

ditions, maximum dissipation ratings are usually given for specific temperature conditions.

In many instances, dissipation ratings for solid-state devices are specified for ambient, case, or mounting-flange temperatures up to 25°C. Such ratings must be reduced linearly for operation of the devices at higher temperatures. Fig. 43 shows a typical power-transistor derating chart that can be used to determine maximum permissible dissipation values at specific temperatures above 25°C. (This chart cannot be assumed to apply to transistor types other than the particular transistors for which it was prepared.) The chart shows the permissible percentage of the maximum dissipation ratings as a function of ambient or case temperature. Individual curves are shown for specific operating temperatures. If the maximum operating temperature of a particular transistor type is some other value, a new curve can be drawn from point A to the desired temperature value on the abscissa, as indicated by the dashed-line curves on the chart.

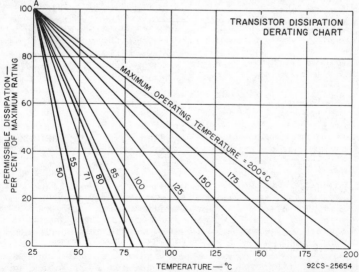

Fig. 43—Chart showing maximum permissible percentage of maximum rated dissipation as a function of temperature.

JUNCTION-TEMPERATURE RATINGS

The temperature of solid-state devices must be closely controlled not only during operation, but also during storage. For this reason, ratings data for these devices usually include maximum and minimum **storage temperatures**, as well as maximum **operating temperatures**.

THERMAL IMPEDANCE

When current flows through a solid-state device, power is dissipated in the semiconductor pellet that is equal to the product of the voltage across the junction and the current through it. As a result, the temperature of the pellet increases. The amount of the increase in temperature depends on the power level and how fast the heat can flow away from the junction through the device structure to the case and the ambient atmosphere. The rate of heat removal depends primarily upon the thermal resistance and capacitance of the materials involved. The temperature of the pellet rises until the rate of heat generated by the power dissipation is equal to the rate of heat flow away from the junction; i.e., until thermal equilibrium has been established.

Thermal resistance can be compared to electrical resistance. Just as electrical resistance is the extent to which a material resists the flow of electric current, thermal resistance is the extent to which a material resists the flow of heat. A material that has a low thermal resistance is said to be a good thermal conductor. In general, materials which are good electrical conductors are good thermal conductors, and vice versa.

The methods of rating solid-state power devices under steady-state conditions are indicated by the following definition of thermal resistance: The thermal resistance of a solid-state device is the ratio of the temperature drop to the heat generated through internal power dissipation under steady-state conditions; the temperature drop is measured between the region of heat generation and some reference point.

The over-all thermal resistance of an assembled device is usually expressed as the rise in junction temperature above the case temperature per unit of power dissipated in the device. This information, together with the maximum junction-temperature rating, enables the user to determine the maximum power level at which the device can be safely operated for a given case temperature. Subtraction of the case temperature from the maximum junction temperature indicates the allowable internal temperature rise. If this value is divided by the specified thermal resistance of the device, the maximum allowable power dissipation is determined.

It should be noted that thermal resistance is defined for steady-state conditions. If a uniform temperature over the entire semiconductor junction is assumed, the power dissipation required to raise the junction temperature to a predetermined value, consistent with reliable operation, can be determined. Under conditions of intermittent or switching loads, however, such a design is unnecessarily conservative and expensive. For such conditions, the effect of **thermal capacitance** should also be considered.

Junction-to-Case Thermal Impedance

The thermal properties of a device may be represented by an electrical analog circuit, such as that shown in Fig. 44, which consists of a current generator connected to a series of resistors that have capacitance to ground distributed along their length. The power P dissipated within the crystal of a solid-state device results in a flow of heat outward from the crystal. This flow of

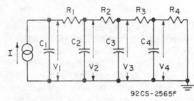

92CS-2565F

Fig. 44—Electrical analog circuit used to describe thermal properties of a solid-state device.

heat (dissipated power P in calories per second or in watts) in a solid-state device is analogous to the flow of charge (electrical current I in coulombs per second or amperes) in such a circuit. Thermal resistances and thermal capacitances of the device are analogous to the electrical resistances and capacitances shown in the circuit. The potential difference or voltage between any two points in the electrical analog circuit is analogous to the temperature difference between the corresponding two points of the device it represents. Table III shows the relationship between various electrical quantities and their corresponding thermal quantities.

Thermal impedance Z_T, like electrical impedance Z, is a complex variable because of the time dependence associated with the thermal capacitance C_T.

In the electrical or thermal-analog circuit shown in Fig. 44, the thermal resistances closest to the heat source are large because the cross section of the semiconductor element is small (all the heat generated flows through a small area). Thermal resistance varies inversely with cross-sectional area. In general, thermal resistances become progressively smaller as distance from the semiconductor element increases.

Thermal capacitance varies directly with both mass and specific heat. Therefore, the small mass of the semiconductor element of a device causes the thermal capacitance to be smallest at the heat source and to become progressively larger as distance from the heat source increases. The final thermal capaci-

Table III—Comparison of Various Electrical Quantities and Corresponding Thermal Quantities

Electrical	Thermal
Current generator	Heat generator (semiconductor crystal)
Resistance R (ohms or volts/ampere)	Thermal Resistance θ (°C/watt)
Capacitance C (ampere-second/volt)	Thermal Capacitance C_T (watt-second/°C)
Potential difference $V_1 - V_2$ (volts)	Temperature difference $T_1 - T_2$ (°C)
Potential above ground $V - V_G$ (volts)	Temperature above ambient $T - T_A$ (°C)
Current I (amperes)	Power dissipation P (watts)
Impedance Z (volts/ampere)	Thermal impedance Z_T (°C/watt)

tance in the series must be considered as an infinite capacitance, which electrically is the same as a direct short across the end of the line.

If a step function of power is applied to a solid-state device (i.e., if the power input at time t_1 increases from P = 0 to P = P_1), the temperature difference between junction and case rises as shown in Fig. 45, and approaches temperature T_1

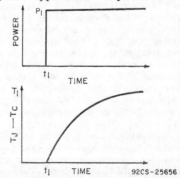

Fig. 45—Temperature-rise curve obtained with step function of power.

asymptotically. This temperature-rise curve is similar to the voltage-rise curve which would be obtained in the analogous resistance-capacitance electrical circuit.

The exact shape of the curve depends upon the magnitudes of the thermal-resistance and thermal-capacitance components of the device. Fig. 46 shows a typical thermal-response curve for a silicon power transistor. This curve indicates that solid-state devices have multiple thermal time constants.

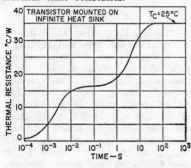

92CS-25657

Fig. 46—Graphical representation of transient thermal response (i.e., thermal-impedance curve).

Case-to-Ambient Thermal Resistance

The thermal equivalent circuits for a transistor discussed in the preceding section considered only the thermal paths from junction to case. For power transistors in which the silicon pellet is mounted directly on the header or pedestal, the total internal thermal resistance from junction to case θ_{J-C} varies from 50°C per watt to less than 1°C per watt. If the transistor is not mounted on a heat sink, the thermal resistance from case to ambient air θ_{C-A} is so large in comparison to that from junction to case that the net over-all thermal resistance from junction to ambient air is primarily the result of the θ_{C-A} term. Table IV lists values of case-to-air thermal resistance for popular JEDEC cases.

Beyond the limit of a few hundred milliwatts, it becomes impractical to increase the size of the case to make the θ_{C-A} term comparable to the θ_{J-C} term. As a result, most power transistors and other solid-state power devices are designed for use on an external heat sink.

Table IV—Case-to-Free-Air Thermal Resistance and Thermal Capacitance for Popular JEDEC Packages

Package Case	Thermal Resistance θ_{C-A} (°C/W)
TO-18	300
TO-46	300
TO-5	150
TO-39	150
TO-8	75
TO-66	60
TO-60	70
TO-3	30
TO-36	25

Package	Thermal Capacitance (Joules/°C)	Thermal Time Constant (Seconds)
TO-5	0.58	69
TO-66 (no button)	2.56	128
TO-8	1.84	110
TO-3 (Cu button)	6.8	204
TO-3 (Mod, 2N5575)	7.8	117

Case-to-Ambient Thermal Capacitance

The thermal capacitance of the over-all package is also an important factor in the thermal circuit of a solid-state power device.

Table IV also lists typical values of thermal capacitances and the thermal time constants for some common types of device packages.

These values can be used to calculate temperature effects of pulses on devices that are not mounted on heat sinks. The thermal time constants can be used to estimate how long units must be cooled between tests to avoid temperature changes. For example, the application of a 150-watt pulse for 1 second results in a temperature rise in the TO-3 package determined as follows:

$$T_{rise} = 150 \text{ watts}/6.8 \text{ joules } °C$$
$$= 22°C$$

The time (t) required to cool the package to within 3°C of room temperature can be determined as follows:

$$T = 22e^{-t/204}$$
$$3 = 22e^{-t/204}$$
$$\ln 3/22 = -t/204$$
$$t = 6.1 \text{ minutes}$$

EFFECT OF EXTERNAL HEAT SINKS

The maximum allowable power dissipation in a solid-state device is limited by the temperature of the semiconductor pellet (i.e., the junction temperature). An important factor that assures that the junction temperature remains below the specified maximum value is the ability of the associated thermal circuit to conduct heat away from the device. For this reason, solid-state power devices should be mounted on a good thermal base (usually copper), and means should be provided for the efficient transfer of heat from this base to the surrounding environment.

Most practical heat sinks used in modern, compact equipment are the result of experiments with heat transfer through convection, radiation, and conduction in a given application. Although there are no set design formulas that provide exact heat-sink specifications for a given application, there are a number of simple rules that reduce the time required to evolve the best design for the job. These simple rules are as follows:

1. The surface area of the heat sink should be as large as possible to provide the greatest possible heat transfer. The area of the surface is dictated by case-temperature requirements and the environment in which the device is to be placed.

2. The heat-sink surface should have an emissivity value near unity for optimum heat transfer by radiation. A value approaching unity can be obtained if the heat-sink surface is painted flat black.

3. The thermal conductivity of the heat-sink material should be such that excessive thermal gradients are not established across the heat sink.

Although these rules are followed in conventional heat-sink systems, the size and cost of such systems often become restrictive in compact, mass-produced power-control and power-switching applications. The use of mass-produced prepunched parts, direct soldering, and batch-soldering techniques eliminates many of the difficulties associated with heat sinks by making possible the use of a variety of simple, efficient, readily fabricated heat-sink configurations that can be easily incorporated into the mechanical design of equipment.

For most efficient heat sinking, intimate contact should exist between the heat sink and at least one-half of the package base. The package can be mounted on the heat sink mechanically, with glue or epoxy adhesive, or by soldering. (Soldering is not recommended for transistors.) If mechanical mounting is employed, silicone grease should be used between the device and the heat sink to eliminate surface voids, prevent insulation buildup due to oxidation, and help conduct heat across the interface. Although glue or epoxy adhesive provides good bonding, a significant amount of resistance may exist at the interface resistance; an adhesive material with low thermal resistance, such as Hysol Epoxy Patch Material No. 6C or Wakefield Delta Bond No. 152, or their equivalent, should be used.

Types of Heat Sinks

Heat sinks are produced in various sizes, shapes, colors, and materials; the manufacturer should be contacted for exact design data. It is convenient for discussion purposes to group heat sinks into three categories as shown below:

1. **Flat vertical-finned types** are normally aluminum extrusions with or without an anodized black finish. They are unexcelled for natural convection cooling and provide reasonable thermal resistance at moderate air-flow rates for forced convection.

2. **Cylindrical or radial vertical-finned types** are normally cast aluminum with an anodized black finish. They are used when maximum cooling in minimum lateral displacement is required, using natural convection.

3. **Cylindrical horizontal-finned types** are normally fabricated from sheet-metal rings and have a painted black matte finish. They are used in confined spaces for maximum cooling in minimum displaced volume.

It is also common practice to use the existing mechanical structure or chassis as a heat sink. The design equations and curves for such heat sinks based upon convection and radiation are shown in Figs. 47, 48, and 49.

A useful nomograph which considers heat removal by both convection and radiation is given in Fig. 50. This nomograph applies for natural bright finish on the copper or aluminum.

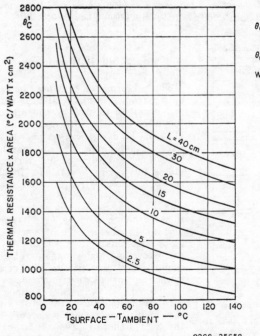

$$\theta_C = \frac{2300}{A} \times \left(\frac{L}{T_s - T_a}\right)^{0.25}$$

$$\theta_C = \frac{\theta_C'}{A} \ {}^\circ C / WATT$$

WHERE θ = CONVECTION THERMAL RESISTANCE °C/WATT
A = AREA IN cm^2, TOTAL EXPOSED SURFACE
L = HEIGHT IN cm

92CS-25658

Fig. 47—Convection thermal resistance as a function of temperature drop from the surface of the heat sink to free air for heat sinks of various heights. (Reprinted from Control Engineering, *October 1956.)*

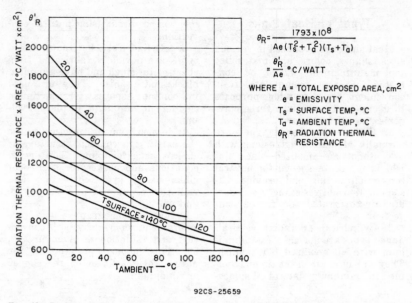

$$\theta_R = \frac{1793 \times 10^8}{Ae(T_s^2 + T_a^2)(T_s + T_a)}$$

$$= \frac{\theta_R'}{Ae} \text{ °C/WATT}$$

WHERE A = TOTAL EXPOSED AREA, cm^2
e = EMISSIVITY
T_s = SURFACE TEMP, °C
T_a = AMBIENT TEMP, °C
θ_R = RADIATION THERMAL
RESISTANCE

92CS-25659

Fig. 48—Radiation thermal resistance as a function of ambient temperature for various heat-sink surface temperatures. (Reprinted from Control Engineering, *October 1956.)*

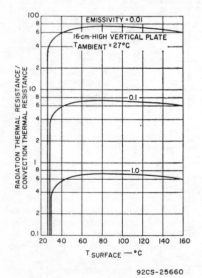

92CS-25660

Fig. 49—Ratio of radiation thermal resistance to convection thermal resistance as a function of heat-sink surface temperature for various surface emissivities. (Reprinted from Control Engineering, *October 1967.)*

Heat-Sink Performance

The performance that may be expected from a commercial heat sink is normally specified by the manufacturer, and the information supplied in the design curves shown in Figs. 47, 48, and 49 provides the basis for the design of flat vertical plates for use as heat sinks. In all cases, it must be remembered that the heat is dissipated from the heat sink by both convection and radiation. Although surface area is important in the design of vertical-plate heat sinks, other factors such as surface and ambient temperature, conductivity, emissivity, thickness, shape, and orientation must also be considered. An excessive temperature gradient can be avoided and the conduction thermal resistance in the heat sink can be minimized by use of a high-conductivity material, such as copper or aluminum, for the heat sink. Radiation losses are increased by an increase in surface

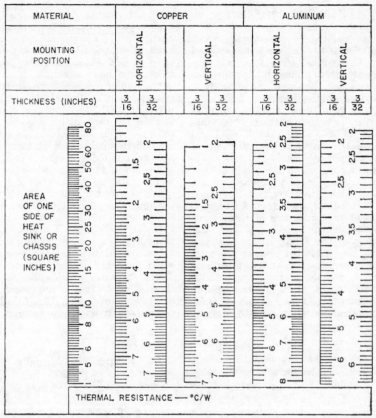

MATERIAL	COPPER					ALUMINUM				
MOUNTING POSITION	HORIZONTAL			VERTICAL		HORIZONTAL			VERTICAL	
THICKNESS (INCHES)	$\frac{3}{16}$	$\frac{3}{32}$		$\frac{3}{16}$	$\frac{3}{32}$	$\frac{3}{16}$	$\frac{3}{32}$		$\frac{3}{16}$	$\frac{3}{32}$

AREA OF ONE SIDE OF HEAT SINK OR CHASSIS (SQUARE INCHES)

THERMAL RESISTANCE — °C/W

INSTRUCTION FOR USE: SELECT THE HEAT-SINK AREA AT LEFT AND DRAW A HORIZONTAL LINE ACROSS THE CHART FROM THIS VALUE. READ THE VALUE OF MAXIMUM THERMAL RESISTANCE DEPENDING ON THE THICKNESS OF THE MATERIAL, TYPE OF MATERIAL, AND MOUNTING POSITION.

92CM-25661

Fig. 50—Thermal resistance as a function of heat-sink dimensions. (Nomograph reprinted from Electronic Design, *August 16, 1961.)*

emissivity, as shown in Fig. 50. Best results are obtained when the heat sink has a black matte finish for which the emissivity is at least 0.9. When free-air convection is used for heat removal, a vertically mounted heat sink provides a thermal resistance that is approximately 30 per cent lower than that obtained with horizontal mounting.

In restricted areas, it may be necessary to use forced-convection cooling to reduce the effective thermal resistance of the heat sink. On the basis of the improved reliability of cooling fans, it can be shown that the over-all reliability of a system may actually be improved by use of forced-convection cooling because the number of components required is reduced.

Economic factors are also important in the selection of heat sinks. It is often more economical to use one heat sink with several properly placed transistors than to use individual heat sinks. It can be shown that the cooling efficiency increases and the unit cost decreases under such conditions.

Heat-Sink Insulators

As pointed out previously, when solid-state devices are to be mounted on heat sinks, some form of electrical isolation must be provided between the case and the heat sink. Unfortunately, however, good electrical insulators usually are also good thermal insulators. It is difficult, therefore, to provide electrical insulation without introduction of significant thermal resistance between case and heat sink. The best materials for this application are mica, beryllium oxide (Beryllia), and anodized aluminum. A comparison of the properties of these three materials for case-to-heat-sink isolation of the TO-3 package is shown in Table V. If the area of the seating plane, the thickness of the material, and the thermal conductivity are known, the case-to-heat-sink thermal resistance θ_{C-S} can be readily calculated by use of the following equation:

$$\theta_{cond} = d/4.186 \, KA \; °C \text{ per watt}$$

where d is the length of the thermal path in centimeters, K is the thermal conductivity in cal/(sec) (cm) (°C), and A is the area perpendicular to the thermal path t in square centimeters. The number 4.186 is a conversion factor used to obtain the result in °C per watt.

Table V—Comparison of Insulating Washers Used for Electrical Isolation of Transistor TO-3 Case from Heat Sink

Material	Thickness (inches)	θ_{C-S} (°C/W)	Capacitace (pF)
Mica	0.002	0.4	90
Anodized Aluminum	0.016	0.35	110
Beryllia	0.063	0.25	15

In all cases, this calculation should be experimentally verified. Irregularities in the bottom of the transistor seating plane or on the face of the heat sink or insulating washer may result in contact over only a very small area unless a filling compound is used. Although silicone grease has been used for years, recently newer compounds with zinc oxide fillers (e.g., Dow Corning #340 or Wakefield #120) have been found to be even more effective.

For small general-purpose transistors, such as the 2N2102, which use a JEDEC TO-5 package, a good method for thermal isolation of the collector from a metal chassis or printed-circuit board is by means of a beryllium-oxide washer. The use of a zinc-oxide-filled silicone compound between the washer and the chassis, together with a moderate amount of pressure from the top of the transistor, helps to decrease thermal resistance. Fin-type heat sinks, which are commercially available, are also suitable, especially when transistors are mounted in Teflon sockets which provide no thermal conduction to the chassis or printed-circuit board. Fig. 51 illustrates both types of mounting.

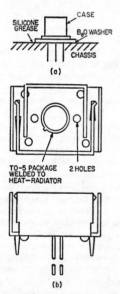

92CS-25662

Fig. 51—Suggested mounting arrangements for transistors having a JEDEC TO-5 package: (a) without heat sink; (b) with fin-type heat sink.

At frequencies of 100 MHz and higher, the effects of stray capacitances and inductances and of ground paths and feedback coupling have a pronounced effect on the gain and power-output capabilities of transistors. As a result, physical aspects such as mechanical layout, shielding, and heat-sink considerations are important in the design of rf amplifiers and oscillators. In particular, it should be noted that the insulating washer necessary for isolation introduces coupling capacitance from collector to chassis which may seriously limit circuit performance. The special techniques used to overcome these effects are explained subsequently in the section **High-Frequency Power Transistors.**

EFFECT OF CYCLIC THERMAL STRESSES

When a solid-state device is alternately heated and allowed to cool, cyclic mechanical stresses are produced within the device because of differences in the thermal expansion of the silicon pellet and the metallic materials to which the pellet is attached. The cyclic stresses may eventually result in physical damage to the semiconductor pellet or the mounting interface.

In most solid-state power devices, a small silicon pellet is bonded to a copper header. The coefficient of thermal expansion for silicon (3×10^{-6}) is much less than that of copper (17.5×10^{-6}). Temperature variations within the device, therefore, result in cyclic stresses at the mounting interface of the silicon pellet and the copper header because of the difference in the thermal expansions of these parts. If a hard solder, such as silicon gold, is used to bond the pellet to the header, these stresses are transmitted to the silicon pellet. Such stresses often result in pellet fractures. In general, however, lead-tin solder is used to bond the silicon pellet to the copper header. The cyclic thermal stresses then are absorbed by non-elastic deformation of the soft lead solder, and very little stress is transmitted to the pellet.

The continuous flexing that results from cyclic temperature changes may eventually cause fatigue failures in a conventional lead solder system. Such failures are a function of the amount of change in temperature at the mounting interface, the difference in the thermal-expansion coefficients of the silicon pellet and the material to which the pellet is attached, and the maximum dimensions of the mounting interface. Fatigue failures occur whenever the cyclic stresses damage the solder to the point at which the transfer of heat between the pellet and the surface to which it is mounted becomes impaired. This condition, which is indicated by a significant rise in junction-to-case thermal resistance, may exist in only a small portion of the pellet. This portion, however, overheats, and device failure results because of regenerative conditions that lead to thermal runaway.

Thermal-fatigue failures in solid-state power devices are accelerated because of dislocation "pile-ups" that result from impurities in the lead solder. RCA has developed a process that substantially reduces the amount of impurities introduced into the solder. Use of this proprietary **"Controlled Solder Process"** makes it possible to avoid the microcracks that propagate to cause fatigue failure in solid-state devices and, therefore, greatly increases the thermal-cycling capability of these devices.

RCA has devised a rating chart that relates the thermal-cycling capablity of a silicon power transistor to total device dissipation and the change in case temperature. A circuit designer may use the rating system to define the limiting value to which the change in case temperature must be restricted to assure that a power transistor is capable of operation at a specified power dissipation over the number of thermal cycles required in a given application. Conversely, if the power dissipation and the change in case temperature are known, the designer may use the rating system to determine whether the thermal-cycling capability of the transistor is adequate for the application. This thermal-cycling rating system is described in the section **Low- and Medium-Frequency Power Transistors.**

Silicon Rectifiers

S ILICON rectifiers are essentially cells containing a simple p-n junction. As a result, they have low resistance to current flow in one (forward) direction, but high resistance to current flow in the opposite (reverse) direction. They can be operated at ambient temperatures up to 200°C and at current levels as high as hundreds of amperes, with voltage levels greater than 1000 volts. In addition, they can be used in parallel or series arrangements to provide higher current or voltage capabilities. The product matrix shown in Table VI indicates the broad range of current and voltage capabilities and the variety of package configurations that can be selected from the extensive line of RCA silicon rectifiers.

Because of their high forward-to-reverse current ratios, silicon rectifiers can achieve rectification efficiencies greater than 99 per cent. When properly used, they have excellent life characteristics which are not affected by aging, moisture, or temperature. They are small and light in weight and can be made impervious to shock and other severe environmental conditions.

ELECTRICAL CHARACTERISTICS

Fig. 52 shows the basic current-voltage characteristics for a silicon rectifier. As explained earlier in the section **Materials, Junctions, and Devices,** the forward current is many times larger than the re-

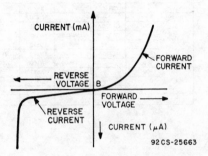

Fig. 52—Current-voltage characteristic of a silicon rectifier.

verse current over the normal operating range of the rectifier. The small reverse (leakage) current gradually rises with an increase in reverse voltage. This increase in reverse current eventually leads to junction breakdown, as indicated by an abrupt increase in reverse current at high reverse voltages. Another important feature of the rectifier characteristic is that the forward voltage drop remains small up to the maximum rated current. The basic characteristic curve shown in Fig. 52 serves as a model in the development of the characteristics data given in the manufacturer's specifications on silicon rectifiers.

Characteristics data given for silicon rectifiers are based on the manufacturer's determination of the inherent qualities and traits of the device. These data, which are usually obtained by direct measurements, provide information that a circuit designer needs to predict the performance capabilities of his circuit and form the basis for the ratings that define the safe operating limits of the rectifier.

Table VI—Rectifier Product Matrix

RCA Rectifiers	Mod. TO-1	DO-1				DO-26			
I_O	0.25A	0.75A	0.75A	1A	1A	0.75A	0.75A Insulated	1A	1A Insulated
I_{FSM}	30A	15A	15A	35A	35A	35A	35A	50A	50A
$V_{RRM(V)}$ 50			1N536		1N2858A				
100	D1300A	1N440B	1N537		1N2859A				
200	D1300B	1N441B	1N538		1N2860A	1N3193	1N3253	1N5211	1N5215
300		1N442B	1N539		1N2861A				
400	D1300D	1N443B	1N540	1N1763A	1N2862A	1N3194	1N3254	1N5212	1N5216
500		1N444B	1N1095	1N1764A	1N2863A				
600		1N445B	1N547		1N2864A	1N3195	1N3255	1N5213	1N5217
800						1N3196	1N3256	1N5214	1N5218
1000							1N3563		
File No.	784	5	3	89	91	41	41	245	245

RCA Rectifiers	DO-15		DO-4		DO-5	
I_O	1A	1.5A	6A	12A	20A	40A
I_{FSM}	30A	50A	160A	240A	350A	800A
$V_{RRM(V)}$ 50	D1201F	1N5391	1N1341B	1N1199A	1N248C	1N1183A
100	D1201A	1N5392	1N1342B	1N1200A	1N249C	1N1184A
200	D1201B	1N5393	1N1344B	1N1202A	1N250C	1N1186A
300		1N5394	1N1345B	1N1203A	1N1195A	1N1187A
400	D1201D	1N5395	1N1346B	1N1204A	1N1196A	1N1188A
500		1N5396	1N1347B	1N1205A	1N1197A	1N1189A
600	D1201M	1N5397	1N1348B	1N1206A	1N1198A	1N1190A
800	D1201N	1N5398				
1000	D1201P	1N5399				
File No.	495	478	58	20	6	38

Table VI—Rectifier Product Matrix (cont'd)

Fast Recovery Types

RCA Rectifiers		DO-26	DO-15	DO-4				DO-5			
I_O		1A	1A	6A	6A	12A	12A	20A	20A	30A	40A
I_{FSM}		35A	50A	75A	125A	150A	250A	225A	300A	300A	700A
$V_{RRM}(V)$	50	D2601F	D2201F	1N3879	D2406F	1N3889	D2412F	1N3899	D2520F	1N3909	D2540F
	100	D2601A	D2201A	1N3880	D2406A	1N3890	D2412A	1N3900	D2520A	1N3910	D2540A
	200	D2601B	D2201B	1N3881	D2406B	1N3891	D2412B	1N3901	D2520B	1N3911	D2540B
	300			1N3882	D2406C	1N3892	D2412C	1N3902	D2520C	1N3912	
	400	D2601D	D2201D	1N3883	D2406D	1N3893	D2412D	1N3903	D2520D	1N3913	D2540D
	500										
	600	D2601M	D2201M		D2406M		D2412M		D2520M		D2540M
	800	D2601N	D2201N								
	1000										
Reverse Recovery Time trr	Typ.	200 ns.	200 ns.	–	200 ns.	–	200 ns.	–	200 ns.	–	200 ns.
	Max.	500 ns.	500 ns.	200 ns.	350 ns.	200 ns.	350 ns.	200 ns.	350 ns.	200 ns.	350 ns.
	File No.	723	629	726	663	727	664	728	665	729	580

For Horizontal-Deflection Circuits

RCA Rectifiers	DO-26			DO-1		DO-15
I_O	0.5A*	1.6A*	1.9A*	–	–	1A
I_{FSM}	30A	70A	70A	70A	30A	50A
Trace			D2601M	D2103SF		D2201M
Commutating		D2601E		D2103S		D2201M
Linearity						D2201B
Regulator						D2201B
Clamp	D2600M				D2101S	
File No.	839	839	839	522	522	629

*$I_{F(RMS)}$ value.

Forward Voltage Drop

The major source of power loss in a silicon rectifier arises from the forward-conduction voltage drop. This characteristic, therefore, is the basis for many of the rectifier ratings.

A silicon rectifier usually requires a forward voltage of 0.4 to 0.8 volt, depending upon the temperature and impurity concentration of the p-n junction, before a significant amount of current flows through the device. As shown in Fig. 53, a slight rise

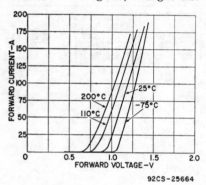

92CS-25664

Fig. 53—Typical forward characteristics of a silicon rectifier.

in the forward voltage beyond this point causes a sharp increase in the forward current. The slope of the voltage-current characteristic at voltages above this threshold value represents the **dynamic resistance** of the rectifier. Losses that result from this resistance characteristic increase as the square of the current and thus increase rapidly at high current levels. The dynamic resistance is dependent upon the construction of the rectifier junction and is inversely proportional to the area of the silicon pellet.

Fig. 53 also shows that, at any reasonable current level, the value of forward voltage required to initiate current flow through the rectifier decreases as the temperature of the rectifier junction increases. This voltage-temperature dependence has

a compensatory effect in rectifiers operated at high currents, but it is a source of difficulty when rectifiers are operated in parallel.

Reverse Current

When a reverse-bias voltage is applied across a silicon rectifier, a limited amount of reverse-blocking current flows through the rectifier. This current is in the order of only a few microamperes, as compared to the milliamperes or amperes of forward current produced when the rectifier is forward-biased. Initially, as shown in Fig. 54, the reverse current increases slightly as the blocking voltage increases, but then tends to remain relatively constant, even though the blocking voltage is increased significantly. Fig. 54 also indicates that an increase in operating temperature causes a substantial increase in reverse current for a given reverse voltage. Reverse-blocking thermal runaway may occur because of this characteristic if the reverse dissipation becomes so large that, as the junction temperature rises, the losses increase faster than the rate of cooling.

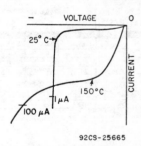

92CS-25665

Fig. 54—Typical reverse characteristics of a silicon rectifier.

If the reverse blocking voltage is continuously increased, it eventually reaches a value (which varies for different types of silicon rectifiers) at which a very sharp increase in reverse current occurs. This voltage is called the **breakdown** or **avalanche** (or **zener**) voltage. Although recti-

fiers can operate safely at the avalanche point, the rectifier may be
destroyed as a result of thermal
runaway if the reverse voltage increases beyond this point or if the
temperature rises sufficiently (e.g.,
a rise in temperature from 25°C to
150°C increases the current by a factor of several hundred).

Reverse Recovery Time

After a silicon rectifier has been
operated under forward-bias conditions, some finite time interval (in
the order of a few microseconds)
must elapse before it can return to
the reverse-bias condition. This reverse-recovery time is a direct consequence of the greatly increased
concentration of charge carriers in
the central region that occurs during forward-bias operation. If the
bias is abruptly reversed, some of
these carriers abruptly change direction and move out in the reverse
direction, and the remainder recombine with opposite-polarity types.
Because there is a finite number of
these carriers in the central region,
and there is no source of additional
charge carriers to replace those
that are removed, the device will
eventually go into the reverse-bias
condition. During the removal
period, however, the charge carriers
constitute a reverse current known
as the **reverse-recovery current.**

Fig. 55 shows the current waveform obtained when a sinusoidal
voltage is applied across a silicon
rectifier. During the positive alternation of the input voltage, the rectifier conducts and accumulates
stored charge. When the supply voltage reverses polarity, the reverse
recovery current flows through the
rectifier until all the stored charge
is removed.

The reverse-recovery time imposes an upper limit on the frequency at which a silicon rectifier
may be used. Any attempt to operate the rectifier at frequencies above
this limit results in a significant decrease in rectification efficiency and

may also cause severe overheating
and resultant destruction of the rectifier because of power losses during
the recovery period.

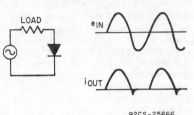

*Fig. 55—Test circuit and output current
waveform obtained when a sinusoidal voltage is applied across a silicon rectifier.*

MAXIMUM RATINGS

Ratings for silicon rectifiers are
determined by the manufacturer on
the basis of extensive testing. These
ratings express the manufacturer's
judgment of the maximum stress
levels to which the rectifiers may be
subjected without endangering the
operating capability of the unit. The
various factors for which silicon
rectifiers must be rated include:
peak reverse voltage, forward current, surge (or fault) current, operating and storage temperatures, amperes squared-seconds, and mounting torque.

Peak Reverse Voltage

Peak reverse voltage (PRV) is
the rating used by the manufacturer
to define the maximum allowable reverse voltage that can be applied
across a rectifier. This rating is less
than the avalanche breakdown level
on the reverse characteristic. With
present-day diffused junctions, the
power dissipation at peak reverse
voltage is a small percentage of the
total losses in the rectifier for operation at the maximum rated current
and temperature levels. The reverse
dissipation may increase sharply,
however, as temperature or blocking
voltage is increased to a point beyond that for which the device is

capable of reliable operation. It is important, therefore, to operate within ratings.

A transient reverse voltage rating may be assigned when it has been determined that increased voltage stress can be withstood for a short time duration provided that the device returns to normal operating conditions when the overvoltage is removed. This condition is illustrated in Fig. 56.

Peak-reverse voltage ratings for single-junction silicon rectifiers range from 50 to 1500 volts and for multiple-junction silicon-rectifier stacks may be as high as several hundred thousands of volts.

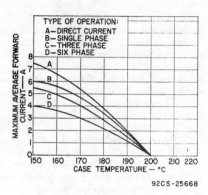

Fig. 57—Current rating chart for a 12-ampere silicon rectifier.

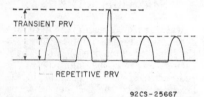

Fig. 56—Typical waveform of repetitive and transient reverse voltages applied across a silicon rectifier.

Forward Current

The current rating assigned to a rectifier is expressed as a maximum value of forward current at a specific case temperature. For these conditions, the power dissipation and internal temperature gradient through the thermal impedance from junction to case are such that the junction is at or near the maximum operating temperature for which the blocking-voltage rating can be maintained. At current levels above this maximum rating, the internal and external leads and terminals of the device may experience excessive temperatures, regardless of the heat sink provided for the pellet itself. The current rating can be described more fully in the form of a curve such as that shown in Fig. 57.

Because the current through a rectifier is not normally a smooth flow, current ratings are usually expressed in terms of average current (I_{avg}), peak current (I_{pk}), and rms current (I_{rms}). Each of these currents may be expressed in terms of the other two currents.

The waveshapes shown in Figs. 58 and 59 help to illustrate the relationships among these ratings. For example, Fig. 58 shows the current variation with time of a sine wave

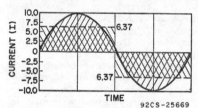

Fig. 58—Variation of current of a sine wave with time.

that has a peak current I_{peak} of 10 amperes. The area under the curve can be translated mathematically into an equivalent rectangle that indicates the average value I_{av} of the sine wave. The relationship between the average and peak values of the total sine-wave current is then given by

$$I_{av} = 0.637\ I_{peak}$$

or

$$I_{peak} = 1.57\ I_{av}$$

However, the power P consumed by a device (and thus the heat generated within it) is equal to the square of the current through it times its finite electrical resistance R (i.e., $P = I^2R$). Therefore, the power is proportional to the square of the current rather than to the peak or average value. Fig. 59 shows the square of the current for the sine wave of Fig. 58. A horizontal line drawn through a point halfway up the I^2 curve indicates the average (or mean) of the squares, and the square root of the I^2 value

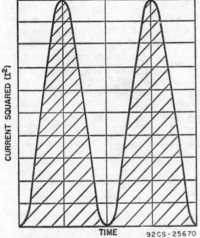

Fig. 59—Variation of the square of sine-wave current with time.

at this point is the root-mean-square (rms) value of the current. The relationship between rms and peak current is given by

$$I_{rms} = 0.707 \, I_{peak}$$

or

$$I_{peak} = 1.414 \, I_{rms}$$

Because a single rectifier cell passes current in one direction only, it conducts for only half of each cycle of an ac sine wave. Therefore, the second half of the curves in Figs. 114 and 115 is eliminated. The average current I_{av} then becomes half of the value determined for full-cycle conduction, and the rms current I_{rms} is equal to the square root of half the

mean-square value for full-cycle conduction. In terms of half-cycle sine-wave conduction (as in a single-phase half-wave circuit), the relationships of the rectifier currents are as shown in Table VII:

Table VII—Rectifier Current Relationships

$$I_{peak} = \pi \times I_{av} = 3.14 \, I_{av}$$
$$I_{av} = (1/\pi) \, I_{peak} = 0.32 \, I_{peak}$$
$$I_{rms} = (\pi/2) \, I_{av} = 1.57 \, I_{av}$$
$$I_{av} = (2/\pi) \, I_{rms} = 0.64 \, I_{rms}$$
$$I_{peak} = 2 \, I_{rms}$$
$$I_{rms} = 0.5 \, I_{peak}$$

For different combinations of rectifier cells and different circuit configurations, these relationships are, of course, changed again. Current (and voltage) relationships have been derived for various types of rectifier applications and are given in the section on **DC Power Supplies**.

Published data for silicon rectifiers usually include maximum ratings for both average and peak forward current. As shown in Fig. 60, the **maximum average forward current** is the maximum average value of current which is allowed to flow in the forward direction during a full ac cycle at a specified ambient or case temperature. Typical average current outputs range from 0.5 ampere to as high as 100 amperes for single silicon diodes. The **peak recurrent forward current** is the maximum repetitive instantaneous forward current permitted under stated conditions.

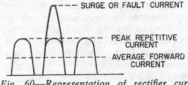

Fig. 60—Representation of rectifier currents.

The dual maximum ratings are required because, under certain conditions (e.g., when a highly capacitive load is used), it is possible for the average current to be low and for the peak current to be high enough to cause overheating of the rectifier. The approximate expression for power losses P in a silicon rectifier, given by the following equation, can be used to explain how this type of operation is possible:

$$P_{watts} = (V_{dc}I_{dc}) + (I_{rms}^2 R_{dyn})$$

where the voltage V_{dc} is 0.4 to 0.8 volt depending upon the junction temperature; the direct current I_{dc} is equivalent to the average current I_{avg}. the current I_{rms} is the true rms current and, for a fixed average current, increases as the peak current increases; and R_{dyn} is the dynamic resistance of the rectifier over the current range considered.

An analysis of the above equation for power losses shows that if the peak current is increased and the conduction time is decreased so that the average current is held constant, the rms current and, therefore, the power dissipated in the rectifier ($I_{rms}^2 R_{dyn}$) are also increased. This behavior explains why the maximum permissible value of average current in multiple-phase circuits is reduced as the number of phases is increased and the conduction period is reduced. Fig. 57 shows the effect of the number of phases on the variation in average current with case temperature.

Surge Current

A third maximum-current limit given in the manufacturer's data on silicon rectifiers is the surge (or **fault**) current rating. During operation, unusually high surges of current may result from inrush current at turn-on, load switching, and short circuits. A rectifier can absorb a limited amount of increased dissipation that results from short-duration high surges of current without any effect except a momentary rise in junction temperature. If the surges become too high, however, the temperature of the junction may be raised beyond the maximum capabil-

ity of the device. The rectifier may then be driven into thermal runaway and, consequently, be destroyed. Fig. 61(a) shows a typical surge-current rating curve for a silicon rectifier.

If the value and duration of anticipated current surges exceed the rating of the rectifier, impedance may be added to the circuit to limit the magnitude of the surge current, or fuses may be used to limit the duration of the surges. In some cases, a rectifier that has an average-current rating higher than that required by the circuit must be used to meet surge requirements of the circuit. This technique eliminates the need for additional circuit impedance elements or special fusing.

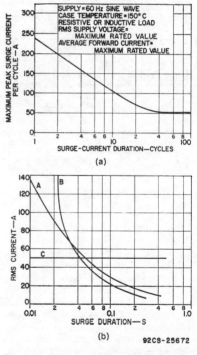

Fig. 61—(a) Peak-surge-current rating chart for a 12-ampere silicon rectifier; (b) coordination chart that relates rectifier surge-current rating (curve A), opening characteristics of circuit fuses (curve B), and maximum available surge current in a circuit (curve C).

If fuses are used to protect the rectifiers, a coordination chart, such as that shown in Fig. 61(b), should be constructed. This chart shows the surge rating of the rectifier (curve A), the opening characteristics of the fuse (curve B), and the maximum surge current available in the circuit (curve C). In the construction of a coordination chart for a particular rectifier, the rms value of the surge current can be obtained from a universal surge-current rating chart, such as that shown in Fig. 62. The opening characteristics of the fuse can be obtained from the manufacturer's published data, and the maximum surge current can be calculated.

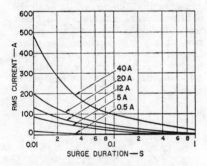

Note: The rms current given by this curve is a partial surge rating and should be added to the normal rms current to determine the total surge rating.

92CS-25673

Fig. 62—Universal surge-current rating chart for RCA silicon rectifiers.

The coordination chart shown in Fig. 61(b) was prepared for a 12-ampere silicon rectifier operated in half-wave service from a 220-volt rms ac source and protected by a fuse having opening characteristics as shown by curve B. If the total short-circuit impedance of all the rectifier elements is determined to be 2.25 ohms, the peak surge current I_s for full-wave operation can be calculated as follows:

$$I_s = \frac{220 \ V_{rms} \times 1.41}{2.25}$$

$$= 137.6 \text{ amperes}$$

For half-wave service, the peak surge current ($I_s = I_{pk}$) can be converted to rms current by use of the relationship given in Table VII, as follows:

$$I_{rms} = \tfrac{1}{2} \ I_{pk}$$

$$= \frac{137.6}{2}, \text{ or } 68.8 \text{ amperes}$$

Curve A of Fig. 61(b), which is merely a reproduction of the 12-ampere curve on the universal rating chart shown in Fig. 62, gives the surge-current rating of the 12-ampere silicon rectifier, but does not consider the normal rms value of current that the rectifier can handle. This normal value of rms current must be subtracted from the total surge current to determine the actual overcurrent of the fault. First, the relationships in Table VII are used to convert the average-current rating of the rectifier to the normal rms value, as follows:

$$I_{rms} = 1.57 \ I_{avg}$$
$$= 1.57 \times 12, \text{ or } 18.8 \text{ amperes}$$

The overcurrent is then determined from the following calculation:

$$I_{surge} - I_{normal} = 68.8 - 18.8,$$
$$\text{or } 50 \text{ amperes}$$

The 50-ampere fault current is represented on the coordination chart in Fig. 61(b) by the straight-line curve C. The 12-ampere rectifier can sustain a fault current of this magnitude for 51 milliseconds, as indicated by the point of intersection of curves A and C. The fuse, however, opens and interrupts the flow of current in the circuit after 43 milliseconds, as indicated by the point of intersection of curves B and C, and the rectifier is protected.

Amperes Squared-Seconds (I²t)

The amperes-squared-seconds (I^2t) rating for silicon rectifiers is a useful figure of merit that provides important information for fuse coordination. This rating indicates the energy, E, required to melt the fusible material of a particular fuse; it is based on the following familiar power relationship:

$$P = I^2R$$

where R is the resistance of the fuse element and I is the rms value of the current in the fuse.

The above equation can be expressed in terms of energy ($E = P t$) as follows:

$$E = I^2Rt$$

The resistance R is constant for a given fuse material; this term is dropped from the equation to obtain the figure of merit I^2t that is directly proportional to the energy required to melt the fuse material.

The I^2t rating for a particular silicon rectifier can be determined directly from the surge-current curves for the device. For example, the I^2t rating for a rectifier operated with a 60-Hz ac input can be calculated from the following relationship:

$$I^2t = \frac{\text{one-cycle surge-current rating}}{2}$$
$$\times 16.67 \times 10^{-3}$$

In this relationship, the factor 16.67×10^{-3} represents the time in seconds for one cycle of operation at 60-Hz, and the peak surge-current value is divided by 2 to obtain the rms value (i.e., $I_{rms} = I_{pk}/2$).

The peak value of surge current that can be sustained by a 12-ampere silicon rectifier is 240 amperes, as indicated by the curves shown in Fig. 61(a). The I^2t rating for the rectifier is calculated as follows:

$$I^2t = \left(\frac{240}{2}\right)^2 \times 16.67 \times 10^{-3}$$

$$= 240 \text{ amperes squared-seconds}$$

For periods of operation less than that for one cycle of a 60-Hz sine wave (i.e., for subcycle time periods), the manufacturer's data does not specify the exact surge-current capability of a rectifier. Under such conditions, a somewhat different procedure is required for fuse coordination. In this procedure, consideration must be given to two important factors.

First, the worst-case condition for fusing results from the application of a square wave of current. If other current waveforms are converted into an equivalent square wave, a conservative fusing rating can be obtained. For example, a half-sine wave of 60-Hz current has a duration of 8.3 milliseconds. The duration of an equivalent square wave having the same peak amplitude is 4.16 milliseconds, as indicated in Fig. 63.

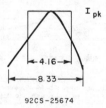

92CS-25674

Fig. 63—Square-wave equivalent of a half sine-wave of 60-Hz current.

Second, because of fundamental differences in thermal and structural characteristics, the surge-current capabilities of solid-state devices and fuses also differ. For fuses, this capability is a constant proportional to I^2t. For a solid-state device this capability is proportional to I^xt, where x is some value between 2 and 3. A safe approximation for a rectifier surge-current failure curve would result from the use of I^3t to define the upper energy limit.

The two factors discussed above are taken into account in the sub-cycle surge-current rating chart shown in Fig. 64. In this chart, the peak surge-current that can be sustained by a rectifier for a subcycle period is normalized by the peak one-cycle surge-current rating (I^2t) and plotted as a cubic function of time. The normalized I^2t value for the single-cycle surge (which is equivalent to a 4.16 millisecond square wave) is assigned the value of 1, and a line that has a slope of 1/3 is drawn through this point (1, 4.16 ms.) on a log-log graph to define the subcycle surge-current capability of a silicon rectifier. Use of this curve is illustrated by the following example, in which the I^2t rating is to be determined for a 1 millisecond duration.

The subcycle surge-current rating curve in Fig. 64 shows that the normalizing factor at 1 millisecond

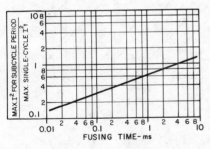

Fig. 64—Subcycle surge-current rating chart.

is 0.62. The 1-millisecond I^2t value is then determined as follows:

$$I^2t_{(1 \text{ ms})} = 0.62 \; I^2t_{(\text{single cycle})}$$

For a 12-ampere silicon rectifier, which has a single-cycle surge capa-

bility of 240 amperes squared-seconds, this value becomes

$$\begin{aligned}
I^2t_{(1 \text{ ms})} &= (0.62) \; 240 \text{ A}^2\text{s} \\
&= 149 \text{ A}^2\text{s}
\end{aligned}$$

For the 12-ampere silicon rectifier, therefore, the I^2t rating for a time period of 1 millisecond is 149 amperes squared-seconds. This value can then be used to calculate the peak square-wave current allowable for a 1-millisecond as follows:

$$\begin{aligned}
I_{pk} \; (1 \text{ ms}) &= \left(\frac{149}{1 \times 10^{-3}} \right)^{1/2} \\
&= \quad 386 \text{ amperes}
\end{aligned}$$

The I^2t fusing rating may be determined by an alternate approach if the peak available short-circuit current is known. Fig. 65 shows a relationship between short-circuit currents and published single-cycle I^2t ratings. This chart, like that shown in Fig. 64, is based on a constant I^2t so that for peak currents that exceed the single-cycle rating, the I^2t capability is inversely proportional to the peak current, as indicated by the following relationship:

$$I^2t = K$$

$$I^2t = K/I$$

As an example of the use of the alternate technique, it is assumed that a device that has a single-cycle peak surge-current rating of 350 amperes is to be used in an application in which a short-circuit current of 400 amperes is available. The 400-ampere fault current is located on the abscissa of Fig. 65. This point is projected vertically upward to intersect the 350-ampere published single-cycle rating. The ordinate of the chart then shows an I^2t rating of 440 amperes squared-seconds. If the same silicon rectifier is used in an application for which a short-circuit current of 700 amperes is available, the fuse should have an I^2t rating of 255 amperes squared-seconds.

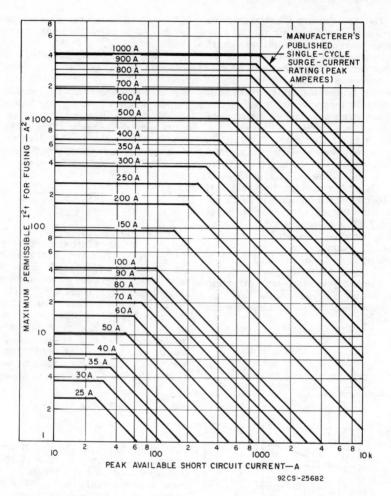

Fig. 65—Maximum permissible I²t as a function of peak available short-circuit current.

SPECIAL TYPES OF SILICON RECTIFIERS

Silicon rectifiers that exhibit unique characteristics can be obtained by special contouring of the basic rectifier junction structure, by use of special processing techniques, or by selective testing. This group includes fast-recovery rectifiers, controlled-avalanche rectifiers, voltage-reference (zener) diodes, and compensating diodes.

Fast-Recovery Rectifiers

In the selection of silicon rectifiers for television high-voltage power supplies, high-speed inverters, switching regulators, and other high-frequency applications, fast recovery characteristics are a major requisite. The actual recovery time of a rectifier is dependent not only upon the structural characteristics of the device, but also upon factors such as the amount of forward current prior to turn-off, the rate of decay of the forward current, the magnitude of the applied reverse voltage, and the junction temperature of the rectifier. A manufacturer's specification for the reverse recovery time of a given rectifier is meaningful, therefore, only if the critical parameters and the actual test circuit used for the measurement are also specified. Table VIII lists and shows significant electrical characteristics for RCA fast-recovery silicon rectifiers.

Table VIII—RCA Fast-Recovery Silicon Rectifiers

RCA TYPE	Forward Current				Voltage V_{RRM} V	Temp. Range Operating °C	Voltage Drop		Rev. Recovery Time		
	RMS $I_{F(RMS)}$ A	Av. I_0 A	Surge I_{FSM} A	Temp.-T_A °C			v_F V	i_F A	t_{rr} µs	I_{FM} A	T_C °C
D2103SF	3	—	70	150●●	750	-30 to 80	1.4	4	0.5	3.14	25
D2103S	3	—	70	150●●	700	-30 to 80	1.4	4	0.5	3.14	25
D2101S	1	—	30	45	700	-30 to 80	1.5	4	0.7	3.14	25
D2201F	1.5	1	50●	100■	50	-40 to 150	1.9	4	0.5	3.14	25
D2201A	1.5	1	50●	100■	100	-40 to 150	1.9	4	0.5	3.14	25
D2201B	1.5	1	50●	100■	200	-40 to 150	1.9	4	0.5	3.14	25
D2201D	1.5	1	50●	100■	400	-40 to 150	1.9	4	0.5	3.14	25
D2201M	1.5	1	50●	100■	600	-40 to 150	1.9	4	0.5	3.14	25
D2201N	1.5	1	50●	100■■	800	-40 to 150	1.9	4	0.5	3.14	25
D2406F	9	6	125	100	50	-40 to 150	1.4	6	0.35	19	25
D2406A	9	6	125	100	100	-40 to 150	1.4	6	0.35	19	25
D2406B	9	6	125	100	200	-40 to 150	1.4	6	0.35	19	25
D2406C	9	6	125	100	300	-40 to 150	1.4	6	0.35	19	25
D2406D	9	6	125	100	400	-40 to 150	1.4	6	0.35	19	25
D2406M	9	6	125	100	600	-40 to 150	1.4	6	0.35	19	25
1N3879	9	6	75	100	50	-65 to 150	1.4	6	0.20	1	25
1N3880	9	6	75	100	100	-65 to 150	1.4	6	0.20	1	25
1N3881	9	6	75	100	200	-65 to 150	1.4	6	0.20	1	25
1N3882	9	6	75	100	300	-65 to 150	1.4	6	0.20	1	25
1N3883	9	6	75	100	400	-65 to 150	1.4	6	0.20	1	25
D2412F	18	12	250	100	50	-40 to 150	1.4	12	0.35	38	25
D2412A	18	12	250	100	100	-40 to 150	1.4	12	0.35	38	25
D2412B	18	12	250	100	200	-40 to 150	1.4	12	0.35	38	25
D2412C	18	12	250	100	300	-40 to 150	1.4	12	0.35	38	35
D2412D	18	12	250	100	400	-40 to 150	1.4	12	0.35	38	25
D2412M	18	12	250	100	600	-40 to 150	1.4	12	0.35	38	25
1N3889	18	12	150	100	50	-65 to 150	1.4	12	0.20	1	25
1N3890	18	12	150	100	100	-65 to 150	1.4	12	0.20	1	25
1N3891	18	12	150	100	200	-65 to 150	1.4	12	0.20	1	25
1N3892	18	12	150	100	300	-65 to 150	1.4	12	0.20	1	25
1N3893	18	12	150	100	400	-65 to 150	1.4	12	0.20	1	25
D2520F	30	20	300	100	50	-40 to 150	1.4	20	0.35	63	25
D2520A	30	20	300	100	100	-40 to 150	1.4	20	0.35	63	25
D2520B	30	20	300	100	200	-40 to 150	1.4	20	0.35	63	25
D2520C	30	20	300	100	300	-40 to 150	1.4	20	0.35	63	25
D2520D	30	20	300	100	400	-40 to 150	1.4	20	0.35	63	25
D2520M	30	20	300	100	600	-40 to 150	1.4	20	0.35	63	25
1N3899	30	20	225	100	50	-65 to 150	1.4	20	0.20	1	25
1N3900	30	20	225	100	100	-65 to 150	1.4	20	0.20	1	25
1N3901	30	20	225	100	200	-65 to 150	1.4	20	0.20	1	25
1N3902	30	20	225	100	300	-65 to 150	1.4	20	0.20	1	25
1N3903	30	20	225	100	400	-65 to 150	1.4	20	0.20	1	25
1N3909	45	30	300	100	50	-65 to 150	1.4	30	0.20	1	25
1N3910	45	30	300	100	100	-65 to 150	1.4	30	0.20	1	25
1N3911	45	30	300	100	200	-65 to 150	1.4	30	0.20	1	25
1N3912	45	30	300	100	300	-65 to 150	1.4	30	0.20	1	25
1N3913	45	30	300	100	400	-65 to 150	1.4	30	0.20	1	25

Table VIII—RCA Fast-Recovery Silicon Rectifiers (cont'd)

RCA TYPE	Forward Current				Voltage V_{RRM} V	Temp. Range Operating °C	Voltage Drop		Rev. Recovery Time		
	RMS $I_{F(RMS)}$ A	Av. I_o A	Surge I_{FSM} A	Temp.-T_C °C			v_F V	i_F A	t_{rr} μs	I_{FM} A	T_C °C
D2540F	60	40	700	165	50	-40 to 150	1.8	100	0.35	125	25
D2540A	60	40	700	165	100	-40 to 150	1.8	100	0.35	125	25
D2540B	60	40	700	165	200	-40 to 150	1.8	100	0.35	125	25
D2540D	60	40	700	165	400	-40 to 150	1.8	100	0.35	125	25
D2540M	60	40	700	165	600	-40 to 150	1.8	100	0.35	125	25
D2601F	1.5	1	35♦	100■	50	-40 to 150	1.9	4	0.5	20	25
D2601A	1.5	1	35♦	100■	100	-40 to 150	1.9	4	0.5	20	25
D2601B	1.5	1	35♦	100■	200	-40 to 150	1.9	4	0.5	20	25
D2601D	1.5	1	35♦	100■	400	-40 to 150	1.9	4	0.5	20	25
D2601M	1.5	1	35♦	100■	600	-40 to 150	1.9	4	0.5	20	25
D2601N	1.5	1	35♦	100■	800	-40 to 150	1.9	4	0.5	20	25

** Junction Temperature

● At Junction Temperature (T_J) = 150 °C

♦ At Junction Temperature (T_J) = 165 °C   ■ Lead Temperature

Recovery-Time Test Circuit—Fig. 66(a) shows a circuit recommended by the JEDEC Committee (JC-22) on Power Rectifiers for use in the measurement of rectifier recovery time. In this circuit, capacitor C is charged during the positive alternation of the input ac voltage. During the negative half-cycle, the silicon controlled rectifier (SCR) is triggered, and capacitor C discharges through inductor L and the rectifier on which the recovery-time measurements is being made. The resultant test-current waveform is shown in Fig. 66(b). Inductor L and capacitor C form a series resonant circuit so that the forward current through the rectifier is very nearly a half sine wave. The peak forward current I_{FM} is specified as π times the average rated value of the half-sine-wave current through the rectifier. The rate of decay of the forward current ($-di/dt$) is specified as the slope of a straight line that passes Regardless of the value of the peak

through the points $I_{FM}/2$ and zero. forward current, the di/dt value is specified as 25 amperes per microsecond for high-power stud-mounted rectifiers and 10 amperes per microsecond for low-power lead-mounted rectifiers. For a true half-sine-wave pulse, the increment from $I_{FM}/2$ to zero represents 30 electrical degrees, or one-sixth the total width of the forward-current pulse. The rate of decay of the forward current, therefore, can be expressed in terms of the over-all pulse width (PW) as follows:

$$\frac{di}{dt} = \frac{I_{FM}/2}{(PW)/6} = \frac{3I_{FM}}{PW}$$

For stud-mounted rectifiers, the di/dt value is specified as 25 amperes per microseconds. The pulse width, therefore, is defined by the following relationship:

$$PW = 3I_{FM}/25 = 0.12I_{FM}$$

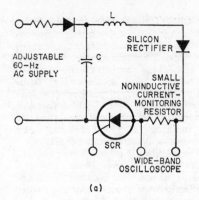

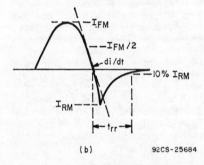

Fig. 66—Test circuit and waveform for rectifier reverse-recovery-time measurement.

The relationships expressed above all assume zero circuit losses. Some adjustment of the calculated values of L and C may be required to compensate for these losses.

For lead-mounted rectifiers, the di/dt value is specified as 10 amperes per microsecond, and the expression for the pulse width becomes

$$PW = 3I_{FM}/10 = 0.3I_{FM}$$

The desired width of the current pulse is obtained by selection of the proper values for L and C in the test circuit. The values of these components are determined from the following relationships:

$$PW = \pi \, (LC)^{1/2}$$
$$C = I_{FM} \, (PW)/\pi \, V_p$$

where V_p is the peak voltage across the capacitor.

A typical practical circuit for measurement of the recovery time of a fast-recovery rectifier is shown in Fig. 67. The diode in parallel with the S6431M SCR in this circuit carries the reverse current through the LC circuit so that the

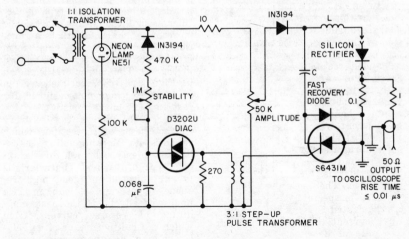

Fig. 67—A typical, practical circuit for measurement of the recovery time of a fast-recovery rectifier.

reverse recovery characteristics of the rectifier are not affected by the reverse recovery characteristic of the SCR.

Many other circuits have been used for measuring reverse recovery time. Fig. 68 shows one method which has been used as the basis for reverse-recovery data by some manufacturers. In this circuit, the forward-current supply and the associated resistors are adjusted to provide a specified value of forward current. The reverse-current supply is adjusted to supply a specified value of reverse-recovery current when switch S is closed. In some cases, the switch S and the reverse-current supply are replaced by a pulse generator.

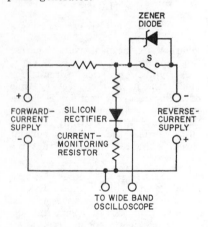

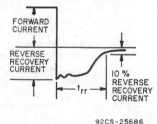

92CS-25686

Fig. 68—Typical circuit used to measure reverse recovery time of a rectifier.

Unfortunately, most of the different methods of measuring reverse recovery time yield widely varying

results. The values obtained depend on many factors, including the magnitude of forward current, the magnitude of reverse-recovery current, the point on the waveform at which recovery time is measured (usually 10 per cent of peak reverse current), and the rate at which forward current decays toward zero (usually a function of circuit layout, stray capacitance, and inductance).

Correlation of reverse-recovery time measurements between equipments at different locations becomes difficult in circuits which produce very rapid rates of change in current (such as the circuit shown in Fig. 68). Seemingly minor changes in circuit layout can have a large effect on the measured reverse-recovery time. The circuit shown in Fig. 66, which uses a half-sine-wave test-current pulse, yields results which are readily reproducible, even with widely differing circuit layouts. For this reason it is considered the most meaningful method of evaluating the reverse-recovery characteristic of a rectifier.

Types of Recovery Characteristics —In a given circuit, three types of recovery characteristics may exist, depending upon the type of rectifier used. Although the exact shape of recovery characteristic is a function of the circuit, the basic form of the characteristic is as shown in Fig. 69. The characteristic shown in Fig. 69(a) is associated with a standard rectifier not designed for fast turn-off characteristics.

Fig. 69(b) shows the recovery characteristic of a rectifier which recovers its blocking-voltage capability suddenly. Alhough this "snap-off" type of turn-off characteristic is an indication of good high-frequency operation, it can produce undesired effects. If the peak magnitude of reverse current from which the rectifier snaps off is relatively high, an appreciable amount of energy is contained in the harmonics generated by the snap-off. If sensitive radio or television re-

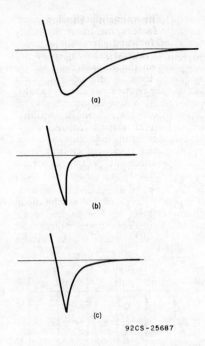

Fig. 69—Rectifier reverse-recovery characteristics: (a) for conventional rectifiers; (b) snap-off fast-recovery characteristic; (c) fast-recovery characteristic without abrupt switch off.

ceiving equipment is in the vicinity, these harmonics can create interference problems. If considerable lead inductance is associated with the rectifier, the snap-off can induce ringing in the lead inductance that may result in circuit malfunction. In some cases, the high di/dt associated with the lead inductance can induce voltages large enough to destroy the rectifier.

The characteristic shown in Fig. 68(c) represents a rectifier which turns off rapidly, but at a somewhat more gradual rate than the rectifier recovery characteristic shown in Fig. 69(b). Reverse-recovery time is only slightly longer than that of the snap-off type, but the generation of harmonics and ringing effects is tremendously reduced. RCA **fast-recovery rectifiers**

are designed to have a reverse-recovery characteristic similar to that shown in Fig. 69(c). This type of characteristic is achieved by use of gold-doping to control the lifetime of minority carriers and a junction geometry designed to prevent abrupt decreases in the peak negative current. Fig. 70 shows the reverse-recovery characteristics of a typical RCA fast-recovery rectifier designed for use in an SCR horizontal-deflection system.

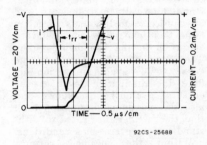

Fig. 70—Reverse-recovery characteristics of an RCA-D2601DF fast-recovery rectifier.

Controlled-Avalanche Rectifiers

Controlled-avalanche types are recommended for silicon-rectifier applications in which the ability to withstand high voltage transients is an important design consideration. In controlled-avalanche rectifiers, the voltage at which avalanching occurs is predetermined during manufacture by precise control of the resistivities (i.e., doping-impurity concentrations) in the junction areas and by careful attention to the geometry of the silicon pellet.

In the manufacture of controlled-avalanche rectifiers, special care is taken to assure exceptional regularity of the silicon pellet and an even distribution of impurities in both the n- and p-type regions of the semiconductor junction. In addition, the edges of the silicon p-n junction are shaped to reduce the intensity of localized electric fields at the junction surface. These conditions assure that breakdown will be uniform across the entire junction area rather than concentrated at weak spots close to the junction surface. Such uniform avalanching, when maintained within acceptable limits, is not destructive.

Extensive tests of controlled-avalanche rectifiers permit precise predictions of the behavior of these devices under high-reverse-voltage conditions. The rectifiers are tested to determine their ability to withstand high-voltage transients and are subjected to life tests to determine their capability for sustained operation in the avalanche region. The slope of the current-voltage curve in the avalanche region defines the dissipation level at the onset of avalanche breakdown and also the maximum dissipation level that the rectifiers are rated to withstand under reverse-bias conditions. Within the specified ratings, controlled-avalanche rectifiers can safely absorb large bursts of energy, such as may result from abrupt switching of inductive circuits.

Zener Diodes

Zener diodes are silicon rectifiers in which the reverse current remains small until the breakdown voltage is reached and then increases rapidly with little further increase in voltage. The breakdown voltage is a function of the diode material and construction, and can be varied from one volt to several hundred volts for various current and power ratings, depending on the junction area and the method of cooling. A stabilized supply can deliver a constant output (voltage or current) unaffected by temperature, output load, or input voltage, within given limits. The stability provided by zener diodes makes them useful as stabilizing devices and as reference sources.

Compensating Diodes

Excellent stabilization of collector current for variations in both supply voltage and temperature can be obtained by the use of a compensating diode operating in the forward direction in the bias network of transistor amplifier or oscillator circuits. Fig. 71(a) shows the transfer characteristics of a transistor; Fig. 71(b) shows the forward characteristics of a compensating diode. In a typical circuit, the diode is biased in the forward direction; the operating point is represented on the diode characteristics by the dashed horizontal line. The diode current at this point determines a bias voltage

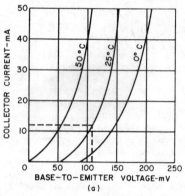

(a)

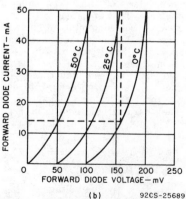

(b)   92CS-25689

Fig. 71—(a) Transfer characteristics of transistor; (b) forward characteristics of compensating diode.

which establishes the transistor idling current. This bias voltage shifts with varying temperature in the same direction and magnitude as the transistor characteristic, and thus provides an idling current that is essentially independent of temperature.

The use of a compensating diode also reduces the variation in transistor idling current as a result of supply-voltage variations. Because the diode current changes in proportion with the supply voltage, the bias voltage to the transistor changes in the same proportion and idling-current changes are minimized. (The use of diode compensation is discussed in more detail under "Biasing" in the section on **Low- and Medium-Frequency Power Transistors.**

OVERLOAD PROTECTION

In the application of silicon rectifiers, it is necessary to guard against both over-voltage and over-current (surge) conditions. A voltage surge in a rectifier arrangement can be caused by dc switching, reverse recovery transients, transformer switching, inductive-load switching, and various other causes. The effects of such surges can be reduced by the use of a capacitor connected across the input or the output of the rectifier. In addition, the magnitude of the voltage surge can be reduced by changes in the switching elements or the sequence of switching, or by a reduction in the speed of current interruption by the switching elements.

In all applications, a rectifier having a more-than-adequate peak reverse voltage rating should be used. The safety margin for reverse voltage usually depends on the application. For a single-phase half-wave application using switching of the transformer primary and having no transient suppression, a rectifier having a peak reverse voltage three or four times the expected working voltage should be used. For a full-wave bridge using load switching and having adequate suppression of transients, a margin of 1.5 to 1 is generally acceptable.

Because of the small size of the silicon rectifier, excessive surge currents are particularly harmful to rectifier operation. Current surges may be caused by short circuits, capacitor inrush, dc overload, or failure of a single rectifier in a multiple arrangement. In the case of low-power units, fuses or circuit breakers are often placed in the ac input circuit to the rectifier to interrupt the fault current before it damages the rectifier. When circuit requirements are such that service must be continued in case of failure of an individual rectifier, a number of rectifiers can be used in parallel, each with its own fuse. Additional fuses should be used in the ac line and in series with the load for protection against dc load faults. In high-power rectifiers, an arrangement of circuit breakers, fuses, and series resistances is often used to reduce the amplitude of the surge current. Fusing requirements can be determined by use of coordination charts for the particular circuits and rectifiers used.

SERIES AND PARALLEL ARRANGEMENTS

Silicon rectifiers can be arranged in series or in parallel to provide higher voltage or current capabilities, respectively, as required for specific applications.

A parallel arrangement of rectifiers can be used when the maximum average forward current required is larger than the maximum current rating of an individual rectifier. In such arrangemnts, however, some means must be provided to assure proper division of current through the parallel rectifiers. Parallel rectifier arrangements are not in general use. Designers normally use a polyphase arrangement to provide higher currents, or simply substitute the readily available higher-current rectifier types.

Series arrangements of silicon rectifiers are used when the applied reverse voltage is expected to be greater than the maximum peak reverse voltage rating of a single silicon rectifier (or cell). For example, four rectifiers having a maximum reverse voltage rating of 200 volts each could be connected in series to handle an applied reverse voltage of 800 volts.

In a series arrangement, the most important consideration is that the applied voltage be divided equally across the individual rectifiers. If the instantaneous voltage is not uniformly divided, one of the rectifiers may be subjected to a voltage greater than its specified maximum reverse voltage, and, as a result, may be destroyed. Uniform voltage division can usually be assured by connection of either resistors or capacitors in parallel with individual rectifiers. Shunt resistors are used in steady-state applications, and shunt capacitors in applications in which transient voltages are expected. Both resistors and capacitors should be usd if the circui is to be exposed to both dc and ac components. When only a few rectifiers are in series, multiple transformer windings may be used, each winding supplying its own assembly consisting of one series diode. The outputs of the diodes are then connected in series for the desired voltage.

Low- and Medium-Frequency Power Transistors

A p-n junction biased in the reverse direction is equivalent to a high-resistance element (low current for a given applied voltage), while a junction biased in the forward direction is equivalent to a low-resistance element (high current for a given applied voltage). Because the power developed by a given current is greater in a high-resistance element than in a low-resistance element ($P = I^2R$), power gain can be obtained in a structure containing two such resistance elements if the current flow is not materially reduced. As explained in the section **Materials, Junctions, and Devices**, a device containing two p-n junctions biased in opposite directions is called a junction or bipolar transistor.

Fig. 72 shows such a two-junction device biased to provide power gain. The thick end layers of the transistor are made of the same type of material (n-type in this case),

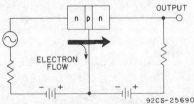

Fig. 72—An n-p-n structure biased for power gain.

and are separated by a very thin layer of the opposite type of material (p-type in the device shown).

By means of the external batteries, the left-hand (n-p) junction is biased in the forward direction to provide a low-resistance input circuit, and the right-hand (p-n) junction is biased in the reverse direction to provide a high-resistance output circuit.

Electrons flow easily from the left-hand n-type region to the center p-type region as a result of the forward biasing. Most of these electrons diffuse through the thin p-type region, however, and are attracted by the positive potential of the external battery across the right-hand junction. In practical devices, approximately 95 to 99.5 percent of the electron current reaches the right-hand n-type region. This high percentage of currrent penetration provides power gain in the righ-resistance output circiut and is the basis for transistor amplification capability.

The operation of p-n-p devices is similar to that shown for the n-p-n device, except that the bias-voltage polarities are reversed, and electron-current flow is in the opposite direction.

DESIGN AND FABRICATION

The ultimate aim of all transistor fabrication techniques is the construction of two parallel p-n junctions with controlled spacing between the junctions and controlled impurity levels on both sides of each junction. A variety of structures has been developed in the course of transistor evolution.

The earliest transistors made were of the point-contact type. In this type of structure, two pointed wires were placed next to each other on an n-type block of semiconductor material. The p-n junctions were formed by electrical pulsing of the wires. This type has been superseded by junction transistors, which are fabricated by various alloy diffusion, and crystal-growth techniques.

In grown-junction transistors, the impurity element of the semiconductor material is changed during the growth of the original crystal ingot to provide the p-n-p or n-p-n regions. The grown crystal is then sliced into a large number of small-area devices, and contacts are made to each region of the devices. Fig. 73(a) shows a cross-section of a grown-junction transistor.

In alloy-junction transistors, two small "dots" of a p-type or n-type impurity element are placed on opposite sides of a thin wafer of n-type or p-type semiconductor material, respectively, as shown in Fig. 73(b). After proper heating, the impurity "dots" alloy with the semiconductor materials to form the regions for the emitter and collector junctions. The base connection in this structure is made to the original semiconductor wafer.

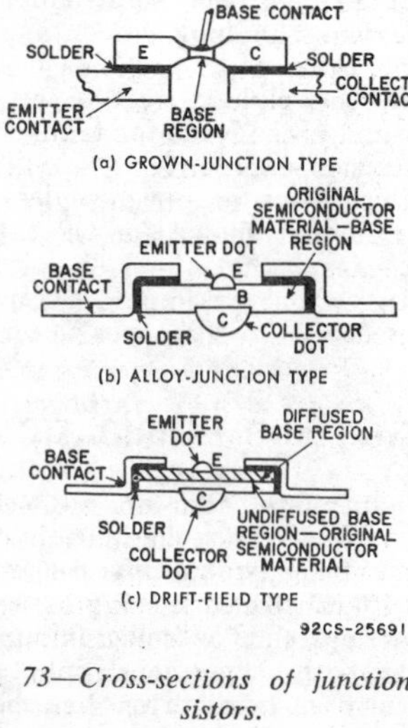

The drift-field transistor is a modified alloy-junction device in which the impurity concentration in the base wafer is diffused or graded, as shown in Fig. 73(c). Two advantages are derived from this structure: (a) the resultant built-in voltage or "drift field" speeds current flow, and (b) the ability to use a heavy impurity concentration in the vicinity of the emitter and a light concentration in the vicinity of the collector makes it possible to minimize capacitive charging times. Both these advantages lead to a substantial extension of the frequency performance over the alloy-junction device.

The **diffused-junction** transistor represents a major advance in transistor technology because increased control over junction spacings and impurity levels makes possible significant improvements in transistor performance capabilities. A cross-section of a **single-diffused** "hometaxial" structure is shown in Fig. 74(a). Hometaxial transistors are fabricated by simultaneous diffusion of impurity from each side of a homogeneously doped base wafer. A **mesa** or flat-topped peak is etched on one side of the wafer in an intricate design to define the transistor emitter and expose the base region for connection of metal contacts. Large amounts of heat can be dissipated from a hometaxial structure through the highly conductive solder joint between the semiconductor material and the device package. This structure provides a very low collector resistance.

Double-diffused transistors have an additional degree of freedom for selection of the impurity levels and junction spacings of the base, emitter, and collector. This structure provides high voltage capability through a lightly doped collector region without compromise of the junction spacings which determine device fre-

(a) GROWN-JUNCTION TYPE

(b) ALLOY-JUNCTION TYPE

(c) DRIFT-FIELD TYPE

92CS-25691

Fig. 73—Cross-sections of junction transistors.

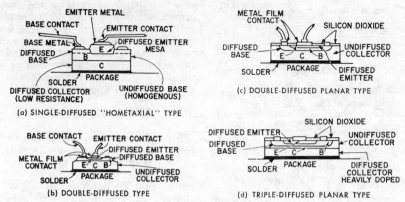

Fig. 74—Cross-sections of diffused transistors.

92CS-25692

quency response and other important characteristics. Fig. 74(b) shows a typical double-diffused transistor; the emitter and base junctions are diffused into the same side of the original semiconductor wafer, which serves as the collector. A mesa is usually etched through the base region to reduce the collector area at the base-to-collector junction and to provide a stable semiconductor surface.

Double-diffused **planar** transistors provide the added advantage of protection or passivation of the emitter-to-base and collector-to-base junction surfaces. Fig. 74(c) shows a typical double-diffused planar transistor. The base and emitter regions terminate at the top surface of the semiconductor wafer under the protection of an insulating layer. Photolithographic and masking techniques are used to provide for diffusion of both base and emitter impurities in selective areas of the semiconductor wafer.

In **triple-diffused** transistors, a heavily doped region diffused from the bottom of the semiconductor wafer effectively reduces the thickness of the lightly doped collector region to a value dictated only by electric-field considerations. Thus, the thickness of the lightly doped or high-resistivity portion of the col-

lector is minimized to obtain a low collector resistance. A section of a triple-diffused planar structure is shown in Fig. 74(d).

Epitaxial transistors differ from diffused structures in the manner in which the various regions are fabricated. Epitaxial structures are grown on top of a semiconductor wafer in a high-temperature reaction chamber. The growth proceeds atom by atom, and is a perfect extension of the crystal lattice of the wafer on which it is grown. In the epitaxial-base transistor shown in Fig. 75(a) a lightly doped base region is deposited by epitaxial techniques on a heavily doped collector wafer of opposite-type dopant. Photolithographic and masking techniques and a single impurity diffusion are used to define the emitter region. This structure offers the advantages of low collector resistance and easy control of impurity spacings and emitter geometry. A variation of this structure uses two epitaxial layers. A thin lightly doped epitaxial layer used for the collector is deposited over the original heavily doped semiconductor wafer prior to the epitaxial deposition of the base region. The collector epitaxial layer is of opposite-type dopant to the epitaxial base layer. This structure, shown in Fig. 75(b), has the added

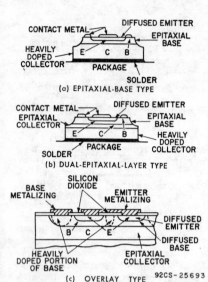

Fig. 75—Cross-sections of epitaxial transistors.

advantage of higher voltage ratings provided by the epitaxial collector layer.

The **overlay** transistor is a double-diffused epitaxial device which employs a unique emitter structure. A large number of separate emitters are tied together by diffused and metalized regions to increase the emitter edge-to-area ratio and reduce the charging-time constants of the transistor without compromise of current- and power-handling capability. Fig. 75(c) shows a section through a typical overlay emitter region.

After fabrication, individual transistor chips are mechanically separated and mounted on individual headers. Connector wires are then bonded to the metalized regions, and each unit is encased in plastic or a hermetically sealed enclosure. In power transistors, the wafer is usually soldered or alloyed to a solid metal header to provide for high thermal conductivity and low-resistance collector contacts, and low-resistance contacts are soldered or metal-bonded from the emitter or

base metalizing contacts to the appropriate package leads. This packaging concept results in a simple structure that can be readily attached to a variety of circuit heat sinks and can safely withstand power dissipations of hundreds of watts and currents of tens of amperes.

BASIC CIRCUITS

Bipolar transistors are ideal current amplifiers. When a small signal current is applied to the input terminals of a bipolar transistor, an amplified reproduction of this signal appears at the output terminals. Although there are six possible ways of connecting the input signal, only three useful circuit configurations exist for current or power amplification: common-base, common-emitter, and common-collector. In the **common-base** (or grounded-base) connection shown in Fig. 76, the signal is introduced into the emitter-base circuit and extracted from the collector-base circuit. (Thus the base element of the transistor is common to both the input and output circuits). Because the input or emitter-base circuit has a low impedance (resistance plus reactance) in the order of 0.5 to 50 ohms, and the output or collector-base circuit has a high impedance in the order of 1000 ohms to one megohm, the voltage or power gain in this type of configuration may be in the order of 1500.

The direction of the arrows in Fig. 21 indicates electron current flow. As stated previously, most of the cur-

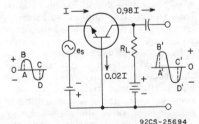

Fig. 76—Common-base circuit configuration.

rent from the emitter flows to the collector; the remainder flows through the base. In practical transistors, from 95 to 99.5 per cent of the emitter current reaches the collector. The current gain of this configuration, therefore, is always less than unity, usually in the order of 0.95 to 0.995.

The waveforms in Fig. 76 represent the input voltage produced by the signal generator e_s and the output voltage developed across the load resistor R_L. When the input voltage is positive, as shown at AB, it opposes the forward bias produced by the base-emitter battery, and thus reduces current flow through the n-p-n transistor. The reduced electron current flow through R_L then causes the top point of the resistor to become less negative (or more positive) with respect to the lower point, as shown at A'B' ont the output waveform. Conversely, when the input signal is negative, as at CD, the output signal is also negative, as at C'D'. Thus, the phase of the signal remains unchanged in this circuit, i.e., there is no voltage phase reversal between the input and the output of a common-base amplifier.

In the **common-emitter** (or grounded-emitter) connection shown in Fig. 77, the signal is introduced into the base-emitter circuit and extracted from the collector-emitter circuit. This configuration has more moderate input and output impedances than the common-base circuit. The input (base-emitter) impedance is in the range of 20 to 5000 ohms, and the output (collector-emitter) impedance is about 50 to 50,000 ohms. Power gains in the order of 10,000 (or approximately 40 dB) can be realized with this circuit because it provides both current gain and voltage gain.

Current gain in the common-emitter configuration is measured between the base and the collector, rather than between the emitter and the collector as in the common-base circuit. Because a very small change in base current produces a relatively large change in collector current, the

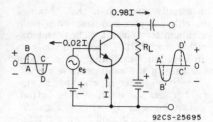

Fig. 77—Common-emitter circuit configuration.

current gain is always greater than unity in a common-emitter circuit; a typical value is about 50.

The input signal voltage undergoes a phase reversal of 180 degrees in a common-emitter amplifier, as shown by the waveforms in Fig. 77. When the input voltage is positive, as shown at AB, it increases the forward bias across the base-emitter junction, and thus increases the total current flow through the transistor. The increased electron flow through R_L then causes the output voltage to become negative, as shown at A'B'. During the second half-cycle of the waveform, the process is reversed, i.e., when the input signal is negative, the output signal is positive (as shown at CD and C'D'.)

The third type of connection, shown in Fig. 78, is the **common-collector** (or grounded-collector) circuit. In this configuration, the signal is introduced into the base-collector circuit and extracted from the emitter-collector circuit. Because the input impedance of the transistor is high and the output impedance low in this connection, the voltage gain is less than unity and the power gain

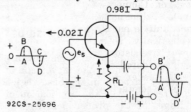

92CS-25696

Fig. 78—Common-collector circuit configuration.

is usually lower than that obtained in either a common-base or a common-emitter circuit. The common-collector circuit is used primarily as an impedance-matching device. As in the case of the common-base circuit, there is no phase reversal of the signal between the input and the output.

The circuits shown in Figs. 76 through 78 are biased for n-p-n transistors. When p-n-p transistors are used, the polarities of the batteries must be reversed. The voltage phase relationships, however, remain the same.

BASIC TRANSISTOR CHARACTERISTICS

THE term "characteristic" is used to identify the distinguishing electrical features and values of a transistor. These values may be shown in curve form or they may be tabulated. When the characteristics values are given in curve form, the curves may be used for the determination of transistor performance and the calculation of additional transistor parameters.

Characteristics values are obtained from electrical measurements of transistors in various circuits under certain definite conditions of current and voltage. Static **characteristics** are obtained with dc potentials applied to the transistor electrodes. **Dynamic characteristics** are obtained with an ac voltage on one electrode under various conditions of dc potentials on all the electrodes. The dynamic characteristics, therefore, are indicative of the performance capabilities of the transistor under actual working conditions.

Current-Voltage Relationships

The currents in a transistor are directly related to the movement of minority carriers in the base region that results from the application of voltages of the proper polarities to the emitter-base and collector-base junctions. A definite mutual relationship exists between the transistor currents and the voltages applied to the transistor terminals. Graphical representations of the variations in transistor currents with the applied voltages provide an excellent indication of the operation of a transistor under different biasing conditions. Transistor manufacturers usually provide curves of current-voltage characteristics to define the operating characteristics of their devices. Such curves are provided for either common-emitter or common-base transistor connections. Fig. 79 shows the bias-voltage polarities and the current components for both common-emitter and common-base connections of a p-n-p transistor. For an n-p-n transistor, the polarities of the voltages and the directions of the currents are reversed.

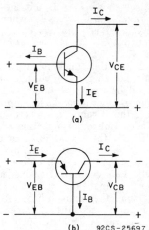

Fig. 79—*Transistor bias-voltage polarities and current components for (a) the common-emitter connection and (b) the common-base connection.*

The common-emitter connection, shown in Fig. 79(a), is the more widely used in practical applications. In this connection, the emitter is the common point between the input (base) and output (collector) circuits, and large current gains are realized by use of a small base current to control a much larger emitter-to-collector current. The common-base connection, shown in Fig. 79(b), differs from the common-emitter connection in that the voltages applied to the transistor are referred to the base rather than to the emitter.

Published data for transistors include both electrode characteristic curves and transfer characteristic curves. These curves present the same information, but in two different forms to provide more useful data. Because transistors are used most often in the common-emitter configuration, characteristic curves are usually shown for the collector or output electrode. The **collector-characteristic curve** is obtained by varying collector-to-emitter voltage and measuring collector current for different values of base current. The **transfer-characteristic curve** is obtained by varying the base-to-emitter (bias) voltage or current at a specified or constant collector voltage, and measuring collector current.

Fig. 80 shows the input (transfer) and output (collector) current-voltage characteristics of a typical p-n-p transistor in a common-base connection. The input characteristic curves, in Fig. 80(a), show the base current as a function of the emitter-to-base voltage for different values of collector-to-base voltage. For any given value of collector voltage, the base current varies with the emitter-to-base voltage in a manner similar to that of any forward-biased narrow p-n junction. After the initial interval required for the forward-bias voltage to overcome the energy barrier at the junction, the current-voltage relationship is almost linear. The slight effect that the collector voltage has on the base current results because variations in the collector voltage change the effective width of the base.

The effective base width is the distance between the edges of the opposing (collector and emitter) depletion layers. This distance is significantly less than the actual thickness of the n-type base material

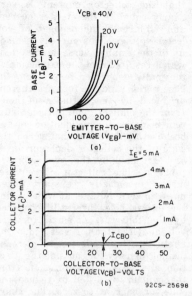

92CS-25698

Fig. 80—Common-base characteristics of a typical p-n-p transistor.

because both depletion layers penetrate into the base region. Because of the small forward bias applied to the emitter-base junction, the emitter depletion layer is much narrower than that of the collector-base junction, and it does not change appreciably with the small range of variation allowed in the emitter-to-base voltage. The collector-junction depletion layer, however, varies markedly with the much larger changes in collector voltage and causes an attendant change in the effective base width. An increase in collector voltage causes a slight increase in base current (for a constant emitter-to-base voltage) because the hole gradient in the p-n-p transistor increases as the effective base width decreases. Many other transistor characteristics are also affected by changes in collector voltage because of the dependence of the effective base width on the collector depletion layer.

The common-base output characteristics, shown in Fig. 80(b), reveal that the collector current is very nearly equal to the emitter current and is largely independent of collector voltage. This current is made up of the small reverse current I_{CBO}, which results from the extraction of thermally generated holes from the base region, and the forward current produced by the diffusion of holes from the emitter. The diffusion current is almost independent of voltage because all the holes that reach the edge of the collector depletion layer are extracted by the reverse-biased junction. The collector current remains essentially constant, even down to zero voltage, at which point the excess holes are still extracted by the collector. A small forward voltage (less than 1 volt) must be applied to the collector to increase the hole density just outside the junction and oppose the diffusion from the emitter in order to reduce the collector current to zero. The characteristic curves show that the collector current varies extremely rapidly with voltage in this region.

The common-emitter characteristics of a p-n-p transistor, shown in Fig. 81, are similar to those of the common-base connection. There are, however, several important differences, the most marked being the small magnitude of the base current which replaces the emitter current in both sets of curves. The base current consists of two components, each of which is in the order of microamperes. These components include the current produced by an inward flow of electrons to replace those lost in the hole injection and diffusion mechanisms, and a current that is largely the result of the extraction of thermally generated holes making up the leakage current of the collector. The total base current, therefore, is just sufficient to make up the losses in the current transfer from emitter to collector. If the base current is increased (by means of the external circuit), more holes can diffuse from emitter to collector, and a considerable increase in the collector current occurs.

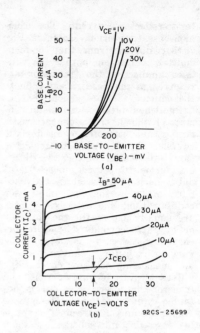

Fig. 81—Common-emitter characteristics of a typical p-n-p transistor.

Physically, this condition is produced by increased hole injection and, therefore, by an increase in the emitter-to-base voltage required to produce an increase in base current. Although changes in base current and voltage must occur together, it is preferable to think of the transistor as a current-controlled device; for this reason, the common-emitter characteristics are usually shown for constant base currents, as indicated in Fig. 81. The curves in this figure confirm that small changes in base current produce much larger changes in collector current.

A final point to note about the common-emitter characteristics is that the current falls rapidly to zero for small collector-to-emitter voltages. The collector-to-emitter voltage is normally divided between the two junctions to provide a small forward bias for the emitter-base junction and a much larger reverse bias for the collector-base junction. If the emitter-to-collector voltage is reduced to the "saturation" value, which is a fraction of a volt, the collector reaches zero bias; below this value, it becomes slightly forward-biased. This condition occurs near the knee of the characteristics. Slightly below the saturated condition, holes continue to flow into the collector, in spite of its forward bias. If the emitter-to-collector voltis reduced further, the collector current falls rapidly to zero.

The input characteristics for the common-emitter connection are similar to those for the common-base connection, except for the greatly reduced magnitude of the base current compared with the emitter current. For the base current to be zero, the emitter junction must have a small forward bias; as a result, all the input characteristics cross the current axis. The collector voltage again affects the characteristics by varying the effective base width, and thus the current gain. With a constant emitter hole density (V_{BE} constant), the base current falls as the collector voltage is raised, and the base transport becomes more efficient.

Current-Gain Parameters

Power gain in transistor circuits is usually obtained by use of a small control signal to produce larger signal variations in the output current. The gain parameter most often specified is the current gain (β) from the base to the collector. The power gain of a transistor operated in a common-emitter configuration is equal to the square of the current gain β times the ratio of the load resistance r_L to the input resistance r_{in}, as indicated in Fig. 82.

Although the input resistance r_{in} affects the power gain, as shown by the equations given in Fig. 82, this parameter is not usually specified directly in the published data on transistors because of the large number of components of which it is

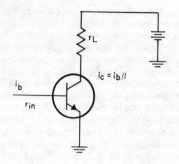

INPUT CURRENT	$= i_b$
INPUT VOLTAGE	$= i_b\, r_{in}$
OUTPUT CURRENT	$= i_c = i_b\beta$
OUTPUT VOLTAGE	$= i_c r_L = i_b\beta r_L$
INPUT POWER	$= i_b^2\, r_{in}$
OUTPUT POWER	$= i_c^2\, r_L = i_b^2\, \beta^2\, r_L$
POWER GAIN	$=$ power output/power input
	$= i_b^2\beta^2 r_L / i_b^2 r_{in}$
	$= \beta^2 r_L / r_{in}$ 92CS-25700

Fig. 82—Test circuit and simplified power-gain calculation for a transistor operated in a common-emitter configuration.

comprised. In general, the input impedance is expressed as a maximum base-to-emitter voltage V_{BE} under specified input-current conditions.

A measure of the current gain of a transistor is its **forward current-transfer ratio**, i.e., the ratio of the current in the output electrode to the current in the input electrode. Because of the different ways in which transistors may be connected in circuits, the forward current-transfer ratio is specified for a particular circuit configuration.

The current gain (or current transfer ratio) of a transistor is expressed by many symbols; the following are some of the most common, together with their particular shades of meaning:

1. beta (β)—general term for current gain from base to collector (i.e., common-emitter current gain)

2. alpha (α)—general term for current gain from emitter to collector (i.e., common-base current gain)

3. h_{fe}—ac gain from base to collector (i.e., ac beta)

4. h_{FE}—dc gain from base to collector. (i.e., dc beta)

Common-base current gain, α, is the ratio of collector current to emitter current (i.e., $\alpha = I_C/I_E$). Although α is slightly less than unity, circuit gain is realized as a result of the large differences of input (emitter-base) and output (collector-base) impedances. The input impedance is small because the emitter-base junction is forward-biased, and the output impedance is large because the collector-base junction is reverse-biased.

Common-emitter current gain, β, is the ratio of collector current to base current (i.e., $\beta = I_C/I_B$). Useful values of β are normally greater than ten.

In the common-base circuit shown in Fig. 76 the emitter is the input electrode and the collector is the output electrode. The dc alpha, therefore, is the ratio of the steady-state collector current I_C to the steady-sate emitter current I_E:

$$\alpha = \frac{I_C}{I_E} = \frac{0.98\ I}{I} = 0.98$$

In the common-emitter circuit shown in Fig. 77, the base is the input electrode and the collector is the output electrode. The dc beta, therefore, is the ratio of the steady-state collector current I_C to the steady-state base current I_B:

$$\beta = \frac{I_C}{I_B} = \frac{0.98\ I}{0.02\ I} = 49$$

Because the ratios given above are based on steady-state currents, they are properly called dc alpha and dc beta. It is more common, however, for the current-transfer ratio to be given in terms of the ratio of signal currents in the input and output electrodes, or the ratio of a change in the output current to the input signal current which causes the change. Fig. 83 shows typical electrode currents in a common-emitter circuit (a) under no-

signal conditions and (b) with a one-microampere signal applied to the base. The signal current of one microampere in the base causes a change of 49 microamperes (147-98) in the collector current. Thus the ac beta for the transistor is 49.

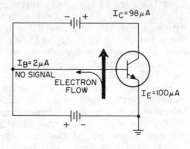

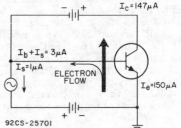

92CS-25701

Fig. 83—Electrode currents under (a) no-signal and (b) signal conditions.

The current-gain parameters α and β are determined by three basic transistor parameters (emitter efficiency γ, collector efficiency α^*, and base transport factor β_0) that are directly dependent upon the physical properties of the device. The following equations express the dependence of the current-gain parameters α and β on the three physical parameters:

$$\alpha = \gamma\beta_0\alpha^*$$

$$\beta = \frac{\gamma\beta_0\alpha^*}{1 - (\gamma\beta_0\alpha^*)}$$

The collector efficiency α^* is generally very near unity and this value is usually assumed for this parameter in the calculations for α and β.

Emitter Efficiency

Emitter Efficiency—In a forward-biased n-p-n transistor, electrons from the emitter diffuse into the base, and holes from the base diffuse into the emitter. The total emitter current I_E is the sum of the hole-current component I_p and the electron-current component I_n and, therefore, may be expressed as follows:

$$I_E = I_p + I_n$$

The potential collector current, I_C, is the difference between the drift currents and is given by

$$I_C = I_n - I_p$$

The holes that diffuse from the base into the emitter originate in the base dc supply and add to the total base current. This hole current I_p, however, does not contribute to the collector current and, in effect, represents a loss in current gain that is directly attributable to poor emitter injection efficiency. The loss in current gain can be held to a minimum if the resistivity of the base is made much greater than that of the emitter so that the number of free holes in the base available to diffuse into the emitter is substantially smaller than the number of free electrons in the emitter available to diffuse into the base.

Base Transport Factor—If high emitter efficiency is assumed, the electrons injected from the emitter into the base diffuse to the collector junction. Some of these electrons, however, recombine with free holes in the base and, in effect, are annihilated. Base current must flow to replenish the free electrons used in the recombination process so that the emitter-to-base forward bias is maintained. In other words, charge neutrality must prevail.

The base transport factor β_0 is an indication of the extent of recombination that takes place in the base region of a transistor. For a high value of the base transport factor β_0, the lifetime of holes in the base (a function of the property of the material) must be long, or the time necessary for the electrons to reach the collector must be short. Any reduction in the time required for the electrons to reach the collector requires a decrease in base width or an increase in the accelerating field used to speed the holes through.

The value of the base transport factor should be in the order of 0.98 for the transistor to provide useful gain.

Transconductance

Extrinsic transconductance may be defined as the quotient of a small change in collector current divided by the small change in emitter-to-base voltage producing it, under the condition that other voltages remain unchanged. Thus, if an emitter-to-base voltage change of 0.1 volt causes a collector-current change of 3 milliamperes (0.003 ampere) with other voltages constant, the transconductance is 0.003 divided by 0.1, or 0.03 mho. (A "mho" is the uint of conductance, and was named by spelling "ohm" backward.) For convenience, a millionth of a mho, or a micromho (μmho), is used to express transconductance. Thus, in the example, 0.03 mho is 30,000 micromhos.

Cutoff Frequencies

For all transistors, there is a frequency f at which the output signal cannot properly follow the input signal because of time delays in the transport of the charge carriers. The three principal cut-off frequencies, shown in Fig. 84, may be defined as follows:

1. The **base cut-off frequency** $f_{\alpha b}$ is that frequency at which alpha (α) is down 3 dB from the low-frequency value.

2. The **emitter cut-off frequency** $f_{\alpha e}$ is that frequency at which beta (β) is down 3 dB from the low-frequency value.

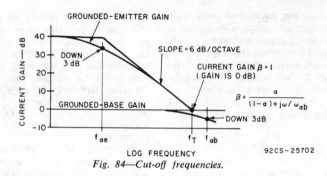

Fig. 84—Cut-off frequencies.

3. The frequency f_T is that frequency at which beta theoretically decreases to unity (i.e., 0-dB gain) with a theoretical 6-dB-per-octave fall off. This term, which is a useful figure of merit for transistors, is referred to as the **gain-bandwidth product**.

The frequency f_T is related to the time delays in a transistor by the following expression:

$$f_T \approx \frac{1}{2\tau \ \Sigma \ t_d}$$

where Σt_d is the sum of the emitter-delay time constant t_e, the base transit time t_b, the collector depletion-layer transit time t_{xm}, and the collector-delay time constant t_c.

The gain-bandwidth product f_T is the term that is generally used to indicate the high-frequency capability of a transistor. Other parameters that critically affect high-frequency performance are the capacitance or resistance which shunts the load and the input impedance, the effect of which is shown by the equations given in Fig. 82.

The base and emitter cut-off frequencies and the gain-bandwidth product of a transistor provide an approximate indication of the useful frequency range of the device, and help to determine the most suitable circuit configuration for a particular application.

The specification of all the characteristics which affect high-frequency performance is so complex that often a manufacturer does not specify all the parameters, but instead specifies transistor performance in an rf-amplifier circuit. This information is very useful when the transistor is operated under conditions very similar to those of the test circuit, but is difficult to apply when the transistor is used in a widely different application. Some manufacturers also specify transistor performance characteristics as a function of frequency, which alleviates these problems. (High-frequency power transistors are discussed in more detail in a later section of this Manual.)

Cutoff Currents

Cutoff currents are small steady-state reverse currents which flow when a transistor is biased into non-conduction. They consist of **leakage currents**, which are related to the surface characteristics of the semiconductor material, and **saturation currents**, which are related to the impurity concentration in the material and which increase with increasing temperatures. Collector-cutoff current is the steady-state current which flows in the reverse-biased collector-to-base circuit when the emitter-to-base circuit is open. Emitter-cutoff current is the current which flows in the reverse-biased emitter-to-base circuit when the collector-to-base circuit is open.

In the common-base configuration, the collector reverse (leakage) current, I_{CBO}, is measured with the emitter circuit open. The presence of the second junction, however, still affects the level of the current because the emitter acquires a small negative bias when it is open-circuited. This bias reduces the hole gradient at the collector and causes the reverse current to decrease. This current, therefore, is much smaller with the emitter open than it is when the emitter-base junction is short-circuited. In the characteristic curves shown in Fig. 80(b), the magnitude of the current I_{CBO} is considerably exaggerated. This current is normally in the order of microamperes or less, although it is often increased by excess surface currents in the same way as the reverse current of any p-n junction.

The reverse current increases with collector voltage, and may lead to avalanche breakdown at high voltages, as described for the silicon rectifier.

The common-emitter reverse collector current I_{CEO}, measured with zero input current ($I_B = O$ in this case), is very much larger than the reverse collector current I_{CBO} in the common-base connection. When the base current is zero, the emitter current adjusts itself so that the losses in the hole-injection and diffusion mechanisms are exactly balanced by the supply of excess electrons left in the vicinity of the collector by hole extraction. For this condition, the collector current is equal to the emitter current.

The common-emitter reverse collector current I_{CEO} increases with collector voltage, unlike the common-base reverse collector current I_{CBO}. This behavior is another consequence of the variation in the effective base width with collector voltage. The narrower the effective base region, the more efficient is the transfer of current from emitter to collector. The more efficient base transport with the higher collector voltage permits a higher emitter current to flow before the losses are again balanced by the supply of elec-

trons from the vicinity of the collector.

Breakdown Voltages

Transistor breakdown voltages define the voltage values between two specified electrodes at which the crystal structure changes and current begins to rise rapidly. The voltage then remains relatively constant over a wide range of electrode currents. Breakdown voltages may be measured with the third electrode open, shorted, or biased in either the forward or the reverse direction. For example, Fig. 85 shows a series of collector-characteristic curves for different base-bias conditions. It can

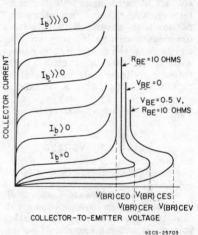

Fig. 85—Typical collector-characteristic curves showing locations of various breakdown voltages.

be seen that the collector-to-emitter breakdown voltage increases as the base-to-emitter bias decreases from the normal forward values through zero to reverse values. The symbols shown on the abscissa are sometimes used to designate collector-to-emitter breakdown voltages with the base open $V_{(BR)CEO}$, with external base-to-emitter resistance $V_{(BR)CER}$, with the base shorted to the emitter $V_{(BR)CES}$, and with a reverse base-to-emitter voltage $V_{(BR)CEV}$.

As the resistance in the base-to-emitter circuit decreases, the collector characteristic develops two breakdown points, as shown in Fig. 85. After the initial breakdown, the collector-to-emitter voltage decreases with increasing collector current until another breakdown occurs at a lower voltage. This minimum collector-to-emitter breakdown voltage is called the **sustaining voltage**.

Punch-Through Voltage

Punch-through (or reach-through) voltage defines the voltage value at which the depletion region in the collector region passes completely through the base region and makes contact at some point with the emitter region. This "reach-through" phenomenon results in a relatively low-resistance path between the emitter and the collector, and causes a sharp increase in current. Punch-through voltage does not result in permanent damage to a transistor, provided there is sufficient impedance in the power-supply source to limit transistor dissipation to safe values.

Saturation Voltage

The curves at the left of Fig. 85 show typical collector characteristics under normal forward-bias conditions. For a given base input current, the collector-to-emitter saturation voltage is the minimum voltage required to maintain the transistor in full conduction (i.e., in the saturation region). Under saturation conditions, a further increase in forward bias produces no corresponding increase in collector current. Saturation voltages are very important in switching applications, and are usually specified for several conditions of electrode currents and ambient temperatures.

Effect of Temperature on Transistor Characteristics

The characteristics of transistors vary with changes in temperature.

In view of the fact that most circuits operate over a wide range of environments, a good circuit design should compensate for such changes so that operation is not adversely affected by the temperature dependence of the transistors.

Current Gain—The effect of temperature on the gain of a silicon transistor is dependent upon the level of the collector current, as shown in Fig. 86. At the lower current levels, the current-gain parameter h_{FE} increases with temperature. At higher currents, however, h_{FE}

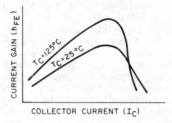

92CS-25705

Fig. 86—Current gain as a function of collector current at different temperatures.

may increase or decrease with a rise in temperature because it is a complex function of many components.

Base-to-Emitter Voltage—Fig. 87 shows the effect of changes in temperature on the base-to-emitter voltage (V_{BE}) of silicon transistors. Two

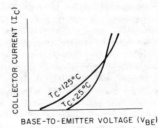

92CS-25706

Fig. 87—Collector current as a function of base-to-emitter voltage at different temperatures.

factors, the base resistance (r_{bb}') and the height of the potential barrier at the base-emitter junction (V_{BE}'), influence the behavior of the base-to-emitter voltage. As the temperature rises, material resistivity increases; as a result, the value of the base resistance r_{bb}' becomes greater. The barrier potential V_{BE}' of the base-emitter junction, however, decreases with temperature. The following equation shows the relationship between the base-to-emitter voltage and the two temperature-dependent factors:

$$V_{BE} = I_B\ r_{bb}' + V_{BE}'$$

$$= \frac{I_C}{h_{FE}}\ r_{bb}' + V_{BE}'$$

As indicated by this equation, the base-to-emitter voltage diminishes with a rise in temperature for low values of collector current, but tends to increase with a rise in temperature for higher values of collector current.

Collector-to-Emitter Saturation Voltage—The collector-to-emitter saturation voltage $V_{CE}(sat)$ is affected primarily by collector resistivity (ρ_C) and the amount by which the natural gain of the device (h_{FE}) exceeds the gain with which the circuit drives the device into saturation. This latter gain is known as the forced gain (h_{FEf}).

At lower collector currents, the natural h_{FE} of a transistor increases with temperature, and the IR drop in the transistor is small. The collector-to-emitter saturation voltage, therefore, diminishes with increasing temperature if the circuit continues to maintain the same forced gain. At higher collector currents, however, the IR drop increases, and gain may decrease. This decrease in gain causes the collector-to-emitter saturation voltage to increase and possibly to exceed the room-temperature (25°C) value. Fig. 88 shows the effect of temperature on the collector-to-emitter saturation voltage.

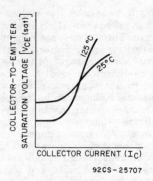

92CS-25707

Fig. 88—Collector current as a function of collector-to-emitter saturation voltage at different temperatures.

Collector Leakage Currents—Reverse collector current I_R is a resultant of three components, as shown by the following equation:

$$I_R = I_D + I_G + I_S$$

Fig. 89 shows the variations of these components with temperature.

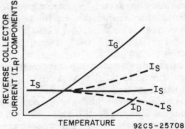

Fig. 89—Reverse collector current as a function of temperature.

The diffusion or saturation current I_D is a result of carriers that diffuse to the collector-base junction and are accelerated across the depletion region. This component is small until temperatures near 175°C are reached. The component I_G results from charge-generated carriers that are created by the flow of diffusion carriers across the depletion region. This component increases rapidly with temperature. I_D and I_G are referred to as bulk leakages. The term I_S represents surface leakage which is caused by local inversion, channeling, ions, and moisture. This leakage component is dependent on many factors, and its variations with changes in temperature are difficult to predict.

At low temperatures, either surface or bulk leakage can be the dominant leakage factor, particularly in transistors that employ a mesa structure. At high temperatures, charge-generated carriers and diffusion current are the major causes of leakage in both mesa and planar transistor structures; the current I_G, therefore, is the dominant leakage component. Because of the dominance of surface leakage I_S at low temperatures and the fact that this leakage may vary either directly or inversely with temperature, it is not possible to define a constant ratio of the leakage current at low temperatures to that at high temperatures. In view of the fact that power transistors are normally operated at high junction temperatures, it is more meaningful to compare the leakage characteristics of both mesa and planar transistors at high temperatures. The relative reliability of different types of power transistors, which is in no way related to the magnitude of low-temperature leakage current, is also best compared at high temperatures.

Leakage currents are important because they affect biasing in amplifier applications and represent the off condition for transistors used in switching applications. The symbol I_R used in the preceding discussion represents any of several different leakage currents commonly specified by transistor manufacturers. The most basic specification is I_{CBO}, which indicates the leakage from collector to base with the emitter open. This leakage is simply the reverse current of the collector-to-base diode.

In addition to the I_{CBO} value, I_{CEV}, I_{CEO}, and I_{CER} specifications are often given for transistors. I_{CEV} is the leakage from the collector to emitter with the base-emitter junction re-

verse-biased. I_{CER} is the leakage current from the collector to the emitter with the base and emitter connected by a specified resistance. I_{CEO} is the leakage current from collector to emitter with the base open. I_{CEV} differs from I_{CBO} only very slightly and in most transistors the two parameters can be considered equal. (This equality is not maintained in symmetrical transistors.) I_{CEO} is simply the product of I_{CBO} at the voltage specified and the h_{FE} of the transistor at a base current equal to I_{CBO}. I_{CEO} is of course the largest leakage current normally specified. I_{CER} is intermediate in value between I_{CEV} and I_{CEO}.

MAXIMUM RATINGS

All solid-state devices undergo irreversible changes if their temperature is increased beyond some critical limit. A number of ratings are given for power transistors, therefore, to assure that this critical temperature limit will not be exceeded on even a very small part of the silicon chip. The ratings for power transistors normally specify the maximum voltages, maximum current, maximum and minimum operating and storage temperatures, and maximum power dissipation that the transistor can safely withstand.

Voltage Ratings

Maximum voltage ratings are normally given for both the collector and the emitter junctions of a transistor. A V_{BEO} rating, which indicates the maximum base-to-emitter voltage with the collector open, is usually specified. The collector-junction voltage capability is usually given with respect to the emitter, which is used as the common terminal in most transistor circuits. This capability may be expressed in several ways. A V_{CEO} rating specifies the maximum collector-to-emitter voltage with the base open; a V_{CER} rating for this voltage implies that the base is returned to the emitter through a specified resistor; a V_{CES} rating gives the maximum voltage when the base is shorted to the emitter; and a V_{CEV} rating indicates the maximum voltage when the base is reverse-biased with respect to the emitter by a specified voltage. A V_{CEX} rating may also be given to indicate the maximum collector-to-emitter voltage when a resistor and voltage are both connected between base and emitter.

If a maximum voltage rating is exceeded, the transistor may "break down" and pass current in the reverse direction. The breakdown across the junction is usually not uniform, and the current may be localized in one or more small areas. The small area becomes overheated unless the current is limited to a low value, and the transistor may then be destroyed.

The collector-to-base or emitter-to-base breakdown (avalanche) voltage is a function of the resistivity or impurity doping concentration at the junction of the transistor and of the characteristics of the circuit in which the transistor is used. When there is a breakdown at the junction, a sudden rise in current (an "avalanche") occurs. In an abruptly changing junction, called a step junction, the avalanche voltage is inversely proportional to the impurity concentration. In a slowly changing junction, called a graded junction, the avalanche voltage is dependent upon the rate of change of the impurity concentration (grade constant) at the physical junction. Fig. 90 shows the two types of junction breakdowns. The basic transistor voltage-breakdown mechanisms and their relationship to external circuits are the basis for the various types of voltage ratings used by transistor manufacturers.

The collector voltage can be limited below its avalanche breakdown value if the depletion layer (space-charge region) associated with the applied collector voltage expands through the thin base width and contacts the emitter junction.

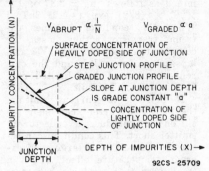

Fig. 90—Step-junction and graded-junction breakdown.

The doping in the base (under the emitter) and the base width in relation to the magnitude of applied voltage govern whether punch-through occurs before avalanche. Higher doping concentrations and wider bases increase punch-through voltage V_{PT} in accordance with the following relationship:

$$V_{PT} = \frac{qNW^2}{2kE_o}$$

where q is the electronic charge, N is the doping-level concentration in the base, W is the base width, k is the dielectric constant, and E_o is the permittivity of free space (kE_o is approximately 1×10^{-12} farad per centimeter for silicon).

Current and Temperature Ratings

The physical mechanisms related to basic transistor action are temperature-sensitive. If the bias is not temperature-compensated, the transistor may develop a regenerative condition, known as **thermal runaway**, in which the thermally generated carrier concentration approaches the impurity carrier concentration. [Experimental data for silicon show that, at temperatures up to 700°K, the thermally generated carrier concentration n_i is deter-

mined as follows: $n_i = 3.87 \times 10^{16}$ $\times$ T $\times$ (3/2) exp $(-1.21/2kT)$.]
When this condition becomes extreme, transistor action ceases, the collector-to-emitter voltage V_{CE} collapses to a low value, and the current increases and is limited only by the external circuit.

If there is no current limiting, the increased current can melt the silicon and produce a collector-to-emitter short. This condition can occur as a result of a large-area average temperature effect, or in a small area that produces hot spots or localized thermal runaway. In either case, if the intrinsic temperature of a semiconductor is defined as the temperature at which the thermally generated carrier concentration is equal to the doped impurity concentration, the absolute maximum temperature for transistor action can be established.

The intrinsic temperature of a semiconductor is a function of the impurity concentration, and the limiting intrinsic temperature for a transistor is determined by the most lightly doped region. It must be emphasized, however, that the intrinsic temperature acts only as an upper limit for transistor action. The maximum operating junction temperature and the maximum current rating are established by additional factors such as the efficiency of heat removal, the yield point and melting point of the solder used in fabrication, and the temperature at which permanent changes in the junction properties occur.

The **maximum current rating** of a transistor indicates the highest current at which, in the manufacturer's judgment, the device is useful. This current limit may be established by setting an arbitrary minimum current gain or may be determined by the fusing current of an internal connecting wire. A current that exceeds the rating, therefore, may result in a low current gain or in the destruction of the transistor.

The basic materials in a silicon transistor allow transistor action at temperatures greater than 300°C. Practical transistors, however, are limited to lower temperatures by mounting systems and surface contamination. If the **maximum rated storage** or **operating temperature** is exceeded, irreversible changes in leakage current and in current-gain characteristics of the transistor result.

Power-Dissipation Ratings

A transistor is heated by the electrical power dissipated in it. A maximum power rating is given, therefore, to assure that the temperature in all parts of a transistor is maintained below a value that will result in detrimental changes in the device. This rating may be given with respect to case temperature (for transistors mounted on heat sinks) or with respect to "free-air ambient" temperature. Case temperature is measured with a small thermocouple or other low-heat-conducting thermometer attached to the outside of the case or preferably inserted in a very small blind hole in the base so that the measurement is taken as close to the transistor chip as possible. Very short pulses of power do not heat the transistor to the temperature which it would attain if the power level was continued indefinitely. Ratings of maximum power consider this factor and allow higher power dissipation for very short pulses.

The dissipation in a transistor is not uniformly distributed across the semiconductor wafer. At higher voltages, the current concentrations become more severe, and hot spots may be developed within the transistor pellet. As a result, the power-handling capability of a transistor is reduced at high voltages. The power rating of a transistor may be presented most easily by a limiting curve that indicates a peak-power safe operating region. This curve shows power-handling capability as a function of voltage for various time durations.

BIASING

For most non-switching applications, the operationg point for a particular transistor is established by the quiescent (dc, no-signal) values of collector voltage and emitter current. In general, a transistor may be considered as a current-operated device, i.e., the current flowing in the emitter-base circuit controls the current flowing in the collector circuit. The voltage and current values selected, as well as the particular biasing arrangement used, depend upon both the transistor characteristics and the specific requirements of the application.

As mentioned previously, biasing of a transistor for most applications consists of forward bias across the emitter-base junction and reverse bias across the collector-base junction. In Figs. 76, 77, and 78, two batteries were used to establish bias of the correct polarity for an n-p-n transistor in the common-base, common-emitter, and common-collector circuits, respectively. Many variations of these basic circuits can also be used. (In these simplified dc circuits, inductors and transformers are represented only by their series resistance.)

Basic Methods

A simplified biasing arrangement for the common-base circuit is shown in Fig. 91. Bias for both the collector-base junction and the emitter-base junction is obtained from the single battery through the voltage-divider network consisting of resistors R_2 and R_3. (For the n-p-n transistor shown in Fig. 91(a) the emitter-base junction is forward-biased because the emitter is negative with respect to the base, and the collector-base junction is reverse-biased because the collector is positive with respect to the base, as shown. For the p-n-p transistor shown in Fig. 91(b), the polarity of the battery and of the electrolyte bypass capacitor C_1 is

reversed.) The electron current I from the battery and through the voltage divider causes a voltage drop across resistor R_2 which biases the base. The proper amount of current then flows through R_1 so that the correct emitter potential is established to provide forward bias relative to the base. This emitter current establishes the amount of collector current which, in turn, causes a voltage drop across R_4. Simply stated, the voltage divider consisting of R_2 and R_3 establishes the base potential; the base potential essentially establishes the emitter potential; the emitter poten-

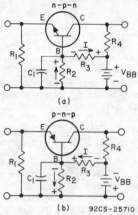

Fig. 91—Biasing network for common-base circuit for (a) n-p-n and (b) p-n-p transistors.

tial and resistor R_1 establish the emitter current; the emitter current establishes the collector current; and the collector current and R_4 establish the collector potential. R_2 is bypassed with capacitor C_1 so that the base is effectively grounded for ac signals.

A single battery can also be used to bias the common-emitter circuit. The simplified arrangement shown in Fig. 92 is commonly called "fixed bias". In this case, both the base and the colelctor are made positive with respect to the emitter by means of the battery. The base resistance R_B is then selected to provide the desired base current I_B for the transistor (which, in turn, establishes the desired emitter current I_E), by means of the following expression:

$$R_B = \frac{V_{BB} - V_{BE}}{I_B}$$

where V_{BB} is the battery supply voltage and V_{BE} is the base-to-emitter voltage of the transistor.

In the circuit shown, for example, the battery voltage is six volts. The

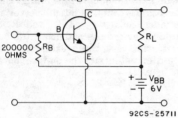

Fig. 92—"Fixed-bias" arrangement for common-emitter circuit.

value of R_B was selected to provide a base current of 27 microamperes, as follows:

$$R_B = \frac{6 - 0.6}{27 \times 10^{-6}} = 20,000 \text{ ohms}$$

The fixed-bias arrangement shown in Fig. 92, however, is not a satisfactory method of biasing the base in a common-emitter circuit. The critical base current in this type of circuit is very difficult to maintain under fixed-bias conditions because of variations between transistors and the sensitivity of these devices to temperature changes. This problem is partially overcome in the "self-bias" arrangement shown in Fig. 93. In this circuit, the base resistor is

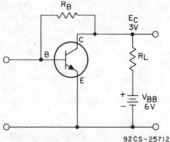

Fig. 93—"Self-bias" arrangement for common-emitter circuit.

tied directly to the collector. This connection helps to stabilize the operating point because an increase or decrease in collector current produces a corresponding decrease or increase in base bias. The value of R_B is then determined as described above, except that the collector voltage V_{CE} is used in place of the supply voltage V_{BB}:

$$R_B = \frac{V_{CE} - V_{BE}}{I_B}$$

$$= \frac{3 - 0.6}{27 \times 10^{-6}} = 90,000 \text{ ohms}$$

The arrangement shown in Fig. 93 overcomes many of the disadvantages of fixed bias, although it reduces the effective gain of the circuit.

In the bias method shown in Fig. 94 the voltage-divider network composed of R_1 and R_2 provides the

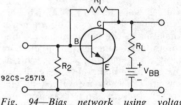

Fig. 94—Bias network using voltage-divider arrangement for increased stability.

required forward bias across the base-emitter junction. The value of the base bias voltage is determined by the current through the voltage divider. This type of circuit provides less gain than the circuit of Fig. 93, but is commonly used because of its inherent stability.

The common-emitter circuits shown in Figs. 95 and 96 may be used to provide stability and yet minimize loss of gain. In Fig. 95, a resistor R_E is added to the emitter circuit, and the base resistor R_2 is returned to the positive terminal of the battery instead of to the collector. The emitter resistor R_E provides additional stability. It is bypassed with capacitor C_E. The value of C_E depends on he lowest frequency to be amplified.

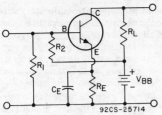

Fig. 95—*Bias network using emitter stabilizing resistor.*

In Fig. 96, the R_2R_3 voltage-divider network is split, and all ac feedback currents through R_3 are shunted to ground (bypassed) by capacitor C_1.

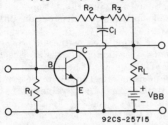

Fig. 96—*Bias network using split voltage-divider network.*

The value of R_3 is usually larger than the value of R_2. The total resistance of R_2 and R_3 should equal the resistance of R_1 in Fig. 94.

In practical circuit applications, any combination of the arrangements shown in Figs. 93, 94, 95, and 96 may be used. However, the stability of Figs. 93, 94, and 96 may be poor unless the voltage drop across the lad resistor R_L is at least one-third the value of the supply voltage. The determining factors in the selection of the biasing circuit are usually gain and bias stability (which is discussed later).

In many cases, the bias network may include special elements to compensate for the effects of variations in ambient temperature or in supply voltage. For example, the **thermistor** (temperature-sensitive resistor) shown in Fig. 97(a) is used to compensate for the rapid increase of collector current with increasing temperature. Because the thermistor resistance decreases as the temperature increases, the emitter-to-base bias voltage is reduced and the collector current tends to remain constant. The addition of the shunt and series resistances provides most effective acompensation over a desired temperature range.

The **diode biasing network** shown in Fig. 97(b) stabilizes collector current for variations in both temperature and supply voltage. The forward-biased diode current determines a bias voltage which establishes the transistor **idling** current (collector current under no-signal conditions). As the temperature increases, this bias voltage decreases. Because the transistor characteristic also shifts in the same direction and magnitude, however, the idling current remains essentially independent of temperature. Temperature stabilization with a properly designed diode network is substantially better than that provided by most thermistor bias networks. Any temperature-stabilizing element should be thermally close to the transistor being stabilized.

In addition, the diode bias current varies in direct proportion with changes in supply voltage. The resultant change in bias voltage is small, however, so that the idling current also changes in direct pro-

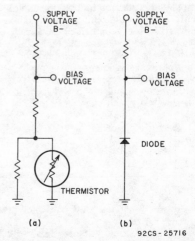

Fig. 97—*Bias networks including (a) a thermistor and (b) a voltage-compensating diode.*

portion to the supply voltage. Supply-voltage stabilization with a diode biasing network reduces current variation to about one-fifth that obtained when resistor or thermistor bias is used for a germanium transistor and one-fifteenth for a silicon transistor.

The bias networks of Figs. 92 through 96 are generally used in class A circuits. Class B circuits normally employ the bias networks shown in Fig. 97. The bias resistor values for class B circuits are generally much lower than those for class A circuits.

Bias Stability

Because transistor currents tend to increase with temperature, it is necessary in the design of transistor circuits to include a "stability factor" to keep the collector-current variation within tolerable values under the expected high-temperature operating conditions. The bias stability factor SF is expressed as the ratio between a change in steady-state collector current and the corresponding change in steady-state collector-cutoff current.

For a given set of operating voltages, the stability factor can be calculated for a maximum permissible rise in steady-state collector current from the room-temperature value, as follows:

$$SF = \frac{I_{Cmax} - I_{C1}}{I_{CBO2} - I_{CBO1}}$$

where I_{C1} and I_{CBO1} are measured at 25°C, I_{CBO2} is measred at the maximum expected ambient (or junction) temperature, and I_{Cmax} is the maximum permissible collector current for the specified collector-to-emitter voltage at the maximum expected ambient (or junction) temperature (to keep transistor dissipation within ratings).

The calculated values of SF can then be used, together with the appropriate values of beta and r_b' (base-connection resistance), to determine suitable resistance values for the transistor circuit. Fig. 98 shows equations for SF in terms of resistance values for three typical circuit configurations. The maximum value which SF can assume is the value of beta. Although this analysis was originally made for germanium transistors, in which the collector saturation current I_{Co} is relatively large, the same type of analysis may be applied to interchangeability with beta for silicon transistors.

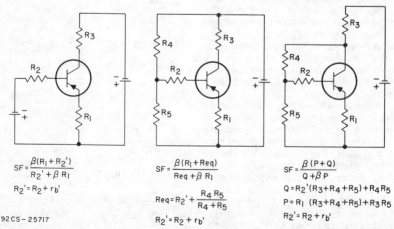

$$SF = \frac{\beta(R_1 + R_2')}{R_2' + \beta R_1}$$
$$R_2' = R_2 + r_b'$$

$$SF = \frac{\beta(R_1 + Req)}{Req + \beta R_1}$$
$$Req = R_2' + \frac{R_4 R_5}{R_4 + R_5}$$
$$R_2' = R_2 + r_b'$$

$$SF = \frac{\beta(P + Q)}{Q + \beta P}$$
$$Q = R_2'(R_3 + R_4 + R_5) + R_4 R_5$$
$$P = R_1(R_3 + R_4 + R_5) + R_3 R_5$$
$$R_2' = R_2 + r_b'$$

92CS–25717

Fig. 98—Bias-stability-factor equations for three typical circuit configurations.

COUPLING METHODS

Three basic methods are used to couple transistor stages: transformer, resistance-capacitance, and direct coupling.

The major advantage of **transformer coupling** is that it permits power to be transferred from one impedance level to another. A transformer-coupled common-emitter n-p-n stage is shown in Fig. 99. The voltage step-down transformer T_1 couples the signal from the collector of the preceding stage to the base of the common-emitter stage. The voltage loss inherent in this transformer is not significant in transistor circuits because, as mentioned previously, the transistor is a current-operated device. Although the voltage is stepped down, the available current is stepped up. The change in base current resulting from the presence of the signal causes an alternating collector current to flow in the primary winding of transformer T_2, and a power gain is obtained between T_1 and T_2.

This use of a voltage step-down transformer is similar to that in the output stage of an audio amplifier, where a **step-down transformer** is normally used to drive the loudspeaker, which is also a current-operated device.

The voltage-divider network consisting of resistors R_1 and R_2 in Fig. 99 provides bias for the transistor. The voltage divider is bypassed by capacitor C_1 to avoid signal attenuation. The stabilizing emitter resistor R_E permits normal variations of the transistor and circuit elements to be compensated for automatically without adverse effects. This resistor R_E is bypassed by capacitor C_2. The voltage supply V_{BB} is also bypassed, by capacitor C_3, to prevent feedback in the event that ac signal voltages are developed across the power supply. Capacitors C_1 and C_2 may normally be replaced by a single capacitor connected between the emitter and the bottom of the secondary winding of transformer T_1, with little change in performance.

The use of **resistance-capacitance coupling** usually permits some economy of circuit costs and reduction of size, with some accompanying sacrifice of gain. This method of coupling is particularly desirable in low-level, low-noise audio amplifier stages to minimize hum pickup from stray magnetic fields. Use of resistance-capacitance (RC) coupling in battery-operated equipment is usually limited to low-power operation. The frequency response of an RC-coupled stage is normally better than that of a transformer-coupled stage.

Fig. 100 shows a two-stage RC-coupled circuit using n-p-n transistors in the common-emitter configuration. The method of bias is similar to that used in the transformer-coupled circuit of Fig. 99. The major additional components are the collector load resistances R_{L1} and R_{L2} and the coupling capacitor C_c. The value of C_c must be made fairly large, in the order of 2 to 10 microfarads, because of the small input and load resistances involved. (It should be noted that electrolyte capacitors are normally used for coupling in transistor audio circuits. Polarity must be observed, therefore, to obtain proper circuit operation. Occasionally, excessive leakage current through an electrolyte coupling capacitor may adversely affect transistor operating currents.)

Impedance coupling is a modified form of resistance-capacitance coupling in which inductances are used

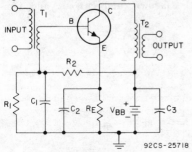

92CS-25718

Fig. 99—Transformer-coupled common-emitter stage.

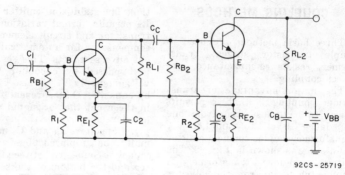

Fig. 100—Two-stage resistance-capacitance coupled circuit.

to replace the load resistors. This type of coupling is rarely used except in special applications where supply voltages are low and cost is not a significant factor.

Direct coupling is used primarily when cost is an important factor. (It should be noted that direct-coupled amplifiers are not inherently dc amplifiers, i.e., that they cannot always amplify dc signals. Low-frequency response is usually limited by other factors than the coupling network.) In the direct-coupled amplifier shown in Fig. 101, resistor R_3 serves as both the collector load resistor for the first stage and the bias resistor for the second stage. Resistors R_1 and R_2 provide circuit stability similar to that of Fig. 94 because the emitter voltage of transistor Q_2 and the collector voltage of transistor Q_1 are within a few tenths of a volt of each other.

Because so few circuit parts are required in the direct-coupled ampli-fier, maximum economy can be achieved. However, the number of stages which can be directly coupled is limited. Temperature variation of the bias current in one stage may be amplified by all the stages, and severe temperature instability may result.

SPECIAL CONSIDERATIONS FOR POWER TRANSISTORS

In power transistors, the main design consideration is power-handling capability. This capability is determined by the maximum junction temperature a transistor can withstand and how quickly the heat can be conducted away from the junction.

In general, the basic physical theory that defines the behavior of any bipolar transistor in relation to charge-carrier interactions, current gain, frequency capabilities, voltage breakdown, and current and tem-

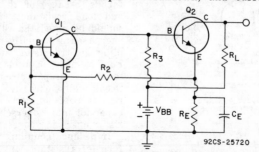

Fig. 101—Two-stage direct-coupled circuit.

perature ratings is not significantly different for power types. Power transistors, however, must be capable of large current densities and are required to sustain large voltage fields. For power types, therefore, the basic transistor theory must be expanded to include the effect that these conditions have on the physical behavior of the devices. In addition, the physical capabilities of power transistors must be defined in terms of factors, such as second-breakdown energy levels, safe operating area, and thermal-cycling stresses, that are not usually considered for small-signal types.

Limiting Phenomena in Power Transistors

Power transistors have many features in common with small-signal transistors, but what distinguishes power transistors is their ability to handle currents up to a few hundred amperes, voltages up to a few thousand volts, large power dissipations, and drastic current and voltage surges.

Several phenomena which are of minor significance in small-signal bipolar transistors become important as currents and voltages are increased, and ultimately they limit performance. These phenomena include: base widening, emitter debiasing, second breakdown, high-voltage surface effects, and thermal fatigue.

Base widening—Base widening occurs in all bipolar transistors having a lightly doped collector region. This effect, shown in Fig. 102, is caused by the mobile carrier flow within the unit, which modifies the electric field distribution so that the effective positions of the base-collector junction and the collector depletion region are different from their positions when there is no current flow.

Consider an n⁺-p-n-n⁺ transistor with no current flowing and a fixed base-collector reverse bias. Under

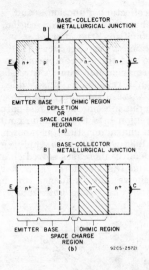

Fig. 102—Base-width modulation in power transistors: (a) no current flow (base narrowing); (b) high current flow (base widening).

these conditions, the depletion region penetrates the base and the lightly doped collector body. Equal fixed total donor and acceptor charges are uncovered on both sides of the metallurgical junction. Initiating current flow (mobile charge) has three effects. First, the mobile charge carriers in the base are of the same sign as the uncovered fixed charge; the depletion region in the base shrinks and the base widens. Second, the mobile carriers in the collector are of the opposite sign and thus subtract from the fixed charge. This condition tends to cause the depletion region in the collector to expand to uncover the requisite charge. Third, the current flow in the transistor introduces an ohmic voltage drop in the lightly doped n part of the collector body. Because this voltage drop subtracts from the applied collector-to-base potential, less voltage is available for the depletion region, and the base becomes wider again.

At high current densities and low collector voltages, the transistor-base widens beyond the metallurgical base-collector junction and approaches the n collector region. In this condition, the current gain and gain-bandwidth product (f_T) decrease and the device approaches the performance of a simple three-layer (n^+-p-n^+) transistor. Base widening with its resultant decreases in current gain introduces a quasi-saturation region into the common emitter static characteristics, as shown in Fig. 103. Operation in this region significantly increases stored charge, rise, and fall times (τ_S, τ_R, τ_F). Fig. 104 shows the influence of base widening on the gain-bandwidth product f_T.

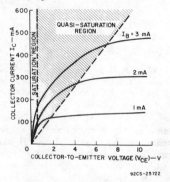

Fig. 103—*Quasi-saturation in power transistors (n^+-p-n-n^+ doping profile).*

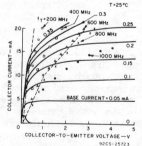

Fig. 104—*Effect of base widening on gain-bandwidth product f_T.*

Debiasing—There are two debiasing phenomena of significance in bipolar power transistors. The first, **resistive debiasing**, shown in Fig. 105, results from the voltage drops along the resistance in emitter and base fingers and in the base region beneath the emitter. These voltage drops cause heavier injection from the edges of the emitter and in emitter regions closest to the emitter contact.

Resistive debiasing beneath the emitter because of the voltage drop in the base resistance can be minimized by making the emitter fingers narrow, the base wide, and the base resistivity beneath the emitter as small as possible. Metallizing the emitters so that the contact is localized toward the finger center inserts a lateral emitter voltage drop which tends to offset the base resistance drop.

Thermal Debiasing, shown in Fig. 106, the second debiasing phenomenon, is a consequence of unintentional temperature differences between various locations of the emitter. Forward current from the emitter is strongly dependent on the temperature of the emitter-base junction, as illustrated in Fig. 106. For example, a temperature of 12°C between two regions of the emitter-base junction results in 3.8 times more current from the hotter region. Because unequal injection results in unequal dissipation, the hot region tends to become hotter and injects more.

Thermal debiasing can be minimized in two ways. First, the emitter can be separated into a number of discrete emitters and a ballast resistance placed in series with each discrete emitter. These resistors insert a voltage drop in each emitter proportional to the current being passed through the emitter, thus inserting current feedback and equalizing the currents between emitters. A second method utilizes a low thermal-resistance coupler between emitters. The emitters are made iso-thermal and, therefore, inject equally. This technique is illustrated in Fig. 107.

Second Breakdown—A bipolar transistor operated at high power densities is subject to a failure mode

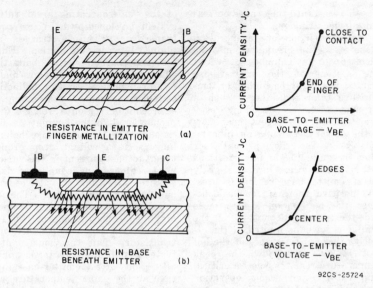

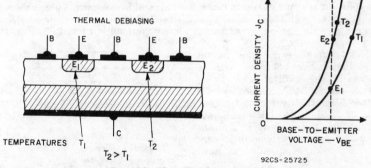

Fig. 105—Resistive debiasing phenomena: (a) along emitter finger; (b) beneath emitter.

Fig. 106—Thermal debiasing.

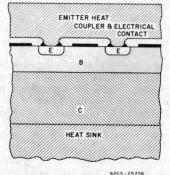

92CS-25726

Fig. 107—Isothermal emitter technique.

termed "second breakdown" in which the emitter-collector voltage suddenly drops, usually 10 to 25 volts. Unless the power is rapidly removed, the transistor is destroyed or materially degraded by overheating. Second breakdown (S/b) is a thermal hot-spot formation within the transistor pellet. It has two phases of development. First is the constriction phase where, because of thermal regeneration, the current tends to concentrate in a small area. The second phase is the destruction phase. In this second phase, local temperatures and temperature gradients increase until they cause permanent device damage.

The constriction or regeneration phase of second breakdown may be initiated in any number of ways. One section of the emitter-base junction need only be higher in temperature than the others. Such a hot spot might be caused by resistive debiasing, divergent heat flow to the device heat sink, an inhomogeneity in the thermal path, or other irregularities or imperfections within the device. Once a slightly hotter emitter-base region is present, positive thermal feedback begins: the hot region injects more and therefore gets hotter. If the available power is limited or the effective thermal resistance of the hot spot is sufficiently low, the peak temperature remains below a critical temperature, and stable operation continues. When the peak temperature reaches a value such that local base-collector leakage currents reach base current magnitude, the device regenerates into second breakdown, often very rapidly.

Second breakdown may occur when the device operates with a forward-biased emitter-base junction or during the application of reverse bias. In the forward-biased form of second breakdown, shown in Fig. 108, the current $I_{s/b}$ above which the device switches into second breakdown is specified as part of the "safe-operating area" rating system developed by RCA for power transistors. (This system is explained later in this section.) Emitter and base resistive ballasting effectively increase forward-biased $I_{s/b}$ of a device. Emitter ballasting equalizes currents by inserting in each emitter region a voltage drop proportional to the current passing through the junction. Base ballasting inserts a voltage drop proportional to base current in the various base regions thus equalizing drive conditions within the device and maintaining uniformity. Thermal coupling between emitter regions may also be used to improve the forward biased $I_{s/b}$ performance of a transistor. This previously illustrated design approach tends to hold all regions of the emitter-base junction at the same temperature and same forward bias, thus maintaining uniform current flow.

Second breakdown is also observed when a transistor operating with an inductive load is turned off. Fig. 109

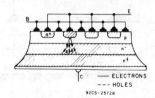

Fig. 109—Reverse-biased second breakdown.

shows this form of second breakdown. When the emitter-base junction is reverse biased, the edges of the emitters are quickly turned-off by the voltage drop caused by the reverse flow of the base current through the base resistance under the emitter. Collector current tends to be rapidly reduced; however, the inductive load responds to the decrease in collector current by driving the collector-emitter voltage to a value at which breakdown can occur in the collector-base space charge region $V_{(BR)CEX}$. The multiplied current resulting from the breakdown

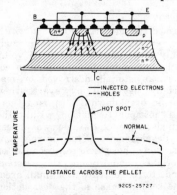

Fig. 108—Forward-biased second breakdown.

is focused towards the emitter centers, keeping the centers on for a longer time. When all center sections of the emitters behave alike, the power is dissipated uniformly by all emitters. If, however, a hot spot exists or develops, the energy stored in the load inductance is dumped into this region. The central region of this "hogging" emitter rapidly rises in temperature, reaching a value where the hot spot sustains itself and second breakdown occurs. Emitter ballasting is not effective in protecting against reverse-biased second breakdown because the hogging portion of the emitter is fed internally from a current source, and this current source is insensitive to the relatively small differences in emitter potential. Ballasting against reverse-biased second breakdown is best done in the collector by the addition of a resistive layer which decreases the internal collector-emitter voltage in the affected region. The maximum energy that may be stored in the load inductance before second breakdown ($E_{s/b}$) is specified for most RCA power transistors intended for switching applications.

High-Voltage Surface Effects—As the voltage ratings of a power transistor are increased, it becomes more difficult to achieve theoretical bulk breakdown values. Furthermore, both the breakdown voltage and junction leakage currents may vary under operating conditions. The problem is usually due to surface phenomena.

High-voltage transistors require large depletion widths in the base-collector junction. This requirement suggests that at least one side of the junction must be lightly doped. Fig. 110 shows what happens in a normal mesa-type device. The external fringing electrical fields terminate on the silicon and modify the depletion regions at the surface. If these fringing fields are large and configured as shown, a local high field condition is established at the surface and premature breakdown

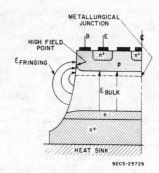

Fig. 110—Electric field distribution in high-voltage "mesa" n-p-n transistor.

occurs. High-intensity fringing fields exist well outside the junction and contribute to the movement of ions external or internal to the applied passivation layers, leading to instabilities.

The state-of-the-art "cures" for these problems are: junction contouring to reduce the magnitude and the shape of the fringing fields; empirical determination of the proper surface etch and the optimum organic encapsulant; or glassing of the junctions to contain the fringing fields. The latter two solutions do not usually yield breakdown voltages equal to the bulk values, but they do lessen the surface instability.

To achieve breakdown voltages approaching the bulk values it is necessary that the fringing field be properly shaped, and once properly shaped it must be kept in this condition. Field electrodes are being investigated to accomplish this objective.

Thermal Fatigue—A power transistor is often used in applications where the power in the device is cycled; the transistor is heated and cooled many times. Because the transistor is constructed of materials that have different thermal expansion coefficients, stress is placed on the chip, the metallurgical bond, and the heat spreader. If the stress is severe enough and sufficient cycles are encountered, the device fails. Usually the chip separates from the heat spreader or one of the contact connections opens. The stress is proportional to the size of the pellet, the temperature variation, elasticity of the connecting members, and the differences in thermal-expansion coefficients. Anything which concentrates the stress, such as voids in the mounting system, aggravates the condition.

The rate of degradation of a metallurgical bond under stressed conditions is also proportional to the average and peak temperature excursions of the bond. The failure-rate dependency of thermal fatigue and other phenomena can be as much as double for every 10°C increase in average and peak temperature. The most economical way to buy reliability in power transistor application is, therefore, to reduce these temperatures by careful consideration of heat flow during equipment design.

Several techniques are used to improve thermal-cycling capability within power transistors. One method is to mount the chip on a metal such as molybdenum, whose thermal expansion coefficient is similar to silicon, and to braze this metal to the package. In this way stresses are evenly distributed, as in a graded glass seal. Another

method, applicable on units using the lead solder mounting technique, uses a controlled solder process in which the thickness and composition of the lead solder are carefully controlled at all times.

An equipment manufacturer should make certain that power-transistor circuit included in his systems are designed so that cyclic thermal stresses are mild enough to assure that no transistor fatigue failures will occur during the required operating life of his equipment.

RCA has developed a **thermal-cycling rating system** that relates the total power dissipation P_T and the change in case temperature $\triangle T_C$ to the total number of thermal cycles N that the transistor is rated to withstand.

Fig. 111 shows a typical **thermal-cycling rating chart** for a power transistor. This chart is provided in the form of a log-log presentation in which total transistor dissipation is denoted by the ordinate and the thermal-cycling capability (number of cycles to failure) is indicated by the abscissa. Rating curves are shown for various magnitudes of changes in case temperature. Use of the thermal-cycling rating charts makes it possible for a circuit designer to avoid transistor thermal-fatigue failures during the

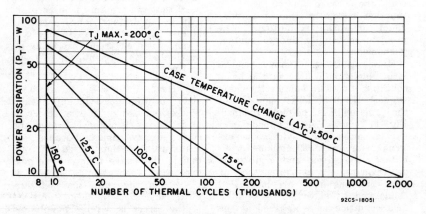

Fig. 111—Thermal-cycling rating chart for an RCA hermetic power transistor.

operating life of his equipment. In general, power dissipation is a fixed system requirement. The design can also readily determine the number of thermal cycles that a power transistor will be subjected to during the minimum required life of the equipment. For these conditions, the charts indicate the maximum allowable change in case temperature. (If the rating point does not lie exactly on one of the rating curves, the allowable change in case temperature can be approximated by linear interpolation. The designer can then determine the minimum size of the heat sink required to restrict the change in case temperature within this maximum value.

RCA thermal-cycling ratings allow a circuit designer to use power transistors with assurance that thermal-fatigue failures of these devices will not occur during the minimum required life of his equipment. These ratings provide valid indications of the thermal-cycling capability of power transistors for all types of operating conditions.

On the basis of these ratings, limiting conditions can be established during circuit design so that the possibility of transistor thermal-fatigue failures are avoided.

Design Tradeoffs

Many design variations are used to reduce the limitations imposed by base widening, emitter debiasing, second breakdown, high voltage surface effects, and thermal fatigue. Each design is replete with compromises, both subtle and obvious.

A power transistor may have its base-collector and emitter-base junctions delineated by two methods, shown in Fig. 112. First, selective etching may be used to create a multi-level "mesa" outline, where the junctions terminate at the edge of the mesa. Second, masked impurity diffusion may be used to convert selected areas to "planar" type junctions. A power transistor may be produced by using one or both of these techniques for the two junctions.

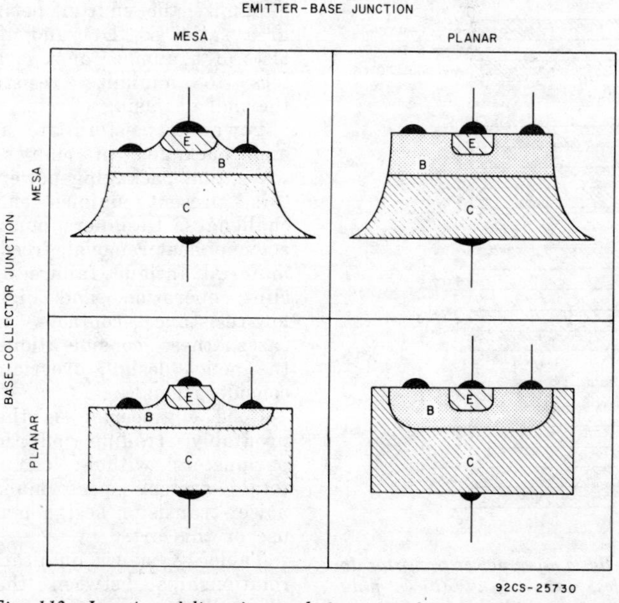

92CS-25730

Fig. 112—Junction delineation techniques used in power transistors.

Planar junctions can be produced with much finer dimensions than mesa junctions while the surface oxide remains covered and ostensibly flat. These characteristics are necessary in high frequency and multi-element types. Mesa junctions are dimensionally inferior but produce junctions with higher breakdown voltages: absence of the radius effect eliminates local internal high electric-field regions. For voltages greater than approximately 300 volts, the mesa junction is the most economical fabrication technique. Mesa type emitter-base junctions also eliminate edge injection and thus result in better high-current current gain.

A power transistor's doping profile may consist of three to six layers obtained by impurity diffusion and/or epitaxial techniques. The basic three-layer profile, shown in Fig. 113, is preferred because of its simplicity, but it necessitates a compromise between voltage and response-time capability. This compromise is required because the collector depletion region forms mostly in the

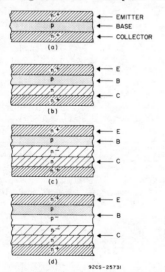

Fig. 113—Basic n-p-n power transistor doping profiles: (a) n^+-p-n^+, (b) n^+-p-n-n^+, (c) n^+-p-n^--n-n^+, (d) n^+-p-p^--n^--n-n^+.

transistor base; consequently, the base must be wide for higher voltage capability. The addition of a lightly doped collector layer, as shown in Fig. 113(b), allows the collector depletion region to expand into the collector rather than into the base, and the base can be kept thin. In this way the voltage capability is increased without reduction in the device response time if base widening is not encountered. The additional layers, as shown in Figs. 113(c) and 113(d), control base widening, improve voltage breakdown through control of the fringing field which surrounds the device, or provide ballasting against second breakdown.

A power transistor may be considered a composite of many unit transistors in parallel. The emitter-base geometry is designed so that each unit transistor is operating efficiently and that, in parallel, all unit transistors operate together, sharing the load current. Fig. 114 shows some of the emitter-base geometries used. In most cases, the emitter-base geometry is complex to maximize the emitter periphery-to-area ratio (E_p/E_A) and often consists of a number of separate emitters to minimize resistive and thermal debiasing effects.

Power transistors are assembled and packaged in many different ways, and packaging power transistors present unique engineering challenges. Designers must achieve efficient heat removal, freedom from material fatigue failure under cycling operation, and high-current low-resistance contacts. In many cases, these considerations dictate the pellet design, overriding other considerations.

Good engineering is the art of profitably trading off individual parameters without compromising total system performance, and power-transistor design makes good use of this art.

Table IX shows qualitatively the relationships between the major physical device design parameters

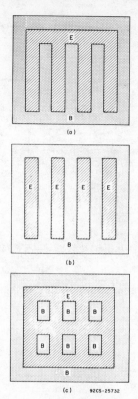

Fig. 114—Basic power transistor emitter geometries: (a) interdigitated, (b) multiple discrete (overlay), and (c) perforated emitter.

built up or decayed within the structure. Obviously, devices having short conduction spaces minimize the transit time; the base must be kept thin. Thin, low-resistivity collector material keeps the space charge region short and limits base widening. These factors also limit the amount of stored charge, thereby reducing switching times. Narrow, highly interdigitated ballasted emitters keep current flow and stored charge more uniform and allow faster charge removal from the base. All of the above factors contribute to a low saturation voltage in the transistor. A compromise in saturation voltage ($V_{CE(sat)}$) must be made, however, if lifetime reduction techniques, such as gold doping, are used during manufacture or if lifetime decreases because of radiation damage. High collector-emitter voltage must be traded against time response, because such devices depend on wide, high resistivity collector and base regions. Furthermore, because lifetime reduction techniques are not completely applicable, switching times are longer.

To achieve high-current capability, the emitter must have a large area, be highly efficient, and inject uniformly. An emitter with many thin fingers meet these conditions and gives a high emitter periphery-to-area ratio. Emitter ballasting also helps. But base widening must be avoided because it limits the current density within a device for any reasonable current gain, and thin, low-resistivity collector material used to prevent base widening is inconsistent with high collector-emitter voltages.

High current gain necessitates high emitter efficiency, narrow base widths, and high minority carrier lifetime. High emitter efficiency requires a lighter doped base. Thin, high-resistivity base regions aggravate emitter debiasing effects; narrow, uniformly injection emitters are desirable to maximize high-current gain. Again, because base widening effects must be avoided,

and the major electrical parameters. Table X shows how the parameters of second breakdown, voltage, radiation resistance, and cost relate to the more conventional device electrical parameters. An examination of these tables indicates that a large number of compromises can be made in the design of the circuit-transistor system to optimize its cost effectiveness. Some of the more salient physical reasons behind the interrelationships are described in the following paragraphs.

The time responses of a power transistor are basically related to the transit time of a free carrier from the emitter to the collector and to the rate at which charge can be

Table IX—Qualitative relationships between physical and electrical parameters.

Physical Parameter	f_T*	τ_d, τ_R τ_f, τ_S	$V_{CE(MAX)}$ $(LV_{(CEO)})$	$V_{CE(SAT)}$	V_{BE}	h_{FE} (peak)	$I_{C(MAX)}$ $(h_{FE}=10)$	$I_{s/b}$	$E_{s/b}$	Radiation Hardness
Base width (W_B) ↑	↓	↑	↑	↑	↑	↓	↓	↓	↑	↓
Base resistivity (ρ_B) ↑	—	↑	—	↓	↑	↑	↓	↑	↓	↓
Collector width (W_C) (n⁻ region) ↑	↓	↑	↑	↑	↑	—	↓	—	↑	↓
Collector resistivity (ρ_C) (n⁻ region) ↑	↓	↑	↑	↑	↑	—	↓	↓	↓	↓
Emitter width (X_E) (finger) ↑	↓	↑	—	—	↑	—	↓	↓	↓	↓
Emitter ballast (R_E) (resistive) ↑	↑	↓	—	↑	↑	—	↑	↑	—	—
Collector ballast (R_C) (resistive) ↑	↓	↑	—	↑	—	—	↓	—	↑	—

Notes:

  * f_T measured at high currents and low voltages (power corner, worst-case point).

Table X—Qualitative relationships between some important parameters.

Specific Electrical Parameters	f_T*	τ_d, τ_R τ_f, τ_S	$V_{CE(SAT)}$	h_{FE} (peak)	$I_{C(MAX)}$ $h_{FE}=10$	Cost
$I_{S/B}$ ↑	↓	↑	↑	↓	↓	↑
$E_{S/B}$ ↑	↓	↑	↑	↓	↓	↑
$V_{CE(MAX)}$ ↑	↓	↑	↑	—	↓	↑
Radiation resistance	↑	↓	↓	↓	↓	↑

Notes:

  * f_T measured at high currents and low voltages (power corner, worst-case point).

Symbols for Tables IX and X

f_T—current gain-bandwidth product
τ_d—delay time
τ_R—rise time
τ_f—fall time
τ_S—storage time
$I_{S/b}$—forward second-breakdown current
h_{FE}—dc current gain

V_{CE}—collector-to-emitter voltage
$V_{CE(sat)}$—collector-to-emitter saturation voltage
I_C—collector current
$E_{s/b}$—reverse second-breakdown energy
↑ Increase
↓ Decrease

low-voltage devices have the advantage.

The ability of a transistor to withstand high transient energy dissipation (second breakdown) is enhanced by keeping the base wide and its resistivity low. This condition conflicts with time response and current gain objectives. A design for uniform injection using narrow emitter fingers and incorporating ballasted emitter sites counters thermal or electrical debiasing effects but tends to increase $V_{CE(sat)}$. Collector regions should be thick and the resistivity as low as possible without losing collector ballasting resistance. Multi-resistive layers in the collector are beneficial but add to the cost. In general, techniques used to assure high second breakdown capability slow time response, increase saturation voltage, and add to cost. For a given volt-ampere product, extremes in voltage or current aggravate the second breakdown problem. In some cases, input regulation and transient control to an entire system is economical because lower-cost power transistors can then be utilized.

Safe-Operating-Area Ratings

During normal circuit operation, power transistors are often required to sustain high current and high voltage simultaneously. The capability of a transistor to withstand such conditions is normally shown by use of a safe-operating-area rating curve. This type of rating curve defines, for both steady-state and pulsed operation, the voltage-current boundaries that result from the combined limitations imposed by voltage and current ratings, the maximum allowable dissipation, and the second-breakdown ($I_{s/b}$) capabilities of the transistor.

If the safe operating area of a power transistor is limited within any portion of the voltage-current characteristics by thermal factors (thermal impedance, maximum junction temperatures, or operating case temperature), this limiting is defined by a constant-power hyperbola ($I = KV^{-1}$) which can be represented on the log-log voltage-current curve by a straight line that has a slope of -1.

The energy level at which second breakdown occurs in a power transistor increases as the time duration of the applied voltage and current decreases. The power-handling capability of the transistor also increases with a decrease in pulse duration because thermal

mass of the power-transistor chip and associated mounting hardware imparts an inherent thermal delay to a rise in junction temperature.

Fig. 115 shows a forward-bias safe-area rating chart for a typical silicon power transistor, the RCA-2N3585. The boundaries defined by the curves in the safe-area chart indicate, for both continuous-wave and nonrepetitive-pulse operation, the maximum current ratings, the maximum collector-to-emitter forward-bias avalanche breakdown-voltage rating [$V\alpha M = 1$, which is usually approximated by $V_{CEO}(sus)$], and the thermal and second-breakdown ratings of the transistors.

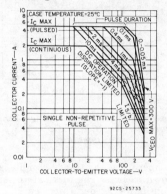

Fig. 115—Safe-area rating chart for the RCA-2N3585 silicon power transistor.

As shown in Fig. 115, the thermal (dissipation) limiting of the 2N3585 ceases when the collector-to-emitter voltage rises above 100 volts during dc operation. Beyond this point, the safe operating area of the transistor is limited by the second-breakdown ratings. During pulsed operation, the thermal limiting extends to higher values of collector-to-emitter voltage before the second-breakdown region is reached, and as the pulse duration decreases, the thermal-limited region increases.

If a transistor is to be operated at a pulse duration that differs from those shown on the safe-area chart, the boundaries provided by the safe-area curve for the next higher pulse duration must be used, or the transistor manufacturer should be consulted. Moreover, as indicated in Fig. 115, safe-area ratings are normally given for single nonrepetitive pulse operation at a case temperature of 25°C and must be derated for operation at higher case temperatures and under repetitive-pulse or continuous-wave conditions.

Fig. 116 shows temperature derating curves for the 2N3585 safe-area chart of Fig. 115. These curves show that thermal ratings are affected far more by increases in case

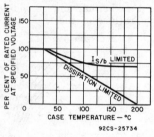

Fig. 116—Safe-area temperature-derating curves for the RCA-2N3585 silicon power transistor.

temperature than are second-breakdown ratings. The thermal (dissipation-limited) derating curve decreases linearly to zero at the maximum junction temperature of the transistor [$T_J(\text{max}) = 200°C$]. The second-breakdown ($I_{s/b}$-limited) temperature derating curve, however, is less severe because the increase in the formation of the high current concentrations that cause second breakdown is less than the increase in dissipation factors as the temperature increases.

Because the thermal and second-breakdown deratings are different, it may be necessary to use both curves to determine the proper derating factor for a voltage-current point that occurs near the breakpoint of the thermal-limited and second-breakdown-limited regions on the safe-area curve. For this condition, a derating factor is read from each

derating curve. For one of the readings, however, either the thermal-limited section of the safe-area curve must be extrapolated upward in voltage or the second-breakdown-limited section must be extrapolated downward in voltage, depending upon which side of the voltage breakpoint the voltage-current point is located. The smaller of the collector-current values obtained from the thermal and second-breakdown deratings must be used as the safe rating.

For pulsed operation, the derating factor shown in Fig. 116 must be applied to the appropriate curve on the safe-area rating chart. For the derating, the effective case temperature $T_C(\text{eff})$ may be approximated by the average junction temperature $T_J(\text{av})$. The average junction temperature is determined as follows:

$$T_J(\text{av}) = T_C + P_{AV} \, (\theta_{J-C})$$

This approach results in a conservative rating for the pulsed capability of the transistor. A more accurate determination can be made by computation of actual instantaneous junction temperatures. (For more detailed information on safe-area ratings and temperature derating the reader should refer to the RCA **Power Circuits Designer's Handbook,** Technical Series SP-52, pp. 126 to 145.)

POWER TRANSISTORS IN SWITCHING SERVICE

An important application of power transistors is power switching. Large amounts of power, at high currents and voltages, can be switched with small losses by use of a power transistor that is alternatively driven from cutoff to saturation by means of a base control signal. The two most important considerations in such switching applications are the speed at which the transistor can change states between saturation and cutoff and the power dissipation.

Basic Operation

Transistor switching applications are usually characterized by large-signal nonlinear operation of the devices. The switching transistor is generally required to operate in either of two states: on or off. In transistor switching circuits, the common-emitter configuration is by far the most widely used.

Typical output characteristics for an n-p-n transistor in the common-emitter configuration are shown in Fig. 117. These characteristics are divided into three regions of operation, i.e., cutoff region, active region, and saturation region.

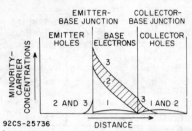

Fig. 118—Minority-carrier concentrations in an n-p-n transistor: (1) in cutoff region, (2) in active region at edge of saturation region, (3) in saturation region.

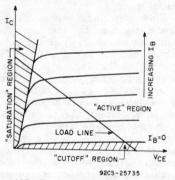

Fig. 117—Typical collector characteristic of an n-p-n transistor showing three principal regions involved in switching.

In the cutoff region, both the emitter-base and collector-base junctions are reverse-biased. Under these conditions, the collector current is very small, and is comparable in magnitude to the leakage current I_{CEO}, I_{CEV}, or I_{CBO}, depending on the type of base-emitter biasing used.

Fig. 118 is a sketch of the minority-carrier concentration in an n-p-n transistor. For the cutoff condition, the concentration is zero at both junctions because both junctions are reverse-biased, as shown by curve 1 in Fig. 118.

In the active region, the emitter-base junction is forward-biased and the collector-base junction is reverse-biased. Switching from the cutoff region to the active region is accomplished along a load line, as indicated in Fig. 117. The speed of transition through the active region is a function of the frequency-response characteristics of the device. The minority-carrier concentration for the active region is shown by curve 2 in Fig. 118.

The remaining region of operation is the saturation region. In this region, the emitter-base and collector-base junctions are both forward-biased. Because the forward voltage drop across the emitter-base junction under this condition $[V_{BE}(sat)]$ is greater than that across the collector-base junction, there is a net collector-to-emitter voltage referred to as $V_{CE}(sat)$. It is evident that any series-resistance effects of the emitter and collector also enter into determining $V_{CE}(sat)$. Because the collector is now forward-biased, additional carriers are injected into the base, and some into the collector. This minority-carrier concentration is shown by curve 3 in Fig. 118.

A basic saturated-transistor switching circuit is shown in Fig. 119. The voltage and current waveforms for this circuit under typical base-drive conditions are shown in Fig. 120. Prior to the application of the positive-going input pulse, the emitter-base junction is reverse-biased by a voltage $-V_{BE}(off) = V_{BB}$. Because the transistor is in the cutoff region, the base current I_B is

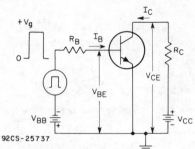

Fig. 119—*Basic saturated transistor switching circuit.*

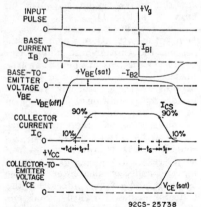

Fig. 120—*Voltage and current waveforms for saturated switching circuit shown in Fig. 119.*

the reverse leakage current I_{BEV}, which is negligible compared with I_{B1}, and the collector current I_C is the reverse leakage current I_{CEV}, which is negligible compared with $V_{CC}/$

R_C. When the positive-going input pulse V_g is applied, the base current I_B immediately goes positive.

The collector current, however, does not begin to increase until some time later. This delay in the flow of collector current (t_d) results because the emitter and collector capacitances do not allow the emitter-base junction to become forward-biased instantaneously. These capacitances must be charged from their original negative potential [$-V_{BE}$(off)] to a forward bias suf-

ficient to cause the transistor to conduct appreciably. After the emitter-base junction is sufficiently forward-biased, there is an additional delay caused by the time required for minority carriers which are injected into the base to diffuse across the base and be collected at the collector. This delay is usually negligible compared with the delay introduced by the capacitive component. The collector and emitter capacitances vary with the collector-base and emitter-base junction voltages, and increase as the voltage V_{BE} goes positive. An accurate determination of total delay time, therefore, requires knowledge of the nonlinear characteristics of these capacitances.

When the collector current I_C begins to increase, the transistor has made the transition from the cutoff region into the active region. The collector current takes a finite time to reach its final value. This time, called rise time (t_r), is determined by the gain-bandwidth product (f_T), the collector-to-emitter capacitance (C_C), and the static forward current-transfer ratio (h_{FE}) of the transistor. At high collector currents and/or low collector voltages, the effect of this capacitance on rise time is negligible, and the rise time of collector current is inversely proportional to f_T. At low currents and/or high voltages, the effect of gain-bandwidth product is negligible, and the rise time of collector current is directly proportional to the product $R_C C_C$. At intermediate currents and voltages, the rise time is proportional to the sum ($1/2\pi f_T$) + $R_C C_C$. Under any of the above conditions, the collector current responds exponentially to a step of base current. If a turn-on base current (I_{B1}) is applied to the device, and the product $I_{B1} h_{FE}$ is less than V_{CC}/R_C, the collector current rises exponentially until it reaches the steady-state value $I_{B1} h_{FE}$. If $I_{B1} h_{FE}$ is greater than V_{CC}/R_C, the collector current rises toward the value $I_{B1} h_{FE}$. The transistor becomes saturated when I_C

reaches the value I_{CS} ($\approx$ V_{CC}/R_C). At this point, I_C is effectively clamped at the value V_{CC}/R_C.

The rise time, therefore, depends on an exponential function of the ratio I_{CS}/I_{B1} : h_{FE}. Because the values of h_{FE}, f_T, and C_C are not constant, but vary with collector voltage and current as the transistor is switching, the rise time as well as the delay time is dependent on nonlinear transistor characteristics.

After the collector curren of the transistor has reached a steady-state value I_{CS}, the minority-charge distribution is that shown by curve 3 in Fig. 44. When the transistor is turned off by returning the input pulse to zero, the collector current does not change immediately. This delay is caused by the excess charge in the base and collector regions, which tends to maintain the collector current at the I_{CS} value until this charge decays to an amount equal to that in the active region at the edge of saturation (curve 2 in Fig. 44). The time required for this charge to decay is called the storage time (t_s). The rate of charge decay is determined by the minority-carrier lifetime in the base and collector regions, on the amount of reverse "turn-off" base current (I_{B2}), and on the overdrive "turn-on" current (I_{B1}) which determined how deeply the transistor was driven into saturation. (In non-saturated switching, there is no excess charge in the base region, so that storage time is negligible.)

When the stored charge (Q_S) has decayed to the point where it is equal to that at the edge of saturation, the transistor again enters the active region and the collector current begins to decrease. This fall-time portion of the collector-current characteristic is similar to the rise-time portion because the transistor is again in the active region. The fall time, however, depends on I_{B2}, whereas the rise time was dependent on I_{B1}. Fall time, like rise time, also depends on f_T and C_C.

The approximate values of I_{B1}, I_{B2}, and I_{CS} for the circuit shown in Fig. 119 are given by:

$$I_{B1} = \frac{V_G - V_{BB} - V_{BE}(sat)}{R_B}$$

$$I_{B2} = \frac{V_{BB} + V_{BE}(sat)}{R_B}$$

$$I_{CS} = \frac{V_{CC} - V_{CE}(sat)}{R_C}$$

Switching Characteristics

The electrical characteristics for a switching transistor, in general, differ from that for a linear-amplifier type of transistor in several respects. The static forward current-transfer ratio h_{FE} and the saturation voltages $V_{CE}(sat)$ and $V_{BE}(sat)$ are of fundamental importance in a switching transistor. The static forward current-transfer ratio determines the maximum amount of current amplification that can be achieved in any given circuit, saturated or non-saturated. The saturation voltages are necessary for the proper dc design of saturated circuits. Consequently, h_{FE} is always specified for a switching transistor, geenrally at two or more values of collector current. $V_{CE}(sat)$ and $V_{BE}(sat)$ are specified at one or more current levels for saturated transistor applications. Control of these three characteristics determines the performance of a given transistor type over a broad range of operating conditions. For non-saturated applications, $V_{CE}(sat)$ and $V_{BE}(sat)$ need not be specified. For such applications, it is important to specify V_{BE} at specific values of collector current and collector-to-emitter voltage in the active region.

Because the collector and emitter capacitances and the gain-bandwidth product influence switching time, these characteristics are specified for most switching transistors. The collector-base and emitter-base junction capacitances are usually measured at some value of reverse bias and are designated C_{ob} and C_{1b}, respectively. The gain-bandwidth product (f_T) of the transistor is the frequency at which the small-signal forward current-transfer ratio (h_{fe}) is unity. Because this characteristic falls off at 6 dB per octave above the corner frequency, f_T is usually controlled by specifying the h_{fe} at a fixed frequency anywhere from 1/2 to 1/10 f_T. Because C_{ob}, C_{ib}, and f_T vary nonlinearly over the operating range, these characteristics are generally more useful as figures of merit than as controls for determining switching speeds. When the switching speeds in a particular application are of major importance, it is preferable to specify the required switching speeds in the desired switching circuit rather than C_{ob}, C_{ib}, and f_T.

The storage time (t_s) of a transistor is dependent on the stored charge (Q_S) and on the driving current employed to switch the transistor between cutoff and saturation. Consequently, either the stored charge or the storage time under heavy overdrive conditions should be specified. Most recent transistor specifications require that storage time be specified.

Because of the dependence of the switching times on current and voltage levels, these times are determined by the voltages and currents employed in circuit operation.

Dissipation, Current, and Voltage Ratings

Up to this point, no mention has been made of dissipation, current, and voltage ratings for a switching transistor. The maximum continuous ratings for dissipation and current are determined in the same manner as for any other transistor. In a switching application, however, the peak dissipation and current may be permitted to exceed these continuous ratings depending on the pulse duration, on the duty factor, and on the thermal time constant of the transistor.

Voltage ratings for switching transistors are more complicated. In the basic switching circuit shown in Fig. 119, three breakdown voltages must be considered. When the transistor is turned off, the emitter-base junction is reverse-biased by the voltage V_{BE}(off), (i.e., V_{BB}), the collector-base junction by $V_{CC} + V_{BB}$, and the emitter-to-collector junction by $+ V_{CC}$. To assure that none of the voltage ratings for the transistor is exceeded under "off" conditions, the following requirements must be met:

The minimum emitter-to-base breakdown voltage $V_{(BR)EBO}$ must be greater than V_{BE}(off).

The minimum collector-to-base breakdown voltage $V_{(BR)CBO}$ must be greater than $V_{CC} + V_{BE}$(off).

The minimum collector-to-emitter breakdown voltage $V_{(BR)CERL}$ must be greater than V_{CC}.

$V_{(BR)EBO}$ and $V_{(BR)CBO}$ are always specified for a switching transistor. The collector-to-emitter breakdown voltage $V_{(BR)CEO}$ is usually specified under open-base conditions. The breakdown voltage $V_{(BR)CERL}$ (the subscript "RL" indicates a resistive load in the collector circuit) is generally higher than $V_{(BR)CEO}$. The requirement that $V_{(BR)CEO}$ be greater than V_{CC} is overly pessimistic. The requirement that $V_{(BR)CERL}$ be greater than V_{CC} should be used wherever applicable.

Coupled with the breakdown voltages are the collector-to-emitter and base-to-emitter transistor leakage currents. These leakage currents (I_{CEV} and I_{BEV}) are particularly important considerations at high operating temperatures. The subscript "V" in these symbols indicates that these leakage currents are specified at a given emitter-to-base voltage (either forward or reverse). In the basic circuit of Fig. 119, these currents are determined by the following conditions:

$$\left.\begin{matrix} I_{CEV} \\ I_{BEV} \end{matrix}\right\} \quad \begin{matrix} V_{CE} = V_{CC} \\ V_{BE} = V_{BE}\text{(off)} = -V_{BB} \end{matrix}$$

In a switching transistor, these leakage currents are usually controlled not only at room temperature, but also at some higher operating temperature near the upper operational limit of the transistor.

Inductive Switching

Most inductive switching circuits can be represented by the basic equivalent circuit shown in Fig. 121. This type of circuit requires a rapid transfer of energy from the switched inductance to the switching mechanism, which may be a relay, a transistor, a commutating diode, or some other device. Often an accurate calculation of the energy to be dissipated in the switching device is required, particularly if that device is a transistor. If the supply voltage is low compared to the sustaining breakdown voltage of the transistor and if the series resistance of the inductor can be ignored, then the energy to be dissipated is $\frac{1}{2} LI^2$. This type of rating for a transistor is called "reverse-bias second breakdown." The energy capability of a

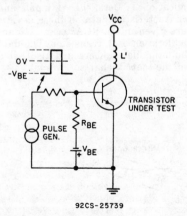

Fig. 121—Basic equivalent circuit for inductive switching circuit.

transistor varies with the load inductance and base-emitter reverse bias. A typical set of ratings which now appears in RCA published data is shown in Fig. 122.

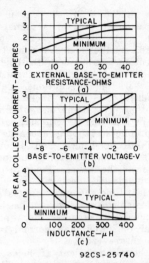

Fig. 122—Typical reverse-bias second-breakdown $(E_{s/b})$ rating curves.

HANDLING CONSIDERATIONS

The generation of static charge in dry weather is harmful to all transistors, and can cause permanent damage or catastrophic failure in the case of high-speed devices. The most obvious precaution against such damage is humidity control in storage and operating areas. In addition, it is desirable that transistors be stored and transported in metal trays rather than in polystyrene foam "snow". During testing and installation, both the equipment and the operator should be grounded, and all power should be turned off when the device is inserted into the socket. Grounded plates may also be used for stockpiling of transistors prior to or after testing, or for use in testing ovens or on operating life racks. Further protection against static charges can be provided by use of partially conducting floor planes and non-insulating footwear for all personnel.

Environmental temperature also affects performance. Variations of as little as 5 per cent can cause changes of as much as 50 per cent in the saturation current of a transistor. Some test operators can cause marked changes in measurements of saturation current because the heat of their hands affects the transistors they work on. Precautions against temperature effects include air-conditioning systems, use of finger cots in handling of transistors (or use of pliers or "plug-in boards" to eliminate handling), and accurate monitoring and control of temperature near the devices. Prior to testing, it is also desirable to allow sufficient time (about 5 minutes) for a transistor to stabilize if it has been subjected to temperature much higher or lower than normal room temperature (25°C).

Although transient rf fields are not usually of sufficient magnitude to cause permanent damage to transistors, they can interfere with accurate measurement of characteristics at very low signal levels or at high frequencies. For this reason, it is desirable to check for such radiation periodically and to eliminate is causes. In addition, sensitive measurements should be made in shielded screen rooms if possible. Care must also be taken to avoid the exposure of transistors to other ac or magnetic fields.

Many transistor characteristics are sensitive to variations in temperature, and may change enough at high operating temperatures to affect circuit performance. Fig. 123 illustrates the effect of increasing temperature on the common-emitter forward current-transfer ratio (beta), the dc collector-cutoff current, and the input and output impedances. To avoid undesired changes in circuit operation, it is recommended that transistors be located away from heat sources in equipment, and also that provisions be made for adequate heat dissipation and, if necessary, for temperature compensation.

RCA POWER-TRANSISTOR PRODUCT MATRICES

RCA offers a very extensive line of low- and medium-frequency power transistors for use in a broad range of linear and switching applications. In the following power-transistor matrices, Tables XI through XVII, these devices are categorized according to current and power ratings, pellet size, structure, and typical applications. Detailed ratings and characteristics data on these devices are given in the RCA technical data bulletin on each transistor. The data bulletin file numbers are also listed in the matrix charts.

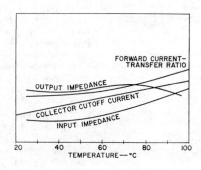

92CS-25741

Fig. 123—Variation of transistor characteristics with temperature.

Table XI—Monolithic Darlington Power Transistors
I_C to 10 A . . . P_T to 100 W . . . h_{FE} to 1000 min.

$I_C = -10$ A max. $P_T = 60$ W max. VERSAWATT (TO-220) 130 x 130▲	$I_C = 8$ A max. $P_T = 90$ W max. (TO-3) 136 x 136	$I_C = 8$ A max. $P_T = 60$ W max. VERSAWATT (TO-220) 136 x 136	$I_C = 10$ A max. $P_T = 100$ W max. (TO-3) 136 x 136	$I_C = 10$ A max. $P_T = 60$ W max. VERSAWATT (TO-220) 136 x 136	$I_C = 10$ A max. $P_T = 70$ W max. (TO-3) 136 x 136
Family Designation					
RCA8203 [P-N-P]	2N6385 [N-P-N]	TA8904 [N-P-N]	2N6385 [N-P-N]	2N6388 [N-P-N]	RCA8350 [P-N-P]
RCA8203 V_{CER}(sus) = -40 V h_{FE} = 1000-20,000 @ -3 A f_T = 20 MHz $I_C = -8$ A File No. 335	**RCA1000** V_{CEO}(sus) = 60 V h_{FE} = 1000 min. @ 3 A t_{on} = 1 μs typ. t_f = 3 μs typ. t_s = 1 μs typ. File No. 594	**RCA122** V_{CER}(sus) = 100 V h_{FE} = 1000 min. @ 3 A f_T = 20 MHz min. File No. 840	**2N6383** V_{CEO}(sus) = 40 V h_{FE} = 1000 min. @ 5 A t_{on} = 1 μs typ. t_f = 3 μs typ. t_s = 1 μs typ. I_C = 10 A CT File No. 609	**2N6386** V_{CEO}(sus) = 40 V h_{FE} = 1000 min. @ 3 A t_{on} = 1 μs typ. t_f = 3 μs typ. t_s = 1 μs typ. I_C = 10 A, P_T = 40 W CT File No. 610	**RCA8350** V_{CER}(sus) = -40 V h_{FE} = 1000-20,000 @ -5 A f_T = 20 MHz min. File No. 861
RCA125 V_{CEO} = -60 V h_{FE} = 1000 min. @ -3 A f_T = 20 MHz min. $I_C = -8$ A File No. 841	**RCA1001** V_{CEO}(sus) = 80 V h_{FE} = 1000 min. @ 3 A t_{on} = 1 μs typ. t_f = 3 μs typ. t_s = 1 μs typ. File No. 594		**2N6055** V_{CER}(sus) = 60 V h_{FE} = 750 min. @ 4 A t_{on} = 1 μs typ. t_f = 3 μs typ. t_s = 1 μs typ. I_C = 8 A File No. 563	**RCA120** V_{CER}(sus) = 60 V h_{FE} = 1000 min. @ 3 A f_T = 20 MHz min. I_C = 8 A P_T = 60 W File No. 840	**RCA8350A** V_{CER}(sus) = -60 V h_{FE} = 1000-20,000 @ -5 A f_T = 20 MHz min. File No. 861
RCA8203A V_{CER}(sus) = -60 V h_{FE} = 1000-20,000 @ -5 A f_T = 20 MHz min. $I_C = -10$ A File No. 835			**2N6384** V_{CER}(sus) = 60 V h_{FE} = 1000 min. @ 5 A t_{on} = 1 μs typ. t_f = 3 μs typ. t_s = 1 μs typ. I_C = 10 A CT File No. 609	**2N6387** V_{CER}(sus) = 60 V h_{FE} = 1000 min. @ 5 A t_{on} = 1 μs typ. t_f = 3 μs typ. t_s = 1 μs typ. I_C = 10 A, P_T = 40 W CT File No. 610	**RCA8350B** V_{CER}(sus) = -80 V h_{FE} = 1000-20,000 @ -5 A f_T = 20 MHz min. File No. 861
RCA126 V_{CEO} = -80 V h_{FE} = 1000 min. @ -3 A f_T = 20 MHz min. $I_C = -8$ A File No. 841			**2N6056** V_{CEO}(sus) = 80 V h_{FE} = 750 min. @ 4 A t_{on} = 1 μs typ. t_f = 3 μs typ. t_s = 1 μs typ. I_C = 8 A File No. 563	**RCA121** V_{CER}(sus) = 80 V h_{FE} = 1000 min. @ 3 A f_T = 20 MHz min. I_C = 8 A P_T = 60 W File No. 840	
RCA8203B V_{CER}(sus) = -80 V h_{FE} = 1000-20,000 @ -5 A f_T = 20 MHz min. $I_C = -10$ A File No. 835			**2N6385** V_{CEO}(sus) = 80 V h_{FE} = 1000 min. @ 5 A t_{on} = 1 μs typ. t_f = 3 μs typ. t_s = 1 μs typ. I_C = 10 A CT File No. 609	**2N6388** V_{CEO}(sus) = 80 V h_{FE} = 1000 min. @ 5 A t_{on} = 1 μs typ. t_f = 3 μs typ. t_s = 1 μs typ. I_C = 10 A, P_T = 40 W CT File No. 610	

▲ Pellet size — values shown are edge dimensions in mils

CT — Complementary Type Available

Table XII—Hometaxial-Base N-P-N Power Transistors
I_C to 80 A . . . P_T to 300 W . . . V_{CE} to 170 V

I_C = 1.5 A max. P_T = 5 W max. (TO-39)*	I_C = 1.5 A max. P_T = 8.75 W max. (TO-39)*	I_C = 1.5 A max. P_T = 8.75 W max. (TO-39)*	I_C = 3.5 A max. P_T = 10 W max. (TO-39)*	I_C = 4 A max. P_T = 50 W max. (TO-66)**	I_C = 4 A max. P_T = 36 W max. VERSAWATT (TO-220)
90 x 90▲	90 x 90	90 x 90	90 x 90	130 x 130	130 x 130
Family Designation					
2N1482	2N1482	40349	2N5786	2N3054	2N5298
2N1479● V_{CEV} = 60 V h_{FE} = 20-60 @ 200 mA ... File No. 135	**40347** V_{CEV}(sus) = 60 V h_{FE} = 25-100 @ 450 mA f_T = 1.5 MHz typ. ... File No. 88	**40349** V_{CEV}(sus) = 160 V h_{FE} = 30-125 @ 150 mA f_T = 1.5 MHz typ. ... File No. 88	**2N5786**● V_{CER}(sus) = 45 V h_{FE} = 20-100 @ 1.6 A f_T = 1 MHz min. ... CT File No. 413	**40250** V_{CER}(sus) = 50 V h_{FE} = 25-100 @ 1.5 A f_T = 1.2 MHz typ. P_T = 29 W ... CT File No. 112	**41504** V_{CER}(sus) = 35 V h_{FE} = 25 min. @ 1 A f_T = 0.8 MHz min. ... File No. 775
2N1481● V_{CEV} = 60 V h_{FE} = 35-100 @ 200 mA ... File No. 135	**40348** V_{CEV}(sus) = 90 V h_{FE} = 30-125 @ 300 mA f_T = 1.5 MHz typ. ... File No. 88		**2N5785**● V_{CER}(sus) = 65 V h_{FE} = 20-100 @ 1.2 A f_T = 1 MHz. min. ... CT File No. 413	**2N6260** V_{CEV}(sus) = 50 V h_{FE} = 20-100 @ 1.5 A f_T = 0.8 MHz min. P_T = 29 W ... File No. 527	**2N5295** **2N5296** V_{CER}(sus) = 50 V h_{FE} = 30-120 @ 1 A f_T = 0.8 MHz min. ... CT File No. 322
2N1480● V_{CEV} = 100 V h_{FE} = 20 min. @ 200 mA ... File No. 135			**2N5784**● V_{CER}(sus) = 80 V h_{FE} = 20-100 @ 1 A f_T = 1 MHz min. ... CT File No. 413	**2N3054** V_{CER}(sus) = 60 V h_{FE} = 25-150 @ 0.5 A f_T = 0.8 MHz min. P_T = 25 W ... CT File No. 527	**2N5297** **2N5298** V_{CER}(sus) = 70 V h_{FE} = 20-80 @ 1.5 A f_T = 0.8 MHz min. ... CT File No. 322
2N1482● V_{CEV} = 100 V h_{FE} = 35-100 @ 200 mA ... File No. 135				**2N6261** V_{CER}(sus) = 85 V h_{FE} = 25-100 @ 1.5 A f_T = 0.8 MHz min. P_T = 50 W ... File No. 527	**2N5293** **2N5294** V_{CER}(sus) = 75 V h_{FE} = 30-120 @ 0.5 A f_T = 0.8 MHz min. ... CT File No. 322

▲ Pellet size — values shown are edge dimensions in mils
* Available with
  a. flange for easy heat sinking $R_{\theta JC}$ = 15° C/W
  b. free-air radiator $R_{\theta JA}$ = 40-50° C/W
** Available with free-air radiator $R_{\theta JA}$ = 30° C/W
● These transistors are also available in TO-5 packages in U.S.A., Canada, Latin America, and Far East
CT — Complementary Type Available

Table XII—Hometaxial-Base N-P-N Power Transistors (cont'd)
I_C to 80 A ... P_T to 300 W ... V_{CE} to 170 V

I_C = 3 A max. P_T = 25 W max. (TO-8)	I_C = 3 A max. P_T = 50 W max. (TO-66)**	I_C = 3A max. P_T = 36 W max. VERSAWATT (TO-220)	I_C = 4 A max. P_T = 50 W max. VERSAWATT (TO-220)	I_C = 4 A max. P_T = 50 W max. VERSAWATT (TO-220)	I_C = 7 A max. P_T = 50 W max. VERSAWATT (TO-220)
130 x 130▲	130 x 130	130 x 130	130 x 130	130 x 130	150 x 150
Family Designation					
2N1486	2N3441	2N6478	2N6478	2N6478	2N5496
2N1483 V_{CEV} = 60 V h_{FE} = 20-60 @ 750 mA File No. 137	**2N6263** V_{CER}(sus) = 130 V h_{FE} = 20-100 @ 0.5 A f_T = 1.2 MHz typ. P_T = 20 W File No. 529	**2N6477** V_{CER}(sus) = 130 V h_{FE} = 25-100 @ 1 A f_T = 0.8 MHz min. File No. 680	**RCA29/SDH** V_{CEO} = 40 V h_{FE} = 40 min. @ 0.2 A f_T = 0.8 MHz min. I_C = 4 A P_T = 36 W File No. 792	**RCA31/SDH** V_{CEO} = 40 V h_{FE} = 25 min. @ 1 A f_T = 0.8 MHz min. I_C = 4 A P_T = 36 W File No. 793	**2N5491** **2N5490** V_{CER}(sus) = 50 V h_{FE} = 20-100 @ 2 A f_T = 0.8 MHz min. CT File No. 353
2N1485 V_{CEV} = 60 V h_{FE} = 35-100 @ 750 mA File No. 137	**2N3441** h_{FE} = 25-100 @ 0.5 A f_T = 1.2 MHz typ. P_T = 25 W CT File No. 529	**2N6478** V_{CER}(sus) = 150 V h_{FE} = 25-100 @ 1 A f_T = 0.8 MHz min. File No. 680	**RCA29A/SDH** V_{CEO} = 40 V h_{FE} = 40 min. @ 0.2 A f_T = 0.8 MHz min. I_C = 4 A P_T = 36 W File No. 792	**RCA31A/SDH** V_{CEO} = 60 V h_{FE} = 25 min. @ 1 A f_T = 0.8 MHz min. I_C = 4 A P_T = 36 W File No. 793	**2N5495** **2N5494** V_{CER}(sus) = 50 V h_{FE} = 20-100 @ 3 A f_T = 0.8 MHz min. CT File No. 353
2N1484 V_{CEV} = 100 V h_{FE} = 20-60 @ 750 mA File No. 137	**2N6264** V_{CER}(sus) = 170 V h_{FE} = 20-60 @ 1 A f_T = 1.2 MHz typ. P_T = 50 W File No. 529		**RCA29B/SDH** V_{CEO} = 60 V h_{FE} = 40 min. @ 0.2 A f_T = 0.8 MHz min. I_C = 4 A P_T = 36 W File No. 792	**RCA31B/SDH** V_{CEO} = 80 V h_{FE} = 25 min. @ 1 A f_T = 0.8 MHz min. I_C = 4 A P_T = 36 W File No. 793	**2N5493** **2N5492** V_{CER}(sus) = 65 V h_{FE} = 20-100 @ 2.5 A f_T = 0.8 MHz min. CT File No. 353
2N1486 V_{CEV} = 100 V h_{FE} = 35-100 @ 750 mA File No. 137			**RCA29C/SDH** V_{CEO} = 100 V h_{FE} = 40 min. @ 0.2 A f_T = 0.8 MHz min. I_C = 2.5 A P_T = 50 W File No. 792	**RCA31C/SDH** V_{CEO} = 100 V h_{FE} = 25 min. @ 1 A f_T = 0.8 MHz min. I_C = 2.5 A P_T = 50 W File No. 793	**2N5497** **2N5496** V_{CER}(sus) = 80 V h_{FE} = 20-100 @ 3.5 A f_T = 0.8 MHz min. CT File No. 353

▲ Pellet size — values shown are edge dimensions in mils
** Available with free-air radiator $R_{\theta\,JA}$ = 30° C/W
CT — Complementary Type Available

Table XII—Hometaxial-Base N-P-N Power Transistors (cont'd)
I_C to 80 A . . . P_T to 300 W . . . V_{CE} to 170 V

I_c = 6 A max. P_T = 75 W max. (TO-3)	I_c = 8 A max. P_T = 83 W VERSAWATT (TO-220)	I_c = 10 A max. P_T = 150 W (TO-36)	I_c = 15 A max. P_T = 150 W max. (TO-3)	I_c = 16 A max. P_T = 75 W max. VERSAWATT (TO-220)
180 x 180▲	180 x 180	180 x 180	180 x 180	180 x 180
Family Designation				
2N1490	2N5037	2N2016	2N3055	2N6103
2N1487 V_{CEV} = 60 V h_{FE} = 15-45 @ 1.5 A	**2N5034** V_{CER}(sus) = 45 V h_{FE} = 20-80 @ 4 A f_T = 800 kHz min. I_c = 6 A	**2N2015** V_{CEO}(sus) = 50 V h_{FE} = 15-50 @ 5 A	**RCS242** V_{CER}(sus) = 50 V h_{FE} = 20 min. @ 3 A f_T = 0.8 MHz min. P_T = 115 W	**RCA41/SDH** V_{CEO} = 40 V h_{FE} = 30 min. @ 0.3 A f_T = 0.8 MHz min. I_c = 16 A
File No. 139	File No. 244	File No. 12	File No. 778	File No. 794
2N1489 V_{CEV} = 60 V h_{FE} = 25-75 @ 1.5 A	**2N5035** V_{CER}(sus) = 45 V h_{FE} = 20-80 @ 4 A f_T = 800 kHz min. I_c = 6 A	**2N2016** V_{CEO}(sus) ≈ 65 V h_{FE} = 15-50 @ 5 A	**2N6371** V_{CEV}(sus) = 50 V h_{FE} = 15-60 @ 8 A f_T = 1 MHz typ. P_T = 117 W CT File No. 607	**RCA41A/SDH** V_{CEO} = 60 V h_{FE} = 30 min. @ 0.3 A f_T = 0.8 MHz min. I_c = 10 A
File No. 139	File No. 244	File No. 12		File No. 794
2N1488 V_{CEV} = 100 V h_{FE} = 15-45 @ 1.5 A	**2N5036** V_{CER}(sus) = 60 V h_{FE} = 20-80 @ 5 A f_T = 800 kHz min. I_c = 8 A		**2N6253** V_{CER}(sus) = 55 V h_{FE} = 20-70 @ 3 A f_T = 0.8 MHz min. P_T = 115 W	**RCA41B/SDH** V_{CEO} = 80 V h_{FE} = 30 min. @ 0.3 A f_T = 0.8 MHz min. I_c = 10 A
File No. 139	File No. 244		File No. 524	File No. 794
2N1490 V_{CEV} = 100 V h_{FE} = 25-75 @ 1.5 A	**2N5037** V_{CER}(sus) = 60 V h_{FE} = 20-80 @ 5 A f_T = 800 kHz min. I_c = 8 A		**2N3055** V_{CER}(sus) = 70 V h_{FE} = 20-70 @ 4 A f_T = 0.8 MHz min. P_T = 115 W CT File No. 524	
File No. 139	File No. 244			
			2N6254 V_{CER}(sus) = 85 V h_{FE} = 20-70 @ 5 A f_T = 0.8 MHz min P_T = 150 W File No. 524	

▲ Pellet size — values shown are edge dimensions in mils
CT — Complementary Type Available

Table XII—Hometaxial-Base N-P-N Power Transistors (cont'd)
I_C to 80 A . . . P_T to 300 W . . . V_{CE} to 170 V

I_C = 16 A max. P_T = 75 W max. VERSAWATT (TO-220)	I_C = 10 A max. P_T = 150 W max. (TO-3)	I_C = 30 A max. P_T = 250 W max. (TO-3)	I_C = 16 A max. P_T = 250 W max. (TO-3)	I_C = 80 A max. P_T = 300 W max. (Modified TO-3)
180 x 180▲	180 x 180	250 x 250	250 x 250	380 x 380
Family Designation				
2N6103	2N3442	2N3771	2N3773	2N5578
2N6102 **2N6103** V_{CER}(sus) = 45 V h_{FE} = 15-60 @ 8 A f_T = 0.8 MHz min. I_C = 16 A max. File No. 485	**2N4347** V_{CEV}(sus) = 140 V h_{FE} = 15-60 @ 2 A f_T = 0.8 MHz typ. P_T = 100 W CT File No. 528	**2N6257** V_{CER}(sus) = 45 V h_{FE} = 15-75 @ 8 A f_T = 0.6 MHz min. P_T = 150 W I_C = 20 A File No. 525	**2N4348** V_{CEV}(sus) = 140 V h_{FE} = 15-60 @ 5 A f_T = 0.7 MHz typ. P_T = 120 W I_C = 10A File No. 526	**2N5575** V_{CEO}(sus) = 50 V h_{FE} = 10-40 @ 60 A f_T = 0.4 MHz min. File No. 359
2N6098 **2N6099** V_{CER}(sus) = 65 V h_{FE} = 20-80 @ 4 A f_T = 0.8 MHz min. I_C = 10 A max. File No. 485	**2N3442** V_{CEV}(sus) = 160 V h_{FE} = 20-70 @ 3 A f_T = 0.8 MHz typ. P_T = 117 W File No. 528	**2N3771** V_{CER}(sus) = 45 V h_{FE} = 15-60 @ 15 A f_T = 0.8 MHz min. P_T = 150 W I_C = 30 A File No. 525	**2N3773** V_{CEV}(sus) = 160 V h_{FE} = 15-60 @ 8 A f_T = 0.7 MHz typ. P_T = 150 W I_C = 16 A File No. 526	**2N5578** V_{CEO}(sus) = 70 V h_{FE} = 10-40 @ 40 A f_T = 0.4 MHz min. File No. 359
2N6100 **2N6101** V_{CER}(sus) = 75 V h_{FE} = 20-80 @ 5 A f_T = 0.8 MHz min. I_C = 10 A max. File No. 485	**2N6262** V_{CEV}(sus) = 170 V h_{FE} 20-70 @ 3 A f_T = 0.8 MHz min. P_T = 150 W File No. 528	**2N3772** V_{CER}(sus) = 70 V h_{FE} = 15-60 @ 10 A f_T = 0.8 MHz min. P_T = 150 W CT File No. 525	**2N6259** V_{CER}(sus) = 160 V h_{FE} = 15-60 @ 8 A f_T = 0.6 MHz typ. P_T = 250 W I_C = 16 A File No. 526	
		2N6258 V_{CER}(sus) = 85 V h_{FE} = 20-60 @ 15 A f_T = 0.6 MHz min. P_T = 250 W I_C = 30 A File No. 525	**43104** V_{CEX} = 160 V h_{FE} = 15-60 @ 8 A f_T = 0.7 MHz typ. I_C = 4 A P_T = 150 W File No. 622	

▲ Pellet size — values shown are edge dimensions in mils
CT — Complementary Type Available

Table XIII—Epitaxial-Base N-P-N and P-N-P Power Transistors I_C to 15 A . . . P_T to 200 W . . . V_{CE} to 125 V

I_C = −2 A max. P_T = 10 W max. RCP Plastic	I_C = 2 A max. P_T = 10 W max. RCP Plastic	I_C = −2 A max. P_T = 10 W max. RCP Plastic	I_C = 2 A max. P_T = 10 W max. RCP Plastic	I_C = −2 A max. P_T = 10 W max. RCP Plastic	I_C = 2 A max. P_T = 10 W max. RCP Plastic
42 x 42▲	42 x 42	42 x 42	42 x 42	42 x 42	42 x 42
Family Designation					
RCP700 [P-N-P]	RCP701 [N-P-N]	RCP700 [P-N-P]	RCP701 [N-P-N]	RCP700 [P-N-P]	RCP701 [N-P-N]
RCP706 V_{CEO}(sus) = −30 V h_{FE} = 20 min. @ −500 mA f_T = 50 MHz min. CT File No. 821	**RCP707** V_{CEO}(sus) = 30 V h_{FE} = 20 min. @ 500 mA f_T = 50 MHz min. CT File No. 820	**RCP702A** V_{CEO}(sus) = −40 V h_{FE} = 30-150 @ −500 mA f_T = 50 MHz min. CT File No. 821	**RCP703A** V_{CEO}(sus) = 40 V h_{FE} = 30-150 @ 500 mA f_T = 50 MHz min. CT File No. 820	**RCP700A** V_{CEO}(sus) = −40V h_{FE} = 50-250 @ −500 mA f_T = 50 MHz min. CT File No. 821	**RCP701A** V_{CEO}(sus) = 40 V h_{FE} = 50-250 @ 500 mA f_T = 50 MHz min. CT File No. 820
RCP704 V_{CEO}(sus) = −30 V h_{FE} = 50 min. @ −500 mA f_T = 50 MHz min. CT File No. 821	**RCP705** V_{CEO}(sus) = 30 V h_{FE} = 50 min. @ 500 mA f_T = 50 MHz min. CT File No. 820	**RCP702B** V_{CEO}(sus) = −60 V h_{FE} = 30-150 @ −500 mA f_T = 50 MHz min. CT File No. 821	**RCP703B** V_{CEO}(sus) = 60 V h_{FE} = 30-150 @ 500 mA f_T = 50 MHz min. CT File No. 820	**RCP700B** V_{CEO}(sus) = −60 V h_{FE} = 50-250 @ −500 mA f_T = 50 MHz min. CT File No. 821	**RCP701B** V_{CEO}(sus) = 60 V h_{FE} = 50-250 @ 500 mA f_T = 50 MHz min. CT File No. 820
RCP706B V_{CEO}(sus) = −60 V h_{FE} = 20 min. @ −500 mA f_T = 50 MHz min. CT File No. 821	**RCP707B** V_{CEO}(sus) = 60 V h_{FE} = 20 min. @ 500 mA f_T = 50 MHz min. CT File No. 820	**RCP702C** V_{CEO}(sus) = −80 V h_{FE} = 30-150 @ −500 mA f_T = 50 MHz min. CT File No. 821	**RCP703C** V_{CEO}(sus) = 80 V h_{FE} = 30-150 @ 500 mA f_T = 50 MHz min. CT File No. 820	**RCP700C** V_{CEO}(sus) = −80 V h_{FE} = 50-250 @ −500 mA f_T = 50 MHz min. CT File No. 821	**RCP701C** V_{CEO}(sus) = 80 V h_{FE} = 50-250 @ 500 mA f_T = 50 MHz min. CT File No. 820
RCP704B V_{CEO}(sus) = −60 V h_{FE} = 50 min. @ −500 mA f_T = 50 MHz min. CT File No. 821	**RCP705B** V_{CEO}(sus) = 60 V h_{FE} = 50 min. @ 500 mA f_T = 50 MHz min. CT File No. 820	**RCP704D** V_{CEO}(sus) = −100 V h_{FE} = 30-150 @ −500 mA f_T = 50 MHz min. CT File No. 821	**RCP703D** V_{CEO}(sus) = 100 V h_{FE} = 30-150 @ 500 mA f_T = 50 MHz min. CT File No. 820	**RCP700D** V_{CEO}(sus) = −100 V h_{FE} = 50-250 @ −500 mA f_T = 50 MHz min. CT File No. 821	**RCP701D** V_{CEO}(sus) = 100 V h_{FE} = 50-250 @ 500 mA f_T = 50 MHz min. CT File No. 820

▲ Pellet size — values shown are edge dimensions in mils

CT — Complementary Type Available

Table XIII—Epitaxial-Base N-P-N and P-N-P Power Transistors (cont'd)
I_C to 15 A . . . P_T to 200 W . . . V_{CE} to 125 V

$I_C = -3.5$ max. $P_T = 10$ W max. (TO-39)*	$I_C = 6$ A max. $P_T = 40$ W max. (TO-66)**	$I_C = -6$ A max. $P_T = 40$ W max. (TO-66)**	$I_C = 7$ A max. $P_T = 40$ W max. VERSAWATT (TO-220)	$I_C = 7$ A max. $P_T = 40$ W max. VERSAWATT (TO-220)
90 x 90▲	90 x 90	90 x 90	90 x 90	90 x 90
Family Designation				
2N5783 [P-N-P]	2N6372 [N-P-N]	2N5954 [P-N-P]	2N6292 [N-P-N]	2N6292 [N-P-N]
2N5783● V_{CER}(sus) = -45 V h_{FE} = 20-100 @ -1.6 A f_T = 8 MHz min. CT File No. 413	**2N6374** V_{CER}(sus) = 45 V h_{FE} = 20-100 @ 3 A f_T = 4 MHz min. CT File No. 675	**2N5956** V_{CER}(sus) = -45 V h_{FE} = 20-100 @ -3 A f_T = 5 MHz min. CT File No. 675	**2N6288** **2N6289** V_{CER}(sus) = 40 V h_{FE} = 30-150 @ 3 A f_T = 4 MHz min. CT File No. 676	**41500** V_{CEX} = 35 V h_{FE} = 25 min. @ 1 A f_T = 4 MHz min. CT File No. 772
2N5782● V_{CER}(sus) = -65 V h_{FE} = 20-100 @ -1.2 A f_T = 8 MHz min. CT File No. 413	**2N6373** V_{CER}(sus) = 65 V h_{FE} = 20-100 @ 2.5 A f_T = 4 MHz min. CT File No. 675	**2N5955** V_{CER}(sus) = -65 V h_{FE} = 20-100 @ -2.5 A f_T = 5 MHz min. CT File No. 675	**2N6290** **2N6291** V_{CER}(sus) = 60 V h_{FE} = 30-150 @ 2.5 A f_T = 4 MHz min. CT File No. 676	
2N5781● V_{CER}(sus) = -80 V h_{FE} = 20-100 @ -1 A f_T = 8 MHz min. CT File No. 413	**2N6372** V_{CER}(sus) = 85 V h_{FE} = 20-100 @ 2 A f_T = 4 MHz min. CT File No. 675	**2N5954** V_{CER}(sus) = -85 V h_{FE} = 20-100 @ -2 A f_T = 5 MHz min. CT File No. 675	**2N6292** **2N6293** V_{CER}(sus) = 80 V h_{FE} = 30-150 @ 2 A f_T = 4 MHz min. CT File No. 676	
		2N6467 V_{CER}(sus) = -105 V h_{FE} = 20-100 @ -1 A f_T = 5 MHz min. File No. 675	**2N6473** V_{CER}(sus) = 110 V h_{FE} = 30-150 @ 1.5 A f_T = 5 MHz typ. CT File No. 676	
		2N6468 V_{CER}(sus) = -125 V h_{FE} = 20-100 @ -1 A f_T = 5 MHz min. File No. 675	**2N6474** V_{CER}(sus) = 130 V h_{FE} = 30-150 @ 1 A f_T = 5 MHz typ. CT File No. 676	

▲ Pellet size — values shown are edge dimensions in mils
· Available with
  a. flange for easy heat sinking $R\theta_{JC} = 15°$ C/W
  b. free-air radiator $R\theta_{JA} = 40$-$50°$ C/W
** Available with free-air radiator $R\theta_{JA} = 30°$ C/W
● These transistors are also available in TO-5 packages in U.S.A., Canada, Latin America, and Far East
CT — Complementary Type Available

Table XIII—Epitaxial-Base N-P-N and P-N-P Power Transistors (cont'd)
I_C to 15 A . . . P_T to 200 W . . . V_{CE} to 125 V

I_C = −7 A max. P_T = 40 W max. VERSAWATT (TO-220)	I_C = −7 A max. P_T = 40 W max. VERSAWATT (TO-220)	I_C = 15 A max. P_T = 125 W max. (TO-3)	I_C = −15 A max. P_C = 125 W max. (TO-3)	I_C = 15 A max. P_T = 75 W max. VERSAWATT (TO-220)	I_C = −15 A max. P_T = 75 W max. VERSAWATT (TO-220)
90 x 90▲	90 x 90	150 x 150	150 x 150	150 x 150	150 x 150
Family Designation					
2N6107 [P-N-P]	2N6107 [P-N-P]	2N6472 [N-P-N]	2N6248 [P-N-P]	2N6488 [N-P-N]	2N6491 [P-N-P]
41501 V_{CER}(sus) = −35 V h_{FE} = 25 min. @ −1 A CT File No. 770	2N6110 2N6111 V_{CER}(sus) = −40 V h_{FE} = 30-150 @ −3 A f_T = 10 MHz min. CT File No. 676	2N6470 V_{CER}(sus) = 45 V h_{FE} = 20-100 @ 5 A f_T = 5 MHz typ. CT File No. 677	2N6469 V_{CER}(sus) = −45 V h_{FE} = 20-100 @ −5 A f_T = 6 MHz min. CT File No. 677	2N6486 V_{CER}(sus) = 50 V h_{FE} = 30-150 @ 6 A f_T = 5 MHz typ. CT File No. 678	2N6489 V_{CER}(sus) = −50 V h_{FE} = 30-150 @ −6 A f_T = 5 MHz typ. CT File No. 678
	2N6108 2N6109 V_{CER}(sus) = −60 V h_{FE} = 30-150 @ −2.5 A f_T = 10 MHz min. CT File No. 676	2N6471 V_{CER}(sus) = 65 V h_{FE} = 20-100 @ 7 A f_T = 5 MHz typ. CT File No. 677	2N6246 V_{CER}(sus) = −65 V h_{FE} = 20-100 @ −7 A f_T = 6 MHz min. CT File No. 677	2N6487 V_{CER}(sus) = 70 V h_{FE} = 30-150 @ 5 A f_T = 5 MHz typ. CT File No. 678	2N6490 V_{CER}(sus) = −70 V h_{FE} = 30-150 @ −5 A f_T = 5 MHz typ. CT File No. 678
	2N6106 2N6107 V_{CER}(sus) = −80 V h_{FE} = 30-150 @ −2 A f_T = 10 MHz min. CT File No. 676	2N6472 V_{CER}(sus) = 85 V h_{FE} = 20-100 @ 6 A f_T = 5 MHz typ. CT File No. 677	2N6247 V_{CER}(sus) = −85 V h_{FE} = 20-100 @ −6 A f_T = 6 MHz min. CT File No. 677	2N6488 V_{CER}(sus) = 90 V h_{FE} = 30-150 @ 4 A f_T = 5 MHz typ CT File No. 678	2N6491 V_{CER}(sus) = −90 V h_{FE} = 30-150 @ −4 A f_T = 5 MHz typ. CT File No. 678
	2N6475 V_{CER}(sus) = −110 V h_{FE} = 30-150 @ −1.5 A f_T = 5 MHz typ. CT File No. 676		2N6248 V_{CER}(sus) = −105 V h_{FE} = 20-100 @ −5 A f_T = 6 MHz min. File No. 677		
	2N6476 V_{CER}(sus) = −130 V h_{FE} = 30-150 @ −1 A f_T = 5 MHz typ. CT File No. 676				

▲ Pellet size — values shown are edge dimensions in mils

CT — Complementary Type Available

Table XIV—High-Voltage N-P-N and P-N-P Power Transistors
I_C to 30 A . . . f_T to 20 MHz . . . P_T to 175 W

I_C = 150 mA max. P_T = 6.25 W max. RCP Plastic	I_C = 150 mA max. P_T = 6.25 W max. RCP Plastic	I_C = 150 mA max. P_T = 6.25 W max. RCP Plastic	I_C = 1 A max. P_T = 20 W max. (Plastic TO-5)	I_C = 1 A max. P_T = 10 W max. (TO-39)*
32 x 32▲	32 x 32	32 x 32	32 x 32	42 x 42
Family Designation				
RCP111 [N-P-N]	RCP111 [N-P-N]	RCP111 [N-P-N]	2N6177 [N-P-N]	2N3439 [N-P-N]
RCP117 V_{CEO}(sus) = 100 V h_{FE} = 20 min. @ 25 mA f_T = 80 MHz typ. File No. 822	**RCP113A** V_{CEO}(sus) = 200 V h_{FE} = 30-150 @ 25 mA f_T = 80 MHz typ. File No. 822	**RCP111A** V_{CEO}(sus) = 200 V h_{FE} = 50-300 @ 25 mA f_T = 80 MHz typ. File No. 822	**41505** V_{CEO}(sus) = 200 V h_{FE} = 20 min. @ 50 mA File No. 771	**2N3440●** V_{CEO}(sus) = 300 V h_{FE} = 40-160 @ 20 mA f_T = 15 MHz min. File No. 64
RCP115 V_{CEO}(sus) = 100 V h_{FE} = 50 min. @ 25 mA f_T = 80 MHz typ. File No. 822	**RCP113B** V_{CEO}(sus) = 250 V h_{FE} = 30-150 @ 25 mA f_T = 80 MHz typ. File No. 822	**RCP111B** V_{CEO}(sus) = 250 V h_{FE} = 50-300 @ 25 mA f_T = 80 MHz typ. File No. 822	**2N6175 40885■** "Plastic 2N3440" V_{CER}(sus) = 300 V h_{FE} = 30-190 @ 20 mA f_T = 20 MHz min. CT File No. 508	**2N3439●** V_{CER}(sus) = 400 V h_{FE} = 40-160 @ 20 mA f_T = 15 MHz min. File No. 64
RCP117B V_{CEO}(sus) = 250 V h_{FE} = 20 min. @ 25 mA f_T = 80 MHz typ. File No. 822	**RCP113C** V_{CEO}(sus) = 300 V h_{FE} = 30-150 @ 25 mA f_T = 80 MHz typ. File No. 822	**RCP111C** V_{CEO}(sus) = 300 V h_{FE} = 50-300 @ 25 mA f_T = 80 MHz typ. File No. 822	**2N6176 40886■** V_{CER}(sus) = 350 V h_{FE} = 30-150 @ 20 mA f_T = 20 MHz min. CT File No. 508	
RCP115B V_{CEO}(sus) = 250 V h_{FE} = 50 min. @ 25 mA f_T = 80 MHz typ. File No. 822	**RCP113D** V_{CEO}(sus) = 350 V h_{FE} = 30-150 @ 25 mA f_T = 80 MHz typ. File No. 822	**RCP111D** V_{CEO}(sus) = 350 V h_{FE} = 50-300 @ 25 mA f_T = 80 MHz typ. File No. 822	**2N6177 40887■** "Plastic 2N3439" V_{CER}(sus) = 400 V h_{FE} = 30-150 @ 50 mA f_T = 20 MHz min. CT File No. 508	

▲ Pellet size — values shown are edge dimensions in mils
• Available with
  a. flange for easy heat sinking $R\theta_{JC}$ = 15° C/W
  b. free-air radiator $R\theta_{JA}$ = 45° C/W
■ Type with a factory-attached heat clip
● These transistors are also available in TO-5 packages in U.S.A., Canada, Latin America, and Far East
CT — Complementary Type Available

Table XIV—High-Voltage N-P-N and P-N-P Power Transistors (cont'd)
I_C to 30 A . . . f_T to 20 MHz . . . P_T to 175 W

$I_C = -1$ A max. $P_T = 10$ W max. (TO-39)*	$I_C = 5$ A max. $P_T = 35$ W max. (TO-66)**	$I_C = -5$ A max. $P_T = 35$ W max. (TO-66)**	$I_C = -2$ A max. $P_T = 20$ W max. (TO-66)**	$I_C = 10$ A peak $P_T = 45$ W max. (TO-66)**
42 x 42▲	103 x 103	124 x 124	124 x 124	130 x 130
Family Designation				
2N5415 [P-N-P]	2N3585 [N-P-N]	2N6213 [P-N-P]	2N6213 [P-N-P]	2N6079 [N-P-N]
RCS880 V_{CER}(sus) = -150 V h_{FE} = 20-150 @ -50 mA P_T = 7.5 W — File No. 777	**2N3583** V_{CER}(sus) = 250 V h_{FE} = 40 min. @ 100 mA h_{FE} = 10 min. @ 1 A f_T = 15 MHz min. — File No. 138	**2N6211** V_{CER}(sus) = -250 V h_{FE} = 10-100 @ -1 A f_T = 20 MHz min. — CT File No. 507	**RCS560** V_{CER}(sus) = -225 V h_{FE} = 7.5 min. @ -0.75 A f_T = 20 MHz min. — File No. 782	**2N6078** V_{CER}(sus) = 275 V h_{FE} = 12-70 @ 1.2 A t_r = 0.3 μs typ. t_f = 0.3 μs typ. — File No. 492
2N5415 V_{CEO}(sus) = -200 V h_{FE} = 30-150 @ -50 mA f_T = 15 MHz min. R_T = 10 W — CT File No. 336	**2N3584** V_{CER}(sus) = 300 V h_{FE} = 40 min. @ 100 mA h_{FE} = 25-100 @ 1 A f_T = 15 MHz min. — CT File No. 138	**2N6212** V_{CER}(sus) = -325 V h_{FE} = 10-100 @ -1 A f_T = 20 MHz min. — CT File No. 507	**RCS559** V_{CER}(sus) = -250 V h_{FE} = 10-100 @ -0.75 A f_T = 20 MHz min. — File No. 782	**2N6077** V_{CER}(sus) = 300 V h_{FE} = 12-70 @ 1.2 A t_r = 0.3 μs typ. t_f = 0.3 μs typ. — CT File No. 492
RCS881 V_{CEO}(sus) = -250 V h_{FE} = 20 min. @ -35 mA f_T = 15 MHz min. P_T = 7.5 W — File No. 780	**2N3585** V_{CER}(sus) = 400 V h_{FE} = 40 min. @ 100 mA h_{FE} = 25-100 @ 1 A f_T = 15 MHz min. — CT File No. 138	**2N6213** V_{CER}(sus) = -375 V h_{FE} = 10-100 @ -1 A f_T = 20 MHz min. — File No. 507		**2N6079** V_{CER}(sus) = 375 V h_{FE} = 12-50 @ 1.2 A t_r = 0.3 μs typ. t_f = 0.3 μs typ. — File No. 492
RCS882 V_{CER}(sus) = -350 V h_{FE} = 20 min. @ -35 mA f_T = 15 MHz min. P_T = 7.5 W — File No. 781	**2N4240** V_{CER}(sus) = 400 V h_{FE} = 40 min. @ 100 mA h_{FE} = 30-150 @ 750 mA f_T = 15 MHz min. — File No. 138	**2N6214** V_{CER}(sus) = -425 V h_{FE} = 10-100 @ -1 A f_T = 20 MHz min. — File No. 507		**40851** V_{CER}(sus) = 375 V h_{FE} = 12 min. @ 1.2 A t_r = 0.3 μs typ. t_f = 0.3 μs typ. — File No. 498
2N5416 V_{CER}(sus) = -350 V h_{FE} = 30-120 @ -50 mA f_T = 15 MHz min. P_T = 10 W — CT File No. 336	**40850** V_{CER}(sus) = 400 V h_{FE} = 25 min. @ 750 mA f_T = 15 MHz min. — File No. 498			

▲ Pellet size — values shown are edge dimensions in mils
* Available with
  a. flange for easy heat sinking $R\theta_{JC} = 15°$ C/W
  b. free-air radiator $R\theta_{JA} = 45°$ C/W
** Available with free-air radiator $R\theta_{JA} = 30°$ C/W
CT — Complementary Type Available

Table XIV—High-Voltage N-P-N and P-N-P Power Transistors (cont'd)
I_C to 30 A . . . f_T to 20 MHz . . . P_T to 175 W

I_C = 10 A peak P_T = 125 W max. (TO-3)	I_C = 5 A max. P_T = 100 W max. (TO-3)		I_C = 7 A max. P_T = 125 W max. (TO-3)	I_C = 15 A peak P_T = 110 W max. (TO-3)	I_C = 30 A peak P_T = 175 W max. (TO-3) Switching
	Switching	Linear			
130 x 130▲	130 x 130	130 x 130	180 x 180	210 x 210	260 x 260
Family Designation					
2N5840 [N-P-N]	2N5840 [N-P-N]	2N5240 [N-P-N]	2N6510 [N-P-N]	2N5804 [N-P-N]	2N6252 [N-P-N]
RCA 410# V_{CEO} = 200 V h_{FE} = 30-90 @ 1 A t_r = 0.35 μs typ. t_f = 0.15 μs typ. File No. 509	**41506** V_{CEO}(sus) = 200 V h_{FE} = 8 min. @ 2 A File No. 776	**2N5239** V_{CER}(sus) = 250 V h_{FE} = 20 min. @ 2 A h_{FE} = 20-80 @ 0.4 A f_T = 5 MHz min. File No. 321	**2N6510** V_{CER}(sus) = 250 V h_{FE} = 10 min. @ 3 A f_T = 3 MHz min. File No. 848	**2N5804** V_{CER}(sus) = 300 V h_{FE} = 25-250 @ 0.5 A h_{FE} = 10-100 @ 5 A t_r = 0.4 μs typ. t_f = 1.2 μs typ. File No. 407	**2N6249** V_{CER}(sus) = 225 V h_{FE} = 10-50 @ 10 A t_r = 0.8 μs typ. t_f = 0.5 μs typ. File No. 523
RCA 411# V_{CEO}(sus) = 300 V h_{FE} = 30-90 @ 1 A t_r = 0.35 μs typ. t_f = 0.15 μs typ. File No. 510	**2N5838** V_{CER}(sus) = 275 V h_{FE} = 20 min. @ 0.5 A h_{FE} = 8-40 @ 3 A t_r = 0.8 μs typ. t_f = 0.4 μs typ. File No. 410	**2N5240** V_{CER}(sus) = 350 V h_{FE} = 20 min. @ 2 A h_{FE} = 20-80 @ 0.4 A f_T = 5 MHz min. File No. 321	**2N6511** V_{CER}(sus) = 300 V h_{FE} = 10 min. @ 4 A f_T = 3 MHz min. File No. 848	**2N5805** V_{CER}(sus) = 375 V h_{FE} = 25-250 @ 0.5 A h_{FE} = 10-100 @ 5 A t_r = 0.4 μs typ. t_f = 1.2 μs typ. File No. 407	**RCS564** V_{CER}(sus) = 225 V h_{FE} = 5 min. @ 10 A t_r = 0.8 μs typ. t_f = 1 μs typ. File No. 782
RCA 413# V_{CEO}(sus) = 325 V h_{FE} = 20-80 @ 0.5 A t_r = 0.35 μs typ. t_f = 0.15 μs typ. File No. 511	**2N5839** V_{CER}(sus) = 300 V h_{FE} = 20 min. @ 0.5 A h_{FE} = 10-50 @ 2 A t_r = 0.6 μs typ. t_f = 0.35 μs typ. File No. 410		**2N6512** V_{CER}(sus) = 350 V h_{FE} = 10 min. @ 4 A f_T = 3 MHz min. File No. 848	**40853** V_{CER}(sus) = 375 V h_{FE} = 10 min. @ 5 A t_r = 0.4 μs typ. t_f = 1.2 μs typ. File No. 498	**2N6250** V_{CER}(sus) = 300 V h_{FE} = 8-50 @ 10 A t_r = 0.8 μs typ. t_f = 0.5 μs typ. File No. 523
RCA 423# V_{CEO}(sus) = 325 V h_{FE} = 30-90 @ 1 A t_r = 0.35 μs typ. t_f = 0.15 μs typ. File No. 512	**2N5840** V_{CER}(sus) = 375 V h_{FE} = 20 min. @ 0.5 A h_{FE} = 10-50 @ 2 A t_r = 0.6 μs typ. t_f = 0.35 μs typ. File No. 410		**2N6514** V_{CER}(sus) = 350 V h_{FE} = 10 min. @ 5 A f_T = MHz min. File No. 848		**2N6251** V_{CER}(sus) = 375 V h_{FE} = 6-50 @ 10 A t_r = 0.8 μs typ. t_f = 0.5 μs typ. File No. 523
RCA 431# V_{CEO}(sus) = 325 V h_{FE} = 15-35 @ 2.5 A t_r = 0.35 μs typ. t_f = 0.15 μs typ. File No. 513	**40852** V_{CER}(sus) = 375 V h_{FE} = 12 min. @ 1.2 A t_r = 0.5 μs typ. t_f = 0.35 μs typ. File No. 498		**2N6513** V_{CER}(sus) = 400 V h_{FE} = 10 min. @ 4 A f_T = 3 MHz min. File No. 848		**40854** V_{CER}(sus) = 325 V h_{FE} = 8 min. @ 10 A t_r = 0.8 μs typ. t_f = 0.5 μs typ. File No. 498

▲ Pellet size — values shown are edge dimensions in mils
\# For new equipment design only—not recommended for retrofit.

Table XV—High-Speed Switching N-P-N and P-N-P Power Transistors
I_C to 60 A . . . f_T to 250 MHz . . . P_T to 140 W

I_c = 1 A max. P_T = 5 W max. (TO-39)*	I_c = −1 A max. P_T = 7 W max. (TO-39)*	I_c = 2 A max. P_T = 5 W max. (Lo Profile TO-39)	I_c = 2 A max. P_T = 10 W max. (TO-39)*	I_c = −2 A max. P_T = 10 W max. (TO-39)*	I_c = 2 A max. P_T = 25 W max. (Plastic TO-5)
30 x 30▲	30 x 30	32 x 32	42 x 42	42 x 42	42 x 42
Family Designation					
2N2102 [N-P-N]	2N4036 [P-N-P]	2N5262 [N-P-N]	2N5320 [N-P-N]	2N5322 [P-N-P]	2N6179 [N-P-N]
41502 V_{CEO}(sus) = 30 V h_{FE} = 20 min. @ 150 mA P_T = 3 W CT File No. 773	**41503** V_{CEO}(sus) = −30 V h_{FE} = 20 min. @ −150 mA CT File No. 774	**2N5189** V_{CEO}(sus) = 35 V h_{FE} = 15 min. @ 1 A f_T = 250 MHz min. t_{on} = 40 ns max. t_{off} = 70 ns max. File No. 296	**2N5321●** V_{CER}(sus) = 65 V h_{FE} = 40-250 @ 500 mA f_T = 50 MHz min. t_{on} = 80 ns max. t_{off} = 800 ns max. CT File No. 325	**2N5323●** V_{CER}(sus) = −65 V h_{FE} = 40-250 @ −500 mA f_T = 50 MHz min. CT File No. 325	**2N6179** "Plastic 2N5321" V_{CER}(sus) = 65 V h_{FE} = 40-250 @ 500 mA f_T = 50 MHz min. t_{on} = 80 ns max. t_{off} = 800 ns max. CT File No. 562
2N3053● V_{CER}(sus) = 50 V h_{FE} = 50-250 @ 150 mA f_T = 100 MHz min. P_T = 5 W CT File No. 432	**2N4037●** V_{CER}(sus) = −60 V h_{FE} = 50-250 @ −150 mA f_T = 60 MHz min. CT File No. 216	**2N5262** V_{CEO}(sus) = 50 V h_{FE} = 25 min. @ 1 A f_T = 250 MHz min. t_{on} = 30 ns max. t_{off} = 60 ns max. CT File No. 313	**2N5320●** V_{CER}(sus) = 90 V h_{FE} = 30-130 @ 500 mA f_T = 50 MHz min. t_{on} = 80 ns max. t_{off} = 800 ns max. CT File No. 325	**2N5322●** V_{CER}(sus) = −90 V h_{FE} = 30-130 @ −500 mA h_{FE} = 10 min. @ −1 A f_T = 50 MHz min. CT File No. 325	**2N6178** "Plastic 2N5320" V_{CER}(sus) = 90 V h_{FE} = 30-130 @ 500 mA f_T = 50 MHz min. t_{on} = 80 ns max. t_{off} = 800 ns. max. CT File No. 562
2N2102● V_{CER}(sus) = 80 V h_{FE} = 40-120 @ 150 mA f_T = 120 MHz min. P_T = 5 W CT File No. 106	**2N4036●** V_{CER}(sus) = −85 V h_{FE} = 40-140 @ −150 mA f_T = 60 MHz min. CT File No. 216				
	2N4314● V_{CER}(sus) = −85 V h_{FE} = 50-250 @ −150 mA f_T = 60 MHz min. File No. 216				

▲ Pellet size — values shown are edge dimensions in mils
· Available with
   a. flange for easy heat sinking $R\theta_{JC}$ = 15° C/W
   b. free-air radiator $R\theta_{JA}$ = 50° C/W
● These transistors are also available in TO-5 packages in U.S.A., Canada, Latin America, and Far East
CT — Complementary Type Available

Table XV—High-Speed Switching N-P-N and P-N-P Power Transistors (cont'd)
I_C to 60 A . . . f_T to 250 MHz . . . P_T to 140 W

I_C = −2 A max. P_T = 25 W max. (Plastic TO-5)	I_C = 7 A max. P_T = 35 W max. (TO-66)**	I_C = 15 A max. P_C = 85 W max. (Radial)	I_C = 20 A max. P_T = 140 W max. (TO-3)	I_C = 25 A max. P_T = 125 W max. (TO-63)	I_C = 30 A max. P_T = 140 W max. (TO-3)	I_C = 60 A max. P_C = 140 W max. (Modified TO-3)
42 x 42▲	103 x 103	155 x 155	146 x 183	215 x 222	220 x 220	220 x 220 [2 CHIPS]
Family Designation			Family Designation			
2N6181 [P-N-P]	2N3879 [N-P-N]	2N6480 [N-P-N]	2N5038 [N-P-N]	2N3263 [N-P-N]	2N5671 [N-P-N]	2N6033 [N-P-N]
2N6181 "Plastic 2N5323" V_{CER}(sus) = −65 V h_{FE} = 40-250 @ −500 mA f_T = 50 MHz min. CT File No. 562	2N3878‡ V_{CER}(sus) = 60 V h_{FE} = 20 min. @ 4 A h_{FE} = 50-200 @ 0.5 A f_T = 60 MHz min. t_r = 400 ns max. t_f = 400 ns max. I_c = 7 A File No. 766	2N6479 (Isolated Collector) 2N6481 (Non-Isolated Coll.) V_{CER}(sus) = 80 V h_{FE} = 20 min. @ 12 A f_T = 100 MHz typ. Radiation Hard File No. 702	2N5039 V_{CER}(sus) = 95 V h_{FE} = 20 min. @ 10 A h_{FE} = 30-150 @ 2 A f_T = 60 MHz min. t_{on} = 0.5 μs max. t_{off} = 2 μs max. File No. 698	2N3266 2N3264■ V_{CER}(sus) = 80 V h_{FE} = 20-80 @ 15 A f_T = 20 MHz min. t_{on} = 0.5 μs max. t_{off} = 2 μs max. File No. 54	2N5671 V_{CER}(sus) = 110 V h_{FE} = 20 min. @ 20 A h_{FE} = 20-100 @ 15 A f_T = 50 MHz min. t_{on} = 0.5 μs max. t_{off} = 2 μs max. File No. 383	2N6032 V_{CER}(sus) = 110 V h_{FE} = 10-50 @ 50 A f_T = 50 MHz min. t_r = 1 μs max. t_f = 0.5μs max. File No. 462
2N6180 "Plastic 2N5322" V_{CER}(sus) = −90 V h_{FE} = 30-130 @ −500 mA @ −1 A f_T = 50 MHz min. CT File No. 562	2N3879 V_{CER}(sus) = 90 V h_{FE} = 40 min. @ 0.4 A h_{FE} = 20-80 @ 4 A f_T = 60 MHz min. t_r = 400 ns max. t_f = 400 ns max. I_c = 7 A File No. 766	2N6480 (Isolated Collector) 2N6482 (Non-Isolated Coll.) V_{CER}(sus) = 80 V h_{FE} = 20 min. @ 12 A f_T = 100 MHz typ. Radiation Hard File No. 702	2N5038 V_{CER}(sus) = 110 V h_{FE} = 20 min. @ 12 A h_{FE} = 50-200 @ 2 A f_T = 60 MHz min. t_{on} = 0.5 μs max. t_{off} = 2 μs max. File No. 698	2N3265 2N3263■ V_{CER}(sus) = 110 V h_{FE} = 25-75 @ 15 A f_T = 20 MHz min. t_{on} = 0.5 μs max. t_{off} = 2 μs max. File No. 54	2N5672 V_{CER}(sus) = 140 V h_{FE} = 20 min. @ 20 A h_{FE} = 20-100 @ 15 A f_T = 50 MHz min. t_{on} = 0.5 μs max. t_{off} = 2 μs max. File No. 383	2N6033 V_{CER}(sus) = 140 V h_{FE} = 10-50 @ 40 A f_T = 50 MHz min. t_r = 1 μs max. t_f = 0.5 μs max. File No. 462
	2N5202 V_{CER}(sus) = 75 V h_{FE} = 10-100 @ 4 A f_T = 60 MHz min. t_r = 400 ns max. t_f = 400 ns max. I_c = 4 A File No. 766		2N6496 V_{CER}(sus) = 130 V h_{FE} = 12-100 @ 8 A f_T = 60 MHz min. t_r = 0.5 μs max. t_s = 1 μs max. t_f = 0.5 μs max. File No. 698			
	2N6500 V_{CER}(sus) = 110 V h_{FE} = 15-60 @ 3 A f_T = 60 MHz min. t_r = 400 ns max. t_f = 500 ns max. I_c = 4 A File No. 766		2N6354 V_{CER}(sus) = 130 V h_{FE} = 20-150 @ 5 A h_{FE} = 10-100 @ 10 A f_T = 80 MHz min. t_r = 0.3 μs max. t_f = 0.2 μs max. I_c = 12 A peak File No. 582			

▲ Pellet size — values shown are edge dimensions in mils
** Available with free-air radiator $R\theta_{JA} = 30°$ C/W
‡ Also available with heat radiator (40375).
■ Flat radial lead version
CT — Complementary Type Available

Table XVI—Power Transistors For Output and Driver Stages in Audio-Frequency Linear Amplifiers

Power Output (8 Ω Imped.)	Bull. File No.	Circuit No.	Output Transistors		Class B Driver Transistors		V_{BE} Mult. (Bias)
			N-P-N	P-N-P	N-P-N	P-N-P	
12 W	642	A012B (True Comp.)	RCA1C10 (2N6292)	RCA1C11 (2N6107)	–	–	–
	642	A012D (IC Driving True Comp.)	RCA1C10 (2N6292)	RCA1C11 (2N6107)	–	–	–
25 W	643	A025C (Quasi-Comp.)	RCA1C14 [2] (2N5496)	–	RCA1A06 (2N2102)	RCA1A05 (2N4036)	–
	644	A025B (Full-Comp.)	RCA1C05 (2N6292)	RCA1C06 (2N6107)	RCA1A06 (2N2102)	RCA1A05 (2N4036)	–
40 W	645	A040C (Quasi-Comp.)	RCA1C09 [2] (2N6103)	–	RCA1A06 (2N2102)	RCA1A05 (2N4036)	–
	646	A040B (Full-Comp.)	RCA1C07 (2N6488)	RCA1C08 (2N6491)	RCA1A06 (2N2102)	RCA1A05 (2N4036)	–
	791	A040D (Full-Comp. Darlington Output)	RCA1B07 (2N6385)	RCA1B08 (TA8925)	–	–	RCA1A18 (2N2102)
70 W	647	A070A (Quasi-Comp. Hom. Output)	RCA1B01 [2] (2N3055)	–	RCA1A03 (2N5320)	RCA1A04 (2N5322)	–
	648	A070C (Quasi-Comp.)	RCA1B06 [2] (2N5840)	–	RCA1C03 (2N6474)	RCA1A04 (2N6476)	RCA1A18 (2N2102)
120 W	649	A120C (Quasi-Comp. Parallel Output)	RCA1B04 [4] (2N5240)	–	RCA1C12 (2N6474)	RCA1C13 (2N6476)	RCA1A18 (2N2102)
200 W	650	A200C (Quasi-Comp. Parallel Output)	RCA1B05 [6] (2N5240)	–	RCA1B05 [2] (2N5240)	–	RCA1A18 (2N2102)

Numbers in brackets indicate number of devices used in the stage.
Type numbers in parentheses indicate the transistor-family designation.

Table XVII—Typical Power Output with 4- and 16-Ohm Load for AF Linear Amplifiers Listed in Table XVI

Amplifier Circuit No.	A012B		A012D		A025A		A025B		A040A		A040B		A040D		A070A		A070C		A120C		A200C	
Impedance — Ω (Load)	4	16	4	16	4	16	4	16	4	16	4	16	4	8	4	16	4	16	4	16	4	16
Typical Power Output — W	12*	6.5	9*	6.5	45	16	45	16	55	25	75	25	40	30	100	40	100	50	180	80	300	130

* Power output limited by driver-circuit capability.
Several of these amplifiers are shown in circuits section of this manual.

High-Frequency Power Transistors

POWER transistors are used in high-frequency amplifiers for military, industrial, and consumer applications. They are operated class A, B, or C, with frequency- or amplitude-modulation, single sideband or double sideband, in environments ranging from airborne to marine.

The increasing number of rf power transistors available today offers the circuit designer a wide selection from which to determine the optimum type for a particular application. The choice is based on factors such as maximum power output, maximum operating frequency, operating efficiency, power gain, reliability, and cost per watt of power generated. The ultimate choice of the transistors produced by any manufacturer, therefore, is dependent upon how well the devices perform in relation to these critical factors. RCA "overlay" silicon power transistors offer significant advantages for rf power applications at frequencies that extend well into the microwave region.

PHYSICAL DESIGN

During the past several years, considerable effort has been expended to improve the quality and reliability of high-frequency power transistors and simultaneously to advance the power-frequency capability of these devices. The technological developments that resulted from this effort have made possible transistor structures that can provide substantial power output and gain and high operating efficiency at frequencies that extend well into the microwave region. Such devices can now be produced with confidence in high-volume productions for use in applications in which high-quality performance and high reliability are primary requirements.

At high current levels, the emitter current of a transistor is concentrated at the emitter-base edge. This current-crowding effect results because the current flow through the base region, between the emitter and the collector, causes a voltage drop that produces the maximum forward bias at the edge of the emitter closest to the base contact. The center of the emitter, therefore, injects very little current. Because of this edge-injection phenomenon, a high emitter periphery-to-area ratio is essential to the achievement of high current-handling capability. This requirement has been a major factor in the evolution of transistor emitters from the circle type, to the line type, to the comb type, and finally to the overlay type of structure.

In addition to the requirement for a high emitter periphery-to-area ratio, power transistors intended for use in high-frequency applications must also exhibit a low capacitance and a short carrier transit time between emitter and collector. These latter factors critically affect the frequency capabilities of the device and, therefore, must be considered in the design of a high-frequency transistor structure.

Overlay Transistor Structure

The exceptional high-frequency power capabilities of the overlay power transistors result from the unique emitter construction used in these devices. In overlay transistors, the size of the emitters is substantially reduced, and a large number (from 16 to several hundred) of separate emitter sites are connected in parallel. This method of construction results in the high emitter periphery-to-area ratios, and makes possible the high current-handling capabilities, low capacitances, and short transit times between emitter and collector, that are required for rf power transistors.

The overlay transistor takes its name from the emitter metallization, shown in Fig. 124, that lies over the base instead of adjacent to it as in the interdigitated structure.

The actual base and emitter areas beneath the metal pattern are insulated from one another by a silicon dioxide layer. The overlay arrangement provides a substantial increase in over-all emitter periphery without increasing the physical area of the device, and thus improves the power-frequency capability of the device.

In addition to the standard base and emitter diffusions, an added diffused region in the base serves as a conductor grid. This p+ region offers three advantages: (1) it distributes base current uniformly over all the separate emitter sites, (2) it reduces the base-contact resistances between the aluminum metallization and the silicon material, and (3) it permits the use of larger emitter sites which makes possible higher design ratios and wider emitter metallizing fingers that result in lower current density.

For lower-power hf/vhf and small-signal uhf RCA transistors, an interdigitated structure is used. In this structure, as shown in Fig. 125, the emitters and bases are built like a set of interlocking combs. The

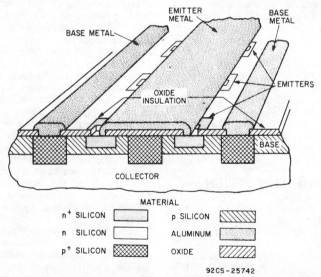

Fig. 124—Top and cross-sectional view of a typical overlay transistor.

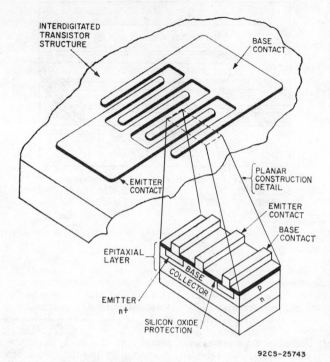

Fig. 125—Top and cross-sectional view of a typical interdigitated transistor.

sizes of the emitter and base areas are controlled by masking and diffusion. The oxide deposit, formed of silicon heated to a high temperature, masks the transistor against either an n- or p-type impurity. This oxide is removed by the usual photoetching techniques in areas where diffusion is required.

Polycrystalline Silicon Layer

The broad emitter fingers and the noncritical metal definition of the overlay transistor structure makes possible the introduction of additional conducting and insulating layers between the aluminum metallization and the shallow diffused emitter sites that are required for good high-frequency performance. RCA has developed a technique in which a polycrystalline silicon layer (PSL) is used to separate these regions. Fig. 126 shows a cross-sectional diagram of an overlay transistor structure in which this interlayer is used.

Use of the polycrystalline silicon layer between the aluminum metallization and the shallow diffused emitter region forms an insulating barrier that substantially reduces the possibility of "alloy spikes" that result from intermetallic formations of silicon and aluminum under severe hot-spot conditions. Such intermetallic formations can cause transistor failures because of emitter-to-base shorts.

As shown in Fig. 126, the polycrystalline silicon layer also forms a barrier between the aluminum emitter fingers and the oxide insulation layer over the base. This barrier minimizes the possibility of dielectric failure, which can also lead to emitter-to-base shorts because of an

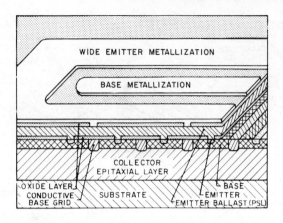

92CS-25744

Fig. 126—Cross-section of an overlay transistor structure that contains the polycrystalline silicon layer (PSL).

interaction between the aluminum and the silicon dioxide.

The addition of the polycrystalline silicon interlayer contributes substantially to the reliability of high-frequency power transistors. Reliability studies of uhf power transistors operated under overstress conditions (i.e., at a junction temperature greater than 200°C) have demonstrated an order-of-magnitude improvement in the mean time between failures of devices that employed the PSL technique over that of devices in which the PSL technique was not used. The PSL technique, therefore, is being used increasingly in RCA high-frequency power transistors.

Emitter-Site Ballasting

In many RCA high-frequency power transistors, the PSL technology is used as the medium for the introduction of emitter-site ballasting. The resistivity and contacting geometry of the aluminum to the polycrystalline silicon interlayer are controlled to form a ballast resistor in series with each emitter site. These resistors function as negative-feedback elements to prevent excessive current in any portion of the transistor and, in this way, minimize the possibility of hot spots in the device. Because the emitter in an overlay transistor is segmented into many separate sites connected in parallel, each hot spot may be isolated and controlled, and the injection of charge carriers across the transistor chip is made more uniform.

Emitter-site ballasting makes possible a more effective use of emitter periphery and, therefore, results in an increased transistor power-output capability. In addition, this ballasting makes high-frequency power transistors more immune to failures caused by high VSWR conditions such as may be encountered in some broadband amplifiers. Transistor failures because of high output VSWR conditions are often related to forward-bias second breakdown. The feedback action of the emitter ballasting resistor tends to minimize the formation of localized concentration of current that is characteristic of forward-bias second breakdown. As a result, transistors that employ emitter-site ballasting have a substantially higher capability to withstand high VSWR conditions.

OPERATING CHARACTERISTICS

At rf and microwave frequencies, the operation of a power transistor is critically dependent upon the high-frequency capability of the device. The ability of the transistor to provide significant power gain, develop useful power output, operate efficiently in circuit applications, and operate over substantial bandwidths, are direct functions of this capability.

Output Power

The power-output capability of a transistor is determined by current- and voltage-handling capabilities of the device in the frequency range of interest. The current-handling capability of the transistor is limited by its emitter periphery and epitaxial-layer resistivity. The voltage-handling capability of the device is limited by the breakdown voltages, which are, in turn, limited by the resistivity of the epitaxial layer and by the penetration of the junction. The breakdown voltage at high frequencies is substantially higher than the dc or static value. In general, the breakdown characteristic increases from the V_{CEO} value under dc conditions to a value approaching V_{CBO} at a frequency f equal to or greater than f_T.

Another parameter that limits the power-handling capability of the transistor is the saturation voltage. The saturation voltage $V_{CE}(sat)$ at high frequencies is significantly greater than the dc value because the active area is smaller than at dc.

In general, the operating voltage restrictions are the same for all high-frequency power transistors; therefore, only current-handling capability differentiates high-power transistors from lower-power units.

At high current levels, the emitter current of a transistor is concentrated at the emitter-base edge; therefore, transistor current-han-

dling capability can be increased by the use of emitter geometries which have high emitter-periphery-to-emitter-area ratios and by the use of improved techniques in the growth of collector substrate material. Transistors for large-signal applications are designed so that the peak currents do not cause base widening which would limit the current-handling capability of the device. Base-width widening is severe in transistors in which the collector side of the collector-base junction has a lower carrier concentration and higher resistivity than the base side of the junction. However, the need for low-resistivity material in the collector to handle high currents without base widening severely limits the breakdown voltages, as discussed previously. As a result, the use of a different-resistivity epitaxial layer for different operating voltages is becoming common.

Power Gain

The power gain of a high-frequency transistor power amplifier is determined by the dynamic f_T, the dynamic input impedance, and the collector load impedance, which depends on the required power output and the collector voltage swing. The power gain, P.G., of a transistor power amplifier can be expressed as follows:

$$P.G. = \frac{(f_T/f)^2 R_L}{Re(Z_{in})}$$

where f_T is the dynamic gain-bandwidth product, f is the frequency of operation, R_L is the real part of the collector parallel equivalent load impedance determined by the required power output, and $Re(Z_{in})$ is the real part of the dynamic input impedance when the collector is loaded with Z_L.

The above equation for power gain shows that for high-gain operation of power transistors, the device should have high current gain at the frequency of operation under large current swing conditions. This performance is achieved with shallow diffusion techniques.

For class B or C operation, the load resistance R_L is defined approximately as follows:

$$R_L \cong K \ \frac{[V_{CC} - V_{CE}(sat)]^2}{2P_O}$$

where K is unity or less, depending on the class of operation. The real part of the dynamic input impedance, $Re(Z_{in})$, varies considerably with signal level, and varies inversely with the power output of the device. The package parasitic inductances also are important in determining the value of $Re(Z_{in})$.

Efficiency

The collector efficiency of a transistor amplifier is defined as the ratio of signal power output at the frequency of interest to the dc input power. It can be calculated as:

$$\eta_c = \eta_v \ \eta_i \ \eta_{ckt}$$

where η_v is the efficiency of conversion of dc collector voltage to signal-frequency collector voltage (determined primarily by the ratio of V_{CC} to V_{CE} and the class of operation), η_i is the efficiency of conversion of dc collector current to signal-frequency collector current (determined primarily by the class of operation and the transit time in the collector depletion region for high-frequency transistors), and η_{ckt} is the circuit efficiency, which is determined by the loaded and unloaded Q's of the collector circuit. (The amplifier collector efficiency η_c, which is a function of both circuit and transistor parameters, should not be confused with the collector efficiency α^*, which is a basic transistor parameter determined solely by the physical structure of the device, as explained in the section **Low- and Medium-Frequency Power Transistors**.)

Bandwidth

The bandwidth of a transistor power amplifier is determined by the intrinsic frequency capability of the transistor (directly related to f_T), package parasitic elements, and the input and output matching circuits.

SPECIAL RATING CONCEPTS

Unlike low-frequency high-power transistors, many rf devices can fail within the dissipation limits set by the classical junction-to-case thermal resistance during operation under conditions of high load VSWR, high collector supply voltage, or linear (Class A or AB) operation. Failure can be caused by hotspotting, which results from local current concentration in the active areas of the device, and may appear as a long-term parameter of degradation. Localized hotspotting can also lead to catastrophic thermal runaway.

Hot-Spot Thermal Resistance

The presence of hotspots can make virtually useless the present method of calculating junction temperature by measurements of average thermal resistance, case temperature, and power dissipation. However, by use of an infrared microscope, the spot temperature of a small portion of an rf transistor pellet can be determined accurately under actual or simulated device operating conditions. The resultant peak-temperature information is used to characterize the device thermally in terms of junction-to-case hot spot thermal resistance, θ_{JS-C}.

The use of hotspot thermal resistance improves the accuracy of junction temperature and related reliability predictions, particularly for devices involved in linear or mismatch service.

DC Safe Area

The safe area determined by infrared techniques represents the locus of all current and voltage combinations within the maximum ratings of a device that produce a specified spot temperature (usually 200°C) at a fixed case temperature. The shape of this safe area is very similar to the conventional safe area in that there are four regions, as shown in Fig. 127: constant current, constant power, derating power, and constant voltage.

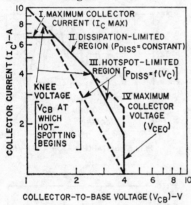

Fig. 127—Safe-area curve for an rf power transistor determined by infrared techniques.

Regions I and IV, the constant-current and constant-voltage regions, respectively, are determined by the maximum collector current and V_{CEO} ratings of the device. Region II is dissipation-limited; in the classical safe area curve, this region is determined by the following relationship:

$$P_{max} = \frac{T_J(max) - T_C}{\theta_{J-C}}$$

where T_C is the case temperature. This relationship holds true for the infrared safe area; P_{max} may be slightly lower because the reference temperature $T_J(max)$ is a peak value rather than an average value. The hotspot thermal resistance (θ_{JS-C}) may be calculated from the infrared safe area by use of the following definition:

$$\theta_{JS-C} = \frac{T_{JS} - T_C}{P}$$

where T_{JS} is highest spot temperature [$T_J(max)$ for the safe area] and P is the dissipated power ($= I \times V$ product in Region II).

The collector voltage at which regions II and III intersect, called the knee voltage V_K, indicates the collector voltage at which power constriction and resulting hotspot formation begins. For voltage levels above V_K, the allowable power decreases. Region III is very similar to the second-breakdown region in the classical safe area curve except for magnitude. For many rf power transistors, the hotspot-limited region can be significantly lower than the second-breakdown locus. Generally V_K decreases as the size of the device is increased.

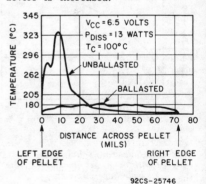

Fig. 128—Thermal profiles of a ballasted and an unballasted power transistor during dc operation.

Fig. 128 shows the temperature profiles of two transistors with identical junction geometries that operate at the same dc power level. If devices are operated on the dissipation-limited line of their classical safe areas, the profiles show that the temperature of the unballasted device rises to values 130°C in excess of the 200°C rating. Temperatures of this magnitude, although not necessarily destructive, seriously reduce the lifetime of the device.

Effect of Emitter Ballasting

The profiles shown in Fig. 128 also demonstrate the effectiveness of emitter ballasting in the reduction of power (current) constriction. In the ballasted device, a biasing resistor is introduced in series with each emitter or small groups of emitters. If one region draws too much current, it will be biased towards cutoff, allowing a redistribution of current to other areas of the device.

The amount of ballasting affects the knee voltage, V_K, as shown in Fig. 129. A point of diminishing returns is reached as V_K approaches V_{CEO}.

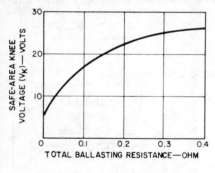

92CS-25747

Fig. 129—Safe-area voltage for an rf power transistor as a function of total ballasting resistance.

RF Operation

In normal class C rf operation, the hotspot thermal resistance is approximately equal to the classical average thermal resistance. If the proper collector loading (match) is maintained, θ_{JS-C} is independent of output power at values below the

saturated- or slumping-power level, and is independent of collector supply voltage at values within +30 per cent of the recommended operating level.

Power constriction in rf service normally occurs only for collector load VSWR's greater than 1.0. A transistor that has a mismatched load experiences temperatures far in excess of device ratings, as shown in Fig. 130(a) for VSWR = 3.0. For comparison, the temperature profile for the matched condition is shown in Fig. 130(b).

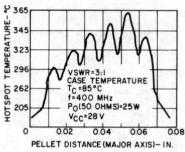

(a)

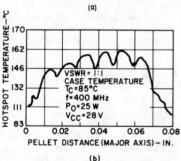

(b)

92CS-25748

Fig. 130—Thermal profile of a power transistor during rf operation: (a) under mismatched conditions; (b) under matched conditions.

Fig. 131 is a typical family of thermal-resistance curves that indicate the response of a device to various levels of VSWR and collector supply voltage. θ_{JS-C} responds to even slight increases in VSWR above 1.0 and saturates at a VSWR in the range of 3 to 6. The saturated

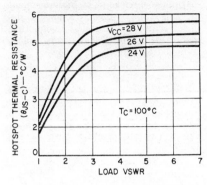

Fig. 131—Mismatch-stress thermal characteristics for the 2N5071.

level increases with increasing supply voltage. Devices with high knee voltages tend to show smaller changes of θ_{JS-C} with VSWR and supply voltage. θ_{JS-C} under mismatch is independent of frequency and power level, and reaches its highest values at load angles that produce maximum collector current. Power level does, however, influence the temperature rise and probability of failure.

Device failure can also occur at a load angle that produces minimum collector current. Under this condition, collector voltage swing is near its maximum, and an avalanche breakdown can result. This mechanism is sensitive to frequency and power level, and becomes predominant at lower frequencies because of the decreasing rf-breakdown capability of the device.

Collector mismatch can be caused by the following conditions:

1. Antenna loading changes in mobile applications when the vehicle passes near a metallic structure.

2. Antenna damage.

3. Transmission-line failure because of line, connector, or switch defects.

4. Variable loading caused by nonlinear input characteristics of a following transistor (particularly broadband) or varactor stage.

5. Supply-voltage changes that reflect different load-line requirements in class C.

6. Tolerance variations in fixed-tuned or stripline circuits.

7. Matching network variations in broadband service.

Case-Temperature Effects

The thermal resistance of both silicon and beryllium oxide, two materials that are commonly used in rf power transistors, increases about 70 per cent as the temperature increases from 25 to 200°C. Other package materials such as steel, kovar, copper, or silver, exhibit only minor increases in thermal resistance (about 5 per cent). The over-all increase in θ_{JS-C} of a device depends on the relative amounts of these materials used in the thermal path of the device; typically the increase of θ_{JS-C} ranges from 5 per cent to 70 per cent. Fig. 132 shows the rf and dc thermal

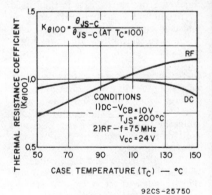

Fig. 132—Thermal-resistance coefficient for the 2N5071.

resistance coefficients for a typical rf transistor. For both cases, the coefficient is referenced to a 100°C case and is defined as follows:

$$K_{\theta_{100}} = \frac{\theta_{JS-C}}{\theta_{JS-C} \text{ at } T_C = 100°C}$$

The rf coefficient changes more than the dc coefficient, because of the power constriction that occurs in rf operation at elevated case temperature.

RELIABILITY CONSIDERATIONS

When the rf and thermal capabilities of a transistor have been established, the next step is to establish the reliability of the device for its actual application. The typical acceptable failure rate for transistors used in commercial equipment is 1 per cent per 1000 hours (100,000 hours MTBF); for transistors used in military and high-reliability equipment, it is 0.01 to 0.1 per cent per 1000 hours. Because it is not practical to test transistors under actual use conditions, dc or other stress tests are normally used to simulate rf stresses encountered in class B or class C circuits at the operating frequencies. Information derived from these tests is then used to predict the failure rate for the end-use equipment. The tests used to assure reliability include high-temperature storage tests, dc and rf operating life tests, dc stress step tests, burn-in, temperature cycling, relative humidity, and high-humidity reverse bias. The end-point measurement for these tests should include collector-to-emitter voltage V_{CEO} and emitter-to-base voltage V_{EBO} in addition to the common end-point collector-to-emitter current I_{CEO}, collector-to-base voltage V_{CBO}, collector-to-emitter saturation voltage $V_{CE}(sat)$, power output, and power gain.

One of the common failure modes in rf power transistors is degradation of the emitter-to-base junction. The high-temperature storage life test and the dc and rf operating life tests can accelerate this failure mode, and it can be detected by measurement of V_{EBO}.

Plastic uhf power transistors are more sensitive to emitter-to-base-junction degradation than similar hermetic devices. The enhancement of this failure mode in plastic devices can be caused by moisture penetration into the very close geometries used in uhf power transistors. Thermal fatigue is also a problem that affects the reliability of uhf plastic power transistors, because large thermal-expansion differences exist between the plastic encapsulant and the fine bonding wires (usually 1 mil) used in the devices.

PACKAGES

The package is an integral part of an rf power transistor. A suitable package for rf applications should have good thermal properties and low parasitic reactances. Package parasitic inductances and resistive losses have significant effects on such circuit performance characteristics as power gain, bandwidth, and stability. The most critical parasitics are the emitter and base lead inductances. Table XVIII gives the inductances of some of the more important commercially available rf power-transistor packages. Photographs of the packages are shown in Fig. 133. The TO-60 and TO-39 packages were first used in devices such as the 2N3375 and the 2N3866. The base and emitter parasitic inductance for both TO-60 and TO-39 packages is in the order of 3 nanohenries; this inductance represents a reactance of 7.5 ohms at 400 MHz. If the emitter is grounded internally in a TO-60 package (as in the RCA-2N5016), the emitter lead inductance is reduced to 0.6 nanohenry.

Hermetic low-inductance radial-lead packages are also available. The HF-19 package introduced by RCA for the 2N5919 utilizes ceramic-to-metal hermetic seals, has isolated electrodes, and has rf performance comparable to an rf plastic package. This package is also available in a studless version (HF-31) for miniaturized or low-power

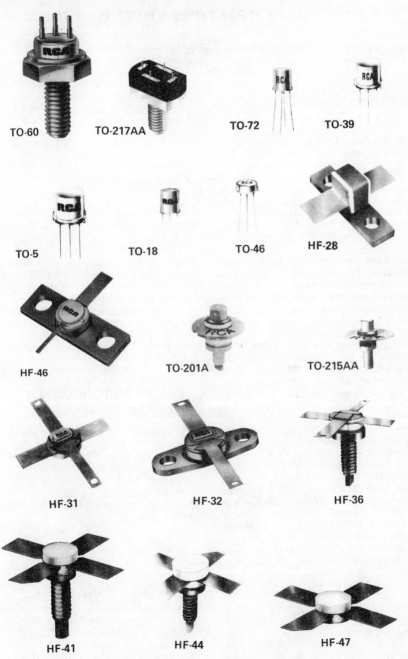

Fig. 133—Packages for RCA rf and microwave power transistors.

Table XVIII—Summary of RCA Packages for RF Power Transistors

PACKAGE	APPROXIMATE INDUCTANCE (nH)	UPPER FREQUENCY OF OPERATION (MHz)
TO-39	3	500
TO-60 (isolated emitter)	3	400
TO-60 (internally grounded emitter)	0.6	500
UHF HERMETIC STRIPLINE		
HF-19 (STUD) = JEDEC TO-216AA	0.5	1000
HF-31 = STUDLESS JEDEC TO-216AA	0.5	1000
HF-32 (FLANGED)	0.5	1000
MICROWAVE HERMETIC STRIPLINE		
HF-28 (FLANGED)	0.2	2500
COAXIAL HERMETIC		
HF-11 = JEDEC TO-215AA	0.1	3000
HF-21 = JEDEC TO-201AA	0.2	2500

applications, and in a grounded-emitter flanged version (HF-32) for compact applications.

Low-parasitic hermetic packages are available for microwave applications. The HF-11, a medium-power hermetic coaxial package first used for the RCA-2N5470, employs ceramic-to-metal construction and has parasitic inductances in the order of 0.1 nanohenry. A larger, higher-power version, the HF-21, uses the same constructional techniques and has parasitic inductances in the range of 0.2 nanohenry. The stripline equivalent of this package is the HF-28 and has approximately the same parasitic reactances as the HF-21.

Table XIX compares the performance of the TO-39 package, the HF-19 hermetic stripline package, and the HF-11 coaxial package with the same transistor chip. At a frequency of 1 GHz and an input power of 0.3 watt, the coaxial package performs significantly better than either the stripline or the TO-39 package. The coaxial package results in twice as much output power as the TO-39 package. In addition, the coaxial-package transistor is capable of delivering an output of more than 1 watt with a gain of 5 dB at 2 GHz.

RCA RF AND MICROWAVE POWER TRANSISTORS

Transistors that can generate tens of watts of power output at frequencies up to and beyond 2.3 GHz are finding applications in a wide variety of new equipment designs. Some of the major applications for these new transistors are in the following types of equipment:

1. Telemetry
2. Microwave relay links
3. Microwave communications

Table XIX—Package Performance with Same Transistor Chip

	f-GHz	P_{in}-W	P_o-W	P.G.-dB	η_C(28V)-%
TO-39	1	0.3	1	5	35
HF-19	1	0.3	1.5	7	45
HF-11	1	0.3	2.2	8.6	50
HF-11	2	0.3	1	5	35

4. Phased-array radars
5. Mobile radio and radio-tele-phones
6. Navigational-aid systems (DME, Collision Avoidance, TACAN)
7. Electronic countermeasures (ECM)
8. Microwave power sources and instrumentation
9. Intrusion-alarm systems

In such equipment, transistors offer the advantages of simplified circuitry, wide bandwidths, and improved reliability, together with reduced size and weight.

The selection of the proper transistor for a specific application is determined by the required power output, gain, and circuit preference. As shown in Table XX, RCA offers the circuit designer a wide variety of rf and microwave power transistors from which to choose the optimum type for a particular application. The selection charts in Figs. 134, 135, and 136 show the power-frequency capabilities, supply-voltage options, and typical applications of these devices.

Table XX—RCA RF and Microwave Power Transistors

Type	Package Type	Collector-Supply Voltage (V)	Frequency (MHz)	Min. Output Power (W) or Noise Figure (dB)	Type	Package Type	Collector-Supply Voltage (V)	Frequency (MHz)	Min. Output Power (W) or Noise Figure (dB)
2N918	TO-72	6-15(V_{CE})	60	NF = 6	2N5179	TO-72	6(V_{CE})	200	NF = 4.5
2N1491	TO-5	20	70	0.01	2N5180	TO-72	10(V_{CE})	200	NF = 4.5
2N1492	TO-5	30	70	0.1	2N5470	TO-215AA	28	2000	1
2N1493	TO-5	50	70	0.5	2N5913	TO-39	12	470	2
2N2631	TO-39	28	50	7.5	2N5914	TO-216AA	12	470	2
2N2857	TO-72	6-15(V_{CE})	450	NF = 4.5	2N5915	TO-216AA	12	470	6
2N2876	TO-60	28	50	10	2N5916	TO-216AA	28	400	2
2N3118	TO-5	28	50	1	2N5917	HF-31	28	400	2
2N3119	TO-5	28	50	1	2N5918	TO-216AA	28	400	10
2N3229	TO-60	50	50	15	2N5919A	TO-216AA	28	400	16
2N3262	TO-39	(High-speed switching)			2N6269	HF-28	22	2300	6.5
2N3375	TO-60	28	400	3	2N6389	TO-72	10	890	NF = 6
2N3478	TO-72	6-15(V_{CE})	200	NF = 4.5	2N6390	HF-46	28	2000	3
2N3553	TO-39	28	175	2.5	2N6391	HF-46	28	2000	5
2N3600	TO-72	6-15(V_{CE})	200	NF = 4.5	2N6392	HF-46	28	2000	10
2N3632	TO-60	28	175	13.5	2N5920	TO-215AA	28	2000	2
2N3733	TO-60	28	400	10	2N5921	TO-201AA	28	2000	5
2N3839	TO-72	6-15(V_{CE})	450	NF = 3.9	2N5995	TO-216AA	12.5	175	7
2N3866	TO-39	28	400	1	2N6093	TO-217AA	28	30	75 (PEP)
2N4012	TO-60	28	1000	2.5	2N6104	HF-32	28	400	30
			(tripler)						
2N4427	TO-39	12	175	1	2N6105	TO-216AA	28	400	30
2N4440	TO-60	28	400	5	2N6265	HF-28	28	2000	2
2N4932	TO-60	13.5	88	12	2N6266	HF-28	28	2000	5
2N4933	TO-60	24	88	20	2N6267	HF-28	28	2000	10
2N5016	TO-60	28	400	15	2N6268	HF-28	22	2300	2
2N5070	TO-60	28	30	25 (PEP)	2N6393	HF-46	28	2000	10
2N5071	TO-60	24	76	24	40080	TO-5	12	27	0.1
2N5090	TO-60	28	400	1.2	40081	TO-5	12	27	0.4
2N5102	TO-60	24	136	15	40082	TO-39	12	27	3
2N5109	TO-39	15	200	NF = 3	40280	TO-39	13.5	175	1

Table XX—RCA RF and Microwave Power Transistors (cont'd)

Type	Package Type	Collector-Supply Voltage (V)	Frequency (MHz)	Min. Output Power (W) or Noise Figure (dB)	Type	Package Type	Collector-Supply Voltage (V)	Frequency (MHz)	Min. Output Power (W) or Noise Figure (dB)
40281	TO-60	13.5	175	4	40967	HF-44	12.5	470	2
40282	TO-60	13.5	175	12	40968	HF-44	12.5	470	6
40290	TO-39	12.5	135	2	40972	TO-39	12.5	175	1.75
40291	TO-60	12.5	135	2	40973	HF-44	12.5	175	10
40292	TO-60	12.5	135	6	40974	HF-44	12.5	175	25
40340	TO-60	13.5	50	25	40975	TO-39	12.5	118	0.05
40341	TO-60	24	50	30	40976	TO-39	12.5	118	0.5
40446	TO-39 + flange	12	27	3	40977	HF-44	12.5	118	6
40581	TO-39	12	27	3.5	41008	HF-47	9	470	0.5
40582	TO-39 + flange	12	27	3.5	41008A	HF-41	9	470	0.5
40608	TO-39	15	200	NF = 3	41009	HF-47	9	470	2
40637A	TO-18	12	175	0.1	41009A	HF-41	9	470	2
40665	TO-60	28	175	13.5	41010	HF-41	9	470	5
40666	TO-60	28	400	3	41024	TO-39	28	1000	1
40836	TO-215AA	21	2000	0.5	41025	HF-41	28	1000	3
40837	TO-215AA	28	2000	1.5	41026	HF-41	28	1000	10
40894	TO-72	12	200	rf amp	41027	HF-41	22	1000	3
40895	TO-72	12	200	mixer	41028	HF-41	22	1000	10
40896	TO-72	12	200	osc	41038	TO-46	20	1680	0.75
40897	TO-72	12	10.7	if amp	41039	TO-39	15(V_{CE})	200	NF = 3.2
40898	TO-215AA	22	2300	2	RCA2001	HF-46	28	2000	1
40899	TO-201AA	22	2300	6	RCA2003	HF-46	28	2000	2.5
40909	TO-201AA	25	2000	2	RCA2005	HF-46	28	2000	5
40915	TO-72	10	450	NF = 2.5	RCA2010	HF-46	28	2000	10
40936	TO-60	28	30	20 (PEP)	RCA3001	HF-46	28	3000	1
40953	TO-39	12.5	156	1.75	RCA3003	HF-46	28	3000	2.5
40954	HF-44	12.5	156	10	RCA3005	HF-46	28	3000	4.5
40955	HF-44	12.5	156	25					
40964	TO-39	12	470	0.4					
40965	TO-39	12	470	0.5					

High-reliability versions of many of these JEDEC and RCA types can be obtained. Such devices are listed and described in the RCA DATABOOK Series SSD-207, "High-Reliability Solid-State Devices."

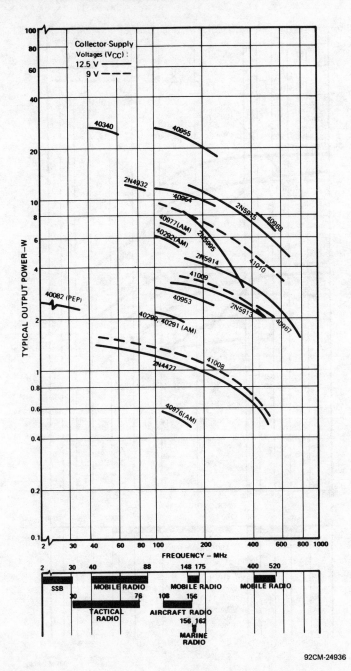

Fig. 134—RF power transistors for operation from a supply voltage of 9 or 12.5 volts.

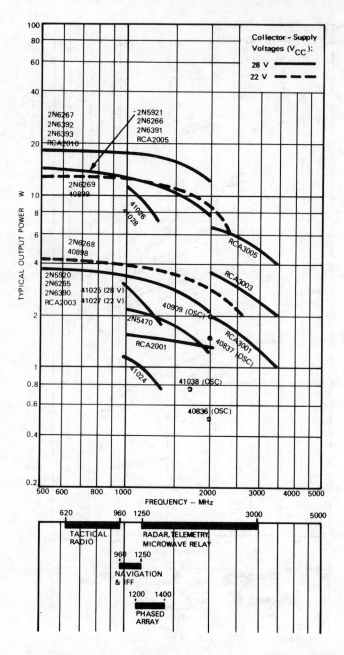

Fig. 135—RF power transistors for operation from a supply voltage of 22 or 28 volts.

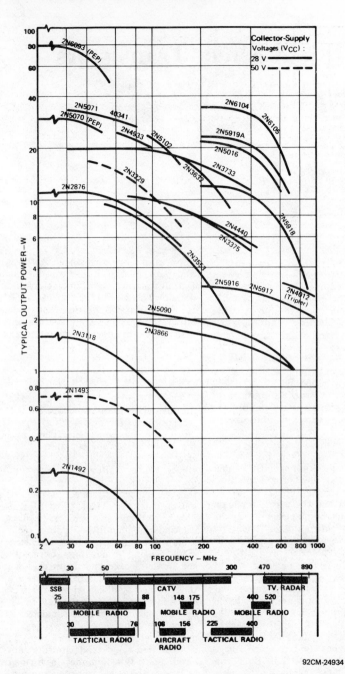

Fig. 136—RF power transistors for operation from a supply voltage of 28 or 50 volts.

MOS Field-Effect Transistors (MOS/FET's)

METAL - OXIDE - SEMICON-DUCTOR (MOS) field-effect transistors represent a unique and important category of solid-state devices. In comparison to bipolar transistors, these devices exhibit a very high input impedance and can handle a very wide dynamic range of input signals. In addition, MOS transistors can provide a square-law transfer characteristic that is especially desirable for amplification of multiple signals in rf amplifiers that are required to exhibit exceptionally low cross-modulation effects. The enhancement type of MOS field-effect transistor, which is essentially a normally open switch is also ideal for switching applications (e.g., digital circuits). In this section, the main emphasis is placed on the depletion type of metal-oxide-semiconductor field-effect transistors, which are becoming increasingly popular in electronic-circuit applications, particularly in receiver rf-amplifier and mixer circuits. The fabrication, electrical characteristics, biasing, and basic circuit configurations of these devices are discussed, and the integral gate-protection system developed for dual-gate types is explained. Fig. 137 shows a chart that lists the variety of MOS transistors available from RCA and indicates typical

applications, performance features, and package type for these devices. Detailed ratings and characteristics data on them are given in the RCA technical data bulletins. The data bulletin File Nos. are listed in the chart below the device type numbers.

FABRICATION

The fabrication techniques used to produce MOS transistors are similar to those used for modern high-speed silicon bipolar transistors. The starting material for an n-channel transistor is a lightly doped p-type silicon wafer. (Reversal of p-type and n-type materials referred to in this description produces a p-channel transistor.) After the wafer is polished on one side and oxidized in a furnace, photolithographic techniques are used to etch away the oxide coating and expose bare silicon in the source and drain regions. The source and drain regions are then formed by diffusion in a furnace containing an n-type impurity (such as phosphorus). If the transistor is to be an enhancement-type device, no channel diffusion is required. If a depletion-type transistor is desired, an n-type channel is formed to bridge the space between the diffused source and drain.

	Industrial Types														Consumer Types													
	Single-Gate								Dual-Gate			Dual-Gate Protected			Single-Gate			Dual-Gate					Dual-Gate Protected					
	3N128	3N138	3N139	3N142	3N143	3N152	3N153	3N154	3N140	3N141	3N159	3N187	3N200	40819	40467A	40468A	40559A	40600	40601	40602	40603	40604	40673	40820	40821	40822	40823	40841
File No.	309	283	284	286	309	314	320	335	285	285	326	436	437	463	324	323	323	333	333	333	334	334	381	464	464	465	465	489
Applications: RF Amplifier, Mixer	■			■	■	■		■	■	■	■	■	■	■	■	■	■	■	■	■	■	■	■	■	■	■	■	■
Applications: Chopper		■					■																					■
Applications: General-Purpose Amplifier			■	■																■								■
Applications: Oscillator	■		■	■	■	■		■	■	■	■	■	■	■	■	■	■	■	■	■	■	■	■	■	■	■	■	■
Features: Low-Noise						■					■	■	■															
Features: Low-Leakage		■					■		■	■	■							■	■	■	■	■						
Features: High-Gain						■			■		■	■	■															
Features: Gain-Controlled									■	■	■	■	■	■				■	■	■	■	■	■	■	■	■	■	■
Features: Premium-Performance						■					■	■	■															

All MOS/FET devices are supplied in the JEDEC TO-72 package

Fig. 137—MOS field-effect (MOS/FET) devices product-classification chart.

The wafer is then oxidized again to cover the bare silicon regions, and a second photolithographic and etching step is performed to remove the oxide in the contact regions. After metal is evaporated over the entire wafer, another photolithographic and etching step removes all metal not needed for the ohmic contacts to the source, drain, and gate. The individual transistor chips are then mechanically separated and mounted on individual headers, connector wires are bonded to the metalized regions, and each unit is hermetically sealed in its case in an inert atmosphere. After testing, the external leads of each device are physically shorted together to prevent electrostatic damage to the gate insulation during branding and shipping.

GATE PROTECTION

One of the most frequent problems encountered during the handling and testing of early MOS/FET's was puncture-failure of the oxide under the metal-gate electrode. The breakdown voltage of the gate-oxide is generally in the order of 100 volts, and the dc resistance is in the order of 10^{12} ohms. Because of the extremely high resistance of the gate oxide, even a very-low-energy source (such as static charge) can have sufficient potential available to puncture the oxide because to the imposition of an over-voltage stress. A single voltage excursion to the breakdown limit can impart sufficient gate-oxide damage (and/or leakage) to adversely effect device operation. Although this problem existed most frequently during handling and testing of devices prior to their installation in a circuit (because normal circuit impedances and voltages make damage of this nature less likely), a solution to the problem was necessary to reduce failure of the MOS/FET during shipping, handling, and assembly into equipment.

The earliest method of gate-oxide protection evolved was the technique in which the device manufacturer provided means to physically short all of the device leads into potentials between the electrodes of the device. Fig. 138 shows a photograph of a commonly used method in which a spiral shorting-spring interconnects the leads during shipping, handling, and assembly. The shorting-spring is not removed until the device has been safely soldered into the circuit. This method of protection continues to be used with devices having so-called "unprotected gates," (e.g., like the devices shown in Figs. 34 and 35).

Fig. 138—Illustration shows shorting spring for RCA MOS field-effect transistors that do not contain the integral gate protection. (Spring should not be removed until after the device is soldered into circuit).

Dual-Gate MOS/FET with Integrated Gate-Protection

With the advent of IC fabrication technology, a new class of dual-gate MOS/FET that features integrated gate protection was introduced. This class of MOS/FET does not require the use of the shorting-spring described above; consequently, the MOS/FET has become a more universally applicable device.

In devices that include this system, a set of back-to-back diodes is diffused directly into the semiconductor pellet and connected between each insulated gate and the source. (The low junction capacitance of the small diodes represents a relatively insignificant addition to the total capacitance that shunts the gate.) Fig. 139 shows a cross-sectional diagram and the schematic symbol for an n-channel **dual-gate-protected depletion-type MOS field-effect transistor.**

The back-to-back diodes do not conduct unless the gate-to-source voltage exceeds $\pm$ 10 volts typically. The transistor, therefore, can handle a very wide dynamic signal swing without significant conductive shunting effects by the diodes (leakage through the "nonconductive" diodes is very low). If the potential on either gate exceeds +10 volts typically, the upper diode [shown in Fig. 139(b)] of the pair associated with that particular gate becomes conductive in the forward direction and the lower diode breaks down in the backward (zener) direction. In this way, the back-to-back diode pair provides a path to shunt excessive positive charge from the gate to the source. Similarly, if the potential on either gate exceeds −10 volts typically, the lower diode becomes conductive in the forward direction and the upper diode breaks down in the reverse direction to provide a shunt path for excessive negative charge from the gate to the source. (The diode gate-protection technique is described in more detail in the following section on **Integral Gate Protection**).

Dual-gate-protected MOS transistors can be connected so that functionally they are directly equivalent to a single-gate type with gate protection. This method of connection is shown in Fig. 140.

Integral Gate-Protection

The advent of an integral system of gate-protection in MOS field-effect transistors has resulted in a class of solid-state devices that exhibits ruggedness on a par with other solid-state devices that provide comparable performance. An

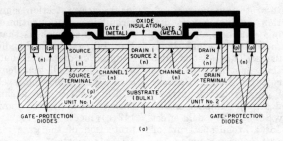

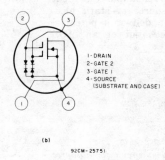

I - DRAIN
2 - GATE 2
3 - GATE 1
4 - SOURCE
(SUBSTRATE AND CASE)

(b)

92CM-25751

Fig. 139—Dual-gate-protected n-channel depletion-type MOS field-effect transistor: (a) sideview cross section; (b) schematic symbol.

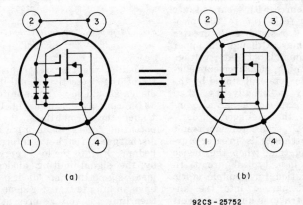

(a)   ≡   (b)

92CS-25752

Fig. 140—Connection of a dual-gate-protected MOS field-effect transistor (a) so that it is functionally equivalent to a single-gate-protected MOS field-effect transistor (b).

integral gate-protection system offers protection against static discharge during handling operations without the need for external shorting mechanisms. This system also guards against potential damage from in-circuit transients. Because the integral gate-protection system has provided a major impact on the acceptability of MOS field-effect transistors for a broad spectrum of applications, it is pertinent to examine the rudiments of this system.

Fig. 141 shows a simple equivalent circuit for a source of static electricity that can deliver a potential e_0 to the gate input of an MOS

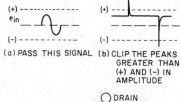

Fig. 141—Equivalent circuit for a source of static electricity.

transistor. The static potential E_s stored in an "equivalent" capacitor C_D must be discharged through an internal generator resistance R_s. Laboratory experiments indicate that the human body acts as a static (storage) source with a capacitance C_D ranging from 100 to 200 picofarads and a resistance R_s greater than 1000 ohms. Although the upper limits of accumulated static voltage can be very high, measurements suggest that the potential stored by the human body is usually less than 1000 volts. Experience has also indicated that the likelihood of damage to an MOS transistor as a result of static discharge is greater during handling than when the device is installed in a typical circuit. In an rf application, for example, static potential discharged into the antenna must traverse an input circuit that normally provides a large degree of attenuation to the static surge before it appears at the gate terminal of the MOS transistor. The

ideal gate-protection signal-limiting circuit is a configuration that allows for a signal, such as that shown in Fig. 142(a), to be handled without clipping or distortion, but limits the amplitude of all transients that exceed a safe operating level, as shown in Fig. 142(b). An arrangement of back-to-back diodes, shown in Fig. 142(c), meets these requirements for protection of the gate insulation in MOS transistors.

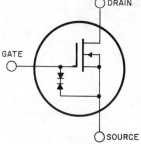

Fig. 142—MOS gate-protection requirements and a solution.

Ideally, the transfer characteristic of the protective signal-limiting diodes should have an infinite slope at limiting, as shown in Fig. 143(a). Under these conditions, the static potential across C_D in Fig. 143(b) discharges through its internal impedance R_s into the load represented by the signal-limiting diodes. The ideal signal-limiting diodes, which have an infinite transfer slope, would then limit the voltage present at the gate terminal to its knee value, e_d. The difference voltage e_s appears as an IR drop across the internal impedance of the source R_s, i.e., $e_s = E_s -$

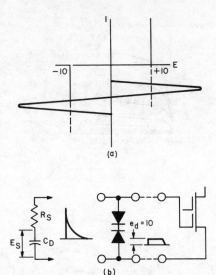

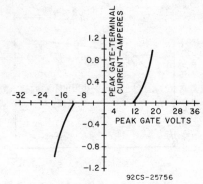

Fig. 144—Typical diode transfer characteristic measured with 1-microsecond pulse width at a duty factor of 4 x 10⁻⁵.

Fig. 143—Transfer characteristic of protective diodes (a), and resulting waveforms in equivalent circuit (b).

ELECTRICAL CHARACTERISTICS

e_d where E_s is the potential in the source of static electricity and e_d is the diode voltage drop. The instantaneous value of the diode current is then equal to e_s/R_s. During physical handling, practical peak values of currents produced by static-electricity discharges range from several milliamperes to several hundred milliamperes.

Fig. 144 shows a typical transfer characteristic curve measured on a typical set of back-to-back diodes used to protect the gate insulation in an MOS field-effect transistor that is nominally rated for a gate-to-source breakdown voltage of 20 volts. The transfer-characteristic curves show that the diodes will constrain a transient impulse to potential values well below the ±20 volt limit, even when the source of the transient surge is capable of delivering several hundred milliamperes of current. (These data were measured with 1-microsecond pulses applied to the protected gate at a duty-factor of 4 x 10⁻³).

The basic current-voltage relationship for an MOS transistor is shown in Fig. 145. With a constant gate-to-source voltage (e.g., $V_{GS} = 0$), the resistance of the channel is essentially constant, and current varies directly with drain-to-source voltage (V_{DS}), as illustrated in region A-B. The flow of drain current (I_D) produces an IR drop along the channel. The polarity of this drop is such as to oppose the field produced within the gate oxide by the gate bias. As the drain voltage is increased, a point is reached at which the IR drop becomes sufficiently high so that the capability of the gate field to attract enough carriers into the channel to sustain a higher drain-current is nullified. When this condition occurs (in the proximity of point B in Fig. 145), the channel is essentially depleted of carriers (i.e., becomes "constricted"), and drain current increases very much more slowly with further increases in drain-to-source voltage V_{DS}. This condition leads to the description of region B-C as the "pinch-off" region because the channel "pinches off" and the drain current (I_{DS}) tends to saturate at a constant value. Beyond point C, the transistor enters the

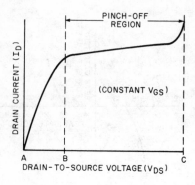

92CS-25757

Fig. 145—Basic current-voltage relationship for an MOS transistor.

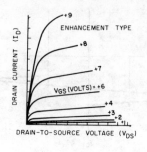

"breakdown" region (also known as the "punch-through" region), in which unrestricted current flow and damage to the transistor result if current flow is not limited by the external circuit.

MOS transistors are especially useful in high-impedance voltage amplifiers when they are operated in the "pinch-off" region. The direct variation in their channel resistance (Region A-B in Fig. 145) makes them very attractive for use in voltage-controlled resistor applications, such as the chopper circuits used in connection with some types of dc amplifiers.

Typical output characteristic curves for n-channel MOS transistors are shown in Fig. 146. The resemblance of these curves to the basic curve shown in Fig. 145 should be noted. (For p-channel transistors, the polarity of the voltages and the direction of the current are reversed.) Typical transfer characteristics for n-channel single-gate MOS transistors are shown in Fig. 147. (Again, voltage polarities and current direction would be reversed for p-channel devices.) The threshold voltage (V_{TH}) shown in connection with the enhancement-type transistor illustrates the "normally-open" source-drain characteristic of the device. In these transistors, conduction does not begin until V_{GS} is in-

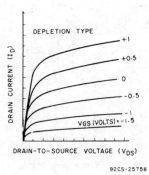

Fig. 146—Typical output-characteristic curves for n-channel MOS transistors.

creased to a particular value. Fig. 148 shows typical drain-current curves for a dual-gate device as a function of gate No. 1-to-source voltage for several values of gate No. 2-to-source voltage.

GENERAL CIRCUIT CONFIGURATIONS

There are three basic single-stage amplifier configurations for MOS transistors: common-source, common-gate, and common-drain. Each of these configurations provides certain advantages in particular applications.

The **common-source** arrangement shown in Fig. 149 is most frequently used. This configuration provides a high input impedance, medium to

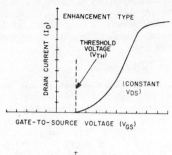

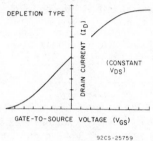

92CS-25759

Fig. 147—Typical transfer characteristics for n-channel MOS transistors.

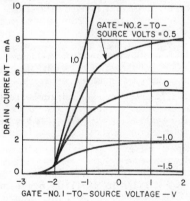

92CS-25760

Fig. 148—Drain current of a dual-gate MOS transistor as a function of gate-No. 1-to-source voltage for several values of gate-No. 2-to-source voltage.

high output impedance, and voltage gain greater than unity. The input signal is applied between gate and source, and the output signal is taken between drain and source. The

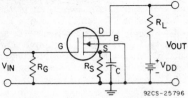

92CS-25796

Fig. 149—Basic common-source circuit for MOS field-effect transistors.

voltage gain without feedback, A, for the common-source circuit may be determined as follows:

$$A = \frac{g_{fs} r_{os} R_L}{r_{os} + R_L}$$

where g_{fs} is the gate-to-drain forward transconductance of the transistor, r_{os} is the common-source output resistance, and R_L is the effective load resistance. The addition of an unbypassed source resistor to the circuit of Fig. 149 produces negative voltage feedback proportional to the output current. The voltage gain with feedback, A', for a common-source circuit is given by

$$A' = \frac{g_{fs} r_{os} R_L}{r_{os} + (g_{fs} r_{os} + 1) R_S + R_L}$$

where R_S is the total unbypassed source resistance in series with the source terminal. The common-source output impedance with feedback, Z_o, is increased by the unbypassed source resistor as follows:

$$Z_o = r_{os} + (g_{fs} r_{os} + 1) R_S$$

The **common-drain** arrangement, shown in Fig. 150, is also frequently referred to as a **source-fol-**

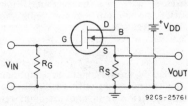

92CS-25761

Fig. 150—Basic common-drain (or source-follower) circuit for MOS transistors.

lower. In this configuration, the input impedance is higher than in the common-source configuration, the output impedance is low, there is no polarity reversal between input and output, the voltage gain is always less than unity, and distortion is low. The source-follower is used in applications which require reduced input-circuit capacitance, downward impedance transformation, or increased input-signal-handling capability. The input signal is effectively injected between gate and drain, and the output is taken between source and drain. The circuit inherently has 100-per-cent negative voltage feedback; its gain A′ is given by

$$A' = \frac{R_S}{\frac{\mu + 1}{\mu} R_S + \frac{1}{g_{fs}}}$$

Because the amplification factor (μ) of an MOS transistor is usually much greater than unity, the equation for gain in the source-follower can be simplified as follows:

$$A' = \frac{g_{fs} R_S}{1 + g_{fs} R_S}$$

For example, if it is assumed that the gate-to-drain forward transconductance g_{fs} is 2000 micromhos (2 x 10^{-3} mho) and the unbypassed source resistance R_S is 500 ohms, the stage gain A′ is 0.5. If the same source resistance is used with a transistor having a transconductance of 10,000 micromhos (1 x 10^{-2} mho), the stage gain increases to 0.83.

When the resistor R_G is returned to ground, as shown in Fig. 150, the input resistance R_1 of the source-follower is equal to R_G. If R_G is returned to the source terminal, however, the effective input resistance R_1' is given by

$$R_1' = \frac{R_G}{1 - A'}$$

where A′ is the voltage amplification of the stage with feedback. For example, if R_G is one megohm and A′ is 0.5, the effective resistance R_1' is two megohms.

If the load is resistive, the effective input capacitance C_1' of the source-follower is reduced by the inherent voltage feedback and is given by

$$C_1' = c_{gd} + (1 - A') c_{gs}$$

where c_{gd} and c_{gs} are the intrinsic gate-to-drain and gate-to-source capacitances, respectively, of the MOS transistor. For example, if a typical MOS transistor having a c_{gd} of 0.3 picofarad and a c_{gs} of 5 picofarads is used, and if A′ is equal to 0.5, then C_1' is reduced to 2.8 picofarads.

The effective output resistance R_o' of the source-follower stage is given by

$$R_o' = \frac{r_{os} R_S}{(g_{fs} r_{os} + 1) R_S + r_{os}}$$

where r_{os} is the transistor common-source output resistance in ohms. For example, if a unit having a gate-to-drain forward transconductance g_{fs} of 2000 micromhos and a common-source output resistance r_{os} of 7500 ohms is used in a source-follower stage with an unbypassed source resistance R_S of 500 ohms, the effective output resistance R_o' of the source-follower stage is 241 ohms.

The source-follower output capacitance C_o' may be expressed as follows:

$$C_o' = c_{ds} + c_{gs} \left(\frac{1 - A'}{A'}\right)$$

where c_{ds} and c_{gs} are the intrinsic drain-to-source and gate-to-source capacitances, respectively, of the MOS transistor. If A′ is equal to 0.5 (as assumed for the sample input-circuit calculations), C_o' is reduced to the sum of c_{ds} and c_{gs}.

The common-gate circuit, shown in Fig. 151, is used to transform from a low input impedance to a high output impedance. The input impedance of this configuration has approximately the same value as the output impedance of the source-follower circuit. The common-gate circuit is also a desirable configuration

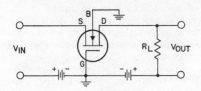

Fig. 151—Basic common-gate circuit for MOS transistors.

for high-frequency applications because its relatively low voltage gain makes neutralization unnecessary in most cases. The common-gate voltage gain, A, is given by

$$A = \frac{(g_{fs} r_{os} + 1) R_L}{(g_{fs} r_{os} + 1) R_G + r_{os} + R_L}$$

where R_G is the resistance of the input-signal source. For a typical MOS transistor ($g_{fs} = 2000$ micromhos, $r_{os} = 7500$ ohms) and with $R_L = 2000$ ohms and $R_G = 500$ ohms, the common-gate voltage gain is 1.8. If the value of R_G is doubled, the voltage gain is reduced to 1.25.

BIASING TECHNIQUES FOR SINGLE-GATE MOS TRANSISTORS

The bias required for operation of a single-gate MOS transistor can be supplied by use of a self-bias (source-bias) arrangement, from a supply of fixed bias, or, preferably, by a combination of these methods. Fig. 152 illustrates each of the three biasing techniques.

The design of a self-bias circuit is relatively simple and straightforward. For example, if a 3N128 MOS transistor is to be operated with a drain-to-source voltage V_{DS} of 15 volts and a small-signal transconductance g_{fs} of 7400 micromhos, the drain current I_D required for the specified value of transconductance is first obtained from published curves, such as those shown in Fig. 153(a). Next, the gate-to-source voltage required for this value of drain current is determined from another published curve, such as the solid-line curve shown in Fig. 153(b). These curves indicate that the drain current should be 5 milliamperes and that the gate-to-source voltage should be -1.1 volts for the specified values of drain-to-source voltage and transconductance. The source voltage V_S, the source resistance R_S, and the dc supply voltage V_{DD} can then be readily calculated, as follows:

$$V_S = V_G - V_{GS} = 1.1 \text{ volts}$$
$$R_S = V_S/I_D = 1.1/5 = 220 \text{ ohms}$$
$$V_{DD} = V_{DS} + V_S = 15 + 1.1$$
$$= 16.1 \text{ volts}$$

The self-bias arrangement is satisfactory for some applications. A particular source resistance, however, must be selected for each device if a specified drain current is required because the drain-current characteristics of individual devices can vary significantly from the typical values.

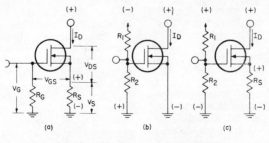

92CS-25763

Fig. 152—Biasing arrangements for single-gate MOS transistors: (a) self-bias circuit; (b) fixed bias supply; (c) combination of self bias and fixed bias.

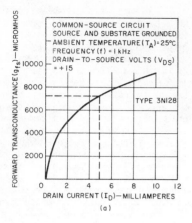

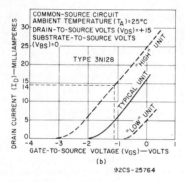

92CS-25764

*Fig. 153—Operating characteristics for the
RCA-3N128 MOS transistor: (a) forward
transconductance as a function of drain
current; (b) drain current as a function of
gate-to-source voltage.*

The dashed-line curves in Fig.
153(b) define the "high" and "low"
limits for the characteristics of the
3N128 MOS transistor. For example,
the zero-bias drain current I_{DSS} can
vary from a low value of 5 milliamperes to a high value of 25 milliamperes, a range of 20 milliamperes.
Use of a source resistor of 220 ohms,
as calculated in the preceding example, reduces the range of the drain
current between "high" and "low"
3N128 transistors operated in self-
bias circuits from 20 milliamperes
to about 4 milliamperes. A reduction

of about 5 to 1 in the range of I_{DSS}
values among individual devices can
be achieved, therefore, by a judicious
choice of the proper value of source
resistance.

Fixed-bias-supply systems, such as
that shown in Fig. 153(b), are
generally unattractive for use with
MOS transistors for two main reasons. First, this type of system is
undesirable because it requires the
use of a separate, negative-voltage
power supply. Second, as shown by
the curves in Fig. 153(b), for a
fixed bias supply of 1.1 volts, drain
current would be 14 milliamperes for
a "high" 3N128 transistor and would
be cut off for a "low" device. Consequently, if an external bias system is used provisions must be
made for adjustment of the bias
voltage if a specific drain current
is required for a particular device.

The combination bias system
shown in Fig. 152(c) is the most
effective arrangement when an application requires a specific drain
current despite the range of drain-
current characteristics encountered
among individual devices. Fig. 154
shows two families of characteristic
curves developed empirically for the
combination bias system shown in
Fig. 152(c). The family of curves on
the left is pertinent for operation at
a drain current of 5 milliamperes.
For operation at a drain current of
10 milliamperes, the family of curves
on the right should be used.

If a drain current of 5 milliamperes is desired, the pertinent curves
in Fig. 154 show that, for a source
resistance of 1000 ohms, a bias system can provide this value of current within 1 milliampere (as indicated by projections of lines a and b
to the abscissa), despite a range of
5 to 25 milliamperes in the value of
I_{DSS} for individual devices. A drain
current I_D of 5 milliamperes, however, develops a self bias of —5 volts
across the 1000-ohm source resistor
R_S, and the transistor will be cut
off unless sufficient positive bias is
applied across the input resistors
(R_1 and R_2) to establish the correct

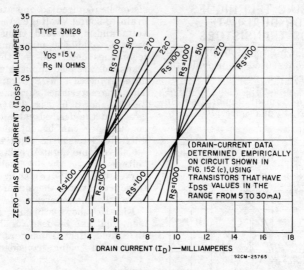

Fig. 154—*Drain current I_D as a function of zero-bias drain current I_{DSS} for several values of source resistance R_S.*

operating point. The positive bias voltage can be obtained from the positive drain supply V_{DD} so that there is no need for a separate bias supply. For a drain-to-source voltage V_{DS} of 15 volts, a drain current I_D of 5 milliamperes, a gate-to-source voltage V_{GS} of —1.1 volts, and a source resistance R_S of 1000 ohms, the circuit parameters for the combination bias system shown in Fig. 152(c) can be calculated as follows:

$$V_S = I_D R_S = (0.005)(1000)$$
$$= 5 \text{ volts}$$
$$V_G = V_{GS} + V_S = -1.1 + 5$$
$$= 3.9 \text{ volts}$$
$$V_{DD} = V_{DS} + V_S = 15 + 5$$
$$= 20 \text{ volts}$$
$$V_{DD}/V_G = (R_1 + R_2)/R_2 = 20/3.9$$
$$= 5.12$$

The lower limits for the values of the input resistors R_1 and R_2 are determined on the basis of the maximum permissible loading of the input circuit. The resistance that corresponds to this value is set equal to the equivalent value of the parallel combination of the two resistors. For example, if the total resistance in shunt with the input circuit is to be no less than 50,000 ohms, the values of R_1 and R_2 are calculated as follows:

$$R_1 R_2 / (R_1 + R_2) = 50,000$$
$$(R_1 + R_2)/R_2 = 5.12$$

Therefore, $R_1 = 256,000$ ohms and $R_2 = 62,000$ ohms.

In rf-circuit applications, the effects of input-circuit loading can be circumvented by use of the circuit arrangement shown in Fig. 155.

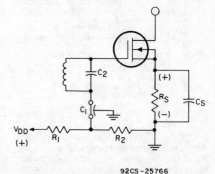

Fig. 155—*Circuit used to eliminate input-circuit loading in rf-amplifier applications.*

BIASING TECHNIQUES
FOR DUAL-GATE
MOS TRANSISTORS

The following example illustrates the techniques used to provide the bias required for operation of a dual-gate MOS transistor. This example assumes a typical application in which a 3N140 dual-gate MOS tran-

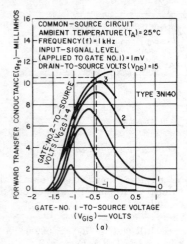

sistor is required to operate with a drain-to-source voltage V_{DS} of 15 volts and a forward transconductance g_{fs} of 10,500 micromhos. (The techniques described for the 3N140 transistor are also applicable to dual-gate-protected MOS transistors.) The characteristic curves for the 3N140, shown in Fig. 156(a), indicate that the desired value of transconductance can be obtained for a gate No. 1-to-source voltage V_{G1S} of −0.45 volt and a gate No. 2-to-source voltage V_{G2S} of +4 volts. The curves in Fig. 156(b) show that for these conditions the drain current I_D is 10 milliamperes.

Fig. 157 shows a biasing arrangement that can be used for dual-gate MOS field-effect transistors. For the application being considered, the

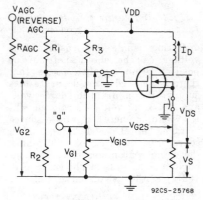

Fig. 157—Typical biasing circuit for dual-gate MOS field-effect transistors.

shunt resistance for gate No. 1 is assumed to be 25,000 ohms. Gate No. 2 is operated at rf ground (by means of adequate bypassing) and is biased with a fixed dc potential. Empirical experience with dual-gate MOS transistors has shown that a source resistance of approximately 270 ohms provides adequate self-bias for the transistor for operation from the proposed dc supply voltage. For this value of source resistance, the remaining parameters of the bias circuit are obtained from the following calculations:

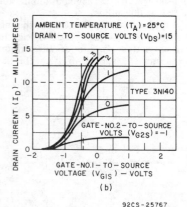

92CS-25767

Fig. 156—Operating characteristics for the RCA-3N140 dual-gate MOS transistor: (a) forward transconductance as a function of gate-No. 1-to-source voltage; (b) drain current as a function of gate-No. 1-to-source voltage.

$V_S = I_D R_S = (0.010)(270)$
         $= +2.7$ volts
$V_{G1} = V_{G1S} + V_S = (-0.45)$
         $+ (+2.7) = +2.25$ volts
$V_{G2} = V_{G2S} + V_S = (+4.0) + (+2.7)$
         $= +6.7$ volts
$V_{DD} = V_{DS} + V_S = (+15) + (+2.7)$
         $= +17.7$ volts

The values of the voltage-divider resistances required to provide the appropriate voltage at each gate are determined in a manner similar to that described for single-gate MOS transistors. The value calculated for R_3 is 197,000 ohms, that for R_4 is 28,600 ohms, and the ratio R_1/R_2 is 11.67.

The circuit shown in Fig. 157 is normally used in rf amplifier applications. In this circuit, the signal voltage is applied at point "a" through appropriate input circuitry. If the agc feature is not employed, (e.g. in mixer circuits), the resistor R_{agc} is disconnected at point "b." In a mixer application, the local oscillator signal is injected at point "b."

TECHNICAL FEATURES

It is apparent from the preceding discussions that MOS field-effect transistors exhibit a number of technical features that result in unique performance advantages in circuit applications such as mixers, product detectors, remote gain-control circuits. balanced modulators, choppers, clippers, and gated amplifiers. These features include:

1. An extremely high input resistance and a low input capacitance—as a result, MO transistors impose virtually no loading on an agc voltage source (i.e., virtually no agc power is required) and have a wide agc range capability.

2. A wide dynamic range—MOS transistors, therefore, can handle positive and negative input-signal excursions without diode-current loading.

3. Cross-modulation effects and spurious response that are substantially less than those of other types of electronic devices—the cross-modulation characteristics of dual-gate transistors actually improve as the device approaches cutoff.

4. Zero offset voltage—this feature is especially desirable for chopper applications.

5. An exceptionally high forward transconductance.

6. Negative temperature coefficient for the drain current—"thermal runaway," therefore, is virtually impossible.

7. A very low gate leakage current that is relatively insensitive to temperature variations.

8. Very low oscillator feedthrough in dual-gate mixer circuits.

9. Dual-gate transistors can provide good gain in common-source amplifiers into the uhf range without neutralization.

HANDLING CONSIDERATIONS

MOS field-effect transistors, like high-frequency bipolar transistors, can be damaged by exposure to excessive voltages. The gate oxide insulation is susceptible to puncture when subjected to voltage in excess of the rated value. The very high resistance of the oxide insulation imposes a negligible load on electrostatically generated potentials and, therefore, provides an ineffective discharge path for sources of static electricity. As discussed earlier, the integral gate-protection system incorporated into some types of dual-gate MOS transistors is highly effective in the protection of these devices against the effects of electrostatic charges. The following special precautions, however, are necessary in handling MOS-transistors which do not contain integral-gate protection systems:

1. Prior to assembly into a circuit, all leads should be kept shorted together by either (a) use of metal shorting springs attached to the device by the

vendor, as shown in Fig. 367, or (b) use of conductive foam such as "ECCOSORB LS26" or equivalent. (ECCOSORB is a Trade Mark of Emerson & Cuming, Inc.). **Note:** Polystyrene insulating "SNOW" can acquire high static charges and should not be used.

2. When devices are removed by hand from their carriers, the hand being used should be at ground potential. Personnel handling MOS transistors during testing should ground themselves, preferably at the hand or wrist.

3. Tips of soldering irons should be grounded.

4. Devices should never be inserted into or removed from circuits with power on.

Thyristors (SCR's and Triacs)

THE term **thyristor** is the generic name for solid-state devices that have characteristics similar to those of thyratron tubes. Basically, this group includes bistable solid-state devices that have two or more junctions (three or more semiconductor layers) and that can be switched between conducting states (from off to on or from on to off) within at least one quadrant of the principal voltage-current characteristic. Reverse-blocking triode thyristors, commonly called silicon controlled rectifiers (SCR's), and bidirectional triode thyristors, usually referred to as triacs, are the most common types. These types have three electrodes and are switched between states by a current pulse applied to the gate terminal.

Thyristors have become increasingly important for use in power-switching and power-control applications at voltages that range from a few volts to more than 1000 volts and current levels from less than one-half ampere to more than 1000 amperes. As mentioned earlier in the section **Materials, Junctions, and Devices**, an SCR is basically a unidirectional device that may be used for both ac and dc functions, and a triac is a bidirectional device that is used mainly for ac functions.

When power control involves conversion of ac voltages and/or currents to dc and control of their magnitudes, SCR's are used because of their inherent rectifying properties. SCR's are also used in dc switching applications such as pulse modulators and inverters, because the currents in the switching device are unidirectional. In addition, SCR's are generally used whenever the desired function can be accomplished adequately by this type of device because of the economics involved.

Triacs, which have symmetrical bidirectional electrical characteristics, were developed specifically for control of ac power. These devices are used primarily to control the power applied to a load from ac power lines.

PELLET STRUCTURES

Fig. 158 shows a cross-sectional diagram of a typical RCA SCR pellet. The shorted-emitter construction used in RCA SCR's can be recognized by the metallic cathode electrode in direct contact with the p-type base layer around the periphery of the pellet. The gate, at the center of the pellet, also makes direct metallic contact to the p-type base so that the portion of this layer under the n-type emitter acts as an ohmic path for current flow between gate and cathode. Because this ohmic path is in parallel with the n-type emitter junction, current preferentially takes the ohmic path until the IR drop in this path reaches the junction threshold voltage of about 0.8 volt. When the gate voltage exceeds this value, the junction current increases rapidly, and injection of electrons by the n-type emitter reaches a level high enough to turn on the device.

In addition to providing a precisely controlled gate current, the

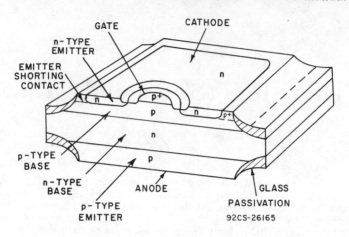

Fig. 158—Cross section of a typical SCR pellet.

shorted-emitter construction also improves the high-temperature and dv/dt (maximum allowable rate of rise of off-state voltage) capabilities of the device. The junction depletion layer acts as a parallel-plate capacitor which must be charged when blocking voltage is applied. Because the charging, or displacement, current (i = Cdv/dt) into this capacitor varies as the rate of rise of forward voltage (dv/dt), a very high dv/dt can result in a high current between anode and cathode. If this current crosses the n-type emitter junction and is of the same order of magnitude as the gate current, it can trigger the device into the conducting state. Such unwanted triggering is minimized by the shorted-emitter construction because the peripheral contact of the p-type base to the cathode electrode provides a large-area parallel path by which the dv/dt current can reach the cathode electrode without crossing the n-type emitter junction. (The critical dv/dt value for thyristors is discussed more fully in a subsequent paragraph.)

The center-gate construction of the SCR pellet provides fast turn-on and high di/dt capabilities (i.e., maximum rate of rise of forward

current). (Thyristor turn-on characteristics and di/dt capabilities are explained later in this section.) In an SCR, conduction is initiated in the cathode region immediately adjacent to the gate contact and must then propagate to the more remote regions of the cathode. Switching losses are influenced by the rate of propagation of conduction and the distance conduction must propagate from the gate. With a central gate, all regions of the cathode are in close proximity to the initially conducting region so that propagation distance is significantly decreased; as a result, switching losses are minimized.

Fig. 159 shows a cross-sectional diagram of a typical RCA triac pellet. In this device, the main-terminal-1 electrode makes ohmic contact to a p-type emitter as well as to an n-type emitter. Similarly, the main-terminal-2 electrode also makes ohmic contact to both types of emitters, but the p-type emitter of the main-terminal-2 side is located opposite the n-type emitter of the main-terminal-1 side, and the main-terminal-2 n-type emitter is opposite the main-terminal-1 p-type emitter. The net result is two four-layer switches in parallel, but oriented in

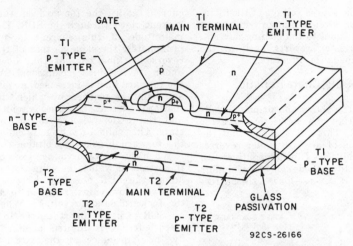

Fig. 159—Cross section of a typical triac pellet.

opposite directions, in one silicon pellet. This type of construction makes it possible for a triac either to block or to conduct current in either direction between main terminal 1 and main terminal 2.

The gate electrodes also makes contact to both n- and p-type regions. As a result, the device can be triggered by either positive or negative gate signals, for either polarity of voltage between the main-terminal electrodes. When the triac is triggered by a positive gate signal, conduction is initiated, as in the SCR, by injection of electrons from the main-terminal-1 n-type emitter, and the gate n-type region is passive. The gate n-type region becomes active when the triac is triggered by a negative gate signal, because it then acts as the n-type emitter of a grounded-base n-p-n transistor. Electrons injected from this region enter the n-type base and cause a forward bias on one of the p-type emitters, depending on which is at the positive end of the voltage between the main-terminal electrodes.

The cathode of an SCR and the main terminal 1 of a triac are fully covered by a relatively heavy metallic electrode. This electrode provides a low-resistance path to distribute current evenly over the cathode or main-terminal-1 area and serves as a thermal capacitor to absorb heat generated by high surge or overload currents. Junction-temperature excursions that result from such conditions are, therefore, held to a minimum.

PRINCIPAL VOLTAGE-CURRENT CHARACTERISTICS

The principal voltage-current characteristics of SCR's and triacs indicate that these devices are ideal for power switching applications. When the voltage across the main terminals of either type of thyristor is below the breakover point, the current through the device is extremely small, and the thyristor is effectively an open switch. When the voltage across the main terminals increases to a value exceeding the breakover point, the thyristor switches to its high-conduction state and is effectively a closed switch. The thyristor remains in the on state until the current through the main terminals drops below a value which is called the **holding current.**

When the source voltage of the main-terminal circuit cannot support a current equal to the holding current, the thyristor reverts back to the high-impedance off state.

SCR Characteristic

Fig. 160 shows the principal voltage-current characteristic curve for an SCR. This curve shows that the operation of an SCR under reverse-bias conditions (anode negative with respect to cathode) is very similar to that of reverse-biased silicon rec-

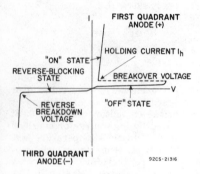

Fig. 160—Principal voltage-current characteristic for an SCR.

tifiers or other solid-state diodes. In this bias mode, the SCR exhibits a very high internal impedance, and only a slight amount of reverse current, called the **reverse blocking current,** flows through the p-n-p-n structure. This current is very small until the reverse voltage exceeds the reverse breakdown voltage; beyond this point, however, the reverse current increases rapidly. The value of the reverse breakdown voltage differs for individual SCR types.

During forward-bias operation (anode positive with respect to cathode), the p-n-p-n structure of the SCR is electrically bistable and may exhibit either a very high impedance (forward-blocking or off state) or a very low impedance (forward-conducting or on state). In the forward-blocking state, a small forward

current, called the forward on-state current, flows through the SCR. The magnitude of this current is approximately the same as that of the reverse-blocking current that flows under reverse-bias conditions. As the forward bias is increased, a voltage point is reached at which the forward current increases rapidly, and the SCR switches to the on state. This value of voltage is called the **forward breakover voltage.**

When the forward voltage exceeds the breakover value, the voltage drop across the SCR abruptly decreases to a very low value, referred to as the **forward on-state voltage.** When an SCR is in the on state, the forward current is limited primarily by the impedance of the external circuit. Increases in forward current are accompanied by only slight increases in forward voltage when the SCR is in the state of high forward conduction.

Triac Characteristic

A triac exhibits the forward-blocking, forward-conducting voltage-current characteristic of a p-n-p-n structure for either direction of applied voltage, as shown in Fig. 161. This bidirectional switching capability results because, as mentioned previously, a triac consists essentially of two p-n-p-n

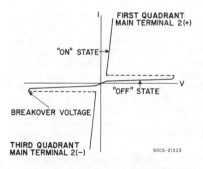

Fig. 161—Principal voltage-current characteristic for a triac.

devices of opposite orientation built into the same crystal. The device, therefore, operates basically as two SCR's connected in parallel, but with the anode and cathode of one SCR connected to the cathode and anode, respectively, of the other SCR. As a result, the operating characteristics of the triac in the first and third quadrants of the voltage-current characteristics are the same, except for the direction of current flow and applied voltage. The triac characteristics in these quadrants are essentially identical to those of an SCR operated in the first quadrant. For the triac, however, the high-impedance state in the third quadrant is referred to as the off state rather than as the reverse-blocking state. Because of the symmetrical construction of the triac, the terms forward and reverse are not used in reference to this device.

Effect of Gate Current on Voltage-Current Characteristics

The breakover voltage of a thyristor can be varied, or controlled, by injection of a signal at the gate terminal. Fig. 162 shows curves of breakover as a function of gate current for first-quadrant operation of an SCR. A similar set of curves can be drawn for both the first and the third quadrant to represent triac operation.

When the gate current I_g is zero, the applied voltage must reach the breakover voltage of the SCR or triac before switching occurs. As the value of gate current is increased, however, the ability of a thyristor to support applied voltage is reduced and there is a certain value of gate current at which the behavior of the thyristor closely resembles that of a rectifier. Because thyristor turn-on, as a result of exceeding the breakover voltage, can produce instantaneous power dissipation during the switching transition, an irreversible condition may exist unless

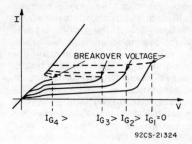

Fig. 162—Thyristor breakover as a function of gate current.

the magnitude and rate of rise of principal current is restricted to tolerable levels. For normal operation, therefore, thyristors are operated at applied voltages lower than the breakover voltage, and are made to switch to the on state by gate signals of sufficient amplitude to assure complete turn-on independent of the applied voltage. Once the thyristor is triggered to the on state, the principal-current flow is independent of gate voltage or gate current, and the device remains in the on state until the principal-current flow is reduced to a value below the holding current required to sustain regeneration.

The gate voltage and current required to switch a thyristor from its high-impedance (off) state to its low-impedance (on) state at maximum rated forward anode current can be determined from the circuit shown in Fig. 163. Resistor R_2 is

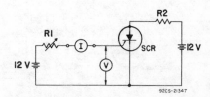

Fig. 163—Circuit used to measure thyristor gate voltage and current switching threshold.

selected so that the anode current specified in the manufacturer's ratings flows when the device latches into its low-impedance or on state. The value of R_1 is gradually decreased until the device under test is switched from its off state to its low-impedance or on state. The values of gate current and gate voltage immediately prior to switching are the values required to trigger the thyristor. For an SCR, there is only one mode of gate firing capable of switching the device into the on state, i.e., a positive gate signal for a positive anode voltage. If the gate polarity is reversed (negative voltage), the reverse current flow is limited by the value of R_2 and the gate-cathode internal shunt. The value of power dissipated for the reverse gate polarity is restricted to the maximum power-dissipation limit imposed by the manufacturer.

Because of its complex structure, a triac can be triggered by either a positive or a negatve gate signal, regardless of the polarity of the voltage across the main terminals of the device. The direction of the principal current, however, influences the gate trigger current; as a result, the magnitude of current required to trigger the triac differs for each triggering mode. The triggering modes in which the principal current is in the same direction as the gate current require less gate current than the triggering modes in which the principal current is in opposition to the gate current. The

directions of the gate current and the principal current for each triggering mode are indicated in the junction diagrams in Fig. 29 in the section **Materials, Junctions, and Devices.** For convenience, Fig. 164 shows these current directions in relation to the schematic symbol of a triac.

RATINGS AND CHARACTERISTICS

Thyristors must be operated within the maximum ratings specified by the manufacturer to assure best results in terms of performance, life, and reliability. These ratings define limiting values, determined on the basis of extensive tests, that represent the best judgment of the manufacturer of the safe operating capability of the device.

Voltage Ratings

The voltage ratings of thyristors are given for both steady-state and transient operation and for both forward- and reverse-blocking conditions. For SCR's, voltages are considered to be in the forward or positive direction when the anode is positive with respect to the cathode. Negative voltages for SCR's are referred to as reverse-blocking voltages. For triacs, voltages are considered to be positive when main terminal 2 is positive with respect to main terminal 1. Alternatively,

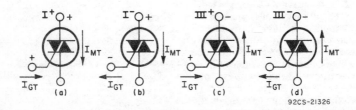

Fig. 164—Gating conditions for each of the four triggering modes of a triac.

this condition may be referred to as operation in the first quadrant.

Off-State Voltages—The repetitive peak **off-state voltage** V_{DRM} is the maximum value of off-state voltage, either transient or steady-state, that the thyristor should be required to block under the stated conditions of temperature and gate-to-cathode resistance. If this voltage is exceeded, the thyristor may switch to the on state. The circuit designer should insure that the V_{DRM} rating is not exceeded to assure proper operation of the thyristor.

Under relaxed conditions of temperature or gate impedance, or when the blocking capability of the thyristor exceeds the specified rating, it may be found that a thyristor can block voltages far in excess of its repetitive off-state voltage rating V_{DRM}. Because the application of an excessive voltage to a thyristor may produce irreversible effects, an absolute upper limit should be imposed on the amount of voltage that may be applied to the main terminals of the device. This voltage rating is referred to as the peak **off-state voltage** V_{DM}. It should be noted that the peak off-state voltage has a single rating irrespective of the voltage grade of the thyristor. This rating is a function of the construction of the thyristor and of the surface properties of the pellet; it should not be exceeded under either continuous or transient conditions.

Reverse Voltages (SCR's only)— Reverse voltage ratings are given for SCR's to provide operating guidance in the third quadrant, or reverse-blocking mode. There are two voltage ratings for SCR's in the reverse-blocking mode: repetitive **peak reverse voltage** (V_{RRM}) and **nonrepetitive peak reverse voltage** (V_{RSM}).

The repetitive peak reverse voltage is the maximum allowable value of reverse voltage, including all repetitive transient voltages, that may be applied to the SCR. Because reverse power dissipation is small at this voltage, the rise in junction temperature because of this reverse dissipation is very slight and is accounted for in the rating of the SCR.

The nonrepetitive peak reverse voltage is the maximum allowable value of any nonrepetitive transient reverse voltage which may be applied to the SCR. These nonrepetitive transient voltages are allowed to exceed the steady-state ratings, even though the instantaneous power dissipation can be significant. While the transient voltage is applied, the junction temperature may increase, but removal of the transient voltage in a specified time allows the junction temperature to return to its steady-state operating temperature before a thermal runaway occurs.

On-State Voltages—When a thyristor is in a high-conduction state, the voltage drop across the device is no different in nature from the forward-conduction voltage drop of a solid-state diode, although the magnitude may be slightly higher. As in diodes, the on-state voltage-drop characteristic is the major source of power losses in the operation of the thyristor, and the temperatures produced become a limiting feature in the rating of the device.

Current Ratings

The current ratings for SCR's and triacs define maximum values for normal or repetitive currents and for surge or nonrepetitive currents. These maximum ratings are determined on the basis of the maximum junction-temperature rating, the junction-to-case thermal resistance, the internal power dissipation that results from the current flow through the thyristor, and the ambient temperature. The effect of these factors in the determination of current ratings is illustrated by the following example.

Fig. 165 shows curves of the maximum average forward power dissipation for the RCA-2N3873 SCR as a function of average forward current for dc operation and for various conduction angles. For the 2N3873, the junction-to-case thermal resistance $\theta_{J\text{-}C}$ is 0.92°C per watt and the maximum operating junction temperature T_J is 100°C. If the maximum case temperature $T_{C(max)}$ is assumed to be 65°C, the maximum average forward power dissipation can be determined as follows:

$$P_{AVG(max)} = \frac{T_{J(max)} - T_{C(max)}}{\theta_{J\text{-}C}}$$

$$= \frac{(100 - 65)\,°C}{0.92\,°C/\text{watt}}$$

$$= 38 \text{ watts}$$

The maximum average forward current rating for the specified conditions can then be determined from the rating curves shown in Fig. 165. For example, if a conduction angle of 180 degrees is assumed, the average forward current rating for a

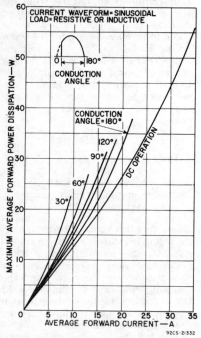

Fig. 165—Power-dissipation rating chart for the 2N3873 SCR.

maximum dissipation of 38 watts is found to be 22 amperes.

These calculations assume that the temperature is uniform throughout the pellet and the case. The junction temperature, however, increases and decreases under conditions of transient loading or periodic currents, depending upon the instantaneous power dissipated within the thyristor. The current rating takes these variations into account.

The **on-state current ratings** for a thyristor indicate the maximum values of average, rms, and peak (surge) current that should be allowed to flow through the main terminals of the device, under stated conditions, when the thyristor is in the on state. For heat-sink-mounted thyristors, these maximum ratings are based on the case temperature; for lead-mounted thyristors, the ratings are based on the ambient temperature.

The **maximum average on-state current rating** for an SCR is usually specified for a half-sine-wave current at a particular frequency. Fig. 166 shows curves of the maximum allowable average on-state current $I_{TF(avg)}$ for the RCA-2N3873 SCR family as a function of case temperature. Because peak and rms currents may be high for small conduction angles, the curves in Fig. 166 also show maximum allowable average currents as a function of conduction angle. The maximum operating junction temperature for the 2N3873 is 100°C. The rating curves indicate, for a given case temperature, the maximum average on-state current for which the average temperature of the pellet will not exceed the maximum allowable value. The rating curves may be used for only resistive or inductive loads. When capacitive loads are used, the currents produced by the charge or discharge of the capacitor through the thyristor may be excessively high, and a resistance should be used in series with the capacitor to limit the current to the rating of the thyristor.

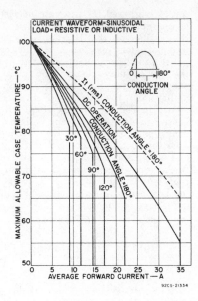

Fig. 166—Current rating chart for the 2N3873 SCR.

The on-state current rating for a triac is given only in rms values because these devices normally conduct alternating current. Fig. 167 shows an rms on-state current rating curve for a typical triac as a function of case temperature. As with the SCR, the triac curve is derated to zero current when the case temperature rises to the maximum operating junction temperature. Triac current rat-

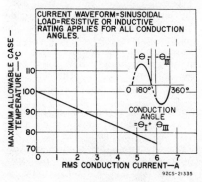

Fig. 167—Current rating chart for a typical RCA triac.

ings are given for full-wave conduction under resistive or inductive loads. Precautions should be taken to limit the peak current to tolerable levels when capacitive loads are used.

The **surge on-state current rating** $I_{TF(surge)}$ indicates the maximum peak value of a short-duration current pulse that should be allowed to flow through a thyristor during one on-state cycle, under stated conditions. This rating is applicable for any rated load condition. During normal operation, the junction temperature of a thyristor may rise to the maximum allowable value; if the surge occurs at this time, the maximum limit is exceeded. For this reason, a thyristor is not rated to block off-state voltage immediately following the occurrence of a current surge. Sufficient time must be allowed to permit the junction temperature to return to the normal operating value before gate control is restored to the thyristor. Fig. 168 shows a surge-current rating curve for the 2N3873

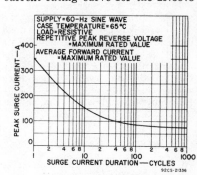

Fig. 168—Surge-current rating curve for the 2N3873 SCR.

SCR. This curve shows peak values of half-sine-wave forward (on-state) current as a function of overload duration measured in cycles of the 60-Hz current. Fig. 169 shows a surge-current rating curve for a typical triac. For triacs, the rating curve shows peak values for a full-sine-wave current as a function of the number of cycles of overload duration. Multicycle surge curves are the basis for the selection of circuit

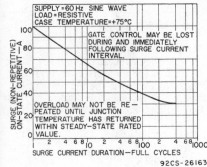

SUPPLY = 60 Hz SINE WAVE
LOAD = RESISTIVE
CASE TEMPERATURE = +75°C

GATE CONTROL MAY BE LOST
DURING AND IMMEDIATELY
FOLLOWING SURGE CURRENT
INTERVAL.

OVERLOAD MAY NOT BE RE-
PEATED UNTIL JUNCTION
TEMPERATURE HAS RETURNED
WITHIN STEADY-STATE RATED
VALUE.

SURGE CURRENT DURATION—FULL CYCLES

92CS-26163

Fig. 169—Surge-current rating curve for a typical triac.

breakers and fuses that are used to prevent damage to the thyristor in the event of accidental short-circuit of the device. The number of surges permitted over the life of the thyristor should be limited to prevent device degradation.

Critical Rate of Rise of On-State Current (di/dt)

In an SCR or triac, the load current is initially concentrated in the small area of the pellet where load current first begins to flow. This small area effectively limits the amount of current that the device can handle and results in a high voltage drop across the pellet in the first microsecond after the thyris-

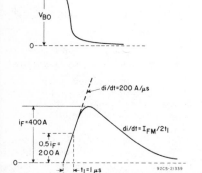

V_{BO}

0

di/dt = 200 A/µs

$i_F = 400 A$

di/dt = I_{FM}/2t_1

0.5i_F =
200 A

$t_1 = 1 µs$

92CS-21339

Fig. 170—Voltage and current waveforms used to determine di/dt rating of the 2N3873 SCR.

tor is triggered. If the rate of rise of current is not maintained within the rating of the thyristor, localized hot spots may occur within the pellet and permanent damage to the device may result. The waveshape for testing the di/dt capability of the RCA 2N3873 is shown in Fig. 170. The critical rate of rise of on-state current is dependent upon the size of the cathode area that begins to conduct initially, and the size of this area is increased for larger values of gate trigger current. For this reason, the di/dt rating is specified for a specific value of gate trigger current.

Holding and Latching Currents

After an SCR or triac has been switched to the on-state condition, a certain minimum value of anode current is required to maintain the thyristor in this low-impedance state. If the anode current is reduced below this critical holding-current value, the thyristor cannot maintain regeneration and reverts to the off or high-impedance state. Because the holding current (I_H) is sensitive to changes in temperature (increases as temperature decreases), this rating is specified at room temperature with the gate open.

The latching-current rating of a thyristor specifies a value of anode current, slightly higher than the holding current, which is the minimum amount required to sustain conduction immediately after the thyristor is switched from the off state to the on state and the gate signal is removed. Once the latching current (I_L) is reached, the thyristor remains in the on, or low-impedance, state until its anode current is decreased below the holding-current value. The latching-current rating is an important consideration when a thyristor is to be used with an inductive load because the inductance limits the rate of rise of the anode current. Precautions should be taken to insure that, under such conditions, the gate signal is present until the anode current rises to the latching value so that complete turn-on of the thyristor is assured.

Critical Rate of Rise of Off-State Voltage (dv/dt)

Because of the internal capacitance of a thyristor, the forward-blocking capability of the device is sensitive to the rate at which the forward voltage is applied. A steep rising voltage impressed across the main terminals of a thyristor causes a capacitive charging current to flow through the device. This charging current $(i = Cdv/dt)$ is a function of the rate of rise of the off-state voltage.

If the rate of rise of the forward voltage exceeds a critical value, the capacitive charging current may become large enough to trigger the thyristor. The steeper the wavefront of applied forward voltage, the smaller the value of the thyristor breakover voltage becomes.

The use of the shorted-emitter construction in SCR's has resulted in a substantial increase in the dv/dt capability of these devices by providing a shunt path around the gate-to-cathode junction. Typical units can withstand rates of voltage rise up to 200 volts per microsecond under worst-case conditions. The dv/dt capability of a thyristor decreases as the temperature rises and is increased by the addition of an external resistance from gate to reference terminal. The dv/dt rating, therefore, is given for the maximum junction temperature with the gate open, i.e., for worst-case conditions.

Turn-on Time

The ratings of thyristors are based primarily upon the amount of heat generated within the device pellet and the ability of the device package to transfer the internal heat to the external case. For high-frequency applications in which the peak-to-average current ratio is high, or for high-performance applications that require large peak values but narrow current pulses, the energy lost during the turn-on process may be the main cause of heat generation within the thyristor. The switching properties of the device must be known, therefore, to determine power dissipation which may limit the device performance.

When a thyristor is triggered by a gate signal, the turn-on time of the device consists of two stages, a delay time t_d and a rise time t_r, as shown in Fig. 171. The total turn-on time t_{gt} is defined as the time interval between the initiation of the gate signal and the time when the resulting current through the thyristor reaches 90 per cent of its maximum value with a resistive load. The **delay time** t_d is defined as the time interval between the 10-per-cent point of the leading edge of the gate-trigger voltage and the 10-per-cent point of the

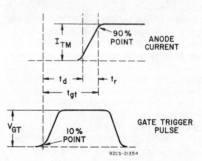

Fig. 171—*Gate-current and voltage turn-on waveforms for a thyristor.*

resulting current with a resistive load. The **rise time** t_r is the time interval required for the principal current to rise from 10 to 90 per cent of its maximum value. The total turn-on time, therefore, is the sum of both the delay and rise times of the thyristor.

Although the turn-on time is affected to some extent by the peak off-state voltage and the peak on-state current level, it is influenced primarily by the magnitude of the gate-trigger current pulse. Fig. 172 shows the variation in turn-on time with gate-trigger current for the RCA-2N3873 SCR.

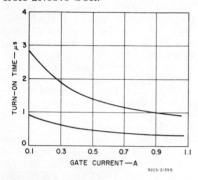

Fig. 172—*Range of turn-on time as a function of gate current for the 2N3873 SCR.*

To guarantee reliable operation and provide guidance for equipment designers in applications having short conduction periods, the voltage drop across RCA thyristors, at a given instantaneous forward current and at a specified time after turn-on from an off-state condition, is given in the published data. The wave-shape for the initial on-state voltage for the RCA-2N3873 SCR is shown in Fig. 173. This initial voltage, together with the time required for reduction of the dynamic forward voltage drop during the spreading time, is an indication of the current-switching capability of the thyristor.

When the entire junction area of a thyristor is not in conduction, the current through that fraction of the pellet area in conduction may result in large instantaneous power losses. These turn-on switching losses are proportional to the current and the voltage from cathode to anode of the device, together with the repetition rate of the gate-trigger pulses. The instantaneous power dissipated in a thyristor under such conditions is shown in Fig. 174. The curves shown in this figure indicate that the peak

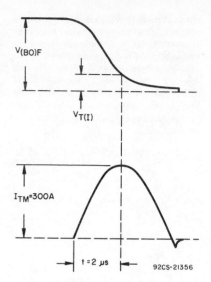

Fig. 173—*Initial on-state voltage and current waveforms for the 2N3873 SCR.*

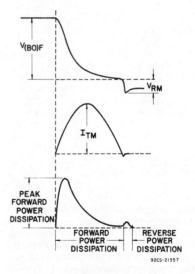

Fig. 174—*Instantaneous power dissipation in a thyristor during turn-on.*

power dissipation occurs in the short interval immediately after the device starts to conduct, usually in the first microsecond. During this time interval, the peak junction temperature may exceed the maximum operating temperature given in the manufacturer's data; in this case, the thyristor should not be required to block voltages immediately after the conduction interval. If the thyristor must block voltages immediately following the conduction interval, the junction-temperature rating must not be exceeded.

Turn-off Time (for SCR's)

The turn-off time of an SCR also consists of two stages, a reverse-recovery time and a gate-recovery time, as shown in Fig. 175. When the

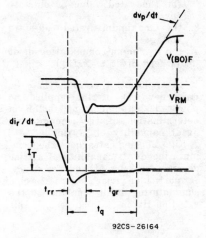

92CS–26164

Fig. 175—Circuit-commutated turn-off voltage and current waveforms for a thyristor.

forward current of an SCR is reduced to zero at the end of a conduction period, application of reverse voltage between the anode and cathode terminals causes reverse current to flow in the SCR until the reverse-blocking junction establishes a depletion region. The time interval between the application of reverse voltage and the time that the reverse current passes its peak value to a steady-state level is called the **reverse-recovery time** t_{rr}. A second recovery period, called the **gate-recovery time**

t_{gr}, must then elapse for the forward-blocking junction to establish a forward-depletion region so that forward-blocking voltage can be re-applied and successfully blocked by the SCR.

The gate-recovery time of an SCR is usually much longer than the reverse-recovery time. The total time from the instant reverse-recovery current begins to flow to the start of the re-applied forward-blocking voltage is referred to as the circuit **commutated turn-off time** t_q. The turn-off time is dependent upon a number of circuit parameters, including the on-state current prior to turn-off, the rate of change of current during the forward-to-reverse transition, the reverse-blocking voltage, the rate of change of the re-applied forward voltage, the gate trigger level, the gate bias, and the junction temperature. The junction temperature and the on-state current, however, have a more significant effect on turn-off time than any of the other factors. Because the turn-off time of an SCR depends upon a number of circuit parameters, the manufacturer's turn-off time specification is meaningful only if these critical parameters are listed and the test circuit used for the measurement is indicated.

Commutating dv/dt Capability (of Triacs)

In ac power-control applications, a triac must switch from the conducting state to the blocking state at each zero-current point, or twice each cycle, of the applied ac power. This action is called commutation. If the triac fails to block the circuit voltage (turn off) following the zero-current point, this action is not damaging to the triac, but control of the load power is lost. Commutation for resistive loading presents no special problems because the voltage and current are essentially in phase. For inductive loading, however, the current lags the voltage so that, following the zero-current point, an applied voltage opposite to the current and equal to the peak of the ac line voltage occurs across the thyristor. The maximum rate of rise of this voltage which can be blocked without the triac reverting to the on state is termed the critical rate of rise of commutation voltage, or the commutating dv/dt capability, of the triac.

SCR's do not experience commutation limitations because turn-on is not possible for the polarity of voltage opposite to current flow.

The commutating dv/dt is a major operating characteristic used to describe the performance capability of a triac. The characteristic can be more easily understood if the triac pellet, shown in Fig. 176, is considered to be divided into two halves.

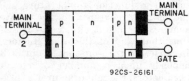

Fig. 176—Junction diagram for a triac pellet.

One half conducts current in one direction, the other half conducts in the opposite direction. The main blocking junctions and a lightly doped n-type base region in which charge can be stored are common to both halves of the triac pellet. (The base region is the section shown between the dotted lines in Fig. 176.)

Charge is stored in the base when current is conducted in either direction. The amount of charge stored at the end of each half-cycle of conduction depends on the commutating di/dt, i.e., the rate of decrease of load current as commutation is approached. The junction capacitance of the triac at commutation is a function of the remaining charge at that time. The greater the di/dt, the more remaining charge, and the greater the junction capacitance.

When the voltage changes direction, the remaining charge diffuses into the opposite half of the triac structure. The rate of rise of this voltage (commutating dv/dt) in conjunction with the junction capacitance results in a current flow which, if large enough, can cause the triac to revert to the conducting state in the absence of a gate signal.

The commutating dv/dt capability is specified in volts per microsecond for the following conditions:

1. the maximum rated on-state current $[I_T(RMS)]$;
2. the maximum case temperature for the rated value of on-state current;
3. the maximum rated off-state voltage (V_{DROM});
4. the maximum commutating di/dt (where $di/dt = 2\pi f I_{pk}$)

It is apparent, therefore, that the frequency (f) of the applied ac power is an important factor in determination of the commutating dv/dt capability of a triac.

Fig. 177 indicates how the commutating dv/dt capability of a triac depends on current and frequency. A particular triac has a specific commutating dv/dt capability at the rated 60-Hz on-state current. If this

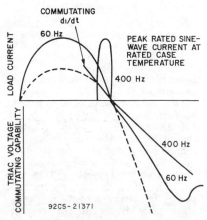

Fig. 177—Dependence of triac commutating capability on current and frequency.

60-Hz on-state current is reduced (dashed-line), then its associated commutating dv/dt capability is increased. It should be noted that as the sine-wave current is decreased in magnitude, the commutating di/dt is also decreased. For a 400-Hz on-state current of the same magnitude, it is evident that the commutating di/dt is much greater than at 60 Hz and, therefore, the commutating dv/dt capability is greatly reduced. These relationships indicate that a triac capable of 400-Hz operation must have an extremely high commutating capability. RCA offers a complete line of triacs rated for 400-Hz operation.

It should be evident that 400 Hz is not an upper limit on frequency capability for triacs; 400 Hz is a characterization point simply because it is a standard operating frequency. Figs. 178 and 179 indicate how the frequency capability of a typical RCA 400-Hz triac can be increased. Fig. 178 shows that reduction of load current increases frequency capability. Maximum rated junction temperature and minimum rated commutating dv/dt are held constant for this test of capability. Fig. 179 shows the effects of junction temperature on frequency capability. For this test, rated current and minimum rated dv/dt are held constant. Therefore, if a typical 400-Hz triac is used at less than its

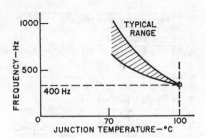

Fig. 179—Frequency capability of a 400-Hz triac as a function of junction temperature.

maximum rated junction temperature and less than its rated current, its frequency capability is greatly enhanced.

One other factor that greatly affects commutating capability is temperature. All commutating characteristic data are specified for maximum operating case temperature at maximum rated steady-state current. If the operating case temperature is below the rated value, the commutating capability is increased.

Gate Characteristics

The manufacturer's specifications indicate the magnitudes of gate current and voltage required to turn on SCR's and triacs. Gate characteristics, however, vary from device to device even among devices within the same family. For this reason, manufacturer's specifications on gating characteristics provide a range of values in the form of characteristic diagrams. A diagram such as that shown in Fig. 180 is given to define the limits of gate currents and voltages that may be used to trigger any given device of a specific family. The boundary lines of maximum and minimum gate impedance on this characteristic diagram represent the loci of all possible triggering points for thyristors in this family. The curve OA represents the gate characteristic of a specific device that is triggered within the shaded area.

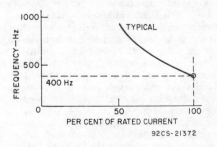

92CS-21372

Fig. 178—Frequency capability of a 400-Hz triac as a function of load current.

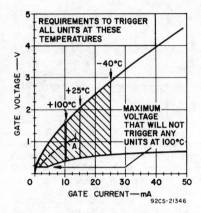

Fig. 180—Gate-characteristic curves for a typical RCA SCR.

Trigger Level—The magnitude of gate current and voltage required to trigger a thyristor varies inversely with junction temperature. As the junction temperature increases, the level of gate signal required to trigger the thyristor becomes smaller. Worst-case triggering conditions occur, therefore, at the minimum operating junction temperature.

The maximum value of gate voltage below the level required to trigger any unit of a specific thyristor family is also an important gate characteristic. At high operating temperatures, the level of gate voltage required to trigger a thyristor approaches the minimum value, and undesirable noise signals may inadvertently trigger the device. The maximum nontriggering gate voltage at the maximum operating junction temperature of the device, therefore, is a measure of the noise-rejection level of a thyristor.

The gate voltage and current required to switch a thyristor to its low-impedance state at maximum rated forward anode current can be determined from the circuit shown in Fig. 181. The value of resistor R_2 is chosen so that maximum anode current, as specified in the manufacturer's current rating,

flows when the device latches into its low-impedance state. The value of resistor R_1 is gradually decreased until the device under test is switched from its high-impedance state to its low-impedance state. The values of gate current and gate voltage immediately prior to switching are the gate voltage and current required to trigger the thyristor.

The **gate nontrigger voltage** V_{GD} is the maximum dc gate voltage that may be applied between gate and cathode of the thyristor for which the device can maintain its maximum rated blocking voltage. This voltage is usually specified at the rated operating temperature (100°C) of the thyristor. Noise signals in the gate circuit should be maintained below this level to prevent unwanted triggering of the thyristor.

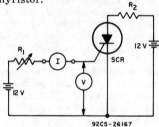

Fig. 181—Test circuit used to determine gate-trigger-pulse requirements of thyristors.

Pulse Triggering—The gate current specified in published data for thyristors is the dc gate trigger current required to switch an SCR or triac into its low-impedance state. For practical purposes, this dc value can be considered equivalent to a pulse current that has a minimum pulse width of 50 microseconds. For gate-current pulse widths smaller than 50 microseconds, the pulse-current curves associated with a particular device should be used to assure turn-on.

When pulse triggering of a thyristor is required, it is always advantageous to provide a gate-current pulse that has a magnitude exceeding the dc value required to trigger the device. The use of large trigger currents reduces variations in turn-on time, increases di/dt capability, minimizes the effect of temperature variation on triggering characteristics, and makes possible very short switching times. When a thyristor is initially triggered into conduction, the current is confined to a small area which is usually the more sensitive part of the cathode. If the anode-current magnitude is great, the localized instantaneous power dissipation may result in irreversible damage unless the rate of rise of principal current is restricted to tolerable levels to allow time for current spreading over a larger area. When a much larger gate signal is applied, a greater part of the cathode is turned on initially; as a result, turn-on time is reduced, and the thyristor can support a much larger peak anode inrush current.

In the past, the maximum value of gate signal that could be used to trigger a thyristor was severely restricted by minimum dc triggering requirements and limitations on maximum gate power. The coaxial gate structure and the "shorted-emitter" construction techniques now used in RCA thyristors, however, has greatly extended the range of limiting gate characteristics. As a result, the gate-dissipation ratings of RCA thyristors are compatible with the power-handling capabilities of elements normally used in the triggering circuits for thyristors. Advantage can be taken of the higher peak-power capability of the gate to improve dynamic performance, increase di/dt capability, minimize interpulse jitter, and reduce switching losses. This higher peak-power capability also allows greater interchangeability of thyristors in high-performance applications.

As explained previously, the "shorted-emitter" technique makes use of the resistance path within the gate layer which is in direct contact with the cathode electrode of the thyristor. When gate current is first initiated, most of the current bypasses the gate-to-cathode junction and flows from the resistive gate layer to the cathode contact. When the IR drop in this gate layer exceeds the threshold voltage of the gate-to-cathode junction, the current across this junction increases until the thyristor is triggered.

When a thyristor is triggered by a gate signal just sufficient to turn on the device, the entire junction area does not start to conduct instantaneously. Instead, as pointed out in the discussion on **Critical Rate of Rise of On-State Current**, the device current is confined to a small area, which is usually the most sensitive part of the cathode. The remaining cathode area turns on as the anode current increases. When a much larger signal is applied to the gate, a greater part of the cathode is turned on initially and the time to complete the turn-on process is reduced. The peak amplitude of gate-trigger currents must be large, therefore, when thyristors have to be turned on completely in a short period of time. Under such conditions, the peak gate power is high, and pulse triggering is required to keep the average gate dissipation within the values given in the manufacturer's specifications. New gate ratings, therefore, are required for this type of application.

The forward gate characteristics for thyristors shown in Figs. 182 and 183, indicate the maximum allowable pulse widths for various peak values of gate input power. The pulse width is determined by the relationship that exists between gate power

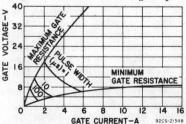

Fig. 182—Forward-gate characteristics for pulse triggering of RCA low-currrent SCR's.

input and the increase in the temperature of the thyristor pellet that results from the application of gate power. The curves shown in Fig. 182 are for RCA SCR's that have relatively small current ratings (2N4101 and 2N4102 types), and the curves shown in Fig. 183 are for RCA SCR's that have larger current

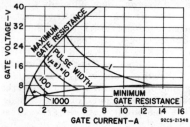

Fig. 183—Forward-gate characteristics for pulse triggering of RCA high-current SCR's.

ratings (2N3670, 2N3873, and 2N3899 types). Because the higher-current thyristors have larger pellets, they also have greater thermal capacities than the smaller-current devices. Wider gate trigger pulses can therefore be used on these devices for the same peak value of gate input power.

Because of the resistive nature of the "shorted-emitter" construction, similar volt-ampere curves can be constructed for reverse gate voltages and currents, with maximum allowable pulse widths for various peak-power values, as shown in Fig. 184. These curves indicate that reverse dissipations do not exceed the maximum allowable power dissipation for the device.

The total average dissipation caused by gate trigger pulses is the sum of the average forward and reverse dissipations. This total dissipation should be less than the **Maximum Gate Power Dissipation** P_{GM} shown in the published data for the selected SCR. If the average gate dissipation exceeds the maximum published value, as the result of high forward gate-trigger pulses and transient or steady-state reverse gate biasing, the maximum al-

lowable forward-conduction current rating of the device must be reduced to compensate for the increased rise of junction temperature caused by the increased gate power dissipation

The gate-trigger requirements of the triac are different in each operating mode. The I$^+$ mode (gate positive with respect to main terminal 1 and main terminal 2 positive with respect to main terminal 1), which is comparable to equivalent SCR operation, is usually the most sensitive. The smallest gate current is required to trigger the triac in this mode. The other three operating modes require larger gate-trigger currents. For RCA triacs, the maximum trigger-current rating in the published data is the largest value of gate current that is required

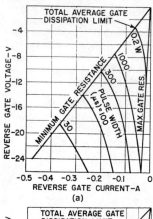

(a)

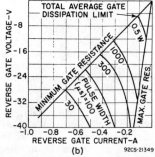

(b)

Fig. 184—Reverse gate characteristics of RCA SCR's: (a) low-current types; (b) high-current types.

to trigger the selected device in any operating mode.

Gate Trigger Circuits—The gate signal used to trigger an SCR or triac must be of sufficient strength to assure sustained forward conduction. Triggering requirements are usually stated in terms of dc voltage and current. Because it is common practice to pulse-fire thyristors, it is also necessary to consider the duration of the firing pulse required. A trigger pulse that has an amplitude just equivalent to the dc requirements must be applied for a relatively long period of time (approximately 30 microseconds) to ensure that the gate signal is provided during the full turn-on period of the thyristor. As the amplitude of the gate-triggering signal is increased, the turn-on time of the thyristor is decreased, and the width of the gate pulse may be reduced. When highly inductive loads are used, the inductance controls the current-rise portion of the turn-on time. For this type of load, the width of the gate pulse must be made long enough to assure that the principal current rises to a value greater than the latching-current level of the device. The latching current of RCA thyristors is always less than twice the holding current.

The application usually determines whether a simple or somewhat sophisticated triggering circuit should be used to trigger a given thyristor. Triggering circuits can be as numerous and as varied as the applications in which they are used; this text discusses the basic types only.

Many applications require that a thyristor be switched full on or full off in a manner similar to the operation of a relay. Although higher currents are handled by the thyristor, only small trigger or gate currents are required from the control circuit or switch. The simplest method of accomplishing this type of triggering is illustrated in Fig. 185.

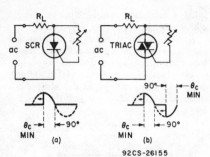

92CS-26155

Fig. 185—Degree of control over conduction angles when ac resistive network is used to trigger SCR's and triacs.

Each circuit shows a variable resistor in the gate circuit to control the conduction angle of the thyristor. The waveforms indicating the degree of control exercised by the variable resistance are also shown in Fig. 185. With maximum resistance in either circuit, the thyristor is off. As the resistance is reduced in the SCR circuit, a point is reached at which sufficient gate trigger current is provided at the positive peak of the voltage wave (90 degrees) to trigger the SCR on. The SCR conducts from the 90-degree point to the 180-degree point for a total conduction angle of $(180 - 90)$, or 90 degrees. In the triac circuit, as the resistance is reduced, the gate current increases until the triac is triggered at both the peak positive (90 degrees) and peak negative (270 degrees) points on the voltage wave. The triac then conducts between 90 degrees and 180 degrees, and between 270 degrees and 360 degrees for a total conduction angle of 180 degrees. The conduction angles of both the SCR and the triac can be increased by further reduction of the resistance in the gate circuits. For the SCR, the firing point is moved back from 90 degrees toward zero for a total conduction angle approaching 180 degrees. The triac firing points can also be moved back from 90 degrees toward zero for the positive half-cycle and from 270 degrees toward 180 degrees for the negative half-cycle to obtain a total conduction angle approaching 360 degrees.

An easier method of obtaining a phase angle greater than 90 degrees for half-wave operation is to use a resistance-capacitance triggering network. Fig. 186 shows the simplest form of such networks for use with

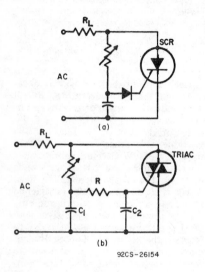

(a)

(b)

92CS-26154

Fig. 186—RC triggering networks used for phase-control triggering of thyristors.

an SCR and a triac. The thyristor is in series with the load and in parallel with the RC network. At the beginning of each half-cycle (positive half-cycle only for the SCR), the thyristor is in the off state. As a result, the ac voltage appears across the thyristor and essentially none appears across the load. Because the thyristor is in parallel with the potentiometer and capacitor, the voltage across the thyristor drives current through the potentiometer and charges the capacitor. When the capacitor voltage reaches the breakover voltage of the thyristor, the capacitor discharges through the gate circuit and turns the thyristor on. At this point, the

ac voltage is transferred from the thyristor to the load R_L for the remainder of the half-cycle. If the potentiometer resistance is reduced, the capacitor charges more rapidly, and the breakover voltage is reached earlier in the cycle; as a result, the power applied to the load is increased.

The gate trigger voltage can be more closely controlled in simple resistance or resistance-capacitance circuits by use of a variety of special triggering devices. Basically, a thyristor triggering device exhibits a negative resistance after a critical voltage is reached, so that the gate-current requirement of the thyristor can be obtained as a pulse from the discharge of the phase-shift capacitor. Because the gate pulse need be only microseconds in duration, the gate-pulse energy and the size of the triggering components are relatively small. Triggering circuits of this type employ elements such as neon bulbs, diacs, unijunction transistors, and two-transistor switches.

Fig. 187 shows a light-dimming circuit in which a diac is used to trigger a triac. The voltage-current characteristic for the diac in this circuit is shown in Fig. 188(a). The

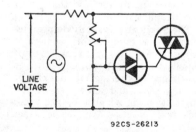

92CS-26213

Fig. 187—A light-dimmer circuit in which a diac is used to trigger a triac.

magnitude and duration of the gate-current pulse are determined by the interaction of the capacitor C_1, the diac characteristics, and the impedance of the thyristor gate.

Fig. 188(b) shows the typical shape of the gate-current pulse that is produced.

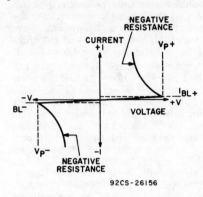

92CS-26156

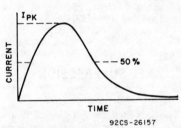

92CS-26157

Fig. 188—(a) Voltage-current characteristic for triggering device (diac) and (b) typical gate-current waveform for circuit shown in Fig. 187.

TRANSIENT PROTECTION

Voltage transients occur in electrical systems when some disturbance disrupts the normal operation of the system. These disturbances may be produced by various sources (such as lighting surges, energizing transformers, and load switching) and may generate voltages which exceed the rating of the thyristors. In addition, transients generally have a fast rate of rise that is usually greater than the critical value for the rate of rise of the thyristor off-state voltage (static dv/dt).

If transient voltages have magnitudes far greater than the device

rating, the thyristor may switch from the off state to the on state, and energy is then transferred from the thyristor to the load. Because the internal resistance of the thyristor is high during the on state, the transients may cause considerable energy to be dissipated in the thyristor before breakover occurs. In such instances, the transient voltage exceeds the maximum allowable voltage rating, and irreversible damage to the thyristor may occur.

Even if the magnitude of a transient voltage is within the maximum allowable voltage rating of the thyristor, the rate of rise of the transient may exceed the static dv/dt capability of the thyristor and cause the device to switch from the off state to the on state. This condition also results in transfer of energy from the thyristor to the load. In this case, thyristor switching from the off state to the on state does not occur because the maximum allowable voltage is exceeded but, instead, occurs because of the fast rate of rise of off-state voltage (dv/dt) and the thyristor capacitance, which result in a turn-on current $i = C dv/dt$. Thyristor switching produced in this way is free from high-energy dissipation, and turn-on is not destructive provided that the current that results from the energy transfer is within the device capability.

In either case, transient suppression techniques are employed to minimize the effects of turn-on because of overvoltage or because the thyristor dv/dt capability is exceeded.

One of the obvious solutions to insure that transients do not exceed the maximum allowable voltage rating is to provide a thyristor with a voltage rating greater than the highest transient voltage expected in a system. This technique, however, does not represent an economical solution because, in most cases, the transient magnitude, which is dependent on the source of transient generation, is not easily defined. Transient voltages as high as 2600 volts have resulted from lighting disturbances on a 120-volt residential power line. Usually, the best solution is to specify devices that can withstand voltage from 2 to 3 times the steady-state value. This technique provides a reasonable safety factor. The effects of voltage transients can further be minimized by use of external circuit elements, such as RC snubber networks across the thyristor terminals, as shown in Fig. 189. The rate at

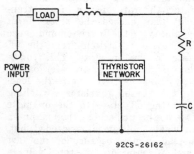

92CS-26162

Fig. 189—Circuit showing use of RC snubber network to minimize effects of voltage transients and use of an inductance in series with the load for suppression of transient rise times.

which the voltage rises at the thyristor terminal is a function of the load impedance and the values of the resistor R and the capacitor C in the snubber network. Because the load impedance is usually variable, the preferred approach is to assume a worst-case condition for the load and, through actual transient measurement, to select a value of C that provides the minimum rate of rise at the thyristor terminals. The snubber resistance should be selected to minimize the capacitor discharge currents through the thyristor during turn-on.

For applications in which it is necessary to minimize false turn-on because of transients, the addition of a coil in series with the load, as shown in Fig. 189, is very

effective for suppression of transient rise times at the thyristor terminals. For example, if a transient of infinite rise time is assumed to occur at the input terminals and if the effects of the load impedance are neglected, the rise time of the transient at the thyristor terminals is approximately equal to $E_{pk}/\sqrt{LC}$. If the value of the added inductor L is 100 microhenries and the value of the snubber capacitor C is 0.1 microfarad, the infinite rate of rise of the transient at the thyristor terminals is reduced by a factor of 3. For a filter network consisting of L = 100 microhenries, C = 22 microfarads, and R = 47 ohms, a 1000-volt-per-microsecond transient that appears at the input terminals is suppressed by a factor of 6 at the thyristor terminals.

RFI SUPPRESSION

The fast switching action of triacs when they turn on into resistive loads causes the current to rise to the instantaneous value determined by the load in a very short period of time. Triacs switch from the high- to the low-impedance state within 1 or 2 microseconds; the current must rise from essentially zero to full-load value during this period. This fast switching action produces a current step which is largely composed of higher-harmonic frequencies of several megahertz that have an amplitude varying inversely as the frequency. In phase-control applications, such as light dimming, this current step is produced on each half-cycle of the input voltage. Because the switching occurs many times a second, a noise pulse is generated into frequency-sensitive devices such as AM radios and causes annoying interference. The amplitude of the higher frequencies in the current step is of such low levels that they do not interfere with television or FM radio. In general, the level of radio-frequency interference (RFI) produced by the triac is well below that produced by most ac/dc brush-type electric motors; however, some type of RFI suppression network is usually added.

There are two basic types of radio-frequency interference (RFI) associated with the switching action of triacs. One form, **radiated RFI**, consists of the high-frequency energy radiated through the air from the equipment. In most cases, this radiated RFI is insignificant unless the radio is located very close to the source of the radiation.

Of more significance is **conducted RFI** which is carried through the power lines and affects equipment attached to the same power lines. Because the composition of the current waveshape consists of higher frequencies, a simple choke placed in series with the load increases the current rise time and reduces the amplitude of the higher harmonics. To be effective, however, such a choke must be quite large. A more effective filter, and one that has been found adequate for most light-dimming applications, is shown in Fig. 190. The LC filter provides

adequate attenuation of the high-frequency harmonics and reduces the noise interference to a low level. The capacitor connected across the entire network bypasses high-frequency signals so that they are not connected to any external circuits through the power lines.

Fig. 191 shows a triac control circuit that includes RFI suppression for the purpose of minimizing high-frequency interference. The values indicated are typical of those used in lamp-dimmer circuits.

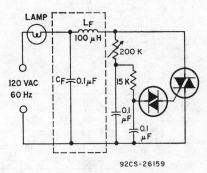

92CS-26159

Fig. 191—Lamp-control circuit incorporating RFI suppression.

RCA PRODUCT MATRICES

The product matrices shown in Tables XXI and XXII indicate the wide range of operating voltages and currents and the variety of package configurations offered by the extensive selection of RCA triacs and SCR's. Detailed ratings and characteristics data on all currently available commercial types of RCA thyristors are given in the RCA Solid State DATABOOK Series SSD-206, "Thyristors, Rectifiers, and Diacs," or in the RCA technical bulletins on each device. The file numbers for the technical bulletins are listed in the product matrices.

92CS-26158

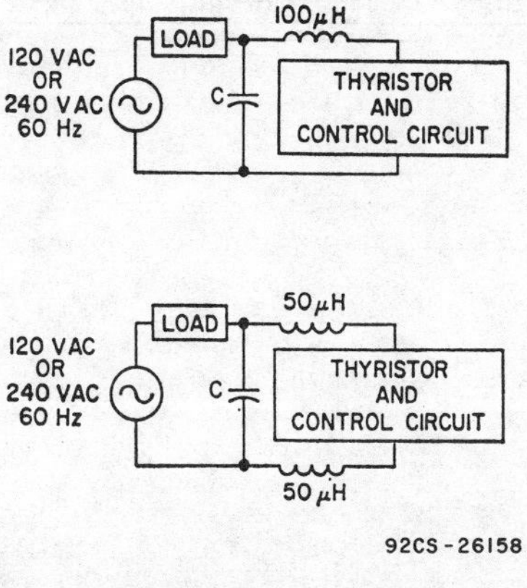

Fig. 190—RFI-suppression networks (C = 0.1 µF, 200 V at 120 V ac; 0.1 µF, 400 V at 240 V ac).

Table XXI—RCA Triac Product Matrix

RCA Triacs			Modified TO-5				Modified TO-5 With Heat Radiator				TO-66	
STANDARD	I_T(RMS)		2.5A	2.5A	2.5A	2.5A	2.5A	2.5A	2.5A	2.5A	6.0A	15.0A
	I_{TSM}		25A	25A	25A	25A	25A	25A	25A	25A	100A	100A
	V_{DROM}(V)	100	T2300A	T2301A	T2302A	2N5754	T2310A	T2311A	T2312A	T2313A		
		200	T2300B	T2301B	T2302B	2N5755	T2310B	T2311B	T2312B	T2313B	T2700B	T4700B
		400	T2300D	T2301D	T2302D	2N5756	T2310D	T2311D	T2312D	T2313D	T2700D	T4700D
		500										
		600				2N5757				T2313M		
		800										
	I_{GT}(mA)	1+, 111−	3	4	10	25	3	4	10	25	25	
		1−, 111+	3	4	10	40	3	4	10	40	40	
	V_{GT}(V)	All Modes	2.2	2.2	2.2	2.2	2.2	2.2	2.2	2.2	2.2	
		File No.	470	431	470	414	470	431	470	414	351	
ZERO VOLTAGE SWITCH	V_{DROM}(V)	100				T2306A				T2316A		
		200				T2306B				T2316B	T2706B	
		400				T2306D				T2316D	T2706D	T4706D
		500										
		600										
	I_{GT}(mA)	1+, 111−				45				45	45	45
	V_{GT}(V)	1+, 111+				1.5				1.5	1.5	1.5
		File No.				406				406	406	406
400-HZ OPERATION	I_T(RMS)				0.5A	0.5A						
	V_{DROM}(V)	200			T2304B	T2305B						
		400			T2304D	T2305D						
	I_{GT}(mA)	1+, 111−			10	25						
		1−, 111+			10	40						
	V_{GT}(V)	All Modes			2.2	2.2						
		File No.			441	441						

Table XXI—RCA Triac Product Matrix (cont'd)

RCA Triacs		TO-66 With Heat Radiator	TO-220AB VERSAWATT				ISOWATT 8A	Press Fit		
STANDARD	I$_T$(RMS)	6.0A	6A	6A	8.0A	8A			10.0A	15.0A
	I$_{TSM}$	100A	60A	80A	100A	100A	100A		100A	100A
	V$_{DROM}$(V) 100						T2850A			
	200	T2710B	T2500B	T2801B	T2800B	T2802B	T2850B		2N5567	2N5571
	300			T2801C	T2800C	T2802C				
	400	T2710D	T2500D	T2801D	T2800D	T2802D	T2850D		2N5568	2N5572
	500			T2801E	T2800E	T2802E				
	600				T2800M	T2802M			T4101M	T4100M
	800									
	I$_{GT}$(mA) 1+, 111–	25	25	80	25	50	25		25	50
	1–, 111+	40	60	–	60	–	60		40	80
	V$_{GT}$(V) All Modes	2.2	2.5	4.0▲	2.5	2.5▲	2.5		2.5	2.5
	File No.	351	615	837	838	838	540		457	458
ZERO VOLTAGE SWITCH	V$_{DROM}$(V) 100									
	200	T2716B			T2806B				T4107B	T4106B
	400	T2716D			T2806D				T4107D	T4106D
	500									
	600									
	I$_{GT}$(mA) 1+, 111–	45			45				45	45
	V$_{GT}$(V) 1+, 111+	1.5			1.5				1.5	1.5
	File No.	406			406				406	406
400-HZ OPERATION	I$_T$(RMS)							6A	10.A	15.0A
	V$_{DROM}$(V) 200							T4105B	T4104B	T4103B
	400							T4105D	T4104D	T4103D
	I$_{GT}$(mA) 1+, 111–							50	50	50
	1–, 111+							80	80	80
	V$_{GT}$(V) All Modes							2.5	2.5	2.5
	File No.							443	443	443

* ISOWATT — Mounting tab electrically isolated from electrodes.   ▲1+, 111– only.

Table XXI—RCA Triac Product Matrix (cont'd)

RCA Triacs			Stud		Isolated Stud		Press Fit		Stud	
	I_T(RMS)		10.0A	15.0A	10.0A	15.0A	30.0A	40.0A	30.0A	40.0A
	I_{TSM}		100A	100A	100A	100A	300A	300A	300A	300A
STANDARD	V_{DROM}(V) 100									
	200		2N5569	2N5573	T4121B	T4120B	T6401B	2N5441	T6411B	2N5444
	400		2N5570	2N5574	T4121D	T4120D	T6401D	2N5442	T6411D	2N5445
	500									
	600		T4111M	T4110M	T4121M	T4120M	T6401M		T6411M	2N5446
	800							T6400N		T6410N
	I_{GT}(mA) 1+, 111−		25	50	25	50	50	50	50	50
	1−, 111		40	80	40	80	80	80	80	80
	V_{GT}(V) All Modes		2.5	2.5	2.5	2.5	2.5	2.5	2.5	2.5
	File No.		457	458	457	458	459	593	459	593
ZERO VOLTAGE SWITCH	V_{DROM}(V) 100									
	200		T4117B	T4116B			T6407B	T6406B	T6417B	T6416B
	400		T4117D	T4116D			T6407D	T6406D	T6417D	T6416D
	500									
	600						T6407M	T6406M	T6417M	T6416M
	I_{GT}(mA) 1+, 111−		45	45			45	45	45	45
	V_{GT}(V) 1+, 111+		1.5	1.5			1.5	1.5	1.5	1.5
	File No.		406	406			406	406	406	406
400-HZ OPERATION	I_T(RMS)	6A	10.0A	15.0A			25.0A	40.0A	25.0A	40.0A
	V_{DROM}(V) 200	T4115B	T4114B	T4113B			T6405B	T6404B	T6415B	T6414B
	400	T4115D	T4114D	T4113D			T6405D	T6404D	T6415D	T6414D
	I_{GT}(mA) 1+, 111−	50	50	50			80	80	80	80
	1−, 111+	80	80	80			120	120	120	120
	V_{GT}(V) All Modes	2.5	2.5	2.5			3.0	3.0	3.0	3.0
	File No.	443	443	443			487	487	487	487

Table XXI—RCA Triac Product Matrix (cont'd)

RCA Triacs			Isolated Stud		Press Fit*		Stud*		Isolated Stud*	
STANDARD	I_T(RMS)		30.0A	40.0A	60A	80A	60A	80A	60A	80A
	I_{TSM}		300A	300A	600A	850A	600A	850A	600A	850A
	V_{DROM}(V)	100								
		200	T6421B	T6420B	T8401B	T8430B	T8411B	T8440B	T8421B	T8450B
		400	T6421D	T6420D	T8401D	T8430D	T8411D	T8440D	T8421D	T8450D
		500								
		600	T6421M	T6420M	T8401M	T8430M	T8411M	T8440M	T8421M	T8450M
		800		T6420N						
	I_{GT}(mA)									
		1+, 111−	50	50	75	75	75	75	75	75
		1−, 111+	80	80	150	150	150	150	150	150
	V_{GT}(V)									
		All Modes	2.5	2.5	2.8	2.5	2.8	2.5	2.8	2.5
		File No.	459	593	725	549	725	549	725	549
ZERO VOLTAGE SWITCH	V_{DROM}(V)	100								
		200								
		400								
		500								
		600								
	I_{GT}(mA)									
		1+, 111−								
	V_{GT}(V)									
		1+, 111+								
		File No.								
400-HZ OPERATION	I_T(RMS)									
	V_{DROM}(V)	200								
		400								
	I_{GT}(mA)									
		1+, 111−								
		1−, 111+								
	V_{GT}(V)									
		All Modes								
		File No.								

* Package has factory-attached flexible leads for main terminals 1 and 2.

Table XXII—RCA SCR Product Matrix

RCA SCR's		TO-8		TO-66						
I_T(RMS)		2.0A	4.5A	5.0A	FTO 5.0A	FTO 5.0A	FTO 5A	FTO 5.0A	FTO 5.0A	FTO 5.0A
I_TSM		60A	200A	60A	80A	80A	80A	75A(I_PM)	50A	50A
V_DROM	15									
V_RROM(V)	25									
	30									
	50									
	100		S2400A				S3704A			
	150									
	200	2N3528	S2400B	2N3228		S3700B	S3704B			
	250									
	300									
	400	2N3529	S2400D	2N3525		S3700D	S3704D			
	500				S3706E					
	600	2N4102	S2400M	2N4101	S3705M	S3700M	S3704M	S3701M		
	700						S3704S		S3702S	
	750									S3703SF
	800									
I_GT(mA)		15	15	15	30	40	40	35	45	40
V_GT(V)		2	2	2	4	3.5	3.5	4	4	4
	File No.	114	567	114	839	306	690	476	522	522

RCA SCR'S		TO-66 With Heat Rad		Low Profile Mod. TO-5	TO-5 With Heat Rad.	TO-5 With Heat Spreader	TO-220AB VERSAWATT			
I_T(RMS)		5.0A	FTO 5A	7.0A	3.3A	7.0A	4.0A	4.0A	4A	8.0A
I_TSM		60A	80A	100A	100A	100A	35A	35A	35A	100A
V_DROM	15						S2060Q	S2061Q	S2062Q	
V_RROM(V)	25									
	30						S2060Y	S2061Y	S2062Y	
	50						S2060F	S2061F	S2062F	
	100		S3714A				S2060A	S2061A	S2062A	S2800A
	150									
	200	S2710B	S3714B	S2600B	S2610B	S2620B	S2060B	S2061B	S2062B	S2800B
	250									
	300						S2060C	S2061C	S2062C	
	400	S2710D	S3714D	S2600D	S2610D	S2620D	S2060D	S2061D	S2062D	S2800D
	500						S2060E	S2061E	S2062E	
	600	S2710M	S3714M	S2600M	S2610M	S2620M	S2060M	S2061M	S2062M	
	700		S3714S							
	750									
	800									
I_GT(mA)		15	40	15	15	15	0.2	0.5	2	15
V_GT(V)		2	3.5	1.5	1.5	1.5	0.8	0.8	0.8	1.5
	File No.	266	690	496	496	496	654	654	654	501

FTO — Fast Turn-Off

Table XXII—RCA SCR Product Matrix (cont'd)

RCA SCR's		Stud	TO-3	Press Fit		Stud		Isolated Stud	
I_T(RMS)		10A	12.5A	20.0A	35.0A	20.0A	35.0A	20.0A	35.0A
I_{TSM}		90A	200A	200A	350A	200A	350A	200A	350A
V_{DROM}	15								
V_{RROM}(V)	25								
	30								
	50								
	100		2N3668	S6200A	2N3870	S6210A	2N3896	S6220A	S6420A
	150								
	200	S5210B	2N3669	S6200B	2N3871	S6210B	2N3897	S6220B	S6420B
	250								
	300								
	400	S5210D	2N3670	S6200D	2N3872	S6210D	2N3898	S6220D	S6420D
	500								
	600	S5210M	2N4103	S6200M	2N3873	S6210M	2N3899	S6220M	S6420M
	700								
	750								
	800				S6400N		S6410N		S6420N
I_{GT}(mA)		40	40	15	40	15	40	15	40
V_{GT}(V)		3.5	2	2	2	2	2	2	2
	File No.	757	116	418	578	418	578	418	578

RCA SCR's		TO-48				
I_T(RMS)		16.0A	25.0A	Pul. Mod. 35.0A	FTO 35.0A	FTO 35A
I_{TSM}		125A	150A	150A	180A	250A
V_{DROM}	15					
V_{RROM}(V)	25	2N1842A	2N681			
	30					
	50	2N1843A	2N682			2N3654
	100	2N1844A	2N683		2N3650	2N3655
	150	2N1845A	2N684			
	200	2N1846A	2N685		2N3651	2N3656
	250	2N1847A	2N686			
	300	2N1848A	2N687		2N3652	2N3657
	400	2N1849A	2N688		2N3653	2N3658
	500	2N1850A	2N689			
	600		2N690	S6431M	S7430M	S7432M
	700					
	750					
	800					
I_{GT}(mA)		45	25	80	180	180
V_{GT}(V)		3.5	3	2	3	3
	File No.	28	96	247	408	724

FTO – Fast Turn-Off

For Horizontal-Deflection Circuits

RCA ITR's*		TO-66	
I_T(RMS)		TRACE 5A	RETRACE 5A
I_{TSM}		50A	50A
V_{DROM}(V)	400		S3800D
	500	S3800E	
	550		S3800EF
	600		S3800M
	650	S3800MF	
	700		S3800S
	750	S3800SF	
I_{GT}(mA)		40	45
V_{GT}(V)		4	4
	File No.	639	639

*Integrated Thyristor/Rectifier

Integrated Circuits for Linear Applications

THE design of linear (analog) circuits involves selection and interconnection of an optimum combination of active and passive components to accomplish a particular signal-processing function with due consideration to factors such as efficiency, reliability, and cost. This general rule is valid whether discrete components and/or integrated circuits are employed.

The advent of integrated circuits has freed the system designer from a considerable amount of detailed work because the integrated-circuit industry now offers him a vast selection of "function blocks" (e.g., amplifiers, voltage regulators, comparators, and a large variety of special-purpose circuits). Because these predesigned electronic function blocks are readily available, the circuit designer can quickly synthesize and test a complex system with the assurance that the integrated-circuit manufacturer has already spent considerable effort in optimizing the detailed design of each function block. For example, the design of a power-supply is simplified considerably by selection of an appropriate integrated-circuit voltage regulator.

In order to "partition" a particular system most effectively, the proficient designer must acquire and maintain a familiarity with the available integrated-circuit types and their characteristics. It is pertinent, therefore, to review the functional classification of integrated circuits for linear circuit applications and describe illustrative examples. The following list indicates the various functional classes of RCA integrated circuits available for use in linear applications:

Active-Device Arrays
Transistor Arrays
Diode Arrays
Thyristor/Transistor Arrays
Transistor/Zener-Diode Arrays

Differential Amplifiers

Operational Amplifiers (Op-Amps)
Operational Voltage Amplifiers
Operational Transconductance
  Amplifiers

Comparators

Power-Control Circuits
Voltage Regulators
Zero-Voltage Switching Circuits
Programmable Power Switches

Special-Purpose Circuits
AM and FM Radio Receiver Circuits (including stereo decoders and audio amplifiers)
TV Receiver Circuits
Analog Multipliers
Phase-Locked Loops
Analog Timing Circuits
Optoelectronic Circuits

INTEGRATED-CIRCUIT ARRAYS

Integrated-circuit transistor and diode arrays provide many of the inherent advantages of monolithic integrated circuits without the limitations sometimes encountered in integrated circuits that contain passive circuit elements (e.g., resistors). They offer the designer a unique class of solid-state devices that feature matched electrical and thermal characteristics, compactness, ease of physical handling, and economy.

Before examining specific designs of integrated-circuit arrays it is instructive to summarize some major advantageous characteristics exhibited by transistors and diodes fabricated within a common silicon substrate.

Matched voltage-transfer characteristics: The base-to-emitter voltage of two integrated-circuit transistors operated at the same emitter current typically are matched to within 2 to 3 per cent over the current range of 50 microamperes to 1 milliampere.

Matched current-transfer characteristics: The current-transfer characteristics of two juxtaposed integrated-circuit transistors typically are matched to within 2 to 3 per cent over the current range of 50 microamperes to 1 milliampere.

Matched thermal characteristics: If no power were dissipated on the integrated-circuit chip, all transistors on the same chip would be at the same temperature. Consequently, even at the outset, the use of integrated-circuit transistors can offer advantages over discrete transistors when a particular circuit requires thermal-tracking between several active components. Thermal-tracking techniques that would be impractical with discrete transistors can be successfully executed by use of integrated-circuit arrays.

Transistor Arrays

Fig. 192 shows the configuration of a very popular integrated-circuit transistor array. This array, the CA3046, contains three independent n-p-n transistors plus two other n-p-n transistors with interconnected emitters. Electrical-match characteristics of the two emitter-coupled transistors (Q_1 and Q_2) have been specified for applications that

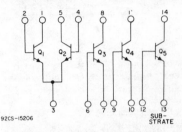

92CS-15206

Fig. 192—CA3046 integrated-circuit transistor array.

require electrical similarity (e.g., differential amplifiers and current-mirrors). For example, the specified maximum magnitude of input offset voltage for this pair of transistors is 5 millivolts (V_{CE} = 3 volts, I_C = 1 milliampere).

The applications for the transistor array shown in Fig. 192 are virtually as varied as is the case for discrete transistors with similar characteristics. A wide variety of bipolar transistor arrays is available for use at supply voltages up to about 40 volts with average-current capabilities in some types of up to 100 milliamperes. Several types have n-p-n and p-n-p transistors on the same monolithic chip. One type, the CA3097E, contains an assortment of devices, viz., a programmable unijunction transistor (PUT), a silicon-controlled rectifier (SCR), a zener diode, a composite combination of n-p-n and p-n-p transistors, and an n-p-n transistor of 100 milliampere capability.

Another example of an integrated-circuit transistor array, the CA3600E, is shown in Fig. 193. This

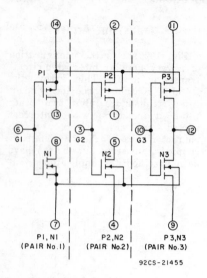

Fig. 193—CA3600E COS/MOS integrated-circuit transistor array.

array consists of three n-channel and three p-channel enhancement-type MOS transistors, i.e., COmplementary-Symmetry MOS (COS/MOS) Field-Effect Transistors. These transistors are uniquely suitable for service in complementary-symmetry circuits at supply voltages in the

range of 3 to 15 volts and are useful at frequencies up to 5 MHz (untuned). Each transistor in the CA3600E can conduct currents up to 10 mA. The transistors feature very high input resistance (10^{11} ohms, typically) and low gate-terminal current (10 picoamperes, typically). Integral gate-protection circuitry is provided in the CA3600E.

Fig. 194 shows the schematic diagram of a single-stage "true-complementary" linear amplifier that uses one pair of the complementary MOS transistors in the CA3600E connected in a common-source circuit. Resistor R_b is used to bias the complementary pair of transistors for class A operation, and resistor R_s represents the source resistance of the signal source. The curves in Fig. 195 show the voltage gain of the amplifier as a function of operating frequency for various supply voltages. This amplifier can produce very high output-swing voltages (V_{out}); for example, its output voltages can swing within several millivolts of either supply voltage "rail". At a supply voltage of 3 volts, the amplifier typically requires only 0.3 microamperes of supply-current; at a supply voltage of 10 volts, it consumes 2.2 milliamperes typically.

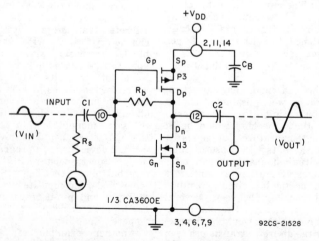

Fig. 194—COS/MOS transistor-pair biased for linear-mode operation.

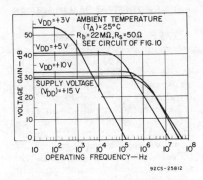

Fig. 195—*Typical voltage gain as a function of frequency characteristics for amplifier circuit shown in Fig. 194.*

Diode Arrays

Fig. 196 shows the configuration of a typical integrated-circuit diode array, the CA3039, together with some of its electrical characteristics. Because all the diodes in the CA3039 array are fabricated simultaneously on a single silicon chip, they have

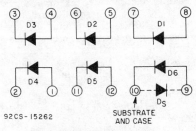

92CS-15262

V_F = 0.90V (max.) at I_F = 10 mA
V_F = 0.69V (max.) at I_F = 50 μA
Diode offset voltage
...... 5 mV (max.) at I_F = 1 mA
PIV for D1 through D5 ...... 5V
DC Leakage Current
...... 100 nA (max.) at V_R = −4V
Reverse Recovery Time
...... 1nS (typ.)
Total Device Dissipation
...... 600 mW (max.)
C_D = 0.65 pF (typ.) at V_R = −2V
10-Lead TO-5 Package

Fig. 196—*Schematic diagram and some characteristics of the CA3039 integrated-circuit diode array.*

nearly identical characteristics, and their parameters track each other with temperature variations as a result of their close physical proximity. The temperature coefficient of the diode offset voltage variation is typically about 1 microvolt per °C. Excellent static and dynamic matching of the diodes makes them particularly useful in communications and switching systems that require balanced or bridge circuitry.

The diodes in the CA3039 are actually diode-connected transistors, connected as shown in Fig. 197. This connection provides a low-voltage diode with low series resistance for general-purpose use. It should be noted that any of the transistors shown in Fig. 192 can be diode-connected in accordance with Fig. 197 to provide a diode which is rea-

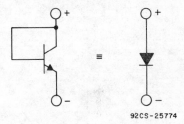

92CS-25774

Fig. 197—*Diagram showing use of a transistor as a diode.*

sonably well matched to transistors on the same chip. Connection of the collector and base of a monolithic transistor to serve as the anode of the diode provides a two-element device which has a good electrical and thermal match to other transistors on the chip for use in thermal-compensation circuitry.

At this juncture, it is also appropriate to observe that the emitter-base junction of a monolithic transistor (such as those shown in Fig. 192) may be biased in the reverse direction to provide a zener diode. The generic connection scheme for such a zener-equivalent element is shown in Fig. 198(a). When transistors like those in the CA3046 (shown in Fig. 192) are used, the zener voltage is approximately 7 volts, and the dynamic impedance is in the range of 60 to 100 ohms. In this connection [Fig. 198(a)], the zener diode exhibits a positive temperature coefficient in the order of 2 millivolts per °C. When a lower dynamic impedance is desired, a transistor may be used as an emitter-

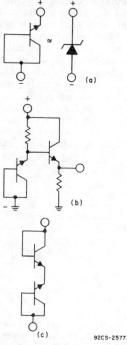

Fig. 198—Diagrams showing (a) use of a transistor emitter-base junction to provide a zener diode, (b) connection of a transistor emitter-follower to provide lower dynamic impedance, and (c) a temperature-compensated zener diode.

follower for the diode, as shown in Fig. 198(b). This configuration provides an equivalent zener voltage of approximately 6.3 volts with a temperature coefficient of +4 millivolts per °C. Fig. 198(c) shows an arrangement for a higher-impedance zener-diode equivalent which is more nearly temperature-compensated; this diode provides a zener voltage of approximately 7.7 volts.

DIFFERENTIAL AMPLIFIERS

The balanced differential amplifier shown in Fig. 199 is the basic configuration used to provide signal gain in linear integrated circuits. This configuration is the fundamental building block for a broad line of RCA all-monolithic-silicon integrated circuits designed for a wide variety of linear applications at frequencies from dc into the vhf region. The differential pair of monolithic transistors Q_1 and Q_2 function in the same way as a pair of discrete transistors operated in a similar circuit configuration. The superior match of the integrated-circuit transistors, however, permits substantially better balanced operation than a discrete-transistor circuit.

Basic Circuit

The currents to the emitter-coupled differential transistors (Q_1 and Q_2) are supplied from a controlled constant-current source (either a transistor or a resistor). Temperature-compensating networks can be readily incorporated as an integral part of the controlled-source circuit to assure that circuit gain, dc operating point, and other important characteristics vary as required over the operating temperature range. The differential amplifier

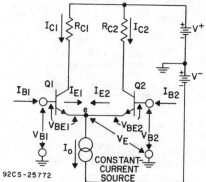

92CS-25772

Fig. 199—Balanced differential-amplifier configuration used as the basic gain stage for RCA linear integrated circuits.

shown in Fig. 199 is connected to operate from symmetrical dual positive and negative supply voltages (V^+ and V^-). Single-supply operation is also feasible, but requires an external voltage-divider network for proper biasing of the bases of the transistors.

The balanced differential amplifier may be considered as two symmetrically arranged "half-circuits," each with a transistor (Q_1 or Q_2) and a load resistor (R_{C1} or R_{C2}). If the characteristics of the transistor-resistor pairs are identical, the two "half-circuits" are perfectly matched and can be mated without introduction of a circuit unbalance when the emitters of Q_1 and Q_2 are joined and the transistors are operated from common dc supplies. If the two input voltages V_{B1} and V_{B2} are either zero or equal in magnitude and of the same polarity, the amplifier does not become unbalanced because the collector currents I_{C1} and I_{C2} remain equal; a zero voltage difference, therefore, is maintained between the collectors of transistors Q_1 and Q_2.

It is apparent from Fig. 199 that the sum of the emitter currents I_{E1} and I_{E2} is always equal to the constant source current I_0. Consequently, an increase in one of the emitter currents is accompanied by an equal decrease in the other emitter current. This current relationship, of course, depends upon the quality of the constant-current source.

When the base of transistor Q_1 is driven positive with respect to the base of transistor Q_2 (i.e., a differential input is applied), the current through Q_1 increases, and the current through Q_2 decreases equally so that the source current I_0 remains constant. For these conditions, I_{C1} is larger than I_{C2} and a voltage difference is developed between the transistor collectors such that the collector voltage of Q_2 is more positive than the collector voltage of Q_1. The differential input voltage, therefore, produces a differential output voltage. This sequence of events describes the operation of the differential amplifier in the **differential-input, differential-output mode.**

The differential amplifier is a versatile configuration that offers a number of circuit-connection modes. For example, if the voltage V_{B1} alone increases in a positive direction with respect to ground, the voltage at the collector of Q_1 decreases with respect to ground. With the output taken from the collector of Q_1, the differential amplifier, then, operates as though it were merely a classical single-stage, phase-inverting transistor amplifier. This type of operation is referred to as the **single-ended-input, single-ended-output inverting mode** of the differential amplifier.

In view of the fact that an increase in the current through Q_1 results in an attendant decrease in the current through Q_2 (to maintain a constant I_0), a positive-going increase in V_{B1} alone also results in an increase in the voltage at the collector of Q_2 with respect to ground. With the output taken from the collector of Q_2, the differential amplifier then operates in the **single-ended-input, single-ended-output noninverting mode,** i.e., merely as a classical single-stage non-inverting transistor amplifier.

The differential amplifier may also be operated in the **differential-input, single-ended output mode.** In this mode of operation, the output voltage is coupled from the collector of either Q_1 or Q_2 in response to a differential input voltage ($V_{B1} - V_{B2}$) applied between the bases of the two transistors.

The ratio of the change in collector voltage to the "difference" in the base voltages is the **differential voltage gain** (A_d). If the collector-to-collector voltage is used in the gain ratio, the result is referred to as the **double-ended differential-voltage gain** (A_{dDE}). If the collector-to-ground change is used, the ratio is called the **single-ended differential-voltage gain** (A_{dSE}).

Transfer Characteristics

The variation in the collector currents I_{C1} and I_{C2} as a function of the differential input voltage ($V_{B1} - V_{B2}$) is the keystone characteristic of the differential amplifier. The transfer curves shown in Fig. 200 provide several important points of information about the differential amplifier:

1. The transfer characteristics are linear in a region about the operating point. For the curves shown, this linear region corresponds to an input-voltage swing of approximately 50 millivolts peak-to-peak.

2. The maximum slope of the curves, which occurs at the operating point, defines the **effective transconductance** of the differential amplifier.

3. The slope of the transfer curves (i.e., the transconductance) is dependent upon the value of the total current I_O supplied to the constant-current source. The slope of the transfer curves can be changed, without changing the linear input region, by varying the value of I_O. This relationship implies that automatic gain control is inherent in the differential amplifier when the current I_O is controlled.

4. The transfer characteristics and the slopes of these characteristics are a function of the alpha of the transistors and of the temperature, both of which are predictable, and of two physical constants.

5. The differential amplifier is a natural limiter; when input excursions exceed approximately ± 100 millivolts (for the curves shown), no further increase in the ouput is obtained.

6. The output current of an amplifier is the product of the input voltage and the transconductance. In the differential amplifier, the transconductance is proportional to the controlled current I_O; this circuit, therefore, may be used for mixing, frequency multiplication, modulation, or product detection

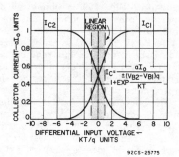

Fig. 200—Transfer curves of the basic differential-amplifier circuit.

when the current I_O is made a multiplicand and the input waveform is the multiplier.

The transconductance at the operating point (at $V_{B1} - V_{B2} = 0$) is the **maximum transconductance** of the differential amplifier. The effective transconductance of the differential amplifier is one-fourth that of a single transistor. This condition results from the fact that, at the operating point, exactly one-half of the total current I_O flows through each transistor of the differential pair and the input voltage must be divided equally between the two transistors.

When the differential amplifier is operated to provide double-ended outputs so that the output voltage is measured between the collectors of the differential pair of transistors, the output currents through the load impedance contribute equally to the output voltage from each transistor. As a result, the output voltage is twice that obtained for single-ended operation. This increase in output voltage results because the load impedance is doubled, not because of any doubling of the transconductance. However, if an impedance is connected between the two collectors and the shunt collector-feed resistors are large compared to this load impedance, the load current is twice as large as can be expected from a single-ended circuit.

Common-Mode Response

If the base voltages of both differential-pair transistors Q_1 and Q_2 are increased or decreased together (i.e., a **common-mode input voltage** is applied), then the emitter currents remain equal and, because their sum must be equal to the constant current I_0, no change in either emitter current occurs. Consequently, no change in collector output voltage results. This absence of output-voltage change in response to a common-mode input-voltage signal constitutes an ability of the differential amplifier to reject common-mode signals. The degree to which rejection is achieved depends on the impedance of the constant-current source (CCS). Because some finite impedance is always associated with the constant-current source, some small output signal is always produced in response to a common-mode signal. The ratio of this small output-voltage change to the change in common-mode input voltage is called the **common-mode voltage gain** (A_c). This gain is normally much less than unity. The ratio of the common-mode voltage gain (A_c) to the differential voltage gain is called the **common-mode rejection ratio** (CMRR),

$$CMRR = A_c/A_d$$

This common-mode rejection ratio is normally expressed in decibels as follows:

$$CMRR \text{ (dB)} = 20 \log A_c/A_d$$

The common-mode rejection ratio is a measure of the ability of the differential amplifier to discriminate between differential and common-mode input signals and is usually of the order of 80 to 120 dB.

A change in common-mode voltage at the input results in a corresponding change in voltage across the constant-current source; however, there are limits to the levels of common-mode voltage that the circuit can handle before its performance is severely degraded. For example, it is apparent from Fig. 199 that if the base voltage of either Q_1 or Q_2 rises above the collector voltage, the transistor will be in saturation, and normal differential-amplifier operation will cease. If the base voltage decreases below some minimum level required by the constant-current source, the impedance of the constant-current source will decrease, and the common-mode rejection ratio will be degraded. This range of common-mode voltages over which the differential amplifier operates is normally called the input **common-mode voltage range**, V_{CMR}, and is usually defined for a specified distortion at the output, or for a specified degradation in the common-mode rejection ratio.

Balanced DC Amplification

The monolithic-integrated-circuit differential amplifier can provide excellent balanced dc amplification because of the close electrical and thermal match of transistors Q_1 and Q_2. If the transistors were perfectly matched in terms of base-to-emitter voltage (V_{BE}), dc beta, and bulk and contact resistances, and if the resistances R_{C1} and R_{C2} were equal, the difference between the collector voltages would be zero when equal voltages were applied to the bases of the two transistors. Unfortunately, small unbalances do exist even with monolithic devices, and a small dc offset between the collectors exists when the bases are at equal potentials. When the bases of transistors Q_1 and Q_2 are driven from low-resistance sources, this offset results primarily from the unbalance in the base-to-emitter voltages of the two transistors. Unequal betas contribute to this offset when the bases are driven from high-resistance sources. For this condition, unequal voltage drops are developed across the resistors because unequal base currents result in a differential voltage drive at the bases.

Any of the following methods may be used to measure and specify the unbalance:

The **output offset voltage** V_{OO} is the difference in dc voltage between the collectors (for equal collector resistors) when the bases are connected to the same dc voltage within the input common-mode voltage range. The measurement can be made with or without resistors in series with the bases, but if resistors are used, they must be of equal value. This measurement is simple to make, but it suffers from a lack of generality because the measured value depends on the dc gain chosen for the test circuit.

The **input offset voltage** V_{IO} is the difference in base voltages ($V_{B1} - V_{B2}$) which must be applied to obtain equal collector dc voltages when the collector resistors R_{C1} and R_{C2} are equal. This offset is relatively independent of beta mismatch and is primarily a function of the mismatch in base-to-emitter voltages. A typical value for the input offset voltage in monolithic transistors is 1 millivolt; consequently, this voltage is more difficult to measure accurately than the output offset voltage. The input offset voltage, however, is more general because it is an offset referred to the input and is independent of circuit gain.

The **input bias-current offset** I_{IO} is the difference in input base-bias currents ($I_{B1} - I_{B2}$) when the collector voltages are equal. This offset is a measure of the beta mismatch at a particular operating current. The input bias-current offset is relatively small for low operating currents and is correspondingly larger for higher operating currents. This current offset is typically 5 to 10 per cent of the input base-bias current.

Offsets are usually specified at +25°C; additional parameters are needed, therefore, to describe the thermal match between the differential-pair transistors Q1 and Q2. The thermal match is indicated by the **offset drift** specifications, which give the change in the particular offset per degree C over a specified temperature range, as follows:

$$\text{Input Offset-Voltage Drift} = \frac{\triangle I_{IO}}{\triangle T}$$

$$\text{Input Offset-Current Drift} = \frac{\triangle I_{IO}}{\triangle T}$$

These parameters are sometimes given as curves of the input offset voltage V_{IO} and the input bias-current offset I_{IO} as a function of temperature.

Constant-Current Source

The preceding discussion of basic differential-amplifier operation indicated the need for a constant-current source (CCS) to assure that the sum of the emitter currents in the differential-amplifier transistor pair would always be constant. This condition was shown to be a requirement for good common-mode rejection. A constant-current source is an infinite-impedance source. The simplest way to approximate a constant-current source is by use of a high-value resistor R_{CCS}, as shown in Fig. 201.

For balanced operation, the resistor R_{CCS} must be very large compared to the emitter impedances of transistors Q_1 and Q_2 to assure that the signal current i_c will not be shunted into R_{CCS} and cause the current i_{c2} to be less than the current i_{c1}. For a given operating current, the value of R_{CCS} is limited by the dc voltage required at the

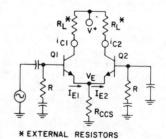

✶ EXTERNAL RESISTORS

92CS-25776

Fig. 201—Balanced differential amplifier that uses a resistor as the constant-current source.

emitter to produce the required value for the constant current I_O. Consequently, the resistance R_{CCS} is determined on the basis of the largest value possible consistent with the minimum permissible value for the current I_O. This compromise normally results in a constant-current source which, although simple, does not provide the best balanced operation or common-mode rejection.

A transistor connected as shown in Fig. 202 provides a much better constant-current source. The impedance, as it appears at the collector of Q_3, is very large, but the voltage required to drive the current I_O through the resistor R_E can be relatively small because a high value of resistance is not required for R_E.

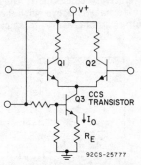

Fig. 202—Balanced differential amplifier that uses a transistor constant-current-source circuit.

The minimum voltage required at the common emitters must be just large enough to sustain the $I_O R_E$ drop plus the collector-to-emitter voltage (approximately 0.5 volt) for Q_3 to prevent saturation of this transistor. The dc voltage would, of course, normally be larger than this minimum limit in order to allow for a large common-mode voltage swing at the input bases. The bias of this constant-current source can be adjusted so that the negative temperature coefficient of the forward base-to-emitter voltage of transistor Q_3 approximately compensates for the positive temperature coefficient of the diffused resistor R_E. This com-

pensation assures that the operating current is maintained nearly constant with changes in temperature. This feature is particularly important for circuits that employ external load resistors because such load resistors do not track with the internal current-setting resistor.

The temperature coefficient for the base-to-emitter voltage is approximately −2 millivolts per degree C, and the temperature coefficient of R_E is approximately +0.2 per cent per degree C. If the current I_O is to remain constant when the change in base-to-emitter voltage with temperature causes a rise in voltage across resistor R_E of 2 millivolts per degree C, the increase of 0.2 per cent per degree C in the voltage drop $I_O R_E$ caused by the change in the value of R_E with temperature must equal 2 millivolts per degree C. The quiescent $I_O R_E$ drop, therefore, must be 1 volt because 0.2 per cent of 1 volt is 2 millivolts.

When diffused load resistors are included on the chip, the emitter voltage of transistor Q_3 must be maintained constant with temperature because changes in the value of R_E will be compensated for by changes in the diffused load resistors R_L. As a result, the operating point and voltage gain remain relatively fixed with changes in temperature. The circuit shown in Fig. 203 illustrates the use of diodes in the base-bias circuit to compensate for the variations in the base-to-emitter voltage of transistor Q_3. The base-to-emitter voltage, therefore, is maintained at a constant value. Two diodes are required to compensate for the temperature-induced changes in the base-to-emitter voltage because the effect of the diodes on the base bias is reduced by the voltage division of the resistances. In RCA differential-amplifier circuits, the base-bias voltage is reduced by a factor of 2. This relationship is clearly illustrated by the Thevenin equivalent circuit of the base-bias network for transistor Q_3 shown in Fig. 204.

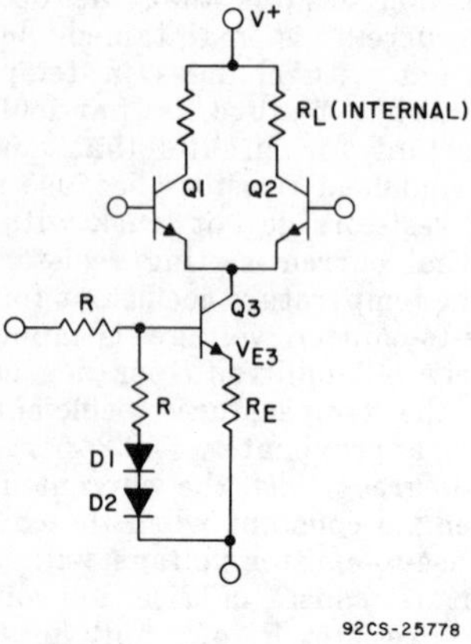

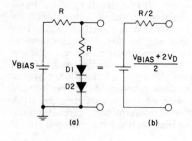

92CS-25778

Fig. 203—Differential amplifier in which temperature-compensating diodes are used in the base-bias network of the constant-current transistor.

Fig. 205 shows a third biasing method for the constant-current-source transistor. This method makes use of the fact that two matched monolithic transistors will conduct equal emitter currents if their bases and emitters are connected in parallel. D_1 is a diode-connected transistor operated in parallel with the base-emitter junction of transistor Q_3. The emitter current of transistor Q_3, therefore, is very nearly equal to the current through D_1. If the dc beta of transistor Q_3 is high enough so that the base current can be neglected, the current through

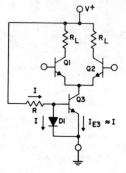

Fig. 204—Base-bias circuit for the constant-current transistor: (a) actual circuit; (b) Thevenin's equivalent circuit.

D_1 can be expressed by the following equation:

$$I = \frac{V^+ - V_{D1}}{R} = I_{E3}$$

This equation shows that the current in the constant-current source can be set by selection of the supply voltage V^+ and the resistance R because the voltage V_{D1} will always be close to 0.7 volt.

The circuit shown in Fig. 205 is similar to the one shown in Fig. 203 in respect to its behavior with variations in temperature. That is, if R and R_L are both diffused resistors, their change with temperature will have compensating effects to keep the operating voltage and voltage gain relatively constant. If they are both external resistors with low or matching temperature coefficients,

92CS-25780

Fig. 205—Differential amplifier in which circuit biasing is based on the match in the characteristics of two monolithic base-emitter junctions connected in parallel.

the constancy will also be maintained.

The arrangement of diode D_1 and transistor Q_3 shown in Fig. 205 is commonly that known as a "current-mirror" because the current I that flows in the diode-connected transistor D_1 is "mirrored" as a current I of equal magnitude flowing in the emitter of Q_3, the base-emitter junctions of D_1 and Q_3 are assumed to be similar. This configuration is frequently used to "scale" current flow in various circuit networks employed in integrated-circuit designs. For example, if the emitter-base junction of Q_3 is twice that of diode-connected transistor D_1, a current-flow of I established by resistor R through D_1 will force a current-flow of 2I in the emitter of Q_3.

Applications of Differential Amplifiers

Differential amplifiers are used in a majority of complex integrated circuits for linear circuit applications. For example, Fig. 38 in the section **Materials, Junctions, and Devices** shows the use of a simple differential amplifier in a voltage regulator. Transistors Q_4, Q_5, Q_6, Q_7, Q_8, Q_9 comprise a differential-amplifier stage used to compare the reference voltage with a sample of the regulated output voltage.

Differential-Amplifier "Building-Block"—The CA3028 integrated circuit shown in Fig. 206 is basically a simple differential-amplifier "building-block" circuit. The CA3028 is intended for general-purpose use at frequencies from dc to 120 MHz. The CA3028B can be operated at supply voltages up to 30 volts, and

the transistors are rated for operation with currents up to 20 milliamperes. This integrated circuit is also ideally suited for use in cascode amplifier configurations with transistors Q_3 and Q_1 in the cascode configuration, and transistor Q_2 optionally applicable as the automatic-gain-control (agc) device.

Differential amplifiers are also very applicable to uses in conjunction with tuned-circuits. Fig. 207

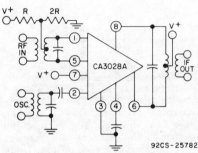

Fig. 207—Schematic of mixer circuit using the CA3028 differential-amplifier integrated circuit.

shows the CA3028 connected in a mixer circuit (e.g., as used in a superheterodyne receiver) with the rf input signal applied differentially at terminals 1 and 5. The local-oscillator signal is injected at terminal 2 and the intermediate-frequency if signal is developed differentially at terminals 6 and 8.

Dual Differential-Amplifier "Building-Block"—A pair of differential amplifiers can readily be fabricated on a monolithic integrated-circuit chip, as shown in Fig. 208. This configuration is ideally suited for use in doubly-balanced circuits, such as those employed in frequency-synthesizing mixers, double-sideband generators, synchronous detectors and the like. In this type of circuit, it is advantageous to balance two differential amplifiers with respect to each other; their match must be maintained despite variations in temperature and the passage of time.

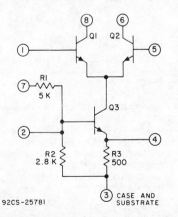

Fig. 206—CA3028A or CA3028B differential-amplifier "building-block" integrated circuit.

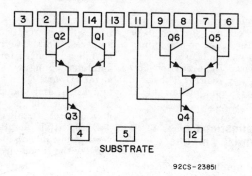

Fig. 208—CA3054 dual differential-amplifier "building-block" integrated circuit.

An example of a doubly-balanced frequency-synthesizing mixer circuit is shown in Fig. 209. The collector outputs for the differential pairs Q_1—Q_2 and Q_5—Q_6 are connected so that the balanced push-pull drive on the emitters and bases essentially cancels each output. Under proper drive conditions, therefore, the output to transformer T_3 can be the desired sum and difference frequencies of f_1 and f_2 with f_1 and f_2 suppressed. When compared to classical mixer circuits, this doubly-balanced circuit is very effective in minimizing the generation of spurious frequencies.

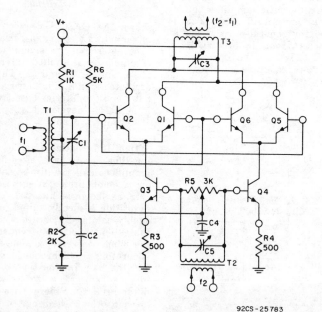

Fig. 209—A frequency-synthesizer mixer using the CA3054 dual differential amplifier.

OPERATIONAL AMPLIFIERS

The term "operational amplifier" was originally intended to denote an amplifier circuit that performed various mathematical operations such as integration, differentiation, summation, and subtraction. The application of the operational amplifier, however, has been so vastly extended that today this term suggests a device that finds the widest use in such applications as signal amplification and wave shaping, servo and process controls, analog instrumentation and system design, impedance transformation, and many other routine functions.

Basic Design Models

An ideal operational amplifier would have infinite open-loop gain and bandwidth, and zero noise, offset, and drift. Although no amplifier has these ideal qualities, practical integrated-circuit operational amplifiers are generally characterized by the following properties:

1. extremely high dc voltage gain, generally in the range from 10^3 to 10^6;

2. wide bandwidth that starts at dc and rolls off to unity gain at from one to several hundred megahertz with a slope of 6 dB per octave or at most 12 dB per octave;

3. positive and negative output voltage over a large dynamic range, preferably to magnitudes that are essentially equal to the supply voltages;

4. very low input dc offset and drift with time and temperature;

5. high input impedance so that amplifier input current can be largely neglected;

6. low output impedance.

The configuration most commonly used for operational voltage amplifiers consists of one or two differential-amplifier circuits, together with an appropriate output stage. The differential-amplifier stages not only fulfill the operational-amplifier requirement for a high-gain direct-coupled amplifier circuit, but also provide significant advantages with respect to the application of the operational amplifier.

Inverting Feedback Configuration —The basic design model for a differential-input operational amplifier operated with an inverting feedback configuration is shown in Fig. 210. The load resistor R_L is assumed to be large enough so that its effect on the transfer characteristic is negligible, i.e., $I_{OUT} = 0$.

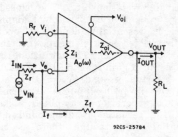

Fig. 210—Inverting operational-amplifier configuration.

Certain differential-input operational amplifiers require a significant flow of bias current at each input. For this condition, the dc paths to ground for each input must be equal so that a minimum dc offset voltage (error) is developed at the output. Thus, for the terminology employed in Fig. 210, R_r must equal the parallel combination of $Z_r(\omega = 0)$ with the series combinaton of $Z_f(\omega = 0)$ and Z_{ol} $(\omega = 0)$.

In the circuit of Fig. 210, the drive-source impedance affects the feedback in the inverting configuration and, therefore, must be considered part of the Z_r term. For brevity, the symbol Z_r is defined to include the source impedance as well as certain feedback design elements. The impedances Z_i and Z_{ol} are the open-loop **intrinsic** input and output impedances of the operational amplifier. Ordinarily, these impedances are assumed in the amplifier symbol. In Fig. 210, however, they are identified to emphasize their importance. The term $A_o(\omega)$ is the **open-loop differential voltage gain** of the operational amplifier; this parameter is frequency-dependent. The terminals on the operational-amplifier symbol labeled minus $(-)$ and plus $(+)$ refer to the inverting and noninverting input, respectively.

The transfer function, or **closed-loop gain,** of an operational voltage amplifier is generally considered to express the relationship between input and output voltages. (It is relatively simple to convert the voltage transfer function to another desired transfer relationship.) The closed-loop gain of the operational amplifier is essentially dependent upon only the feedback elements (i.e., for the ideal inverting configuration, the closed-loop gain is equal to $-Z_f/Z_r$).

A virtual ground $(V_e = 0)$ exists at the inverting-input $(-)$ terminal of operational amplifier. That is, the terminal is at ground potential even though there is no electrical connection between this nodal point and ground. Moreover, no current flows into the negative terminal of the amplifier when $A_o(\omega)$ is infinite (ideal amplifier), as indicated by the fact that the nodal-assigned voltage V_e is zero and the impedance at the negative terminal $(Z_i + Z_r)$ is not zero.

The **closed-loop output impedance** for an inverting feedback operational amplifier is defined as the ratio of the unloaded output voltage to the short-circuit output current. Although this impedance is in no way indicative of the output-current capabilities of the amplifier, it is a useful small-signal parameter that can be employed to determine the gain reduction that results when the operational amplifier is operated into a finite load impedance.

Noninverting Feedback Configuration—Fig. 211 shows the general-circuit model for a differential-input

operational amplifier operated with a noninverting feedback configuration.

The model for the noninverting feedback configuration, as did that for the inverting circuit, assumes that the load resistance R_L is large enough so that its effect is negligible, i.e., $R_L \to \infty$ and $I_{OUT} = 0$.

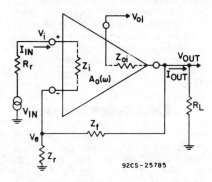

Fig. 211—Noninverting operational-amplifier configuration.

A noninverting operational amplifier, unlike the inverting type, requires a differential-input arrangement because it uses the common-mode effect in its feedback scheme. The following basic requirements and definitions that apply to the inverting circuit shown in Fig. 210 are also valid for the general noninverting circuit shown in Fig. 211:

1. The dc return paths to ground for the two inputs must be equal and finite for amplifiers that require a significant amount of input bias current.

2. The input and output impedances Z_i and Z_{o1}, are inherent in the basic amplifier unit and are shown on the diagram to emphasize their importance in the determination of the classical design relationships.

3. The open-loop gain is frequency-dependent and is represented by the symbol $A_o(\omega)$.

4. The plus and minus labels on the input terminals designate the

noninverting and inverting terminals respectively.

In the noninverting circuit, however, the source impedance is included in the passive element R_r rather than the frequency-dependent parameter Z_r, as in the inverting circuit.

As with the inverting-feedback configuration, the **closed-loop gain** for the noninverting configuration expresses the relationship between input and output voltage. The term $1 + (Z_t/Z_r)$ represents the closed-loop gain for the ideal noninverting configuration. This term, which is referred to as the **ideal feedback characteristic**, is basic to operational-amplifier phase-compensation theory.

As with the inverting configuration, the **closed-loop output impedance** for the noninverting configura-

tion is defined as the ratio of the open-circuit output voltage V_{OUT} to the short-circuit output current I_{out}.

General-Purpose Operational Voltage Amplifier (OVA)

Fig. 212 shows the basic circuit configuration for a general-purpose integrated-circuit operational amplifier. This circuit consists basically of two differential amplifiers and a single-ended output circuit in cascade. The pair of cascaded differential amplifiers are responsible for virtually all the gain provided by the operational-amplifier circuit.

Circuit Operation—The inputs to the operational amplifier are applied to the bases of the pair of emitter-coupled differential input transistors Q_1 and Q_2 in the first differential

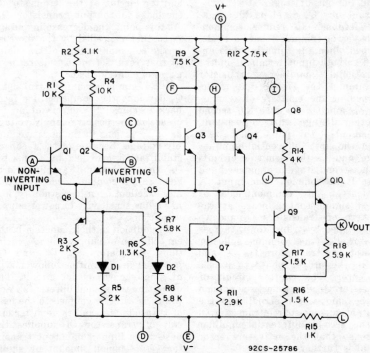

Fig. 212—Circuit configuration for RCA general-purpose integrated-circuit operational amplifiers (CA3008, CA3008A, CA3010, CA3010A, CA3015, CA3015A, CA3016, CA3016A, CA3029, CA3029A, CA3030, CA3030A, CA3037, CA3037A, CA3038, and CA3038A).

amplifier. The inverting input is applied to the base of transistor Q_2, and the noninverting input is applied to the base of transistor Q_1. These transistors develop the driving signals for the second differential amplifier. The constant-current source transistor Q_6 provides bias stabilization for transistors Q_1 and Q_2. Diode D_1 provides thermal compensation for the first differential-amplifier stage.

The emitter-coupled transistors Q_3 and Q_4 in the second differential amplifier are driven push-pull by the outputs from the first differential amplifier. Bias stabilization for the second differential amplifier is provided by the constant-currrent-source transistor Q_7. Compensating diode D_2 provides the thermal stabilization for the second differential-amplifier and also for the transistor Q_9, in the output stage.

Transistor Q_5 develops the negative feedback to reduce common-mode error signals that are developed when the same input is applied to both input terminals of the operational amplifier. Transistor Q_5 samples the signal that is developed at the emitters of transistors Q_3 and Q_4. Because the second differential stage is driven push-pull, the signal at this point will be zero when the first differential-amplifier stage and the base-emitter circuits of the second stage are matched and there is no common-mode input. A portion of any common-mode, or error, signal that appears at the emitters of transistors Q_3 and Q_4 is developed by transistor Q_5 across resistor R_2 (the common collector resistor for transistors Q_1, Q_2, and Q_5) in the proper phase to reduce the error. The emitter circuit of transistor Q_5 also reflects a portion of the same error signal into the constant-current-source transistor Q_7 in the second differential-amplifier stage so that the activating error signal is further reduced.

Transistor Q_5 also develops feedback signals to compensate for dc common-mode effects produced by variations in the supply voltages.

For example, a decrease in the dc voltage from the positive supply results in a decrease in the voltage at the emitters of transistors Q_3 and Q_4. This negative-going change in voltage is reflected by the emitter circuit of transistor Q_5 to the bases of transistors Q_7 and Q_9. Less current then flows through these transistors. The decrease in the collector current of transistor Q_7 results in a reduction of the current through transistors Q_3 and Q_4, and the collector voltages of these transistors tend to increase. This tendency to increase on the part of the collector voltages partially cancels the decrease that occurs with the reduction in the positive supply voltage. The partially cancelled decrease in the collector voltage of transistor Q_4 is coupled directly to the base of transistor Q_8 and is transmitted by the emitter circuit of this transistor to the base of output transistor Q_{10}. At this point, the decrease in voltage is further cancelled by the increase in the collector voltage of transistor Q_9 that results from the decrease in current mentioned above.

In a similar manner, transistor Q_5 develops the compensating feedback to cancel the effects of an increase in the positive supply voltage or of variations in the negative supply voltage. Because of the feedback stabilization provided by transistor Q_5, the operational amplifier provides high common-mode rejection and excellent open-loop stability, and has a low sensitivity to power-supply variations.

In addition to their function in the cancellation of supply-voltage variations, transistors Q_8, Q_9, and Q_{10} are used in an emitter-follower type of single-ended output circuit. The output of the second differential amplifier is directly coupled to the base of transistor Q_8, and the emitter circuit of transistor Q_8 supplies the base-drive input for output transistor Q_{10}. A small amount of signal gain in the output circuit is made possible by the bootstrap connection from the emitter of output transistor Q_{10} to the emitter circuit of transistor Q_9. If this bootstrap connection were neglected, transistor Q_9 could be considered as merely a dc constant-current source for drive transistor Q_8. Because of the bootstrap arrangement, however, the output circuit can provide a signal gain of 1.5 from the collector of differential-amplifier transistor Q_4 to the output. Although this small amount of gain may seem insignificant, it does increase the output-swing capabilities of the operational amplifiers.

The output from the operational-amplifier circuit is taken from the emitter of output transistor Q_{10} so that the dc level of the output signal is substantially lower than that of the differential-amplifier output at the collector of transistor Q_4. In this way, the output circuit shifts the dc level at the output so that it is effectively the same as that at the input when no signal is applied.

Resistor R_{15} increases the negative-going-signal capability of the operational amplifier, when terminal L is shorted to terminal K so that the resistor is connected between the output and the negative supply.

AC Characteristics—Fig. 213 shows the open-loop gain-frequency characteristics of the operational-amplifier circuit shown in Fig. 212. For operation from ±6-volt supplies, the

amplifier provides an open-loop gain of 60 dB. The first "break" in the frequency response occurs at about 300 Hz and then begins to drop sharply at about 3 MHz. The unity-gain frequency is 18 MHz. Fig. 214

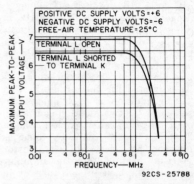

Fig. 214—Output-swing capabilities as a function of frequency for operational-amplifier circuit shown in Fig. 212.

shows the unloaded output-voltage as a function of frequency for operation from ±6-volt supplies.

Phase-Compensation—The principal usefulness of an operational amplifier is generally realized with large amounts of negative feedback (i.e., closed-loop operation). It is imperative, therefore, that the inevitable phase-shifts be adequately "compensated" so that the desired low-frequency negative feedback cannot become positive at higher frequencies and cause amplifier instabilities (i.e., oscillations). The open-loop gain-frequency characteristics of an operational amplifier are useful guidelines for prediction of circuit stability in a great majority of closed-loop (feedback) applications. A particular amplifier will usually be stable in closed-loop operation if its open-loop gain-frequency characteristic decreases at a rate no greater than about 6 dB per octave (20 dB per frequency-decade) for all frequencies into the proximity of the unity-gain ("cut-off frequency") point. Otherwise, some auxiliary circuit network (commonly called a "phase-compensation"

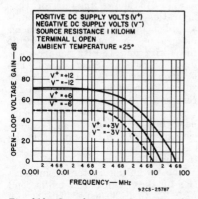

Fig. 213—Open-loop gain-frequency characteristics of the operational-amplifier circuit shown in Fig. 212.

network) must be provided to establish the requisite open-loop frequency-roll-off characteristic.

Fig. 213 shows that under open-loop conditions the operational-amplifier circuit shown in Fig. 212 suffers a gain loss of about 35 dB (50 dB − 15 dB) over the frequency-decade between 1 and 10 MHz in ±6-volt operation. Consequently, it can be predicted with absolute certainty that the amplifier would be unstable in many closed-loop (feedback) circuits. This situation is easily corrected for the circuit in Fig. 212 by addition of two "phase-compensation" networks, each consisting of a 27-picofarad capacitor and a 2000-ohm resistor connected in series between the collector of a transistor in the first differential-amplifier stage to the base of a transistor in the second differential-amplifier stage (i.e., one network is connected between terminals C and F and the other network between terminals H and I). Fig. 215 shows the response of the amplifier, with and without the compensating networks. When the "compensation" networks are connected, the open-loop gain-frequency response of the amplifier is rolled-off at a rate of somewhat less than 20 dB per frequency-decade over about three frequency-decades. Consequently, the amplifier will operate with stability in most applications that require closed-loop networks.

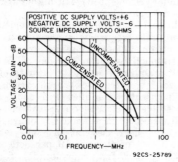

Fig. 215—Open-loop gain as a function of frequency for both phase-compensated and uncompensated operation for operational-amplifier circuit shown in Fig. 212.

Monolithic integrated-circuit operational amplifiers are usually classified, insofar as phase-compensation networks are concerned, as "uncompensated" (i.e., external phase-compensaton networks are employed as required in particular applications) or "compensated" (i.e., integral on-chip compensation circuitry is provided) types. The circuit shown in Fig. 212 is of the "uncompensated" type. The "uncompensated" operational-amplifier circuit provides a degree of flexibility for the designer in using a particular circuit over a broad spectrum of applications. On the other hand, an operational amplifier with internal phase compensation is simpler to apply in many circuits, at the expense of some performance restrictions in certain applications.

Integrator Application—The versatility of the operational-amplifier circuit shown in Fig. 212 can be illustrated in connection with its use in an integrator circuit. Integrators have diverse functions, e.g., ranging from their ability to integrate accurately in the solution of differential equations to their use as an electronic low-pass filter with current-gain capability. An important design consideration when an operational-amplifier is to be used as an integrator is that dc feedback be provided to meet the stringent requirements for stability.

Fig. 216 shows the circuit configuration when the operational-amplifier in Fig. 212 is used as an integrator; responses of the circuit for 1-kHz and 10-kHz square-wave input signals are also shown. The dc gain of the circuit is limited to 20 dB by the 390,000-ohm feedback resistor (Z_f). The effect of this resistor on the gain, however, becomes negligible for ac signals at frequencies above 13 Hz because of the 0.03-microfarad capacitor connected in parallel with Z_f.

The weighting factor of integration for the circuit is 1.17-millisecond $(T = Z_r\ C_1 = 39,000 \times 0.03 \times 10^{-6} = 1.17$ milliseconds$)$. The functions

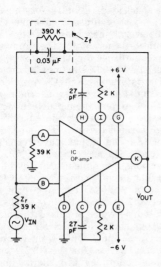

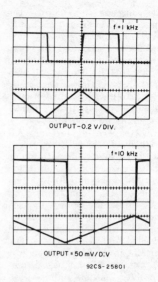

* Circuit shown in Fig. 212.

Fig. 216—Circuit diagram and the input and output waveforms for an operational amplifier used as an integrator.

of the compensation networks connected at terminals H-I and C-F have been described above.

The integrator shown in Fig. 216 provides an output that is proportional to the time integral of the input signal (V_{IN}). The gain function for the integrator is given by:

$$V_{OUT} = \frac{1}{Z_r \, C_1} \int V_{IN} \; dt$$

For example, if a symmetrical square-wave input signal with an average value of zero volts, a peak amplitude of A volts, and a period T is applied to the integrator, then the peak-to-peak output can be calculated by integrating over one-half the input period as follows:

$$-V_{OUT \; (p-p)} = \frac{1}{Z_r \, C_1} \int^{T/2} A \; dt$$

$$= \frac{AT}{2Z_r \, C_1} \; (volts)$$

The waveshape of the output will be triangular, corresponding to the in-

tegral of the square wave. The circuit will perform as an integrator at frequencies above those given by:

$$f = \frac{1}{2\pi \, Z_f \, C_1} \; = 13.6 \; Hz$$

The excellent linearity of response by the complete integrator circuit is illustrated by the waveforms shown in Fig. 216.

An Operational Amplifier (OVA) with Internal Phase Compensation

The "741" type of integrated-circuit operational amplifier has achieved universal popularity among users. One of its attractive features is the incorporation of an internal phase-compensation capacitor on the monolithic chip to provide unity-gain compensation so that it is unnecessary to add any additional stabilization components for closed-loop operation. Consequently, the user need not master the subtleties of phase compensation to use the "741" in a wide variety of applications.

A simplified schematic of the "741" operational amplifier is shown in Fig. 217. Essentially, the circuit consists of a differential input stage (Q_1 and Q_2) followed by a high-gain driver stage (Q_{16}), with a complementary-symmetry output stage (Q_{13} and Q_{17}) operating in Class AB service. Diodes D_a and D_b represent the circuit used to appropriately bias the output stage. The combination of D_c and Q_7 represents the active-device collector loads for transistors Q_1 and Q_2 and additionally is used for conversion from the differential configuration into the single-ended output needed to drive transistor Q_{16}. Capacitor C_1 is a 30-pF capacitor used to provide sufficient phase compensation for stable unity-gain closed-loop operation.

The detailed schematic diagram for the CA741 operational-amplifier is shown in Fig. 218. Low input bias currents are achieved by connection of transistor pairs Q_1, Q_3 and Q_2, Q_4 as shown. Transistors Q_6 and Q_7 are used as load resistors for the input stage; they present effective values of several megohms. The operating collector current for the input stage is established by transistor Q_9 and resistor R_4. Darlington-connected transistors Q_{15} and Q_{16} provide the second-stage gain, with their operating current being determined by resistor R_5 driving the current-mirror formed by diode D_2 and transistor Q_{10}. The output-stage transistors Q_{13} and Q_{17} are connected in a conventional complementary-symmetry arrangement, with transistor Q_{12} in the output short-circuit protection network. Transistor Q_{11} and associated network are used to bias the output stage for class AB operation.

The open-loop gain-frequency characteristics for the CA741 operational amplifier are shown in Fig. 219. Use of internal compensation in the device rolls-off the amplifier frequency response at a rate of 20 dB per frequency-decade over the range from 10 Hz to 10 MHz. This characteristic insures stability in closed-loop applications.

Internally compensated integrated-

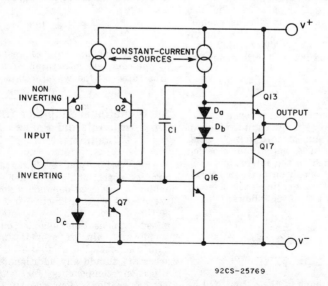

92CS-25769

Fig. 217—Simplified schematic of a "741" integrated-circuit operational amplifier.

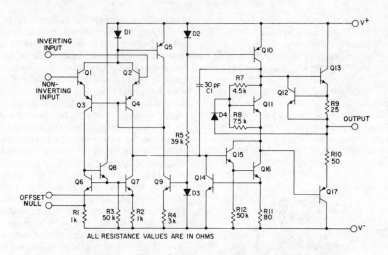

Fig. 218—Schematic diagram of the RCA-CA741 internally compensated integrated-circuit operational amplifier.

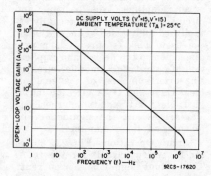

Fig. 219—Open-loop voltage gain-frequency characteristics for the CA741 integrated-circuit operational amplifier.

circuit operational amplifiers, such as the "741", offer the user the simplest solution to applications in which feedback is employed to establish accurately and maintain a fixed amount of gain. Fig. 220 shows that inverting and noninverting amplifiers can be designed with the use of only three external resistors to establish the gain desired.

The use of operational amplifiers with internal compensation is not, however, a universal panacea for all applications in which such circuits can be used. Fig. 219 shows that compensation for unconditional stability (100-per-cent feedback), such as practiced with the "741", sharply restricts the frequency response of an amplifier and also acutely reduces response speed of the circuit under transient-signal conditions. Consequently, many operational-amplifier

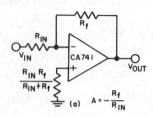

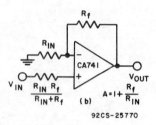

92CS-25770

Fig. 220—Connection of the CA741 integrated-circuit operational amplifier in (a) inverting-amplifier and (b) noninverting-amplifier configurations.

circuits that feature internal phase compensation are also offered in "uncompensated" versions for use in specialized applications for which phase compensation either is not required (e.g., comparators)or can be minimized. For example, the configuration of the CA748 integrated-circuit operational amplifier is precisely similar to that of the CA741 except for the fact that the internal phase-compensation capacitor has been omitted, and two terminals have been provided for the connection of an appropriate phase-compensation capacitor when required in particular applications.

Operational Transconductance Amplifier (OTA)

The RCA operational transconductance amplifier (OTA) is a unique type of operational amplifier

that offers numerous distinctive advantages to the circuit designer. It has all the generic characteristics of the classical operational voltage amplifier (OVA) described in the preceding section except that the forward transfer characteristic is best described by transconductance rather than voltage gain. The output of the operational transconductance amplifier is a current, the magnitude of which is equal to the product of transconductance and the input voltage (i.e., $i_{out} = g_m e_{in}$). The output circuit of this amplifier, therefore, may be characterized by an infinite-impedance current generator, rather than the zero-impedance voltage generator used to represent the output circuit of an operational voltage amplifier. The low output conductance of the operational transconductance amplifier permits the circuit to approach the ideal current generator.

When the operational transconductance amplifier is terminated in a suitable resistive load impedance and provisions are included for feedback, its performance is essentially identical in all respects to that of an equivalent operational voltage amplifier. The electrical characteristics of the OTA circuits, however, are functions of the amplifier bias current. In the OTA, therefore, access is provided to bias the amplifier by means of an externally applied current. As a result, the transconductance, amplifier, dissipation, and circuit loading may be externally established and varied at the option of the user. This feature adds a new dimension to the design and application of operational-amplifier circuits. The brief summary of the goals of ideal OTA and OVA circuits shown in Table XXIII points out the basic similarities and the significant differences of these amplifiers.

Circuit Operation—Fig. 221 shows the equivalent circuit for the OTA. The output signal is a "current" which is proportional to the transconductance (g_m) of the OTA established by the amplifier bias current (I_{ABC}) and the differential in-

Table XXIII—Comparison of Idealized OTA and OVA Circuits

Characteristic	OTA		OVA
INPUT IMPEDANCE		HIGH	
INPUT BIAS CURRENT		LOW	
OFFSET VOLTAGE		0	
GAIN	HIGH TRANSCONDUCTANCE		HIGH VOLTAGE GAIN
BANDWIDTH		INFINITE	
SLEW RATE		INFINITE	
OUTPUT VOLTAGE		LIMITED BY SUPPLIES	
OUTPUT CURRENT		LIMITED BY SUPPLIES	
OUTPUT IMPEDANCE	INFINITE		0
OPERATING CURRENT (AT FIXED SUPPLY VOLTAGE)	ADJUSTABLE		FIXED

put voltage. The OTA can either source or sink current at the output terminals, depending on the polarity of the input signal.

The availability of the amplifier bias current (I_{ABC}) terminal significantly increases the flexibility of the OTA and permits the circuit designer to exercise his creativity in the utilization of this device in many unique applications not possible with the conventional operational amplifier.

A simplified block diagram of the OTA is shown in Fig. 222. Transistors Q_1 and Q_2 comprise the differential input amplifier found in most operational amplifiers, and the lettered-circles (with arrows leading either into or out of the circles) denote "current-mirrors". Fig. 223 shows a simple current-mirror which is

comprised of two transistors, one of which is diode-connected. In a cur-

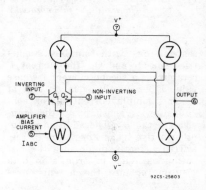

Fig. 222—Simplified diagram of OTA.

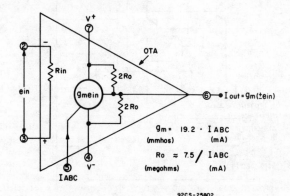

$$g_m = 19.2 \cdot I_{ABC}$$
(mmhos)   (mA)

$$R_o \approx 7.5 / I_{ABC}$$
(megohms)   (mA)

92CS-25802

Fig. 221—Basic equivalent circuit of the OTA.

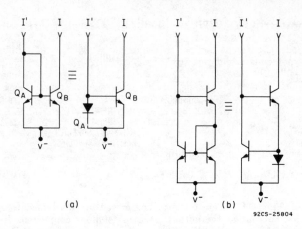

Fig. 223—*Representative types of current mirrors; (a) diode-connected transistor paired with transistor; (b) improved version: employs an extra transistor.*

rent-mirror, with similar geometries for Q_A and Q_B, the current I′ establishes a second current I whose value is essentially equal to that of I′. This simple current-mirror configuration is sensitive to the transistor beta (β). The addition of another active transistor shown in Fig. 223(b), greatly diminishes the circuit sensitivity to transistor beta (β) and increases the current-source output impedance in direct proportion to the transistor beta (β). Current-mirror W (shown in Fig. 222) uses the configuration shown in Fig. 223(a), while mirrors X, Y, and Z are basically the version shown in Fig. 223(b). Mirrors Y and Z employ p-n-p transistors, as depicted by the arrows pointing outward from the mirrors.

Fig. 224 is the complete schematic

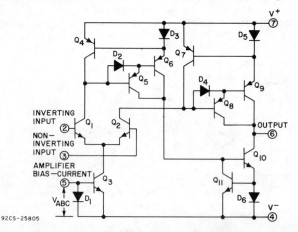

Fig. 224—*Schematic diagram of an operational transconductance amplifier (CA3080 series).*

diagram of an OTA. This OTA employs only active devices (transistors and diodes). Current applied to the amplifier-bias-current terminal, I_{ABC}, establishes the emitter current of the input differential amplifier Q_1 and Q_2. Hence, effective control of the differential transconductance (g_m) is achieved. At room temperature, $g_m = 19.2 \times I_{ABC}$, where g_m is in millimhos and I_{ABC} is in milliamperes.

Transistor Q_3 and diode D_1 (shown in Fig. 224) comprise the current mirror "W" of Fig. 222. Similarly, transistors Q_7, Q_8, and Q_9 and diode D_5 of Fig. 224 comprise the generic current mirror "Z" of Fig. 224. Darlington-connected transistors are employed in mirrors "Y" and "Z" to reduce the voltage sensitivity of the mirror, by the increase of the mirror output impedance. Transistors Q_{10} and Q_{11} and diode D_6 of Fig. 224 comprise the current-mirror "X" of Fig. 222. Diodes D_2 and D_4 are connected across the base-emitter junctions of Q_5 and Q_8, respectively, to improve the circuit speed. The amplifier output signal is derived from the collectors of the "Z" and "X" current-mirror of Fig. 222, providing push-pull Class A output stage that produces full differential g_m.

Typical Applications—In addition to the classical applications in which operational amplifiers are ordinarily used, OTA's can be used advantageously in a variety of applications not easily accomplished with effectiveness by ordinary operational amplifiers, such as amplitude modulators, time-division multiplexers, sample-and-hold circuits, analog multipliers, mixers, and gain control. These types of circuit applications are more easily executed by using an OTA because it contains circuitry by which the amplifier gain can be linearly or digitally programmed, independently of the differential input signal.

The uniqueness of the OTA from an applications standpoint can be illustrated by describing its use in an amplitude-modulator circuit. Effective gain control of a signal may be obtained by controlled variation of the amplifier-bias-current (I_{ABC}) in the OTA because its g_m is directly proportional to the amplifier-bias-current (I_{ABC}). For a specified value of amplifier-bias-current, the output current (I_0) is equal to the product of g_m and the input signal magnitude. The output voltage swing is the product of output current (I_0) and the load resistance (R_L).

Fig. 225 shows the configuration for this form of basic gain control (a modulation system). The output signal current (I_0) is equal to -$g_m \times V_x$; the sign of the output signal is negative because the input signal is applied to the inverting input terminal of the OTA. The transconductance of the OTA is controlled

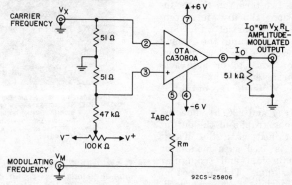

Fig. 225—Amplitude modulator circuit using the OTA.

by adjustment of the amplifier bias current, I_{ABC}. In this circuit, the level of the unmodulated carrier output is established by a particular amplifier-bias-current (I_{ABC}) through resistor Rm. Amplitude modulation of the carrier frequency occurs because variation of the voltage Vm forces a change in the amplifier-bias-current (I_{ABC}) supplied via resistor Rm. When Vm goes positive the bias current increases which causes a corresponding increase in the g_m of the OTA. When the Vm goes in the negative direction (toward the amplifier-bias-current terminal potential), the amplifier-bias-current decreases, and reduces the g_m of the OTA.

As discussed earlier, $g_m = 19.2 \times I_{ABC}$, where g_m is in millimhos when I_{ABC} is in milliamperes.. In this case, I_{ABC} is approximately equal to:

$$\frac{Vm - (V^-)}{Rm} = I_{ABC}$$

$$(I_0) = -g_m Vx$$

$$g_m Vx = (19.2)(I_{ABC})(Vx)$$

$$I_0 = \frac{-19.2\,[Vm - (V^-)]\,Vx}{Rm}$$

$$I_0 = \frac{19.2\,(Vx)\,(V^-)}{Rm} - \frac{19.2\,(Vx)\,(Vm)}{Rm}$$

(Modulation Equation).

There are two terms in the modulation equation: the first term represents the fixed carrier input, independent of Vm, and the second term represents the modulation, which either adds to or subtracts from the first term. When Vm is equal to the V^- term, the output is reduced to zero.

Figs. 226(a) and 226(b) show oscilloscope photographs of the output voltages obtained when the circuit of Fig. 225 is used as a modulator for both sinusoidal and triangular modulating signals. This method of modulation permits a range exceeding 1000:1 in the gain, and thus

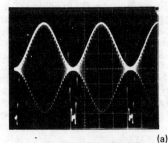

TOP TRACE: MODULATION FREQUENCY INPUT
≈ 20 VOLTS P-P & 50μsec / DIV
CENTER TRACE: AMPLITUDE MODULATE OUTPUT
500mV/DIV & 50 μsec /DIV
BOTTOM TRACE: EXPANDED OUTPUT TO SHOW
DEPTH OF MODULATION 20mV/DIV
& 50μsec /DIV   92CS-25807

(a)

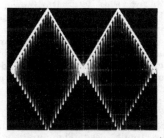

TOP TRACE: MODULATION FREQUENCY INPUT
20 VOLTS & 50μsec / DIV
BOTTOM TRACE: AMPLITUDE MODULATED OUTPUT
500mV/DIV & 50μsec /DIV

92CS-25808

(b)

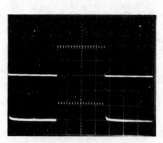

TOP TRACE: GATED OUTPUT I V/DIV AND 50μsec /DIV
BOTTOM TRACE: VOLTAGE EXPANSION OF ABOVE
SIGNAL—SHOWING NO RESIDUAL
I mV/DIV AND 50μsec/DIV— AT
LEAST 80 db OF ISOLATION
fq = 100 kHz

92CS-25809

(c)

Fig. 226—(a) Oscilloscope photograph of amplitude modulator circuit waveforms with Rm = 40 kΩ, V+ = 10 V, and V− = −10 V.

provides modulation of the carrier input in excess of 99%. The photo in Fig. 226(c) shows the excellent isolation achieved in this modulator during the "gated-off" condition.

COMPARATORS

Many electronics applications require that a comparison be made between two voltages, and that an output is supplied in accordance with which of the two is the greater. Examples of such applications for comparators include use as a threshold detector, a voltage comparator in analog-to-digital (A/D) converters, a pulse-height discriminator, pulse-width modulator, a zero-crossing detector, and the like. A dual comparator can be used as a core-memory sense amplifier, a window discriminator in pulse-height analyzers, or a double-ended limit detector for automatic go/no-go test equipment. In many instances, a comparator is used to interface directly with digital circuits by producing a digital one or zero at the output when one input is higher than the other. Many industrial control applications also employ comparators. For example, the RCA-CA3059 integrated-circuit zero-voltage switch (described later in the section **AC Power Control**), for example, contains an integral comparator circuit function to actuate a control function when the output voltage from a sensor [e.g., a thermistor] exceeds a predetermined reference voltage. The comparator toggles the output voltage to its original state when voltage from the sensor drops below the reference voltage.

Most integrated-circuit comparators employ a differential-amplifier input stage coupled to a single-ended output amplifier. A reference voltage is applied to one terminal of the differential input stage, and the voltage to be compared is applied to the other terminal. The output stage is toggled to a "high" or "low" output in accordance with the voltage comparison. When the voltage to be compared has a slowly varying characteristic, "snap action" can be added to the circuit by means of positive feedback to reduce the "uncertainty region" in which toggle-switching occurs.

Although a number of integrated circuits have been specifically designed for comparator service, operational amplifiers are also frequently used as comparators. Because the input-stage of a comparator circuit is usually a differential amplifier, the same basic input-circuit parameters are specified for them as for operational amplifiers, namely, input bias current, input offset current, and input offset voltage. In the comparator application, however, the effect of input offset voltage is not "amplified" as is the case in closed-loop operational-amplifier circuits. In a comparator, errors introduced by the input offset voltage are directly related to the reference voltage. For example, an offset voltage of 5 millivolts represents an error of only 1 per cent if the reference voltage is 5 volts dc. Because comparators do not employ linear feedback circuitry, they can be operated without the frequency-compensation capacitor required in closed-loop operational-amplifier circuits. Consequently, a comparator can change its output voltage state more rapidly than a compensated operational amplifier because there is no compensation capacitor to be charged and discharged. Finally, the output stage of a comparator is usually required to either source or sink current, as is the case for an operational amplifier. Following is a description of a unique operational-amplifier circuit which is also very attractive for use in comparator circuits.

COS/MOS Operational-Amplifier for Comparator Applications

The CA3130 COS/MOS integrated circuit shown in Fig. 227 has many distinctive op-amp characteristics, but additionally has features (e.g.,

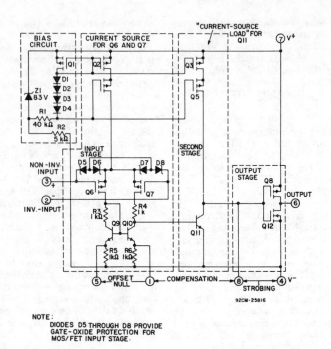

Fig. 227—CA3130 COS/MOS integrated-circuit operational amplifier.

NOTE:
DIODES D5 THROUGH D8 PROVIDE
GATE-OXIDE PROTECTION FOR
MOS/FET INPUT STAGE.

strobing capability and ideal characteristics for interfacing with COS/MOS digital circuits) of great utility in comparator applications.

Circuit Description—The CA3130 is basically a three-stage amplifier; it consists of a differential-input stage that uses P-MOS transistors (Q_6 and Q_7), an intermediate gain-stage that uses a single bipolar transistor (Q_{11}) and a drain-loaded output stage that uses COmplementary-Symmetry MOS transistors (Q_8 and Q_{12}). Cascode-connected P-MOS transistors (Q_2 and Q_4) are used in the constant-current source for the differential input-stage; the current-flow is established by the diode-connected P-MOS transistor Q_1 "mirrored" to the P-MOS transistor Q_2. Zener diode Z_1 establishes an 8.3-volt source across the series-

connected circuit consisting of resistor R_1, diodes D_1 through D_4, and transistor Q_1. A tap at the R_1-D_4 junction provides bias-voltage for transistor Q_4. The small diodes (D_5 through D_8) provide gate-oxide protection against high-voltage transients (e.g., static electricity) for transistors Q_6 and Q_7. The combination of the diode-connected P-MOS transistor Q_9 and transistor Q_{10} represents the active-device collector-loads for transistors Q_6 and Q_7 and additionally is used for conversion from the differential configuration into the single-ended output needed to drive transistor Q_{11}. The cascode-connected P-MOS transistors Q_3 and Q_5, which are biased in the manner described for transistors Q_2 and Q_4, respectively, are used as a collector load for transistor Q_{11}. The output stage uses drain-loaded COS/MOS transistors (P-MOS type Q_8 and N-MOS type Q_{12}) that are normally operated under class A conditions. Offset-nulling is accomplished by connection of a potentiometer between terminals 1 and 5, with the slider arm connected to terminal 4. A single external capacitor, connected between terminals 1 and 8, is used to "compensate" the amplifier. The open-loop gain as a function of frequency for both phase-compensated and uncompensated operation of the COS/MOS operational amplifier is shown in Fig. 228.

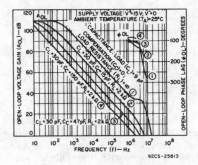

Fig. 228—Open-loop gain as a function of frequency for the CA3130 COS/MOS integrated-circuit operational amplifier in phase-compensated and uncompensated configurations.

Some of the major advantages of the CA3130 COS/MOS operational amplifier are as follows:

1. Inputs can operate with signals down to the negative-supply terminal.

2. Output can be swung to within millivolts of either supply-terminal under lightly loaded conditions; [e.g., interfacing with COS/MOS digital circuits].

3. Very high input impedance (Z_I), 1000 Megohms (typical).

4. Low input-bias current, 2 picoamperes (typical).

5. Wide bandwidth, unity-gain crossover at 15 MHz.

6. High open-loop gain, 100 dB (typical).

7. High slew-rate, 8 volts per microsecond (typical) as unity-gain follower.

8. High output-current capability, 20 milliamperes (typical).

9. Output can be controlled by strobing (e.g., for use in comparators and power-control circuits).

CA3130 Comparator Circuit

The foregoing advantages are likewise applicable when the CA3130-series circuits are used in comparator applications. Additionally, as previously noted, comparators are usually operated in the uncompensated mode. Under these conditions, the CA3130 has an unusually high slew-rate, about 10 volts per microsecond.

Fig. 229 shows the schematic of a strobed comparator using the CA3130 COS/MOS operational amplifier operating from a single supply. In its quiescent state, (logic "0" on strobing-terminal 8) the output is "high", i.e., essentially equal to the V^+ supply-voltage. The comparator is strobed "on" by the application of a $+10$-volt signal at terminal 8. If the signal voltage at terminal 2 is more positive than the positive reference voltage at terminal 3, the output voltage at terminal 6 swings down to within a few millivolts of ground (logic "0"). If the signal voltage at terminal 2 is less positive than the reference voltage, the output voltage remains "high" (logic "1"), i.e., essentially equal to the V^+ supply-voltage. The "sense" of the output voltage can be reversed by interchanging the signal and reference voltages. This type of single-supply comparator can be used to compare potentials whose magnitudes are in the proximity of ground potential; it can also compare potentials with values up to about 12 volts.

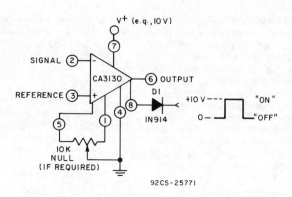

Fig. 229—Strobed comparator using the CA3130 COS/MOS integrated-circuit operational amplifier.

Programmable Comparator with Memory

Fig. 230 shows the block diagram of an integrated-circuit programmable comparator. This integrated circuit, the CA3099E, is particularly useful in the direct control of high-operating-current loads such as thyristors, lamps, and relays, that require currents up to 150 milli-

amperes. Nevertheless, it operates with microwatt standby power dissipation when the current to be controlled is less than 30 milliamperes. As indicated in Fig. 230, the CA3099E contains the following six major circuit-function features:

1. **Differential amplifiers and summer.** The circuit uses two differential amplifiers, one to compare the input voltage with the "high" ref-

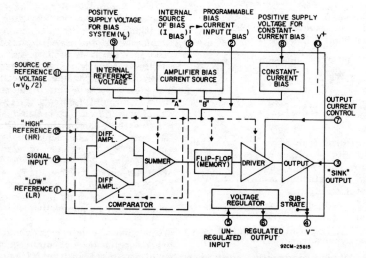

Fig. 230—Block diagram of CA3099E programmable comparator.

erence, and the other to compare the input with the "low" reference. The resultant output of the differential amplifiers actuates a summer circuit which delivers a trigger that initiates a change in state of a flip-flop.

2. **Flip-flop.** The flip-flop functions as a bistable "memory" element that changes state in response to each trigger command.

3. **Driver and output stages.** These stages permit the circuit to "sink" maximum peak load currents up to 150 milliamperes at terminal 3.

4. **Programmable operating current.** The circuit incorporates a separate terminal to permit programming the desired quiescent operating current and performance parameters.

5. **Internal sources of reference voltage and programmable bias current.** An integral circuit supplies a temperature-compensated reference voltage ($V_b/2$) which is about one-half the externally applied bias voltage (V_b). Additionally, integral circuitry can optionally be used to supply an uncompensated constant-current source of bias (I_{bias}).

6. **Voltage regulator.** Optional on-chip voltage regulation is available when power for the CA3099E is provided by an unregulated supply.

Fig. 231 shows the simplicity with which the CA3099E can be used for off/on control of a triac in response to the output of a sensor (thermal, light, or other types).

Fig. 232 shows the schematic diagram for a unique **integrated-circuit programmable power/switch amplifier**, the CA3094. In this circuit, a high-gain OTA type of preamplifier is used to drive the power-amplifier output stage. The preamplifier in this circuit is identical to the CA3080 operational transconductance amplifier shown in Fig. 224. The output stage employs a Darlington-connected pair of output transistors, Q_{12} and Q_{13} that can deliver average power of 3 watts or peak power of 10 watts to an external load. Although the CA3094 integrated circuit can be operated under quiescent conditions with only a few microwatts of power, when "keyed" on, it can control currents of 100 milliamperes average (or 300 milliamperes peak).

POWER-CONTROL CIRCUITS

Linear integrated circuits play an increasingly important role in power-control applications, such as voltage regulators, zero-voltage switching circuits, and programmable power switches.

Integrated-circuit voltage regulators range in complexity from comparatively simple three-terminal types that provide a fixed-voltage output to more sophisticated types that can be adjusted for operation over a wide range of output voltages. An example of this latter type was described earlier in the section **Materials, Junctions, and Devices.** Other examples of the application of integrated-circuit voltage regulators are described in the section **DC Power Supplies.**

The principles and advantages of **zero-voltage switching** in power control applications are explained later in the section **AC Power Controls.** This section also describes the operation and application of an integrated circuit, the CA3059, that is an excellent and widely used trigger cir-

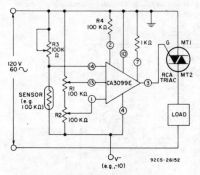

RI FOR SETTING "HIGH" REFERENCE VOLTAGE
R2 FOR SETTING "LOW" REFERENCE VOLTAGE
R3 FOR VARIATION OF HYSTERESIS

Fig. 231—Off/on control of triac with programmable hysteresis.

cuit for thyristors in zero-voltage power-switching and control applications.

Fig. 233 shows a line-operated thyristor-firing circuit controlled by a CA3094 that operates from an ac-

bridge sensor. This circuit is particularly suited to certain classes of sensors that cannot be operated from dc. The CA3094 is inoperative when the high side of the ac line is negative because there is no I_{ABC}

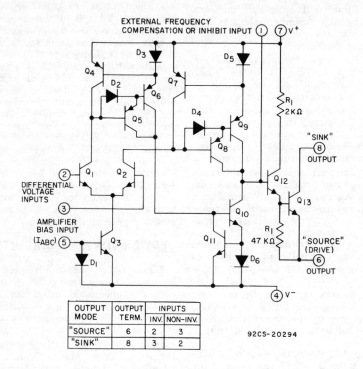

OUTPUT MODE	OUTPUT TERM.	INPUTS	
		INV.	NON-INV.
"SOURCE"	6	2	3
"SINK"	8	3	2

92CS-20294

Fig. 232—Circuit diagram for CA3094-series integrated-circuit programmable power switch/amplifier.

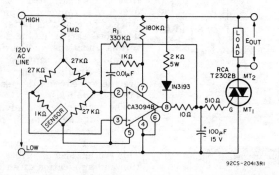

92CS-20413R1

Fig. 233—Line-operated thyristor-firing circuit controlled by ac-bridge sensor.

supply to terminal 5. When the sensor bridge is unbalanced so that terminal 2 is more positive than terminal 3, the output stage of the CA3094 is cut off when the ac line swings positive, and the output level at terminal 8 of the CA3094 goes "high". Current from the line flows through the 1N3193 diode to charge the 100-microfarad reservoir capacitor, and also provides current to drive the triac into conduction. During the succeeding negative swing of the ac line, there is sufficient remanent energy in the reservoir capacitor to maintain conduction in the triac.

When the bridge is unbalanced in the opposite direction so that terminal 3 is more positive than terminal 2, the output of the CA3094 at terminal 8 is driven sufficiently "low" to "sink" the current supplied through the 1N3193 diode so that the triac gate cannot be triggered. Resistor R_1 supplies the hysteresis feedback to prevent rapid cycling between turn-on and turn-off.

The CA3094 integrated circuit can directly control lights and relays; it is very useful in motor-speed controllers. In addition to the power-switching applications, it is also an excellent comparator and is an ideal driver for audio power transistors. This later application of the CA3094 is discussed in the section **Audio Power Amplifiers.**

SPECIAL-PURPOSE CIRCUITS

The functional classifications of integrated circuits for linear applications reviewed in the preceding paragraphs have dealt with categories of devices that have a broad, general spectrum of applicability. There are, however, a multitude of special-purpose types of linear integrated circuits being produced, some types in very high volume. Some typical special-purpose types of integrated circuits for linear application are as follows:

AM and FM radio-receiving circuits

TV receiver circuits
Analog multipliers
Phase-locked loops
Analog timing circuits
Optoelectronic circuits
Analog-to-digital (A/D) converters
Digital-to-analog (D/A) converters
Sense amplifiers for memory cores
Thermal sensors

The following paragraphs describe RCA integrated circuits used in analog multipliers, phase-locked loops, analog timing circuits, and optoelectronic circuits. RCA integrated circuits for AM, FM, and TV receiver applications are described in a later section of this Manual.

Analog Multipliers

An analog multiplier is, essentially, a gain-controlled amplifier that multiplies the input signal (V_x) with the external gain controlling signal (V_y) to produce the resultant output (V_o). The gain is externally adjustable by a coefficient (k). Stated simply, a multiplier produces an output voltage that is the linear product of two input voltages. Fig. 234 shows the symbolic diagram for an analog multiplier.

The basic multiplier, shown in Fig. 235(a), is a two-quadrant multiplier. The input signal (V_x) may have either a positive or negative polarity; however, the external gain-controlling signal (V_y) must be positive and greater than the base-

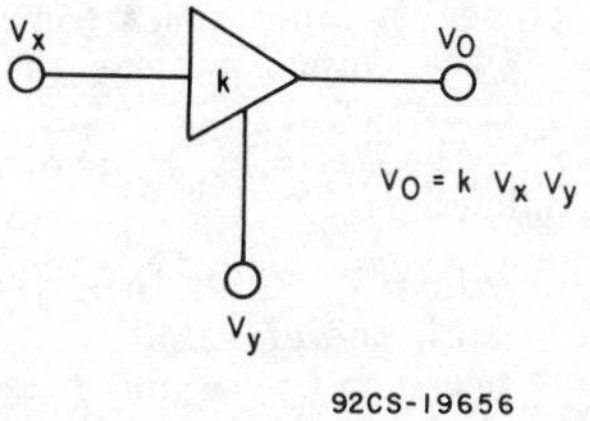

$$V_O = k\ V_x\ V_y$$

92CS-19656

Fig. 234—Gain-controlled amplifier.

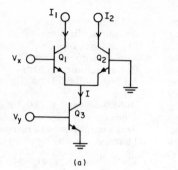

92CS-19657

92CS-19658

Fig. 235—Two-quadrant multiplier: (a) basic circuit; (b) multiplier functional only in shaded region.

to-emitter voltage, as indicated in Fig. 235(b). The output current $(I_1 - I_2)$ of the differential amplifier, comprised of transistors Q_1 and Q_2, is related to both the input signal (V_x) and the source current (I).

Since the course current (I) is related to the gain-controlling signal (V_y), the output current $(I_1 - I_2)$, therefore, is related to both V_x and V_y. This relationship is essentially nonlinear; thus an appropriate linearization circuit must be provided in the input stage to achieve the following linear relationship:

$$I_1 - I_2 = k' \, V_x \, V_y$$

where k' is a constant.

Fig. 236 shows a typical arrangement of three differential amplifiers to form a four-quadrant multiplier. This arrangement incorporates the operating principles of the two-quadrant multiplier, but, in addition, it permits both of the input signals $(V_x$ and $V_y)$ to have positive or negative polarities (or zero). When either input is zero, the output current $(I_1 - I_2)$ must, theoretically, be zero as is shown by the following:

1. Assume $V_x = 0$, then $i_1 = i_2$ and $i_3 = i_4$ therefore $i_1 + i_4 = i_2 + i_3$. Since $I_1 = i_1 + i_4$ and $I_2 = i_2 + i_3$, then $I_1 = I_2$. This equality is independent of V_y.

2. Now assume $V_y = 0$, then $i_5 = i_6$. Since $i_5 = i_1 + i_2$ and $i_6 = i_3 + i_4$, then $i_1 + i_2 = i_3 + i_4$. Since $i_1 = i_3$ and $i_2 = i_4$ then $i_1 + i_4 = i_3 + i_2$. Therefore $I_1 = I_2$. This equality is independent of V_x.

The multiplying operation discussed in the previous section applies when neither V_x nor V_y is zero. The output current $(I_1 - I_2)$ then satisfies the following equation:

$$I_1 - I_2 = k'V_xV_y.$$

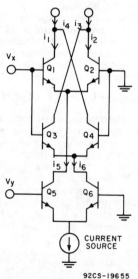

92CS-19655

Fig. 236—Basic four-quadrant multiplier.

The multiplying action of the four-quadrant multiplier is dependent on current unbalance in the three differential amplifiers. Ideally, the multiplying operation should not occur if either V_x or V_y is 0. However, in practical applications, slight current unbalances do exist. It is necessary, therefore, to null out such unbalances with external potentiometers prior to operation.

The functional block diagram of a monolithic integrated circuit analog multiplier, the CA3091D, is shown in Fig. 237. With the circuit values shown, the circuit can provide an output current at terminal 10 in accordance with the product of V_x (terminal 2) and V_y (terminal 13). Appropriate external peripheral circuit components are shown to provide adjustment for linearity, x-y balance, k-factor, and zero output.

In addition to functioning as a multiplier, the CA3091 can be used as a divider, squarer, square rooter, and power-series approximator. This integrated circuit is also useful in applications such as ideal full-wave rectifiers, automatic level controllers, rms converters, frequency discriminators, and voltage-controlled filters.

Phase-Locked Loops

Phase-locked-loops (PLL's), especially in monolithic integrated-circuit form, are finding significantly increased usage in signal-processing and digital systems. FM demodulation, FSK demodulation, tone decoding, frequency multiplication, signal conditioning, clock synchronization, and frequency synthesis are some of the many applications of a PLL.

Basic System—The basic phase-locked-loop system is shown in Fig. 238; it consists of three parts: phase comparator, low-pass filter, and voltage-controlled oscillator (VCO); all are connected to form a closed-loop frequency-feedback system.

With no signal input applied to the PLL system, the error voltage

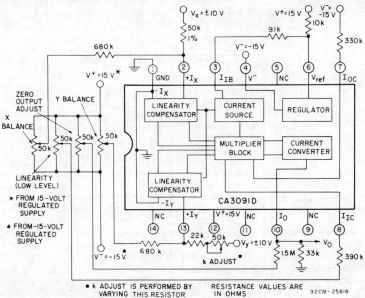

Fig. 237—Functional block diagram of CA3091D with typical multiplier outboard (peripheral) circuitry.

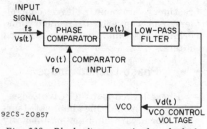

Fig. 238—Block diagram of phase-locked loop (PLL).

at the output of the phase comparator is zero. The voltage, $V_d(t)$, from the low-pass filter is also zero, which causes the VCO to operate at a set frequency, f_o, called the center frequency. When an input signal is applied to the PLL, the phase comparator compares the phase and frequency of the signal input with the VCO frequency and generates an error voltage proportional to the phase and frequency difference of the input signal and the VCO. The error voltage, $V_e(t)$, is filtered and applied to the control input of the VCO; $V_d(t)$ varies in a direction that reduces the frequency difference

between the VCO and signal-input frequency. When the input frequency is sufficiently close to the VCO frequency, the closed-loop nature of the PLL forces the VCO to lock in frequency with the signal input; i.e., when the PLL is in lock, the VCO frequency is identical to the signal input except for a finite phase difference. The range of frequencies over which the PLL can maintain this locked condition is defined as the **lock range** of the system. The lock range is always larger than the band of frequencies over which the PLL can acquire a locked condition with the signal input. This latter band of frequencies is defined as the **capture range** of the PLL system.

COS/MOS Phase-Locked Loop— Fig. 239 shows the block diagram of an integrated-circuit PLL implemented with COS/MOS transistors. This circuit, type CD4046A, consists of a low-power, linear voltage-controlled oscillator (VCO) and two different phase comparators having a common signal-input amplifier and a common comparator input. A 5.2-volt zener diode is provided for sup-

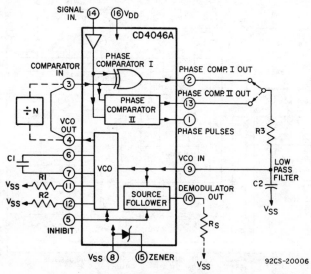

Fig. 239—COS/MOS phase-locked loop block diagram.

ply regulation, if necessary. The CD4046A operates at frequencies up to 1.2 MHz with a supply voltage V_{DD} of 10 volts; it consumes only 70 microwatts of power with the VCO operated at $f_o = 10$ kHz and $V_{DD} = 5$ V; VCO linearity is 1 per cent.

The VCO requires one external capacitor C_1 and one or two external resistors (R_1 or R_1 and R_2). Resistor R_1 and capacitor C_1 determine the frequency range of the VCO and resistor R_2 enables the VCO to have a frequency offset if required. The high input impedance (10^{12} ohms) of the VCO simplifies the design of low-pass filters by permitting the designer a wide choice of resistor-to-capacitor ratios. In order not to load the low-pass filter, a source-follower output of the VCO input voltage is provided at terminal 10 (DEMODULATED OUTPUT). If this terminal is used, a load resistor (R_S) of 10 kilohms or more should be connected from this terminal to V_{SS}. If unused, this terminal should be left open. The VCO can be connected either directly or through frequency dividers to the comparator input of the phase comparators. A full COS/MOS logic swing is available at the output of the VCO and allows direct coupling to COS/MOS frequency dividers such as the CD4024A, CD4018A, CD4020A, CD4022A, or CD4029A COS/MOS integrated-circuit counters. (These counter circuits are described later in the section **Counters and Registers.**) One or more CD4018A (Presettable Divide-by-N Counter) or CD4029A (Presettable Up/Down Counter), together with the CD4046A (Phase-Locked Loop) can be used to build a micro-power low-frequency synthesizer. A logic 0 on the INHIBIT input "enables" the VCO and the source follower, while a logic 1 "turns off" both to minimize standby power consumption.

Analog Timing Circuits

Timing functions have long been performed by using analog electronic circuitry, such as the monostable ("one-shot") multivibrator circuit. Their functioning has been predicated on the practical use of the classic resistance-capacitance (RC) time-constant relationship to control the switching of an electronic circuit. With the advent of integrated circuits, however, the performance of electronic analog timers has become more sophisticated, and the user is being provided with additional versatility in their application. Consequently, they have begun to pervade the role of thermal relays and electromechanical timers which have long dominated this field. Integrated circuits are now being used economically to provide accurate control of timing over periods ranging from microseconds to hours.

Operation of an analog timing circuit is contingent upon the period required to charge (or discharge) a capacitor to some predetermined level. Some timers employ the classic exponential charge (or discharge) characteristic of an RC network. A number of integrated circuit timers, however, employ a source of constant-current flow to charge (or discharge) the timing capacitor linearly with respect to time. The following examples illustrate timing circuits that use constant-current sources. The first example is predicated on linearly discharging a capacitor; the second example is predicated on charging a capacitor linearly through a constant-current source.

Presettable Timer—The programmability feature inherent in the operational transconductance amplifier (OTA) simplifies the design of presettable analog timers such as the one shown in Fig. 240. Long timing intervals (e.g., up to 4 hours) are achieved by discharging a timing capacitor C_1 into the signal-input terminal (e.g., No. 3) of the CA3094 integrated circuit. (The schematic diagram for the CA3094 is shown in Fig. 232.) This discharge current is controlled precisely by the magni-

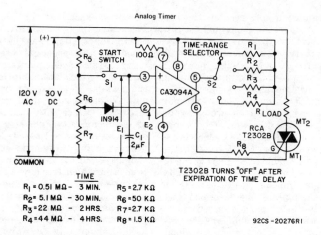

Fig. 240—*Presettable analog timer using the CA3094A COS/MOS integrated-circuit transistor array.*

tude of the amplifier bias current I_{ABC} programmed into terminal 5 through a resistor selected by switch S_2. Operation of the circuit is initiated by charging capacitor C_1 through the momentary closing of switch S_1. Capacitor C_1 starts discharging and continues discharging until voltage E_1 is less than voltage E_2. The differential input transistors in the CA3094 then change state, and terminal 2 draws sufficient current to reverse the polarity of the output voltage (terminal 6). Thus, the CA3094 not only has provision for readily presetting the time delay, but also provides significant output current to drive control devices such as thyristors. Resistor R_5 limits the initial charging current for C_1. Resistor R_7 establishes a minimum voltage of at least 1 volt at terminal 2 to insure operation within the common-mode-input range of the device. The diode limits the maximum differential input voltage to 5 volts. Gross changes in time-range selection are made with switch S_2, and vernier trimming adjustments are made with potentiometer R_6.

Analog Timer Using Constant-Current Capacitor-Charging—Fig. 241 shows the schematic diagram of

a timer circuit that uses a CA3600E COS/MOS transistor-array (Fig. 193 shows the schematic diagram of the CA3600E). For purposes of explanation, it is assumed that capacitor C_1 initially is in a completely discharged condition. Terminal 10, therefore, is initially at ground potential, and transistor n_3 is nonconductive. The circuitry at the left of terminal 10 provides a source of constant-current flow through p_1 to charge capacitor C_1 increasingly positive with respect to ground. After the passage of time (T), capacitor C_1 is charged sufficiently in the positive direction so that transistor n_3 is driven into conduction by its gate, and the lamp is lighted to signify the end of the time-delay period. The circuit is reset by momentarily closing switch S_1 to discharge capacitor C_1 through R_4.

Resistor-divider network R_1, R_2 establishes the supply voltage to a constant-current network comprised of resistor R_3 and the series-connected COS/MOS pair n_2 and p_2, biased for linear operation by resistor R_b as previously described. This combination is connected to the gate terminal (No. 6) of transistor p_1 to form a current mirror, i.e., the current flowing through p_1 to charge

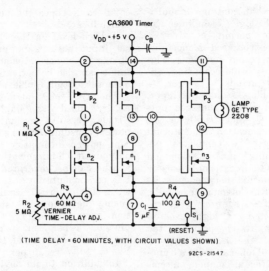

Fig. 241—Analog timer using CA3600E COS/MOS integrated-circuit transistor array.

C_1 will be essentially equal to the constant-current flow established through R_3, n_2, and p_2.

Low-Power COS/MOS Monostable/Astable Multivibrator—As already noted, the monostable (one-shot) multivibrator has long been used in electronic timing circuitry. Its output changes state in response to the control enforced by its servility to a time-determining RC network. The practicality of integrated-circuit technology has led to more popular usage of multivibrators. In digital applications, for example, an integrated circuit is an attractive means of interfacing a multivibrator circuit with appropriate digital-control circuitry.

Fig. 242 shows the block diagram of a COS/MOS integrated circuit,

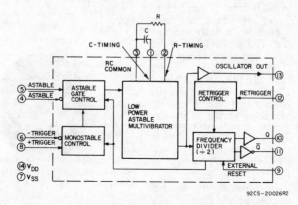

Fig. 242—Block diagram of CD4047 COS/MOS integrated-circuit monostable/astable multivibrator.

the CD4047, that contains a multivibrator with logic techniques incorporated to permit positive or negative edge-triggered monostable multivibrator action having retriggering and external counting options. Inputs include +Trigger, − Trigger, Astable, $\overline{\text{Astable}}$, Retrigger, and External Reset. Buffered outputs are Q, $\overline{\text{Q}}$, and Oscillator. In all modes of operation, an external capacitor must be connected between C-Timing and RC-Common terminals, and an external resistor must be connected between the R-Timing and RC-Common terminals.

Astable ("free running") operation is enabled by a high level on the Astable input. The period of the square wave at the Q and $\overline{\text{Q}}$ outputs in this mode of operation is a function of the external components employed. "True" input pulses on the Astable input or "Complement" pulses on the $\overline{\text{Astable}}$ input allow the circuit to be used as a gatable multivibrator. An output whose period is half of that which appears at the Q terminal is available at the Oscillator Output terminal.

In the monostable mode, positive-edge triggering is accomplished by application of a leading-edge pulse to the "+Trigger" input and a low level to the "−Trigger" input. For negative-edge triggering, a trailing-edge pulse is applied to the "−Trigger" and a high level is applied to the "+Trigger". Input pulses may be of any duration relative to the output pulse. The multivibrator can be retriggered (on the leading edge only) by application of a common pulse to both the "Retrigger" and "+Trigger" inputs. In this mode, the output pulse remains "high" as long as the input pulse period is shorter than the period determined by the RC components.

An external countdown option can be implemented by coupling "Q" to an external "N" counter (e.g., CD4017A) and resetting the counter with the trigger pulse. The counter output pulse is fed back to the $\overline{\text{Astable}}$ input and has a duration

equal to N times the period of the multivibrator.

A high level on the External Reset input assures no output pulse during a power-on condition. This input can also be activated to terminate the output pulse at any time.

The COS/MOS transistor technology used to implement the CD4047 results in unusually low power consumption, e.g., its quiescent power dissipation is typically only about 0.5 microwatt. Long pulse widths can be generated with small RC components by using external COS/MOS digital counters. Frequency deviation in astable operation is typically only ± 0.5 per cent plus 0.015 per cent per °C at an operating frequency of 10 kHz, with $V_{DD} = 10$ V ± 10 per cent.

In addition to its obvious uses in timing and time-delay circuits, the CD4047 is applicable in frequency generation, frequency multiplication, frequency division, frequency discrimination, and envelope detection applications.

Optoelectronic Circuits

Integrated circuit technology permits the fabrication of photosensitive elements on a monolithic chip, together with amplifiers and/or switching circuits. The photograph of a simple device in this category of integrated circuits is shown in Fig. 243. The monolithic chip can be

Fig. 243—Photograph of CA3062 optoelectronic integrated circuit.

seen in the center of a modified 12-lead TO-5 style package containing a glass window. A block diagram of the integrated circuit, the CA3062, is shown in Fig. 244. The photosensitive section consists of Darlington pairs. The power amplifier has a differential configuration which provides complementing output in response to a light input—normally "on" and normally "off." The separate photodetector, amplifier, and high-current switch provide a flexible circuit arrangement. This feature, together with the high current capability of the output section, provides the user with a complete system particularly useful in photoelectric control applications that use IR emitters and visible-light sources. The output stages can switch currents up to 100 milliamperes.

This type of device provides the designer with a simple "building block" for use in applications such as counters, sorting, level controls, position sensing, edge monitoring, isolators, and the like. The compactness of the device and the ease with which it can be employed are advantageous.

PRODUCT CLASSIFICATION CHARTS FOR RCA LINEAR INTEGRATED CIRCUITS

The product-classification charts shown in Tables XXIV through XXVIII lists the more than 100 types of linear integrated circuits available from RCA and indicate typical applications, performance features, and package for each of them. Detailed ratings and characteristics for these circuits are given in the RCA technical data bulletins. The data-bulletin File numbers are included in the charts below the circuit type numbers.

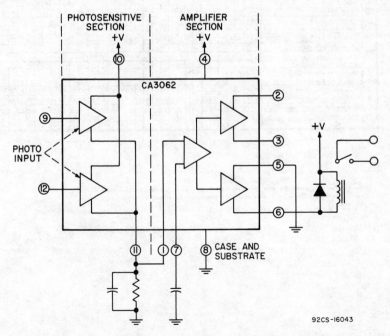

Fig. 244—Light-operated relay using CA3062 optoelectronic integrated circuit.

Table XXIV—Operational Amplifiers

Column groupings:

- **Micropower** — Single OTA •: CA3080, CA3080A (File No. 475); Triple OTA •: CA3060A, CA3060B, CA3060 (File No. 537); Single OP-AMP: CA3078A, CA3078 (File No. 535)
- **High-Current** — CA3033, CA3033A, CA3047, CA3047A (File No. 360); CA3094, CA3094A, CA3094B (File No. 598)
- **General Purpose — Single Unit** — Low Noise: CA6741, CA6078A ▲ (File No. 592); CA101, CA101A, CA201, CA201A, CA301A (File No. 786); CA107, CA207, CA307 (File No. 785); CA741C, CA741, CA748C, CA748 (File No. 531); COS/MOS Output, MOS/FET Input: CA3130, CA3130A, CA3130B (File No. 817)

	CA3080	CA3080A	CA3060A	CA3060B	CA3060	CA3078A	CA3078	CA3033	CA3033A	CA3047	CA3047A	CA3094	CA3094A	CA3094B	CA6741	CA6078A ▲	CA101	CA101A	CA201	CA201A	CA301A	CA107	CA207	CA307	CA741C	CA741	CA748C	CA748	CA3130	CA3130A	CA3130B
APPLICATIONS																															
Sample and Hold																													■	■	■
Switching	■	■	■	■	■							■	■	■															■	■	■
Schmitt Trigger	■	■	■	■	■			■	■	■	■	■	■	■															■	■	■
Multivibrator	■	■	■	■	■	■	■	■	■	■	■	■	■	■	■		■	■	■	■	■	■	■	■	■	■	■	■	■	■	■
Modulator	■	■	■	■	■	■	■					■	■	■	■																
Mixer	■	■	■	■	■	■	■					■	■	■	■																
Detector	■	■	■	■	■	■	■					■	■	■	■																
Comparator	■	■	■	■	■	■	■	■	■	■	■	■	■	■	■		■	■	■	■	■				■	■	■	■	■	■	■
DC Amplifier	■	■	■	■	■			■	■	■	■	■	■	■	■							■	■	■	■	■	■	■	■	■	■
Timer												■	■	■	■		■	■	■	■	■	■	■	■					■	■	■
Wideband Large Signal																															
FEATURES																															
Multiple Unit			■	■	■																										
AGC Capability	■	■	■	■	■							■	■	■																	
Balanced Input	■	■	■	■	■	■	■	■	■	■	■	■	■	■	■		■	■	■	■	■	■	■	■	■	■	■	■	■	■	■
Short-Circuit Protection	■	■	■	■	■							■	■	■	■										■	■	■	■	■	■	■
Internal Frequency Compensation															■							■	■	■	■	■					
Offset Adjustment						■	■	■	■	■	■																	■	■	■	■
Negative VV_{ICR} near V⁻	■	■	■	■	■							■	■	■															■	■	■
Low Power Supply Current (<1 mA)	■	■	■	■	■	■	■					■	■	■																	
Ultra-Low I_{IB}	■	■	■	■	■							■	■	■															■	■	■
Very Low V_{IO} & I_{IO}																															
PACKAGE																															
Flat Pack Ceramic																															
Dual In-Line Ceramic (DIC)			D	D	D			■	■																						
Dual In-Line Plastic (DIP)	E*				E					■	■	E*	E*											E*		E*	E*	E*			
TO-5 Style Straight Lead	■	■				T	T			T	T				T	T	T	T	T	T	T	T	T	T	T	T	T	T	T	T	T
TO-5 Style Dual-In-Line (DIL-CAN)	S	S				S	S			S	S	S	S	S	S	S	S	S	S	S	S	S	S	S	S	S	S	S	S	S	S
Frit Seal Dual In-Line Ceramic	F	F				F	F																								
Beam Lead																									L						
Chip						H																			H	H					

Note 1: The indicated suffix letter identifies the package type for the device type number having a suffix letter; a black square is shown for a type number with no suffix letter.

● Operational Transconductance Amplifier   ▲ Micropower Type   *Available in 8-lead DIP (MINI-DIP)

Table XXIV—Operational Amplifiers (cont'd)

	General Purpose				Quad	Wideband																	Precision					
	Multiple Unit																											
	CA747C	CA747	CA1458	CA1558	CA3401	CA3008	CA3008A	CA3010	CA3010A	CA3015	CA3015A	CA3016	CA3016A	CA3029	CA3029A	CA3030	CA3030A	CA3037	CA3037A	CA3038	CA3038A	CA3100	CA108	CA108A	CA208	CA208A	CA308	CA308A
File No.		531			630	316	310	316	310	316	310	316	310	316	310	316	310	316	310	316	310	625			621			
APPLICATIONS																												
Switching																												
Schmitt Trigger																												
Multivibrator	■	■	■	■	■	■	■	■	■	■	■	■	■	■	■	■	■	■	■	■	■		■	■	■	■	■	■
Modulator																												
Mixer																												
Detector																												
Comparator	■	■	■	■																		■	■	■	■	■	■	■
DC Amplifier	■	■	■	■	■	■	■	■	■	■	■	■	■	■	■	■	■	■	■	■	■	■	■	■	■	■	■	■
Timer																							■	■	■	■	■	■
Wideband Large Signal																						■						
FEATURES																												
Multiple Unit	■	■	■	■	■																							
AGC Capability																												
Balanced Input	■	■	■	■		■	■	■	■	■	■	■	■	■	■	■	■	■	■	■	■	■	■	■	■	■	■	■
Short-Circuit Protection	■	■	■	■																		■	■	■	■	■	■	■
Internal Frequency Compensation	■	■	■	■																			■	■	■	■	■	■
Offset Adjustment	■	■				■	■	■	■	■	■	■	■	■	■	■	■	■	■	■	■	■						
Negative V$_{ICR}$ near V⁻					■																							
Low Power Supply Current (<1 mA)																							■	■	■	■	■	■
Ultra-Low I$_B$																							■	■	■	■	■	■
Very Low V$_{IO}$ & I$_{IO}$																						■						
PACKAGE	TYPE DESIGNATION SUFFIX LETTER (See Note 1)																											
Flat Pack Ceramic						■	■							■	■													
Dual In-Line Ceramic (DIC)																							■	■	■	■		
Dual In-Line Plastic (DIP)	E		E*		E									■	■	■	■											
TO-5 Style Straight Lead	T	T	T	T				■	■	■	■												T	T	T	T	T	T
TO-5 Style Dual In-Line (DIL-CAN)			S	S																			S	S	S	S	S	S
Frit Seal Dual In-Line Ceramic																							F					
Beam Lead																												
Chip	H				H																		H					H

Note 1   The indicated suffix letter identifies the package type for the device type number having a suffix letter, a black square is shown for a type number with no suffix letter.

* Available in 8-lead DIP (MINI-DIP)

Table XXV—Arrays

	Diode Arrays		Transistor Arrays															
	Individual	Quad Plus Two	General-Purpose								2 Transistors, 2 Zener Diodes, 1 Diode	Dual Darlington Connected			Darlington Connected Pair Plus Two Individual			
			n-p-n					p-n-p	p-n-p & n-p-n									
	CA3019	CA3039	CA3081	CA3082	CA3083	CA3183A	CA3183	CA3084	CA3096	CA3096A	CA3093	CA3036	CA3050	CA3051	CA3018	CA3018A	CA3118A	CA3118
File No.	236	343	480	481	532			482		595	533	275		361	338		532	
Applications																		
Comparator				■	■	■					■							
Detector	■	■		■	■	■			■	■				■	■	■	■	■
Differential Amplifier				■	■	■		■	■	■	■	■	■	■	■	■	■	■
Limiter	■	■		■	■	■		■							■	■	■	■
Mixer	■	■		■	■	■		■								■	■	■
Modulator	■	■		■	■	■		■							■	■	■	■
Multivibrator	■	■		■	■	■		■							■	■	■	■
Oscillator			■	■	■	■		■							■	■	■	■
Schmitt Trigger				■	■	■												
Sense Amplifier				■	■	■			■	■								
Switching	■		■	■	■	■	■	■										
Thyristor & SCR Control			■	■	■	■	■	■	■	■					■	■	■	■
Timer																		
VHF																		
Regulator											■							
Features																		
High Input Resistance																		
Balanced Input				■	■	■		■	■	■		■	■	■	■	■	■	■
Balanced Output				■	■	■		■	■	■		■	■	■	■	■	■	■
Low Noise									■	■		■						
AGC Capability															■	■	■	■
Multiple Unit												■	■	■	■	■	■	■
Wide Band															■	■	■	■
TYPE DESIGNATION SUFFIX LETTER (SEE NOTE 1)																		
Package																		
Flat Pack Ceramic																		
Dual In-Line Ceramic													■					
Dual In-Line Plastic			■	■	■	E	E	■	E	E	E			■				
TO-5 Style Straight Lead	■	■													■	■	T	T
TO-5 Style Formed Lead																		
Frit Seal Dual-In-Line Ceramic			F	F	F													
Chip	H	H	H	H	H			H	H	H	H					H		H
Beam-Lead					L			L								L		

NOTE 1: The indicated suffix letter identifies the package type for the device type number having a suffix letter; a black square is shown for a type number with no suffix letter.

Table XXV—Arrays (cont'd)

	Transistor Arrays									Amplifier Arrays					
	Differentially Connected Pair Plus Three Individual					Super β Amp. Plus 3 n-p-n Trans.	1 n-p-n & 1 p-n-p/n-p-n transistors, 1 zener diode, 1 PUT*, 1 SCR▲ (Thyristor)	COS/MOS Array 3 n-channel & 3 p-channel transistors	High-Freq. n-p-n	Dual Independent (Differential)				Three Ampl.	Four Ampl.
	CA3045	CA3046	CA3086	CA3146A	CA3146	CA3095	CA3097	CA3600	CA3127	CA3026	CA3049	CA3102	CA3054	CA3035	CA3048
File No.	341	483		532		591	633	619	662	388	611		388	274	377
Applications															
Comparator							■	■							
Detector	■	■	■	■	■	■			■	■	■	■	■		
Differential Amplifier	■	■	■			■		■	■	■	■	■			
Limiter			■					■							
Mixer			■			■		■	■		■	■			■
Modulator			■					■	■		■	■			
Multivibrator			■			■		■	■		■	■			■
Oscillator			■			■	■	■	■		■	■			
Schmitt Trigger			■						■						
Sense Amplifier			■						■						
Switching			■				■	■	■		■	■			
Thyristor & SCR Control			■				■	■							
Timer							■	■							
VHF							■		■						
Regulator							■								
Features															
High Input Resistance								■							
Balanced Input	■	■	■	■	■	■				■	■	■			
Balanced Output	■	■	■	■	■	■				■	■	■			
Low Noise						■								■	■
AGC Capability						■			■		■	■			
Multiple Unit	■	■	■	■	■	■		■	■	■	■	■	■	■	■
Wide Band	■	■	■	■	■	■			■	■	■	■		■	■
	TYPE DESIGNATION SUFFIX LETTER									(SEE NOTE 1)					
Package															
Flat Pack Ceramic															
Dual In-Line Ceramic	■	■	■												
Dual In-Line Plastic				E	E	E	E	E	E			E	■		
TO-5 Style Straight Lead										■	■			■	
TO-5 Style Formed Lead														VI	
Frit Seal Dual-In-Line Ceramic	F														
Chip	H			H		H	H			H	H	H	H	H	H
Beam-Lead	L										L	L			

* Programmable Unijunction Transistor   ▲ Silicon Controlled Rectifier

Table XXVI—Broadband (Video) and Differential Amplifiers and AM/FM Communications Circuits

	Broadband (Video) Amplifiers								Differential Amplifiers															AM/FM Communications Circuits									
	CA3002	CA1352	CA3020	CA3020A	CA3021	CA3022	CA3023	CA3040†	CA3000	CA3001*	CA3004	CA3005	CA3006	CA3007	CA3026	CA3028A	CA3028B	CA3049	CA3050	CA3051	CA3053	CA3054	CA3102E	CA3011	CA3012	CA3013	CA3014	CA3043	CA3075	CA3076	CA3088	CA2111A	CA3123
File No.	123	Prel.		339		243		363	121	122	124	125		126	388	382		611	361		382	388	611	128	129		331		424	430	560	612	631
Applications																																	
Voltage Regulator																																	
Comparator									■	■	■	■	■		■	■	■	■						■									
Comparator – High Current Output																																	
Control – Relays, Heaters, LED's Lamps, etc.																																	
Detector	■											■	■	■	■										■		■	■			■	■	■
Differential Amplifier	■	■	■	■					■	■	■	■	■	■	■	■	■	■	■	■	■	■	■	■	■	■	■	■	■	■		■	■
Limiter			■	■	■	■	■					■	■	■		■	■	■				■		■	■	■	■		■	■		■	■
Mixer	■		■	■							■	■	■	■	■			■	■	■		■	■										
Modulator	■										■	■	■	■	■			■	■	■	■	■	■										
Multivibrator									■	■					■				■	■	■												
Oscillator			■	■					■						■				■	■	■	■											■
Schmitt Trigger	■								■	■	■				■	■	■	■	■	■	■	■											
Sense Amplifier	■								■	■	■				■	■	■	■	■	■	■		■	■	■								
Switching			■	■							■	■	■	■	■	■	■	■	■				■										
Thyristor & SCR Control																																	
Freq. Doubler, Mult., Divide, Sq. Root, Squarer																																	
Display Decoder-Driver																																	
Timer																																	
Features																																	
Balanced Input	■		■	■					■	■	■	■	■	■	■	■	■	■	■	■	■	■	■	■	■	■	■	■				■	■
Balanced Output			■	■					■	■	■	■	■	■	■	■	■	■	■	■	■	■		■	■	■							
Low Noise (1/f)																																	
Regulated Power Supply			■	■																								■	■	■	■	■	■
Class B Output			■	■																													
AGC Capability	■	■			■	■	■	■	■	■	■	■	■		■	■															■	■	■
Multiple Unit															■				■	■	■	■		■	■								
Wide Band	■	■	■			■	■		■	■	■	■												■	■	■	■		■	■	■	■	■
Micropower				■																													
Decimal Pt. Output																																	
Ripple Blanking																																	
Type Designation Suffix Letter ■ = No Suffix Letter																																	
Package																																	
Flat Pack (FP)																											`,`						
Dual-In-Line Ceramic (DIC)																					■												
Dual-In-Line Plastic (DIP)		E																						■	■	E'				■†	E	E	E‡ E
TO-5	■		■	■	■	■	■	■	■	■	■	■	■	■	■	■	■	T					■		■	■	■	■	■				
Chip	H		H			H			H H		H				H H		H				H H		H		H				H H	H			
Beam Lead															L		L					L											
Frit Seal															F	F				F									F				
TO-5 Style Dual-In-Line (DIL-CAN)															S	S																	

NOTE 1: The indicated suffix letter identifies the package type for the device type number having a suffix letter; a black square is shown for a type number with no suffix letter.

* CA3001 is also useful as a Broadband (Video) Amplifier.

† CA3040 is also useful as a Differential Amplifier.

‡ In quad-in-line package (QUIP)

Table XXVII—Power-Control and Voltage-Regulator Circuits, Analog Multiplier, and Computer Interface Circuits

	CA3068	CA3059	CA3079	CA3097:	CA3094A	CA3094B	CA3094	CA3099	CA555, C	CA3062	CA3085	CA3085A	CA3085B	CA3091	CA1541	CD2500E	CD2501E	CD2502E	CD2503E
Power-control group	Thyristor Control			Power Control Switch/Ampl.				Programmable Comparator	Timers	Photo Det.	Voltage Regulators			Analog Multiplier	Sense Ampl.	Decoder Drivers			
File No.	490	490	490	633	598	598	598	620	834	421	491	491	491	534	536	392	392	392	392
Applications																			
Voltage Regulator					■	■	■				■	■	■						
Comparator	■	■	■	■	■	■	■	■											
Comparator – High Current Output	■	■	■	■	■	■	■	■			■	■	■						
Control – Relays, Heaters, LED's, Lamps, Etc.	■	■	■	■	■	■	■	■											
Detector	■	■	■							■				■					
Differential Amplifier	■	■	■		■	■	■	■						■					
Limiter																			
Mixer					■	■	■							■					
Modulator					■	■	■	■		■									
Multivibrator				■	■	■	■		■										
Oscillator	■	■	■	■	■	■	■	■											
Schmitt Trigger					■	■	■	■											
Sense Amplifier																			
Switching	■	■	■		■	■	■	■			■								
Thyristor & SCR Control	■	■	■	■	■	■	■	■			■	■	■						
Freq. Doubler, Mult., Divide, Sq. Root, Squarer														■					
Display Decoder-Driver																■	■	■	■
Timer	■	■	■	■	■	■	■	■	■										
Features																			
Balanced Input					■	■	■												
Balanced Output									■										
Low Noise (1/f)					■	■	■				■	■	■						
Regulated Power Supply	■	■	■					■											
Class B Output																			
AGC Capability					■	■	■												
Multiple Unit																			
Wide Band																			
Micropower					■	■	■	■											
Decimal Pt. Output																■		■	
Ripple Blanking																■		■	■
Package — Type Designation Suffix Letter ■ = No Suffix Letter																			
Flat Pack (FP)																			
Dual-In-Line Ceramic (DIC)	■													D	D				
Dual-In-Line Plastic (DIP)		■	■	E				E	E		E					E	E	E	E
TO-5					T	T	T			T	■	■	■	■					
Chip	H			H			H	H			H			H	H				
Beam Lead														L					
Frit Seal																			
TO-5 Style Dual-In-Line (DIL-CAN)					S	S	S		S		S	S	S						

NOTE 1: The indicated suffix letter identifies the package type for the device type number having a suffix letter; a black square is shown for a type number with no suffix letter.

CA555, CA555C, CA3085, CA3085A available in 8-lead DIP (MINI-DIP) package.

Table XXVIII—Consumer Circuits

Circuit Functions	Audio Circuits — Pre-Amp. CA3036	Audio Circuits — Pre-Amp. CA3052	Audio Circuits — Drivers CA3094	Audio Circuits — Drivers CA3094A, B	Audio Circuits — Power Amplifiers CA810	Audio Circuits — Power Amplifiers CA3131, CA3132	Multiplex Decoders CA758	Multiplex Decoders CA1310	Multiplex Decoders CA3090A	AM Rcvr. Ckts. CA3088	AM Rcvr. Ckts. CA3123	FM Receiver — FM IF Subsystems CA2111A	FM Receiver — FM IF Subsystems CA3089	FM Receiver — FM IF Subsystems CA3075	FM Receiver — FM IF Subsystems CA3043	FM Receiver — FM IF Subsystems CA3013, CA3014	FM Receiver — FM IF Gain Blocks CA3011	FM Receiver — FM IF Gain Blocks CA3012	FM Receiver — FM IF Gain Blocks CA3076
File No.	387		598		Prel.	Prel.	760	761	684	560	631	612	561	424	331	129	128		430
Audio Driver			■	■															
Audio Preamplifier	■	■	■	■						■			■	■	■	■			
ACC																			
AFC/AFT													■						
AFPC																			
AGC			■	■						■			■						
Chroma Amplifier																			
Chroma Demodulator																			
Chroma Signal Processor																			
Converter										■	■								
Detector										■		■	■	■	■	■			
Video Amplifier																	■	■	■
Sync Processor																			
IF Amplifier										■		■	■	■	■	■	■	■	■
Limiter												■	■	■	■	■	■	■	■
Oscillator										■									
Audio Power Amplifier	·				■	■													
Tint Control																			

Multiplex Decoders columns (CA758, CA1310, CA3090A): **Stereo Multiplex Decoder**

Package — (TYPE DESIGNATION SUFFIX LETTER (See Note 1))

Package	CA3036	CA3052	CA3094	CA3094A, B	CA810	CA3131, CA3132	CA758	CA1310	CA3090A	CA3088	CA3123	CA2111A	CA3089	CA3075	CA3043	CA3013, CA3014	CA3011	CA3012	CA3076
Dual-In-Line Plastic		■				EM	E	E		E	E	E	E						
Quad-In-Line Plastic					Q,QM				Q			Q	■						
TO-5 Standard Lead	■		T	T											■	■	■	■	■
TO-5 Formed Lead																			

Note 1: Where a code letter is shown (E, EM, Q, T, V1), add the code letter as a suffix to the type number to identify the package (and lead configuration) option. A black square indicates no suffix code is added to the type number for that package option.

Table XXVIII—Consumer Circuits (cont'd)

TV Receiver Circuits

Circuit Functions	Remote Control	Automatic Fine-Tuning (AFT)		IF Systems — Sound					IF Systems — Pix	Chroma Systems — 2 Package									Chroma Systems — 3 Package			"Jungle" Circuit
Type	CA3035	CA3044	CA3064	CA3134	CA3041	CA3042	CA3065	CA2111A	CA3068	CA1352	CA3066	CA3067	CA3070	CA3121	CA3067	CA3126	CA1398	CA3125	CA3128	CA3070	CA3071	CA3072	CA3120
File No.	274	340	396	Prel.	318	319	412	612	467	Prel.	466	468	688	466	860	686	685	Prel.		468			691
Audio Driver						■	■	■															
Audio Preamplifier	■					■	■	■															
ACC											■		■	■		■	■		■	■	■		
AFC/AFT		■	■																				
AFPC													■					■	■				
AGC									■	■													■
Chroma Amplifier											■		■	■	■	■			■	■			
Chroma Demodulator												■	■		■	■		■				■	
Chroma Processor: PAL Systems											■		■	■	■	■			■	■			
Converter																							
Detector		■	■		■	■	■	■	■											■			
Video Amplifier																				■			■
Sync Processor																							■
IF Amplifier		■	■		■	■	■	■	■														
Limiter		■		■	■	■	■	■															
Oscillator											■		■	■	■	■			■	■			
Audio Power Amplifier				■																			
Tint Control											■	■				■							

(TYPE DESIGNATION SUFFIX LETTER (See Note 1))

Package																							
Dual-In-Line Plastic			E	E					E	E				■	■		E	E	E	■	■	■	E
Quad-In-Line Plastic					■	■	■	Q	■		■	■				■	Q		Q				
TO-5 Standard Lead	■	■																					
TO-5 Formed Lead	VI	VI	■																				

Note 1: Where a code letter is shown (E, EM, Q, T, V1), add the code letter as a suffix to the type number to identify the package (and lead configuration) option. A black square indicates no suffix code is added to the type number for that package option.

COS/MOS Integrated Circuits for Digital Applications

A COS/MOS digital integrated circuit is a monolithic circuit that uses both n-channel and p-channel COmplementary-Symmetry MOS transistors. The use of these complementary transistors on a single monolithic chip provides a circuit flexibility that makes COS/MOS a versatile building block for a wide range of applications.

COS/MOS DESIGN AND LAYOUT

COS/MOS integrated circuits are normally fabricated on an n-type substrate which serves as the substrate material for all p-channel MOS devices as well as for p^+ tunnels, diodes, and resistors. A p-type substrate is provided for the complementary n-channel MOS devices, n^+ tunnels, diodes, and resistors by diffusion of a lightly doped p-well region into the original n-type substrate. The n-channel units exhibit the higher carrier mobility associated with electrons; they have approximately twice the transconductance of p-channel units with identical geometry. Therefore, the matching of a p-channel with an n-channel unit requires that a p-channel unit with a given channel length L have approximately twice the channel width W of the n-channel unit with which it is to be matched.

Guard Bands

Protective **guard bands** surround separate MOS devices, tunnels, wells, and diodes or combinations of MOS devices which are interconnected through common diffused regions for the purpose of preventing leakage between the entities named. All p-channel devices, tunnels, and diodes must be surrounded by a continuous n^+ guard band which also serves as a tunnel to help conduct current from the external supply voltage V_{DD} across the n-type substrate to every p-channel device tied to the external supply. Similar heavily doped p^+ guard bands surround all n-channel devices, tunnels, and diodes to help conduct current from the external ground supply V_{SS} across the p-well to every n-channel device tied to ground. Contact to the n-type substrate may be made through the n^+ guard band and returned to the V_{DD} pad; contact to the p-well substrate pad may be made through the p^+ guard band and returned to the ground pad. Fig. 245 illustrates a typical p- and n-channel MOS complementary pair with appropriate connections to V_{DD} and V_{SS}, respectively. Fig. 246 is a cross section of the complementary pair of Fig. 245 showing the use of n^+ and p^+ guard bands for interconnecting the source regions to the V_{DD} and V_{SS} pads, respectively. Guard bands

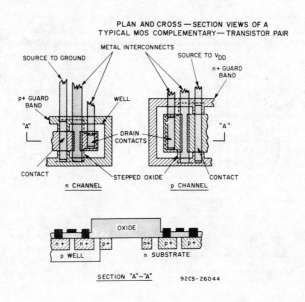

Fig. 245—Plan and cross-sectional views of a typical MOS complementary-transistor pair.

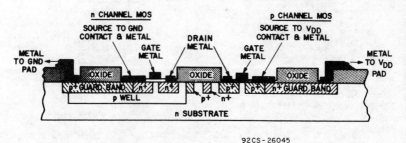

Fig. 246—A cross section of the complementary pair of Fig. 245 showing the use of guard bands.

may be narrow strips, or, where space permits, large diffused areas that minimize resistance in the V_{DD} and V_{SS} supply lines. Guard bands are also used to assure positive device cutoff; this cutoff is accomplished by having the gate metal, as it leaves the end of the channel, cross a guard band prior to stepping up over the thick oxide as shown in Fig. 246.

Typical Inverter and Logic Gate Layout

A typical layout for the basic COS/MOS inverter circuit is shown in Fig. 247; the schematic diagram for this basic circuit is shown in Fig. 248. The figure shows the single n-channel MOS transistor and its complementary p-channel coun-

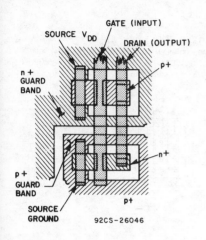

Fig. 247—Typical layout for the basic COS/MOS inverter circuit.

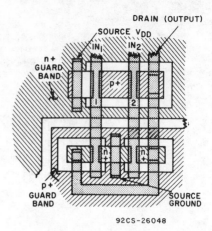

92CS-26048

Fig. 249—Typical layout for a 2-input NOR gate.

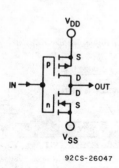

92CS-26047

Fig. 248—Schematic diagram for the basic COS/MOS inverter circuit.

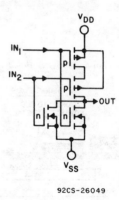

92CS-26049

Fig. 250—Schematic diagram for a 2-input NOR gate.

terpart with the gates and drains tied to each other and with metal interconnects between source and V_{SS} and source and V_{DD}, respectively.

The basic inverter may be extended, as described in the section on **Basic Building Blocks,** to true complementary logic. The logical NOR gate is formed by placing the n-channel units in parallel and the p-channel units in series. Fig. 249 shows a typical layout and Fig. 250 shows the schematic diagram for a 2-input NOR gate. The source-

to-V_{DD} and source-to-V_{SS} connections are made to the n^+ and p^+ guard bands because these guard bands are tied directly to V_{DD} and ground, respectively. Variations of this layout are used in the CD4000A dual 3-input gate and CD4001A quad 2-input gate. If the above arrangement is reversed, i.e., if the p-channel units are placed in parallel and the n-channel units in series, NAND gates result, as in the CD4011A quad 2 and CD4012A dual 4-input NAND gate circuits.

Packages

RCA COS/MOS integrated circuits are currently packaged in three distinct configurations: the TO-5-style glass-metal package, the ceramic flat pack, and the dual-in-line package. The dual-in-line package may be either ceramic or plastic. The TO-5-style package has 12 leads; the flat pack and the dual-in-line packages have 14, 16, or 24 leads. The various integrated-circuit packages are shown in the section **Guide to RCA Solid-State Devices.**

The letters D, E, F, K, and T appended to a device identification identify the type of package in which the device is enclosed.

COS/MOS integrated circuits are also provided in chip form. Suffix letter H designates the unpackaged circuit chip.

Desig-nation	Meaning	Standard Pins Available
E	Dual-in-Line Plastic	14, 16
D	Dual-In-Line Weld-Seal Ceramic	14, 16, 24, 28
K	Flat Pack	14, 16, 24, 28
T	TO-5-Style Package	12
H	Chip	—

For example, a CD4006AK is a CD4006A circuit in a ceramic flat pack, a CD4006AD is a CD4006A circuit in a dual in-line ceramic package, and the CD4006AE is the CD4006A circuit in a dual in-line plastic package.

BASIC BUILDING BLOCKS

This section describes several circuits (building-block units) that form the basis for the more complex circuits discussed later in the Man-ual. Because COS/MOS IC's are particularly well suited for digital applications, the basic circuits are described from that point of view.

Inverter

The most basic of all the COS/MOS circuits, the **inverter** circuit, consists of one p-channel and one n-channel enhancement-type MOS transistor, as shown in Fig. 251.

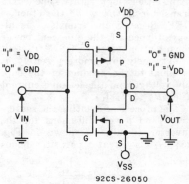

Fig. 251—COS/MOS inverter circuit.

The substrate of the p-channel device is at $+V_{DD}$, and the substrate of the n-channel device is at ground. Consequently, when the voltage at the input of the inverter is zero (logic 0), the input is at $-V_{DD}$ relative to the substrate of the p-channel device and at zero volts relative to the substrate of the n-channel device. The result is that the p-channel device is turned on and the n-channel device is turned off. Under these conditions, there is a low-impedance path from the output to V_{DD}, and a very-high-impedance path from the output to ground; therefore, the output voltage approaches V_{DD} (logic 1) under normal loading conditions. When the input voltage is $+V_{DD}$ (logic 1), the situation is reversed: the p-channel unit is OFF and the n-channel unit is ON, with the result that the output voltage approaches zero (logic 0). Fig. 252 shows the logic diagrams for a COS/MOS inverter.

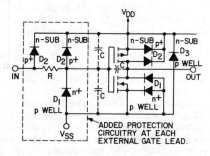

92CS-26051

Fig. 252—Logic diagrams for a COS/MOS inverter.

In either logic state, one MOS transistor is ON while the other is OFF. Because one transistor is always turned off, the quiescent power consumption of the COS/MOS unit is extremely low; more precisely, it is equal to the product of the supply voltage and the leakage current (i.e., $P_D = V_{DD}I_L$).

The process of creating the source-drain and p-well diffusions for the inverter circuit, as shown in Fig. 253(a), also creates parasitic diodes

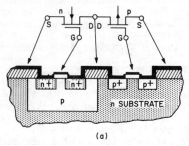

(a)

(b)

92CS-26052

Fig. 253—(a) Cross section of inverter-circuit chip showing parasitic diodes; (b) circuit diagram showing connection of the diodes to basic inverter nodes.

which are connected to the basic inverter nodes, as shown in Fig. 253(b). These parasitic elements (D_1, D_2, and D_3) are back-biased across the power supply and contribute in part to the device leakage current and thus to the quiescent dissipation. Additional diodes are also diffused into the over-all structure, as shown in Fig. 253(b), to form an input limiter circuit that provides protection against static voltages. Any gate input node that is brought out to a package terminal is protected by this network. The operation of the input protection circuit is described in the next section.

RCA COS/MOS products range from circuits as simple as 2-input **logic gates** through the complexity of a 200-stage shift register. These devices are composed of varying numbers of interconnected inverter circuits placed on silicon chips of varying area. Therefore, the leakage current ranges widely because it depends on the number of interconnected circuits and the parasitic diode area associated with each circuit. For example, the dissipation of a logic gate (the CD4001) is typically 0.01 microwatt at 10 volts, while the dissipation of the 64-stage shift register (CD4031) is typically 10 microwatts at 10 volts, even though these device types are processed similarly.

Device Switching Characteristics —Because of the complementary nature of the interconnections of the series p- and n-type devices in the basic inverter, the **transfer characteristic** of a COS/MOS logic gate is as shown in Fig. 254. The high input impedance of the gate results in no dc loading on the output so that the input and output signals are allowed to swing completely from zero volts (logic 0) to V_{DD} (logic 1) when sufficient time settling is allowed. The switching point is shown to be typically 45 to 50 per cent of the magnitude of the power-supply voltage, and varies directly with that voltage over the entire range of supply volt-

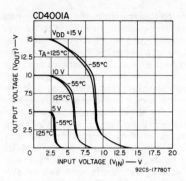

Fig. 254—Transfer characteristic of a COS/MOS logic gate.

age specified for COS/MOS devices. The COS/MOS transfer characteristic of Fig. 254 illustrates the high noise immunity of COS/MOS devices; i.e., typically 45 per cent of the supply voltage. Fig. 254 also shows the negligible change in operating point as temperature ranges from −55°C to +125°C. Because of the ideal nature of these switching characteristics, COS/MOS devices operate reliably over a much wider range of voltage than other forms of logic circuits.

AC Dissipation Characteristics— All significant COS/MOS power dissipation is ac in nature and is a direct function of load capacitance C, operating supply voltage V, and switching rate f.

During the transition from a logical 0 to a logical 1, both transistors in the COS/MOS inverter are momentarily ON with the result that, instantaneously, a pulse of current is drawn from the power supply. The magnitude of this current depends on the impedance and threshold voltage of the inverter transistors, as well as the magnitude of the power-supply voltage and the length of time spent in transition (e.g., the input rise or fall time). Current is also required to charge and discharge the output load capacitance. The dissipation that results

from the current components described above is directly proportional to the frequency of operation and amount of capacitive loading and may be expressed as follows:

$$P_{ac} = CV^2f$$

The more often the circuit switches, the greater the current; the heavier the capacitive loading, the greater the resultant dissipation.

AC Performance Characteristics— As indicated above, the node capacitances located within or external to a given circuit are charged and discharged during switching through the channel resistance of the p- or n-type device. As the magnitude of V_{DD} increases, the impedance of the conducting channel decreases; therefore, the maximum switching speed of COS/MOS devices increases with increasing supply voltage, as shown in Fig. 255. The effect of increasing external load capacitance is shown in Fig. 256.

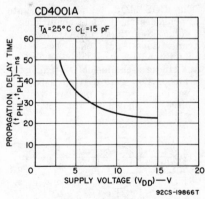

Fig. 255—Curve illustrating increase in maximum switching speed with increase in supply voltage.

Transmission Gates

The COS/MOS transmission gate is a single-pole single-throw switch formed by the parallel connection of a p-type and an n-type device. This switch expands the versatility of

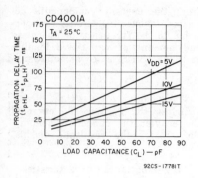

CD4001A

$T_A = 25\,°C$

PROPAGATION DELAY TIME ($t_{pHL} = t_{pLH}$) — ns

$V_{DD} = 5V$

10V

15V

LOAD CAPACITANCE (C_L) — pF

92CS-17781T

Fig. 256—Curve showing effect on switching speed of increasing external load capacitance.

COS/MOS circuits in both digital and linear applications.

The perfect transmission gate or switch may be characterized as having zero forward and reverse resistance when closed and infinite resistance when open; i.e., it has an infinite OFF/ON impedance ratio. The COS/MOS transmission gate approaches these ideal conditions.

The advantages of a COS/MOS transmission gate can be better understood by consideration first of the single n-channel MOS-transistor transmission gate driving a capacitive load from a positive voltage source, as shown in Fig. 257. With

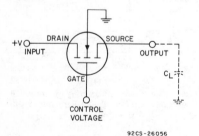

92CS-26056

Fig. 257—Single n-channel MOS-transistor transmission gate.

0 volts applied to the gate of the n-channel device, no current can flow, and the load capacitance C_L remains uncharged. As the gate volt-

age to the transmission gate is made positive enough to turn the transmission gate on, however, the load capacitance begins to charge. However, the load capacitance can only charge to a level equal to the gate voltage minus the threshold voltage of the n-channel transistor because the single n-channel transmission gate operates as a source-follower circuit in which premature gate cutoff occurs. Another aspect of MOS transmission gates is that they are bilateral; i.e., drain and source are interchangeable. This type of transmission gate also operates with slow speed in large-signal applications; i.e., as the device begins to turn on, the RC time constant is large. A COS/MOS transmission gate which overcomes these disadvantages is made by paralleling n- and p-channel MOS transistors as shown in Fig. 258. This arrangement overcomes

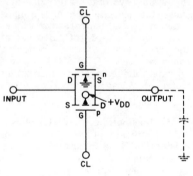

92CS-26057

Fig. 258—COS/MOS transmission gate.

the premature-cutoff problem associated with the single-channel transmission gate because one of the two channels is always being operated as a drain-loaded stage regardless of what the input or output voltage may be. If each MOS channel of the COS/MOS circuit has a 2-volt threshold voltage and if 0 volts is applied to the gate of the p-channel unit and 10 volts to the gate of the n-channel unit, an increase in input voltage in excess of 8 volts (10 volts — 2 volts) cannot be switched through the n-channel unit. However, proper switching can occur in the p-channel unit because the magnitude of the voltage from gate to source (0 volts − 8 volts = −8 volts) is greater than the p-channel threshold (−2 volts). As a result, the switch does not turn off prematurely because the gate-to-source voltages of both the n- and p-channel units never equal the threshold voltages of these devices. The full 10-volt supply voltage (V_{DD} - V_{SS} = 10 volts) can, therefore, be switched. The COS/MOS transmission gate is also considerably faster than the single-channel MOS transmission gate; the RC time constant is always smaller.

Transmission Gate and Inverter Applications

At present, all COS/MOS integrated circuits are constructed of two basic building blocks, the inverter and the transmission gate. The basic inverter forms all the NOR and NAND gates. Combined with transmission gates, the inverter forms more complicated circuits, such as flip-flops, counters, shift registers, arithmetic blocks, and memories.

NOR Gate—A two-input NOR gate is an inverter with two n-type units in parallel and two p-type units in series, as shown in Fig.

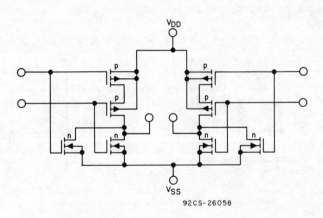

92CS-26058

Fig. 259—Schematic diagram for a pair of two-input NOR gates.

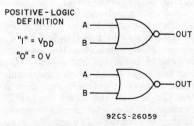

POSITIVE - LOGIC DEFINITION

"1" = V_{DD}

"0" = 0 V

92CS-26059

Fig. 260—Logic diagram for the pair of two-input NOR gates shown in Fig. 259.

259. Fig. 260 shows the logic diagram for a two-input NOR gate. Each of the two inputs is connected to the gate of one n- and one p-channel transistor. A negative output is obtained when either the A or the B input is positive because the positive input turns off the associated p-channel transistor, disconnecting the output from the V_{DD} supply, and turns on the associated n-channel transistor, connecting it to ground and causing a low output.

When both of the input signals are at ground potential, both p-channel units are on and both n-channel units off. In this case the output is coupled to the V_{DD} terminal and provides a high output. Three- and four-input NOR gates may be formed by placing three or four n-channel transistors in parallel and three or four p-channel transistors in series in an arrangement similar to that shown in Fig. 259.

NAND Gate—A NAND gate is an inverter with two p-channel transistors in parallel, and two n-channel transistors in series, as shown in Fig. 261. The logic diagram for a two-input NAND gate is shown in Fig. 262. The output goes negative only if both inputs are positive, in which case the p-channel transistors are turned off and the n-channel

transistors on. This condition couples the output to ground. If either input is negative, the associated n-channel transistor is turned off, and the associated p-channel transistor on; thus the output is coupled to V_{DD} and goes high. Again, three- or four-input NAND gates may be formed by placing three or four p- and n-channel transistors in parallel and in series, respectively, in an arrangement similar to that shown in Fig. 261.

Set-Reset Flip-Flop—Two NOR gates may be connected as shown in Fig. 263 to form a set-reset flip-flop. Fig. 264 shows the logic diagram and truth table for this circuit. When the set and reset inputs are low, one amplifier output is low, and the other high, and there is a stable condition. If the set input is

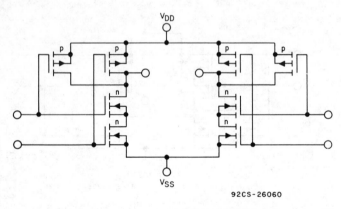

92CS-26060

Fig. 261—Schematic diagram for a pair of NAND gates.

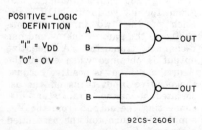

92CS-26061

Fig. 262—Logic diagrams for NAND gates shown in Fig. 261.

raised to a higher level, the associated n-channel unit is turned on so that the output of the set stage goes high and becomes logic 1 or Q. Under these conditions the flip-flop is said to be in the set state. Raising the reset input level causes the other output to go high, and places the flip-flop in the reset state (Q̄ represents the low output state). Thus, the circuit of Fig. 263 represents a static flip-flop with set and reset capability.

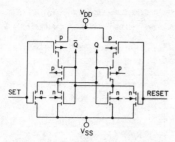

ALL p – UNIT SUBSTRATES CONNECTED TO V_{DD};
ALL n – UNIT SUBSTRATES TO V_{SS}.

92CS-26062

Fig. 263—Schematic diagram of a set-reset flip-flop.

NOR-SET/RESET = 8 DEVICES

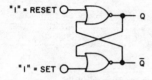

TRUTH TABLE

SET	RESET	Q	$\bar{Q}$	
1	0	1	0	
0	1	0	1	
1	1	0	0	
0	0	Q	$\bar{Q}$	NO CHANGE

92CS-26063

Fig. 264—Logic diagram and truth table for set-reset flip-flop shown in Fig. 263.

D Flip-Flop—A block diagram of a D-type flip-flop is shown in Fig. 265; the schematic diagram for the flip-flop is shown in Fig. 266(a). The block diagram shows a master flip-flop formed from two inverters and two transmission gates (shown as switches) that feeds a slave flip-

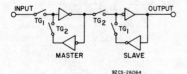

Fig. 265—Block diagram for a D-type flip-flop.

flop having a similar configuration. When the input signal is at a low level, the TG_1 transmission gates are closed and the TG_2 gates open. This configuration allows the master flip-flop to sample incoming data, and the slave to hold the data from the previous input and feed it to the output. When the clock is high, the TG_1 transmission gates open and the TG_2 transmission gates close, so that the master holds the data entered and feeds it to the slave. The D flip-flop is static and holds its state indefinitely if no clock pulses are applied, i.e., it stores the state of the input prior to the last clocked input pulse. A **clock pulse** is the pulse applied to the logical elements of a sequential digital system to initiate logical operations. Both the "clock", CL, and "inverted clock", $\overline{CL}$, as shown in Fig. 266(b), are required; clock inversion is accomplished by an inverter internal to each D flip-flop.

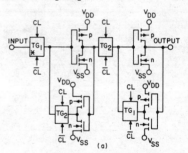

ALL p – UNIT SUBSTRATES CONNECTED TO V_{DD};
ALL n – UNIT SUBSTRATES TO V_{SS}.

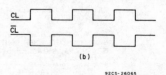

(b)

92CS-26065

Fig. 266—(a) Schematic diagram and (b) clock-pulse waveforms for a D-type flip-flop.

Fig. 267 shows the logic diagram and truth table for a D-type flip-flop.

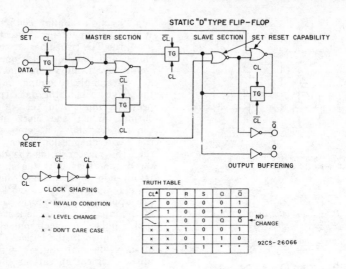

STATIC "D" TYPE FLIP-FLOP

TRUTH TABLE

CL▲	D	R	S	Q	Q̄	
╱	0	0	0	0	1	
╱	1	0	0	1	0	
╲	x	0	0	Q	Q̄	NO CHANGE
x	x	1	0	0	1	
x	x	0	1	1	0	
x	x	1	1	*	*	

* = INVALID CONDITION
▲ = LEVEL CHANGE
x = DON'T CARE CASE

92CS-26066

Fig. 267—Logic diagram and truth table for a D-type flip-flop.

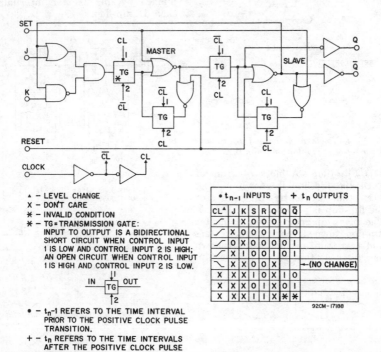

▲ – LEVEL CHANGE
X – DON'T CARE
∗ – INVALID CONDITION
∗ – TG = TRANSMISSION GATE:
INPUT TO OUTPUT IS A BIDIRECTIONAL
SHORT CIRCUIT WHEN CONTROL INPUT
1 IS LOW AND CONTROL INPUT 2 IS HIGH;
AN OPEN CIRCUIT WHEN CONTROL INPUT
1 IS HIGH AND CONTROL INPUT 2 IS LOW.

• – t_{n-1} REFERS TO THE TIME INTERVAL
PRIOR TO THE POSITIVE CLOCK PULSE
TRANSITION.
+ – t_n REFERS TO THE TIME INTERVALS
AFTER THE POSITIVE CLOCK PULSE
TRANSITION.

• t_{n-1} INPUTS						+ t_n OUTPUTS		
CL▲	J	K	S	R	Q	Q	Q̄	
╱	1	X	0	0	0	1	0	
╱	X	0	0	0	1	1	0	
╱	0	X	0	0	0	0	1	
╱	X	1	0	0	1	0	1	
╲	X	X	0	0	X			←(NO CHANGE)
X	X	X	1	0	X	1	0	
X	X	X	0	1	X	0	1	
X	X	X	1	1	X	∗	∗	

92CM-17188

Fig. 268—Logic diagram and truth table for a J-K flip-flop.

J-K Flip-Flop—The logic diagram and truth table for a J-K flip-flop are shown in Fig. 268. The J-K flip-flop is similar in some respects to the D flip-flop, but has some additional circuitry to accommodate the J and K inputs. The J and K inputs provide separate clocked set and reset inputs, and allow the flip-flop to change state on successive clock pulses.

The J-K flip-flop circuit also has set and reset capability; the inverters in the master and slave flip-flop each have an added OR input for direct (unclocked) setting and resetting of the flip-flop.

Memory Cell—The basic storage element common to all RCA COS/MOS memories consists of two COS/MOS inverters cross-coupled to form a flip-flop as shown in Fig. 269. Single-transistor transmission

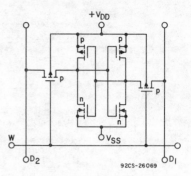

Fig. 270—A word-organized storage cell: W is word line, D_1 and D_2 are data lines.

six transistors; one word line, W; and two digit-sense lines, D_1 and D_2. Addressing is accomplished by energizing a word line; this action turns on the transmission gates on both sides of the selected flip-flop. Because the cell in Fig. 270 has p-channel transmission gates, a ground-level voltage is required for selection.

Fig. 271 shows an 8-transistor bit-organized memory cell employing X-Y selection. A modification of

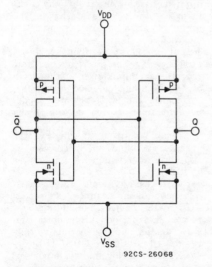

Fig. 269—The basic storage element common to all RCA COS/MOS memories.

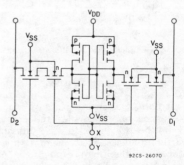

Fig. 271—Eight-transistor bit-organized memory cell with X-Y selection.

gates are employed as a simple and efficient means of performing the logic functions associated with storage-cell selection; i.e., the sensing and storing operations. The resulting word-organized storage cell, shown in Fig. 270, is composed of

this circuit in which the Y-select transistors are common for each column of storage elements in used in large memory arrays.

Dynamic Shift Register—Fig. 272 shows a two-stage shift register; each stage consists of two

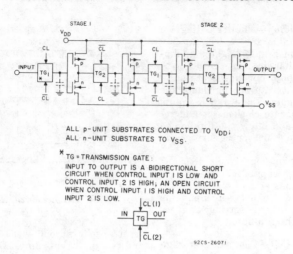

Fig. 272—Schematic diagram for a two-stage shift register.

inverters and two transmission gates. Each transmission gate is driven by two out-of-phase clock signals arranged, as shown in Fig. 273, so that when alternate trans-

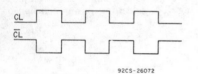

Fig. 273—Clock-pulse waveforms for two-stage shift register shown in Fig. 272.

mission gates are turned on, the others are turned off. When the first transmission gate in each stage is turned on, it couples the signal from the previous stage to the inverter, and causes the signal to be stored on the input capacitance of the inverter. The shift register utilizes the input of the inverter for temporary storage.

When the transmission gate is turned off on the next half cycle of the clock, the signal is stored on this input capacitance, and the signal remains at the output of the inverter where it is available to the next transmission gate, which is now

turned on. Again, this signal is applied to the input of the next inverter where it is stored on the input capacitance of the inverter, making the signal available at the output of the stage. Thus a signal progresses to the right by one half stage on each half cycle of the clock, or by one stage per clock cycle.

Because the shift register is dependent upon stored charge which is subject to slow decay, there is a minimum frequency at which it will operate; reliable operation can be expected at frequencies as low as 5 kHz.

COS/MOS dynamic shift registers have all the advantages of other COS/MOS devices, including low power dissipation, high noise immunity, and wide operating voltage range. In addition, COS/MOS devices are superior in two important ways to the single-channel (p-MOS and n-MOS) dynamic shift registers. First, the COS/MOS device easily generates the two-phase clock signals required internal to itself with just one supply voltage. Second, TTL and DTL logic compatibility is maintained on all inputs and outputs with one supply voltage.

PROTECTION CIRCUIT

The standard input protection device used in RCA COS/MOS integrated circuits is shown in Fig. 274. Protection is required to prevent damage to the MOS input gates that could result from careless handling and/or testing prior to final installation. Fig. 274 illustrates the posi-

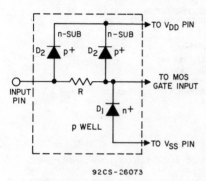

92CS-26073

Fig. 274—COS/MOS IC protection circuit showing diode clamps.

tive, built-in protection afforded by the diode clamps in a COS/MOS circuit; this approach is in contrast to the widely varying zener-diode breakdown protection used in bipolar circuits.

The breakdown voltage of an MOS gate oxide is in the order of 100 volts; the dc resistance is in the order of 10^{12} ohms. In contrast to semiconductor diodes in which the breakdown limit can be tested any number of times without damaging the device, the MOS gate oxide is shorted as a result of only one voltage excursion to the breakdown limit. Because of the extremely high resistance of the gate oxide, even a very-low-energy source (such as a static charge) is capable of developing this breakdown voltage.

The input resistance R, as shown in Fig. 274, is nominally between 1 and 3 kilohms. This value, in conjunction with the capacitances of the gate and the associated protective diodes, integrates and clamps the

device voltages to a safe level. Input circuits can be designed to limit extraneous voltages to safe levels under all operating conditions. Because of the low RC time constants of these circuits, they have no noticeable effect on circuit speed and do not interfere with logic operation.

In circuits that contain gate-protection circuits, the power-supply voltage V_{DD} should not be turned off while a signal from a low-impedance pulse generator is applied at any of the inputs to the COS/MOS IC. Therefore, if, in any system design, any input excursion is expected to exceed $+V_{DD}$ or fall below $-V_{SS}$, the current through the input diodes should be limited to 10 milliamperes to assure safe operation.

Fig. 275 shows the over-all protection circuit (interconnected with a COS/MOS inverter) that is incorporated into all RCA COS/MOS integrated circuits. In addition to the basic input protection discussed in the preceding paragraphs, all inverter outputs and all transmission-gate inputs and outputs are fully protected by substrate diodes (D3, D4, and D5), as shown in Fig. 275.

CLASSIFICATION OF RCA COS/MOS INTEGRATED CIRCUITS

The RCA CD4000A series of COS/MOS integrated circuits provides the equipment designer with a very comprehensive line of circuits for a wide variety of logic-system applications. This series of circuits includes arithmetic devices, counter/dividers, decoders, flip-flops, gates, hex-buffers, multiplexers, shift registers, and latches. These circuits feature low power requirements, wide operating voltage range, high noise immunity, fully protected inputs, excellent temperature stability, and high fanout capabilities. Table XXIX lists COS/MOS circuits

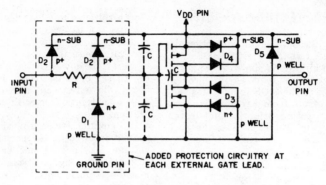

DIODE BREAKDOWNS

D_1 = n$^+$ TO p WELL   25 V MAX
D_2 = p$^+$ TO n SUB   50 V
D_3 = n SUB TO p WELL   100 V
R = NORMAL p$^+$ DIFFUSION IN n SUB ISOLATION

92CS-26074

Fig. 275—Gate-oxide protection circuit used in COS/MOS integrated circuits.

Table XXIX—RCA CD4000A-Series COS/MOS Integrated Circuits

Gates		Data Bulletin File No.
CD4000A	Dual 3-Input NOR Gate Plus Inverter	479
CD4001A	Quad 2-Input NOR Gate	479
CD4002A	Dual 4-Input NOR Gate	479
CD4011A	Quad 2-Input NAND Gate	479
CD4012A	Dual 4-Input NAND Gate	479
CD4019A	Quad AND-OR Select Gate	479
CD4023A	Triple 3-Input NAND Gate	479
CD4025A	Triple 3-Input NOR Gate	479
CD4030A	Quad Exclusive-OR Gate	503
CD4037A	Triple AND-OR Bi-Phase Pairs	576
CD4048A	Expandable 8-Input Gate	636
Flip-Flops		
CD4013A	Dual D with Set/Reset Capability	479
CD4027A	Dual J-K with Set/Reset Capability	503
CD4047A	Monostable/Astable Multivibrator	623
Latches		
CD4042A	Quad Clocked D Latch	589
CD4043A	NOR R/S Latch (3 Output States)	590
CD4044A	NAND R/S Latch (3 Output States)	590
Arithmetic Devices		
CD4008A	Four-Bit Adder, Parallel Carry-Out	479
CD4032A	Triple Serial Adder, Internal Carry (Neg. Logic)	503
CD4038A	Triple Serial Adder, Internal Carry (Pos. Logic)	503
CD4057A	LSI 4-Bit Arithmetic Logic Unit	635

Table XXIX—RCA CD4000A-Series COS/MOS Integrated Circuits (cont'd)

currently included in the CD4000A series and indicates the logic function for which each circuit is normally used. Complete data on these circuits can be found in RCA Solid-State DATABOOK SSD-203, or in the data bulletins listed in Table XXIX.

RATINGS AND CHARACTERISTICS

The RCA family of COS/MOS digital integrated circuits includes a standard line of devices (the CD4000A series) designed to operate from voltage supplies of 3 to 15 volts. Each pellet in the series is supplied in both ceramic and plastic packages so that all devices included in the series are available in two different operating temperature ranges. Devices supplied in the ceramic flat-pack (CD4000AK-series), dual-in-line ceramic (CD4000AF-series), dual-in-line weld-seal ceramic (CD4000AD-series), and TO-5-style (CD4000AT-series) packages operate over the temperature range of $-55°C$ to $+125°C$. Devices supplied in the dual-in-line plastic (CD4000AE-series) packages operate over a temperature range of $-40°C$ to $+85°C$. The storage-temperature range for all packages is from $-65°C$ to $+150°C$. Table **XXX** lists the maximum ratings for the RCA CD4000A series of COS/MOS digital integrated circuits.

The ratings shown in Table **XXX** are based on the **Absolute Maximum System** and are limiting values of operating and environmental conditions that should not be exceeded by any circuit of a specified type under any conditions of operation. Effective use of these ratings requires close control of supply-voltage variations, component variations, equipment-control adjustment, load variations, and environmental conditions.

COS/MOS devices have many performance characteristics that are not found in other integrated-circuit technologies. These unique characteristics, which are given in the COS/MOS data bulletins listed in Table **XXIX**, include the following:

Quiescent power dissipation. The quiescent (dc) power dissipation of COS/MOS devices is extremely low; the static or standby power is normally in the microwatt range.

Output drive current. COS/MOS devices will normally drive inputs to other COS/MOS devices; these other devices appear as purely capacitive loads. Therefore, except during switching, no output **source** or **sink current** flows. In some instances, however, COS/MOS devices are directly interfaced with other logic forms. Typical output sink and source currents are a few milliamperes.

Table XXX—Maximum Ratings (Absolute Maximum Values)

	CD4000AK,AD,AT,AF	CD4000AE
Recommended Operating Voltage Range (V_{DD}-V_{SS}) (Volts)	3 to 15	
DC Supply Voltage (Volts)	-0.5 to $+15$	
Dissipation per package (milliwatts)	200	
Operating Temperature Range (°C)	-55 to $+125$	-40 to $+85$
All Inputs*	$V_{SS} \leq V_{IN} \leq V_{DD}$	
Storage Temperature Range (°C)	-65 to $+150$	

* For types CD4009A and CD4010A, $V_{SS} \leq V_{CC} \leq V_{DD}$

Noise immunity. Noise immunity data are expressed in terms of absolute values referenced above or below the normal-state level for which the logic level will change state. For example, if the device input is at a logic level of $+10$ volts (V_{DD} of 10 volts) and the guaranteed noise immunity for the device is 3 volts, the device will not change state when the input level drops from $+10$ to 7 volts because of noise. Similarly, if the input is at 0 volts (logic 0) under the same conditions, the device will not change state when noise signals of 3 volts appear at the input. The interfering or noise voltage may be of either a slow drift variety (dc), transient in nature (ac), or a combination of both. Typical noise immunity is $\pm$ 4.5 volts at $V_{DD} = 10$ V, and $\pm$ 2.25 volts at $V_{DD} = 5$ V.

Propagation delay. The propagation delay in COS/MOS devices depends upon the load that is driven and upon the supply voltage. Typical values are tenths of microseconds.

Clock frequency. For sequential circuits such as flip-flops, counters, and registers, a maximum clock frequency (**toggle rate**) is given; values range from 1 to 10 MHz. Clock rise-time and fall-time are also listed. Maximum values are on the order of 5 to 15 microseconds.

Dynamic power consumption. Power dissipation in COS/MOS devices is a function of load capacitance (ac fanout), supply voltage, and switching frequency. Typical values range from less than a microwatt to a fraction of a watt.

USE OF COS/MOS DEVICES IN EQUIPMENT DESIGN

The following paragraphs outline some guidelines for the design of equipment that uses RCA COS/MOS integrated circuits. The equipment operating conditions have an important bearing on the reliability and life of an integrated circuit;

therefore, the equipment designer must make sure that the system does not impose excessive electrical stresses that may adversely affect the life or performance of any device and thereby degrade the reliability of the system.

Maximum Ratings

The first rule in the design of a COS/MOS logic system is to make certain that no maximum rating of a COS/MOS integrated circuit will be exceeded under any condition of operation. The power-supply voltage should never be applied to a COS/MOS integrated circuit in the reverse polarity. Application of a reverse voltage greater than 0.5 volt may damage the integrated circuit. Input signals should not exceed the power-supply range unless special precautions are taken to limit current through the input protective diodes. The V_{CC} terminal of the CD4009A and CD4010A circuits must never be more positive than the V_{DD} voltage, although this terminal can be connected to the same voltage value if these circuits are to be used as non-level-shifting buffers.

Power supply regulation and turn-on and turn-off transients must not exceed the maximum supply rating of 15 volts. Reliable designs should normally not use power supply voltages greater than 12 to 13 volts.

A worse-case design for high-reliability equipment should be based on the maximum or minimum rating specified for a device, rather than on typical values.

Unused Inputs

All unused input leads must be connected to either V_{SS} or V_{DD}, whichever is appropriate for the logic circuit involved. A floating input on a high-current type, such as the CD4009A, CD4010A, CD4041A, CD4049A, and CD4050A, not only can result in faulty logic operation, but can cause the maximum power dissipation of 200 milliwatts to be exceeded and may result in damage to the device. Inputs to these types on printed-circuit boards that may temporarily become unterminated should have a pull-up resistor to V_{SS} or V_{DD}. A useful range of resistor values is from 0.2 to 1 megohm.

Input Signals

Signals must not be applied to the inputs while the device power supply is off unless the input current source is limited to a steady-state value of less than 5 milliamperes.

Handling and Input/Output Protection

All COS/MOS gate inputs have a resistor/diode gate protection network. All transmission gate inputs and all outputs have diode protection provided by inherent p-n junction diodes that limit input signals to worst-case values of less than 20 to 30 volts. These diode networks at input and output interfaces fully protect COS/MOS devices from gate-oxide failure (80- to 100-volt limit) for static discharges or signal voltages up to 1 to 2 kilovolts under most transient or low-current conditions.

Output Short Circuits

Shorting of outputs to V_{SS} or V_{DD} can damage many of the higher-output-current COS/MOS types, such as the CD4007A, CD4009A, CD4010A, CD4041A, CD4049A, and CD4050A. In general, these types can all be safely shorted for supplies up to 5 volts, but will be damaged (depending on type) at higher power-supply voltages. For cases in which a short-circuit load, such as the base of a p-n-p or n-p-n bipolar transistor, is directly driven, the device output characteristics given in the published data should be consulted to determine the requirements for safe operation below 200 milliwatts.

DC Power Supplies

A dc power supply converts the power from an ac line to direct current and a steady voltage of a desired value. The ac input power is first rectified to provide a pulsating dc, and then filtered to produce a smooth voltage. Finally, the voltage may be regulated to maintain a constant output level despite fluctuations in the power-line voltage or circuit loading. The rectification, filtering, and regulation steps in a dc power supply are illustrated in Fig. 276.

A dc power supply need not include all three of the elements shown in Fig. 276. Electroplating supplies and battery chargers require only rectification of the ac, and broadcast receivers and phonograph amplifiers need only the rectifier and filter steps. However circuits such as oscillators, high-gain amplifiers, and low-voltage logic, which have exacting frequency, stability, or output requirements, can be critically affected by variations in dc supply voltages. Therefore, some type of regulation is frequently required to prevent significant changes in the output of a dc power supply as a result of line-voltage fluctuations or variations in circuit loading.

RECTIFICATION

The most suitable type of rectifier circuit for a particular application depends on the dc voltage and current requirements, the amount of rectifier "ripple" (undesired fluctuation in the dc output caused by an ac component) that can be tolerated in the circuit, and the type of ac power available.

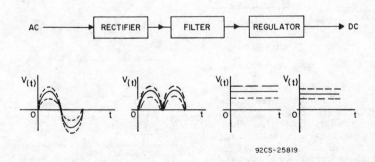

92CS-25819

Fig. 276—Simplified block diagram of a regulated dc power supply. Waveforms show effects of rectification, filtering, and regulation. (Dashed lines indicate voltage fluctuations as a result of input variations.)

Figs. 277 through 283 show seven basic rectifier configurations. (Filters used to smooth the rectifier output are not shown for these circuits, but are discussed later.) These illustrations also include the output-voltage waveforms for the various circuits and the current waveforms for each individual rectifier in the circuits. Ideally, the voltage waveform should be as flat as possible (i.e., approaching almost pure dc). A flat curve indicates a peak-to-average voltage ratio of one.

The single-phase half-wave circuit shown in Fig. 277 delivers only one phase of current for each cycle of ac input voltage. As shown by the current waveform, the single rectifier conducts the entire current flow. This type of circuit contains a very high percentage of output ripple.

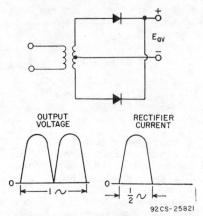

Fig. 278—Single-phase full-wave circuit with center-tapped power transformer.

The single-phase full-wave bridge circuit shown in Fig. 279 uses four rectifiers, and does not require the use of a transformer center-tap. It can be used to supply twice as much output voltage as the circuit of Fig. 278 for the same transformer voltage, or to expose the individual rectifiers to only half as much peak

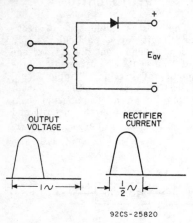

92CS-25820

Fig. 277—Single-phase half-wave circuit.

Fig. 278 shows a single-phase full-wave circuit that operates from a center-tapped high-voltage transformer winding. This circuit has a lower peak-to-average voltage ratio than the circuit of Fig. 277 and about 65 per cent less ripple. Only 50 per cent of the total current flows through each rectifier. This type of circuit is widely used in television receivers and large audio amplifiers.

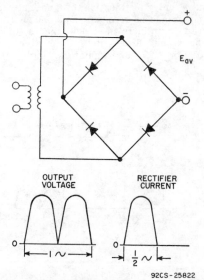

Fig. 279—Single-phase full-wave circuit without center-tapped power transformer (i.e., bridge-rectifier circuit).

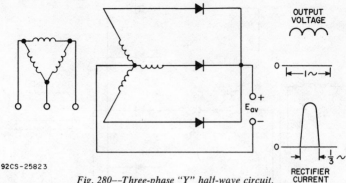

Fig. 280—Three-phase "Y" half-wave circuit.

92CS-25823

reverse voltage for the same output voltage. Only 50 per cent of the total current flows through each rectifier. This type of circuit is popular in amateur transmitter use.

The three-phase circuits shown in Figs. 280 through 283 are usually found in heavy industrial equipment such as high-power transmitters. The three-phase Y half-wave circuit shown in Fig. 280 uses three rectifiers. This circuit has considerably less ripple than the circuits discussed above. In addition, only one-third of the total output current flows through each rectifier.

Fig. 281 shows a three-phase full-wave bridge circuit which uses six rectifiers. This circuit delivers twice as much voltage output as the circuit of Fig. 280 for the same trans-

former conditions. In addition, this circuit, as well as those shown in Figs. 282 and 283, has an extremely small percentage of ripple.

In the six-phase "star" circuit shown in Fig. 282, which also uses six rectifiers, the least amount of the total output current (one-sixth) flows through each output rectifier. The three-phase double-Y and interphase transformer circuit shown in Fig. 283 uses six half-wave rectifiers in parallel. This arrangement delivers six current pulses per cycle and twice as much output current as the circuit shown in Fig. 280.

Table **XXXI** lists voltage and current ratios for the circuits shown in Figs. 277 through 283 for resistive or inductive loads. These ratios apply for sinusoidal ac input volt-

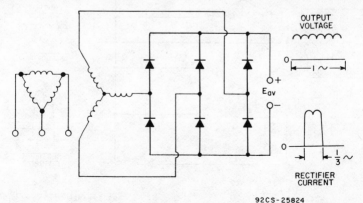

92CS-25824

Fig. 281—Three-phase "Y" full-wave circuit.

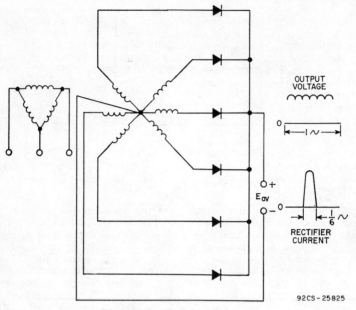

OUTPUT
VOLTAGE

RECTIFIER
CURRENT

92CS-25825

Fig. 282—Six-phase "star" circuit.

ages. It is generally recommended that inductive loads rather than re-sistive loads be used for filtering of rectifier current, except for the cir-

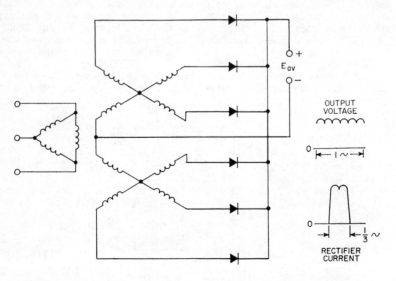

OUTPUT
VOLTAGE

RECTIFIER
CURRENT

92CS-25826

Fig. 283—Three-phase "double-Y" and interphase-transformer circuit.

cuit of Fig. 277. Current ratios given for inductive loads apply only when a filter choke is used between the output of the rectifier and any capacitor in the filter circuit. Values shown do not take into consideration voltage drops which occur in the power transformer, the silicon rectifiers, or the filter components under load conditions. When a particular rectifier type has been selected for use in a specific circuit, Table XXXI can be used to determine the parameters and characteristics of the circuit.

In Table XXXI, all ratios are shown as functions of either the average output voltage E_{av} or the average dc output current I_{av}, both of which are expressed as unity for each circuit. In practical applications, the magnitudes of these average values will, of course, vary for the different circuit configurations.

FILTERING

Filter circuits are generally used to smooth out the ac ripple in the output of a rectifier circuit. Filters consist of two basic types, inductive "choke" input and capacitive input.

Table XXXI—Voltage and Current Ratios for Rectifier Circuits Shown in Figs. 277 Through 283. Fig. 277 Uses a Resistive Load, and Figs. 278 Through 283 Use Resistive or Inductive Loads

CIRCUIT RATIOS	Fig. 277	Fig. 278	Fig. 279	Fig. 280	Fig. 281	Fig. 282	Fig. 283
Output Voltage:							
Average	E_{av}	E_{av}	E_{av}	E_{av}	E_{av}	E_{av}	E_{av}
Peak (x E_{av})	3.14	1.57	1.57	1.21	1.05	1.05	1.05
RMS (x E_{av})	1.57	1.11	1.11	1.02	1.00	1.00	1.00
Ripple (%)	121	48	48	18.3	4.3	4.3	4.3
Input Voltage (RMS):							
Phase (x E_{av})	2.22	1.11*	1.11	0.855•	0.428•	0.74•	0.855•
Line-to-Line (x E_{av})	2.22	2.22	1.11	1.48	0.74	1.48†	1.71‡
Average Output (Load)							
Current	I_{av}	I_{av}	I_{av}	I_{av}	I_{av}	I_{av}	I_{av}
RECTIFIER CELL RATIOS							
Forward Current:							
Average (x I_{av})	1.00	0.5	0.5	0.333	0.333	0.167	0.167
RMS (x I_{av}):							
resistive load	1.57	0.785	0.785	0.587	0.579	0.409	0.293
inductive load	—	0.707	0.707	0.578	0.578	0.408	0.289
Peak (x I_{av}):							
resistive load	3.14	1.57	1.57	1.21	1.05	1.05	0.525
inductive load	—	1.00	1.00	1.00	1.00	1.00	0.500
Ratio peak to average:							
resistive load	3.14	3.14	3.14	3.63	3.15	6.30	3.15
inductive load	—	2.00	2.00	3.00	3.00	6.00	3.00
Peak Reverse Voltage:							
x E_{av}	3.14	3.14	1.57	2.09	1.05	2.42	2.09
x E_{rms}	1.41	2.82	1.41	2.45	2.45	2.83	2.45

* to center tap   • to netural   † maximum value   ‡ maximum value, no load

Combinations and variations of these types are often used; some typical filter circuits are shown in Fig. 284.

The simplest of these filtering circuits is the capacitive input. This type of filtering is most often used in low-current circuits in which a

fairly large amount of ripple can be tolerated. Such circuits are usually single-phase, half-wave or full-wave. In this type of filter, the capacitor charges up to approximately the peak of the input voltage on each half-cycle that a rectifier conducts. The current into the load

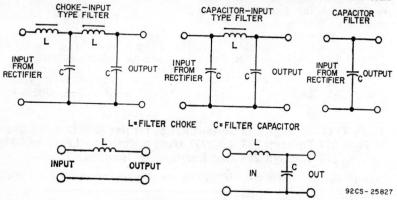

Fig. 284—Typical filter circuits.

is then supplied from the capacitor rather than from the power supply until the point in the next half-cycle when the input voltage again equals the voltage across the capacitor. A rectifier circuit that uses a smoothing capacitor and the voltages involved are shown in Fig. 285.

Higher average dc output voltages and currents can be obtained from this type of circuit by the use of larger capacitors. A larger capacitor also tends to reduce the ripple. However, care must be taken that the capacitor is not so large that excessive peak and rms currents cause overheating of the rectifier. The effects of capacitor loading on a rectifier circuit are discussed in detail in the section on **Capacitive-Load Circuits.**

The next simplest filter is the inductive input filter. This filter performs the same function as a capacitive input filter in that it smooths the load current by storing energy during one part of the cycle and releasing it to the load during another part of the cycle.

However, the inductor acts in a different way by extending the time during which current is drawn from a rectifier. When a smoothing inductor is used in series with a full-

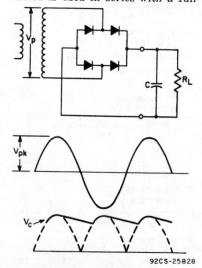

Fig. 285—Bridge-rectifier circuit with capacitor input filter.

wave rectifier circuit, the conduction period of each rectifier may be extended so that conduction does not stop in one rectifier until the other rectifier starts conducting. As a result of this spreading action, any increase in inductance to reduce ripple results in a decrease in the average output voltage and current.

The smoothing capabilities of capacitors and inductors can be combined as shown in the other filters of Fig. 284 to take advantage of the best feature of each. Filters which provide maximum output and minimum ripple and use reasonably small components can thus be designed.

CAPACITIVE-LOAD CIRCUITS

When rectifiers are used in circuits with capacitive loads, the rectifier current waveforms may deviate considerably from their true sinusoidal shape. This deviation is most evident for the peak-to-average-current ratio, which is somewhat higher than that for a resistive load. For this reason, capacitive-rating calculations are generally more complicated and time-consuming than those for resistive-load rectifier circuits. However, the simplified rating system described below allows the designer to calculate the characteristics of capacitive-load rectifier circuits quickly and accurately.

Fig. 286 shows typical half-wave and voltage-doubling rectifier circuits that use capacitive loads. In such circuits, the low forward voltage drop of the silicon rectifiers may result in a very high surge of current when the capacitive load is first energized. Although the generator or source impedance may be high enough to protect the rectifier, additional resistance must be added in some cases. The sum of this resistance plus the source resistance is referred to as the total limiting resistance R_S. The magnitude of R_S required for protection of the rectifier may be calculated from surge

rating charts such as those shown in Figs. 287 and 288. Each point of these curves defines a surge rating by indicating the maximum time for which the device can safely carry a specific value of rms current.

With a capacitive load, maximum surge current occurs if the circuit is switched on when the input voltage is near its peak value. When the time constant R_SC of the surge loop

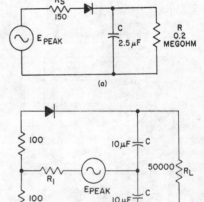

Fig. 286—Typical rectifier circuits using capacitive loads: (a) half-wave rectifier circuit; (b) voltage doubler.

is much smaller than the period of the input voltage, the peak current I_{peak} is equal to the peak voltage E_{peak} divided by the limiting resistance R_S, and the resulting surge approximates an exponentially decaying current with the time constant R_SC.

Surge-current ratings for rectifiers are often given in terms of the rms value of the surge current and the time duration t of the surge. For rating purposes, the surge duration t is defined by the time constant R_SC. The rms surge current

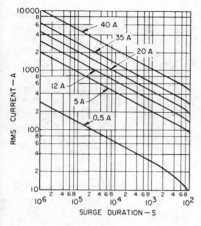

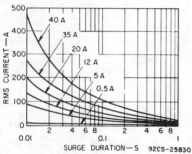

Fig. 287—Universal surge rating charts for RCA rectifiers.

I_{rms} is then approximated by the following equations:

$$I_{rms} = 0.7 \ (E_{peak}C/R_SC)$$

$$= 0.7 \ (E_{peak}C/t)$$

and

$$I_{rms}t = 0.7 \ E_{peak} \ C$$

where E_{peak} and C are the values specified by the circuit design. This equation may then be plotted on the surge-rating chart, which has axes labeled I_{rms} and t. Because R_SC is equal to t, any given value of R_S defines a specific time t, and hence a specific point on the plot of the equation for $I_{rms}t$. However, R_S must be large enough to make this point fall below the rating curve for the rectifier used.

The following example illustrates the use of this simplified procedure for the half-wave rectifier circuit shown in Fig. 286(a), which has a frequency f of 60 Hz and a peak input voltage E_{peak} of 4950 volts. The values shown for E_{peak} and C are substituted in the equation for $I_{rms}t$ as follows:

$$I_{rms}t = 0.7 \ (4950) \ (2.5 \times 10^{-6})$$

$$= 0.0086$$

When this value is plotted on the surge-rating chart of Fig. 289, the resulting line intersects the rectifier rating curve at 3.3×10^{-4} second. The minimum limiting resistance which affords adequate surge proction is then calculated as follows:

$$R_SC \geq 3.3 \times 10^{-4}$$

$$R_S \geq \frac{3.3 \times 10^{-4}}{2.5 \times 10^{-6}} = 132 \ \text{ohms}$$

Therefore the value of 150 ohms shown for R_s in Fig. 286(a) provides adequate surge-current protection for the rectifier.

The design of rectifier circuits having capacitive loads often requires the determination of rectifier current waveforms in terms of average, rms, and peak currents. These

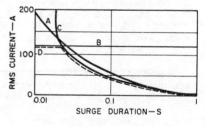

92CS-25831

Fig. 288—Typical coordination chart for determination of fusing requirements: Curve A—surge rating for 20-ampere rectifier; Curve B—expected surge current in half-wave circuit; Curve C—opening characteristics of protective device; Curve D—resulting surge current in modified circuit.

waveforms are needed for calculation of circuit parameters, selection of components, and matching of circuit parameters with rectifier ratings. Although actual calculation of rectifier current is a rather lengthy process, the current-relationship charts shown in Figs. 290 and 291

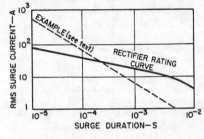

Fig. 289—Surge rating chart for stack rectifier CR210.

can be used to determine peak or rms current if the average current is known, or vice versa.

The ratios of peak-to-average current (I_{peak}/I_{av}) and rms-to-average current (I_{rms}/I_{av}) are shown in Fig. 289 as functions of the circuit constants $n\omega CR_L$ and R_S/nR_L. The quantity ωCR_L is the ratio of resistive-to-capacitive reactance in the load, and the quantity R_S/R_L is the ratio of the limiting resistance to the load resistance. The factor n, referred to as the "charge factor," is simply a multiplier which allows the chart to be used for various circuit configurations. The value of n is equal to unity for half-wave circuits, to 0.5 for doubler circuits, and to 2 for full-wave circuits. (These values actually represent the relative quantity of charge delivered to the capacitor on each cycle.)

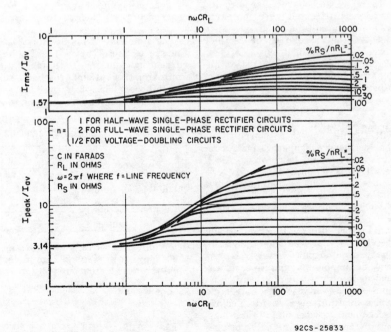

Fig. 290—Relationship of peak, average, and rms rectifier currents in capacitor-input circuits.

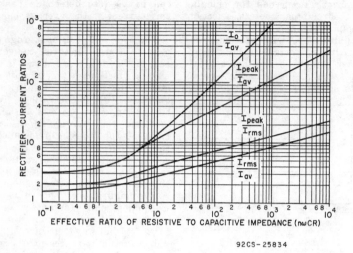

92CS-25834

Fig. 291—Forward-current ratios for rectifiers in capacitor-input circuits in which R_s is much less than R_L.

In many silicon rectifier circuits, R_S may be neglected when compared with the magnitude of R_L. In such circuits, the calculation of rectifier currents is simplified by use of Fig. 291, which gives current ratios under the limitation that R_S/R_L approaches zero. Even if this condition is not fully satisfied, the use of Fig. 291 merely indicates a higher peak and higher rms current than will actually flow in the circuit, i.e., the rectifiers will operate more conservatively than calculated. As a result, this simplified solution can be used whenever a rough approximation or a quick check is needed on whether a particular rectifier will fit a specific application. When more exact information is needed, the chart of Fig. 290 should be used.

Average output voltage E_{av} is another important quantity in capacitor-input rectifier circuits because it can be used to determine average output current I_{av}. The relationships between input and output voltages for half-wave, voltage-doubler, and full-wave circuits are shown in Figs. 292, 293, and 294, respectively. Fig. 295 shows curves

of output ripple voltage (as a percentage of E_{av}) for all three types of circuits.

The following example illustrates the use of these curves in rectifier-current calculations. Both exact and approximate solutions are given. For the half-wave circuit of Fig. 286(a), the resistive-to-capacitive reactance ωCR_L is given by

$$\omega CR_L = 2\pi \times 60 \times 2.5 \times 10^{-6} \times 200{,}000$$
$$= 189$$

For an exact solution using Fig. 290, the ratio of R_S to R_L is first calculated as follows:

$$\frac{R_S}{R_L} = \frac{150}{200{,}000} = 0.075$$

The values for ωCR_L and R_S/R_L are then plotted in Fig. 292 to determine the average output voltage E_{av} and the average output current I_{av} as follows:

$E_{av}/E_{peak} = 98$ per cent
$E_{av} = 0.98 \times 4950 = 4850$ volts
$I_{av} = E_{av}/R_L$
$I_{av} = 4850$ volts$/200{,}000$ ohms
$\quad\ = 24.2$ milliamperes

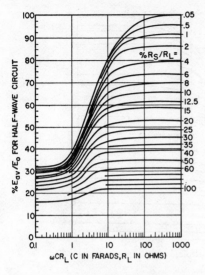

Fig. 292—Relationship of applied ac peak
voltage to dc output voltage in half-wave
capacitor-input circuit.

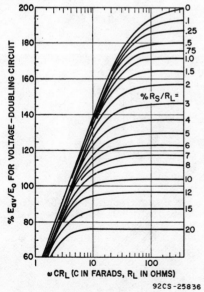

Fig. 293—Relationship of applied ac peak
voltage to dc output voltage in capacitor-
input voltage-doubler circuit.

This value of I_{av} is then substituted
in the ratio of I_{rms}/I_{av} obtained
from Fig. 290, and the exact value
of rms current I_{rms} in the rectifier
is determined as follows:

$$I_{rms}/I_{av} = 4.4$$
$$I_{rms} = 4.4 \times 24.2$$
$$= 107 \text{ milliamperes}$$

For a simplified solution using
Fig. 291, it is assumed that the average output current I_{av} is approximately equal to the peak input
voltage E_{peak} divided by the load resistance R_L, as follows:

$$I_{av} = E_{peak}/R_L$$
$$I_{av} = 4950/200,000$$
$$= 24.7 \text{ milliamperes}$$

This value of I_{av} is then substituted
in the ratio of I_{rms}/I_{av} obtained
from Fig. 291, and the approximate
rms current is determined, as follows:

$$I_{rms}/I_{av} = 5.7$$
$$I_{rms} = 5.7 \times 24.7$$
$$= 141 \text{ milliamperes}$$

Current-versus-temperature ratings for rectifiers are usually given
in terms of average current for a resistive load with 60-Hz sinusoidal
input voltage. When the ratio of
peak-to-average current becomes
higher (as with capacitive loads),
however, junction heating effects become more and more dependent on
rms current rather than average
current. Therefore, capacitive-load
ratings should be obtained from a
curve of rms current as a function
of temperature. Because the ratio
of rms-to-average current for the
rated service is 1.57 (as shown by
I_{rms}/I_{av} at low ωCR on Figs. 290 and
291), the current axis of the average-current rating curves for a
sinusoidal source and resistive
load can be multiplied by 1.57 to
convert the curves to rms rating
curves.

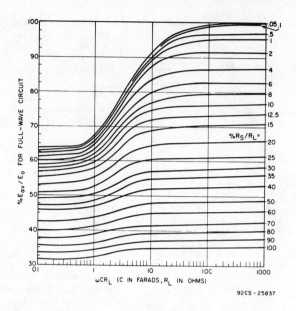

Fig. 294—Relationship of applied ac peak voltage to dc output voltage in full-wave capacitor-input circuit.

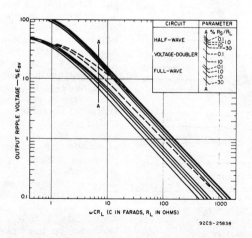

Fig. 295—RMS ripple voltage in capacitor-input circuits.

REGULATION

The regulation of a dc power supply is usually accomplished by some type of feedback circuit that senses any change in the dc output and develops a control signal to cancel this change. As a result, the output is maintained essentially constant. The nature of the control exercised by the feedback circuit (regulator) is determined by the type of circuit arrangement (series or shunt) and the mode of operation of the variable-resistance **pass element,** which is a transistor or an SCR. In a transistor regulator, the output voltage from the dc power supply is compared with a reference voltage, and the difference signal is amplified and fed back to the base of a pass transistor. In response to the feedback signal, the conduction of the pass transistor is varied, either linearly or as a switch, to regulate the output voltage. When the pass transistor can be operated at any point between cutoff and saturation, the regulator circuit is referred to as a **linear voltage regulator.** When the pass transistor operates only at cutoff or at saturation, the circuit is referred to as a **switching regulator.** All SCR regulators are by nature of SCR operation switching regulators.

All linear voltage regulators can be classified as either **series** or **shunt** types, as determined by the arrangement of the pass element with respect to the load. In a series regulator, as the name implies, the pass transistor is connected in series with the load. Regulation is accomplished by variation of the current through the series pass transistor in response to a change in the line voltage or circuit loading. In this way, the voltage drop across the pass transistor is varied and that delivered to the load circuit is maintained essentially constant. In the shunt regulator, the pass transistor is connected in parallel with the load circuit, and a voltage-dropping resistor is connected in series with this parallel network. If the load current tends to fluctuate, the current through the pass transistor is increased or decreased as required to maintain an essentially constant current through the dropping resistor.

Series Regulators

Fig. 296 shows a basic configuration for a linear series regulator which is representative of the type used in **voltage-regulating power supplies.** In this type of regulator, the series pass transistor is usually operated as an emitter-follower, and the control (error) signal is used to initiate the regulating action is applied to the base. The base control is developed by a dc amplifier. This amplifier, which is included in the feedback loop from the load circuit to the pass transistor, senses any change in the output voltage by comparison of this voltage with a known reference voltage. If an error exists, the error voltage is amplified and applied to the base of the pass transistor. The conduction of the pass transistor is then increased or decreased in response to the error signal input as required to maintain the output voltage at the desired value.

Fig. 297 shows the basic configuration for a linear regulator circuit used in **current-regulating power supplies.** This regulator senses the voltage across a resistor in series with the load, rather than the voltage across the load circuit as in the linear voltage regulator. Because the voltage across the series resistor is directly proportional to the load current, a detected error signal can be used to cancel any tendency for a change in load current from the desired value. Ideally, the linear current regulator has an infinite output impedance.

Performance Parameters—Most voltage-regulated power supplies are required to provide voltage

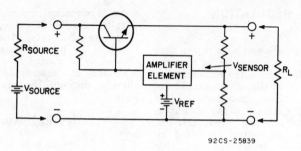

92CS-25839

Fig. 296—Basic series voltage regulator.

regulation for wide variation in load current. It is important, therefore, to specify the output impedance of the supply, $\triangle V_{out}/\triangle I_{out}$, over a large band of frequencies. This parameter indicates the ability of the power supply to maintain a constant output voltage during rapid changes in load. The output impedance of a typical voltage-regulated supply is normally less than 0.1 ohm at all frequencies below 2 kHz. Above this frequency, the impedance increases and may be as much as several ohms.

A power supply must continue to supply a constant voltage (or current) regardless of variations in line voltage. An index of its ability to maintain a constant output voltage or current during input variation is called the **line regulation** of the supply, which is defined as 100 (V_o'/V_o), or as the change in output voltage $\triangle V_o$, for a specified change in input voltage, expressed in per cent. Typical values of line regulation are less than 0.01 per cent.

Another important power-supply parameter is **load regulation,** which specifies the amount that the regulated output quantity (voltage or current) changes for a given change in the unregulated quantity. Load regulation is mainly a function of the stability of the reference source and the gain of the feedback network.

Foldback Current Limiting—Foldback current limiting is a form of protection against excessive current. If the load impedance is reduced to

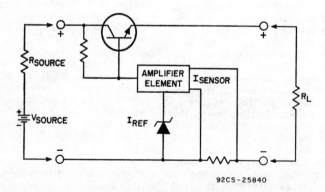

92CS-25840

Fig. 297—Basic series regulator modified for current sensing.

a value that would draw more than the predetermined maximum current, the foldback circuit reduces output voltage and thus reduces the current. Further reduction of load impedance causes further decrease of output voltage and current; therefore a regulated power supply that includes a foldback current-limiting circuit has the voltage-current characteristic shown in Fig. 298. The

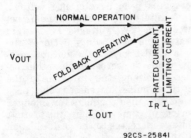

Fig. 298—*Output characteristic of a regulated power supply with foldback current-limiting protection for pass transistor.*

foldback process is reversible; if the load impedance is increased while the circuit is in the limiting mode, the output voltage and cur-

rent increase. When the current reaches the threshold level, the regulator is re-activated, and the power supply returns to normal operation.

A foldback current-limiting circuit is shown in Fig. 299. At low output current, transistor Q_5 is cut off; the value of resistor R_5 is selected so that Q_5 has zero bias when the output current reaches its rated value, I_R. When the load current I_{OUT} reaches the limiting value, I_X, Q_5 begins to conduct; current flows through resistor R_2, transistor Q_4 turns on, and the base-to-emitter voltage of transistor Q_3 is reduced. Therefore, the base-to-emitter voltage of transistor Q_2 decreases, and the output voltage of the power supply decreases. This decrease in the output voltage V_{OUT} reduces the output current, so that Q_5 continues to conduct at the same emitter current. If the load impedance is reduced further, Q_5 is driven even harder, and the output voltage and current decrease even further.

Shunt Regulators

Although shunt regulators are not as efficient as series regulators for

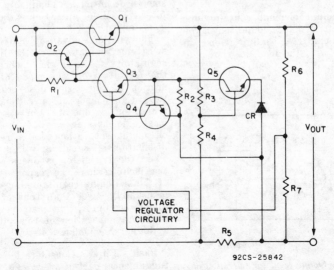

Fig. 299—*Foldback current-limiting circuitry in a series voltage regulator.*

most applications, they have the advantage of greater simplicity. The shunt regulator includes a shunt element and a reference-voltage element. The output voltage remains constant because the shunt-element current changes as the load current or input voltage changes. This current change is reflected in a change of voltage across the resistance R_1 in series with the load. A typical shunt regulator is shown in Fig. 300.

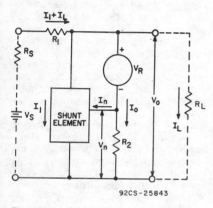

92CS-25843

Fig. 300—Basic configuration for a typical shunt regulator.

The shunt element contains one or more transistors connected in the common-emitter configuration in parallel with the load, as shown in Fig. 301.

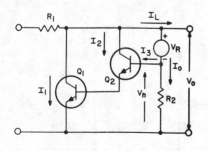

92CS-25844

Fig. 301—Shunt regulator circuit using two transistors as the shunt pass element.

Switching Regulator

Fig. 302 shows the basic configuration for a switching type of transistor voltage regulator. In this circuit, the pass transistor is connected in series with the load, and regulation of the output voltage is accomplished by on-off switching of the pass transistor through a feedback circuit. The feedback circuit samples the output voltage and compares it to a reference voltage. The difference (error signal) between the two voltages is used to control the on-off duty cycle of the pass transistor. If the output voltage tends to decrease below the reference voltage, the duration of the on-time pulse increases. The pass transistor then conducts for a longer period of time so that the output voltage increases to the desired level. If the output voltage tends to rise above the reference voltage, the duration of the on-time pulse decreases. The shorter conduction period of the pass transistor then results in a compensating decrease in output voltage. Some type of filter is required between the pass transistor and the load to obtain a smooth dc output. A commonly used filter consists of an LC network and a commutating diode.

The major advantage of the switching regulator over the linear regulator is the higher efficiency that results from the mode of operation of the series pass transistor. In this mode of operation, the transistor is operated in its two most efficient stages, either at cutoff or at saturation. As a result, dissipation is considerably less than when the transistor is operated in the linear region. The response time of the switching regulator, however, is usually slower than that of the linear regulator, but can be improved by operation of this circuit at higher frequencies.

Basic Filter Considerations—A fundamental part of every switching regulator is the filter. Fig. 303

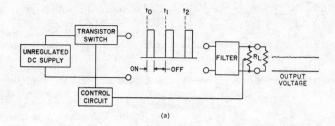

(a)

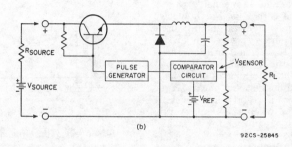

(b)

92CS-25845

Fig. 302—Basic configuration of switching type of transistor voltage regulator: (a) block diagram; (b) schematic diagram.

shows the various types of filters that can be used. Selection of the optimum filter for a power supply is based on the load requirements of the particular circuit and consideration of the basic disadvantages of the various types of filters.

A capacitive filter, shown in Fig. 303(a), has two primary disadvantages: (1) because large peak currents exist, R must be made large enough to limit peak transistor current to a safe value; and (2) the resistance in this circuit introduces loss.

An inductive filter, shown in Fig. 303(b), has three disadvantages: (1) The inductance may produce a destructive voltage spike when the transistor turns off. This problem, however, can be solved effectively by the addition of a commutating diode, as shown in Fig. 304. This diode commutates the current flowing through the inductor I_L when the transistor switches off.

(2) An abrupt change in the load resistance R_L produces an abrupt change in the output voltage because the current through the load I_L can-

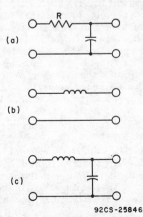

(a)

(b)

(c)

92CS-25846

Fig. 303—Typical filter circuits for use between pass element and load in a switching regulator: (a) capacitive filter; (b) inductive filter; (c) inductive-capacitive filter.

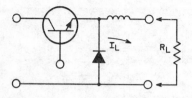

92CS-25847

Fig. 304—Use of inductance and commutating diode as filter network between pass transistor and load in switching voltage regulator.

not change instantaneously. (3) A third disadvantage of the inductive filter becomes evident during light loads. The energy stored in an inductor is proportional to the inductance and the square of the load current. Under light load conditions, the inductor must be much larger to provide a relatively constant current flow when the transistor is off than is required for a heavy load.

Most of the problems associated with either a capacitive filter or an inductive filter can be solved by use of a combination of the two as shown in Fig. 303(c). Because the energy stored in an inductor varies directly as current squared, whereas the energy output at constant voltage varies directly with current, it is not usually practical to design the inductor for continuous current at low current outputs. The addition of a capacitor eliminates the need for a continuous flow of current through the inductor. With the addition of a commutating diode, this filter has the following advantages.

(1) No "lossy" elements are required.

(2) The inductive element need not be oversized for light loads because the capacitance maintains the proper output voltage V_{out} if the inductive current becomes discontinuous.

(3) High peak currents through the transistor are eliminated by the use of the inductive element.

The means by which the switching regulator removes the line-frequency ripple component is illustrated in Fig. 305. The "on" time increases under the valley points of the unregulated supply and decreases under the peaks. The net result is to remove the 60-Hz component of ripple and introduce only ripple at the switching frequency, which is a relatively high frequency and easily filtered out.

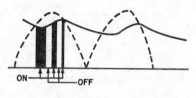

92CS-25848

Fig. 305—Effect of high-frequency switching of the switching regulator on power-supply ripple component.

Step-Down Switching Regulator— A transistor switching regulator can be used as a dc step-down transformer. This circuit is a very efficient means of obtaining a low dc voltage directly from a high-voltage ac line without the need for a step-down transformer. Fig. 306 shows a typical step-down transistor switching regulator. This regulator utilizes the dc voltage obtained from a rectified 117-volt line to provide a constant 60-volt supply.

Phase-Controlled SCR Regulated Power Supply

In a different type of pulse-width-modulated switching regulator, the pass element is switched at the line frequency and the conduction angle is varied to obtain the desired pulse width. This type of control is generally used with SCR's because turn-on of an SCR is simple and turn-off is accomplished automatically when the line voltage reverses.

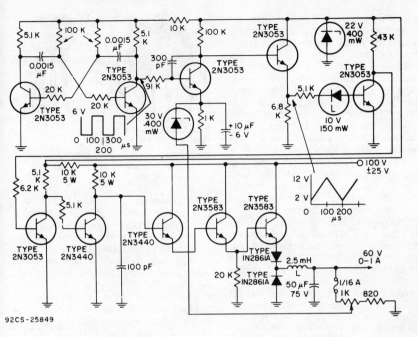

Fig. 306—Typical step-down transistor switching regulator.

Fig. 307 shows the circuit configuration for a regulated dc power supply that uses an SCR as a series pass element. This type of circuit is designed to provide approximately 125 volts, regulated to ± 3 per cent for both line and load. Ripple is less than 0.5 per cent rms.

The power supply is basically a half-wave phase-controlled rectifier. The 5-microfarad capacitor between the cathode and gate of the SCR charges up during half of each cycle and is discharged by the firing of the SCR. The firing angle of the SCR is advanced or retarded by the

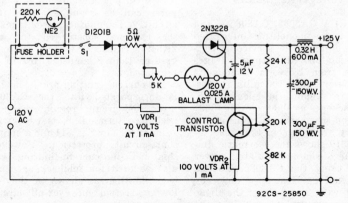

Fig. 307—SCR regulated power supply.

charging current flowing into the capacitor. Some of the current which would normally charge this capacitor is shunted by the collector of the control transistor. As the current in the control transistor increases, current is shunted around the capacitor, through the ballast lamp, so that the capacitor charging time is increased. As a result, the firing angle of the SCR is retarded, and a lower output voltage results.

The controlling voltage on the control transistor is derived from both the dc output and the line voltage in such a manner as to provide load and line regulation respectively. The voltage-dependent resistor (VDR_1) in the base circuit of the control transistor decreases resistance for an increase in line voltage and thus increases base current (and collector current) as line voltage is increased. In addition, the ballast lamp exhibits an increase in resistance with increasing line voltage, and thus tends to retard the firing angle of the SCR. Changes in dc output voltage that result from variations in load current are fed back to the base of the control transistor by a voltage divider at the input to the filter in the proper polarity to adjust collector current in a direction to compensate for changes in dc output voltage.

Integrated-Circuit Voltage Regulators

The RCA CA3085, CA3085A, and CA3085B are monolithic integrated circuits designed for service as voltage regulators at output voltages ranging from 1.7 volts to 46 volts with currents up to 100 milliamperes. They are supplied in 8-lead TO-5-style packages as shown in Fig. 308; block and schematic diagrams of the circuitry are shown in Fig. 309.

Fig. 310 shows the schematic diagram of a simple regulated power supply using the CA3085. The ac supply voltage is stepped down by T_1, full-wave rectified by the diode bridge circuit, and smoothed by the

**8-LEAD TO-5 with
Dual-In-Line Formed Leads**

**8-LEAD
TO-5**

Fig. 308—Packages for CA3085-series integrated-circuit voltage regulators.

large electrolytic capacitor C_1 to provide unregulated dc to the CA3085 regulator circuit. Frequency compensation of the error-amplifier is provided by capacitor C_2. Capacitor C_3 bypasses residual noise in the reference-voltage source, and thus decreases the incremental noise-voltage in the regulator circuit output.

The CA3085A and CA3085B have output current capabilities up to 100 milliamperes and the CA3085 up to 12 milliamperes without the use of external pass transistors. However, all the devices can provide voltage regulation at load currents greater than 100 milliamperes with the use of suitable external n-p-n transistors as shown in Fig. 311. In this circuit, the output current available from the regulator is increased in accordance with the h_{FE} of the external n-p-n pass transistor. Output currents up to 8 amperes can be regulated with this circuit. A Darlington power transistor can be substituted for the 2N5497 transistor when currents greater than 8 amperes are to be regulated.

The current-limiting provision that is incorporated into the CA3085 integrated-circuit regulators, connected to terminal 8, is used with resistor R_{SCP} in Fig. 311 for a simple short-circuit protection. However, this built-in current-limiting can also be used for foldback or snapback limiting, as shown in the following section on over-all circuit configurations.

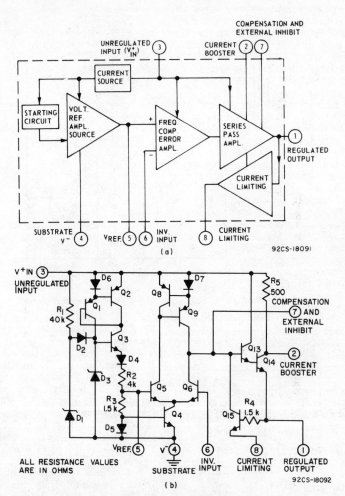

Fig. 309—Block diagram (a) and schematic diagram (b) of the CA3085-series integrated-circuit voltage regulator.

OVER-ALL CIRCUIT CONFIGURATIONS

Some representative dc power supply circuits have been described in the previous sections to illustrate the material under discussion. Several more supplies are described here.

60-Watt, 20-Volt Supply

A 60-watt 20-volt regulated power supply that uses integrated circuits and a single pass transistor is shown in Fig. 312. A 2N3055 transistor, driven by a 2N5781, extends the current capability of the CA3085 integrated-circuit voltage regulator; the overload protection provided by foldback current limiting permits

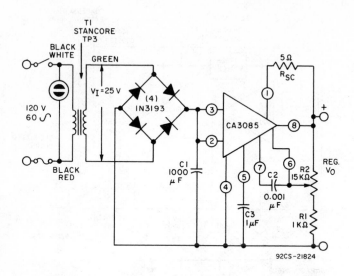

Fig. 310—Basic power supply that uses CA3085 integrated circuit.

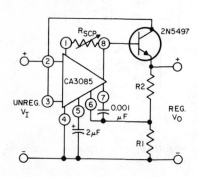

Fig. 311—High-current voltage regulator using CA3085 with an external n-p-n pass transistor.

operation of the pass transistor at a dissipation level close to its limit. The foldback circuit achieves high efficiency by use of a CA3030 integrated-circuit operational amplifier.

In this power supply, transformer T_1 and rectifiers D_1 through D_4 supply the raw dc power that is regu-

lated by pass transistor Q_1; this pass transistor is driven by driver Q_2, which is driven by the control circuit IC_1. Transformer T_2, rectifiers D_6, D_7, D_9, and D_{10}, and shunt regulator Q_4, provide positive and negative supplies for operational amplifier IC_2; this operational amplifier drives the current-limiting control Q_3. Output voltage is sensed at resistance string $(R_6 + R_{13})$, and load current is sensed by R_8.

For voltage regulation, the output voltage is sampled by the voltage divider $(R_6 + R_{13})$, and a portion is fed to terminal No. 6 (the inverting input) of the CA3085. (This portion is less than the 3.3-volt breakdown voltage of zener diode D_5; the zener is present only to protect the integrated circuit from accidental overvoltages.) If the output voltage decreases, the base-to-emitter voltage of Q_2 increases, as explained in the next paragraph. Therefore the pass transistor Q_1 is driven harder, and as a result the output voltage increases to its original value (minus the error dictated by the system gain).

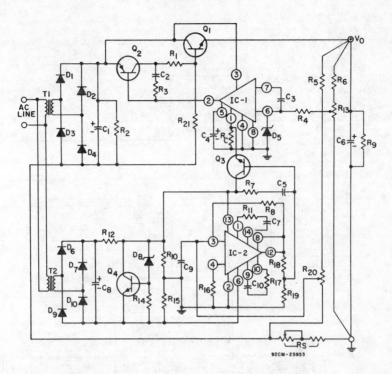

Fig. 312—Schematic diagram of 60-watt, 20-volt regulated power supply with foldback current limiting.

Parts List for Fig. 312

T_1	Signal Transformer Co., Part No. 24-4 or equivalent
T_2	Signal Transformer Co., Part No. 12.8-0.25 or equivalent
D_1-D_4	RCA-1N1344B
D_5	Zener Diode, 1N5225 (3.3 V)
D_6, D_7 $\}$	Power Rectifier, D1201B
D_9, D_{10}	Power Rectifier, D1201B
D_8	Zener Diode, 1N5242 (12 V)
C_1	5900 μF, 75 V, Sprague type 36D592-F075BC or equivalent
C_2	0.005 μF, ceramic disc, Sprague TGD50 or equivalent
C_3, C_7, C_{10}	50 pF, ceramic disc, Sprague 30GA-Q50 or equivalent
C_4	2 μF, 25 V, electrolytic, Sprague 500D G025BA7 or equivalent
C_5	0.01 μF, ceramic disc, Sprague TG510 or equivalent
C_6	500 μF, 50 V, Cornell-Dubilier No. BR500-50 or equivalent
C_8	250 μF, 25 V, Cornell-Dubilier BR 250-25 or equivalent
C_9	0.47 μF, film type, Sprague type 220P or equivalent
R_1	5 ohms, 1 watt, IRC type BWH or equivalent
R_2	1000 ohms, 5 watts, Ohmite type 200-5 1/4 or equivalent
R_3	1200 ohms, 1/2 watt, carbon, IRC type RC 1/2 or equivalent
R_4	100 ohms, 1/2 watt, carbon, IRC type RC 1/2 or equivalent
R_5	430 ohms, 2 watts, wire wound, IRC type BWH or equivalent
R_6	9100 ohms, 2 watts, wire wound, IRC type BWH or equivalent
R_7	470 ohms, 1/2 watt, carbon, IRC type RC 1/2 or equivalent
R_8	5100 ohms, 1/2 watt, carbon, IRC type RC 1/2 or equivalent
R_9, R_{14}	1000 ohms, 2 watts, wire wound, IRC type BWH or equivalent
R_{10}, R_{15}	250 ohms, 2 watts, 1%, wire wound, IRC type AS-2 or equivalent
R_{11}, R_{17}	1000 ohms, 1/2 watt, carbon, IRC type RC 1/2 or equivalent
R_{12}	82 ohms, 2 watts, IRC type BWH or equivalent
R_{13}	1000 ohms, potentiometer, Clarostat series U39 or equivalent
R_{16}	1200 ohms, 2 watts, wire wound, IRC type BWH or equivalent
R_{18}	510 ohms, 1/2 watt, carbon, IRC type RC 1/2 or equivalent
R_{19}	10,000 ohms, 1/2 watt, carbon, IRC type RC 1/2 or equivalent
R_{20}	300 ohms, potentiometer, Clarostat series U39 or equivalent
R_{21}	510 ohms, 3 watts, wire wound, Ohmite type 200-3 or equivalent

Parts list continued on page 274.

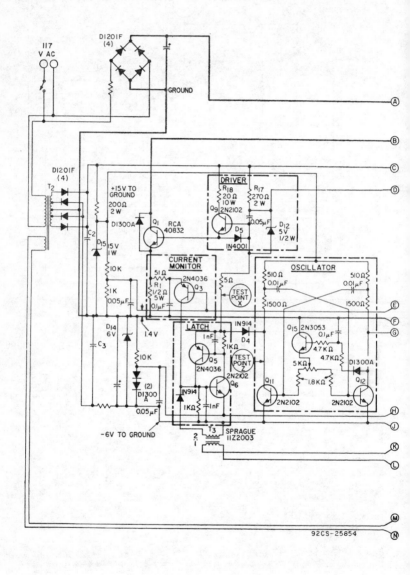

Fig. 313—Diagram of switching-regulator power supply (continued on page 273).

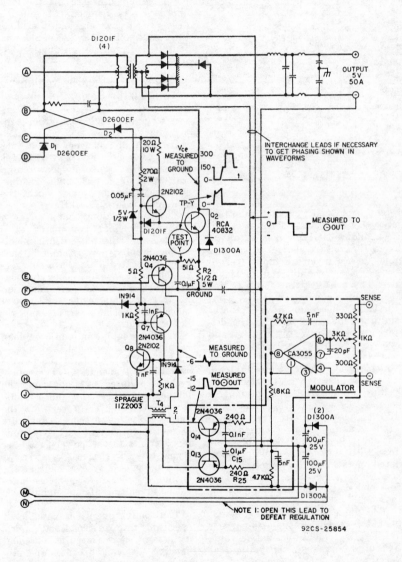

Fig. 313—Diagram of switching-regulator power supply (continued from page 272).

Parts List for Fig. 312 (cont'd from page 271)

R_C	240 ohms, 1%, wire wound, IRC type AS-2 or equivalent
R_S	(See text for fixed portion); 1 ohm, 25 watts, Ohmite type H or equivalent
IC_1	RCA-CA3085
IC_2	RCA-CA3030
Q_1	RCA-2N3055
Q_2	RCA-2N5781
Q_3, Q_4	RCA-40347

Miscellaneous

(1 Req'd)	Heat Sink, Delta Division Wakefield Engineering NC-423 or equivalent
(3 Req'd)	Heat Sink, Thermalloy #2207 PR-10 or equivalent
(1 Req'd)	8-pin socket Cinch #8-ICS or equivalent
(1 Req'd)	14-pin DiL socket, T.I., #IC 014ST-7528 or equivalent
(2 Req'd)	TO-5 socket ELCO #05-3304 or equivalent
	Vector Board #838AWE-1 or equivalent
	Vector Receptacle R644 or equivalent
	Chassis — As required
	Cabinet — As required
	Dow Corning DC340 filled grease

The foldback current limiting uses the CA3030 integrated circuit as a differential amplifier. A signal from the voltage divider R_{20} is applied to the inverting input (terminal No. 3) of the differential amplifier. The non-inverting input is tied to system ground through R_{16}. Thus the base-to-base signal that actuates the differential amplifier is the difference between V_{RS} ($= I_0R_S$) and the voltage drop across the lower end of R_{20}. The CA3030 output, which is the voltage at terminal No. 12, varies linearly with the actuating voltage. When the load current is zero, V_{RS} is zero; therefore terminal 12 is negative with respect to ground, and Q_3 is back-biased (i.e., cut off). Therefore Q_3 does not interfere with the normal voltage-regulated operation of the supply. As the load current increases, V_{RS} increases and the voltage at terminal 12 increases.

The value of resistor R_S is adjusted so that when the load current reaches the foldback-activation value (about 3 amperes) the voltage at terminal No. 12 of the CA3030 becomes positive. At about 0.7 volt, transistor Q_3 begins to conduct; current flows through the current-limiting resistor R_C, with the result that terminal No. 1 of the CA3055 con-

trol circuit is driven positive, the base-to-emitter voltage of Q_2 is reduced, and the output voltage of the power supply decreases. If the load impedance is reduced, Q_3 will be driven even harder, and therefore the output voltage and the load current will decrease even further.

250-Watt, 5-Volt Switching-Regulator Power Supply

A switching-regulator power supply that produces 250 watts at 5 volts with an efficiency of 70 per cent is shown in Fig. 313 on pages 272 and 273. It uses two switching transistors in a push-pull arrangement with variable pulse width; the switching rate is 200 kHz.

The power supply can deliver a load current of 50 amperes at 5 volts. All of the pulse-width modulation circuits, drivers, and latches are duplicated for each power-switching transistor. This duplication uses more than the minimum number of components, but it provides wide design margins and reliable operation.

Voltage regulation and over-load regulation are accomplished by reducing the duty cycle of the power-switching transistors. The duty cycle is reduced by triggering the latches on, either from pulse transformers T_3 and T_4 to regulate the output voltage, or from transistors Q_3 and Q_4 to prevent excessive emitter currents in the power-switching transistors. The excessive currents could be caused by overloads at the output or by transformer core saturation resulting from unbalanced duty cycles.

This power supply is capable of operating into any load impedance, including short circuits, without damage. It can operate at duty cycles from less than 10 per cent to 100 per cent. With a duty cycle of 100 per cent, the supply operates as a straight inverter at the full capacity of the transistors, transformers, and rectifiers.

Inverters and Converters

IN many applications the optimum value of voltage is not available from the primary power source. In such instances, dc-to-dc converters or dc-to-ac inverters may be used to provide the desired value of voltage. An inverter is used to transform dc power to ac power. If the ac output is rectified and filtered to provide dc again, the over-all circuit is referred to as a converter. The purpose of the converter is to change the magnitude of the available dc voltage.

BASIC CIRCUIT ELEMENTS

Power-conversion circuits, both inverters and converters, consist basically of some type of "chopper". Fig. 314(a) shows a simple chopper circuit. In this circuit, a switch S is connected between the load and a dc voltage source E. If the switch is alternately closed and opened, the output voltage across the load will be as shown in Fig. 314(b). If the on-off intervals are equal, the average voltage across the load is equal to E/2. The average voltage across the load can be varied by varying the ratio of the on-to-off time of the switch, by periodically varying the repetition rate, or by a combination of these factors. If a filter is added between the switch and the load, the fluctuations in the output can be suppressed, and the circuit becomes a true dc-to-dc stepdown transformer (or converter).

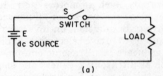

(a)

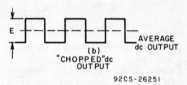

(b)

92CS-26251

Fig. 314—Simple chopper circuit and output-voltage waveform.

In practice, the switch shown in Fig. 314 may be replaced by a power transistor or a silicon controlled rectifier (SCR). When a power transistor is used, the switch is opened or closed by application of the appropriate polarity signal to the transistor base. The SCR switch can easily be closed by application of a positive pulse to its gate. Once conduction has been initiated, however, the gate loses control, and some means must be provided to stop conduction and open the switch.

The design of the transformer is an important consideration because this component determines the size and frequency of the converter (or inverter), influences the amount of regulation required after the conversion or inversion is completed, and provides the transformation ratio necessary to assure that the desired value of output voltage is delivered to the load circuit.

Inverters may be used to drive any equipment which requires an ac supply, such as motors, ac radios, television receivers, or fluorescent lighting. In addition, an inverter can be used to drive electromechanical transducers in ultrasonic equipment, such as ultrasonic cleaners and sonar detection devices. Similarly, converters may be used to provide the operating voltages for equipment that requires a dc supply.

Transistor and SCR inverters can be made very light in weight and small in size. They are also highly efficient circuits and, unlike their mechanical counterparts, have no moving components.

TRANSISTOR INVERTERS AND CONVERTERS

Several types of transistor circuits may be used to convert a steady-state dc voltage into either an ac voltage (inversion) or another dc voltage (conversion). The simplest converter circuit is the blocking-oscillator, or ringing-choke, power converter which consists of one transistor and one transformer. More complex circuits use two transistors and one or two transformers.

Basic Design Considerations

The design of practical inverter (or converter) circuits involves, essentially, selection of the proper transistors and design of the transformers to be used. The particular requirements for the transistors and transformers to be used are specified by the individual circuit design. Basic transistor and transformer requirements of ringing-choke and push-pull transformer-coupled inverter circuits for a variety of voltage and power levels are given in Tables XXXII and XXXIII.

Table XXXII—Typical Design Parameters For Ringing-Choke-Type DC-To-DC Converters That Have Output Ratings Up To 50 Watts.

	APPLICATION REQUIREMENTS		TRANSISTOR REQUIREMENTS				TRANSFORMER-CORE PARAMETERS		CIRCUIT EFFICIENCY
P_{out} (W)	Max. V_{out} (V)	DC V_{in} (V)	Max. R_{sat} (Ω)	Min. V_{CB}(max) (V)	Min. I_C(pk) (A)	Min. P_D* (W)	Area A (cm²)	Length l_1 (cm)	Factor η
	250	6-10	5	25	0.5	0.1			
1	500	10-15	10	35	0.4	0.08	0.5-1.5	2.5-10	0.75
	750	15-20	20	45	0.3	0.07			
	250	6-12	1	30	3	1.5	0.5-5	2.5-12	0.75
5	500	12-20	2	45	2	1			
	750	20-28	8	60	1	0.5	0.5-5	2.5-12	0.7
	300	6-12	0.8	30	6	3			
10	500	12-18	1	45	4	2	1-7.5	2.5-15	0.7
	750	18-28	1.2	60	2	1			
	400	10-18	0.5	45	10	10			
25	600	18-26	0.8	60	6	5	1-10	5-15	0.65
	750	26-36	1	80	3	2			
50	500	12-24	0.5	60	15	20	2-15	7.5-20	0.6
	750	24-36	0.5	80	8	7.5			

* Case or Flange Temperature = 55°C.

Table XXXIII—Design Data for Push-Pull, Transformer-Coupled DC-to-DC Converters.

	APPLICATION REQUIREMENTS			TRANSISTOR REQUIREMENTS				TRANSFORMER-CORE PARAMETERS		CIRCUIT EFFICIENCY FACTOR
P_{out} (W)	Max. V_{out} (V)	DC V_{in} (V)	Max. R_{sat} (Ω)	Min. $V_{CB(max)}$ (V)	Min. $I_C(pk)$ (A)	Min. P_D^* (W)	Area A (cm^2)	Length l_1 (cm)	η
2	250	6-12	2	30	0.5	0.1	0.5-4	2.5-10	0.85
	500	12-20	4	45	0.4	0.075			
10	400	12-18	1.5	45	2	1	0.5-5	2.5-10	0.85
	600	18-28	3	60	1	0.5			
25	400	12-18	1	45	5	3	1.5	5-15	0.85
	600	18-28	2	60	3	1.5			
50	250	8-18	0.5	45	12	10	2-7.5	7.5-20	0.85
	500	18-28	0.8	60	8	5			
	800	28-38	1	80	5	2			
100	400	12-18	0.5	45	18	15	3-12	10-25	0.85
	600	18-28	0.5	60	10	10			
	800	28-38	0.5	80	7.5	5			
200	400	12-24	0.2	60	20	25			
	600	24-36	0.2	80	15	15	5-15	15-35	0.8
	800	36-48	0.5	100	10	10	7.5-20	20-40	0.75

* Case or Flange Temperature = 55° C.

Special Transistor Requirements— The type of transistor selected for use in a high-speed converter circuit is dictated by the following conditions:

1. In a high-speed converter, the peak value of the collector-to-emitter voltage of each transistor is equal to twice the supply voltage plus the amplitude of the voltage spikes generated by transient elements. Therefore, the collector-to-emitter breakdown voltage V_{CEO} of the transistors should be slightly greater than twice the supply voltage (usually an additional 20 per cent is sufficient).

2. The transistors must be capable of handling the currents necessary to produce the required output power at the given supply voltage, and their saturation voltage at these currents must be low enough so that the high efficiency desired can be obtained.

3. The junction-to-case thermal resistance of the transistors θ_{J-C} must be low enough so that the manufacturer's maximum ratings, for the given ambient temperature and the available heat sink and cooling apparatus, are not exceeded.

Table XXXIV indicates the operating frequency range and the peak current and voltage capabilities of RCA power transistors recommended for use in inverter or converter applications.

Current and dissipation ratings: The maximum collector current, the dissipation, and the heat-sink thermal resistance of the transistors can be approximated on the basis of these limiting conditions as follows:

The maximum collector current I_C is approximately given by

$$I_C = P_{out}\, \eta\, / [V_S - V_{CE(sat)}]$$

where V_S is the supply voltage, $V_{CE(sat)}$ is the transistor collector-to-emitter saturation voltage (for a specific I_C), P_{out} is the required power output, and η is the desired efficiency of the output transformer (usually 90 to 95 per cent).

The transistor dissipation can be approximated as follows (because the base dissipation is very small, it is neglected in this approximation):

$$P_D = (T_1/T)\, (V_{CE(sat)}\, I_C + 2I_{CEX}\, V_S) + [(t_{on} + t_f)/T]\, (V_S I_C/3)$$

where V_S is the supply voltage,

Table XXXIV—RCA Power Transistors (N-P-N and P-N-P Types) for Inverter or Converter Service

Frequency Range	Peak Voltage Required	Up to 0.2 A	0.2 to 1 A	1 to 4 A	4 to 20 A	> 20 A
	10 to 60 V	2N3053 2N4037•	2N5321• 2N5323 2N6179 2N6181•	2N3054 2N5497▲	2N3055 2N3772	—
60 Hz to 50 kHz	60 to 150 V	2N1486 2N2102 2N4036•	2N1486 2N3441 2N5298 2N5781• 2N5784	2N3442 2N3879 2N5293▲ 2N5954•	2N3265 2N3773 2N5039 2N5672 2N6248•	2N5671 2N6032
	150 to 450V	2N3440 2N5416• 2N6177▲	—	2N3585 2N6212• 40850 40851	2N5805 2N5840 2N6251 410 411 413 423 40852 431 40853 40854	—

• P-N-P types $V_{PEAK} = V_{CEX}$ value $V_{PEAK} = 2.2$ V (V_{CC}) for push-pull inverters
▲ Plastic-packaged types $= 1.1$ V (V_{CC}) for bridge inverters

$V_{CE(sat)}$ is the transistor saturation voltage (for a specific I_C); I_C is the collector current, as given in the preceding equation; I_{CEX} is the collector current with the base reverse-biased (for $V_{CE} = 2V_S$); t_{on} is the transistor "turn-on" time [at I_C given by Eq. (354) and h'_{FE} given in step 4 of the general procedure]; t_f is the transistor "fall" time; T is the period reciprocal of the operating frequency; and $T_1 = \frac{1}{2} [T - (t_{on} + t_f)]$.

The equation for P_D given above is used as a guide for the first stages of design; the exact dissipation is determined experimentally. The transistor saturated-switching characteristics must be fast enough to prevent the transient dissipation from becoming excessive.

The required heat-sink thermal resistance may be approximated by the following equation:

$$\theta_{C-A} = (\Delta T/P_D) - \theta_{J-C}$$

where ΔT is the permissible junction temperature rise ($\Delta T = T_{J(max)} - T_A$); P_D is the transistor dissipation; and θ_{C-A} is the case-to-air thermal resistance, including mount-ing, interface, any insulation material, and heat sink.

The estimate of the required heat-sink thermal resistance, together with the manufacturer's maximum rating curve or safe operating region, completes the determination of transistor requirements.

Second-breakdown considerations: A high-speed, high-power inverter requires transistors that have high power-handling capabilities and very fast saturated-switching speeds. Reverse-bias second breakdown (which is discussed in an earlier section of this Manual) is a factor that must also be considered in the design of these circuits.

Reverse-bias second breakdown can be analyzed as follows: During the turn-off time t_{off}, the transistor is subjected to high energy as a result of energy stored in the output-transformer leakage inductance. This leakage inductance can be made small by careful winding of the transformer to obtain close coupling. An approximation of the value of leakage inductance can be obtained by measuring the inductance of one-half the primary with the other half of the primary short-circuited.

Transformer Considerations—The selection of the proper core material in the design of a transformer to be used in an inverter depends on the power-handling requirements, operating frequency, and operating temperature of the inverter. For high-frequency applications, the ferrite core is superior to the iron type in both performance and economy. Even at low frequencies, ferrite cores may be more economical because the iron type must be made in thin laminations or in the form of a tape-wound toroid.

Power loss in ferrite is approximately a linear function of frequency up to 40 kHz. Above this frequency, eddy-current losses decrease the efficiency of most ferrites. Laminated iron cores are normally restricted to frequencies below 10 kHz. Table XXXV indicates optimum core materials for different operating frequencies.

The operating temperature of the transformer is an important consideration in the choice of the particu-

Table XXXV—Optimum Core Materials For Different Operating Frequencies.

Transformer Material	Operating Frequency (kHz)
Ferrite	1-20
Silicon Iron (Grain-Oriented)	0.1-1
Silicon Steel	0.1-1

lar ferrite core. For many ferrite cores, the Curie temperature is low. The manufacturer's data on ferrite material indicate the maximum operating temperature which, together with the variation in flux density as a function of temperature and the desired flux density (B), must be considered to select the proper core.

Another important consideration is the efficiency of the transformer. The transformer efficiency desired can be used to obtain an approxi-

mation of allowable magnetic power, P_M, dissipated by the transformer. When P_M and the core loss factor are known, the maximum volume of core material which can be used is estimated. The core loss factor at the operating frequency is obtained from the manufacturer's data.

The remaining design considerations follow the conventional rules of transformer design. The size of the wire must be large enough to assure that copper losses are low. The selection is made on the basis of a 50-per-cent duty cycle. If the wire size is too small, copper losses will be appreciable and cause an increase in core temperatures. In high-power, high-frequency inverters, a large number of turns in the primary should be avoided to minimize copper losses and maintain a low value of leakage inductance. Moreover, because of the relatively small size of the core and the large size of wire that must be used, a large number of turns may be physically impossible. Good balance and close coupling between primaries is normally achieved by the use of bifilar windings.

Additional Considerations—Other factors, such as starting-bias methods, the use of voltage-multiplication techniques, and maximum operating temperature, may also require consideration in the design of practical inverters or converters. Excellent starting under heavy load conditions may be obtained by the use of a transistor-type switch which will provide a large starting bias and then be cut off by the buildup of the output voltage. It is also possible to obtain satisfactory starting by the use of a fixed bias resistance, provided the value of this resistance is high enough so that it does not materially affect normal switching. Such techniques are explained subsequently in the discussion of the specific circuit types in which they are required.

For dc output voltages higher than those given in the particular design, a voltage-multiplier-type rectifier circuit may be used to avoid use of larger transformer step-up ratios. Although the use of a voltage-multiplier circuit results in a reduction in over-all efficiency, this condition may be more acceptable than one which results in higher copper losses, magnetic-coupling problems, and higher core losses that may result from the use of higher transformer step-up ratios.

The transistor requirements given in Tables XXXII through XXXIII are for operation at a case or flange temperature of 55°C. To relate case or flange temperature to ambient temperature, it is necessary to know the thermal resistance between the transistor and free air. This resistance is a function of the contact resistance between the transistor case or flange and the chassis; the thermal resistance of any insulating washer used; the size, thickness, and material of the chassis; and the method used to cool the chassis (for example, forced-air cooling, water cooling, or simple convection cooling).

To assure reliable operation at any permissible ambient temperature, care must be taken that the collector-junction temperature of the transistor is not greater than that specified by the manufacturer. The average temperature of the junction $T_{J(av)}$ is equal to the ambient temperature plus the product of the average power dissipated in the transistor and the thermal resistance between junction and case plus the case-to-air thermal resistance as indicated by the following equation:

$$T_{J(AV)} = T_A + P_{AV}\ \theta_{J-C} + \theta_{C-A}$$

The average junction temperature calculated by use of the above equation is equivalent to the effective case temperature $T_{C(eff)}$ usually given on transistor safe-area-rating charts. The effects of switching on the instantaneous temperature must be evaluated by use of standard safe-area techniques, as described earlier

in the section on **Low- and Medium-Frequency Power Transistors.**

Ringing-Choke Converter

In the ringing-choke type of dc-to-dc converter, a blocking oscillator (chopper circuit) is transformer-coupled to a half-wave rectifier type of output circuit. The rectifier converts the pulsating oscillator output into a fixed-value dc output voltage.

When the oscillator transistor conducts (as a result of either a forward bias or external drive), energy is transferred to the collector inductance presented by the primary winding of the transformer. The voltage induced across the transformer is fed back (from a separate feedback winding) to the transistor base through a resistor. This voltage increases the conduction of the transistor until it is driven into saturation. A rectifier diode in series with the secondary winding of the transformer is oriented so that no power is delivered to the load circuit during this portion of the oscillator cycle.

Fig. 315(a) shows the basic configuration for a practical ringing-choke converter, which is basically a one-transistor, one-transformer circuit. Fig. 315(b) shows the waveforms obtained during an operating cycle.

During the "on" or conduction period of the transistor (t_{on}), energy is drawn from the battery and stored in the inductance of the transformer. When the transistor switches off, this energy is delivered to the load. At the start of t_{on}, the transistor is driven into saturation, and a substantially constant voltage, waveform A in Fig. 315(b), is impressed across the primary by the battery. This primary voltage produces a linearly increasing current in the collector-primary circuit, waveform B. This increasing current induces substantially constant voltages in the base windings, shown by waveform C, and in the secondary winding.

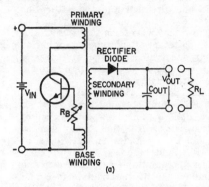

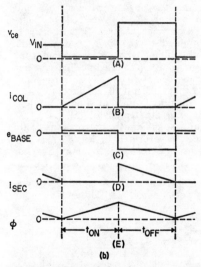

Fig. 315—Ringing-choke converter circuit: (a) Schematic diagram; (b) Typical operating waveforms in a ringing-choke converter —(A) primary voltage; (B) primary current; (C) base-to-emitter voltage; (D) secondary current; (E) magnetic flux in transformer core.

The resulting base current is substantially constant and has a maximum value determined by the base-winding voltage, the external base resistance R_B, and the input conductance of the transistor. Because the polarity of the secondary volt-age does not permit the rectifier diode to conduct, the secondary is open-circuited. Therefore, during the conduction period of the transistor t_{on}, the load is supplied only by energy stored in the output capacitor C_{OUT}.

The collector-primary current increases until it reaches a maximum value I_P which is determined by the maximum base current and base voltage supplied to the transistor. At this instant, the transistor starts to move out of its saturated condition with the result that the collector-primary current and the voltage across the transformer windings rapidly decrease, and "switch-off" occurs.

After the transistor has switched off, the circuit starts to "ring", i.e., the energy stored in the transformer inductance starts to discharge into the stray capacitance of the circuit, with the result that the voltages across the primary, base, and secondary windings reverse polarity. These reverse voltages rapidly increase until the voltage across the secondary winding exceeds the voltage across the output capacitor. At this instant the diode rectifier starts to conduct and to transfer the energy stored in the inductance of the transformer to the output capacitor and load. Because the output capacitor tends to hold the secondary voltage substantially constant, the secondary current decreases at a substantially constant rate, as shown by waveform D in Fig. 315(b). When this current reaches zero the transistor switches on again, and the cycle of operation repeats.

The operating efficiency of the ringing-choke inverter is low, and the circuit, therefore, is used primarily in low-power applications. In addition, because power is delivered to the output circuit for only a small fraction of the oscillator cycle (i.e., when the transistor is not conducting), the circuit has a relatively high ripple factor which substantially increases output filtering requirements. This converter, however, provides definite advantages to the system designer in terms of design simplicity and compactness.

Push-Pull Transformer-Coupled Inverters and Converters

The **push-pull switching inverter** is probably the most widely used type of power-conversion circuit. For inverter applications, the circuit provides a square-wave ac output. When the inverter is used to provide dc-to-dc conversion, the square-wave voltage is usually applied to a full-wave bridge rectifier and filter. A single saturable transformer controls circuit switching and provides the desired voltage transformation to the bridge rectifier. The rectifier for the square-wave output delivered and filter convert the square-wave voltage into a smooth, fixed-amplitude dc output voltage.

Two-Transistor, One-Transformer Converter—Fig. 316 shows a push-pull, transformer-coupled, dc-to-dc converter that uses one transformer and two transistors. Fig. 317 shows the waveforms obtained from this circuit during one complete operating cycle.

During a complete cycle, the flux density in the transformer core varies between the saturation value in one direction and the saturation value in the opposite direction, as shown by waveform A in Fig. 317. At the start of the conduction period for one transistor, the flux density in the core is at either its maximum negative value $(-B_{sat})$ or the maximum positive value $(+B_{sat})$.

For example, transistor A switches on at $-B_{sat}$. During conduction of transistor A, the flux density changes from its initial level of $-B_{sat}$ and becomes positive as energy is simultaneously stored in the inductance of the transformer and supplied to the load by the battery. When the flux density reaches $+B_{sat}$, transistor A is switched off and transistor B is switched on. The transformer assures that energy is supplied to the load at a constant rate during the entire period that transistor A conducts. This energy-transformation cycle is repeated when transistor B conducts.

Initially, sufficient bias is applied to saturate transistor A. As a result, a substantially constant voltage, waveform B in Fig. 317, is impressed across the upper half of the primary winding by the dc source

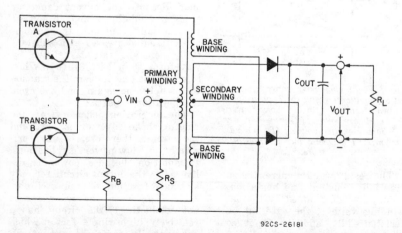

92CS-26181

Fig. 316—Two-transistor, one-transformer push-pull switching converter.

V_{IN}. This bias voltage can be a temporary bias, a small fixed bias, or even a small forward bias developed across the bias winding as a result of leakage and saturation current flowing in the transformer primary. The constant primary voltage causes a dc component and a linearly increasing component of current, waveform C in Fig. 317, to flow

through transistor A. As in the ringing-choke converter, the linearly increasing primary current induces substantially constant voltages, waveform D in Fig. 317, in the base winding and secondary winding. The induced voltage in the base winding limits the maximum value of the base current and, therefore, of the collector current.

In the push-pull transformer-coupled converter, the transition to switch-off is initiated when the transformer begins to saturate. As long as the transistor is not saturated, the product of the transformer inductance and the time rate of change of the collector current remains constant. When the transformer core saturates, however, the inductance decreases rapidly toward zero, with the result that the time rate of change of the collector current increases towards infinity. When the collector current reaches its maximum value, transistor A moves out of saturation and the winding voltages decrease and then reverse and thereby cause transistor A to switch off. The reversal of the winding voltages switches transistor B on, and the switching operation is repeated.

Fig. 318 shows the circuit sche-

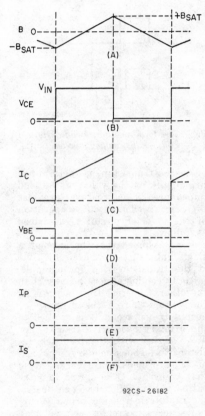

92CS-26182

Fig. 317—Typical operating waveforms for a two-transistor, one-transformer switching converter: (A) flux density in transformer core; (B) collector voltage of one transistor; (C) collector current of one transistor; (D) base voltage of one transistor; (E) primary current; (F) secondary current.

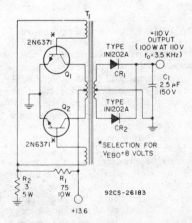

92CS-26183

Fig. 318—Schematic diagram of 13.6-to-110-volt transformer-coupled push-pull converter.

matic and Fig. 319 shows the performance curves for the 13.6-to-110-volt converter.

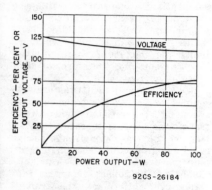

92CS-26184

Fig. 319—Output voltage and efficiency as a function of power output for the converter circuit shown in Fig. 318.

Two-Transistor, Two-Transformer Inverters—There are three basic disadvantages associated with the two-transistor, one-transformer inverter. First, the peak collector current is independent of the load. This current, therefore, depends on the available base voltage, the gain of the transistor, and the input characteristic of the transistor. Second, because of the dependence of the peak current on transistor characteristics, the circuit performance depends on the particular transistor used because there is a wide spread in transistor characteristics. Third, the transformer, which is relatively large, must use expensive square-loop material and must have a high value of flux density at saturation. These disadvantages can be overcome by the use of two transformers in various circuit arrangements, such as that shown in Fig. 320.

In this type of circuit, a saturable base-drive transformer T_1 controls the inverter switching operation at base-circuit power levels. The linearly operating output transformer transfers the output power to the load. Because the output transformer T_2 is not allowed to saturate, the peak collector current

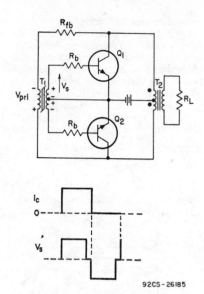

92CS-26185

Fig. 320—Two-transistor, two-transformer push-pull switching inverter.

of each transistor is determined principally by the value of the load impedance. This feature provides high circuit efficiency. The operation of the inverter circuit is described as follows:

It is assumed that, because of a small unbalance in the circuit, one of the transistors, Q_1 for example, initially conducts more heavily than the other. The resulting increase in the voltage across the primary of output transformer T_2 is applied to the primary base-drive transformer T_1 in series with the feedback resistor R_{fb}. The secondary windings of transformer T_1 are arranged so that transistor Q_1 is driven to saturation. As transformer T_1 saturates, the rapidly increasing primary current causes a greater voltage drop across feedback resistor R_{fb}. This increased voltage reduces the voltage applied to the primary of transformer T_1; thus, the drive input and ultimately the collector current of transistor Q_1 are decreased.

In the circuit arrangement shown in Fig. 320, the base is driven hard compared to the expected peak collector current (forced beta of ten, for example). If the storage time of the transistor used is much longer than one-tenth of the total period of oscillation T, the transistors begin to have an appreciable effect on the frequency of operation. In Fig. 320, the storage time could conceivably be quite long because there is no turn-off bias (the drive voltage only decreases to zero) for Q_1 until the collector current of Q_1 begins to decrease.

Two methods of overcoming this problem by decreasing the storage time are shown in Fig. 321. In Fig. 321(a), a capacitor is placed in parallel with each base resistor R_B. When V_s is positive, the capacitor charges with the polarity shown. When V_s decreases to zero, this capacitor provides turn-off current for the transistor. In Fig. 321(b), a feedback winding from the output transformer is placed in series with each base. The base-to-emitter voltage V_{BE} is then expressed as follows:

$$V_{BE} = V_S - V_{rb} - V_T$$

If V_s decreases to zero and the collector current does not begin to decrease, then the base-to-emitter voltage is expressed simply by

$$V_{BE} = V_{rb} - V_T$$

A turn-off bias is thus provided to decrease the collector current.

The energy stored in the output transformer by its magnetizing current is sufficient to assure a smooth changeover from one transistor to the other. The release of this stored energy allows the inverter-circuit switching to be accomplished without any possibility of a "hang-up" in the crossover region during the short period when neither transistor is conducting.

The operation of the high-speed converter is relatively insensitive to small system variations that may cause slight overloading of the cir-

cuit. Under such conditions, the base power decreases; however, this loss is so small that it does not noticeably affect circuit performance. At the same time, the amount of energy stored in the output transformer also increases. Although this increase results in a greater transient dissipation, the inverter switching is still effected smoothly.

A practical design of the high-speed converter should include some means of initially biasing the transistors into conduction to assure that the circuit will always start. Such starting circuits, as described later, can be added readily to the converter, and are much more reliable than one which depends on circuit imbalance to shock the converter into oscillation.

Feedback Resistance—The value of feedback resistance R_{fb} is computed as the resistance required to produce the difference in voltage that should exist between the collector-to-collector voltage of the two transistors and the voltage applied to the primary of transformer T_1 at a given primary current I_{pri1}. The optimum value of the feedback resistor is then determined experimentally. A decrease in the value of R_{fb} increases the loss that results from the circuit resistance and that in the transformer core because the magnetizing current increases. The voltage across the primary of the transformer then increases and the operating frequency increases. An increase in the value of R_{fb} causes a greater voltage drop across this resistance, and less voltage is then available to the primary of transformer T_1; therefore, the frequency decreases. Thus, R_{fb} can be used to control frequency over a limited range only.

Starting Circuits—The circuits shown in Figs. 320 and 321 will not necessarily begin to oscillate, especially under a heavy load. As a result, a starting bias must be applied

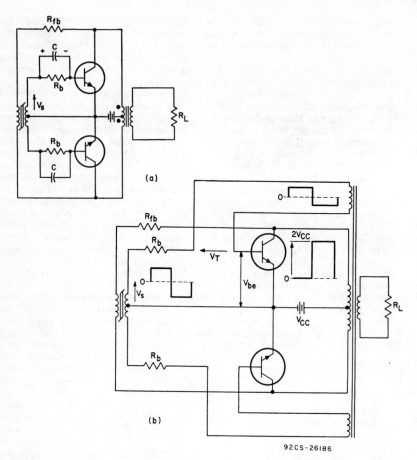

*Fig. 321—Two-transistor, two-transformer **push-pull** switching inverters in which transistor-storage times are reduced: (a) Capacitor in parallel with each base resistor assures rapid turn-off of associated transistor; (b) Feedback winding from output transformer in series with base of each transistor assures sharp cutoff characteristics.*

so that the circuit has a loop gain greater than unity and is always capable of initiating oscillation. This bias arrangement can be such that it is connected only during starting, or can be connected permanently within the circuit. Two practical starting circuits are described in the following paragraphs.

Fig. 322 shows an inverter that uses a resistive voltage-divider network to supply the necessary starting bias. With this circuit, a compromise of reliable starting and

tolerable bleeder current must be reached.

Fig. 323 shows a diode starting circuit in which the bases of the two inverter transistors are supplied by a resistance R_1, which is determined as follows:

$$R_1 = V_{CC}/2I_B$$

As the inverter begins to oscillate, the base current is conducted through the base-emitter diode and through the forward direction of the starting diode. Usually, additional drive is needed to compensate for the diode voltage drop. Low-voltage silicon diodes capable of carrying the base current continuously are normally used.

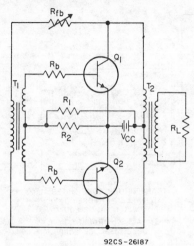

92CS-26187

Fig. 322—Two-transistor, two-transformer push-pull inverter that uses a resistive voltage-divider network to provide starting bias.

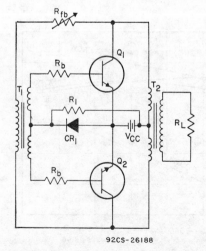

92CS-26188

Fig. 323—Two-transistor, two-transformer push-pull inverter that uses a diode starting circuit.

250-Watt, 50-kHz Converter—Fig. 324 shows a schematic diagram for a 250-watt, 50-kHz two-transistor, two-transformer push-pull converter. The leakage inductance of the output transformer, as measured on a Q meter, is about 0.5 microhenry. The peak collector current is calculated to be about 10 amperes, and the reverse base-to-emitter bias voltage is about −2 volts. The 2N3265 transistor has an assured capability to withstand second breakdown at currents in excess of 10 amperes for a collector inductance of 90 microhenries and a reverse bias of 6 volts. The published data on the 2N3265 indicate that a reduction in bias voltage or in collector inductance allows the transistor to handle larger amounts of reverse-bias energy. The operating conditions for the output transformer are well within the safe area. Both transformers should be constructed with a minimum of tape to provide as much surface area as possible to ensure a low core temperature.

Fig. 325 shows the output characteristics of the converter as a function of the load. The output characteristics were measured at the load at the output terminals of the rectifier bridge. Thus, the efficiency shown represents the total circuit efficiency. The range of values indicated on the efficiency curve (i.e., 82 to 88 per cent) takes into account the transistor dissipation, transformer losses, rectifier-bridge losses, and all other circuit IR losses.

Fig. 326 shows the experimental transistor load line for a load resistance of 25.6 ohms and a supply voltage of 28 volts. The area enclosed by the load line shows that high dissipation occurs during switching. This area is decreased somewhat when loads having a small capacitive reactance are used.

Fig. 327 shows the collector current and voltage waveforms. The collector-current waveform exhibits the transformer saturation current. The collector-voltage waveform exhibits the voltage spikes resulting from the transformer leakage inductance. Fig. 328 shows the collector current on an expanded time scale to illustrate the current rise and fall times.

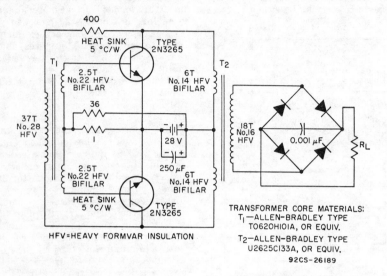

Fig. 324—Schematic diagram of 250-watt, 50-kHz push-pull dc-to-dc converter.

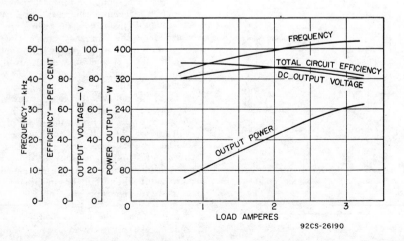

Fig. 325—Output characteristics (i.e., frequency, efficiency, voltage, and power) of the 250-watt converter as a function of the load.

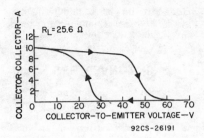

Fig. 326—*Experimental load line for the 2N3265 transistor using a load impedance of 25.6 ohms and a supply voltage of 28 volts.*

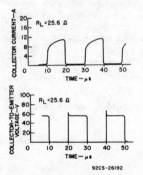

Fig. 327—*Collector-current and voltage waveforms for the 2N3265 transistors used in the 250-watt converter.*

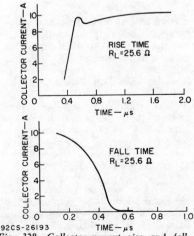

Fig. 328—*Collector-current rise and fall times.*

Four-Transistor Bridge Inverters —Fig. 329 shows a **four-transistor, single-transformer bridge configuration** that is often used in inverter or converter applications. In this type

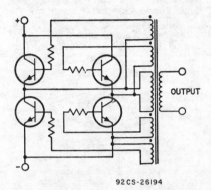

Fig. 329—*Basic circuit configuration of a four-transistor, single-transformer bridge inverter.*

of circuit, the primary winding of the output transformer is simpler and the breakdown-voltage requirements of the transistors are reduced to one-half those of the transistors in the push-pull converter shown in Fig. 316.

The separate saturable-transformer technique may also be applied in the design of a bridge converter, as shown in Fig. 330.

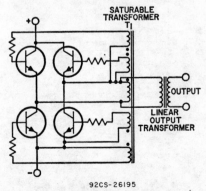

Fig. 330—*Basic configuration of a four-transistor bridge inverter that uses a saturable output transformer.*

Three-phase bridge inverters for induction motors are usually used to convert dc, 60-Hz, or 400-Hz input to a much higher frequency, possibly as high as 10 kHz. Increasing frequency reduces the motor size and increases the horsepower-to-weight ratio, desirable features in military, aviation, and portable industrial power-tool markets. Fig. 331 shows a typical three-phase

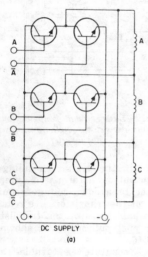

DC SUPPLY

(a)

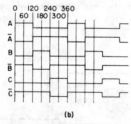

(b)

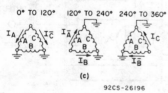

(c)

92CS-26196

Fig. 331—Three-phase bridge inverter: (a) circuit configuration; (b) base driving signals; (c) transformer primary current switching.

bridge circuit with base driving signals and transformer primary currents.

SCR INVERTERS

SCR inverters offer an efficient and economical method for conversion of direct current to alternating current. In the design of an SCR inverter, the fact that the SCR is basically a "latching" device must be considered. Anode current can be initiated at any time by application of a signal of the proper polarity to the gate. However, the gate loses control as soon as conduction begins, and current continues to flow, regardless of any gate signal which may be applied, as long as the anode remains positive. Special circuitry is required to turn off the SCR at the proper time.

Commutation of Inverter SCR's

Circuits that are used to turn off an SCR are commonly called commutating circuits. The commutating circuit normally bypasses the load current around the SCR for a time sufficiently long to permit the SCR to recover its forward-blocking capability.

Parallel-Capacitor Commutation— One of the simplest methods for commutation of an SCR in an inverter circuit is by use of a capacitor connected in parallel with the load, as shown in Fig. 332. If the switch S is open at the time the SCR is triggered by a pulse applied to its gate, the negative terminal of capacitor C will be connected to the negative terminal of the supply voltage. Capacitor C then charges at an exponential rate through resistor R toward the dc supply voltage. When the switch is closed, capacitor C will be connected between anode and cathode of the SCR in a polarity such that it provides a negative anode voltage for the SCR. Load current is then diverted through the capacitor, and the SCR turns off. The capacitance and the voltage to which it is charged must be sufficient to bypass the load current around the SCR for the time required for the SCR to recover its forward-blocking capability.

In an actual circuit, the switch S shown in Fig. 332 is replaced by a second SCR, as shown in Fig. 333. The resultant circuit is the familiar single-phase capacitor-commutated inverter. Capacitor C alternately commutates SCR₁ and SCR₂. The inductor limits the current that flows into C during the charging intervals in a manner similar to that of resistor R in Fig. 332, but without the I^2R losses that occur in a resistor.

Series-Capacitor Commutation—A basic series-capacitor-commutated circuit is shown in Fig. 334. In

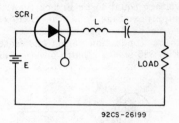

Fig. 334—Basic series-capacitor type of SCR commutation circuit.

this circuit, commutation is achieved through the action of a capacitor connected in series with the load. When the SCR is turned on by a positive pulse applied to its gate, inductor L and capacitor C are connected in series with the load to the supply voltage. The inductor L and the capacitor C form a series resonant circuit. Current through this resonant circuit (and the SCR) builds up sinusoidally to a maximum, decreases to zero, and then attempts to reverse. When the current attempts to reverse, the SCR turns off, and reverse voltage is main-

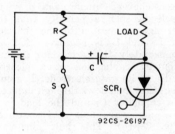

Fig. 332—Parallel-capacitor type of SCR commutation circuit.

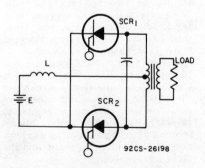

Fig. 333—Single-phase capacitor-commutated SCR inverter.

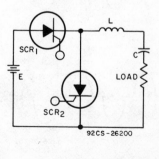

Fig. 335—Basic series-capacitor-commutated SCR inverter.

tained across it by the charge on capacitor C. With the configuration shown in Fig. 334, a recurrent waveform cannot be maintained because capacitor C will be charged to a voltage equal to or higher than the supply voltage, and the SCR can never be forward-biased after the initial conduction period has ceased. Fig. 335 shows a more practical circuit for use as an inverter.

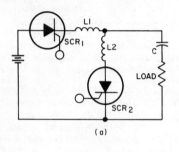

(a)

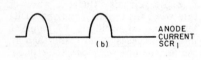

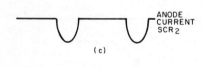

ANODE CURRENT SCR₁

(b)

ANODE CURRENT SCR 2

(c)

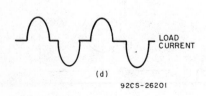

LOAD CURRENT

(d)

92CS-26201

Fig. 336—Series-capacitor-commutated SCR inverter that uses an inductor to reduce the triggering delay.

In this circuit, SCR₁ and SCR₂ are gated on alternately, with sufficient time between gating pulses to permit the LC circuit to commutate the conducting SCR. When SCR₁ conducts, the voltage across capacitor C rises to some positive value above the supply voltage. When SCR₂ conducts, the voltage across capacitor C falls to some negative value below the supply voltage. When one SCR is triggered on, forward voltage is immediately applied to the other SCR. Therefore, sufficient delay must be allowed between the end of one half-cycle of load current and the start of another. This delay must be at least equal to the turnoff time (t_q) of the SCR. The required delay in triggering can be reduced by the use of an additional inductor, as shown in Fig. 336.

If both SCR₁ and SCR₂ are in a blocking state and capacitor C has been charged during the previous cycle in such a polarity that the terminal connected to the junction of L_1 and L_2 is negative, SCR₁ is immediately triggered into conduction by application of a positive pulse to its gate electrode. The current that then flows through inductor L_1, capacitor C, and the load is a half sine-wave pulse, as shown in Fig. 335(b). This half sine-wave pulse resonantly charges capacitor C in the opposite polarity. By the end of the current pulse, capacitor C is charged to a value higher than the supply voltage. Current then attempts to reverse through SCR₁. Some reverse-recovery current flows for a few microseconds, and SCR₁ then reverts to its reverse-blocking state (i.e., turns off). Current remains at zero until a gating pulse is applied to SCR₂. When SCR₂ is triggered on, capacitor C discharges through SCR₂ in a half sine-wave pulse. When current attempts to reverse, SCR₂, after a few microseconds of reverse-recovery current, reverts to its blocking state, and the cycle is repeated.

In the circuit shown in Fig. 336, the current carried by SCR₂ is sup-

plied completely by energy stored
in capacitor C during the period
when SCR_1 is conducting. Because
the objective of the inverter is to
supply energy to the load, an arrangement that supplies energy from
the supply on each half-cycle, such
as shown in Fig. 337 is desirable.

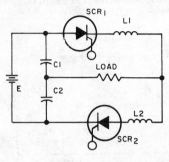

92CS-26202

*Fig. 337—Series-capacitor-commutated SCR
inverter in which energy from the power
source is supplied to the load on each half-
cycle of inverter operation.*

In the circuit shown in Fig. 337,
capacitor C is divided into two equal
parts. During the half-cycle that
SCR_1 conducts, capacitor C_2 is
charged from the supply, and capacitor C_1 is discharged through the
load. Both the charging current of
capacitor C_2 and the discharging
current of capacitor C_1 flow through
SCR_1, inductor L_1, and the load. On
the next half-cycle, a similar action
takes place, except that SCR_2 conducts, capacitor C_1 charges from the
supply, and capacitor C_2 discharges.
If $C_1 = C_2$ and $L_1 = L_2$, one-half
the load current on each half cycle
is supplied by the power supply, and
the other half of the load current
is supplied by the capacitor being
discharged.

If separate inductors are used for
L_1 and L_2, the maximum operating
frequency of the inverter must always be slightly less than the series
resonant frequency of L_1C_1 and L_2C_2.
However, if L_1 and L_2 are closely
coupled on a common core, the operating range of the inverter can be
extended slightly above the resonant
frequency of the series LC circuit.
If L_1 and L_2 are coupled, when one
SCR is turned on, the current that
flows in the reactor associated with
that SCR induces a voltage in the
other reactor that reverses the voltage on the other SCR and causes it
to cease conduction.

Impulse Commutation—If a very
short pulse is used to briefly reverse
the voltage on an SCR, the SCR is
said to be impulse-commutated. The
pulse must be of sufficient magnitude to cause the SCR to stop conduction and must be of sufficient
duration to permit the SCR to recover its forward-blocking capability (must provide sufficient turn-off
time).

There are three general types of
impulse commutation. In one type,
the commutating pulse is generated
by an SCR that is complementary
to the SCR being turned off. The
parallel-capacitor commutation circuit shown in Fig. 332 is an example
of **complementary impulse-commu-
tation**. One disadvantage of such a
circuit is that if the gating signals
are lost for any reason, the conducting SCR will not be turned off, i.e.,
will fail to commutate. In the event
of such a failure, the conducting
SCR may be destroyed by excessive
forward dissipation, unless a fuse
or circuit breaker opens to remove
the dc power from the circuit. The
commutating impulse may also be
generated automatically through
some action of the circuit which
causes it to automatically turn itself off after a certain time interval.
The series-capacitor-commutated circuits shown in Figs. 335, 336, and
337 are examples of **self impulse-
commutation**. A third type of impulse commutation is one in which
the commutating impulse is generated by some auxiliary means, separate from the power-generating portion of the inverter. This method is
called **auxiliary-impulse commutation**.

Self-impulse commutation: Fig. 338 shows a basic "chopped" circuit in which the commutating impulse is generated by action of the circuit.

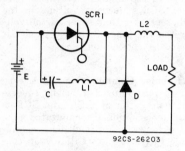

92CS-26203

Fig. 338—Basic self-impulse-commutated SCR inverter.

When the SCR is off, but load current has been established during a previous cycle, load current will be maintained by L_2. Because the SCR is off, the return path for the load current is through the rectifier D. During the off time of the SCR, capacitor C charges through inductors L_1 and L_2 and the load to the supply voltage E. When the SCR is gated on, the voltage at the junction of inductors L_1 and L_2 rises immediately to the supply voltage E. Rectifier D then becomes reverse-biased, and the load current flows through the SCR. Capacitor C also discharges through inductor L_1 and the SCR in an oscillatory manner and reverses its charge within one-cycle. After the first half-cycle, the current through inductor L_1 and capacitor C reverses and causes current to flow through the SCR which is in a reverse direction from the load current. The net current through the SCR is then the difference between the load current and the current through L_1 and C. When the current through L_1 and C equals the load current, the SCR current becomes zero, and the SCR turns off. The voltage remaining on capacitor C then appears as inverse voltage across the SCR. Load current continues to flow through C, L_1, and L_2 until C has been charged

back to the supply voltage E. At this time, current through C stops, and load current is transferred to the rectifier D. The cycle is then repeated.

Fig. 339 shows a circuit that uses another form of self-impulse commutation. This configuration is the well-known **Morgan circuit.** During

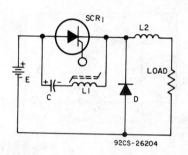

92CS-26204

Fig. 339—Basic Morgan type of inverter circuit.

the off-time, load current is maintained by L_2, as in the circuit shown in Fig. 338. However, the current that charges capacitor C during the off time saturates the core of L_1 (positively, it is assumed for this explanation) so that the inductance of this inductor is small. When the SCR is triggered on, the capacitor will be connected directly across L_1 and will drive its core towards saturation in the reverse direction (negative saturation). A finite time interval is required for the core to "switch" from positive saturation to negative saturation. During this time interval, current flows through the SCR, inductor L_2, and the load. Rectifier D is reverse-biased and therefore does not conduct. When the core of L_1 switches to negative saturation, capacitor C discharges in an oscillatory manner through the SCR. Because the inductance of inductor L_1 is very small at this time, the resonant discharge is very rapid, and the voltage on capacitor C very quickly falls from $+E$ to $-E$. The core of L_1 remains in the negatively saturated condition for a certain time interval and then switches rapidly to the positively saturated condition. During this interval, the SCR continues to conduct current which flows through L_2 and the load. When the core of L_1 becomes positively saturated again, the capacitor C once more discharges in a resonant manner, and current is driven in the reverse direction through the SCR. When the reverse current through the SCR equals the load current, the SCR turns off. In practice, a rectifier is usually connected across the SCR in the circuit shown in Fig. 338 and Fig. 339 to carry the excess reverse current once the reverse current becomes equal to the load current.

Auxiliary impulse commutation: An example of auxiliary impulse commutation is shown in Fig. 340.

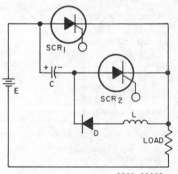

92CS-26205

Fig. 340—SCR inverter circuit that employs auxiliary type of impulse commutation of SCR's.

In this circuit, if SCR_2 has been triggered on, capacitor C is permitted to charge from the supply in the polarity shown. SCR_2 turns off automatically once capacitor C becomes fully charged because of the lack of current. When SCR_1 is triggered on, load current flows through SCR_1 and the load. In addition, capacitor C discharges through inductor L, rectifier D, and SCR_1 until it has reversed its charge. The hold-off rectifier D then prevents

the current through capacitor C from reversing again. Load current continues to flow through SCR_1 until SCR_2 is triggered on, capacitor C is then allowed to discharge through SCR_2, and, in a reverse direction, through SCR_1. When the reverse current through SCR_1 equals the load current, SCR_1 turns off.

Commutating Capacitors—In all the commutation arrangements discussed in the preceding paragraphs, the commutation current for the conducting SCR must be carried by a capacitor. This fact imposes rather severe requirements on the commutating capacitor. The capacitor must be able to carry the high peak currents necessary for commutation without excessive losses that would reduce circuit efficiency, cause excessive temperature rise, and premature capacitor failure. Also, because the series inductance of the capacitor may limit the width of the commutating pulse and the initial rate of rise of current in the conducting SCR, capacitor manufacturers should be consulted as to the current ratings of the commutating capacitors at the operating frequency of the inverter. In addition, temperature life tests should be conducted to assure that operating-temperature ratings of the capacitor are not exceeded after long periods of operation.

High-Frequency SCR Switching Inverter

Fig. 341 shows a typical high-frequency SCR switching inverter; Fig. 342 shows the waveshapes across each SCR and the output of the transformer. For resistive loads, this inverter is capable of delivering 500 watts of output power at an operating frequency of 8 kHz, and is provided with regulation from a no-load condition to full load. With proper output derating, this circuit can also accommodate inductive and capacitive loads. Under a capacitive load the power dissipation of the SCR's is increased; under an inductive load the turn-off time is decreased.

The inverter can be operated at any optional frequency up to 10 kHz provided that a suitable output transformer is used and the timing capacitors are changed in the gate-trigger-pulse generator. A change in

operating frequency, however, does not require any change in the commutating components C_1 and L_1.

Circuit Operation—Fig. 341 shows the two thyristors SCR_1 and SCR_2 connected to the output transformer T_1. These thyristors are alternately triggered into conduction by the gate-trigger-pulse generator shown

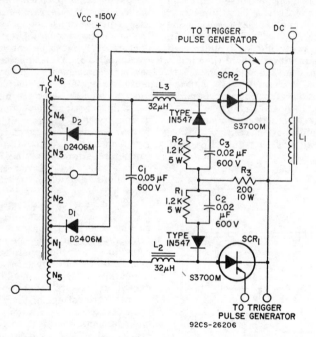

Fig. 341—High-frequency (10-kHz) SCR push-pull switching inverter.

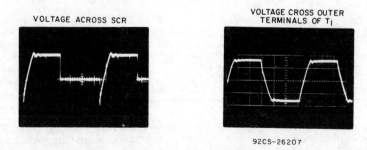

Fig. 342—Typical operating waveforms for SCR inverter shown in Fig. 341.

in Fig. 343 to produce an alternating current in the primary of the power transformer.

The thyristors are commutated by capacitor C_1, which is connected between the anodes of SCR_1 and SCR_2. The flow of current through the circuit can be traced more easily if it is assumed that initially SCR_1 is conducting and SCR_2 is cut off and that the common cathode connection of the SCR's is the reference point. For this condition, the voltage at the anode of SCR_2 is twice the voltage of the dc power supply, i.e., $2E_{CC}$. The load current flows from the dc power supply through one-half the primary winding of transformer T_1, inductor L_2, SCR_1, and inductor L_1. When the firing current is applied to the gate of SCR_2, this SCR turns on and conducts.

During the on period of SCR_2, the capacitor C_1 begins to discharge through L_3, SCR_2, SCR_1, and L_2. In-ductors L_2 and L_3 function to limit the rate of rise of the discharge current di/dt so that the associated stresses are maintained within the capablity of the device during the turn-on of the SCR. The effect of this control is to decrease the turn-on dissipation, which becomes a significant portion of the total device dissipation at high repetition rates.

The discharge current through SCR_1 flows in a reverse direction, and after the carriers are swept out (and recombined) the SCR_1 switch opens (i.e., SCR_1 switches to the off state). At this time, the voltage across the capacitor C_1, which is approximately equal to $-2E_{CC}$, appears across SCR_1 as reverse voltage. This voltage remains long enough to allow the device to recover its forward blocking capability. Simultaneously during this interval, the conducting SCR_2 establishes another discharge path for capacitor C_1 through transformer

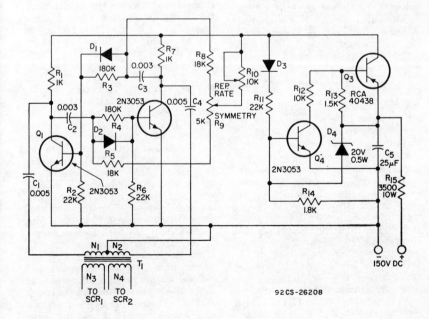

Fig. 343—Gate-trigger pulse generator for SCR inverter shown in Fig. 341.

92CS-26208

T_1 and inductors L_1 and L_3. The role of inductor L_1 is to control the rate of discharge of the capacitor to allow sufficient time for turn-off.

After capacitor C_1 is discharged from $-2E_{cc}$ to zero, it starts to charge in the opposite direction to $+2E_{cc}$. When C_1 is charged to $+2E_{cc}$, because of the phase shift between voltage and current the flux at that time in the inductor L_1 is a maximum. This reactive energy stored in the inductor is normally transferred to the capacitor and causes an "overvoltage" or "overcharge", which in this particular case is undesirable. Voltages on the capacitor higher than $2E_{cc}$ produce a negative voltage at the anode of SCR_2 with respect to the negative terminal of the dc power supply. This condition is prevented by use of a clamping diode D_2 connected to an extra tap on the transformer oriented close to the anode of SCR_2. As a result, the amount of "overcharge" of the capacitor is considerably reduced. The energy stored in inductor L_1 causes current to flow through diode D_2, the N_4 transformer winding inductor L_3 and SCR_2. Transformer windings N_4 and N_3 act as an autotransformer through which the energy stored in the inductor is fed back to the power supply.

When the firing current is applied to the gate of SCR_1, this device conducts and the process described above is repeated.

Each time the SCR's turn off to interrupt the reverse recovery current, a certain amount of energy remains in the inductor. This energy is transferred to the device capacitance, which is relatively small, and thus a high-voltage transient is generated. This high-voltage transient may exceed the rating of the device, produce undesirable stresses, and increase the switching dissipation. A transient-suppressor network consisting of two 1N547 diodes, resistors R_1, R_2 and R_3, and capacitors C_2 and C_3 prevents this transient voltage from exceeding the maximum rating of the SCR's.

Gate-Trigger-Pulse Generator— The gate-trigger-pulse generator, as shown in Fig. 343, is a conventional astable (free-running) multivibrator, combined with a threshold-sensitive switch consisting of transistors Q_3 and Q_4 which turns the generator on and off. The square-wave output of the generator is differentiated and fed to the gates of SCR_1 and SCR_2 through the N_3 and N_4 windings of pulse transformer T_1. The threshold-sensitive switch holds the generator off until the required dc level is achieved in the power supply. This minimum level is necessary to maintain a nominal repetition rate and to supply sufficient current to trigger both SCR's. As dc power is applied through resistor R_{15} to charge capacitor C_5, the gradually increasing voltage at the emitter of transistor Q_3 eventually rises to a value above the zener voltage of the zener diode D_4 connected between the emitter of transistor Q_3 and the base of transistor Q_4. So long as this voltage is not exceeded, the base current of transistor Q_4 is zero. Because transistor Q_4 is cut off, transistor Q_3 also remains cut off. As the voltage of the power supply increases and exceeds the zener voltage of D_4, the zener diode conducts current to the base of transistor Q_4 and causes the transistor to conduct. The collector current of Q_4 then flows into the base of Q_3 and causes this transistor to conduct. The collector current of Q_3 is then applied to the astable multivibrator. A polarity-sensitive positive feedback loop consisting of diode D_3 and resistor R_{11} provides regenerative feedback to transistors Q_4 and Q_3 when the zener diode D_4 is conducting. In the event that the power-supply voltage decreases and current ceases to flow through the zener diode, this feedback network maintains transistor Q_3 in saturation until the voltage in the circuit drops to a few volts.

The collector current through transistors Q_1 and Q_2 does not maintain perfect balance as the base currents of transistors Q_1 and Q_2 increase. Any slight unbalance in collector current is amplified through the positive feedback loops. As a result, one transistor is cut off and the other is turned on at the extreme limit of unbalance. If transistor Q_1 is assumed turned on, the base of transistor Q_2 is driven negative by capacitor C_2, which is connected to the collector of Q_1. The negative bias on the base of Q_2 drives the transistor into the cut-off state. Capacitor C_3 connected to the base of Q_1 is then charged through the load resistor R_7 of transistor Q_2, and the base drive on transistor Q_1 increases until the capacitor is fully charged. Capacitor C_2, with its negatively charged plate connected to the base of transistor Q_2 through a resistor divider consisting of R_4 and R_6, is discharged through resistor R_5. Resistor R_5 is connected to a potentiometer R_9 which controls the waveshape symmetry and another potentiometer R_{10} which is connected to the positive supply voltage and serves as the repetition-rate control.

When the negative bias decreases to zero and the base of Q_3 becomes positive, transistor Q_2 turns on and causes Q_1 to turn off. The capacitor C_4 which was charged through load resistor R_7 starts to discharge through the N_2 primary windings of the pulse transformer T_1 after Q_2 is turned on. This discharge current is fed to the gate of the SCR, in the appropriate direction to fire the device. During the alternate half-cycle of multivibrator operation, capacitor C_1 discharges through the N_1 primary windings of the pulse transformer to trigger SCR_1.

Applications—Some of the applications of the SCR inverter are as follows:

1. DC-to-dc converter. Conversion can be accomplished by the use of small, light-weight, low-cost transformers, inductors, and capacitors. This circuit is suitable for use in computer power supplies, telephone equipment, radio transmitters, battery chargers, and similar equipment.

2. High-frequency fluorescent-lighting supply. Because of the high frequency of the inverter circuit, the size and weight of the inductive ballast is considerably reduced; in addition, half of the inductive components can be replaced with low-cost capacitors to maintain a unity power factor in the circuit. The over-all system efficiency can also be improved; for example, the 20- to 26-per-cent power dissipation as a result of the low-efficiency ballast at 60 Hz can be reduced to a few per cent by use of high-frequency, high-efficiency inductors at moderate cost. This decrease in power dissipation in a large industrial building can mean less burden on the air-conditioning system.

Audio Amplifiers

AUDIO amplifier circuits are used in radio and television receivers, public address systems, sound recorders and reproducers, and similar applications to amplify signals in the frequency range from 20 to 20,000 Hz. Each transistor in an audio amplifier can be considered as either a current amplifier or a power amplifier. The type of circuit configuration selected is dictated by the requirements of the given application. The output power to be supplied, the required sensitivity and frequency response, and the maximum distortion limits, together with the capabilities and limitations of available devices, are the main criteria used to determine the circuit that will provide the desired performance most efficiently and economically.

In addition to the consideration that must be given to the achievement of performance objectives and the selection of the optimum circuit configuration, the circuit designer must also take steps to assure reliable operation of the audio amplifier under varying conditions of signal level, frequency, ambient temperature, load impedance, line voltage, and other factors which may subject the transistors to either transient or steady-state high stress levels.

The input to an audio amplifier is a low-power-level audio signal from the phonograph or magnetic-tape pickup head or, in a radio receiver, from the detector stage as indicated in the section **Tuned Amplifiers, Frequency Converters, and Detectors for AM, FM, and TV Receivers**. This signal is usually amplified through a preamplifier stage, one or more low-level (predriver or driver) audio stages, and an audio power amplifier. The system may also include frequency-selective circuits which act as equalization networks and/or tone controls.

CLASSES OF OPERATION

A circuit designer may select any one of three classes of operation for transistors used in linear-amplifier applications. This selection is made on the basis of a combination of such factors as required power output, dissipation capability, efficiency, gain, and distortion characteristics.

The three basic classes of operation (class A, class B, and class C) for linear transistor amplifiers are defined by the operating point of the transistor. In class A operation, the active element conducts for the entire input cycle. In class B operation, the active element conducts for 180 degrees of an input cycle and is cut off during the remainder of the time. In class C operation, the active element conducts for some amount less than 180 degrees of an input cycle. The following paragraphs discuss the distinguishing features of class A and class B operation. In general, because of the high harmonic distortion introduced as a result of the short conduction angle, class C operation is used primarily in rf-amplifier applications in which it is practical to use tuned output circuits to eliminate the harmonic components. For this reason, class C operation is not discussed further.

Class A Operation

Class A amplifiers are used for linear service at low power levels. When power amplifiers are used in this class of operation, the amplifier output is usually transformer-coupled to the load circuit, as shown in Fig. 344. At low power levels, the class A amplifier can also be coupled to the load by resistor, capacitor, or direct coupling techiques.

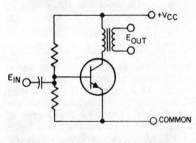

92CS-25855

Fig. 344—Basic class A, transformer-coupled amplifier.

There is some distortion in a class A stage because of the nonlinearity of the active device and circuit components. The maximum efficiency of a class A amplifier is 50 per cent; in practice, however, this efficiency is not realized. The class A transistor amplifier is usually biased so that the quiescent collector current is midway between

the maximum and minimum values of the output-current swing. Collector current, therefore, flows at all times and imposes a constant drain on the power supply. The consistent drain is a distinct disadvantage when higher power levels are required or operation from a battery is desired.

Class B Operation

Class B power amplifiers are usually used in pairs in a push-pull circuit because conduction is not maintained over the complete cycle. A circuit of this type is shown in Fig. 345. If conduction in each de-

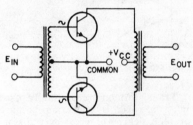

92CS-25856

Fig. 345—Basic class B, push-pull transformer-coupled amplifier.

vice occurs during approximately 180 degrees of a cycle and the driving wave is split in phase, the class B stage can be used as a linear power amplifier. The maximum efficiency of the class B stage at full power output is 78.5 per cent when two transistors are used. In a class B amplifier, the maximum power dissipation is 0.203 times the maximum power output and occurs at 42 per cent of the maximum output.

Ideally, transistors used in class B push-pull service should be biased to collector cutoff so that no power is dissipated under zero-signal conditions. At low signal inputs, however, the resulting signal would be distorted, as shown in Fig. 346, because of the low forward current-transfer ratio of the transistor at very low currents. This type of distortion, called **cross-over distortion,** can be suppressed by the use of a bias voltage which permits a small collector current flow at zero signal level. Any residual distortion can be further reduced by the use of negative feedback.

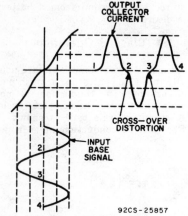

Fig. 346—Waveforms showing cause of cross-over distortion.

Transistors are not usually used in true class B operation because of high degree of cross-over distortion at low power levels. Most power stages operate in a biased condition somewhat between class A and class B. This intermediate class is defined as class AB. Class AB transistor amplifiers operate with a small forward bias on the transistor to minimize the nonlinearity. The quiescent current level, however, is still low enough so that class AB amplifiers provide good efficiency. This advantage makes class AB amplifiers an almost universal choice for high-power linear amplification, especially in battery-operated equipment.

PREAMPLIFIERS

Simple class A amplifier circuits are normally used in low-level audio stages such as **preamplifiers** and **drivers.** Preamplifiers usually follow low-level output transducers such as microphones, hearing-aid and phono-

graph pickup devices, and recorder-reproducer heads.

Basic Design Features

In the design of an audio preamplifier, special consideration must be given to the circuit noise figure, equalization networks, input impedance, and volume and tone controls.

Noise Figure—One of the important characteristics of a low-level amplifier circuit is its **signal-to-noise ratio,** or **noise figure.** The input circuit of an amplifier inherently contains some thermal noise contributed by the resistive elements in the input device. All resistors generate a predictable quantity of noise power as a result of thermal activity. This power is about 160 dB below one watt for a bandwidth of 10 kHz.

When an input signal is amplified, therefore, the thermal noise generated in the input circuit is also amplified. If the ratio of signal power to noise power (S/N) is the same in the output circuit as in the input circuit, the amplifier is considered to be "noiseless" and is said to have a noise figure of unity, or zero dB.

In practical circuits, however, the ratio of signal power to noise power is inevitably impaired during amplification as a result of the generation of additional noise in the circuit elements. A measure of the degree of impairment is called the noise figure (NF) of the amplifier, and is expressed as the ratio of signal power to noise power at the input (S_i/N_i) divided by the ratio of signal power to noise power at the output (S_o/N_o), as follows:

$$NF = \frac{S_i/N_i}{S_o/N_o}$$

The noise figure in dB is equal to ten times the logarithm of this power ratio. For example, an amplifier with a 1-dB noise figure decreases the signal-to-noise ratio by a factor of 1.26, a 3-dB noise figure by a factor of 2, a 10-dB noise figure by a factor of 10, and a 20-dB noise figure by a factor of 100.

In audio amplifiers, it is desirable that the noise figure be kept low. In general, the lowest value of NF is obtained by use of an emitter current of less than one milliampere and a collector voltage of less than two volts for a signal-source resistance between 300 and 3000 ohms. If the input impedance of the transistor is matched to the impedance of the signal source, the lowest value of NF that can be attained is 3 dB. Generally, the best noise figure is obtained by use of a transistor input impedance approximately 1.5 times the source impedance. However, this condition is often not realizable in practice because many transducers are reactive rather than resistive. In addition, other requirements such as circuit gain, signal-handling capability, and reliability may not permit optimization for noise.

In the simple low-level amplifier stage shown in Fig. 347, resistor R_1 determines the base bias for the transistor. The output signal is de-

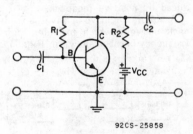

92CS-25858

Fig. 347—Simple low-level class A amplifier.

veloped across the load resistor R_2. The collector voltage and the emitter current are kept relatively low to reduce the noise figure. If the load impedance across the capacitor C_2 is low compared to R_2, very little voltage swing results on the collector. Therefore, ac feedback through R_1 does not cause much reduction in gain.

Equalization—In many cases, low-level amplifier stages used as preamplifiers include some type of **frequency-compensation network** to enhance either the low-frequency or the high-frequency components of the input signal. The frequency range and dynamic range* which can be recorded on a phonograph record or on magnetic tape depend on several factors, including the composition, mechanical characteristics, and speed of the record or tape, and the electrical and mechanical characteristics of the recording equipment. To achieve wide frequency and dynamic range, manufacturers of commercial recordings use equipment which introduces a nonuniform relationship between amplitude and frequency. This relationship is known as a "recording characteristic". To assure proper reproduction of a high-fidelity recording, therefore, some part of the reproducing system must have a frequency-response characteristic which is the inverse of the recording characteristic. Most manufacturers of high-fidelity recordings use the RIAA characteristic for discs and the NARTB characteristic for magnetic tape.

The simplest type of equalization network is shown in Fig. 348. Because the capacitor C is effectively an open circuit at low frequencies, the low frequencies must be passed through the resistor R and are attenuated. The capacitor has a lower reactance at high frequencies, however, and bypasses high-frequency components around R so that they

---

* The dynamic range of an amplifier is a measure of its signal-handling capability. The dynamic range expresses in dB the ratio of the maximum usable output signal (generally for a distortion of about 10 per cent) to the minimum usable output signal (generally for a signal-to-noise ratio of about 20 dB). A dynamic range of 40 dB is usually acceptable; a value of 70 dB is exceptional for any audio system.

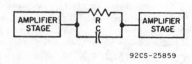

92CS-25859

Fig. 348—Simple RC frequency-compensation network.

receive negligible attenuation. Thus the network effectively "boosts" the high frequencies. This type of equalization is called "attenuative."

The location of the frequency-compensation network or "equalizer" in the reproducing system depends on the types of recordings which are to be reproduced and on the pickup devices used. All commercial pickup devices provide very low power levels to a transistor preamplifier stage.

A ceramic high-fidelity phonograph pickup is usually designed to provide proper compensation for the RIAA recording characteristic when the pickup is operated into the load resistance specified by its manufacturer. Usually, a "matching" resistor is inserted in series with the input of the preamplifier transistor. However, this arrangement produces a fairly small signal current which must then be amplified. If the matching resistor is not used, equalization is required, but some improvement can be obtained in dynamic range and gain.

A magnetic high-fidelity phonograph pickup, on the other hand, usually has an essentially flat frequency-response characteristic. Because a pickup of this type merely reproduces the recording characteristic, it must be followed by an equalizer network, as well as by a preamplifier having sufficient gain to satisfy the input requirements of the tone-control amplifier and/or power amplifier. Many designs include both the equalizing and amplifying circuits in a single unit.

A high-fidelity magnetic-tape pickup head, like a magnetic phonograph pickup, reproduces the recording characteristic. This type of pickup device, therefore, must also be followed by an equalizing network and preamplifier to provide equalization for the NARTB characteristic.

Feedback networks may also be used for frequency compensation and for reduction of distortion. Basically, a feedback network returns a portion of the output signal to the input circuit of an amplifier. The feedback signal may be returned in phase with the input signal (**positive** or **regenerative** feedback) or 180 degrees out of phase with the input signal (**negative, inverse,** or **degenerative** feedback). In either case, the feedback can be made proportional to either the output voltage or the output current, and can be applied to either the input voltage or the input current. A negative feedback signal proportional to the output current raises the output impedance of the amplifier; negative feedback proportional to the output voltage reduces the output impedance. A negative feedback signal applied to the input current decreases the input impedance; negative feedback applied to the input voltage increases the input impedance. Opposite effects are produced by positive feedback.

A simple negative or inverse feedback network which provides high-frequency boost is shown in Fig. 349.

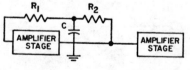

92CS-25860

Fig. 349—Negative-feedback frequency-compensation network.

This network provides equalization comparable to that obtained with Fig. 348, but is more suitable for low-level amplifier stages because it does not require the first amplifier stage to provide high-level low frequencies. In addition, the inverse feedback improves the distortion characteristics of the amplifier.

Input Impedance—As mentioned previously, it is undesirable to use a high-resistance signal source for a transistor audio amplifier because the extreme impedance mismatch results in high noise figure. High source resistance cannot be avoided, however, if an input device such as a ceramic pickup is used. In such cases, the use of negative feedback to raise the input impedance of the amplifier circuit (to avoid mismatch loss) is no solution because feedback cannot improve the signal-to-noise ratio of the amplifier. A more practical method is to increase the input impedance somewhat by operating the transistor at the lowest practical current level and by using a transistor which has a high forward current-transfer ratio.

Volume and Tone Controls—Some preamplifier or low-level audio amplifier circuits include variable resistors or potentiometers which function as volume or tone controls. Such circuits should be designed to minimize the flow of dc currents through these controls so that little or no noise will be developed by the movable contact during the life of the circuit. Volume controls and their associated circuits should permit variation of gain from zero to maximum, and should attenuate all frequencies equally for all positions of the variable arm of the control.

A tone control is a variable filter (or one in which at least one element is adjustable) by means of which the user may vary the frequency response of an amplifier to suit his own taste. In radio receivers and home amplifiers, the tone control usually consists of a resistance-capacitance network in which the resistance is the variable element.

The simplest form of tone control is a "treble cut" network such as that shown in Fig. 350. As R_1 is made smaller, the capacitor C_2 bypasses more of the high audio frequencies; therefore, the output of

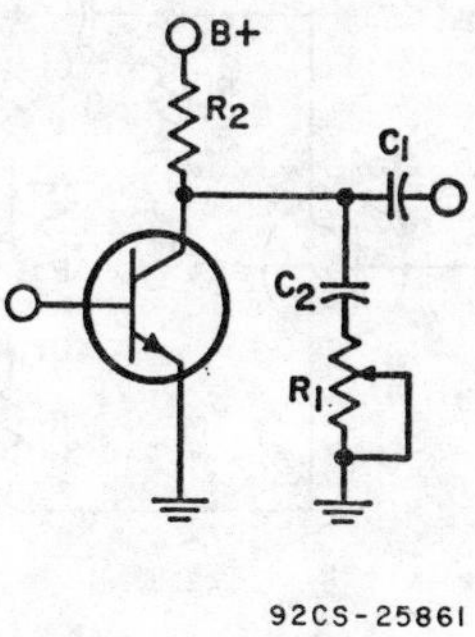

92CS-25861

Fig. 350—Simple tone-control network for fixed tone compensation or equalization.

the network is decreased by an amount dependent upon the value of R_1. The resistance of R_1 should be very large in comparison to the reactance of C_2 at the highest audio frequency.

The tone-control network shown in Fig. 351 has two stages with completely separate bass and treble controls. Fig. 352 shows simplified representations of the bass control when the potentiometer is turned to its extreme variations (labeled BOOST and CUT). At very high frequencies, C_1 and C_2 are effectively short circuits and the network becomes the simple voltage divider R_1 and R_2. In the bass-boost position, R_3 is inserted in series with R_2 so that there is less attenuation to very low frequencies than to very high frequencies. Therefore, the bass is said to be "boosted". In the bass-cut position, R_3 is inserted in series with R_1 so that there is more attenuation to very low frequencies.

Fig. 353 shows extreme positions of the treble control. R_6 is generally much larger than R_4 or R_5 and may be treated as an open circuit in the extreme positions. In both the boost and cut positions, very low frequencies are controlled by the voltage divider R_4 and R_5. In the boost position, R_4 is bypassed by the high frequencies, and the voltage-divider point D is placed closer to C. In the cut position, R_5 is bypassed, and there

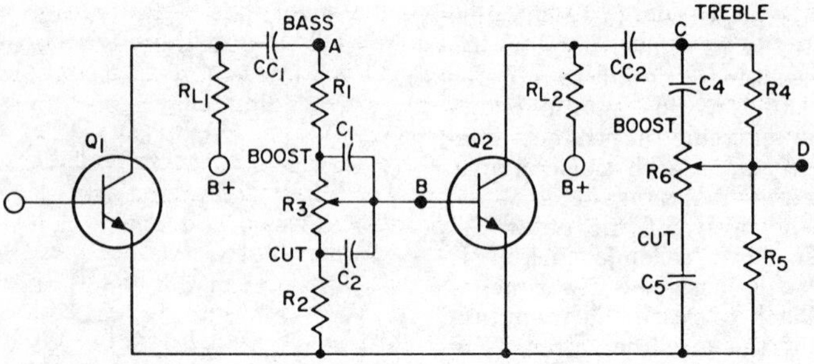

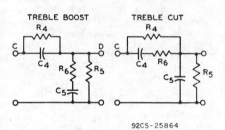

92CS-25862

Fig. 351—Two-stage tone-control circuit incorporating separate bass and treble controls.

is greater attenuation of the high frequencies.

The frequencies at which boost and cut occur in the circuit of Fig. 351 are controlled by the values of C_1, C_2, C_4, and C_5. Both the output impedance of the driving stage (generally R_{L1}) and the loading of the driven stage affect the response curves and must be considered. This tone-control circuit, like the one in Fig. 350, is attenuative. Feedback tone controls may also be employed.

The location of a tone-control network is of considerable importance. In a typical preamplifier, it may be in the collector circuit of the final low-level stage or in the input circuit of the first stage. If the amplifier incorporates negative feedback, the tone control must be inserted in a part of the amplifier which is external to the feedback loop, or must be made a part of the feedback network. The over-all gain of a well designed tone-control network should be approximately unity. The system

92CS-25864

Fig. 353—Simplified representations of treble-control circuit at extreme ends of potentiometer.

dynamic range should be adequate for all frequencies anticipated with the tone controls in any position. The high-frequency gain should not be materially affected as the bass control is varied, nor should the low-frequency gain be sensitive to the treble control.

92CS-25863

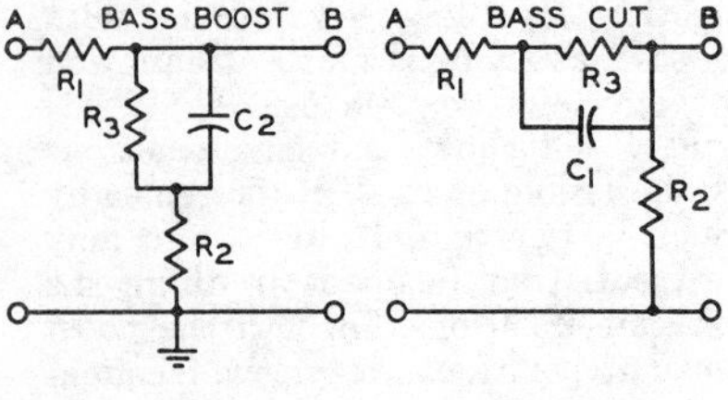

Fig. 352—Simplified representations of bass-control circuit at extreme ends of potentiometer.

Integrated-Circuit Stereo Preamplifier

The RCA CA3052 silicon monolithic integrated circuit is designed specifically for stereo preamplifier service. This circuit consists of four identical independent amplifiers that can be connected to provide all the amplification necessary in a dual-channel preamplifier for a high-quality phonograph system. When a signal source is connected to the inputs of a CA3052 unit operated in this mode, the output of each channel may be used to drive a high-quality, high-power audio amplifier; all intermediate functions are accomplished by interconnection with the CA3052 preamplifier circuit.

The CA3052 is specified using RIAA* test methods for equivalent input noise using one test method for amplifiers 1 and 4, and an appropriately different method for amplifiers 2 and 3. These circiuts are supplied in 16-terminal dual-in-line plastic packages and may be operated over a temperature range of $-25°C$ to $+85°C$.

Circuit Description—Figs. 354 and 355 show a block diagram and a detailed schematic, respectively, of the CA3052 integrated-circuit amplifier array. Each of the amplifiers A_1 through A_4 provides two stages of voltage gain. The input stage is basically a differential amplifier with a Darlington transistor added on

one side. The output stage uses a combination of three transistors connected in an inverting configuration.

Input signals to the amplifiers in the CA3052 are normally applied to the noninverting input terminal (terminal 9 for amplifier A_3) to the base of the Darlington input transistor (Q_{19}). The 0.1-megohm resistor R_{37} supplies bias current for this transistor. The voltage drop across the resistor is small because the base current of the Darlington transistor is very small.

Each amplifier in the CA3052 array may be viewed as an ac operational amplifier in which a fixed resistance is permanently connected between the output and the inverting input. In amplifier A_3, this resistance is provided by the series combination of resistors R_{41} and R_{42}.

The amplifiers in the CA3052 arrays are normally operated in the noninverting configuration; it is important, therefore, to minimize the capacitance from output to input. Excessive capacitance between output and input can result in a peaked response, instability, or even oscillation in extreme cases. In general, however, if sound design practices, careful layout, and typical terminations are used, a stable predictable circuit results.

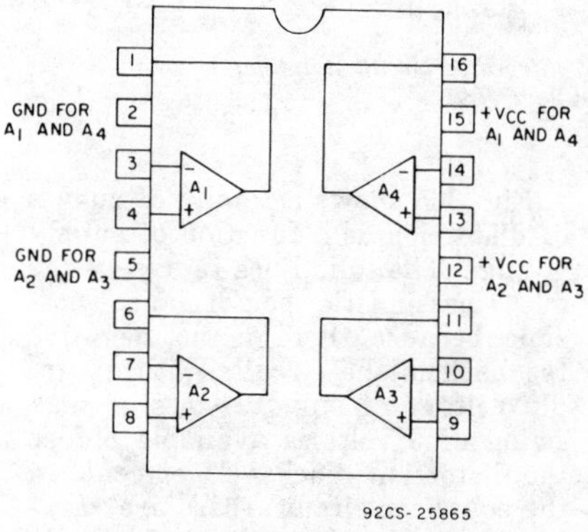

Operating Characteristics—Figs. 356 through 360 show typical dc and ac operating characteristics for the CA3052 as measured in the test circuits shown in Figs. 361 and 362.

Fig. 363 shows the curve of total harmonic distortion as a function of ambient temperature.

Fig. 354—Block diagram of CA3052 integrated-circuit amplifier array.

First Amplifier Section of Preamplifier—In order that the signal applied to the second amplifier section is not degraded in signal-to-noise ratio, the gain of the first amplifier section of the preamplifier should be sufficient to raise the signal more than 40 dB. The gain of the first amplifier, however, should not be so

* Record Industry Association of America

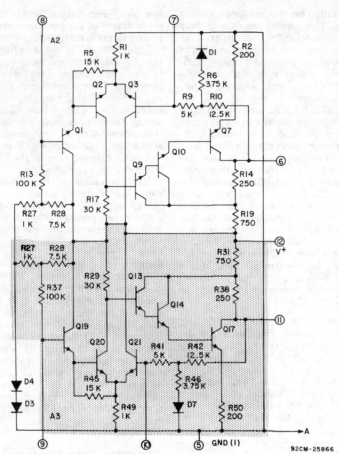

Fig. 355—Schematic of the CA3052 integrated-circuit amplifier array.
(continued on page 309)

high that the amplifier overloads at maximum signal levels (i.e., at a recorded velocity of 25 centimeters per second).

Each amplifier of the CA3052 array has an open-loop gain of 58 dB which is sufficient to allow the design described above to be realized. A schematic of the first-section amplifier is shown in Fig. 364. The breakpoints ω_2 and ω_3 are determined as indicated in the figure. The low-frequency breakpoint ω_2 is caused mainly by C_3 and the input impedance at the inverting input terminal.

Fig. 365 shows a family of curves of distortion as a function of supply voltage. The solid lines represent the most pessimistic performance possible because there is no negative feedback in the circuit. With a supply voltage of only 10 volts, a peak swing of 3 volts is available before the distortion reaches 2 per cent. In the equalizer circuit, there are varying amounts of negative feedback, depending on the frequency; the poorest situation occurs at low frequencies where there is maximum boost.

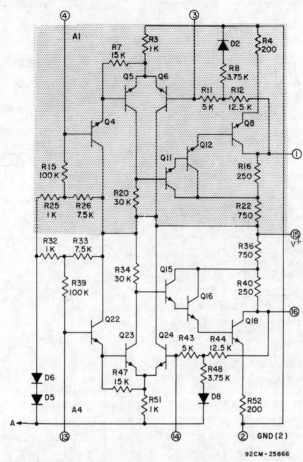

Fig. 355—Schematic of the CA3052 integrated-circuit amplifier array.
(continued from page 308)

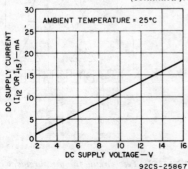

Fig. 356—Typical dc supply current as a
function of supply voltage.

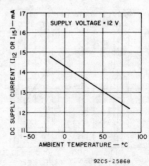

Fig. 357—Typical dc supply current as a
function of ambient temperature.

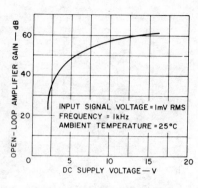

92CS-25869

Fig. 358—Typical amplifier gain as a function of supply voltage.

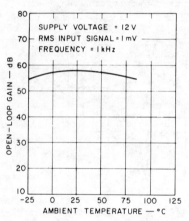

92CS-25870

Fig. 359—Typical open-loop gain as a function of ambient temperature.

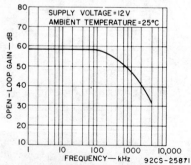

Fig. 360—Typical open-loop gain as a function of frequency.

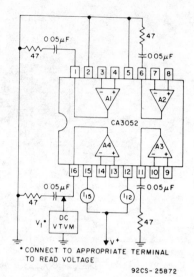

92CS-25872

Fig. 361—Test circuit for measurement of collector supply voltage and currents.

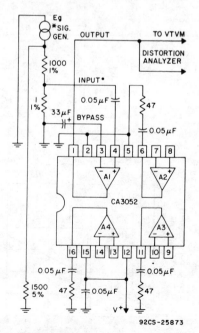

92CS-25873

Fig. 362—Test circuit for measurement of distortion, open-loop gain, and bandwidth characteristics.

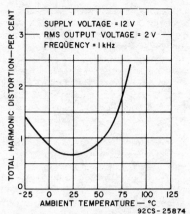

Fig. 363—Typical total harmonic distortion as a function of ambient temperature.

Because each amplifier of the CA3052 has an independent feedback point, it is possible to vary the gain for balancing. With this arrangement, the gain of one channel is increased while that of the other is decreased as the balance control is varied. The result is negligible change in level throughout the range of the control.

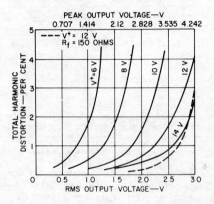

92CS-25876

Fig. 365—Total harmonic distortion of circuit shown in Fig. 364 as a function of output voltage for different dc supply voltages.

Complete Preamplifiers—A complete schematic diagram of a conventional single-channel preamplifier is shown in Fig. 366. The first amplifier/equalizer is the same as that shown in Fig. 364. Tone controls and a second amplifier have been added. The second amplifier provides a flat response in the audio range, but is rolled off at about 20 kHz by capacitor C_1.

The resistor R_1 in Fig. 366 acts in parallel with the feedback resistance already on the integrated-circuit chip to reduce the gain. The resistors R_2 and R_3 reduce the negative feedback supplied by R_1. This same amount of gain could have been programmed by omission of R_1 and a corresponding increase in the values of R_2 and R_3. Analysis of differential-amplifier stages has shown that the equivalent noise source resistance comes from both inputs. Therefore, the low resistor values achieved by the use of R_1 result in a decrease in the noise output by about 4 dB. This noise reduction is important when the level control is at the minimum setting and the signal-to-noise ratio is 0 dB. A schematic diagram of a second preamplifier (one channel only) is shown in Fig. 367. In this circuit,

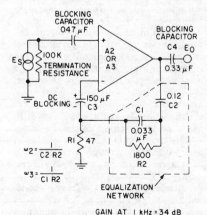

92CS-25875

Fig. 364—First-amplifier section of the CA3052 preamplifier (single channel) with RIAA equalization network.

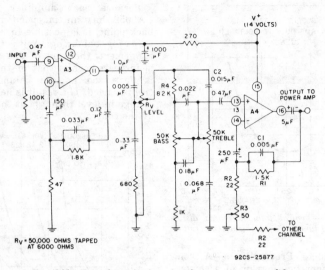

Fig. 366—One channel of a complete stereo preamplifier.

the equalizer and second-amplifier stages have been kept intact, but the positions of the level and tone controls have been reversed. In some cases, the design of complete systems is complicated by the fact that many loudspeakers tend to overload at low frequencies. In addition, there may be acoustic coupling between the input and output that causes an unstable "microphonic" condition at full gain.

With the arrangement shown in Fig. 367, it is possible to realize a system which has a great deal of bass boost at normal listening levels, but in which the gain at the bass end is restricted as the maximum level is approached. The value of resistor R_4 in the tone-control network is made smaller in this system. At average listening levels, the difference is made up by the reduced series impedance presented by R_4. The treble-boost capacitor must be increased so that there is little apparent loss in treble at low settings. The gain of the second amplifier is reduced to make the system gain equivalent at the reference frequency, but in so doing the net gain at the bass end is less.

As the level control is advanced, the sensation is one of a small change in emphasis from lows to highs.

Feedback Level Control—Fig. 368 shows the use of the CA3052 in a preamplifier that employs feedback volume control. In a feedback-volume-control circuit, the gain, rather than the input level, of the amplifier is varied. The level control is located between the output and the inverting input of the CA3052. At minimum volume, the entire output is fed back to the input. With this amount of feedback, some external stabilization is required; C_3 and R_5 are used to provide this stabilization.

The maximum gain level of the second amplifier stage is determined by the ratio of R_v and R_6. Adjustment of R_v also varies the ratio of feedback resistance to source resistance. The input impedance to the second stage varies from R_6 at maximum volume to $R_6 + R_v$ at minimum volume. Adjustment of R_v, therefore, varies the loading on the preceding tone-control circuit. The circuit shown in Fig. 368 exhibits

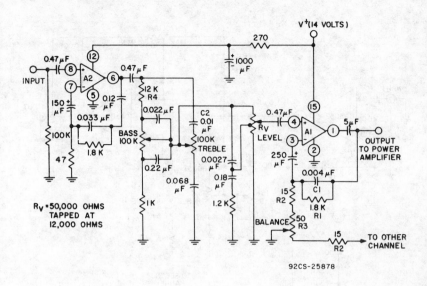

Fig. 367—One channel of a complete stereo preamplifier in which the locations of the tone and level controls have been reversed from those in the preamplifier shown in Fig. 366.

less bass boost at maximum volume than at lower levels, as does the circuit in Fig. 367. To maintain bass boost at higher levels, it is necessary to scale the impedances of the tone-control circuit to lower values.

At minimum volume, the feedback-volume-control circuit effectively places the noise source for the second stage at the output of the preamplifier. Under these conditions, the source resistance seen by the power amplifier is reduced.

The feedback-volume-control circuit requires a special taper on the volume-control potentiometer. A linear taper acts rather like a switch in that it provides very little volume as the control is rotated up to about 90 per cent of its rotation. The level then rises very quickly to maximum. The correct taper is a counterclockwise logarithmic type, i.e., one in which the rate of change of resist-ance is very fast at first, and then slows down as maximum rotation is approached.

Musical-Instrument Applications— The four independent amplifiers of the CA3052 make possible simple designs for circuits normally used in electronic musical instruments.

The **tone-generation system** for musical instruments may be a system of oscillators and frequency dividers designed for particular characteristics. The oscillators should be stable and tunable; each frequency divider is then phased-locked to its synchronizing source so that its output is exactly one-half the synchronizing frequency. As each divider is added, another tone is generated which is one octave below the input frequency. It is desirable for each element of the string to exhibit an

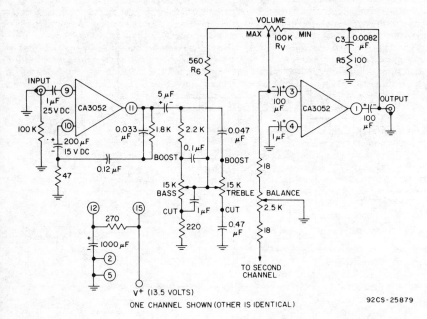

Fig. 368—One channel of a complete stereo preamplifier that employs feedback volume control.

identical waveform which is rich in harmonics. The tones may then be mixed and passed through various filters so that different frequency bands are passed to the power amplifier to provide a variety of tonal colors.

Fig. 369 shows the CA3052 connected as a tone generator for an electronic organ. Amplifier A_3 is connected as a Hartley oscillator, with its frequency governed by the relationship $\omega_0 = 1/(L_1 C_1)^{1/2}$.

Amplifiers A_1, A_2, and A_4 are set up as astable multivibrators. The period of A_4 is made somewhat longer than one-half the period of A_3 by timing capacitor C_3. Capacitor C_2 provides a synchronizing signal to amplifier A_4 by triggering A_4 before its natural period is completed. In similar fashion, amplifier A_4 provides a synchronizing signal to amplifier A_1 by means of capacitor C_4, and to amplifier A_2 by means of capacitor C_5.

Outputs are taken from the timing capacitors C_3, C_6, and C_7 in the multivibrators, and from the integrating capacitor C_8 in the master oscillator. In each case, a triangular waveform, approximately 200 millivolts peak-to-peak, is obtained.

Twelve generators of this type would constitute the tone-generation system for a four octave electronic organ.

Tremolo is widely used in musical-instrument amplifiers to create effects which add to the capability of the instrument itself. It is a musical effect characterized by a sub-audio modulation of the musical tone. When amplitude modulation is used, it is called tremolo; a frequency-modulation effect is termed **vibrato**.

If a variable impedance is connected in the feedback loop of the CA3052, it is possible to vary its gain by a modulating signal. This feature is used in the circuit shown in Fig. 370.

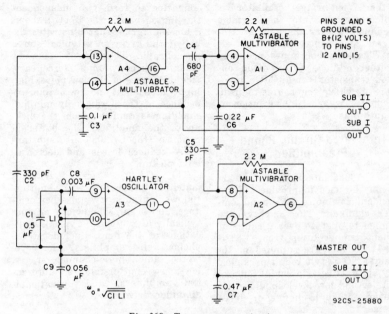

Fig. 369—Tone-generator circuit.

$$\omega_0 = \frac{1}{\sqrt{C1 \ L1}}$$

92CS-25880

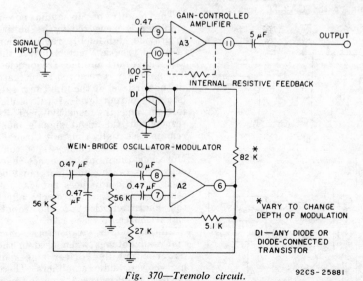

Fig. 370—Tremolo circuit.

92CS-25881

Amplifier A_2 is connected as a Wien-bridge oscillator that operates at about 6 Hz. Amplifier A_3 is operated as a voltage amplifier with a variable impedance between the inverting input and ground. The resistors internal to the CA3052 chip, in combination with the variable impedance, control the gain. The output of the Wien-bridge oscillator is coupled directly into the variable impedance to vary the gain at a 6-Hz rate to produce the tremolo effect. The variable impedance may be either a diode (or a diode-connected transistor) or an MOS transistor in which the source-to-drain impedance is varied by variation of the gate bias.

Integrated-Circuit Phono Preamplifier

The CA3036 integrated-circuit array may be used to provide two independent low-noise wide-band amplifier channels. These arrays are designed to operate over a range of ambient temperatures from $-55°C$ to $+125°C$ and are supplied in 10-terminal TO-5-style metal packages. They are particularly useful for preamplifier and low-level amplifier applications in single-channel or stereo systems.

Fig. 371(a) shows the schematic diagram of the CA3036 array. The array consists of four transistors connected to form two independent Darlington pairs. Fig. 371(b) shows a block diagram that illustrates the use of the array in a typical stereo phonograph. The CA3036 can be mounted directly on a stereo cartridge. Because of the low noise, high input impedance, and low output impedance of the array, only minimal shielding is required from the pickup to the amplifier. The buffering action of the CA3036 also substantially reduces losses and decreases hum pickup.

The CA3036 array features matched transistors with emitter-follower outputs, low-noise performance, and a gain-bandwidth product typically of 200 MHz. Typical applications of the array include stereo phonograph amplifiers, low-level stereo and single-channel stages, low-noise emitter-follower differential amplifiers, and operational-amplifier drivers.

POWER AMPLIFIERS

The quality of an audio power amplifier is measured by its ability to provide high-fidelity reproduction of audio program material over the full range of audible frequencies. The amplifier is required to increase the power level of the input to a satisfactory output level with little distortion, and the sensitivity of its response to the input signals must remain essentially constant throughout the audio-frequency spectrum. Moreover, the input-impedance characteristics of the amplifier must be such that the unit does not load excessively and thus adversely affect the characteristics of the input-signal source.

Silicon power transistors offer many advantages when used in the power-output and driver stages of high-power audio amplifiers. These devices may be used over a wide range of ambient temperatures to develop tens of watts of audio-frequency power to drive a loudspeaker system.

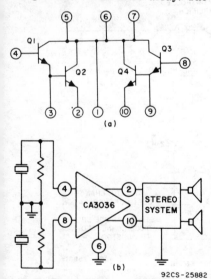

92CS-25882

Fig. 371—(a) Circuit diagram for CA3036 and (b) block diagram of stereo system using a CA3036 in a phono preamplifier.

Audio Amplifiers

Basic Circuit Configurations

The selection of the basic circuit configuration for an audio power amplifier is dictated by the particular requirements of the intended application. The selection of the basic circuit configuration that provides the desired performance most efficiently and economically is based primarily upon the following factors: power output to be supplied, required sensitivity and frequency-response characteristics, maximum allowable distortion, and capabilities of available devices.

Class A Transformer-Coupled Amplifiers—Fig. 372 shows a three-stage class A transformer-coupled audio amplifier that uses dc feedback (coupled by R_1, R_2, R_3, R_4, and

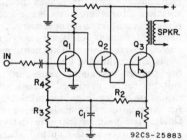

Fig. 372—Three-stage transformer-coupled, class A amplifier.

C_1) from the emitter of the output transistor to the base of the input transistor to obtain a stable operating point. An output capability of 5 watts with a total harmonic distortion of 3 per cent is typical for this type of circuit. In general, this output level is the upper limit for class A amplifiers because the power dissipated by the output transistor in such circuits is more than twice the output power. For this reason, it is economically impractical to use class A audio amplifiers to develop higher levels of output power. A circuit such as the one shown in Fig. 372 usually requires no over-all feedback unless extremely low distortion is required. Local feedback

in each stage is adequate; amplifiers of this type, therefore, are usually very stable.

Class AB Push-Pull Transformer-Coupled Amplifiers—At power-output levels above 5 watts, the operating efficiency of the circuit becomes an important factor in the design of audio power amplifiers. The circuit designer may then consider a class AB push-pull amplifier for use as the audio-output stage.

Fig. 373 shows a class AB push-pull transformer-coupled audio-output stage. Resistors R_1, R_2, and R_3 form a voltage divider that provides the small amount of transistor forward bias required for class AB operation. The transformer type of output coupling used in the circuit is advantageous in that a suitable output transformer can be selected to match the audio system to any desired load impedance. This feature assures maximum transfer of the audio-output power to the load circuit, which is especially important in sound-distribution systems that use high-impedance transmission lines to reduce losses. A major disadvantage of transformer output coupling is that it tends to limit the amplifier frequency response, particularly at the low-frequency end.

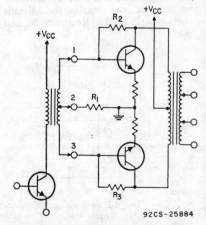

Fig. 373—Class AB, push-pull transformer-coupled audio output stage.

Variations in transformer impedance with frequency may produce significant phase shifts in the signal at both frequency extremes of the amplifier response. Such phase shifts are potential causes of amplifier instability if they occur within the feedback loop. Open-circuit stability is always a problem in designs that use output transformers because the gain increases sharply when the load is removed. If too much over-all feedback is employed, the amplifier may oscillate. The local feedback caused by the bias arrangement of R_2 and R_3 helps to eliminate this problem.

Push-pull output stages, which use identical output transistors, require some form of phase inversion in the driver stage. In the circuit shown in Fig. 373, a center-tapped driver transformer is used for this purpose. The requirements of this transformer depend upon the power levels involved, the bandwidth required, and the distortion that can be tolerated. This transformer also introduces phase-shift problems that tend to cause instabilities in the circuit when high levels of feedback are employed. Phase-shift problems are substantially reduced when the output stage is designed to operate at low drive requirements. The reduced drive requirements can be achieved by use of the Darlington circuit shown in Fig. 374. Resistors R_1 and

R_2 shunt the leakage of the driver and also permit the output transistors to turn off more rapidly. Impedance levels between the class A driver and the output stage can be easily matched by the use of an appropriate transformer turns ratio.

An alternative method of phase inversion is to use a transistor in a phase-splitter circuit, such as described later in the discussion on **Drive Requirements.** Unlike the center-tapped transformer method, impedance matching may be a problem because the collector of the driver, which has a relatively high impedance, operates into the low input impedance of the output stage. One solution is to reduce the output impedance of the driver stage by the use of smaller resistors. The resultant increase in collector current, however, also increases the dissipation. Moreover, very large coupling capacitors are necessary for the achievement of good low-frequency performance. The nonlinear impedance exhibited by the input of the output transistor causes a dc voltage to be produced across the capacitor under high signal levels. An alternate solution is to use a Darlington pair to increase the input impedance of the output stage.

Class AB Series-Output Amplifiers —For applications in which low distortion and wide frequency response are major requirements, a transformerless approach is usually employed in the design of audio power amplifiers. With this approach, the common type of circuit configuration used is the series-output amplifier.

The class-AB-operated n-p-n transistors used in the series-output circuits shown in Fig. 375 require some form of phase inversion of the drive signal for push-pull operation. A common approach is to use a driver transformer that has split secondary windings, as shown in Fig. 376. The split secondary windings are required because of the mode in which

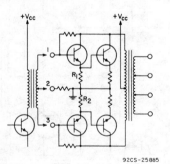

92CS-25885

Fig. 374—Class AB, push-pull transformer-coupled audio output stage in which Darlington pairs are used to reduce drive requirements of output transistors.

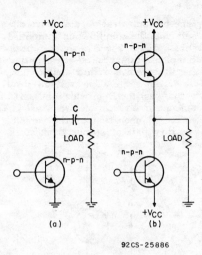

Fig. 375—*Circuit arrangements for operation of series output circuit from (a) a single dc supply and (b) symmetrical dual supplies.*

each of the series output transistors operates.

If ground were used as the drive reference for both secondary windings of the circuit shown in Fig. 376, transistor Q_1 would operate as an

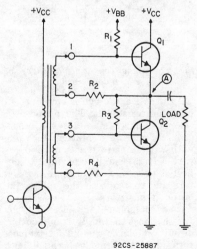

Fig. 376—*Circuit using a driver transformer that has split secondary windings to provide phase inversion for push-pull operation of a series-output circuit.*

emitter-follower and would provide gain of somewhat less than unity. Transistor Q_2, however, is connected in a common-emitter configuration which can provide substantial voltage gain. For equal output-voltage swings in both directions, the drive input to transistor Q_1 is applied directly across the base and emitter terminals. Transistor Q_1 is then effectively operated in a common-emitter configuration (although there is no phase reversal from input to output) and has a voltage gain equal to that of transistor Q_2.

The disadvantages of a driver transformer discussed previously also apply to the circuit shown in Fig. 376. In addition, coupling through interwinding capacitances can adversely affect the performance of the circuit. Such coupling is particularly serious because at both ends of the upper secondary (terminals 1 and 2) the ac voltage with respect to ground is approximately equal to the output voltage. During signal conditions, when output transistor Q_1 is turned on, this coupling provides an unwanted drive to Q_1. The forward transistor bias required to maintain class AB circuit operation is provided by the resistive voltage divider R_1, R_2, R_3, and R_4. These resistors also assure that the output point between the two transistors (point A) is maintained at one-half the dc supply voltage V_{CC}.

As in the case of the transformer-coupled output, phase inversion can be accomplished by use of an additional transistor. Fig. 377 shows a circuit in which the transistor phase inverter is used, together with a Darlington output stage to minimize loading on the phase inverter. It should be noted that capacitor C provides a drive reference back to the emitter of the upper output transistor. In effect, this arrangement duplicates the drive conditions of the split-winding transformer approach. A disadvantage of this circuit is the high quiescent dissipation of the phase inverter Q_1 which is

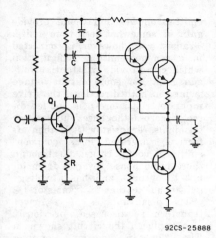

92CS-25888

Fig. 377—Push-pull series-output amplifier in which driver and output transistors are connected as Darlington pairs and drive-signal phase inversion is provided by phase-splitter stage Q_1.

type of configuration, the drive circuit for the amplifier is simplified substantially. Fig. 378 shows a basic complementary type of series-output circuit together with a simple class A driver stage. The voltage drop across resistor R provides the small amount of forward bias required for class AB operation of the complementary pair of output transistors.

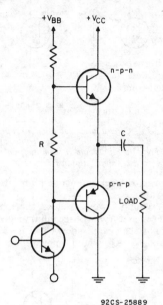

92CS-25889

Fig. 378—Basic complementary type of series-output circuit.

necessary to obtain adequate drive at full power output. An unbypassed emitter resistor R is necessary because a signal is derived from this point to drive the lower output transistor. When transistor Q_1 is driven into saturation, the minimum collector-to-ground voltage that can be obtained is limited primarily by the peak emitter voltage under these conditions. To obtain the necessary voltage swing at this collector (a voltage swing that is also approximately equal to the output voltage swing), it is necessary to use a quiescent collector-to-emitter voltage higher than that required in a stage that uses a bypassed emitter resistor.

Complementary-Symmetry Amplifiers—When a complementary pair of output transistors (n-p-n and p-n-p) is used, it is possible to design a series-output type of audio power amplifier which does not require push-pull drive. Because phase inversion is unnecessary with this

In practice, a diode is employed in place of resistor R. The purpose of the diode is to maintain the quiescent current at a reasonable value with variations in junction temperatures. It is usually thermally connected to one of the output transistors and tracks with the V_{BE} of the output transistors.

The complementary circuit is by far the most thermally stable output circuit. It places the output transistors in a V_{CES} mode because both transistors are operated with a low impedance between base and emitter. Therefore, the I_{CBO} leakage is the only component of concern in the stability criteria. At power-output levels from 3 to 20 watts, a complementary-symmetry amplifier offers advantages in terms of circuit simplicity. At higher power levels, however, the class A driver transistor is required to dissipate considerable heat, the quiescent power-supply current drain becomes significant, and excessively large filter capacitors are required to maintain a low hum level. For these reasons, the maximum practical output for a true complementary-symmetry amplifier is considered to be about 20 watts; at higher power levels, this type of amplifier is usually replaced by the quasi-complementary circuit.

Q u a s i-Complementary-Symmetry Amplifiers—In the quasi-complementary amplifier, shown in Fig. 379, the driver transistors provide the necessary phase inversion. A single but descriptive way to analyze the operation of a quasi-complementary

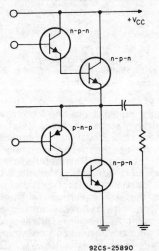

Fig. 379—Basic quasi-complementary type of series-output circuit.

amplifier is to consider the result of connecting a p-n-p transistor to a high-power n-p-n output transistor, as shown in Fig. 380. The collector current of the p-n-p transistor becomes the base current of the n-p-n transistor. The n-p-n transistor, which is operated as an emitter-follower, provides additional current gain without inversion. If the emitter of the n-p-n transistor is considered as the "effective" collector of the composite circuit, it becomes apparent that the circuit is equivalent to a high-gain, high-power p-n-p transistor. The output characteristics of the p-n-p circuit shown in Fig. 380 and of a high-gain, high-

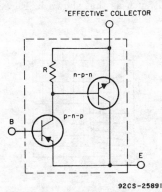

Fig. 380—Connection of n-p-n driver transistor to n-p-n output transistor.

power n-p-n circuit formed by the connection of the same type of n-p-n output transistor and an n-p-n driver transistor in a Darlington configuration, such as shown in Fig. 381, are compared in Fig. 382.

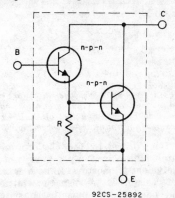

Fig. 381—Darlington connection of n-p-n driver transistor to n-p-n output transistor.

The saturation characteristics of the over-all circuit in both cases are the combination of the base-to-emitter voltage V_{BE} of the output transistor and the collector saturation voltage of the driver transistor. Moreover, in both cases the current gain is the product of the individual betas of the transistors used. A quasi-complementary amplifier, therefore, is effectively the same as a simple complementary output circuit such as that shown in Fig. 378,

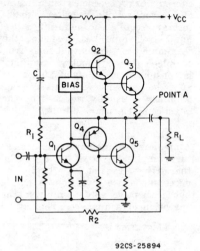

92CS-25894

Fig. 383—Quasi-complementary audio power amplifier that operates from a single dc supply.

the complementay amplifier, but presents no problem for silicon transistors.

A typical quasi-complementary amplifier is shown in Fig. 383. Capacitor C performs two functions essential to the successful operation of the circuit. First, it acts as

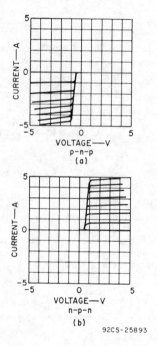

92CS-25893

Fig. 382—Output characteristics for (a) p-n-p/n-p-n driver-output transistor pair shown in Fig. 380 and for (b) Darlington pair of n-p-n transistors shown in Fig. 381.

and is formed by the use of high-gain, high-power n-p-n and p-n-p equivalent transistors. In both cases, the resistor R between the emitter and base of the output transistor places the device in a V_{CER} mode. This mode is not as stable as that of

a bypass to decouple any power-supply ripple from the driver and predriver stages. Second, it is connected as a "boot-strap" capacitor to provide the drive necessary to pull the upper Darlington pair of transistors into saturation. This latter function results from the fact that the stored voltage of the capacitor, with reference to the output point A, provides a higher voltage than the normal collector-supply voltage to drive transistor Q_2. This higher voltage is necessary during the signal conditions that exist when the upper transistors are being turned on because the emitter voltage of transistor Q_2 then approaches the normal supply voltage. An increase in the base voltage to a point above this level is required to drive the transistor into saturation. Resistor R_1 provides the necessary dc feedback to maintain point A at approximately one-half the nominal supply voltage. Over-all ac feedback from output to input is coupled by resistor R_2 to reduce distortion and to improve low-frequency performance.

Series-output circuits can be employed with separate positive and negative supplies; no series output capacitor is then required. The elimination of this capacitor may result in an economic advantage, even though an additional power supply is used, because of the size of the series output capacitor necessary in the single-supply case to obtain good low-frequency performance (e.g., a 2000-microfarad capacitor is required to provide a 3-dB point at 20 Hz for a 4-ohm load impedance). Split supplies, however, pose certain problems which do not exist in the single-supply case. The output of the amplifier must be maintained at zero potential under quiescent conditions for all environmental conditions and device parameter variations. Also, the input ground reference can no longer be at the same point as that indicated in Fig. 383, because this point is at the negative supply potential in a split-supply system.

If the ground-point reference for the input signal were a common point between the split supplies, any ripple present on the negative supply would effectively drive the amplifier through transistor Q_1, with the result that this stage would operate as a common-base amplifier with its base grounded through the effective impedance of the input signal source. To avoid this condition, the amplifier must include an additional p-n-p transistor as shown in Fig. 384. This transistor (Q_6) reduces the drive effects of the negative supply ripple because of the high collector impedance (1 megohm or more) that it presents to the base of transistor Q_1, and effectively isolates the input source impedance from transistor Q_1. In practice,

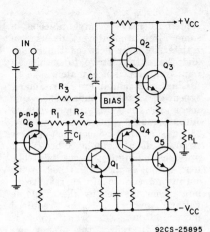

92CS-25895

Fig. 384—Quasi-complementary audio power amplifier that operates from symmetrical dual dc power supplies. The p-n-p transistor input stage is required to prevent ripple component from driving amplifier.

transistor Q_1 may be replaced by a Darlington pair to reduce the loading effects on the p-n-p predriver.

Negative dc feedback is applied from the output to the input stage by R_1, R_2, and C_1 so that the output is maintained at about zero potential. Actually, the output is maintained at approximately the forward-biased base-emitter voltage of transistor Q_6, which may be objectionable in a few cases, but which can be eliminated by a method discussed later. Capacitor C_1 effectively bypasses the negative dc feedback at all signal frequencies. Resistor R_3 provides ac feedback to reduce distortion in the amplifier.

Drive Requirements

In class A amplifiers, the output stage is usually connected in a common-emitter configuration. The relatively low input impedance that generally characterizes this type of configuration may result in a severe mismatch with the output impedance of the driver transistor. Usually, at low power levels, RC coupling is used and the loss is accepted. It may be advantageous in some circuits, however, to use an emitter-follower between the driver and the output stage to obtain an improved impedance match.

Class AB amplifiers have many types of output connections. One form is the **transformer-coupled output stage** illustrated in Fig. 385(a). Again, the common-emitter circuit is usually employed because it provides the highest power gain. The load circuit is never matched to the output impedance of the transistor, but rather is fixed by the available voltage swing and the required power output. The transformer is designed to reflect the proper im-

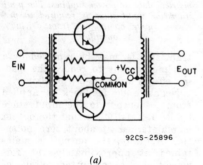

92CS-25896

(a)

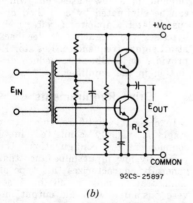

92CS-25897

(b)

Fig. 385—Class AB, push-pull amplifiers (a) with transformer-coupled output; (b) with series output connection.

pedance to the output transistors so that the desired power output can be achieved with a specific supply voltage.

The use of transformer coupling from the driver to the input of the power transistor assures that the phase split required for push-pull operation of the output stages and any necessary impedance transformation can be readily achieved. Output transformer coupling provides an easy method for matching several values of load impedance, including those encountered in sound-distribution systems. For paging service, servo motor drive, or other applications requiring a limited bandwidth, the transformer-coupled output stage is very useful. However, there are disadvantages to the use of tranformer coupling. One disadvantage is the phase shift encountered at low- and high-frequency extremes, which may lead to unstable operation. In addition, the output transistors must be capable of handling twice the supply voltage because of the transformer requirements.

In the **series-connected output stage,** the transistors are connected in series across the supply and the load circuit is coupled to the midpoint through a capacitor. There must be a 180-degree phase shift between the driving signals for the upper and lower transistors. A transformer can be used in this application provided that the secondary consists of two separate windings, as shown in Fig. 385(b). Other forms of phase splitting can be used; all have problems such as insufficient swing or poor impedance matching. Capacitor output coupling also has disadvantages. A low-frequency phase shift is usually associated with the capacitor, and it is difficult to obtain a capacitor that is large enough to produce an acceptable low-frequency output. These disadvantages can be alleviated by use of a split supply and by connection of the load between the transistor midpoint and the supply midpoint with the return path through the power-supply capacitors. The power-supply capacitors must be large enough to prevent excessive ripple.

Complementary amplifiers are produced when p-n-p and n-p-n transistors are used in series. A capacitor can be used to couple the amplifier output when a single supply is used, or direct coupling can be employed when a split power supply is used, as shown in Fig. 386. Because no

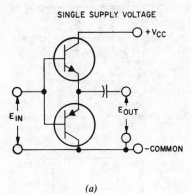

SINGLE SUPPLY VOLTAGE

(a)

SPLIT SUPPLY VOLTAGE

92CS-25898

(b)

Fig. 386—Circuit arrangements for operation of complementary output stages (a) from single dc supply; (b) from symmetrical dual (positive and negative) supplies.

phase inversion is needed in the driving circuit for this output configuration, there are definite advan-

tages in the simplicity of the design. One disadvantage of this type of amplifier is that the driver must be a class A stage which may have a high dissipation. This dissipation can be reduced, however, by use of a Darlington compound connection for the output stage. This compound connection reduces the driving-stage requirement. A method of overcoming this disadvantage completely is to use a quasi-complementary configuration. In this configuration, the output transistors are a pair of p-n-p or n-p-n transistors driven by a complementary pair in the driver. In this manner the n-p-n/p-n-p drivers provide the necessary phase inversion. The availability of both n-p-n and p-n-p silicon driving transistors that have the same electrical characteristics is good. The driving transistors are connected directly to the bases of the output transistors, as illustrated in Fig. 387.

Adequate drive may be a problem with the transistor pair shown in the upper part of the quasi-complementary amplifier unless suitable techniques are used to assure that this pair saturates. Care must also be taken when split supplies are used to assure that any ripple on the lower supply is not introduced into the predriving stages by this technique. The advantage of a split supply is that it makes possible direct connection to the load and thus improves low-frequency response.

To this point, phase inversion has been mentioned but not discussed. Phase inversion may be accomplished in many ways. The simplest electronic phase inverter is the single-stage configuration. This configuration can be used at low power levels or with high-gain devices when the limited drive capability is not a drawback. At higher power levels, some impedance transformation and gain may be required to supply the drive needed. There are several complex phase-splitting circuits; a few of them are shown in Fig. 388.

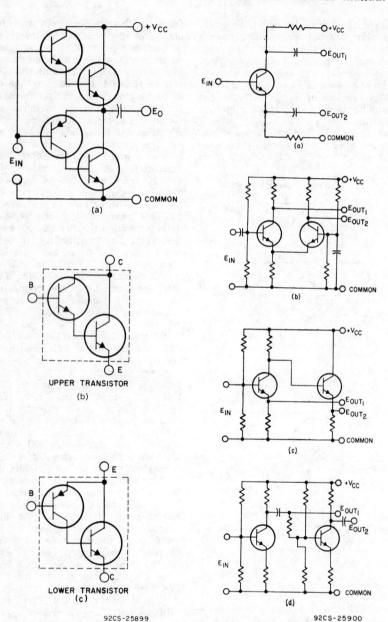

Fig. 387—Compound output stage in which output transistors are driven by complementary driver transistors: (a) overall circuit; (b) upper transistor pair; (c) lower transistor pair.

Fig. 388—Basic phase-inverter circuits: (a) single-stage phase-splitter type; (b) two-stage emitter-coupled type; (c) two-stage low-impedance type; (d) two-stage similar-amplifier type.

92CS-25899

92CS-25900

Power Output in Class B Audio Amplifiers

For all cases of practical interest, the power output (P_o) of an audio amplifier is given by the following equation:

$$P_o = I(\text{rms}) \times E(\text{rms}) = (I_p E_p)/2$$
$$= (I_p{}^2 R_L)/2 = E_p{}^2/2R_L \quad (1)$$

where I_p and E_p are the peak load current and voltage, respectively, and R_L is the load impedance presented to the transistor. Fig. 389 shows the relationship among these various factors in graphic form. Ob-

viously, the peak load current is the peak transistor current, and the transistor breakdown-voltage rating must be at least twice the peak load voltage. The vertical lines that denote 4-ohm, 8-ohm, and 16-ohm resistances are particularly useful for transformerless designs in which the transistor operates directly into the loudspeaker.

Rating Methods—The **Institute of High Fidelity** (IHF) and the **Electronic Industries Association** (EIA) have attempted to standardize power-output ratings to establish a common reference of comparison and to provide a solid definition of

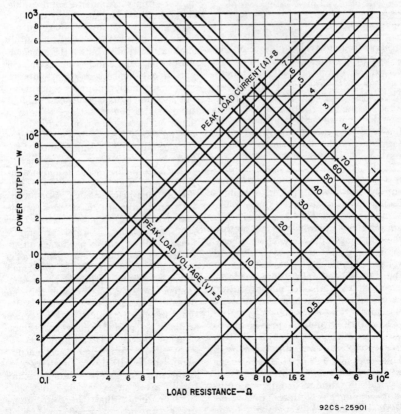

Fig. 389—Peak transistor currents and load voltages for various output powers and load resistances.

92CS-25901

the capabilities of audio power amplifiers. Obviously, an audio power amplifier using an unregulated supply can deliver more output power under transient conditions than under steady-state conditions. The rating methods which have been standardized for this type of operation are the **IHF Dynamic Output Rating (IHF-A-201)** and the **EIA Music Power Rating (EIA RS-234-A)**.

Both of these measurement methods allow the use of regulated supply voltage to simulate transient conditions. Because the regulated supply has no source impedance or ripple, the results do not completely represent the transient conditions, as will be explained later.

Measurement techniques: The EIA standard is used primarily by manufacturers of packaged equipment, such as portable phonographs, packaged stereo hi-fi consoles, and packaged home-entertainment consoles. The **EIA music power output** is defined as the power obtained at a total harmonic distortion of 5 per cent or less, measured after the "sudden application of a signal during a time interval so short that supply voltages have not changed from their no-signal values." The supply voltages are bypassed voltages. These definitions mean that the internal supply may be replaced with a regulated supply equal in voltage to the no-signal voltage of the internal supply. For a stereo amplifier, the music power rating is the sum of both channels, or twice the single-channel rating.

The IHF standard provides two methods to measure dynamic output. One is the **constant-supply method.** This method assumes that under music conditions the amplifier supply voltages undergo only insignificant changes. Unlike the EIA method, this measurement is made at a reference distortion. The constant-supply method is used by most

high-fidelity component manufacturers. The reference distortion chosen is normally less than one per cent, or considerably lower than the EIA value of 5 per cent used by packaged-equipment manufacturers.

A second IHF method is called the **"transient distortion" test.** This method requires a complex setup including a low-distortion modulator with a prescribed output rise time and other equipment. The modulator output is required to have a rise time of 10 to 20 milliseconds to simulate the envelope rise time of music and speech. This measurement is made using the internal supply of the amplifier and, consequently, includes distortion caused by voltage decay, power-supply transients, and ripple. This method tends to be more realistic and to yield lower power-output ratings than the constant-supply method. Actually, both IHF methods should be used, and the lowest power rating obtained at reference distortion with both channels operating, both in and out of phase, should be used as the power rating. (There is some question concerning unanimity among high-fidelity manufacturers on actually performing both IHF tests.)

Because music is not a continuous sine wave, and has average power levels much below peak power levels, it would appear that the music power or dynamic power ratings are true indications of a power amplifier's ability to reproduce music program material. The problem is that all three methods described have a common flaw. Even the transient-distortion method fails to account for the ability of the audio amplifier to reproduce power peaks while it is already delivering some average power. The amplifier is almost never delivering zero output when it is called on to deliver a transient. For every transient that occurs after an extremely quiet passage or zero signal, there are hundreds that are imposed on top of some low but non-zero average power level.

This condition can best be clarified by consideration of the power supply. Many amplifiers have regulated supplies for the front-end or low-level stages, but almost none provides a regulated supply for the power-output stages because regulation requires extra transistors or other devices; it becomes costly, especially at high power levels. The power supply for the output stages of power amplifiers is commonly a nonregulated rectifier supply having a capacitive input filter. The output voltage of such a supply is a function of the output current and, consequently, of the power output of the amplifier.

Effect of power-supply regulation:
Power-supply regulation is dependent on the amount of effective internal series resistance present in the power supply. The effective series resistance includes such things as the dc resistance of the transformer windings, the amount and type of iron used in the transformer, the amount of surge resistance present, the resistance of the rectifiers, and the amount of filtering. The internal series resistance causes the supply voltage to drop as current is drawn from the supply.

Fig. 390 shows a typical regulation curve for a rectifier power supply that has a capacitive input filter.

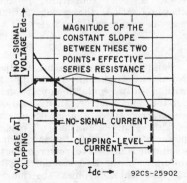

Fig. 390—*Regulation curve for capacitive rectifier power supply.*

The voltage is a linear function of the average supply current over most of the useful range of the supply. However, a rapid change in slope occurs in the regions of both very small and very large currents. In class B amplifiers, the no-signal supply current normally occurs beyond the low-current knee, and the current required for the amplifier at the clipping level occurs before the high-current knee. The slope between these points is nearly linear and may be used as an approximation of the equivalent series resistance of the supply.

The amount of power lost depends on the quality of the power supply used in the amplifier. Accordingly, rating amplifier power output with a superb **external** power supply (that is, not using the built-in amplifier power supply) provides false music power outputs. Under actual usage, the output is lower.

It should be emphasized that, while there is a discrepancy between the actual power available and the power measured under the EIA Music Power or the IHF Dynamic Power methods, these methods are not without merit. The IHF dynamic power rating, in conjunction with the continuous power rating, produces an excellent indication of how the amplifier will perform. The EIA music power rating, which is measured at a total harmonic distortion of 5 per cent with a regulated power supply, provides a less adequate indication of amplifier performance because there is no indication of how the amplifier power-supply voltage reacts to power output.

Some important factors considered by packaged-equipment manufacturers, the primary users of the EIA music power rating, are mostly economic in nature and affect many aspects of the amplifier performance. Because there is no continuous power output rating required, two amplifiers may receive the same EIA music power rating but have different continuous power ratings.

The ratio of music power to continuous power is, of course, a function of the regulation and effective series resistance of the supply.

One reason for the difference between ratings used by the console or the packaged-equipment manufacturer and those used by the hi-fi component manufacturer is that the latter does not always know just what will be required of the amplifier. The console manufacturer always designs an amplifier as part of a system, and consequently knows the speaker impedances and the power required for adequate sound output. The console manufacturer may use high-efficiency speakers requiring only a fraction of the power needed to drive many component-type acoustic-suspension systems. The difference may be such that the console may produce the same sound pressure level with an amplifier having one-tenth of the power output. High ratios of music-power to continuous-power capability are common in these consoles. A typical ratio of IHF music power to continuous power may be 1.2 to 1 in component amplifiers, whereas a typical ratio of EIA music power to continuous power in a console system may be 2 to 1. Console manufacturers use the EIA music power rating to economic advantage as a result of the reduced regulation requirement of the power supply. A high ratio of music power to continuous power means higher effective series resistance in the power supply. This resistance, in turn, means less continuous dissipation on the output transistors, smaller heat sinks, and a lower-cost power supply.

Basic Power-Dissipation Relationships—Under ideal conditions (i.e., with a perfectly regulated dc power supply), maximum transistor power dissipation in a class B audio output stage, such as the complementary circuits shown in Fig. 391, is approximately 20 per cent of the

maximum unclipped sine-wave power output and occurs when the output stage is delivering approximately 40

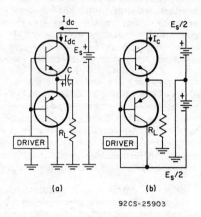

Fig. 391—Typical complementary-symmetry circuits.

per cent of the maximum output power to the load. Clipping begins at the point where the peak collector current I_{pk} is given by

$$I_{pk} = \frac{E_o \ \pi}{R_S + 2 \ \pi \ R_L}$$

Power output at clipping can then be expressed as follows:

$$P_o \text{ (clipping)} = \frac{E_o{}^2 \ \pi^2 \ R_L}{2 \ (R_S + 2\pi \ R_L)^2}$$

If $R_S = 0$ is substituted in the above equation, the power output may be expressed as follows:

$$P_o = E_o{}^2/8 \ R_L$$

This value is equivalent to the power output just prior to clipping with a fully regulated supply and, for the remainder of this discussion, is referred to as the music power output. [This definition of music power output, i.e., as the maximum unclippued sine-wave power output, differs from the EIA standard (RS-234-A), which definies the music power output as the point at which the total harmonic distortion is 5 per cent when a regulated supply is used. The EIA value is about 10 per cent greater.]

Maximum average transistor dissipation is related to the music power output by the following expression:

$$\frac{P_T \ (max)}{P_o \ (music)} = \left[\frac{\pi^2}{2} + \frac{R_S}{R_L} \right] - 1$$

The power output at which maximum average transistor dissipation occurs P_o (max diss.) is related to the music power output as follows:

$$\frac{P_o \ (max \ diss.)}{P_o \ (music)} =$$

$$\left[\frac{\pi^2}{4} + \frac{R_S}{R_L} + \frac{R_S{}^2}{\pi^2 \ R_L{}^2} \right] - 1$$

The continuous power output at the clipping level, P_o (clipping), is related to the music power output by the following expression:

$$\frac{P_o \ (clipping)}{P_o \ (music)} =$$

$$\left[1 + \frac{R_S}{\pi \ R_L} + \frac{R_S{}^2}{4 \ \pi^2 \ R_L{}^2} \right] - 1$$

The three equations above are plotted in Fig. 392. Power levels are normalized with respect to the music power output and are plotted as a function of R_S/R_L.

The equations plotted in Fig. 392 suggest some interesting possibilities. Transistor power dissipation is only a small fraction of the clipping power output for higher ratios of R_S/R_L. For example, a 100-watt amplifier could be built using transistors and associated heat sinks capable of only about 7 watts of maximum dissipation each.

The equations presented, however, do not consider high line voltage or effects of ripple voltage. Calculations for average transistor dissipation should also include no-signal bias dissipation and the increase in

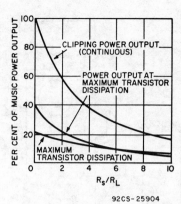

92CS-25904

Fig. 392—Power output and dissipation as functions of R_S and R_L.

bias dissipation with increasing ambient and junction temperatures in class AB circuits. Storage effects, phase shift, and thermal tracking should also be considered.

Of the above factors, bias dissipation probably contributes the greatest percentage of average worst-case transistor dissipation. The output stage is usually biased on slightly (class AB) to reduce cross-over distortion.

It is possible, however, to design amplifiers for which bias dissipation is not a problem. One such amplifier is shown in Fig. 393. The bias dissipation in this amplifier is negligible at all practical temperatures. One side is cut off and the other conducts less than one milliampere. Thermal runaway cannot be initiated in the output stage at any junction temperature below its maximum rating. Consequently, thermal tracking may also be neglected so long as the ambient temperature plus the product of the instantaneous dissipation times the junction-to-ambient thermal resistance is less than the maximum junction temperature rating.

Storage effects are also reduced as a result of the reverse bias provided for the off-transistor by the

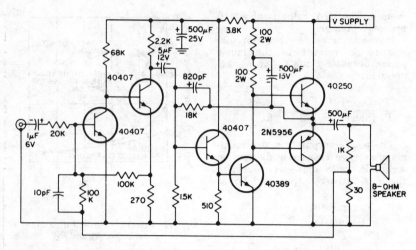

92CS-25905

Fig. 393—Class B complementary-symmetry power amplifier.

on-transistor in complementary symmetry. This circuit, then, is one practical example of an amplifier capable of achieving the characteristics shown in Fig. 392.

Ratio of Music Power to Continuous Power—Some advantages of high values of the ratio R_s/R_L and correspondingly high ratios of music power output to transistor dissipation are as follows:

1. Reduced heat sink or transistor cost: Because the volt-ampere capacity of the transistor is determined by the music power output, it is not likely that reduced thermal-resistance requirements will produce significant cost reductions. Alternatively, the heat-sink requirements may be reduced.

2. Reduced power supply costs: Transformer and/or filter-capacitor specifications may be relaxed.

3. Reduced speaker cost: Continuous power-handling capability may be relaxed.

These cost reductions may be passed along to the consumer in the form of more music power per dollar.

The question arises as to how high the ratio R_s/R_L and the corre-

sponding ratio of music power output to continuous power output may go before the capability of the amplifier to reproduce program material is impaired.

The objective is to provide the listener with a close approximation of an original live performance. Achievement of this objective requires the subjective equivalents of sound pressure levels that approach those of a concert hall. Although the peak sound pressure level of a live performance is about 100 dB, the average listener prefers to operate an audio system a a peak sound pressure level of about 80 dB. The amplifier, however, should also accommodate listeners who desire higher-than-average levels, perhaps to peaks of 100 dB.

A sound pressure level of 100 dB corresponds to about 0.4 watt of acoustic power for an average room of about 3,000 cubic feet. If speaker efficiencies are considered to be in the order of 1 per cent, a stereophonic amplifier must be capable of delivering about 20 watts per channel. Higher power outputs are required for lower-efficiency speakers. The peak-to-average level for most program material is between 20 and

23 dB. A system capable of providing a continuous level of 77 dB and peaks of 100 dB would satisfy the power requirements of nearly all listeners. For this performance to be attained, the power-supply voltage cannot drop below the voltage required for 100 dB of acoustic power while delivering the average current required for 77 dB. Moreover, because sustained passages that are as much as 10 dB above the average may occur, the power-supply voltage cannot drop below the value required for 100 dB of acoustic power while delivering 87 dB of acoustic power (87 dB of acoustic power corresponds to about 1 watt per channel). This performance means that for 8-ohm loads, with output-circuit losses neglected, the power-supply voltage must not decrease to a value less than 36 volts, while delivering the average current required for 1 watt per channel (0.225 ampere dc).

It should be noted that the power-output capability for peaks while the amplifier is delivering a total of 2 watts is not the music power rating of the amplifier because the power-supply voltage is below its no-signal value by an amount depending on its effective series resistance.

Maximum Effective Series Resistance—There is a relationship between the maximum effective series resistance of the power supply and the music power rating of the amplifier if it is to perform to the standards as outlined above.

The power-supply series resistance R_s may be expressed as a function of music-power output, as follows:

$$R_s = \left[\frac{(8R_L) \; P_o \; \text{(music)}}{\overline{I^2}} \right]^{1/2} - \frac{E_s \; \text{(min)}}{\overline{I}}$$

where E_s(min) is the minimum volt-

age required for 100 dB of acoustical power output and I is the current required for 87 dB of acoustical power output, less the idle current. [E_s(min) should, in practice, be increased by peak output-circuit voltage losses.]

The equation for R_s given above is plotted in Fig. 394. The value of R_s is the absolute maximum value of effective supply resistance for each music-power value that will allow the amplifier to deliver a mini-

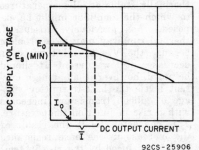

Fig. 394—*Power-supply regulation curve.*

mum of 100 dB of acoustical power output as described above.

Use of Fig. 395 in conjunction with Fig. 392 shows that very high ratios of music power output to continuous power output may be employed without sacrifice of the subjective ability of the amplifier to reproduce program material. This

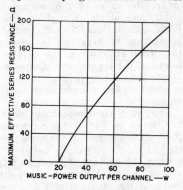

Fig. 395—*Maximum effective series resistance as a function of music power output.*

technique provides economic advantages while adhering to a minimum "power margin" for the faithful reproduction of program material, even at loud levels.

Thermal-Stability Requirements

One serious problem that confronts the design engineer is the achievement of a circuit which is thermally stable at all temperatures to which the amplifier might be exposed. As previously discussed, thermal runaway may be a problem because the V_{BE} of all transistors decreases at low current. It should be noted, however, that at high current levels the base-to-emitter voltage of silicon transistors increases with a rise in junction temperature. This characteristic is the result of the increase in the base resistance that is produced by the rise in temperature. The increase in base resistance helps to stabilize the transistor against thermal runaway. In high-power amplifiers, the emitter resistors employed usually have a value of about 1 ohm or less. The size of the capacitor required to bypass the emitter adequately at all frequencies of interest makes this approach economically impractical. A more practical solution is to increase the value of the emitter resistor and shunt it with a diode. With this technique, sufficient degeneration is provided to improve the circuit stability; at low currents, however, the maximum voltage drop across the emitter resistor is limited to the forward voltage drop of the diode.

The quasi-complementary amplifier shown in Fig. 396 incorporates the stabilization techniques described. A resistor-diode network is used in the emitter of transistor Q_3, and another such network is used in the collector of transistor Q_5, with the emitter of transistor Q_4 returned to the collector of transistor Q_5. Previous discussion regarding the p-n-p

driver and n-p-n output combination (Q_4 and Q_5) showed that the collector of the output device becomes the "effective" emitter of the high-gain, high-power p-n-p equivalent, and vice versa. For maximum operating-point stability, therefore, the diode-resistor network should be in the "effective" emitter of the p-n-p equivalent. Quasi-complementary circuits employing the stabilization resistor in the emitter of the lower output transistor, as shown in Fig. 383, do not improve the operating-point stability of the over-all circuit.

The circuit shown in Fig. 396 is biased for class AB operation by the voltage obtained from the forward drop of two diodes, D_1 and D_2, plus the voltage drop across potentiometer R, which affords a means for a slight adjustment in the value of the quiescent current. The current necessary to provide this voltage reference is the collector current of driver transistor Q_1.

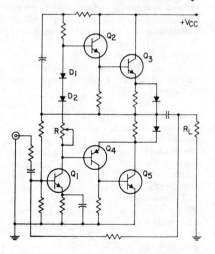

92CS-25908

Fig. 396—Quasi-complementary amplifier that incorporates two stabilization networks.

The diodes may be thermally connected to the heat sink of the output transistors so that thermal feedback is provided for improvement of further thermal stability. Because the forward voltage of the reference diodes decreases with increasing temperature, these diodes compensate for the decreasing V_{BE} of the output transistors by reducing the external bias applied. In this way, the quiescent current of the output stage can be held relatively constant over a wide range of operating temperatures.

Effects of Large Signal Phase Shifts

The amplifier frequency-response characteristic is an important factor with respect to the ability of the amplifier to withstand unusually severe electrical stress conditions. For example, under certain conditions of input-signal amplitude and frequency, the amplifier may break into high-frequency oscillations which can lead to destruction of the output transistors, the drivers, or both. This problem becomes quite acute in transformer-coupled amplifiers because the characteristics of transformers depart from the ideal at both low and high frequencies. The departure occurs at low frequencies because the inductive reactance of the transformer decreases, and at high frequencies because the effects of leakage inductance and transformer winding capacitance become appreciable. At both frequency extremes, the effect is to introduce a phase shift between input and output voltage.

Negative feedback is used almost universally in audio amplifiers; the voltage coupled back to the input through the feedback loop may cause the amplifier to be potentially unstable at some frequencies, especially if the additonal phase shift is sufficient to make the feedback positive. Similar effects can occur in transformerless amplifiers because reactive elements (such as coupling and bypass capacitors, transistor junction capacitance, stray wiring ca-

pacitance, and inductance of the loudspeaker voice coil) are always present. The values of some of the reactive elements (e.g., transistor junction capacitance and transformer inductance as the core nears saturation) are functions of the signal level; coupling through wiring capacitance and unavoidable ground loops may also vary with the signal level. As a result, an amplifier that is stable under normal listening levels may break into oscillations when subjected to high-level signal transients.

A large phase shift is not only a potential cause of amplifier instability, but also results in additional transistor power dissipation and increases the susceptibility of the transistor to forward-bias second-breakdown failures. The effects of large-signal phase shifts at low frequencies are illustrated in Fig. 397, which shows the load-line characteristics of a transistor in a class AB push-pull circuit similar to that shown in Fig. 396 for signal frequencies of 1000 Hz and 10 Hz. The phase shift is caused primarily by the output capacitor. In both cases, the amplifier is driven very strongly into saturation by a 5-volt input signal. The increased dissipation at 10 Hz, compared to that obtained at 1000 Hz, results from simultaneous

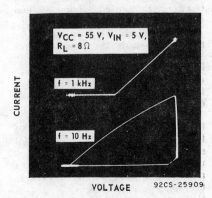

Fig. 397—Effect of large signal phase shift on the load-line characteristics of a transistor at low frequencies.

high-current high-voltage operation. The transistor is required to handle safely a current of 0.75 ampere at a collector voltage of 40 volts for an equivalent pulse duration of about 10 milliseconds; it must be free from second-breakdown failures under these conditions.

Effect of Excessive Drive

Simultaneous high-current high-voltage operation may also occur in class B amplifiers at high frequencies when the amplifier is overdriven to the point that the output signals are clipped. For example, if the input signal applied to the series-output push-pull circuit shown in Fig. 398(a) is large enough to drive the transistors into both saturation and cutoff, transistor A is drive into saturation, and transistor B is cut off during a portion of the input cycle. Fig. 398(b) shows the collector-current waveform for transistor A under these conditions.

During the interval of time from t_2 to t_3, transistor A operates in the saturation region, and the output voltage is clipped. The effective negative feedback is then reduced because the output voltage does not follow the sinusoidal input signal. Transistor A, therefore, is driven even further into saturation by the unattenuated input signal. When transistor B starts to conduct, transistor A cannot be turned off immediately because the excessive drive results in a large storage time. As a result, transistor B is required to support almost the full supply voltage (less only the saturation voltage of transistor A and the voltage drop across the emitter resistors, if used) as its current is increased by the drive signal. For this condition to occur, a large input signal is required at a frequency high enough so that the storage time is greater than one-quarter cycle.

Because of the charging current through the output coupling capacitor, transistor A in Fig. 398(a) is

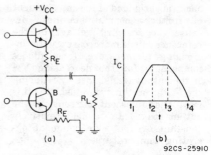

92CS-25910

Fig. 398—(a) Class B series-output stage, (b) collector-current waveform under over-drive (clipping) conditions.

also subject to forward-bias second-breakdown failure if the dc supply voltage and a large input signal are applied simultaneously.

All of these conditions point to the need for a good "safe area" of operation. Fig. 399 shows the safe area for the RCA-2N3055. In all cases, the load lines fall within the area guaranteed safe for this transistor.

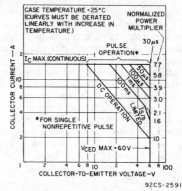

92CS-25911

Fig. 399—Safe-area-of-operation rating chart for the RCA-2N3055 homotaxial-base transistor.

Short-Circuit Protection

Another important consideration in the design of high-power audio amplifiers is the ability of the circuit to withstand short-circuit conditions. As previously discussed, overdrive conditions may result in disastrously high currents and excessive dissipation in both driver and output stages. Obviously, some form of short-circuit protection is necessary. One such technique is shown in Fig. 400. A current-sampling resistor R is placed in the ground leg of the load. If any condition (including a short) exists such that higher-than-normal load current flows, diodes D_1 and D_2 conduct on alternate half-cycles and thus provide a high negative feedback which effectively reduces the drive of the amplifiers. This feedback should not exceed the stability margin of the amplifier. This technique in no way affects the normal operation of the amplifier.

A second approach to current limiting is illustrated by the circuit shown in Fig. 401. In this circuit, a diode biasing network is used to establish a fixed current limit on the driver and output transistors. Under sustained short-circuit conditions, however, the output transistors are required to support this current limit and one-half the dc supply voltage.

The circuit shown in Fig. 402 illustrates a dissipation-limiting technique that provides positive protection under all loading conditions. The limiting action of this circuit is

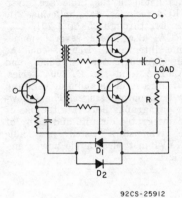

92CS-25912

Fig. 400—Push-pull power amplifier with short-circuit protection.

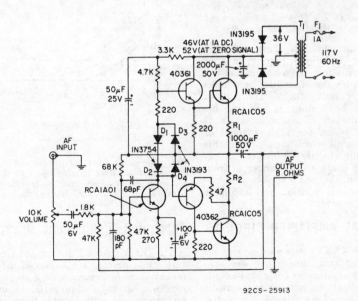

92CS-25913

Fig. 401—25-watt (rms) quasi-complementary audio amplifier using current-limiting diodes (D_3 and D_4).

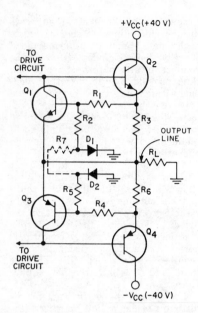

+V_CC (+40 V)

TO DRIVE CIRCUIT

Q₁ R₁ Q₂

R₂ R₃

R₇ D₁ OUTPUT LINE

R_L

D₂

Q₃ R₅ R₆

R₄

TO DRIVE CIRCUIT

Q₄

-V_CC (-40 V)

92CS-25914

Fig. 402—Quasi-complementary audio output in which diode-resistor biasing network is used to prevent complementary transistors Q₁ and Q₂ from being forward-biased by the output voltage swing.

shown in Fig. 403. This safe-area limiting technique permits use of low-dissipation driver and output transistors and of smaller heat sinks in the output stages. The use of smaller heat sinks is possible because the worst-case dissipation is normal 4-ohm operation instead of short-circuit conditions. With this technique, highly inductive or capacitive loads are no longer a problem, and thermal cut-outs are unnecessary. In addtion, the technique is inexpensive.

TYPICAL AMPLIFIER CIRCUITS

Table XXXVI lists a broad variety of power transistors that are specially characterized for audio-amplifier applications at rms power-output levels from 12 watts to 200 watts. Even higher power outputs can be obtained by connections of several

output transistors in parallel. The transistors listed in Table XXXVI are classified according to function in the circuit for which each type is recommended. Ratings and characteristics data for these transistors and the circuit diagram and performance data for the amplifier circuits listed in the table are given in RCA technical data bulletins. File Nos. 642 through 653 and 791.

RCA also offers several monolithic integrated circuits, such as the CA3007, the CA3020 and CA3020A, and the CA3094 series, that are useful as audio drivers and low-level output stages. The following paragraphs illustrate the use of the integrated circuits and many of the transistors listed in Table XXXVI in typical audio power amplifiers.

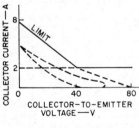

92CS-25915

Fig. 403—Load lines for the circuit of Fig. 402. Load lines showing effect of the inclusion of high-resistance diode-resistor network in the forward-biasing path of Q₁ are shown dotted.

Integrated-Circuit Audio Driver

The RCA-CA3007 integrated-circuit audio driver is a balanced differential configuration with either a single-ended or a differential input and two push-pull emitter-follower outputs. The circuit is intended for use as a direct-coupled driver in a class B audio amplifier which exhibits both gain and operating-point stability over the temperature range from −55 to 125°C. Because of its circuit configuration (a balanced differential pair fed by a constant-current transistor), the CA3007 is an excellent controlled-gain audio driver for systems requiring audio squelching. This circuit is also usable as a servo driver. The audio driver circuit is available in a 12-terminal TO-5 low-silhouette package.

Dual-Supply Audio Driver in a Direct-Coupled Audio Amplifier—Fig. 404 shows the CA3007 used as a dual-supply audio driver in a direct-coupled audio amplifier. This amplifier provides a power output of 300 milliwatts for an audio input of 0.3 volt rms (V_{CC} = 6 volts, V_{EE} = −6 volts, V = 30 volts). For a voltage V of 6 volts, the output power is 10 milliwatts without transformer optimization; the use of a lower-impedance transformer would permit power outputs in the order of 100 milliwatts.

The external resistor R connected between terminals 3 and 4 is used to set the class B output-stage standby current as required for a particular application. If the standby current is too low, crossover distortion will result; if it is too high, standby power drain will be excessive. Decreasing the value of resistor R reduces the standby current; for a standby current of 10 milliamperes, R is typically 10,000 ohms.

Terminal 2 must be grounded or, if an audio squelch is desired, must be connected to a positive voltage supply of 5 volts minimum. When terminal 2 is near ground, the audio amplifier functions normally. When terminal 2 is at 5 volts, the differential pair of the audio driver saturates, and the push-pull output stage is cut off. The squelch source must be capable of supplying a current of 1.5 milliamperes in the 5-volt condition, and 0.75 milliampere in the near-ground condition.

Table **XXXVII** shows values of harmonic distortion and intermodulation distortion for the amplifier.

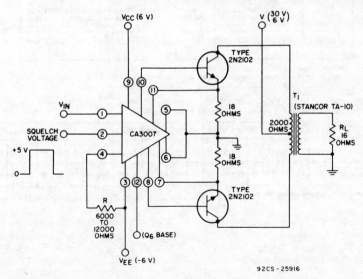

Fig. 404—CA3007 used as an audio driver for a direct-coupled 300-milliwatt audio amplifier.

Table XXXVI—Types for Audio-Frequency Linear Amplifiers

Power Output (8 Ω Imped.)	Bull. File No.	Circuit No.	Output Transistors N-P-N	Output Transistors P-N-P	Class B Driver Transistors N-P-N	Class B Driver Transistors P-N-P	V_{BE} Mult. (Bias)
12 W	642	A012B (True Comp.)	RCA1C10 (2N6292)	RCA1C11 (2N6107)	–	–	–
	642	A012D (IC Driving True Comp.)	RCA1C10 (2N6292)	RCA1C11 (2N6107)	–	–	–
25 W	643	A025C (Quasi-Comp.)	RCA1C14 [2] (2N5496)	–	RCA1A06 (2N2102)	RCA1A05 (2N4036)	–
	644	A025B (Full-Comp.)	RCA1C05 (2N6292)	RCA1C06 (2N6107)	RCA1A06 (2N2102)	RCA1A05 (2N4036)	–
40 W	645	A040C (Quasi-Comp.)	RCA1C09 [2] (2N6103)	–	RCA1A06 (2N2102)	RCA1A05 (2N4036)	–
	646	A040B (Full-Comp.)	RCA1C07 (2N6488)	RCA1C08 (2N6491)	RCA1A06 (2N2102)	RCA1A05 (2N4036)	–
	791	A040D (Full-Comp. Darlington Output)	RCA1B07 (2N6385)	RCA1B08 (TA8925)	–	–	RCA1A18 (2N2102)
70 W	647	A070A (Quasi-Comp. Hom. Output)	RCA1B01 [2] (2N3055)	–	RCA1A03 (2N5320)	RCA1A04 (2N5322)	–
	648	A070C (Quasi-Comp.)	RCA1B06 [2] (2N5840)	–	RCA1C03 (2N6474)	RCA1A04 (2N6476)	RCA1A18 (2N2102)
120 W	649	A120C (Quasi-Comp. Parallel Output)	RCA1B04 [4] (2N5240)	–	RCA1C12 (2N6474)	RCA1C13 (2N6476)	RCA1A18 (2N2102)
200 W	650	A200C (Quasi-Comp. Parallel Output)	RCA1B05 [6] (2N5240)	–	RCA1B05 [2] (2N5240)	–	RCA1A18 (2N2102)

Numbers in brackets indicate number of devices used in the stage.
Type numbers in parentheses indicate the transistor-family designation.

Typical Power Output for 4 Ω and 16 Ω Load for AF Linear Amplifiers

Amplifier Circuit No.	A012B		A012D		A025A		A025B		A040A		A040B		A040D		A070A		A070C		A120C		A200C	
Impedance – Ω (Load)	4	16	4	16	4	16	4	16	4	16	4	16	4	8	4	16	4	16	4	16	4	16
Typical Power Output – W	12*	6.5	9*	6.5	45	16	45	16	55	25	75	25	40	30	100	40	100	50	180	80	300	130

*Power output limited by driver-circuit capability.

Table XXXVI—Types for Audio-Frequency Linear Amplifiers (cont'd)

Class B Pre-Driver Transistors		Protection Circuit		Class A Pre-Driver Transistors		Input Devices
N-P-N	P-N-P	N-P-N	P-N-P	N-P-N	P-N-P	
–	–	–	–	–	RCA1A08 (2N4036)	RCA1A07 [2] (2N2102)
–	–	–	–	–	–	CA3094AT
–	–	RCA1A18 (2N2102)	RCA1A19 (2N4036)	RCA1A01 (2N2102)	–	RCA1A02 [2] (2N4036)
–	–	RCA1A18 (2N2102)	RCA1A19 (2N4036)	RCA1A01 (2N2102)	–	RCA1A02 [2] (2N4036)
–	–	RCA1A18 (2N2102)	RCA1A19 (2N4036)	RCA1A01 (2N2102)	–	RCA1A02 [2] (2N4036)
–	–	RCA1A18 (2N2102)	RCA1A19 (2N4036)	RCA1A01 (2N2102)	–	RCA1A02 [2] (2N4036)
–	–	RCA1A18 (2N2102)	RCA1A19 (2N4036)	RCA1A15 [2] (2N3440)	RCA1A16 [2] (2N5416)	RCA1A17 [2] (2N2102)
–	–	RCA1A18 (2N2102)	RCA1A19 (2N4036)	RCA1A17 (2N2102)	–	RCA1A02 [2] (2N4036)
–	–	RCA1A18 (2N2102)	RCA1A19 (2N4036)	RCA1A15■ [2] (2N3439)	RCA1A16 [2] (2N5415)	RCA1A17 [2] (2N2102)
–	–	RCA1A18 (2N2102)	RCA1A19 (2N4036)	RCA1A09■ [2] (2N3439)	RCA1A10 [2] (2N5415)	RCA1A11 [2] (2N3439)
RCA1E02 (1N3585)	RCA1E03 (2N6211)	RCA1A18 (2N2102)	RCA1A19 (2N4036)	RCA1A09■ [2] (2N3439)	RCA1A10 [2] (2N5415)	RCA1A11 [2] (2N3439)

■ Current Source

Other applications for the types above . . .
Audio Power Amplifiers–Linear Modulators–Servo Amplifiers–Operational Amplifiers

Single-Supply Audio Driver in a Capacitor-Coupled Audio Amplifier —Fig. 405 shows the CA3007 used as a single-supply audio driver in a capacitor-coupled audio amplifier. This amplifier provides a power output of 30 milliwatts for an audio input of 6.5 millivolts rms (V_{cc} =

Table XXXVII—Distortion Measurements for Direct-Coupled Amplifier Shown in Fig. 404

HARMONIC DISTORTION

Power Output (mW)	Output-Signal Level (mVrms) with 2-kHz Input Signal						Harmonic Distortion (%)
	2 kHz	4 kHz	6 kHz	8 kHz	10 kHz	12 kHz	
62.5	1000	9	3.0	—	—	—	0.95
140	1500	18	4.0	2.0	1.4	1.0	1.24
250	2000	25	4.2	5.0	1.0	1.5	1.30
330	2300	27	6.0	9.0	3.0	2.0	1.27

INTERMODULATION DISTORTION

Output-Signal Level:

at f_1 (2 kHz) .... 1000 mV rms
at f_2 (3 kHz) .... 1000 mV rms
at $2f_2-f_1$ (4 kHz) .... 0.7 mV rms
3rd-Order IMD .... 0.07 %

9 volts) with the transformer shown.

The connection shown in Fig. 405 still represents a differential-pair phase splitter fed from a constant-current transistor. The two output signals from the phase splitter are direct-coupled through two emitter-

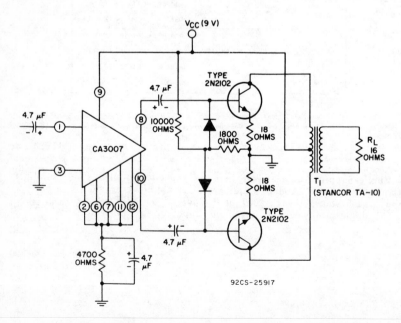

Fig. 405—CA3007 used as an audio driver for a 30-milliwatt audio amplifier.

followers which are capacitor-coupled to the push-pull output stage. Because of the ac coupling, there is no longer a dc dependence between the driver and the output stage, and any desired audio output design or drive source may be used. As a single stage, the CA3007 audio driver provides a voltage gain of 24 dB for a dc power dissipation of 20 milliwatts with the harmonic distortion reaching 3 per cent for outputs of 0.6 volt rms at terminals 8 and 10 (without feedback).

Both dc and ac feedback loops are eliminated in the circuit of Fig. 405. Although the dc feedback loop is no longer required because of the ac coupling, removal of the ac feedback loop causes the output power gain to decrease about 1 dB for a 50°C rise in temperature.

Integrated-Circuit Audio Amplifiers

The RCA CA3020 and CA3020A integrated circuits are multipurpose, multifunction power amplifiers designed for use as power-output amplifiers and driver stages in portable and fixed communications equipment and in ac servo control systems.

The CA3020 and CA3020A are designed to operate from a single supply voltage which may be as low as +3 volts. The maximum supply voltage is dictated by the type of circuit operation. For transformer-loaded class B amplifier service, the maximum supply voltages are +9 and +12 volts for the CA3020 and the CA3020A, respectively. When operated as a class B amplifier, either circuit can deliver a typical output of 150 milliwatts from a +3-volt supply or 400 milliwatts from a +6-volt supply. At +9 volts, the idling dissipation can be as low as 190 milliwatts, and either circuit can deliver an output of 550 milliwatts. An output of slightly more than 1 watt is available from the CA3020A when a +12-volt supply is used.

The CA3020 and the CA3020A integrated-circuits are well suited for use in audio-amplifier applications. These circuits may be used without transformers to drive a high-impedance speaker directly, or they may be used with power transformers to deliver power to a low-impedance speaker. They may also be used as transformer-coupled drivers for one or more power transistors to develop up to 10 watts of audio output power.

Basic Class B Amplifier—Fig. 406 shows a typical audio-amplifier cir-

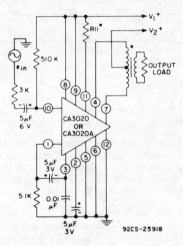

Fig. 406—Basic class B audio amplifier circuit using the CA3020 or CA3020A.

cuit in which the CA3020 or CA3020A can provide a power output of 0.5 or 1 watt, respectively. Table XXXVIII shows performance data for both types in this amplifier. The circuit can be used at all voltage and power-output levels applicable to the CA3020 and CA3020A.

The input transistor in the CA3020 or CA3020A is connected as an emitter-follower stage at the input of the amplifier in Fig. 406 to provide a high input impedance. Although many variations of biasing may be applied to this stage, the

Table XXXVIII—Typical Performance of CA3020 and CA3020A in Circuit of Fig. 406*

Characteristics	CA3020	CA3020A	
Power Supply — V_1^+	9	9	V
V_2^+	9	12	V
Zero-Signal Idling Current — I_{CC1}	15	15	mA
I_{CC2}	24	24	mA
Maximum-Signal Current — I_{CC1}	16	16.6	mA
I_{CC2}	125	140	mA
Maximum Power Output at 10% THD	550	1000	mW
Sensitivity	35	45	mV
Power Gain	75	75	dB
Input Resistance	55	55	kΩ
Efficiency	45	55	%
Signal-to-Noise Ratio	70	66	dB
% Total Harmonic Distortion at 150 mW	3.1	3.3	%
Test Signal	1000 Hz/600Ω generator		
Equivalent Collector-to-Collector Load	130	200	Ω
Idling-Current Adjust Resistor (R11)	1000	1000	Ω

* Integrated circuit mounted on a heat sink, Wakefield 209 Alum. or equiv.

method shown is efficient and economical. The output of this buffer stage is applied to terminal 3 of the differential amplifier for proper balance of the push-pull drive to the output stages. Terminals 2 and 3 must be bypassed for approximately 1000 ohms at the desired low-frequency roll-off point.

At low power levels, the cross-over distortion of the class B amplifier can be high if the idling current is low. For low cross-over distortion, the idling current should be approximately 12 to 24 milliamperes, depending on the efficiency, idling dissipation, and distortion requirements of the particular application. The idling current may be increased by connection of a jumper between terminals 8 and 9. If higher levels of operating idling current are desired, a resistor (R_{11}) may be used to increase the regulated voltage at terminal 11 by a slight amount with additional current injection from the power supply V_1^+.

In some applications, it may be desirable to use the input transistor Q1 of the CA3020 or CA3020A for other purposes than the basic buffer amplifier shown in Fig. 406. In such cases, the input ac signal can be applied directly to terminal 3.

The extended frequency range of the CA3020 and CA3020A requires that a high-frequency ac bypass capacitor be used at the input terminal 3. Otherwise, oscillation could occur at the stray resonant frequencies of the external components, particularly those of the transformers. Lead inductance may be sufficient to cause oscalliation if long power-supply leads are not properly ac bypassed at the CA3020 or CA3020A common ground point. Even the bypassing shown may be insufficient unless good high-frequency construction practices are followed.

Fig. 407 shows typical power output of the CA3020A at supply voltages of +3, +6, +9, and +12 volts, and of the CA3020 at +6 and +9 volts, as measured in the basic class B amplifier circuit of Fig. 406. The CA3020A has higher power output for all voltage-supply conditions because of its higher peak-output-current capability.

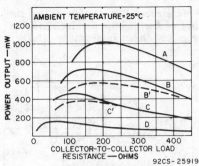

Fig. 407—Power output of the CA3020 or CA3020A as a function of collector-to-collector load resistance R_{cc}.

Fig. 408 shows total harmonic distortion (THD) as a function of power output for each of the voltage conditions shown in Fig. 407. The values of the collector-to-collector load resistance (R_{cc}) and the idling-

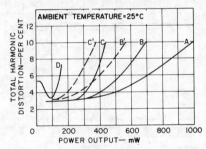

Fig. 408—Total harmonic distortion of the CA3020 or CA3020A as a function of power output.

current adjust resistor (R_{11}) shown in the figure are given merely as a fixed reference; they are not necessarily optimum values. Higher idling-current drain may be desired for low cross-over distortion, or a higher value of R_{cc} may be used for better sensitivity with less power-output capability. Because the maximum power output occurs at the same conditions of peak-current limitations, the sensitivities at maximum power output for the curves of Figs. 407 and 408 are approximately the same. Increasing the idling-current drain by reducing the value of resistor

R_{11} also improves the sensitivity.

Fig. 409 illustrates the improvement in cross-over distortion at low power levels. Distortion at 100 milliwatts is shown as a function of

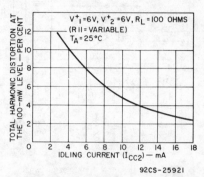

Fig. 409—Total harmonic distortion as a function of idling current for a supply voltage of 6 volts and an output of 100 milliwatts.

idling current I_{CC2} (output stages only). There is a small improvement in total harmonic distortion for a large increase in idling current as the current level exceeds 15 milliamperes.

The circuit shown in Fig. 406 may be used as a highly efficient class B audio power-output circuit in such applications as communications systems, AM or FM radios, tape recorders, phonographs, intercom sets, and linear mixers.

545-Milliwatt Amplifier Driving a Low-Impedance Speaker—Fig. 410 shows a circuit configuration that has the required characteristics for driving a conventional low-impedance speaker. The circuit shown uses a transformer capable of driving a 3.2-ohm speaker; other transformers may be used to drive 8-ohm and 16-ohm speakers. This circuit has the following characteristics:

Input voltage for full
  power output .... 45 mV
Maximum power output .... 545 mV
Idling current .... 22 mA
Input resistance ....50,000 ohms
Total harmonic distortion at P_{out} =
  135 mW 3.3%

Signal-to-noise ratio (input voltage reference of 20 mV) .... 77 dB

Intercom—Fig. 412 illustrates the use of the audio amplifier shown in Fig. 406 in an intercom in which a

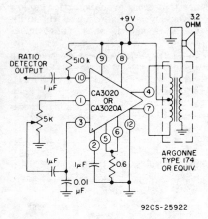

Fig. 410—545-milliwatt amplifier driving a low-impedance speaker.

4-Watt Class A Audio Amplifier—Fig. 411 shows a class A audio amplifier in which the CA3020 or CA3020A is used with a driver transformer, a 2N2148 power transistor, and an output transformer. This circuit can deliver a power output of 4 watts to an 8-ohm speaker for an input voltage of 18 millivolts, or 0.45 watt for an input of 5.5 millivolts.

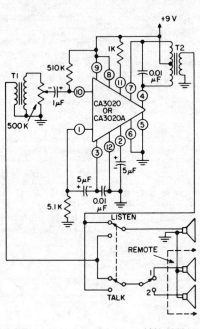

Fig. 412—Intercom using CA3020 or CA3020A.

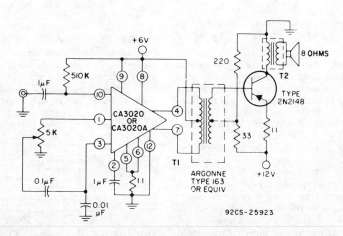

Fig. 411—4-watt audio amplifier.

listen-talk position switch controls two or more remote positions. Only the speakers, the switch, and the input transformer are added to the basic audio amplifier circuit. A suitable power supply for the intercom could be a 9-volt battery used intermittently rather than continuously.

12-Watt True-Complementary-Symmetry Audio Amplifiers

Fig. 413 shows a block diagram and Fig. 414 shows the detailed circuit diagram of a true-complementary-symmetry audio power ampli-

terized for audio-output service. They are provided in the JEDEC TO-220AB version of the VERSA-WATT plastic package.

Table **XXXIX** lists typical performance characteristics for the 12-watt amplifier, and Figs. 415 and 416 show the distortion and response curves for the circuit.

Fig. 417 shows a block diagram and Fig. 418 shows the circuit diagram of a 12-watt true-complementary-symmetry audio amplifier that uses RCA1C10 and RCA1C11 discrete transistors, an integrated circuit, one diode, and a 36-volt split power supply; the amplifier output is

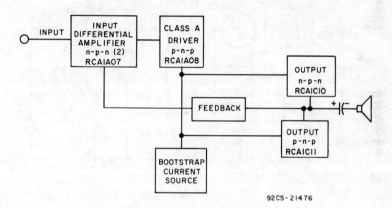

92CS-21476

Fig. 413—Block diagram and transistor complement for 12-watt true-complementary-symmetry audio amplifier.

fier that can supply 12 watts of audio output power.

The audio-amplifier circuit uses RCA1C10 and RCA1C11 output transistors in conjunction with three other (input and driver) transistors, two diodes, and a single 36-volt power supply; the amplifier output is capacitively coupled to an 8-ohm speaker. The choice of a true-complementary-symmetry output stage provides excellent fidelity for a low-cost system.

The RCA1C10 and RCA1C11 are n-p-n and p-n-p epitaxial-base silicon power transistors, respectively. These devices are especially charac-

directly coupled to an 8-ohm speaker. The integrated circuit-true-complementary-symmetry combination provides a high-quality, low-cost amplifier. This circuit is essentially the same as the amplifier shown in Fig. 414, except that the low-level transistor stages have been replaced by the RCA CA3094AT integrated circuit.

The CA3094AT integrated circuit provides sufficient drive current for the complementary-symmetry output stage. Tone controls, bass and treble, with functions of "boost" and "cut" are incorporated into the feedback loop of the amplifier, resulting in excellent signal-to-noise ratio and freedom from distortion.

Table XL shows typical performance data for this amplifier, and Figs. 419 and 420 shows the distortion and frequency-response curves.

40-Watt Full-Complementary-Symmetry Audio Amplifier

Fig. 421 shows a block diagram and Fig. 422 shows the circuit diagram for a full-complementary-sym-

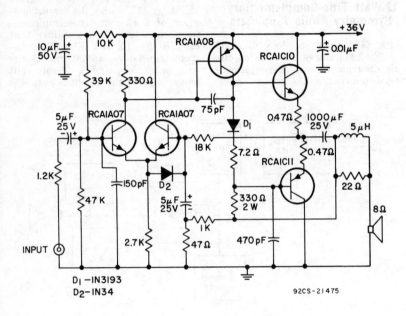

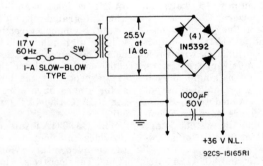

Fig. 414—12-watt amplifier circuit featuring complementary-symmetry output.

Table XXXIX—Typical Performance Data For 12-Watt Audio Amplifier Circuit in Fig. 414.

Measured at a line voltage of 120 V, $T_A = 25°C$, and a frequency of 1 kHz, unless otherwise specified.

Power:
   Rated power (8-Ω load, at rated distortion) 12W
   Typical power (4-Ω load) 12W
   Typical power (16-Ω load) .. 6.5W
   Music power (8-Ω load, at 5% THD with regulated supply) 15W
   Dynamic power (8-Ω load, at 1% THD with regulated supply) 13W
Total Harmonic Distortion:
   Rated distortion 1.0%

IM Distortion:
   10 db below continuous power output at 60 Hz and 7 kHz (4:1) 1.5%
Sensitivity:
   At continuous power-output rating .... 600 mV
Hum and Noise:
   Below continuous power output:
     Input shorted .... 90 dB
     Input open .... 70 dB
Input Resistance .... 23 kΩ

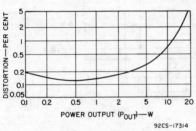

Fig. 415—Distortion as a function of power output.

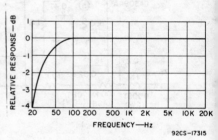

Fig. 416—Response curve.

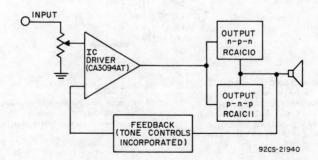

Fig. 417—Block diagram and transistor complement for 12-watt true-complementary-symmetry audio amplifier with integrated-circuit driver.

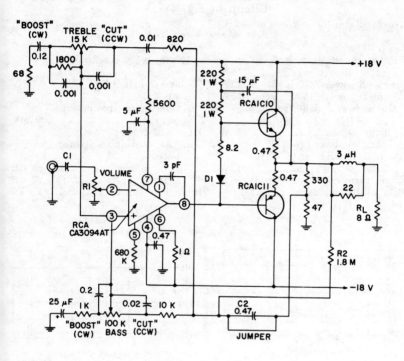

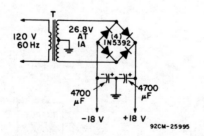

92CM-25995

NOTES:

1. T: Stancor No.P-8609 (120 V AC to 26.8 V CT @ 1 A) or equivalent

2. FOR STANDARD INPUT: Short C_2; R_1 = 250 K; C_1 = 0.047 μF; Remove R_2

3. FOR CERAMIC-CARTRIDGE INPUT: C_1 = 0.0047 μF; R_1 = 2.5 MΩ; Remove Jumper from C_2; Leave R_2.

4. D1 1N5392

5. Resistors are 1/2-watt unless otherwise specified; values are in ohms.

6. Capacitances are in μF unless otherwise specified.

7. Non-inductive resistors.

Fig. 418—12-watt amplifier circuit featuring an integrated-circuit driver and a true-complementary-symmetry output stage.

Table XL—Typical Performance Data For 12-Watt Audio Amplifier Circuit in Fig. 418.

Measured at a line voltage of 120 V, $T_A = 25°C$, and a frequency of 1 kHz, unless otherwise specified.

Power:
    Rated power (8-Ω load, at
      rated distortion) 12W
    Typical power (4-Ω load) 9W
    Typical power (16-Ω load) 6.5W
    Music power (8-Ω load, at
      5% THD with regu-
      lated supply) 15W
Total Harmonic Distortion:
    Rated distortion 1.0%
    Typical at 1 W 0.05%
IM Distortion:
    10 dB below continuous
      power output at 60 Hz
      and 2 kHz (4:1) 0.2%

Sensitivity:
    At continuous power-out-
      put rating (tone con-
      trols flat) .... 100 mV
Hum and Noise:
    Below continuous power
      output:
        Input open .... 83 dB
Input resistance .... 250 kΩ
Voltage Gain .... 40 dB
Tone Control Range .... See Fig. 420

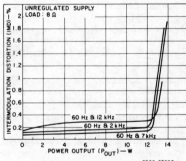

Fig. 419—Intermodulation distortion as a function of power output.

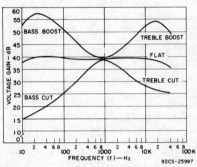

Fig. 420—Voltage gain as a function of frequency.

metry audio amplifier (i.e., both driver and output stage employ complementary n-p-n and p-n-p transistor pairs). This circuit uses RCA1C07 n-p-n and RCA1C08 p-n-p output transistors in conjunction with seven other low-level-stage transistors, ten diodes, and a 64-volt split power supply to develop up to 40 watts of audio output power. The amplifier output is directly coupled to an 8-ohm speaker. The high-frequency performance of this

40-watt amplifier will provide excellent reproduction for the most critical listener. The amplifier also features an overload protection circuit that prevents damage to the output transistors in the event that output should be inadvertently short-circuited.

Table XLI shows typical performance data for the 40-watt amplifier, and Figs. 423 and 424 show frequency-response and distortion curves.

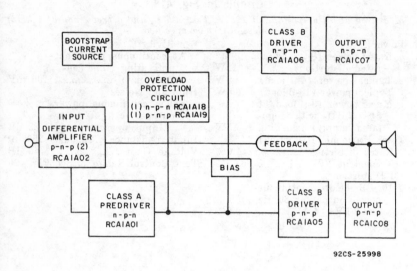

92CS-25998

Fig. 421—Block diagram and transistor complement for 40-watt full-complementary-symmetry audio amplifier.

Table XLI—Typical Performance Data For 40-Watt Audio Amplifier Circuit in Fig. 422.

Measured at a line voltage of 120 V, $T_A = 25°C$, and a frequency of 1 kHz, unless otherwise specified.

Power:
    Rated power (8-Ω load, at rated distortion) 40W
    Typical power (4-Ω load)   75W
    Typical power (16-Ω load)   25W
Total Harmonic Distortion:
    Rated distortion 1.0%
    Typical at 20 W 0.05%
IM Distortion:
    10 dB below continuous power output at 60 Hz and 7 kHz (4:1) 0.1%

IHF Power Bandwidth:
    3 dB below rated continuous power at rated distortion 80kHz
Sensitivity:
    At continuous power-output rating 600mV
Hum and Noise:
    Below continuous power output:
        Input shorted .... 80 dB
        Input open .... 75 dB
Input Resistance .... 20 kΩ

70-Watt Quasi-Complementary-Symmetry Audio Amplifiers

Fig 425 shows a block diagram and Fig. 426 shows the circuit diagram of a quasi-complementary-symmetry audio amplifier that can supply 70 watts of audio output power. This 70-watt audio amplifier uses RCA 1B06 n-p-n output transistors in conjunction with eleven other transistors, thirteen diodes, and a 90-volt split power supply. The amplifier output is directly coupled to an 8-ohm speaker. The high-fre-

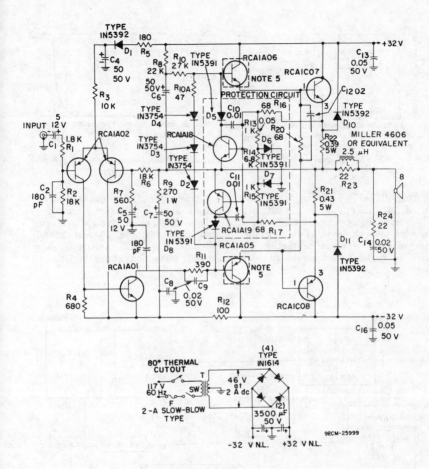

NOTES:

1. T: Signal 88-2 (parallel secondary)●, Signal Transformer Co., 1 Junius St., Brooklyn, N.Y. 11212

2. Resistors are 1/2-watt unless otherwise specified; values are in ohms.

3. Capacitances are in μF unless otherwise specified.

4. Non-inductive resistors.

● Or equivalent.

Fig. 422—40-Watt amplifier circuit featuring full-complementary-symmetry output using load line limiting.

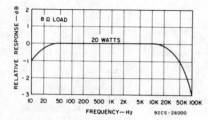

Fig. 423—Response curve.

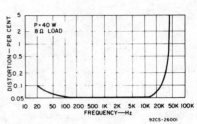

Fig. 424—Typical distortion as a function of frequency.

quency RCA1B06 output transistors used in the amplifier circuit produce excellent transient response at a high power level. The amplifier includes an overload protection circuit to prevent damage to the out-put transistors in the event that the output should be short-circuited.

Table XLII shows typical performance data for the 70-watt amplifier and Figs. 427 and 428 show typical distortion characteristics.

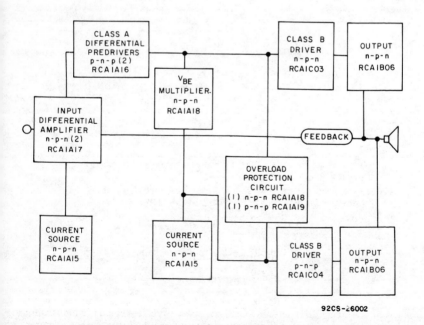

92CS-26002

Fig. 425—Block diagram and transistor complement for 70-watt quasi-complementary-symmetry audio amplifier with pi-nu output transistors.

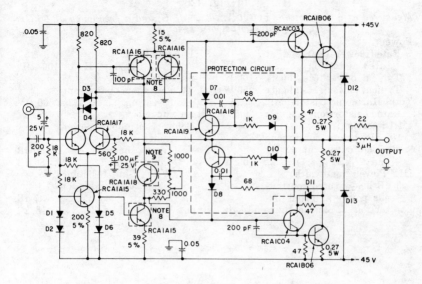

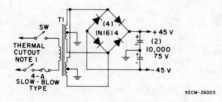

92CM-26003

NOTES:

1. 100°C thermal cutout attached to heat sink for output transistors (Elmwood Sensor part No. 2455-88-4).●

2. Power transformer: Signal 120-2 (parallel secondary),● Signal Transformer Co., 1 Junius St., Brooklyn, N.Y. 11212.

3. Resistors are 1/2-watt unless otherwise specified; values are in ohms.

4. Capacitances are in μF unless otherwise specified.

5. Non-inductive resistors.

6. D1-D8, D11-1N5391
   D9, D10, D12, D13-1N5393   ● Or equivalent.

Fig. 426—70-Watt amplifier circuit featuring quasi-complementary-symmetry output employing pi-nu construction output transistors.

Table XLII—Typical Performance Data For 70-Watt Audio Amplifier in Fig. 426.

Measured at a line voltage of 120 V, $T_A = 25°C$, and a frequency of 1 kHz, unless otherwise specified.

Power:
  Rated power (8-Ω load, at rated distortion) .... 70 W
  Typical power (4-Ω load) 100 W
  Typical power (16-Ω load) 50 W
Total Harmonic Distortion:
  Rated distortion 0.5%
IM Distortion:
  10 dB below continuous power output at 60 Hz and 7 kKz (4:1) .... <0.2%
IHF Power Bandwidth:
  3 dB below rated continuous power at rated distortion .... 5 Hz to 50 kHz

Bandwidth at
  1 W .... 5 Hz to 100 kHz
Sensitivity:
  At continuous power-output rating .... 600 mV
Hum and Noise:
  Below continuous power output:
    Input shorted .... 100 dB
    Input open .... 85 dB
    With 2-kΩ resistance on 20-ft. cable on input .... 97 dB
Input Resistance .... 18 kΩ

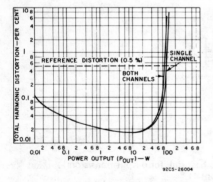

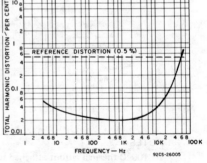

Fig. 427—Typical total harmonic distortion as a function of power output at 1 kHz.

Fig. 428—Typical total harmonic distortion as a function of frequency at 35 watts.

Tuned Amplifiers, Frequency Converters, and Detectors for AM, FM, and TV Receivers

WHEN speech, music, or video information is transmitted from a radio or television station, the station radiates a modulated radio-frequency (rf) carrier. The function of a radio or television receiver is simply to reproduce the modulating wave from the modulated carrier.

As shown in Fig. 429, a super-heterodyne radio receiver picks up the transmitted modulated rf signal, amplifies it and converts it to a modulated intermediate-frequency (if) signal, amplifies the modulated if signal, separates the modulating signal from the basic carrier wave, and amplifies the resulting audio signal to a level sufficient to produce the desired volume in a speaker. In addition, the receiver usually includes some means of producing automatic gain control (agc) of the modulated signal before the audio information is separated from the carrier.

The transmitted rf signal picked up by the radio receiver may contain either amplitude modulation (AM) or frequency modulation (FM). (These modulation techniques are described later under the heading **Detection.**) In either case, amplification prior to the detector stage is performed by tuned amplifier circuits designed for the proper frequency and bandwidth. Frequency conversion is performed by mixer and oscillator circuits or by a single converter stage which performs both mixer and oscillator functions. Separation of the modulating signal is normally accomplished by one or more diodes in a detector or discriminator circuit. Amplification of the audio signal is then performed by one or more audio amplifier stages. The portion of the receiver before the audio amplifier is called the **tuner.** (Audio amplifiers are discussed in a later section of this Manual.)

The operation of a television receiver (shown in block-diagram form in Fig. 430) is more complex

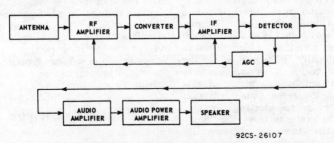

92CS-26107

Fig. 429—Simplified block diagram for a broadcast-band receiver.

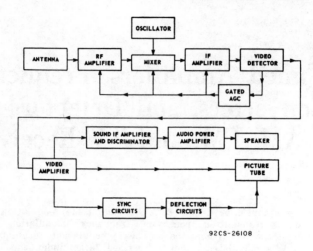

Fig. 430—Simplified block diagram for a television receiver.

than that of a radio receiver, as shown by a comparison of Figs. 429 and 430.

The tuner section of the television receiver selects the proper rf signals for the desired channel frequency, amplifies them, and converts them to a lower intermediate frequency. As in a radio receiver, these functions are accomplished in rf-amplifier, mixer, and local-oscillator stages. The if signal is then amplified in if-amplifier stages which provide the additional gain required to bring the signal level to an amplitude suitable for detection.

After if amplification, the detected signal is separated into sound and picture information. The sound signal is amplified and processed to provide an audio signal which is fed to an audio amplifier system. The picture (video) signal is passed through a video amplifier which conveys beam-intensity information to the television picture tube and thus controls instantaneous "spot" brightness. At the same time, deflection circuits cause the electron beam of the picture tube to move the "spot" across the faceplate horizontally and vertically. Special "sync" signals derived from the video signal

assure that the horizontal and vertical scanning are timed so that the picture produced on the receiver exactly duplicates the picture being viewed by the camera or pickup tube. (The sync and deflection circuits are described in the section on **TV Deflection**.)

In a television receiver, the video signal contains a dc component, and therefore the average carrier level varies with signal information. As a result, the agc circuit is designed to provide a control voltage proportional to the peak modulated carrier level rather than the average modulated carrier level. The time constant of the agc detector circuit is made large enough so that the picture content of the composite video signal does not influence the magnitude of the agc voltage. In addition, an electronic switch is often included in the circuit so that it can be operated only during the retrace portion of the scanning cycle. This "gated agc" technique prevents noise peaks from affecting agc operation.

TUNED AMPLIFIERS

In radio-frequency (rf) and intermediate-frequency (if) amplifiers, the bandwidth of frequencies to be amplified is usually only a small percentage of the center frequency. Tuned amplifiers are used in these applications to select the desired bandwidth of frequencies and to suppress unwanted frequencies. The selectivity of the amplifier is obtained by means of tuned interstage coupling networks.

Resonant-Circuit Characteristics

The properties of tuned amplifiers depend upon the characteristics of **resonant circuits**. A simple parallel resonant circuit (sometimes called a "tank" because it stores energy) is shown in Fig. 431. For practical purposes, the resonant frequency of such a circuit may be considered independent of the resistance R, provided R is small compared to the inductive reactance X_L. The resonant frequency f_r is then given by

$$f_r = \frac{1}{2\pi\sqrt{LC}}$$

For any given resonant frequency, the product of L and C is a constant; at low frequencies LC is large; at high frequencies it is small.

The **Q (selectivity)** of a parallel resonant circuit alone is the ratio of the current in the tank (I_L or I_C) to the current in the line (I). This un-

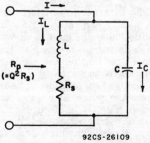

92CS-26109

Fig. 431—Simple parallel resonant circuit.

loaded Q, or Q_o, may be expressed in various ways, for example:

$$Q_o = \frac{I_C}{I} = \frac{X_L}{R_S} = \frac{R_p}{X_C}$$

where X_L is the inductive reactance ($= 2\pi fL$), X_c is the capacitive reactance ($= 1/[2\pi fC]$), and R_p is the total impedance of the parallel resonant circuit (tank) at resonance. The Q varies inversely with the resistance of the inductor R_S. The lower the resistance, the higher the Q and the greater the difference between tank impedance at frequencies off resonance compared to the tank impedance at the resonant frequency.

The Q of a tuned interstage coupling network also depends upon the impedances of the preceding and following stages. The output impedance of a transistor can be considered as consisting of a resistance R_o in parallel with a capacitance C_o, as shown in Fig. 432. Similarly, the input impedance can be considered as consisting of a resistance R_i in parallel with a capacitance C_i. Because the

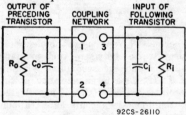

92CS-26110

Fig. 432—Equivalent output and input circuits of transistors connected by a coupling network.

tuned circuit is shunted by both the output impedance of the preceding transistor and the input impedance of the following transistor, the effective selectivity of the circuit is the loaded Q (or Q_L) based upon the total impedance of the coupled network, as follows:

$$Q_L = \frac{\left\{\begin{array}{c}\text{total loading on}\\\text{coil at resonance}\end{array}\right\}}{X_L \text{ or } X_C}$$

The capacitances C_o and C_i in Fig. 432 are usually considered as part of the coupling network. For example, if the required capacitance between terminals 1 and 2 of the coupling network is calculated to be 500 picofarads and the value of C_o is 10 picofarads, a capacitor of 490 picofarads is used between terminals 1 and 2 so that the total capacitance is 500 picofarads. The same method is used to allow for the capacitance C_i at terminals 3 and 4.

When a tuned resonant circuit in the primary winding of a transformer is coupled to the nonresonant secondary winding of the transformer, as shown in Fig. 433(a), the effect of the input impedance of the following stage on the Q of the tuned circuit can be determined by considering the values reflected (or referred) to the primary circuit by transformer action. The reflected resistance r_i is equal to the resistance R_i in the secondary circuit times the square of the effective turns ratio between the primary and secondary windings of the transformer T:

$$r_i = R_i (N_1/N_2)^2$$

where N_1/N_2 represents the electrical turns ratio between the primary winding and the secondary winding of T. If there is capacitance in the secondary circuit (C_s), it is reflected to the primary circuit as a capacitance C_{sp}, and is given by

$$C_{sp} = C_p \div (N_1/N_2)^2$$

The loaded Q, or Q_L, is then calculated on the basis of the inductance L_p, the total shunt resistance (R_o plus r_i plus the tuned-circuit impedance $Z_t = Q_o X_c = Q_o X_L$), and the total capacitance ($C_p + C_{sp}$) in the tuned circuit.

Fig. 433(b) shows a coupling network which consists of a single-tuned circuit using mutual inductive coupling. The capacitance C_t includes the effects of both the output capacitance of the preceding transistor and the input capacitance of the following transistor (referred to the primary of transformer T_1).

The bandwidth of a single-tuned transformer is determined by the half-power points on the resonance curve (-3 dB or 0.707 down from

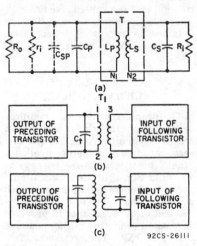

(a)

(b)

(c)

92CS-26111

Fig. 433—Equivalent circuits for transformer-coupling networks: (a) having tuned primary winding; (b) using inductive coupling; (c) using tap on primary winding.

the maximum). Under these conditions, the band pass $\triangle f$ is equal to the ratio of the center or resonant frequency f_r divided by the loaded (effective) Q of the circuit, as follows:

$$\triangle f = f_r/Q_L$$

The inherent internal feedback in transistors can cause instability and oscillation as the gain of an amplifier stage is increased (i.e., as the load and source impedances are increased from zero to matched conditions). At low radio frequencies, therefore, where the potential gain of transistors is high, it is often desirable to keep the transistor load impedance low. Relatively high capacitance values in the tuned collector circuit can then be avoided by use of a tap on the primary winding of the coupling transformer, as shown in Fig. 433(c). At higher frequencies, the gain potential of the transistor decreases, and impedance matching is permissible. However, lead inductance becomes significant at higher frequencies, particularly in the emitter circuit. All lead lengths should be kept short, therefore, and especially the emitter lead, which not only degrades performance but is also a mutual coupling to the output circuit.

Gain and Noise Figure

In the design of low-level tuned rf amplifiers, careful consideration must be given to the transistor and circuit parameters which control circuit stability, as well as those which maintain adequate power gain. The power gain of an rf transistor must be sufficient to provide a signal that will overcome the noise level of succeeding stages. In addition, if the signals to be amplified are relatively weak, it is important that the transistor and its associated circuit provide low noise figure at the operating frequency. In communication receivers, the noise figure of the rf stage determines the absolute sensitivity of the receiver and is, therefore, one of the most important characteristics of the device used in the rf stage.

The relative power-gain capabilities of transistors at high frequencies are indicated by their theoretical maximum frequency of oscillation f_{max}. At this frequency, the unilateralized matched power gain, or **maximum available gain MAG**, is 0 dB. MAG as a function of frequency for a typical rf transistor rises approximately 6 dB per octave below f_{max}.

Because most practical rf amplifiers are not individually unilateralized, the power gain that can be obtained is somewhat less than the MAG because of internal feedback in the circuit. This feedback is greater in unneutralized circuits than in neutralized circuits, and therefore gain is lower when neutralization is not used. From a practical consideration, the feedback capacitance which must be considered is the total feedback capacitance between collector and base, including both stray and socket capacitances. In neutralized circuits, stray capacitances, socket capacitance, and the typical value of device capacitance can generally be neutralized. At a given frequency, therefore, the **maximum usable power gain MUG** of a neutralized circuit depends on the transconductance g_m and the amount of internal feedback capacitance C_f. In unneutralized circuits, however, both socket and stray capacitances are involved in the determination of gain and must be included in the value of C_f. The ratio of g_m to C_f should be high to provide high power gain.

Cross-Modulation Distortion

Cross-modulation, an important consideration in the evaluation of transistorized tuner circuits, is produced when an undesired signal within the pass band of the receiver input circuit modulates the carrier of the desired signal. Such distortion occurs when third- and higher-odd-order nonlinearities are present in an rf-amplifier stage. In general, the severity of cross-modulation is independent of both the semiconductor material and the construction of the transistor (provided gain and noise factor are not sacrificed). At low frequencies, cross-modulation is also independent of the amplitude of the desired carrier, but varies as the square of the amplitude of the interfering signal.

In most rf circuits, the undesirable effects of cross-modulation can be minimized by good selectivity in the antenna and rf interstage coils. Minimum cross-modulation can best be achieved by use of the optimum circuit Q with respect to bandwidth and tracking considerations, which implies minimum loading of the tank circuits.

Cross-modulation may occur in the mixer or rf amplifier, or both. Accordingly, it is important to analyze the entire tuner as well as the individual stages. Cross-modulation is also a function of agc. At sensitivity conditions where the rf stage is operating at maximum gain and the interfering signal is far removed from the desired signal, cross-modulation occurs primarily in the rf stage. As the desired signal level increases and agc is applied to the rf stage, the rf transistor gain decreases and provides improved cross-modulation. If the interfering signal is close to the desired signal, it is the rf gain at the undesired signal frequency which determines whether the rf stage or mixer stage is the prime contributor of cross-modulation. For example, it is possible that the rf stage gain (including selectivity of tuned circuits) at the undesired frequency is greater than unity. In this case, the undesired signal at the mixer input is larger than that at the rf input; thus the contribution of the mixer is appreciable. Intermediate and high signal conditions may be analyzed similarly by considering rf agc.

If adequate limiting is employed, cross-modulation does not occur in an FM signal.

Limiting

A **limiter** circuit is essentially an if-amplifier stage designed to provide clipping at a desired signal level. Such circuits are used in FM receivers to remove AM components from the if signal prior to FM detection. The limiter stage is normally the last stage prior to detection, and is similar to preceding if stages. At low input rf signal levels, it amplifies the if signal in the same manner as preceding stages. As the signal level increases, however, a point is reached at which the limiter stage is driven into saturation (i.e., the peak currents and voltages are limited by the supply voltage and load impedances and increases in signal produce very little increase in collector current). At this point, the if signal is "clipped" (or flattened) and further increases in rf signal level

produce no further output in if signal to the detector.

Limiter stages may be designed to provide clipping at various input-signal levels. A high-gain FM tuner is usually designed to limit at very low rf input signal levels, and possibly even on noise signals.

Integrated-Circuit RF and IF Amplifiers

Integrated circuits, such as the CA3028 described in the section **Integrated Circuits for Linear Applications**, can be used advantageously for rf and if amplification. The CA3028B integrated circuit, used with external tuned-circuit, transformer, or resistive loads, can provide high gain, low noise figure, and effective limiting at frequencies from dc to 120 MHz. It can be operated in either the differential mode or the cascode mode, as shown in Figs. 434(a) and 434(b). The cascode mode of operation is recommended for maximum gain; the differential mode is best if good limiting is required. Figs. 435 and 436 show the gain and noise characteristics of the two modes.

The 10.7-MHz if strip shown in Fig. 437 uses two CA3028A or CA3028B integrated circuits. The first integrated circuit is connected as a cascode amplifier and yields voltage gain of 50 dB; the second integrated circuit is connected as a differential amplifier and yields voltage gain of 42 dB.

When a practical interstage transformer having a voltage insertion loss of 9 dB is used, over-all gain is 83 dB and the sensitivity at the base of the first integrated circuit is 140 microvolts. A less sophisticated converter filter (double-tuned) could be employed at the expense of about 26 dB of second-channel attenuation. If the voltage insertion loss of the converter filter is assumed to be 18 dB and the front-end voltage gain (antenna to mixer collector) is 50 dB, this receiver would have an

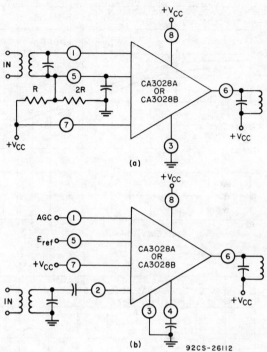

Fig. 434—Connection of CA3028 as (a) differential amplifier; (b) cascode amplifier.

IHFM* sensitivity of approximately 8 microvolts.

MOS-Transistor RF and IF Amplifiers

Dual-gate MOS field-effect transistors (MOS/FET's) are useful in a wide variety of rf and if amplifier applications. These devices provide excellent power gain, low noise figures, and wide dynamic range. Fig. 438 shows a typical circuit using a dual-gate MOS/FET as the rf stage of a TV tuner. The two serially-connected channels with independent control gates make possible a greater dynamic range and lower cross-modulation than is normally achieved using devices having only a single control element.

The two-gate arrangement also makes possible a desirable reduction in feedback capacitance by operating in the common-source configuration and ac-grounding Gate No. 2. The reduced capacitance allows operation at maximum gain without neutralization; and, of special importance in rf-amplifiers, it reduces local oscillator feedthrough to the antenna.

Although maximum theoretical power gain cannot be achieved in practical circuits, the gain of MOS transistors at high frequencies closely approximates the theoretical limit except for some losses in the input and output matching circuits.

In practical rf-amplifier circuits using MOS transistors, the best possible noise figures are obtained when the input impedance of the transistor is slightly mismatched to that of the source. With this technique,

---

* Institute of High-Fidelity Manufacturers.

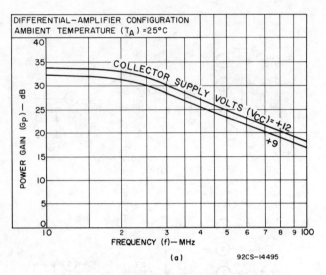

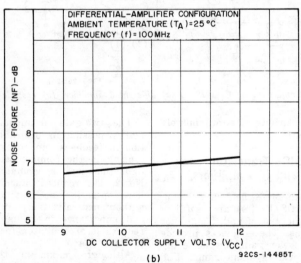

Fig. 435—Gain (a) and noise characteristics (b) for the CA3028 in a differential amplifier configuration.

noise figures as low as 1.9 dB have been obtained. Dual-gate MOS transistors typically exhibit a noise figure of 3.5 dB in the vhf range and of 4.5 dB in the uhf range.

The dynamic-range capability of MOS field-effect transistors is about 25 times greater than that of bipolar transistors. In an actual tuner circuit, this large intrinsic dynamic range is reduced by a factor proportional to the square of the circuit source impedances. The net result is a practical dynamic range for MOS tuner circuits about five times that for bipolar types.

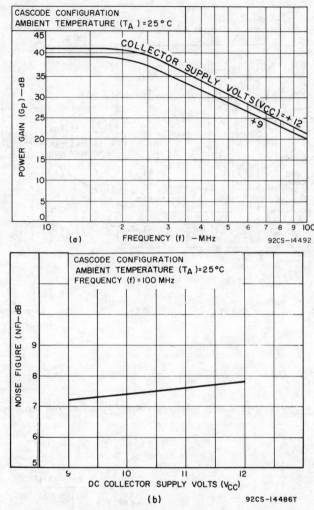

Fig. 436—Gain and noise characteristics for CA3028 in cascode configuration.

The cross-modulation characteristics of MOS transistors are as good as those of bipolar transistors in the high-attenuation region, and are as much as ten times better in the low-attenuation region (when the incoming signal is weak). This low cross-modulation distortion should ultimately lead to extensive use of MOS transistors in the rf stages of all types of communication receivers.

A typical application of a MOS transistor for if amplification is shown in Fig. 439.

OSCILLATORS

Bipolar and field-effect transistor oscillator circuits are similar in many respects to the tuned amplifiers discussed previously, except that a portion of the output power is returned to the input network in

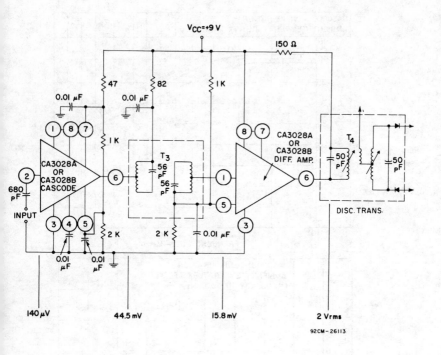

Fig. 437—10.7-MHz if strip using two CA3028A or CA3028B integrated circuits.

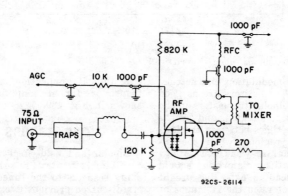

Fig. 438—A dual-gate MOS/FET used as an rf amplifier.

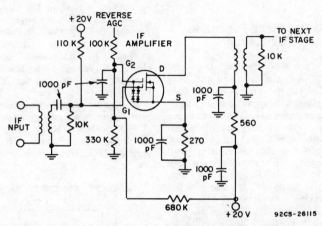

Fig. 439—TV if amplifier stage utilizing an MOS transistor.

phase with the starting power (regenerative or positive feedback) to sustain oscillation. DC bias-voltage requirements for oscillators are similar to those discussed for amplifiers.

The maximum operating frequency of an oscillator circuit is limited by the frequency capability of the transistor used. The maximum frequency of oscillation of a transistor is defined as the frequency at which the power gain is unity. Because some power gain is required in an oscillator circuit to overcome losses in the feedback network, the operating frequency must be some value below the transistor maximum frequency of oscillation.

For sustained oscillation in a transistor oscillator, the power gain of the amplifier network must be equal to or greater than unity. When the amplifier power gain becomes less than unity, oscillations become smaller with time (are "damped") until they cease to exist. In practical oscillator circuits, power gains greater than unity are required because the power output is divided between the load and the feedback network, as shown in Fig. 440. The feedback power must be equal to the input power plus the losses in the

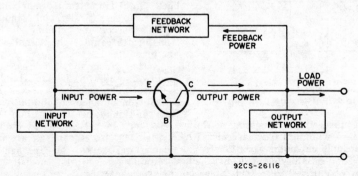

Fig. 440—Block diagram of transistor oscillator showing division of output power.

feedback network to sustain oscillation.

LC Resonant Feedback Oscillators

The frequency-determining elements of an oscillator circuit may consist of an inductance-capacitance (LC) network, a crystal, or a resistance-capacitance (RC) network. An LC tuned circuit may be placed in either the base circuit or the collector circuit of a common-emitter transistor oscillator. In the tuned-base oscillator shown in Fig. 441, one battery is used to provide all the dc operating voltages for the transistor. Resistors R_1, R_3, and R_4 provide the necessary bias conditions. Resistor R_2 is the emitter stabilizing resistor. The components within the dotted lines comprise the transistor amplifier. The collector shunt-feed arrangement prevents dc current flow through the tickler (primary) winding of transformer T. Feedback is accomplished by the mutual inductance between the transformer windings.

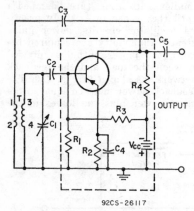

Fig. 441—Tuned-base oscillator.

The tuned circuit consisting of the secondary winding of transformer T and variable capacitor C_1 is the frequency-determining element of the oscillator. Variable capacitor C_1 permits tuning through a range of fre-

quencies. Capacitor C_2 couples the oscillation signal to the base of the transistor, and also blocks dc. Capacitor C_4 bypasses the ac signal around the emitter resistor R_3 and prevents degeneration. The output signal is coupled from the collector through coupling capacitor C_5 to the load.

A tuned-collector transistor oscillator is shown in Fig. 442. In this circuit, resistors R_1 and R_3 establish the base bias. Resistor R_2 is the emitter stabilizing resistor. Capaci-

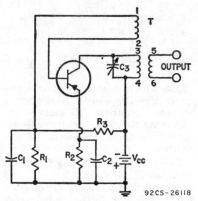

Fig. 442—Tuned-collector oscillator.

tors C_1 and C_2 bypass ac around resistors R_1 and R_2, respectively. The tuned circuit consists of the primary winding of transformer T and the variable capacitor C_3. Regeneration is accomplished by coupling the feedback signal from transformer winding 3-4 to the tickler coil winding 1-2. The secondary winding of the transformer couples the signal output to the load.

Another form of LC resonant feedback oscillator is the Hartley oscillator. This oscillator makes use of split inductance to obtain feedback and may be either shunt or series fed. In the shunt-fed circuit of Fig. 443, R_1, R_2, and R_3 are the biasing resistors; the frequency-determining network consists of variable capacitor C_1 in series with the windings of T_1. The frequency

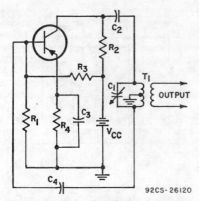

Fig. 443—Shunt-fed Hartley oscillator.

of the oscillator is varied by C_1; C_2 is the dc blocking capacitor and C_3 is an ac bypass capacitor.

The circuit inductance functions in the manner of an auto transformer and provides the regenerative feedback signal obtained from the voltage induced in the lower half of the transformer winding and coupled through C_4 to the transistor base. No dc current flows through the primary of T_1 because the collector is shunt fed through R_2.

In the series-fed Hartley circuit shown in Fig. 444, the base-emitter circuit is biased through R_1 and R_2;

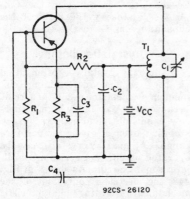

Fig. 444—Series-fed Hartley oscillator.

the collector is biased through the upper half of the transformer windings. Again, as in the shunt-fed

circuit, C_3 provides an ac bypass. Feedback in the series-fed Hartley circuit is obtained from the lower-half of the transformer winding and is coupled through C_4 to the base of the transistor. The center-tap of the transformer winding is maintained at ac ground potential by C_2.

Fig. 445 shows two arrangements of a Hartley oscillator circuit using MOS field-effect transistors. Circuit (a) uses a bypassed source resistor to provide proper operating conditions; circuit (b) uses a gate-leak resistor and biasing diode. The amount of feedback in either

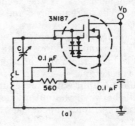

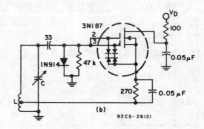

Fig. 445—Hartley oscillator circuits using MOS transistors.

circuit is dependent on the position of the tap on the coil. Too little feedback results in a feedback signal voltage at the gate insufficient to sustain oscillation; too much feedback causes the impedance between source and drain to become so low that the circuit becomes unstable. Output from these circuits can be obtained through inductive coupling to the coil or through capacitive coupling to the gate.

Another form of LC resonant feedback oscillator is the transistor version of the Colpitts oscillator, shown in Fig. 446. Regenerative feedback is obtained from the tuned circuit consisting of capacitors C_2 and C_3 in parallel with the primary winding of the transformer, and is applied to the emitter of the transistor. Base bias is provided by resistors R_2 and R_3. Resistor R_4 is the collector load resistor. Resistor R_1 develops the emitter input signal and also acts as the emitter stabilizing resistor. Capacitors C_2 and C_3 form a voltage divider; the voltage developed across C_3 is the feedback voltage. The frequency and the amount of feedback voltage can be controlled by adjustment of either or both capacitors. For minimum feedback loss, the ratio of the capacitive reactance between C_2 and C_3 should be approximately equal to the ratio between the output impedance and the input impedance of the transistor.

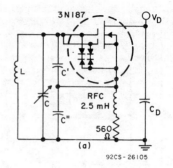

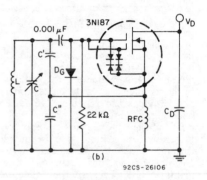

Fig. 447—Colpitts oscillator circuits using MOS transistors.

frequencies by the Hartley circuits. Feedback is controlled in the Colpitts oscillator by the ratio of the capacitance of C' to C".

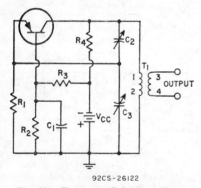

Fig. 446—Transistor Colpitts oscillator.

Fig. 447 shows the field-effect transistor in use in two forms of the Colpitts oscillator circuit. These circuits are more commonly used in vhf and uhf equipment than the Hartley circuits because of the mechanical difficulty involved in making the tapped coils required at these

Crystal Oscillators

A quartz crystal is often used as the frequency-determining element in a transistor oscillator circuit because of its extremely high Q (narrow bandwidth) and good frequency stability over a given temperature range. A quartz crystal may be operated as either a series or parallel resonant circuit. As shown in Fig. 448, the electrical equivalent of the mechanical vibrating characteristic of the crystal can be represented by a resistance R, an inductance L, and a capacitance C_s in series. The lowest impedance of the crystal occurs at the series resonant frequency of C_s and L; the resonant frequency of the circuit is then determined only by the mechanical vibrating characteristics of the crystal.

The parallel capacitance C_p shown in Fig. 448 represents the electrostatic capacitance between the crystal electrodes. At frequencies above the

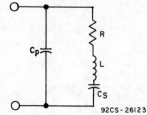

Fig. 448—Equivalent circuit of quartz crystal.

series resonant frequency, the combination of L and C_s has the effect of a net inductance because the inductive reactance of L is greater than the capacitive reactance of C_s. This net inductance forms a parallel resonant circuit with C_p and any circuit capacitance across the crystal. The impedance of the crystal is highest at the parallel resonant frequency; the resonant frequency of the circuit is then determined by both the crystal and externally connected circuit elements.

Increased frequency stability can be obtained in the tuned-collector and tuned-base oscillators discussed previously if a crystal is used in the feedback path. The oscillation frequency is then fixed by the crystal. At frequencies above and below the series resonant frequency of the crystal, the impedance of the crystal increases and the feedback is reduced. Thus, oscillation is prevented at frequencies other than the series resonant frequency.

The parallel mode of crystal resonance is used in the Pierce oscillator shown in Fig. 449. (If the crystal were replaced by its equivalent circuit, the functioning of the oscillator would be analogous to that of the Colpitts oscillator shown in Fig. 446.)

The resistances shown in Fig. 449 provide the proper bias and stabilizing conditions for the common-emitter circuit. Capacitor C_1 is the emitter bypass capacitor. The required 180-degree phase inversion of the feedback signal is accomplished through the arrangement of the voltage-divider network C_2 and C_3. The connection between the capacitors is grounded so that the voltage developed across C_3 is applied between base and ground and a 180-degree phase reversal is obtained. The oscillating frequency of the circuit is determined by the crystal and the capacitors connected in parallel with it.

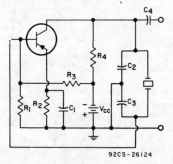

Fig. 449—Pierce-type transistor crystal oscillator.

The field-effect transistor also operates well in crystal oscillator circuits such as the Pierce-type oscillators shown in Fig. 450. Pierce

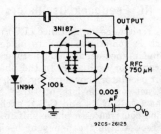

Fig. 450—Pierce-type crystal oscillator circuit using MOS transistor.

oscillators are extremely popular because of their simplicity and minimum number of components.

A crystal oscillator can also be implemented with a COS/MOS inverter-amplifier and feedback network as shown in Fig. 451.

the common-emitter oscillator occurs only at one frequency; thus, the output frequency of the oscillator is fixed. Phase-shift oscillators may be made variable over particular frequency ranges by the use of ganged variable capacitors or resistors in the

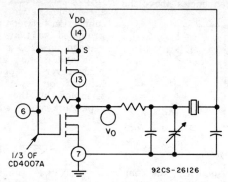

Fig. 451—A COS/MOS crystal oscillator.

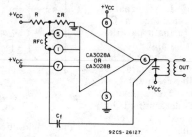

Fig. 452—A linear integrated circuit connected as an oscillator.

Fig. 452 shows a CA3028 linear integrated circuit connected as an oscillator for use in an rf tuner.

RC Feedback Oscillators

A resistance-capacitance (RC) network is sometimes used in place of an inductance-capacitance network in a transistor oscillator. In the phase-shift oscillator shown in Fig. 453, the RC network consists of three sections (C_1R_1, C_2R_2, and C_3R_3), each of which contributes a phase shift of 60 degrees at the frequency of oscillation. Because the capacitive reactance of the network increases or decreases at other frequencies, the 180-degree phase shift required for

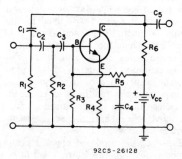

Fig. 453—Transistor RC phase-shift oscillator.

RC networks. Three or more sections must be used in the phase-shifting networks to reduce feedback losses. The use of more sections contributes to increased stability.

FREQUENCY CONVERTERS

Transistors can be used in various types of circuits to change the frequency of an incoming signal. In radio and television receivers, frequency conversion is used to change the frequency of the rf signal to an intermediate frequency. In communications transmitters, frequency multiplication is often used to raise the frequency of the developed rf signal.

In a radio or television receiver, the oscillating and mixing functions are performed by a nonlinear device such as a diode or a transistor. As shown in the diagram of Fig. 454,

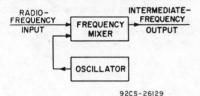

92CS-26129

Fig. 454—Block diagram of simple frequency-converter circuit.

two voltages of different frequencies, the rf signal voltage and the voltage generated by the oscillator, are ap-

plied to the input of the mixer. These voltages "beat," or heterodyne, within the mixer transistor to produce a current having, in addition to the frequencies of the input voltages, numerous sum and difference frequencies.

The output circuit of the mixer stage is provided with a tuned circuit which is adjusted to select only one beat frequency, i.e., the frequency equal to the difference between the signal frequency and the oscillator frequency. The selected output frequency is known as the intermediate frequency, or if. The output frequency of the mixer transistor is kept constant for all values of signal frequency by tuning of the oscillator circuit.

In AM broadcast-band receivers, the oscillator and mixer functions are often accomplished by use of a single transistor called an "autodyne converter". In FM receivers, stable oscillator operation is more readily obtained when a separate transistor is used for the oscillator function. In such a circuit, the oscillator voltage is applied to the mixer by inductive coupling, capacitive coupling, or a combination of the two.

Fig. 455 shows the basic circuit

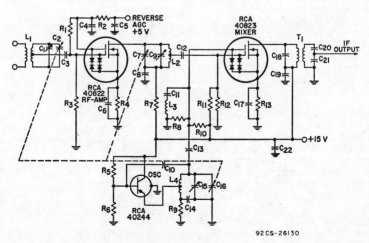

92CS-26130

Fig. 455—Circuit diagram of FM tuner using dual-gate MOS transistors in the rf amplifier and mixer stages.

configuration for the "front-end" stages of an FM tuner that uses dual-gate-protected MOS field-effect transistors in both the rf-amplifier and mixer stages. A bipolar transistor is used in the local-oscillator stage.

The dual-gate MOS transistor is very attractive for use in mixer service because the two signals to be mixed are applied to separate gate terminals. This arrangement is an effective technique for reduction of oscillator radiation. In the circuit shown in Fig. 455, the signal frequency is applied to gate No. 1 of the mixer transistor and the local-oscillator input to gate No. 2.

Connection of a CA3028 integrated circuit as a converter is shown in Fig. 456(a). A mixer using the CA3028 integrated circuit is shown in Fig. 456(b).

DETECTORS

The circuit of a radio, television, or communications receiver in which the modulation is separated from the carrier is called the demodulator or detector stage. Transmitted rf signals may be modulated in either of two ways. If the frequency of the carrier remains constant and its amplitude is varied, the carrier is called an amplitude-modulated (AM) signal. If the amplitude remains essentially constant and the frequency is varied, the carrier is called a frequency-modulated (FM) signal.

The effect of **amplitude modulation (AM)** on an rf carrier wave is shown in Fig. 457. The audio-frequency (af) modulation can be extracted from the amplitude-modulated carrier by means of a simple **diode detector** such as that shown

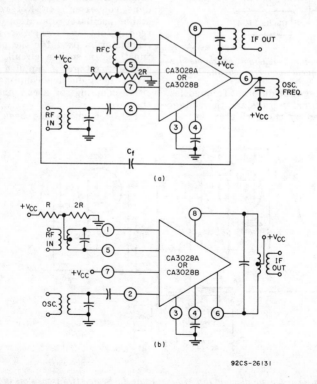

92CS-26131

Fig. 456—Connection of a CA3028 as (a) a converter; (b) a mixer.

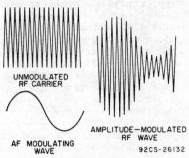

UNMODULATED
RF CARRIER

AF MODULATING
WAVE

AMPLITUDE—MODULATED
RF WAVE

92CS-26132

Fig. 457—Waveforms showing effect of amplitude modulation on an rf wave.

in Fig. 458(a). This circuit eliminates alternate half-cycles of the waveform, and detects the peaks of the remaining half-cycles to produce the output voltage shown in Fig. 458(b). In this figure, the rf voltage applied to the circuit is shown in light line; the output voltage across the capacitor C is shown in heavy line.

Between points a and b of Fig. 458(b), capacitor C charges up to the peak value of the rf voltage. Then, as the applied rf voltage falls

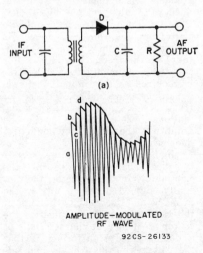

D

IF
INPUT

C   R   AF
OUTPUT

(a)

d
b
c
a

AMPLITUDE—MODULATED
RF WAVE

92CS-26133

Fig. 458—(a) Basic diode detector circuit and (b) waveform showing modulated rf input (light line) and output voltage (heavy line) of diode-detector circuit.

away from its peak value, the capacitor holds the cathode of the diode at a potential more positive than the voltage applied to the anode. The capacitor thus temporarily cuts off current through the diode. While the diode current is cut off, the capacitor discharges from b to c through the diode load resistor R.

When the rf voltage on the anode rises high enough to exceed the potential at which the capacitor holds the cathode, current flows again, and the capacitor charges up to the peak value of the second positive half-cycle at d. In this way, the voltage across the capacitor follows the peak value of the applied rf voltage and reproduces the af modulating signal. The jaggedness of the curve in Fig. 458(b), which represents an rf component in the voltage across the capacitor, is exaggerated in the drawing. In an actual circuit, the rf component of the voltage across the capacitor is small. When the voltage across the capacitor is amplified, the output of the amplifier reproduces the speech or music that originated at the transmitting station.

Another way to describe the action of a diode detector is to consider the circuit as a half-wave rectifier. When the signal on the anode swings positive, the diode conducts and the rectified current flows. The dc voltage across the capacitor C varies in accordance with the rectified amplitude of the carrier and thus reproduces the af signal. Capacitor C should be large enough to smooth out rf or if variations, but should not be so large as to affect the audio variations. (Although two diodes can be connected in a circuit similar to a full-wave rectifier to produce full-wave detection, in practice the advantages of this connection generally do not justify the extra circuit cost and complication.)

In the circuit shown in Fig. 458(a), it is often desirable to forward-bias the diode almost to the point of conduction to improve performance for weak signal levels. It is also desirable that the resistance of the ac load which follows the detector be considerably larger than the diode load resistor to avoid severe distortion of the audio waveform at high modulation levels.

The basic diode detector may also be adapted to provide video-signal detection in black-and-white and color television receivers. Fig. 459 shows an example of a diode type of video detector for a color television receiver.

The **video detector** demodulates the if signal so that the luminance, chrominance, and sync signals are available at the output of the detector circuit. A crystal diode with an if filter is commonly used for this purpose. The video detector in a color receiver may employ a sound-carrier trap in its input. This trap attenuates the sound carrier and insures against the development of an undesirable 920-kHz beat frequency which is the frequency difference between the sound carrier and the color subcarrier. When the sound carrier is attenuated in this manner, the sound take-off point is located ahead of the video detector.

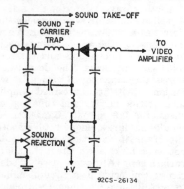

Fig. 459—Video detector for a color television receiver.

The effect of **frequency modulation (FM)** on the waveform of an rf carrier wave is shown in Fig. 460. In this type of transmission, the frequency of the rf carrier deviates from the mean value at a rate pro-

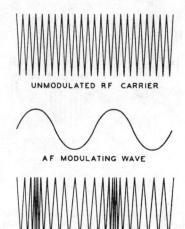

Fig. 460—Waveforms showing effect of frequency modulation on an rf wave.

portional to the audio-frequency modulation and by an amount (determined in the transmitter) proportional to the amplitude of the af modulating signal. That is, the number of times the carrier frequency deviates above and below the center frequency is a measure of the frequency of the modulating signal; the amount of frequency deviation from the center frequency is a measure of the loudness (amplitude) of the modulating signal. For this type of modulation, a detector is required to discriminate between deviations above and below the center frequency and to translate these deviations into a voltage having an amplitude that varies at audio frequencies.

The FM detector shown in Fig. 461 is called a **balanced phase-shift discriminator**. In this detector, the mutually coupled tuned circuits in the primary and secondary windings of the transformer T are tuned to the center frequency. A characteristic of a double-tuned transformer is that the voltages in the primary and secondary windings are 90 degrees out of phase at resonance, and that the phase shift changes as the frequency changes from resonance. Therefore,

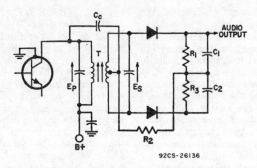

92CS-26136

Fig. 461—Balanced phase-shift discriminator circuit.

the signal applied to the diodes and the RC combinations for peak detection also changes with frequency.

Because the secondary winding of the transformer T is center-tapped, the applied primary voltage E_p is added to one-half the secondary voltage E_s through the capacitor C_c. The addition of these voltages at resonance can be represented by the diagram in Fig. 462; the resultant volt-

92CS-26137

Fig. 462—Diagram illustrating phase shift in double-tuned transformer at resonance.

age E_1 is the signal applied to one peak-detector network consisting of one diode and its RC load. When the signal frequency decreases (from resonance), the phase shift of $E_s/2$ becomes greater than 90 degrees, as shown at (a) in Fig. 463, and E_1 becomes smaller. When the signal frequency increases (above

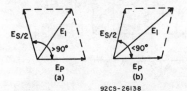

92CS-26138

Fig. 463—Diagrams illustrating phase shift in double-tuned transformer (a) below resonance and (b) above resonance.

resonance), the phase shift of $E_s/2$ is less than 90 degrees, as shown at (b), and E_1 becomes larger. The curve of E_1 as a function of frequency in Fig. 464 is readily identified as the response curve of an FM detector.

Because the discriminator circuit shown in Fig. 461 uses a push-pull configuration, the diodes conduct on alternate half-cycles of the signal frequency and produce a plus-and-minus output with respect to zero rather than with respect to E_1. The primary advantage of this arrangement is that there is no output at resonance. When an FM signal is applied to the input, the audio output voltage varies above and below zero as the instantaneous frequency varies above and below resonance.

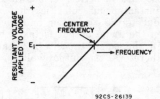

92CS-26139

Fig. 464—Diagram showing resultant voltage E_1 in Fig. 462 as a function of frequency.

The frequency of this audio voltage is determined by the modulation frequency of the FM signal, and the amplitude of the voltage is proportional to the frequency excursion from resonance. (The resistor R_2 in the circuit provides a dc return for the diodes, and also maintains a load impedance across the primary winding of the transformer.)

One disadvantage of the balanced phase-shift discriminator shown in Fig. 461 is that it detects amplitude modulation (AM) as well as frequency modulation (FM) in the if signal because the circuit is balanced only at the center frequency. At frequencies off resonance, any variation in amplitude of the if signal is reproduced to some extent in the audio output.

The **ratio-detector** circuit shown in Fig. 465 is a discriminator circuit which has the advantage of being relatively insensitive to amplitude variations in the FM signal. In this circuit, E_p is added to $E_s/2$ through the mutual coupling M_2 (this voltage addition may be made by either mutual or capacitive coupling). Because of the phase-shift relationship of these voltages, the resultant detected signals vary with frequency variations in the same manner as described for the phase-discriminator circuit shown in Fig. 461. However, the diodes in the ratio detector are placed "back-to-back" (in series, rather than in push-pull) so that both halves of the circuit operate simultaneously during one-half of the signal frequency cycle (and are cut off on the other half-cycle). As a result, the detected voltages E_1 and E_2 are in series, as shown for the instantaneous polarities that occur during the conduction half-cycle. When the audio output is taken between the equal capacitors C_1 and C_2, therefore, the output voltage is

equal to $(E_2—E_1)/2$ (for equal resistors R_1 and R_2).

The dc circuit of the ratio detector consists of a path through the secondary winding of the transformer, both diodes (which are in series), and resistors R_1 and R_2. The value of the electrolytic capacitor C_3 is selected so that the time constant of R_1, R_2, and C_3 is very long compared to the detected audio signal. As a result, the sum of the detected voltages $(E_1 + E_2)$ is a constant, and the AM components on the signal frequency are suppressed. This feature of the ratio detector provides improved AM rejection as compared to the phase-shift discriminator circuit shown in Fig. 461.

AUTOMATIC GAIN CONTROL

Automatic gain control (agc) is often used in rf and if amplifiers in AM radio and television receivers to provide lower gain for strong signals and higher gain for weak signals. (In radio receivers, this gain-compensation network may also be called **automatic volume control** or avc.) When the signal strength at the antenna changes, the agc circuit modifies the receiver gain so that the output of the last if-amplifier stage remains nearly constant and consequently maintains a nearly constant speaker volume or picture contrast.

The agc circuit usually reduces the rf and if gain for a strong signal by varying the bias on the rf-amplifier and if-amplifier stages when the sig-

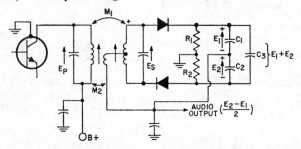

92CS-26140

Fig. 465—Ratio-detector circuit.

nal increases. A simple **reverse agc** circuit is shown in Fig. 466. On each positive half-cycle of the signal voltage, when the diode anode is positive with respect to the cathode, the diode passes current. Because of the flow of diode current through R_1, there is a voltage drop across R_1 which makes the upper end of the resistor negative with respect to ground. This voltage drop across R_1 is applied, through the filter R_2 and C, as reverse bias on the preceding stages. When the signal strength at the antenna increases, therefore, the signal applied to the agc diode increases, the voltage drop across R_1 increases, the reverse bias applied to the rf and if stages increases, and the gain of the rf and if stages is decreased. As a result, the increase in signal strength at the antenna does not produce as much increase in the output of the last if-amplifier stage as it would without agc.

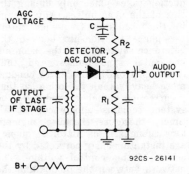

Fig. 466—Simple reverse agc circuit.

When the signal strength at the antenna decreases from a previous steady value, the agc circuit acts in the opposite direction, applying less reverse bias and thus permitting the rf and if gain to increase.

The filter composed of C and R_2 prevents the agc voltage from varying at an audio frequency. This filter is necessary because the voltage drop across R_1 varies with the modulation of the carrier being received. If agc voltage were taken directly from R_1 without filtering, the audio variations in agc voltage would vary

the receiver gain so as to smooth out the modulation of the carrier. To avoid this effect, the agc voltage is taken from the capacitor C. Because of the resistance R_2 in series with C, the capacitor can charge and discharge at only a comparatively slow rate. The agc voltage therefore cannot vary at frequencies as high as the audio range, but can vary rapidly at frequencies high enough to compensate for most changes in signal strength.

There are two ways in which automatic gain control can be applied to a transistor. In the reverse agc method shown in Fig. 466, agc action is obtained by decreasing the collector or emitter current of the transistor, and thus its transconductance and gain. The use of **forward agc** provides improved cross-modulation characteristics and better signal-handling capability than reverse agc. For forward agc operation, however, the transistor used must be specially designed so that transconductance decreases with increasing emitter current. In such transistors, the current-cutoff characteristics are designed to be more remote than the typical sharp-cutoff characteristics of conventional transistors. (All transistors can be used with reverse agc, but only specially designed types with forward agc.)

Reverse agc is simpler to use, and provides less bandpass shift and tilt with signal-strength variations. The input and output resistances of a transistor increase when reverse agc is applied, but the input and output capacitances are not appreciably changed. The change in the loading of tuned circuits is minimal, however, because considerable mismatch already exists and the additional mismatch caused by agc has little effect.

In forward agc, however, the input and output resistances of the transistor are reduced when the collector or emitter current is increased, and thus the tuned circuits are damped. In addition, the input and output capacitances change drastically, and alter the resonant frequency of the tuned circuits. In a practical circuit, the bandpass shift and tilt caused by forward agc can be compensated to a large extent by the use of passive coupling circuits.

AUTOMATIC FREQUENCY CONTROL

An automatic frequency control (afc) circuit is often used to provide automatic correction of the oscillator frequency of a superheterodyne receiver when, for any reason, it drifts from the frequency which produces the proper if center frequency. This correction is made by adjustment of the frequency of the oscillator. Such a circuit automatically compensates for slight changes in rf carrier or oscillator frequency, as well as for inaccurate manual or push-button tuning.

An afc system requires two sections: a frequency detector and a variable reactance. The detector section may be essentially the same as the FM detector illustrated in Fig. 461. In the afc system, however, the output is a dc control voltage, the magnitude of which is proportional to the amount of frequency shift. This dc control voltage is used to control the bias on a transistor or diode which comprises the variable reactance.

Automatic frequency control is also used in television receivers to keep the horizontal oscillator in step with the horizontal-scanning frequency at the transmitter. A widely used horizontal afc circuit is shown in Fig. 467. This circuit, which is often referred to as a balanced-phase-detector or phase-discriminator circuit, is usually employed to control the frequency of the horizontal-oscillator circuit. The detector diodes supply a dc control voltage to the horizontal-oscillator circuit which counteracts changes in its operating frequency. The magnitude and polarity of the control voltages are determined by phase relationships in the afc circuit.

The horizontal sync pulses obtained from the sync-separator circuit are fed through a phase-inverter or phase-splitter circuit to the two diode detectors. Because of the action of the phase-inverter circuit, the signals applied to the two diode units are equal in amplitude but 180 degrees out of phase. A reference sawtooth voltage obtained from the horizontal output circuit is also applied simultaneously to both units. The diodes are biased so that conduction takes place only during the tips of the sync pulses. Any change in the oscillator frequency alters the phase relationship betwen the reference sawtooth and the incoming horizontal sync pulses, and thus causes one of the diodes to conduct more heavily than the other so that a correction signal is produced. The system remains unbalanced at all times, therefore, because momentary changes in oscillator frequency are instantaneously corrected by the action of this control voltage. The network between the diodes and the horizontal-oscillator circuit is essentially a low-pass filter which prevents the horizontal sync pulses

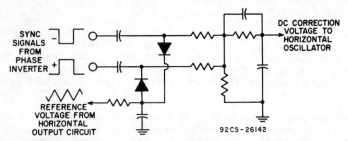

Fig. 467—Balanced-phase-detector or phase-discriminator circuit for horizontal afc.

from affecting the horizontal-oscillator performance.

AM RADIO RECEIVERS USING INTEGRATED CIRCUITS

Modern radio receivers often use a single multistage integrated circuit to provide several major circuit functions. The CA3123E is an example of an integrated circuit used to provide multiple circuit functions in AM superheterodyne receiver applications. With the exception of tuning elements, this integrated circuit contains the basic electronic circuitry required for the rf-ampli-

(diode D_1 and associated components) to the 262-kHz intermediate frequency commonly used in auto AM radio receivers. Audio signals in the order of 10 millivolts can be delivered from the volume-control potentiometer to an audio amplifier.

The CA3088E is another example of an integrated-circuit sub-system for AM receiver applications. This integrated circuit includes the electronic circuitry for the converter, if-amplifier, detector, agc, and audio-preamplifier functions. Fig. 470 shows the circuit diagram for an AM home-radio receiver that uses this integrated circuit together with

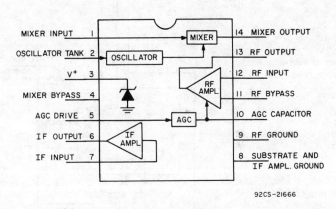

Fig. 468—Block diagram of the CA3123E integrated circuit for AM radio receivers.

fier, mixer, local-oscillator, if-amplifier, and agc functions. Fig. 468 shows the block diagram of the CA3123E integrated circuit. Fig. 469 shows a circuit diagram that illustrates the use of this circuit in an automobile AM radio receiver. Mechanically ganged tuning coils are used to adjust the operating frequencies of the antenna circuit and the rf-amplifier and local-oscillator sections of the CA3123E to cover the AM broadcast band of 550 to 1600 kHz. Transformers T_2 and T_3 tune the if-amplifier section of the CA3123E and the detector circuit

external tuning circuitry and an optional MOS-transistor rf amplifier.

FM TUNERS USING INTEGRATED CIRCUITS

Fig. 471 shows the circuit diagram of an FM tuner that can provide stereo-output audio signals. The "front-end" tuner circuits employ a classical design that uses MOS transistors in the rf-amplifier and mixer stages to produce the 10.7-MHz intermediate frequency. The if-amplifier/limiter and detector functions in this tuner are provided by a CA3089E integrated circuit.

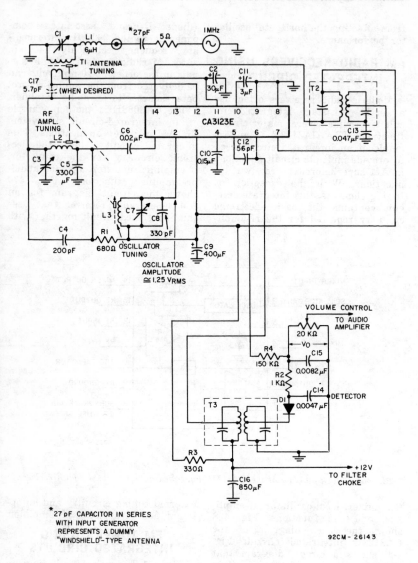

Fig. 469—Schematic diagram of an automobile AM radio receiver using type CA3123E integrated circuit.

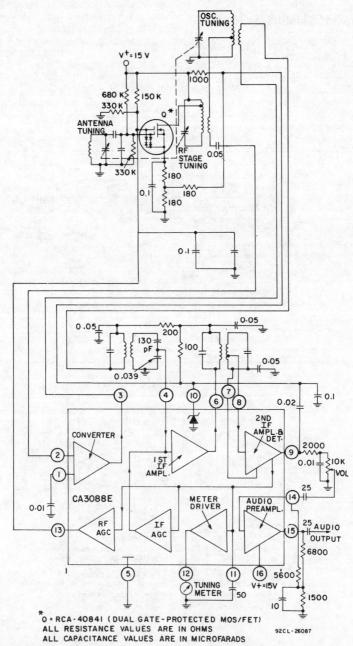

Fig. 470—Typical AM broadcast receiver using the CA3088E integrated circuit with an optional MOS-transistor rf amplifier stage.

FM-IF
Amplifier/Limiter/Detector

As shown in Fig. 471, the CA3089E integrated-circuit subassembly includes a three-stage FM-if amplifier/limiter with level detectors at each stage, a doubly balanced FM quadrature detector, and a low-level audio amplifier that features an optional muting (squelch) circuit. This latter feature permits the audio output to be muted during tuning and/or to reject stations for which the signal strength is low.

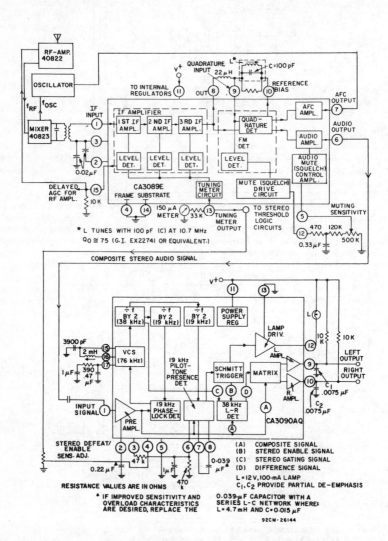

Fig. 471—FM stereo broadcast receiver using integrated circuits for FM-if system (CA3089E) and stereo multiplex decoding (CA3090AQ).

The if-amplifier section includes several desirable deluxe features such as delayed agc for the rf amplifier, an afc drive circuit, and an output signal to drive a tuning meter and/or to provide the switching logic potential that maintains the stereo multiplex decoder in the monaural mode unless the received signal is sufficiently strong to assure noise-free stereo reception.

A ceramic filter is commonly used to provide the immediate-frequency selectivity between the mixer stage and the input stage of the CA3089E integrated circuit. The CA3089E quadrature detector can be tuned with a single tuned circuit, which simplifies receiver alignment procedures. The audio output may be applied directly to an audio amplifier if stereo multiplex reception is not desired.

Integrated-Circuit Stereo Multiplex Decoder

Fig. 471 shows the functional elements and external circuit connections for an integrated-circuit FM stereo multiplex decoder, the CA3090AQ, as used in a typical FM receiver. For stereo operation, a minimum input signal voltage (composite) of 40 mV is required for a proper operation of the multiplex decoder. This stereo multiplex decoder requires only one low-inductance tuning coil (requires only one adjustment for complete alignment), provides automatic stereo switching, energizes a stereo indicator lamp, and operates from a wide range of voltage supplies.

The input signal from the FM-receiver detector is applied at terminal 1 of the CA3090Q. This signal is amplified by a low-distortion preamplifier and applied simultaneously to both the 19-kHz and 38-kHz synchronous detectors. A 76-kHz signal, generated by a local voltage-controlled oscillator (VCO), is counted down by two frequency dividers to a 38-kHz signal and to two 19-kHz signals in phase quadrature. The 19-

kHz pilot-tone supplied by the FM detector is compared to the locally generated 19-kHz signal in a synchronous detector. The resultant signal controls the voltage-controlled oscillator (VCO) so that it produces an output signal to phase-lock the stereo decoder with the pilot tone. A second synchronous detector compares the locally generated 19-kHz signal with the 19-kHz pilot tone. If the pilot tone exceeds an externally adjustable threshold voltage, a Schmitt trigger circuit is energized. The signal from the Schmitt trigger lights the stereo indicator, enables the 38-kHz synchronous detector, and automatically switches the CA3090AQ from monaural to stereo operation. The output signal from the 38-kHz detector and the composite signal from the preamplifier are applied to a matrixing circuit from which emerge the resultant left-and-right-channel audio signals. The signals are applied to their respective left and right post-amplifiers to a level sufficient to drive most audio amplifiers. The CA3090AQ may be used without the stereo defeat/enable function if a control voltage for this function is not readily available. In this mode of operation, terminal 4 should be grounded.

The CA3090AQ integrated-circuit FM stereo-broadcast multiplex-decoder employs sophisticated phase-locked-loop (PLL) circuitry. The input signal contains a 19-kHz pilot-tone when stereo signals are being broadcast. This signal is used to "lock" a 76-kHz VCO. Appropriate frequency-dividers are used to provide the 19-kHz and 38-kHz signals required for the operation of the three phase-detectors used in the circuit.

TV RECEIVER CIRCUITS

A wide variety of integrated circuit types is used in TV receivers, particularly in color-TV receivers. Fig. 472 shows the block diagram of a color-TV receiver that uses six types of integrated circuits.

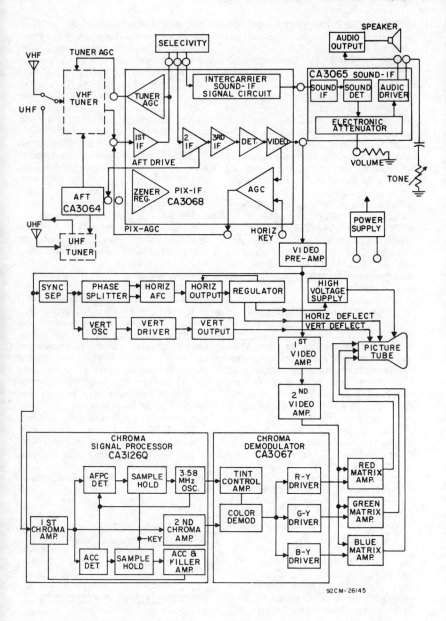

Fig. 472—Block diagram of solid-state color TV receiver using integrated circuits.

AFT Circuit

Accurate tuning is a prerequisite for optimized reception of color-TV signals; the use of automatic-fine-tuning (AFT) circuitry, therefore, is a valuable asset. The CA3064 integrated circuit operates in coujunction with a phase-detector transformer tuned to the picture intermediate frequency (e.g., 45.75 MHz). A sample of the PIX-if signal is applied to the input of the CA3064. Any deviation from the desired fre-

sired 45.75 MHz signal to the picture if section of the receiver. Fig. 473 shows a typical AFT circuit that uses the CA3064.

Picture IF Circuit

The CA3068 is a comprehensive integrated circuit that performs all the subsystem functions for the TV-if section, e.g., video if amplification, linear detection, video output amplification, agc from a keyed supply, agc delay for the tuner, sound-

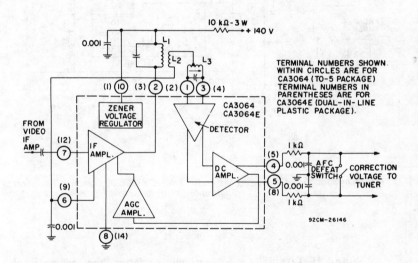

Fig. 473—Block diagram of typical operating circuit utilizing the CA3064.

quency results in a dc error-signal output from the CA3064. This error signal is applied to varactor tuning diodes in the vhf and uhf local-oscillator circuits in the tuners. The local-oscillator frequencies are then "pulled" in the appropriate direction until the error-signal correction goes to zero, indicating that the mixer in the vhf tuner is delivering the de-

carrier detection, sound-carrier amplification, and a buffered output to drive the AFT circuitry. This integrated circuit also includes secondary functions for improved noise immunity and a zener diode to control the power-supply regulator. A typical circuit arrangement that uses the CA3068 is shown in Fig. 474.

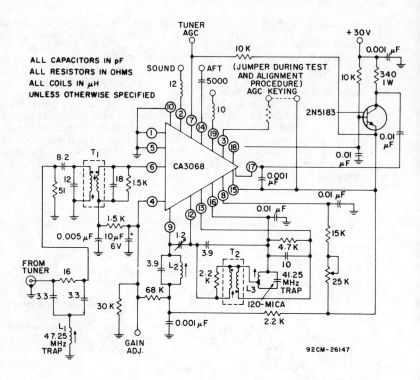

ALL CAPACITORS IN pF
ALL RESISTORS IN OHMS
ALL COILS IN μH
UNLESS OTHERWISE SPECIFIED

92CM-26147

Fig. 474—Schematic diagram showing connections for CA3068 in a picture-if-amplifier application.

Sound IF Circuit

The CA3065 integrated-circuit TV-sound subsystem contains a multistage intercarrier-sound (e.g., 4.5 MHz) if-amplifier/limiter, an FM detector, an electronic audio attenuator, an integral voltage regulator, and an audio amplifier-driver that can directly drive an n-p-n transistor audio-power stage. The FM signal is detected by a differential peak-detector which requires only one tuned circuit for ease of alignment. A unique feature of the CA3065 is the electronic attenuator which performs the conventional volume control function by means of a simple external rheostat. Because no audio signal is present in this control, hum or noise pickup can be by-passed. Fig. 475 shows a typical TV-sound output circuit that uses the CA3065.

Color Demodulation Circuit

Two integrated circuits, the CA3126Q and CA3067, perform the entire chroma processing and demodulation functions. The CA3126Q uses phase-locked-loop techniques in

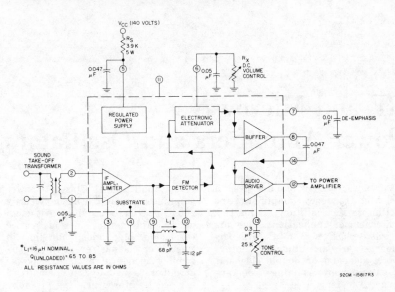

Fig. 475—Block diagram of CA3065 in a typical TV sound-if application.

regenerating the chroma sub-carrier and employs sample-and-hold techniques in the automatic frequency-phase control (AFPC) servo loop. The automatic chrominance control (ACC)/color-killer detector circuity also employs sample-and-hold techniques. Supplementary ACC is provided to prevent oversaturation of the picture tube. Few external components are required and only initial crystal-filter tuning is required for alignment. The CA3067 performs the demodulation and tint control functions. The chroma demodulation components are matrixed and dc-shifted in voltage to give the R-Y, G-Y, and B-Y color difference components with close dc balance and the proper amplitude ratios for the matrix amplifiers which drive

the picture tube. A more detailed discussion of the use of integrated circuits for color demodulation is given in the section **TV Deflection and Chroma Circuits**.

Remote Control Circuit

Many TV receivers use remote-control systems as an option. Many of these systems employ ultrasonic signals to effect remote control. A microphone in the receiver is used to pick up the comparatively weak ultrasonic signals (e.g., in the proximity of 40 kHz). The CA3035 integrated circuit can be used to provide the high voltage gain (e.g., 129 dB) required between the pickup microphone and the control system.

RF and Microwave
Power Amplifiers and Oscillators

RF power transistors are used in high-frequency amplifiers and oscillators for military, industrial, and consumer applications. They are operated class A, B, or C, with frequency- or amplitude-modulation, single sideband or double sideband, in environments ranging from airborne to marine.

BASIC DESIGN CONSIDERATIONS

In the design of silicon-transistor rf power amplifiers for use in transmitting systems, several fundamental factors must be considered. As with any power amplifier, the class of operation has an important bearing on the power output, linearity, and operating efficiency. The matching characteristics of input and output terminations significantly affect power output and frequency stability and, therefore, are particularly important considerations in the design of transistor rf power amplifiers. The selection of the proper transistor for a given circuit applications is also a major consideration, and the circuit designer must realize the significance of the various transistor parameters to make a valid evaluation of different types.

Class of Operation

The class of operation of an rf amplifier is determined by the circuit performance required in the given applications. Class A power amplifiers are used when extremely good linearity is required. Although power gain in this class of service is considerably higher than that in class B or class C service, the operating efficiency of a class A power amplifier is usually only about 25 per cent. Moreover, the standby drain and thermal dissipation of a class A stage are high, and care must be exercised to assure thermal stability.

In applications that require good linearity, such as single-sideband transmitters, class B push-pull operation is usually employed because the transistor dissipation and standby drain are usually much smaller and operating efficiency is higher. Class B operation is characterized by a collector conduction angle of 180 degrees. This conduction is obtained by use of only a slight amount of forward bias in the transistor stage. In this class of service, care must be taken to avoid thermal runaway.

In a class C transistor stage, the collector conduction angle is less than 180 degrees. The gain of the class C stage is less than that of a class A or class B stage, but is entirely usable. In addition, in the class C stage, standby drain is virtually zero, and circuit efficiency is the highest of the three classes. Because of the high efficiency, low collector dissipation, and negligible standby drain, class C operation is the most commonly used mode in rf power-transistor applications.

For class C operation, the base-to-emitter junction of the transistor must be reverse-biased so that the collector quiescent current is zero during zero-signal conditions. Fig. 476 shows four methods that may be used to reverse-bias a transistor stage.

developed across the base spreading resistance. The magnitude of this bias is small and uncontrollable because of the variation in r_{bb}' among different transistors. A better approach, shown in Fig. 476(c), is to develop the bias across an external resistor R_B. Although the bias level

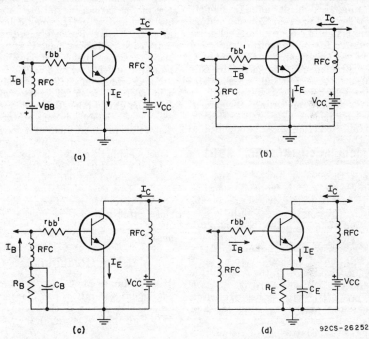

(a)

(b)

(c)

(d)

92CS-26252

Fig. 476—Methods for obtaining class C reverse bias: (a) by use of fixed dc supply V_{BB}; (b) by use of dc base current through the base spreading resistance r_{bb}'; (c) by use of dc base current through an external base resistance R_B; (d) by use of self bias developed across an emitter resistor R_E.

Fig. 476(a) shows the use of a dc supply to establish the reverse bias. This method, although effective, requires a separate supply, which may not be available or may be difficult to obtain in many applications. In addition, the bypass elements required for the separate supply increase the circuit complexity.

Figs. 476(b) and 476(c) show methods in which reverse bias is developed by the flow of dc base current through a resistance. In the case shown in Fig. 476(b), bias is

is predicable and repeatable, the size of R_B must be carefully chosen to avoid reduction of the collector-to-emitter breakdown voltage.

The best reverse-bias method is illustrated in Fig. 476(d). In this method, self-bias is developed across an emitter resistor R_E. Because no external base resistance is added, the collector-to-emitter breakdown voltage is not affected. An additional advantage of this approach is that stage current may be monitored by measurement of the voltage drop across R_E. This technique is very helpful in balancing the shared power in paralleled stages. The bias resistor R_E must be bypassed to provide a very-low-impedance rf path to ground at the operating frequency to prevent degeneration of stage gain. In practice, emitter bypassing is difficult and frequently requires the use of a few capacitors in parallel to reduce the series inductance in the capacitor leads and body. Alternatively, the lead-inductance problem may be solved by formation of a self-resonant series circuit between the capacitor and its leads at the operating frequency. This method is extremely effective, but may restrict stage bandwidth.

Modulation (AM, FM, SSB)

Amplitude modulation of the collector supply of a transistor output stage does not result in full modulation. During down-modulation, a portion of the rf drive feeds through the transistor. Better modulation characteristics can be obtained by modulation of the supply to at least the last two stages in the transmitter chain. On the downward modulation swing, drive from the preceding modulated stages is reduced, and less feed-through power in the output results. Flattening of the rf output during up-modulation is reduced because of the increased drive from the modulated lower-level stages.

The modulated stages must be operated at half their normal voltage levels to avoid high collector-voltage swings that may exceed transistor collector-to-emitter breakdown ratings. RF stability of the modulated stages should be checked for the entire excursion of the modulating signal.

Amplitude modulation of transistor transmitters may also be obtained by modulation of the lower-level stages and operation of the higher-level stages in a linear mode. The lower efficiencies and higher heat dissipation of the linear stages override any advantages that are derived from the reduced audio-drive requirements; as a result, this approach is not economically practical.

Frequency modulation involves a shift of carrier frequency only. Carrier deviations are usually very small and present no problems in amplifier bandwidth. For example, maximum carrier deviations in the 50-MHz and 150-MHz mobile bands are only 5 kHz. Because there is no amplitude variation, class C rf transistor stages have no problems handling frequency modulation.

Single-sideband (SSB) modulation requires that all stages after the modulator operate in a linear mode to avoid intermodulation-distortion products near the carrier frequency. In many SSB applications, channel spacing is close, and excessive distortion results in adjacent-channel interference. Distortion is effectively reduced by class B operation of the rf stages, with close attention to biasing the transistor base-to-emitter junction in a near-linear region.

Matching Requirements

A simplified high-frequency equivalent circuit of an "overlay" type of transistor is shown in Fig. 477. This circuit is similar to the

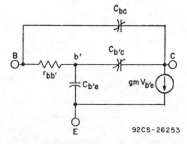

Fig. 477—Simplified high-frequency equivalent circuit for an "overlay" transistor.

hybrid-pi equivalent circuit of a transistor except for the addition of the capacitance C_{bc}. This capacitance represents the high collector-to-base capacitance in the overlay transistor which is created by the large area of the collector-to-base junction together with the active area under the emitter. This capacitance and the capacitance $C_b'_c$ vary nonlinearly with the collector-to-emitter voltage.

Maximum performance in a transistor rf amplifier can be obtained only if the base and collector terminals are properly terminated. The input network generally is required to match a 50-ohm source to the relatively low base-to-emitter impedance, which includes approximately 1 to 10 ohms of resistance and some series reactance. The output network must match a resistive component and the transistor output capacitance to a load impedance, which is generally about 50 ohms. In most applications, the output network also acts as a band-rejection filter to eliminate unwanted frequency components that may be included in the collector waveform. The filter presents a high impedance to these unwanted frequencies and also increases collector efficiency. The power output and collector-voltage swing determine the resistive components to be presented to the collector. The design and form of the output networks (resonant circuits for narrow-band operation or transmission lines for broad-band operation) are discussed in a later section.

Matching networks for rf amplifiers perform two important functions. First, they transform impedance levels as required by the active and fixed elements (e.g., transistor output to antenna impedance). Second, they provide frequency discriminaton by virtue of the "quality factor" (Q) of the resonant circuit, transform harmonic energy into desired output-frequency energy, and prevent the presence of undesired frequency components in the output.

The design of matching circuits is based on the following requirements:

(1) desired or actual network output impedance specified by the series reactance X_s or shunt conductance G_p and shunt susceptance B_p;

(2) desired or actual network input impedance specified by R_s and X_s or G_p and B_p;

(3) loaded circuit Q calculated with input and output terminations connected.

The usual approach is to use L, T, or twin-T matching pads or tuned-transformer networks. More sophisticated systems may use exponential lines and balun transformers.

Input-Circuit Requirements—In practically all power-transistor stages, the input circuit must provide a match between a source impedance that is high compared to the transistor input impedance and the transistor input. When several stages are used, both the input and output impedance of a driver stage are usually higher than those of the following stage.

In most good rf transistors, the real part of the input impedance is usually low, in the order of a few tenths of an ohm to several ohms. In a given transistor family, the resistive part of the common-emitter input impedance is always inversely proportional to the area of the transistor and, therefore, is inversely proportional to the power-output capability of the transistor, if equal emitter inductances are assumed.

The reactive part of the input impedance is a function of the transistor package inductance, as well as the input capacitance of the transistor itself. When the capacitive reactance is smaller than the inductive reactance, low-frequency feedback to the base may be excessive. It is not uncommon to use an inductive input for high-power large-area transistors because the input reactance is a series combination of the package lead inductance and the input capacitance of the transistor itself. Thus, at low frequencies, the input is capacitive, and at higher frequencies, it becomes inductive. At some single frequency, it is entirely resistive.

Output-Circuit Requirements—Although maximum power gain is obtained under matched conditions, a mismatch may be required to meet other requirements. Under some conditions, a mismatch may be necessary to obtain the required selectivity. In power amplifiers, the load impedance presented to the collector, R_L, is not made equal to the output resistance of the transistor. Instead, the value of R_L is dictated by the required power output and the peak dc collector voltage. The peak ac voltage is always less than the supply voltage because of the rf saturation voltage. The collector load resistance R_L may be expressed as follows:

$$R_L = (V_{CC})^2 / 2P_o$$

Designs for tuned, untuned, narrow-band high-Q, and broad-band coupling networks are considered later under specific applications. In some cases, particularly mobile and aircraft transmitters, considerations for safe operation must include variations in the load, both in magnitude and phase. Safe-operation considerations may include protective circuits or actual test specifications imposed on the transistor to assure safe operation under the worst-load conditions.

Transistor Selection

In selection of a transistor and circuit configuration for an rf power amplifier, the designer should be familiar with the following transistor and circuit characteristics:

(1) maximum transistor dissipation and derating,
(2) maximum collector current,
(3) maximum collector voltage,
(4) input and output impedance characteristics,
(5) high-frequency current-gain figure of merit (f_T),

(6) operational parameters such as efficiency, usable power output, power gain, and load-pulling capability.

Proper cooling must be provided to prevent destruction of the transistor because of overheating. Transistor dissipation and derating information reflect how well the heat generated within the transistor can be removed. This factor is determined by the junction-to-case thermal resistance of the transistor. A good rf power transistor is characterized by a low junction-to-case thermal resistance.

The current gain of an rf transistor varies approximately inversely with emitter current at high emitter-current levels. Peak collector current may be determined by the allowable amount of gain degradation at high frequencies. For applications in which amplitude modulation or low supply voltages are involved, peak current-handling capabilities are very important criteria to good performance.

The maximum collector-voltage rating must be high enough so that junction breakdown does not occur under conditions of large collector voltage swing. The large voltage swing is produced under conditions of amplitude modulation or reactive loading because of load mismatch and circuit tuning operations.

Before the proper matching networks of an rf amplifier can be designed, transistor impedance (or admittance) characteristics at the expected operating conditions of the circuit must be known. It is important that the value and dependence of transistor impedances on collector current, supply voltage, and operating frequency be defined.

The term f_T defines the frequency at which the current gain of a device is unity. This parameter is essential to the determination of the power-gain performance of an rf transistor at a particular frequency. Because f_T is current-dependent, it normally decreases at very high emitter currents. Therefore, it should be determined at the operating current levels of the circuit. A high f_T at high emitter or collector current levels characterizes a good rf transistor.

The operational parameters of an rf transistor can be considered to be those measured during the performance of a given circuit in which this type of transistor is used. The information displayed by these parameters is of a direct and practical interest. Operating efficiencies can normally be expected to vary between 30 and 80 per cent. Whenever possible, a circuit should employ transistors that have operational parameters specified at or near the operating conditions of the circuit so that comparisons can be made.

In some rf power applications, such as mobile radio, the transistors must withstand adverse conditions because high SWR's are produced by faulty transmission cables or antennas. The ability of a transistor to survive these faults is sometimes referred to as load-pulling or mismatch capability, and depends on transistor breakdown characteristics as well as circuit design. The load-pulling effects that the transistor may be subjected to can be determined by replacement of the rf load with a shorted stub and movement of the short through a half wavelength at the operating frequency. Dissipation capabilities of a transistor subjected to load pulling must be higher than normal to handle the additional device dissipation created by the mismatch.

Multiple Connection of Power Transistors

Many applications require more rf power than a single transistor can supply. The parallel approach is the most widely used method for multiple connection of power transistors.

In parallel operation of transistors, steps must be taken to assure equal rf and thermal load sharing. In one approach, the transistors are connected directly in parallel. This approach, however, is not very practical from an economic standpoint because it requires the use of transistors that are exactly matched in efficiencies, power gains, terminal impedances, and thermal resistances. A more practical approach is to employ signal splitting in the input matching network. By use of adjustable components in each leg, adequate compensation can be made for variations in power gains and input impedances to assure equal load sharing between the transistors. For applications in which low supply voltages are used and high power outputs are desired, the output impedance of the rf amplifier is very low. For this reason, it is beneficial, in the interest of paralleling efficiency, to split the collector loads. By use of separate collector coils, the power outputs may be combined at higher impedance levels at which the effect of any asymmetry introduced by lead inductances is insignificant and resistive losses are less. The use of separate collector coils also permits individual collector currents to be monitored.

Circuit Stability

Frequency stability is an important consideration in the design of high-frequency transistor circuits. Most instabilities occur at frequencies well below the frequency of operation because of the increased gain at lower frequencies. With the gain increasing at 6 dB per octave any parasitic low-frequency resonant loop can set the circuit into oscillation. Such parasitic oscillations can result in possible destruction of the transistor. These low-frequency loops can usually be traced to inadequate bypassing of power-supply leads, circuit component self-resonances, or rf choke resonances with circuit or transistor capacitances. Supply bypassing can be effected by use of two capacitors, one for the operating frequency and another for the lower frequencies. For amplifiers operated in the 25-to-70-MHz range, sintered-electrode tantalum capacitors can provide excellent bypassing at all frequencies of concern. At uhf and higher frequencies, these capacitors may be lossy and therefore not effective for bypassing. High-Q ceramic bypass capacitors are better suited for uhf use. RF chokes, when used, should be low-Q types and should be kept as small as possible to reduce circuit gain at lower frequencies. Chokes of the ferrite-bead variety have been used very successfully as base chokes. Collector rf chokes can be avoided by use of a coil in the matching network to apply dc to the collector.

Because of the variation of transistor parameters with changes in collector voltage and current, the stability of an rf transistor stage should be checked under all expected conditions of supply voltage, drive level, source mismatch, load mismatch, and, in the case of amplitude modulation, modulation swing.

Parametric oscillation is another form of instability that can occur in rf circuits that use power transistors. The transistor collector-to-base capacitance, as stated previously, is nonlinear and can cause oscillations that appear as low-level spurious frequencies not related to the carrier frequency.

Careful selection of components is necessary to obtain good performance in an rf transistor circuit. The components should be checked with an impedance bridge for parasitic impedances and self-resonances. When parasitic elements are encountered, their possible detrimental effects on circuit performance should be determined. This procedure helps the designer select coils and capacitances with low losses and high self-resonances (capacitors of the "bypass feed-through" or "mica postage stamp" variety can have very high self-resonances). Resistors used in rf current paths should have low series inductance and shunt capacitance (generally, low-wattage carbon resistors are quite acceptable).

Circuit layout and construction are also important for good performance. Chassis should be of a high-conductivity material such as copper or aluminum. Copper is sometimes preferable because of its higher conductivity and the fact that components can be soldered directly to the chassis. Another chassis approach now becoming popular is the use of double-side laminated printed-circuit boards. The circuit, in this approach, may be arranged so that all the conductors are on one side of the board. The opposite-side foil is then employed as an additional shield. Whenever possible, the chassis should be designed on a single plane to reduce chassis inductance and to minimize unwanted ground currents.

It must be remembered that, at rf frequencies, any conductor has an inductive and resistive impedance that can be significant when compared to other circuit impedances in a transistor amplifier. It follows, therefore, that wiring should be as direct and short as possible. It is also helpful to connect all grounds in a small area to prevent chassis inductance from causing common-impedance gain degeneration in the emitter circuit. Busses or straps may be used, but it should be remembered that these items have some inductance and that the point at which a component is connected to a buss can affect the circuit.

Coils used in input and output matching networks should be oriented to prevent unwanted coupling. In some applications, such as high-gain stages, coil orientation alone is not enough to prevent instability or strange tuning characteristics, and additional shielding between base and collector circuits must be used.

In common-emitter circuits, stage gain is very dependent on the impedance in series with the emitter. Even very small amounts of inductive degeneration can drastically reduce circuit gain at high frequencies. Although emitter degeneration results in better stability, it should be kept as low as possible to provide good gain and to reduce tuning interaction and feedback between output and input circuits. The emitters of many rf power transistors are internally connected to the case so that the lowest possible emitter-lead inductance is achieved. This technique substantially reduces the problems encountered when the transistor is fastened directly to the chassis. If a transistor with a separate emitter lead is used, every attempt should be made to provide a low-inductance connection to the chassis, even to the point of connecting the chassis directly to the lead (or pin) as close to the transistor body as practicable. In extreme cases, emitter tuning by series resonating of the emitter-lead inductance is employed.

Another important area of concern involves the removal of heat generated by the transistor. Adequate thermal-dissipation capabilities must be provided; in the case of low-power devices, the chassis itself may be used. Finned heat sinks and other means of increasing radiator area are used with higher-power devices. Consideration must also be given to ambient variations and mismatch conditions during tuning operations or load pulling, when transistor dissipation can increase. Under such conditions, the thermal resistance of the transistor may be the limiting factor, and may dictate either a change to another device of lower thermal resistance or a parallel mode of operation using the existing transistor.

MOBILE AND MARINE RADIO

In the United States, three frequency bands have been asigned to two-way mobile radio communications by the Federal Communications Commission. These frequency bands are 25 to 50 MHz, 148 to 174 MHz, and 450 to 470 MHz. The low-frequency band for overseas mobile

Frequency modulation (FM) is used for mobile radio communications in the Untied States and most overseas countries. The modulation is achieved by phase-modulation of the oscillator frequencies (usually the 12th or 18th submultiple of the operating frequency). In vhf bands, the frequency deviation is ± 5 kHz and channel spacing is 25 kHz. In uhf bands, at present, the modulation deviation is ± 15 kHz and channel spacing is 50 kHz. In the United Kingdom, AM as well as FM is used in mobile communications.

The minimum mobile-transmitter power-output levels in the United States are 50 watts in the 50-MHz band, 30 watts in the 174-MHz band, and 15 watts in the 470-MHz band. Some of the transmitters used in the United States have power-output ratings well in excess of 100 watts. Overseas power requirements are more moderate and are often regulated by law; the most common power output level is 12 watts at the antenna.

Most mobile-radio transmitters are designed to operate directly from the 12-volt electrical system of a vehicle. Tables XLIII and XLIV list the variety of RCA rf power transistors characterized for mobile-radio applications. The power output and gain, operating frequency, supply voltage, and package for each type are also indicated. The RCA data bulletin for each type provides detailed ratings and characteristics data for each device.

Table XLIII—Types for UHF Mobile-Radio Applications

Type	Operating Frequency (MHz)	Min. Output Power (W)	Collector- Supply Voltage (V)	Min. Power Gain (dB)	Package Type
41008	470	0.5	9	5.2	HF-47
41008A	470	0.5	9	5.2	HF-41
41009	470	2	9	6	HF-47
41009A	470	2	9	6	HF-41
41010	470	5	9	4	HF-41
40964	470	0.4	12	6	TO-39
40965	470	0.5	12	7	TO-39
2N5914	470	2	12.5	7	TO-216AA
40967	470	2	12.5	7	HF-44
40968	470	6	12.5	4.8	HF-44
2N5915	470	6	12.5	4.8	TO-216AA

Table XLIV—Types for VHF Mobile-Radio Applications

Type	Operating Frequency (MHz)	Min. Output Power (W)	Collector- Supply Voltage (V)	Min. Power Gain (dB)	Package Type
2N4427	175	1	12	10	TO-39
40280	175	1	13.5	9	TO-39
2N5913	175	1.75	12.5	12.4	TO-39
40972	175	1.75	12.5	12.4	TO-39
40281	175	4	13.5	6	TO-60
2N5995	175	7	12.5	9.7	TO-216AA
40973	175	10	12.5	7.6	HF-44
40282	175	12	13.5	4.8	TO-60
40974	175	25	12.5	4.5	HF-44

SINGLE-SIDEBAND TRANSMITTERS

The increase in communication traffic, especially in the hf and vhf ranges, necessitates more effective use of the frequency spectrum so that more channels can be assigned to a given spectrum. It has been shown that one of the more efficient methods of communication is through the use of single-sideband (SSB) techniques. In the past, the power-amplifier stages of an SSB transmitter invariably employed tubes because of the lack of suitable high-frequency power transistors. Recent transistor developments, however, have made it feasible and practical to design and construct all-solid-state single-sideband equipment for both portable and vehicular applications.

Unlike most commercially available rf power transistors, which are normally designed primarily for class C operation, an SSB transistor is designed for linear applications and should have a flat beta curve for low distortion, and emitter ballast resistance for stability and degeneration. In high-power amplifiers, transistor junctions experience wide excursions in temperature and a means must be provided to sense the collector-junction temperature so that an external circuit can be used to provide bias compensation to prevent an excessive shift in operating point and to avoid catastrophic device failure as a result of thermal runaway. Table XLV lists several rf power transistors for single-side-band applications. The power output, frequency, supply voltage, and package for these types are also indicated. More detailed ratings and characteristic data are contained in the RCA data bulletin for each device.

Advantages of SSB Transmission

Single-sideband communication systems have many advantages over AM and FM systems. In areas in which reliability of transmission as well as power conservation are of prime concern, SSB transmitters are usually employed. The main advantages of SSB operation include reduced power consumption for effective transmission, reduced channel width to permit more transmitters to be operated within a given frequency range, and improved signal-to-noise ratio.

In a conventional 100-per-cent modulated AM transmitter, two-thirds of the total power delivered by the power amplifier is at the carrier frequency, and contributes nothing to the transmission of intelligence. The remaining third of the total radiated power is distributed equally between the two sidebands. Because both sidebands are identical in intelligence content, the transmission of one sideband would be sufficient. In AM, therefore, only one-sixth of the total rf power is fully utilized. In an SSB system, no power is transmitted in the suppressed sideband, and power in the carrier is greatly reduced or eliminated; as a

Table XLV—Types for Single-Sideband Applications and for Military Communications

Type	Operating Frequency (MHz)	Min. Output Power (W)	Collector-Supply Voltage (V)	Min. Power Gain (dB)	Package Type
40082	30	2.5(PEP)	12.5	10	TO-39
40936	30	20(PEP)	28	13	TO-60
2N5070	30	25(PEP)	28	13	TO-60
2N6093	30	75(PEP)	28	13	TO-217AA
2N5071	76	24	24	9	TO-60

result, the dc power requirement is substantially reduced. In other words, for the same dc input power, the peak useful output power of an SSB transmitter in which the carrier is completely suppressed is theoretically six times that of a conventional AM transmitter.

Another advantage of SSB transmission is that elimination of one sideband reduces the channel width required for transmission to one-half that required for AM transmission. Theoretically, therefore, two SSB transmitters can be operated within a frequency spectrum that is normally required for one AM transmitter.

In a single-sideband system, the signal-to-noise power ratio is eight times as great as that of a fully modulated double-sideband system for the same peak power.

Linearity Test

For an amplifier to be linear, a relationship must exist such that the output voltage is directly proportional to the input voltage for all signal amplitudes. Because a single-frequency signal in a perfectly linear single-sideband system remains unchanged at all points in the signal path, the signal cannot be distinguished from a cw signal or from an unmodulated carrier of an AM transmitter. To measure the linearity of an amplifier, it is necessary to use a signal that varies in amplitude. In the method commonly used to measure nonlinear distortion, two sine-wave voltages of different frequencies are applied to the amplifier input simultaneously, and the sum, difference, and various combination frequencies that are produced by nonlinearities of the amplifier are observed. A frequency difference of 1 to 2 kHz is used widely for this purpose. A typicla two-tone signal without distortion, as displayed on a spectrum analyzer, is shown in Fig. 478. The resultant signal envelope varies continuously between zero and maxi-

mum at an audio-frequency rate. When the signals are in phase, the peak of the two-frequency envelope is limited by the voltage and current ratings of the transistor to the same power rating as that for the single-frequency case. Because the amplitude of each two-tone frequency is equal to one-half the cw amplitude under peak power condition, the average power of one tone of a two-tone signal is one-fourth the single-frequency power. For two tones, conversely, the PEP rating of a single-sideband system is two times the average power rating.

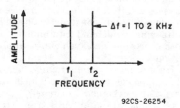

Fig. 478—Frequency spectrum for a typical two-tone signal without distortion.

Intermodulation Distortion

Nonlinearities in an amplifier generate intermodulation (IM) distortion. The important IM products are those close to the desired output frequency, which occur within the pass band and cannot be filtered out by normal tuned circuits. If f_1 and f_2 are the two desired output signals, third-order IM products take the form of $2f_1 - f_2$ and $2f_2 - f_1$. The matching third-order terms are $2f_1 + f_2$ and $2f_2 + f_1$, but these matching terms correspond to frequencies near the third harmonic output of the amplifier and are greatly attenuated by tuned circuits. It is important to note that only odd-order distortion products appear near the fundamental frequency. The frequency spectrum shown in Fig. 479 illustrates the frequency relationship of some distortion products to the test signals f_1 and f_2. All such products are either in the difference-frequency region or in the harmonic regions of the original frequencies. Tuned circuits or filters following the nonlinear elements can effectively remove all products generated by the even-order components of curvature. Therefore, the second-order component that produces the second harmonic does not produce any distortion in a narrow-band SSB linear amplifier. This factor explains why class AB and class B rf amplifiers can be used as linear amplifiers in SSB equipment even through the collector-current pulses contain large amounts of second-harmonic current.

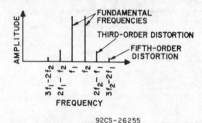

Fig. 479—Frequency spectrum showing the frequency relationship of some distortion products to two test signals f_1 and f_2.

In a wideband linear application, however, it is possible for harmonics of the operating frequency to occur within the pass band of the output circuit. Biasing the output transistor further into class AB can greatly reduce the undesired harmonics. Operation of two transistors in the push-pull configuration can also result in cancellation of even harmnics in the output.

The signal-to-distortion ratio (in dB) is the ratio of the amplitude of one test frequency to the amplitude of the strongest distortion product. A signal-to-distortion specification of -30 dB means that no distortion product will exceed this value for a two-tone signal level up to the PEP rating of the ampli-

fier. A typical presentation of IM distortion for an RCA-2N6093 transistor at various output-power levels is shown in Fig. 480.

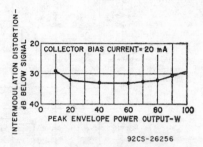

Fig. 480—Typical intermodulation distortion in an RCA-2N6093 transistor at various output power levels.

Typical Linear Amplifier

The common-emitter configuration should be used for the power amplifier because of its stability and high power gain. Tuning is less critical, and the amplifier is less sensitive to variations in parameters among transistors. The class AB mode is used to obtain low intermodulation distortion. Neither resistive loading nor neutralization is used to improve linearity because of the resulting drastic reduction in power gain; furthermore, neutralization is difficult for large signals because parameters such as output capacitance and output and input impedances vary nonlinearly over the limits of signal swing.

The RCA-2N5070 transistor is specified for SSB applications without temperature compensation as follows:

Frequency = 30 MHz
P_o (PEP) at 28 V = 25 W
Power Gain = 13 dB (min.)
Collector Efficiency =
   40% (min.)

Fig. 481 shows a 25-watt 2-to-30-MHz wideband linear amplifier

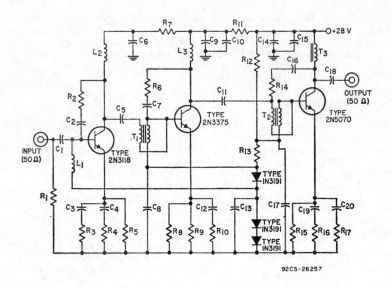

92CS-26257

Fig. 481—2-to-30-MHz linear power amplifier.

that uses RCA rf transistors. At 5 watts (PEP) output, IM distortion products are more than 40 dB below one tone of a two-tone signal. Power gain is greater than 40 dB.

AIRCRAFT RADIO

The aircraft radios discussed in this section are of the type used for communication between the pilot and the airport tower. The transmitter operates in an AM mode on specific channels between 118 and 136 MHz. Radios of this type are regulated by both the FCC (Federal Communications Commission) and the FAA (Federal Aeronautics Administration). The FCC assigns frequencies to airports and places some requirements on the transmitters, particularly as regards spurious radiation and interference. The FAA sets minimum requirements on radio performance which are based on the maximum authorized altitudes for the plane, whether paying passen-

gers are carried, and on the authorization for instrument flying. The FAA gives a desirable TSO certification to radio equipment that satisfies their standards of airworthiness.

The FCC checks aircraft-radio transmitter designs for interference and other electrical characteristics (as it does all transmitters). Additional requirements are specified for radios intended for use by scheduled airlines by a corporation supported by the airlines themselves. This corporation is ARINC (Aeronautical Radio, Inc., 2552 Riva Road, Annapolis, Maryland 21401).

All these specifications combine to generate radio-transmitter requirements for different types of aircraft, as indicated in Table XLVI. RCA rf power transistors intended for use in such applications are listed and briefly described in Table XLVII. The RCA data bulletin for each device provides definitive ratings and characteristics.

Table XLVI—Popular Aircraft-Radio Transmitters
(Designs by Aircraft Type*)

TYPICAL OWNER	NO. OF ENGINES IN AIRCRAFT	FAA & ARINC CLASS	VOLTAGE AVAIL-ABLE	TRANS-MITTER POWER (MIN.)	TYPICAL POWER RANGE	TRANSMITTER FEATURES
Private Planes	1	I	13 V	1 W	>1.5 W	Low cost, few channels, may be portable
Owner/Pilot	1	I	13 V	4 W	>6.0 W	Panel mounted, 90 or 360 channels.
Private/Business	2	II	28 V	4 W	6 to >20 W	
Chartered & Cargo	2-4	III	28 V	16 W	>20 W	Remote Operation, 360 channels.
Scheduled Air Lines	2-4 Jets	III & ARINC	28 V	25 W	30 W	Maximum reliability

* This chart is not complete or exact and is not intended to show actual requirements, but merely what is typical. Consult FAA for complete requirements.

Table XLVII—Types for Aircraft-Radio Applications

Type	Operating Frequency (MHz)	Min. Output Power (W)	Collector-Supply Voltage (V)	Min. Power Gain (dB)	Package Type
40975	118–136	0.05	12.5	10	TO-39
40976	118–136	0.5	12.5	10	TO-39
40977	118–136	6	12.5	10.8	HF-44
40290	118–136	2	12.5	6	TO-39
40291	118–136	2	12.5	6	TO-60
40292	118–136	6	12.5	4.8	TO-60
2N5102	118–136	15	24	4	TO-60

VHF AND UHF MILITARY RADIO

Military radios, which operate in the vhf and uhf ranges, vary greatly in requirements. Telemetering devices may operate with as little output as 0.25 watt, while communication systems may require outputs of 50 watts and more. Modulation may be AM, FM, PM (pulse modulation), or PCM (pulse-code modulation). Equipment may be designed for fixed, mobile, airborne, or even space applications. Although the circuits described in this section apply only to specific military applications, they are representative of the general design techniques used in all military vhf and uhf radio equipment. RCA transistor types specified for use in such applications are listed in Table XLVIII.

Sonobuoy Transmitter

A sonobuoy is a floating submarine-detecting device that incorporates an underwater sound detector (hydrophone). The audio signals received are converted to a frequency-modulated rf signal which is transmitted to patrolling aircraft or surface vessels. The buoy is battery-operated and is designed to have a very limited active life.

Typical requirements for the rf-transmitter section of the sonobuoy are as follows:

Frequency = 165 MHz
Supply Voltage = 8 to 15 volts
CW Output = 0.25 to 1.5 watts
Over-all Efficiency = 50 per cent
Harmonic Output = 40 dB down
  from carrier

Fig. 482 shows the circuit configuration of an experimental sonobuoy transmitter designed to produce a power output of 2 watts at 160 MHz. Only three stages, including the crystal-controlled oscillator section, are required. Efficiency is greater than 50 per cent (overall) with a battery supply of 12 to 15 volts.

The 2N3866 or 2N4427 transistor can be used in a class A oscillator-quadrupler circuit which is capable of delivering 40 milliwatts of rf power at 80 MHz. Narrow-band frequency modulation is accomplished by "pulling" of the crystal oscillator. The crystal is operated in its fundamental mode at 20 MHz. The oscillator is broadly tuned to 20 MHz in the emitter circuit and is sharply tuned to 80 MHz in the collector circuit. The supply voltage to the oscillator section is regulated at 12 volts by means of a zener diode. Spectrum-analyzer tests indicate that this stage is highly stable even though rather high operating levels are used.

The oscillator-quadrupler section is followed by a 2N3553 class C doubler stage. This stage delivers

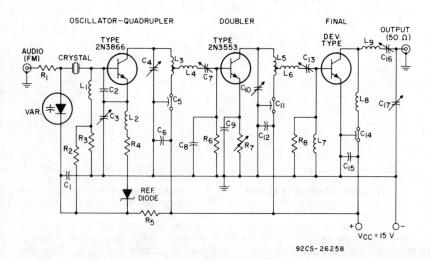

Fig. 482—2-watt (rf power output) sonobuoy transmitter.

92CS-26258

Table XLVIII—Types for UHF Military Applications

Type	Operating Frequency (MHz)	Min. Output Power (W)	Collector-Supply Voltage (V)	Min. Power Gain (dB)	Package Type
2N3866	400	1	28	10	TO-39
2N5916	400	2	28	10	TO-216AA
2N5917	400	2	28	10	HF-31
2N5918	400	10	28	8	TO-216AA
2N5919A	400	16	28	6	TO-216AA
2N6104	400	30	28	5	HF-32
2N6105	400	30	28	5	TO-216AA

a power output of 250 milliwatts at 160 MHz from a 12- to 15-volt supply. The over-all output of the sonobuoy can be adjusted by varying the emitter resistance of this stage.

The final power output is developed by an RCA-2N2711 transistor which operates as a straight-through class C amplifier at 160 MHz. A pi network matches this output to the 50-ohm line. The spurious output (measured directly at the output port) is more than 35 dB down from the carrier. This suppression is achieved by means of series resonant trap circuits between stages and the use of the pi network in the output.

Sonobuoy circuits, in general, must be reliable, simple, and low in cost. The three-stage transmitter circuit shown in Fig. 482 is intended to be representative of the general design techniques used in these systems. However, four-stage sonobuoy trans-

mitter systems are also in common use at the present time. Typically, a four-stage arrangement consists of an oscillator-tripler stage, a second tripler stage, a buffer stage, and a final amplifier stage. Most present-day sonobuoy applications require cw power output between 0.25 and 0.5 watt.

Air-Rescue Beacon

The air-rescue beacon is intended to aid rescue teams in locating airplane crew members forced down on land or at sea. The beacons are amplitude-modulated or continuous-tone line-of-sight transmitters. They are battery-operated and small enough to be included in survival gear.

Typical requirements for rescue beacons are as follows:

Frequency = 243 MHz (fixed)
Power Output = 300 milliwatts
                    (carrier)
Efficiency = greater than 50 per
                    cent
Supply Voltage = 6 to 12 volts
Modulation = AM, up to ±100
                    per cent

The 2N4427 transistor is especially suited for this service. A general circuit for the driver and output stages is shown in Fig. 483. Collector modulation, as well as some driver modulation, is used to achieve good down-modulation of the final amplifier. Conventional transformer-coupled modulation is used; however, a separate power supply and resistor network in the driver circuit are provided to adjust the modulation level of this stage independently of the output stage.

The rf-amplifier design is conventional; pi- and T-matching networks are used; simpler circuits (e.g., device-resonated tapped coils), however, could be used. The T-matching

network at the driver input is used to match the amplifier to a 50-ohm source for test purposes. A 10-to-20-milliwatt input signal is needed to develop a 300-to-400-milliwatt carrier output level.

Communication Systems

The frequency range from 225 to 400 MHz is used in a large variety of relatively-high-power military communication systems. Equipments are usually amplitude-modulated and used for voice-communication purposes. The circuits discussed in this section are class B and class C amplifiers for use in driver or final output stages that provide power outputs in the range from 5 to 30 watts from a single transistor. Higher power can be obtained from combinations of transistors.

These amplifiers make extensive use of power combiners and broadband impedance matching.

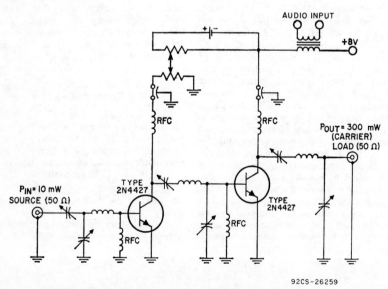

92CS-26259

Fig. 483—Driver and output stage for a 243-MHz beacon transmitter.

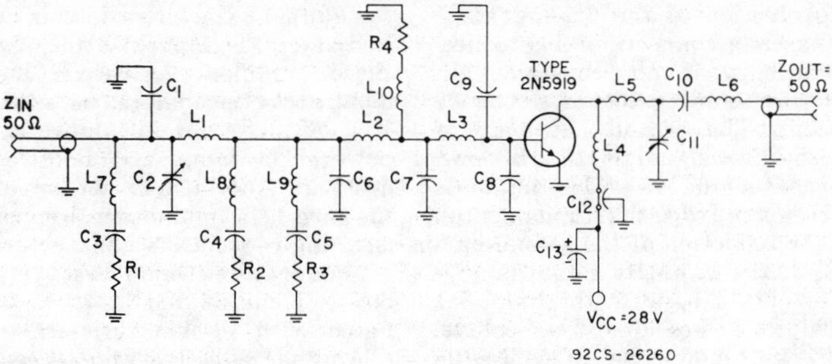

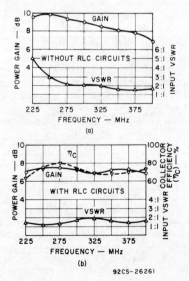

Fig. 484 circuit schematic

92CS-26260

$C_1 = 10$ pF silver mica
$C_2 = 0.8$–10 pF, Johanson 3957*
$C_3 = 2.2$ pF, Quality Components type 10% QC, "gimmick"*
$C_4 = 1.0$ pF, Quality Components type 10% QC, "gimmick"*
$C_5 = 1.5$ pF, Quality Components type 10% QC, "gimmick"*
$C_6 = 36$ pF, ATC-100*
$C_7 = 51$ pF, ATC-100*
$C_8 = 47$ pF, ATC-100*
$C_9 = 68$ pF, ATC-100*
$C_{10} = 12$ pF, silver mica
$C_{11} = 0.8$–20 pF, Johanson 4802*
$C_{12} = 1000$ pF feedthrough type, Allen-Bradley FA5C*
$C_{13} = 1$ μF electrolytic
$L_1 = 1\frac{1}{2}$ turns▲
$L_2 =$ Copper strip $\frac{5}{8}$ in. (15.875 mm) L; 5/32 in. (3.96 mm) W

$L_3 =$ Transistor base lead, 3/6 in. (4.74 mm) L
$L_4, L_6 = 3$ turns▲
$L_5 = 2$ turns▲
$L_7, L_8, L_9 = 0.18$ μH RFC, Nytronics, P.#DD-0.18
$L_{10} = 0.1$ μH RFC, Nytronics, P.#DD-0.10
$R_1 = 100$ Ω, 1 W, carbon
$R_2, R_3 = 100$ Ω, $\frac{1}{2}$ W, carbon
$R_4 = 5.1$ Ω, $\frac{1}{2}$ W, carbon
* Or equivalent
▲ All coils are 5/32 in. (3.96 mm) I. D., # 18 wire, 12 turns per inch.
Allen-Bradley Co., Milwaukee, Wis.
American Technical Ceramics, Huntington Station, N. Y. 11746
Johanson Mfg. Corp., Boonton, N. J. 07005
Nytronics, Inc., Berkeley Heights, N. J.

Fig. 484—16-watt broadband amplifier circuit using the RCA-2N5919A.

Fig. 484 shows a schematic diagram of an amplifier that uses the RCA-2N5919A. The circuit utilizes a lumped-element approach to broadband design. Typical amplifier performance is shown in Fig. 485. For a constant power output of 16 watts, response is fairly flat; the gain variation is within 1 dB across the band. Maximum input VSWR is 2:1. Such flatness of response and low input VSWR are obtained by designing for the best possible match across the band and then dissipating some of the power at the low end of the band through dissipative RLC networks. The effectiveness of this technique can be evaluated by comparison of the gain and input VSWR curves in Fig. 485(a) with those in Fig. 485(b). The flatter the response, the smaller the dynamic range required in the output leveling system. Low input VSWR is necessary for effective cascading

Fig. 485—Typical performance for circuit of Fig. 484 from 225 to 400 MHz.

92CS-26261

and protection of the driving stage in a cascade connection. The collector efficiency is not constant, but has a minimum value of about 63 per cent. The second harmonic of the 225-MHz signal is 12 dB down and that of the 400-MHz signal is 30 dB down from the fundamental. Further reduction of the second harmonic of the 225-MHz signal is difficult to obtain because the amplifier bandwidth covers almost an octave.

Fig. 486 shows the RCA-2N6105 high-power transistor in a 30-watt 225-to-400-MHz broadband amplifier. This circuit utilizes lumped-circuit-element broadbanding. No

special care is evident in this circuit to reduce the input VSWR. Two of these amplifiers can be combined by quadrature combiners, as shown in Fig. 487, to obtain higher output power. The input VSWR of the individual amplifier is not important in such a combination because of the high isolation characteristics of quadrature combiners; reflected power is dissipated in ports terminated with 50-ohm resistors.

Another effective way to combine transistors is push-pull operation utilizing transmission-line techniques. The low second harmonic in the push-pull configuration is es-

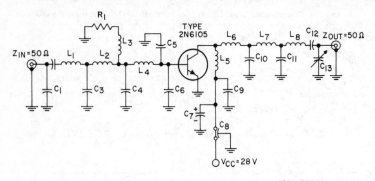

92CS-26262

$C_1 = 8.2$ pF chip, Allen-Bradley*
$C_2 = 18$ pF silver mica
$C_3 = 33$ pF chip, Allen-Bradley*
$C_4 = 47$ pF chip, Allen-Bradley*
$C_5 = 68$ pF chip, ATC-100*
$C_6 = 62$ pF chip, ATC-100*
$C_7, C_8 = 1000$ pF, Feedthrough
$C_9, C_{12} = 1000$ pF chip, Allen-Bradley*
$C_{10} = 22$ pF chip, Allen-Bradley*
$C_{11} = 6.9$ pF chip, Allen-Bradley*
$C_{13} = 0.8$-10 pF variable air, Johanson No. 3957*

$L_1 = 2$ turns, 5/32 in. I.D. coil
$L_2 = 17/32$-in. long wire
$L_3 = $ RFC, 0.1 μH, Nytronics*
$L_4 = 5/32$-in. long transistor base lead
$L_5, L_7 = 13/16$-in. long wire
$L_6 = 9/16$-in. long wire
$L_8 = 7/8$-in. long wire
$R_1 = 5.0$ Ω, 1/4 W

All wire is No. 20 AWG
* Or equivalent

Fig. 486—30-watt 225-to-400-MHz broadband amplifier using RCA-2N6105.

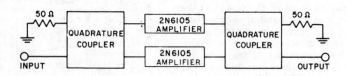

92CS-26263

Fig. 487—Two RCA-2N6105 amplifiers connected in parallel by use of quadrature couplers.

pecially important in the 225-to-400 MHz frequency band; because the second harmonic of the low frequency falls just outside the band, filtering it out presents considerable difficulty. The input VSWR in a push-pull amplifier is very high at the low end of the band, so this type of circuit is especially suitable for use with quadrature combiners. Fig. 488 shows two push-pull amplifiers combined by quadrature combiners to make up a 100-watt broadband module.

In such equipment, transistors offer the advantages of simplified circuitry, wide bandwidths, and improved reliability, together with reduced size and weight.

The selection of the proper transistor for a specific application is determined by the required power output, gain, and circuit preference. RCA offers the circuit designer a wide variety of microwave power transistors from which to choose the optimum type for a particular application. Such transistors are sup-

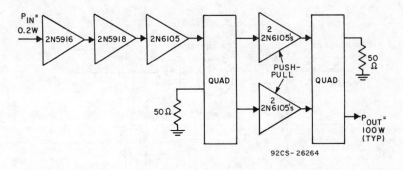

Fig. 488—Block diagram of a 100-watt 225-400-MHz amplifier employing 2N6105's in the driver and output stages.

MICROWAVE POWER AMPLIFIERS

Transistors that can generate tens of watts of power output at frequencies up to and beyond 3 GHz are finding applications in a wide variety of new equipment designs. Some of the major applications for these new transistors are in the following types of equipment:

Telemetry
Microwave relay links
Microwave communications
Phased-array radars
Mobile radio and radio-telephones
Navigational aid systems (DME, Collision Avoidance, TACAN)
Electronic countermeasure (ECM)
Microwave power sources and instrumentation
Intrusion-alarm systems

plied in either coaxial or stripline packages. In addition, transistor types that are optimized for operation from either 22 or 28 volts and that are packaged specifically for either amplifier or oscillator applications are also readily available.

At 2 GHz, a single transistor can supply an output power greater than 10 watts with a gain of 7 dB. If higher output power is required, two or more transistors can be connected in parallel, either directly or by use of hybrid combiners. Higher gain can be achieved by connection of several stages in cascade. The transistor supply voltage is usually dictated by the application. In telemetry systems and microwave relay links, the supply voltage normally ranges from 20 to 23 volts; most other systems use 28 volts. The type of package selected, either coaxial or stripline, usually depends on the type of circuit in which the designer plans to use the transistor.

Coaxial Amplifier Circuits

Fig. 489 shows a typical 2N5921 coaxial-circuit amplifier for 2-GHz

Lower-power amplifiers can be made with the RCA-2N5920 or RCA-2N5470 transistors.

Microstripline Amplifier Circuits

Fig. 492 shows the circuit diagram for a typical microstripline power amplifier that uses an RCA-2N6267

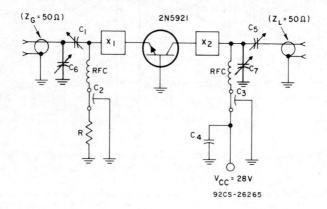

92CS-26265

C₁ = 1 to 10 pF, Johanson 4581, or equiv.*
C₂, C₃ = 470 pF.
C₄ = 0.01 μF.
C₅, C₆, C₇ = 0.3 to 3.5 pF, Johanson 4700, or equiv.*

RFC = 3 turns No. 32 wire 1/16 in. (1.59 mm) ID, 3/16 in. (4.76 mm) long.
R = 0.43 ohm.
X₁, X₂; Coaxial-line circuits (see Fig. 490).
* Johanson Mfg. Corp., Boonton, N. J. 07005

Fig. 489—2-GHz coaxial-line amplifier circuit.

operation, and Fig. 490 shows the mechanical construction of the circuit. The performance data in Fig. 491 show that the 2N5921 can provide an output of 6 watts at 2 GHz with a gain of 7.8 dB and a collector efficiency greater than 40 per cent.

transistor to develop a power output of 10 watts at 2 GHz when operated from a 28-volt supply. This amplifier is tunable by adjustment of variable capacitors C1 and C4. Similar circuits can be designed using the 2N6265 and 2N6266 transistors.

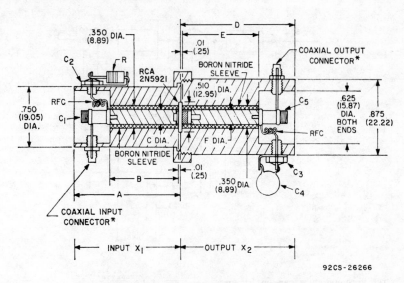

92CS-26266

DIMENSIONS OF COAXIAL LINES X_1 AND X_2

INPUT (X_1)				OUTPUT (X_2)			
A	B	C	Center Conductor	D	E	F	Center Conductor
0.860 (21.84)	0.350 (8.89)	0.265 (6.73)	0.300 (7.62)	1.06 (26.92)	0.550 (13.97)	0.270 (6.86)	0.385 (9.78)

Dimensions in inches (and millimeters)
MATERIAL: Center conductor—copper
          Outer conductor for input and output—brass
* Conhex 50-045-0000 Sealectro Corp., or equiv.

Fig. 490—Constructional details for 2-GHz coaxial-line circuit.

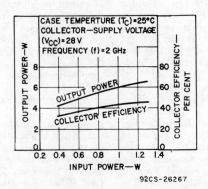

92CS-26267

Fig. 491—Typical power output and collector efficiency as a function of power input at 2 GHz for circuit shown in Fig. 489.

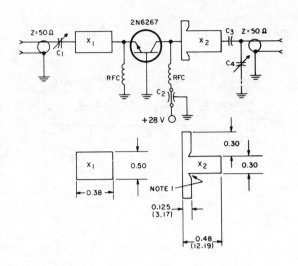

NOTE 1: Shunt stubs can be trimmed.
To shorten, cut overall length.
To lengthen, cut taper in stubs
as shown.

C₁, C₃, C₄ = 0.3 to 3.5 pF, Johanson 1700 or
equivalent
C₂ = Filtercon, Allen Bradley SMFB A1, or
equivalent
RFC = No. 32 wire, 0.4 in (1.02 mm) long

Dielectric material = 1/32 in. (0.79 mm)
thick Teflon-Fiberglas double-clad circuit board (ε = 2.6). Lines X₁ and X₂
are produced by removing upper copper
layer to dimensions shown.

Fig. 492—2-GHz power amplifier using an RCA-2N6267 microwave power transistor.

MICROWAVE POWER OSCILLATORS

The design of microwave power oscillators is usually a complex process because of the many diverse requirements that must be met in these circuits. The design of the power oscillator shown in Fig. 493 however, is made relatively simple by application of power-amplifier considerations. This oscillator uses an RCA-2N6267 microwave power transistor to develop a typical power output of 4.0 watts at 1.7 GHz when operated from a collector supply of 20 volts. A power output of 2.3 watts at 1.6 GHz can be obtained when the transistor is operated from a collector supply of 12.5 volts. The collector efficiency in each case is approximately 37 per cent.

In the design of the oscillator, the values of the line section L_1 and the variable capacitor C_2 are chosen so that the resonant frequency of these elements is slightly less than the desired circuit operating frequency. At frequencies above resonance (e.g., the oscillator operating frequency), the combination of L_1 and C_2, in essence, becomes a variable inductance L. The transistor input and output reactances, the inductance L, and the transistor collector-to-emitter capacitance C_{CE} form a resonant circuit that establishes the oscillation frequency of the over-all circuit and also determines the correct level of the in-phase signal fed back to the input to sustain oscillation. The operating frequency of the oscillator is controlled by adjustment of the variable capacitor C_2.

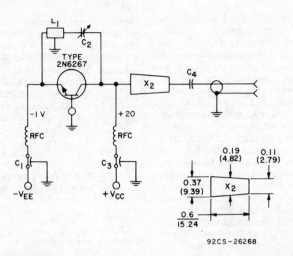

92CS-26268

C₁, C₃ = Filtercon, Allen Bradley SMFB-A1, or equivalent
C₂ = 0.3-3.5 pF, Johanson 4700, or equiv.
C₄ = 300 pF, ATC 100 or equivalent
L₁ = 1.0 in. section miniature 50-ohm cable, or microstripline equivalent
RFC = 3 turns, No. 32 wire 1/16 in. ID, 3/16 in. long

X₂ = 13-mil Teflon-Kapton double-clad circuit board (Grade PE-1243 as supplied by Budd Polychem Division, Newark, Delaware) or equivalent
Line X₂ is exponentially tapered
NOTE: Oscillator is single-screw tunable from 1.6 GHz to 1.8 GHz

Fig. 493—Typical 1.7-GHz power oscillator using the RCA-2N6267 microwave power transistor.

The real part of the collector load impedance, $Re(Z_{CL})$, determined on the basis of large-signal class C load conditions, is transformed to 50 ohms by use of a quarter-wavelength transformer X_2. Because of the wide range of frequency control provided by the variable capacitor C_2, a quarter-wavelength tapered line is used for transformer X_2.

MICROWAVE FREQUENCY MULTIPLIERS

Operation of a transistor in the harmonic-frequency mode can extend the upper limit of the frequency range far beyond that possible from the same transistor operating in the fndamental-frequency mode. A further advantage of the harmonic mode of operation is that frequency multiplication and power amplification can be realized simultaneously.

A transistor operating in this mode provides power amplification at the fundamental frequency of the input-drive power, and the nonlinear capacitance of the collector-to-base junction, acting as a varactor, generates harmonics of the input-drive frequency.

The design of transistor frequency-multiplier circuits generally consists of the selection of a suitable transistor and the design of proper filtering and matching networks for optimum circuit performance.

In the design of such circuits, the input impedance at the fundamental frequency that exists at the emitter-to-base junction of the transistor as well as the load impedance presented to the collector at both the fundamental and harmonic frequencies must be known. Knowledge of the collector load impedance at the harmonic frequency is required for design of the output circuit. Knowledge of the collector impedance at the fundamental frequency is needed to determine the input impedance of the transistor at that frequency so that matching networks can be designed between the driving source and the transistor. The three impedances, of course, are interrelated and are functons of operating power level (i.e., are determined by voltage and current swings). Once the impedances are established, the design of the matching networks is straightforward. For the input circuit, a matching section having low-pass characteristics is preferred; for the output circuit, a matching section having high-pass or band-pass characteristics is preferred. Such arrangements assure good isolation between input and output circuits. As the frequency of operation increases above 800 MHz, the design of transistor multiplier circuits requires the use of distributed circuit techniques.

The 2N4012 power transistor is characterized for frequency-multiplication applications and can provide a minimum power output of 2.5 watts as a frequency tripler at an output frequency of 1 GHz and a collector efficiency of 25 per cent. This overlay transistor is designed to operate in military and industrial communications equipment as a frequency multiplier in the uhf or L-band range. It can be operated as a doubler, tripler, or quadrupler to supply a power output of several watts at frequencies in the low gigahertz range.

Fig. 494 shows the power-output capabilities as a function of output frequency for a typical 2N4012 transistor used in common-emitter circuit configurations for frequency doubling, tripling, and quadrupling. In a common-emitter doubler circuit, the transistor delivers power output of 3.3 watts at 800 MHz with a conversion gain of 5 dB. In a common-emitter tripler circuit, it can supply power output of 2.8 watts at 1 GHz with a conversion gain of 4.5 dB.

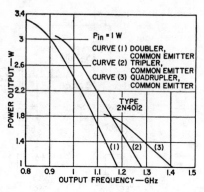

92CS-26269

Fig. 494—Power output of the RCA-2N4012 overlay transistor as a function of frequency when operated in common-emitter doubler, tripler and quadrupler circuits.

In a common-emitter quadrupler circuit, it can provide power output of 1.7 watts at 1.2 GHz with a conversion gain of 2.3 dB.

Fig. 494 shows that the amount of power output that can be supplied by a transistor frequency multiplier depends upon the order of multiplication. For a given multiplier circuit, the highest output power is obtained at the frequency for which the product of power gain and conversion efficiency has the largest value. When a 2N4012 overlay transistor is used, maximum power output is obtained at 800 MHz from a doubler circuit, at 1 GHz from a tripler circuit, and at 1.3 GHz from a quadrupler circuit.

The circuit arrangements and performance data shown in the following paragraphs illustrate several practical frequency-multiplier circuits that use the 2N4012 and other RCA overlay transistors. These circuits include a 400-to-800-MHz doubler, a 367-to-1100-MHz tripler, and a 420-to-1680-MHz quadrupler. As mentioned previously, the design of multiplier circuits that have an output frequency of 800 MHz or higher requires the use of distributed-circuit techniques. All such high-frequency circuits described use coaxial-cavity output circuits. These circuits are discussed first. The low-frequency circuits, which use lumped-element output circuits, are then described.

400-to-800-MHz Doubler

Fig. 495 shows the complete circuit diagram of a 400-to-800-MHz doubler that uses the 2N4012 transistor. This circuit uses lumped-element input and idler circuits and a coaxial-cavity output circuit. The transistor is placed inside the cavity with its emitter properly grounded to the chassis. A pi section (C_1, C_2, L_1, L_2, and C_3) is used in the input to match the impedances, at 400 MHz, of the driving source and the base-emitter junction of the transistor. L_2 and C_3 provide the necessary ground return for the nonlinear ca-

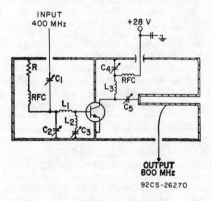

Fig. 495—*400-to-800-MHz common-emitter transistor frequency multiplier.*

a point near the shorted end of the cavity.

Fig. 496 shows the power output at 800 MHz as a function of the power input at 400 MHz for the doubler circuit, which uses a typical

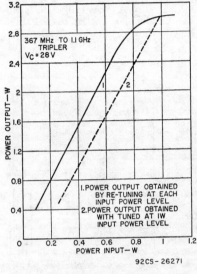

Fig. 496—*Output power and collector efficiency as a function of input power for the 400-to-800-MHz frequency doubler.*

2N4012 operated at a collector supply voltage of 28 volts. The curve is nearly linear at a power output level between 0.9 and 2.7 watts. The power output is 3.3 watts at 800 MHz for an input drive of 1 watt at 400 MHz, and rises to 3.9 watts as the input drive increases to 1.7 watts. The collector efficiency, which is defined as the ratio of the rf power output to the dc power input at a supply voltage of 28 volts, is also shown in Fig. 496. The efficiency is 43 per cent measured at an input power of 1 watt. The 3-dB bandwidth of this circuit measured at a power output of 3.3 watts is 2.5 per cent. The fundamental-frequency component measured at a power-

pacitance of the transistor. L_3 and C_4 form the idler loop for the collector at 400 MHz. The output circuit consists of an open-ended 1¼-inch-square coaxial cavity. A lumped capacitance C_5 is added in series with a ¼-inch hollow center conductor of the cavity near the open end to provide adjustment for the electrical length. Power output at 800 MHz is obtained by direct coupling from output level of 3.3 watts is 22 dB down from the output carrier. Higher attenuations of spurious components can be achieved if more filtering sections are used.

367-to-1100-MHz Tripler

The 367-to-1100-MHz tripler shown in Fig. 497 is essentially the same as the doubler shown in Fig. 495 except that an additional idler loop (L_t and C_6) is added in shunt with

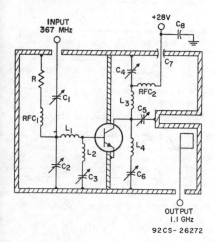

Fig. 497—367-MHz-to-1.1-GHz common-emitter transistor frequency tripler.

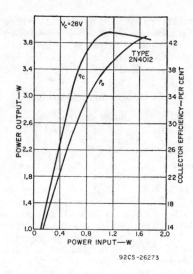

92CS-26273

Fig. 498—Power output as a function of power input for the 367-MHz-to-1.1-GHz frequency tripler.

level. A power output of 2.9 watts at 1.1 GHz is obtained with drive of 1 watt at 367 MHz.

A similar tripler circuit that uses a selected 2N3866 and that is operated from 500 MHz to 1.5 GHz can delived a power output of 0.5 watt at 1.5 GHz with an input drive of 0.25 watt at 500 MHz.

Common-Emitter and Common-Base Circuits

The performance data in this section are given for amplifier-multipliers in which the transistor is connected in a common-emitter configuration. When transistors are used in common-base circuit configurations, different results are obtained. Fig. 499 shows curves of power output and efficiency for a common-base and a common-emitter tripler circuit using a 2N4012 transistor. At low power levels, the common-base tripler provides higher gain and collector efficiency; at high power levels, higher gain and collector efficiecy are provided by the common-emitter circuit. At a power

the collector of the transistor. This idler loop is resonant with the transistor junction capacitance at the second harmonic frequency (734 MHz) of the input drive.

Fig. 498 shows the power output of the tripler at 1.1 GHz as a function of the power input at 367 MHz. This circuit also uses a typical 2N4012 transistor operated at a collector supply voltage of 28 volts. The solid-line curve shows the power output obtained when the circuit is retuned at each power-input level. The dashed-line curve shows the power output obtained with the circuit tuned at the 2.9-watt output

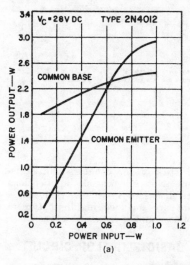

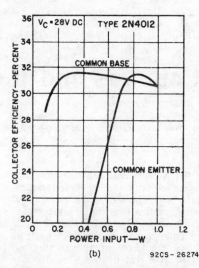

92CS- 26274

Fig. 499—Comparison of performance characteristics of common-base and common-emitter tripler circuits using the RCA-2N4012 transistor: (a) power output as a function of power input; (b) collector efficiency as a function of power input.

input of 1 watt at 367 MHz, the common-emitter tripler delivers a power output of 2.9 watts at 1.1 GHz and the common-base circuit an output of 2.4 watts. The collector efficiencies for both cirlcuits are approximately the same and are better than 30 per cent. The 3-dB bandwidth measured in the common-emitter tripler is 2.3 per cent, as compared to 2.5 per cent in a common-base tripler. The major difference between the two circuits is that the power output of the common-emitter tripler saturates at a much higher power-input level than that of the common-base circuit. This effect has also been observed in a straight-through amplifier. In addition, the common-emitter circuit is less sensitive to hysteresis and high-frequency oscillations.

420-MHz-to-1.68-GHz Oscillator-Quadrupler

The inherent varactor frequency-multiplication ability in overlay transistors also permits use of these devices as oscillator-multipliers. Fig. 500 shows an oscillator-quad-

rupler circuit that uses a selected 2N3866 transistor. This circuit can deliver a power output of more than 300 milliwatts at 1.68 GHz. The first two rf chokes and the resistors R_1 and R_2 form the bias circuit. The fundamental frequency of the oscillator is 420 MHz, as determined by C_0, L_1, and C_1. L_2 and C_2 form the second-harmonic idler. The second-harmonic component produced by this idler circuit beats with the fundamental-frequency component to generate additional fourth-harmonic components. A series-tuned circuit consisting of L_3 and C_3 completes the output circuit.

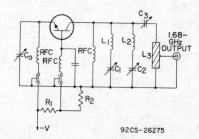

92CS-26275

Fig. 500—Oscillator-quadrupler circuit.

Counters and Registers

COUNTERS and registers are digital circuits used in computers, industrial controls, timepieces, and other applications that require repetitive steps. These circuits are made up from arrays of flip-flops such as those discussed in the section on COS/MOS Integrated Circuits for Digital Applications; COS/MOS counters and registers consist of between 100 and 300 MOS devices, supporting interconnect tunnels, metal runs, and bond pads, all on a single monolithic pellet. Table XLIX summarizes the major operating characteristics of the counters and registers. Not all possible applications are discussed in this section; those that are included highlight typical counter and register uses.

BASIC FLIP-FLOP CIRCUIT

The D-type flip-flop that was shown in Fig. 267 is the basic building-block for RCA COS/MOS counters and registers. In this basic static master-slave flip-flop, redrawn here as Fig. 501, the logic level present at the D (data) input is transferred

Table XLIX—Typical Features and Characteristics of COS/MOS Counters and Registers

Features		
Operating-Temperature Range	—55 to +125°C (Ceramic Package) —40 to +85°C (Plastic Package)	
Operating-Voltage Range (V)	3 to 15	
Full MOS Gate-Oxide Protection at all Terminals.		
Output Buffers Provided		
Full Static Operation		
Characteristics (T$_A$ = +25°C)	**V$_{DD}$ = 10 Volts**	**V$_{DD}$ = 5 Volts**
Clock-Pulse Frequency (MHz)	5	2.5
Clock Rise and Fall Times (μs)	5	5
Quiescent Power Dissipation per Package (μW)	5	1.5
Noise Immunity—All Inputs	45% of V$_{DD}$	
Drive Capability	I$_D$ = 0.5 to 3 mA at V$_1$ = 7 V	I$_D$ = 0.1 to 1 mA at V$_1$ = 4 V
Sink Capability	I$_D$ = 0.5 to 3 mA at V$_0$ = 3 V	I$_D$ = 0.1 to 1 mA at V$_0$ = 1 V

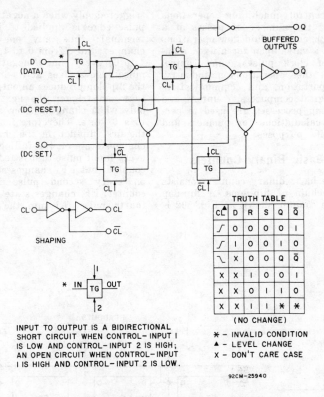

BUFFERED
OUTPUTS

D (DATA)

R (DC RESET)

S (DC SET)

CL

SHAPING

INPUT TO OUTPUT IS A BIDIRECTIONAL
SHORT CIRCUIT WHEN CONTROL-INPUT I
IS LOW AND CONTROL-INPUT 2 IS HIGH;
AN OPEN CIRCUIT WHEN CONTROL-INPUT
I IS HIGH AND CONTROL-INPUT 2 IS LOW.

TRUTH TABLE

CL▲	D	R	S	Q	Q̄
⌐/	0	0	0	0	1
⌐/	1	0	0	1	0
⌐\	X	0	0	Q	Q̄
X	X	1	0	0	1
X	X	0	1	1	0
X	X	1	1	*	*

(NO CHANGE)

* - INVALID CONDITION
▲ - LEVEL CHANGE
X - DON'T CARE CASE

92CM-25940

Fig. 501—Basic COS/MOS master-slave flip-flop stage.

to the Q output during the positive-going transition of the clock pulse. DC reset or set is accomplished by a high level at the respective input. Reset and/or set functions are easily omitted, as shown in some of the designs. Output lead isolation at the Q and/or Q outputs is realized by use of inverters. These inverters eliminate all possibility of MOS gate-oxide damage at the output leads, and also improve circuit speed, noise immunity, and drive capability. The sizes of the p-channel and n-channel MOS devices in the output inverters are tailored to meet the desired drive-current and sink-current requirements. The internal clock-shaping permits a loosely specified input waveshape requirement, and the basic flip-flop configuration operates

from a non-critical single-phase clock input signal. Both 1 and 0 clock-pulse durations can go to infinity, and rise and fall times up to 5 microseconds are permissible.

COUNTERS

Counters are among the most important circuits used in digital applications. They count electrical pulses, but these pulses can be generated by optical, mechanical, acoustical, or other means. Therefore counters can be used to count objects passing in front of a photocell, or to measure time by counting pulses of a known frequency, or to monitor motor speed by counting revolutions in one second. By extension of these techniques, counters can control machining operations (count the number of passes of a cutter), control fluid flow operations (hold a valve open for a fixed number of clock pulses), and control many other operations in industry, transportation, and communication. In digital computers, counters and counting processes are used to perform calculations for business and scientific purposes.

Basic Binary Counters

The basic binary counter consists of a chain of flip-flops. A flip-flop in the counter shown as Fig. 502 is triggered only when a negative-going pulse edge is applied to its input terminal; thus any one flip-flop changes state (from 0 to 1, or from 1 to 0) every time a negative-going pulse edge arrives at its input, but the flip-flop produces an output pulse that can trigger the next flip-flop only when changing from a 1 state to a 0 state. Therefore in Fig. 502 the first flip-flop in the chain, FF_1, changes state on the falling edge of every input pulse that is fed into the counter; but FF_2 changes state only on every **second** pulse into the counter. FF_3 changes state on every **fourth** pulse fed into the counter,

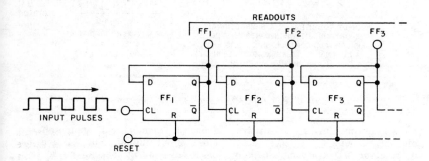

INPUT PULSE	STATE OF FLIP-FLOPS AFTER INPUT PULSE			
NUMBER	FF_1	FF_2	FF_3	FF_4
1	1	0	0	0
2	0	1	0	0
3	1	1	0	0
4	0	0	1	0
5	1	0	1	0
6	0	1	1	0
7	1	1	1	0
8	0	0	0	1
9	1	0	0	0
etc.	etc.	etc.	etc.	etc.

92CS-25941

Fig. 502—Basic binary counter.

and each successive FF would react to the next higher power of 2 (i.e., 16 pulses, 32 pulses, and so forth). The readout for each FF shows its state, and thus the readouts provide a running total of the pulses that have entered the counter since it was reset to all zeros.

The basic counter circuit can be modified to perform various specialized functions. Some of these functions are described below in the discussions of particular RCA COS/MOS circuits. In simplest terms, they include:

Up-Down Counter—An Up-Down Counter shows the difference between two different running counts, such as the difference between the number of nuts and the number of bolts packed in an assembly kit. This circuit requires more elements than the basic counter; after each flip-flop, the true output of the FF and the up-count pulse enter one AND gate while the complementary output of the FF and the down-count pulse enter another AND gate. The outputs from the two AND gates then enter an OR gate, which feeds the next flip-flop.

Decade Counter—For applications that require a decimal counter rather than a binary counter, there are special binary counter circuits that count ten pulses before starting over. These counters are called Decade Counters. A decade counter is basically a four-stage binary counter,

which would normally count up to 16 before recycling. To convert the circuit to a decimal count, therefore, six counts must be removed. One way to accomplish this removal is by arranging the reset so that counting starts at six instead of zero. Then the first pulse produces a setting of seven, the second pulse sets the circuit to eight, and so forth until the sixteenth pulse again recycles to six.

Divide-By-N Counter—A divide-by-N counter is one that can be preset so that N input pulses will cause reset to zero and will produce an output pulse. The value of N is limited by the number of stages in the counter; for example, an 8-stage counter could be set to divide by numbers less than 256. The presetting of the divide-by-N counter consists of providing feedback paths that inject extra pulses into the counter before the actual count begins. Thus a divide-by-225 counter would use an 8-stage assembly preset to $(256 - 225)$, i.e., with a preset count of 31.

Seven-Stage Ripple-Carry Binary Counter (CD4024A)

Fig. 503 shows the logic diagram of the CD4024A, a seven-stage ripple-carry binary counter with buffered reset. Figs. 504 and 505

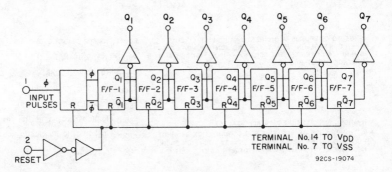

Fig. 503—Logic diagram for CD4024AE seven-stage binary counter with buffered reset.

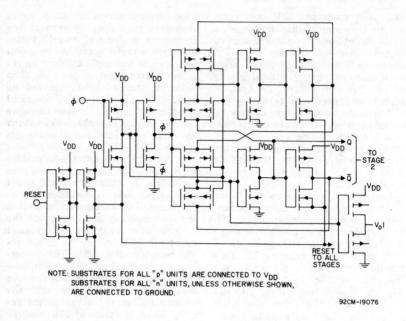

NOTE: SUBSTRATES FOR ALL "p" UNITS ARE CONNECTED TO V_{DD}
SUBSTRATES FOR ALL "n" UNITS, UNLESS OTHERWISE SHOWN,
ARE CONNECTED TO GROUND.

92CM-19076

Fig. 504—Schematic diagram for pulse shaper and one binary stage of the CD4024A seven-stage binary counter.

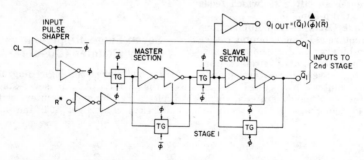

INPUT TO OUTPUT IS A BIDIRECTIONAL SHORT CIRCUIT
WHEN CONTROL–INPUT 1 IS LOW AND CONTROL–INPUT 2
IS HIGH; AN OPEN CIRCUIT CONTROL–INPUT 1 IS HIGH AND
CONTROL–INPUT 2 IS LOW.

* R = HIGH DOMINATES (RESETS ALL STAGES)

▲ ACTION OCCURS ON NEGATIVE GOING TRANSITION OF INPUT
PULSE. COUNTER ADVANCES ONE BINARY COUNT ON EACH
NEGATIVE CL TRANSITION (128 TOTAL BINARY COUNT).

EQUATIONS FOR STAGES 3 TO 7

$Q_{2OUT} = (\bar{Q}_2)(Q_1)(\overset{\blacktriangle}{\overline{CL}})(\bar{R})$

$Q_{3OUT} = (\bar{Q}_3)(Q_1)(Q_2)(\overset{\blacktriangle}{\overline{CL}})(\bar{R})$

$Q_{4OUT} = (\bar{Q}_4)(Q_1)(Q_2)(Q_3)(\overset{\blacktriangle}{\overline{CL}})(\bar{R})$

$Q_{5OUT} = (\bar{Q}_5)(Q_1)(Q_2)(Q_3)(Q_4)(\overset{\blacktriangle}{\overline{CL}})(\bar{R})$

$Q_{6OUT} = (\bar{Q}_6)(Q_1)(Q_2)(Q_3)(Q_4)(Q_5)(\overset{\blacktriangle}{\overline{CL}})(\bar{R})$

$Q_{7OUT} = (\bar{Q}_7)(Q_1)(Q_2)(Q_3)(Q_4)(Q_5)(Q_6)(\overset{\blacktriangle}{\overline{CL}})(\bar{R})$

92CM-25942

(b)

Fig. 505—Logic diagram for pulse shaper and one binary stage of the CD4024A seven-stage binary counter.

show the schematic and logic diagrams for one of the seven counter stages. Operation is similar to that of the basic master-slave flip-flop, with the following exceptions: The D-line connection is derived from the Q output of that stage so that it complements the stage at the negative clock-pulse transition. The clocking of a stage is derived from the previous counter stage (ripple carry). The dc reset function is realized by raising the ground return path of the Q outputs of all seven stages (both master and slave sections). This mode of resetting saves four devices and associated interconnections per counter stage. An internal reset driver is provided so that the reset input presents only one COS/MOS load.

Decade Counter Plus 10 Decoded Decimal Outputs (CD4017A)

Fig. 506 shows the logic diagram of the CD4017A, a decade counter plus 10 decoded decimal ouputs.

Five flip-flop stages are used to implement the decade counter. (These stages are connected in the Johnson counter configuration; an n-stage Johnson counter counts up to 2n rather than to 2^n.) The flip-flop stages are similar to that described in Fig. 501. Clock, reset, inhibit, and carry-out signals are provided. The decade counter advances one count at the positive clock-signal transition provided the inhibit signal is low. Counter advancement by way of the clock line is inhibited when the inhibit signal is high. A high reset signal clears the decade counter to its zero count. Use of the Johnson decade-counter configuration permits high-speed operation, two-input decimal-decode gating, and spike-free decoded outputs. Anti-lock gating is provided to permit only the proper counting sequence. The ten decimal outputs are normally low and go high only at their respective decoded-decimal time slot. Each decimal output remains high for one full clock cycle. The carry-out signal completes one cycle for every ten clock input cycles, and is used to

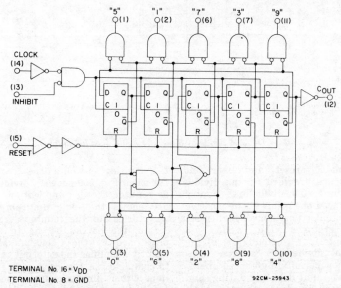

TERMINAL No. 16 = V_{DD}
TERMINAL No. 8 = GND

92CM-25943

Fig. 506—CD4017A decade counter plus 10 decoded decimal outputs.

clock the following decade directly in any multi-decade application.

Divide-by-8 Counter and 8 Decoded Outputs (CD4022A)

Fig. 507 shows the logic diagram of the CD4022A, a divide-by-8 counter and 8 decoded outputs. A four-stage Johnson counter is used to implement the divide-by-8 counter. The basic flip-flop stages are similar to that described in Fig. 501. Clock,

only the proper counting sequence. The eight decoded outputs are normally low and go high only at their respective decoded time slot. Each decoded output remains high for one full clock cycle. The carry-out signal completes one cycle for every eight clock input cycles, and is used to clock the following counter stage directly in multi-stage applications.

Fig. 508 shows the CD4022A in a divide-by-8 counter/decoder application. One CD4022A unit provides

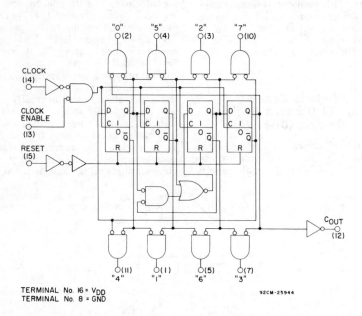

TERMINAL No. 16 = V_DD
TERMINAL No. 8 = GND   92CM-25944

Fig. 507—CD4022A divide-by-8 counter and 8 decoded outputs.

clock-enable, reset and carry-out signals are provided on the divide-by-8 counter. The divide-by-8 counter is advanced one count at the positive clock-signal transition. A high reset signal returns the divide-by-8 counter to its zero count. Use of the Johnson divide-by-8 counter configuration permits high-speed operation, two-input decoding gating, and spike-free decoded outputs. Anti-lock gating is provided to permit

the counting as well as the decoding function. Fig. 509 shows a divide-by-64 counter/decoder application. Fig. 510 shows the logic diagram and waveforms for the counter/decoder. Again, the partial decode function performed by the CD4022A significantly simplifies the external gating to complete the 1-in-64 decode function. Other binary counter/decoder applications can be realized similarly.

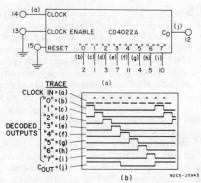

Fig. 508—(a) CD4022A divide-by-8 counter/decoder; (b) circuit waveforms.

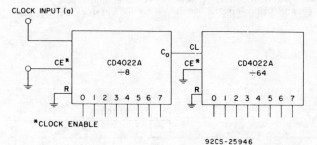

Fig. 509—CD4022A divide-by-64 counter/decoder.

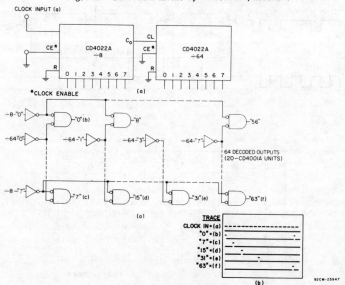

Fig. 510—(a) Logic diagram and (b) circuit waveforms for CD4022A divide-by-64, counter/decoder.

REGISTERS

A shift register is a chain of flip-flops used for storing a set of binary digits. The basic shift register circuit, shown in Fig. 511, resembles a counter. Every time that a clock pulse is applied, the 1 or 0 in each flip-flop is shifted to the next flip-flop in the chain; a new 1 or 0 enters FF_1, and the 1 or 0 in the last flip-flop is shifted out of the register.

In addition to the one-at-a-time (serial) readout from the register, parallel readout is possible. In parallel readout, the data in all of the flip-flops are simultaneously transferred into some other circuit, such

as a memory bank. Conversely, the shift register can be operated in a parallel-input, serial-output mode.

Shift registers are used for temporary storage of information, for signal delay, for multiplication and division of binary numbers, and for numerous other applications in computation, communication, and control. For example:

• Data for storage in a parallel-feed memory can be entered serially into a register, and then dumped (parallel-shifted) into the memory.

• To retrieve data from a particu-

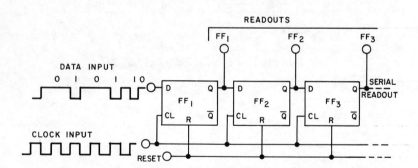

CLOCK PULSE NO.	DATA INPUT	STATE OF FLIP-FLOPS AFTER CLOCK PULSE			
		FF_1	FF_2	FF_3	FF_4
1	1	1	0	0	0
2	1	1	1	0	0
3	0	0	1	1	0
4	1	1	0	1	1
5	0	0	1	0	1
6	0	0	0	1	0
7	1	1	0	0	1
8	1	1	1	0	0
9	0	0	1	1	0
etc.	etc.	etc.	etc.	etc.	etc.

92CS-25948

Fig. 511—Basic shift register.

lar memory location, the address number for that location is stored in a register. A counter counts location pulses until it reaches the number in the register, and then signals for readout from the memory.

Multiplication of binary numbers can be performed by shifting the digits and adding. Therefore shift registers are used for binary multiplication in computers.

Several RCA COS/MOS shift registers are described in the following paragraphs.

18-Stage Static Shift Register (CD4006A)

Fig. 512 shows the logic diagram of the CD4006A, an 18-stage static shift register. The register stages are similar to those shown in Fig. 501. The CD4006A consists of four separate shift-register sections: two four-stage sections and two five-stage sections with output taps at the fourth stage. Each register section has independent data inputs to the first stage. The clock input is common to all 18 register stages. Through appropriate connection of inputs and outputs, multiple register

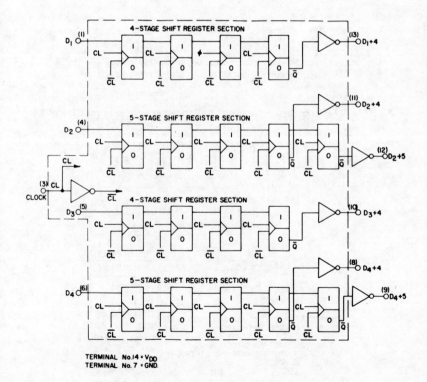

Fig. 512—CD4006A 18-stage static shift register.

sections of 4, 5, 8, and 9 stages or single register sections of 10, 12, 13, 14, 16, 17, and 18 stages can be implemented using one CD4006A. Longer shift-register sections can be assembled by use of more than one

CD4006A. Figs. 513 and 514 show the schematic and logic diagrams for one of the 18 register stages. Register shifting occurs on the negative clock-pulse transition.

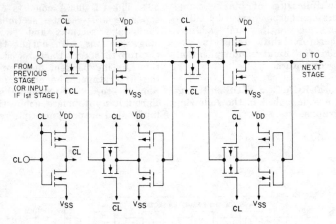

NOTE:
ALL "P"-UNIT SUBSTRATES ARE CONNECTED TO V_{DD}
ALL "N"-UNIT SUBSTRATES ARE CONNECTED TO V_{SS}

92CS-25950

Fig. 513—Schematic diagram for one of the 18 register sections of the CD4006A.

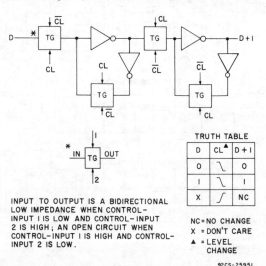

INPUT TO OUTPUT IS A BIDIRECTIONAL
LOW IMPEDANCE WHEN CONTROL-
INPUT I IS LOW AND CONTROL-INPUT
2 IS HIGH; AN OPEN CIRCUIT WHEN
CONTROL-INPUT I IS HIGH AND CONTROL-
INPUT 2 IS LOW.

TRUTH TABLE

D	CL▲	D+1
O	\	O
I	\	I
X	/	NC

NC = NO CHANGE
X = DON'T CARE
▲ = LEVEL
    CHANGE

92CS-25951

Fig. 514—Logic diagram and truth table for one of the 18 register sections of the CD4006A.

Dual 4-Stage Serial-Input/Parallel-Output Register (CD4015A)

Fig. 515 shows the logic diagram of the CD4015A, which consists of two identical, independent, four-stage serial-input/parallel-output registers. Each register has independent clock and reset inputs, as well as a serial data input. Q outputs are available from each of the four stages on both registers. All register stages are similar to that shown in Fig. 501. The logic level present at

of the CD4014A, an 8-stage synchronous parallel-input/serial-output register. A clock input and a single serial data input together with individual parallel jam inputs to each register stage and a common parallel/serial control signal are provided. Q outputs from the 6th, 7th, and 8th stages are available. All register stages are similar to that shown in Fig. 501 except that extra transmission gates permit parallel or serial entry. Parallel or serial entry is made into the register synchronously with the positive clock

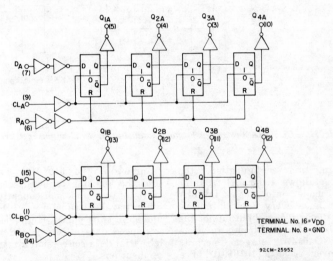

Fig. 515—CD4015A dual 4-stage serial-input/parallel-output register.

the data input is transferred into the first register stage and shifted over one stage at each positive clock transition. Reset of a four-stage section is accomplished by a high level on the reset line. Register expansion to 8 stages with one CD4015A or to more than 8 stages with multiple CD4015A package is permitted.

8-Stage Synchronous Parallel-Input/Serial-Output Register (CD4014A)

Fig. 516 shows the logic diagram

transition and under control of the parallel/serial input. When the parallel/serial input is low, data is serially shifted into the 8-stage register synchronously with the clock positive transition. When the parallel/serial input is high, data is jammed into the 8-stage register by way of the parallel input lines and synchronously with the positive clock transition. Register expansion with multiple CD4014A packages is permitted.

Fig. 517 shows the CD4014A in an 8-stage synchronous parallel-input/serial-output register applica-

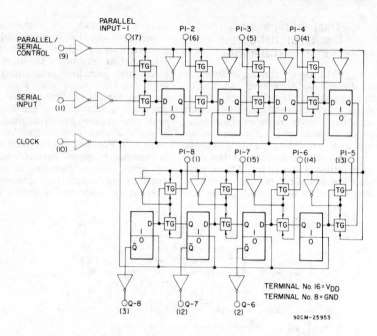

Fig. 516—CD4014A 8-stage synchronous parallel-input/serial-output register.

tion. In this configuration, the CD4013A allows a parallel transfer to be made into the CD4014A register once every 8 clock pulses. Use of the divide-by-2 outputs of the CD4013A as parallel inputs to alternate CD4014A stages permits changeover from a 10101010 to a 01010101 parallel input pattern every 8 pulses. A scope trace of the changeover would show the parallel transfer of the 01010101 pattern into the register followed by eight shift pulses and subsequently another parallel transfer of 10101010 and eight shift pulses. High-speed-to-low-speed data queueing and parallel/serial data conversion are typical applications.

The CD4014A can be utilized in pseudo-random code-generation applications by means of combined

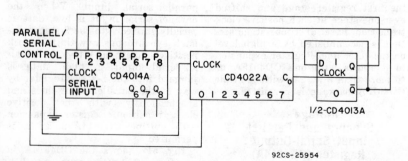

Fig. 517—CD4014A 8-stage synchronous parallel-input/serial-output register application.

control of the parallel input conditions and gating of the feedback of the 6th, 7th, and 8th stages to the serial input.

3-Stage Asynchronous Parallel-Input/Serial-Output Register (CD4021A)

Fig. 518 shows the logic diagram of the CD4021A 8-stage asynchronous parallel-input/serial-output register. Operation is basically the same as for the CD4014A except that

parallel transfers are made as soon as the parallel/serial control input goes high. Parallel transfers are thus made asynchronously with the clock input. Serial shifting is still performed synchronously with the clock input. The CD4014A thus permits the parallel transfer to be synchronized with a different clocking signal than the serial transfer. In high-speed-to-low-speed data queueing, for example, an externally gated high-speed clock may control the parallel transfer while the low-speed clock controls the serial shifting.

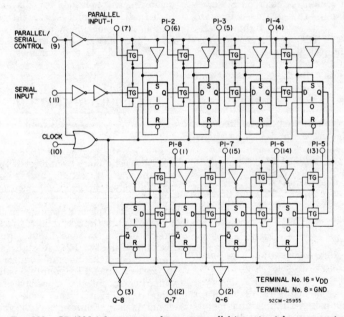

Fig. 518—CD4021A 8-stage asynchronous parallel-input/serial-output register.

Digital Display Systems

DIGITAL display devices are widely used to present information in a clear, interesting, attention-catching manner. For reasons of speed and convenience, electronic seven-segment numerals are used in many of these displays; light-emitting diodes (LED's), incandescent-segment lamps, fluorescent displays, and liquid crystals are examples of such devices. In these units, a combination of two or more of the seven segments can be activated to produce any numeral from 0 through 9. The seven segments are identified by the letters a, b, . . . g, as shown in Fig. 519.

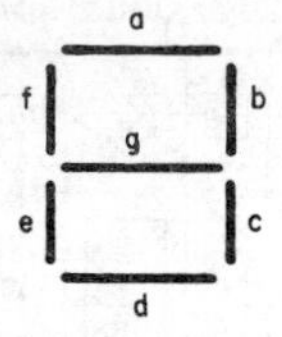

92CS-25968

Fig. 519—Designations of the segments in a seven-segment digit.

Seven-segment digital displays for use in counting, timing, and metering applications can be driven by the RCA CD4026A or CD4033A COS/MOS decade counter/dividers. Each of these circuits consists of a 5-stage Johnson decade counter and an output decoder that converts the Johnson code into a 7-segment decoded output to drive a seven-segment digital display. The CD4033A has ripple-blanking input and output (RBI and RBO) and lamp-test capability. The CD4026A has display-enable capability.

This section describes the CD4033A and CD4026A and their use with various 7-segment display units presently available. Interface packages and methods are discussed to help the designer select the best system to meet his needs. Also included are battery-operated systems for digital clocks and watches.

CIRCUIT OPERATION AND PERFORMANCE CHARACTERISTICS

The inputs to the CD4033A are clock, reset, clock enable, ripple-blanking input (RBI), and lamp test as shown in Fig. 520. The outputs are carry out, ripple-blanking output (RBO), and the seven decoded outputs (a, b, c, d, e, f, g). The effects of the various input signals are shown in Fig. 521. A high reset signal clears the decade counter to its zero count. The counter is advanced one count at the positive clock signal transition if the clock-enable signal is low. Counter advancement by way of the clock line is inhibited when the clock-enable signal is high.

The carry-out (Cout) signal completes one cycle every ten clock input cycles and is used to clock the succeeding decade directly in a multidecade counting chain. The seven decoded outputs (a, b, c, d, e, f, g) illuminate the proper segments in a 7-segment display device used to present the decimal number 0 to 9. The 7-segment outputs go "high" on selection.

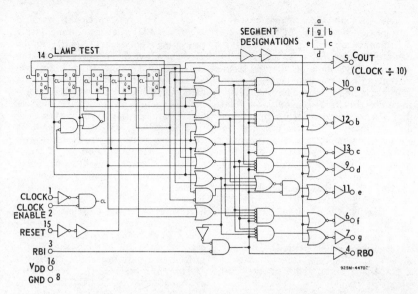

Fig. 520—Logic diagram for CD4033A decade counter/divider with decoded 7-segment display outputs.

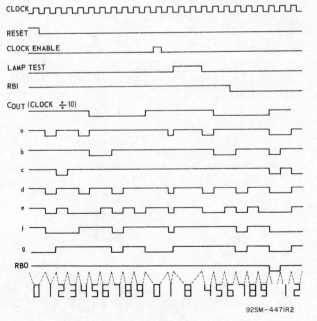

Fig. 521—CD4033A timing diagram.

The CD4033A has provisions for automatic blanking of the non-significant zeros in a multi-digit decimal number. This feature results in an easily readable display consistent with normal writing practice. For example, the number 0050.0700 in an eight-digit display would be displayed as 50.07. Zero suppression on the integer side is obtained by the connection of the RBI terminal of the CD4033A associated with the most significant digit in the display to a low-level voltage and the connection of the RBO terminal of the same stage to the RBI terminal of the CD4033A in the next-lower-significant position in the display. This procedure is continued for each succeeding CD4033A on the integer side of the display.

On the fraction side of the display, the RBI of the CD4033A associated with the least significant bit is connected to a low-level voltage and the RBO of the same CD4033A is connected to the RBI terminal of the CD4033A in the next more significant-bit position. This procedure is continued for each CD4033A on the fraction side of the display.

In a purely fractional number (e.g. 0.7346), the zero immediately preceding the decimal point can be displayed by the connection of the RBI of that stage to a high-level voltage (instead of the RBO of the next more significant stage). Similarly, the zero in a number such as 763.0 can be displayed by the connection of the RBI of the CD4033A associated with it to a high-level voltage. Ripple blanking of non-significant zeros provides an appreciable savings of display power.

The CD4033A has a lamp-test input which, when connected to a high-level voltage, overrides normal decoder operation and enables a check to be made of possible display malfunctions by putting the 7 outputs in the high state.

Fig. 522 shows the logic diagram of the RCA CD4026A. The CD4026A is identical to the CD4033A except that the ripple-blanking input (RBI) and the ripple-blanking output (RBO) and lamp-test capabilities are replaced by a "display enable" control. An extra c-segment output (not gated with the display enable) is available to retain the

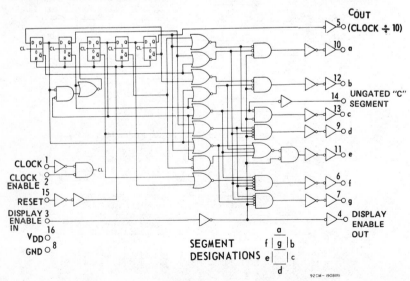

Fig. 522—Logic diagram of CD4026A.

ability to implement the divide-by-12 function. The power dissipation and output characteristics of the CD4026A and CD4033A are identical.

DISPLAY DRIVERS

The decoded outputs of the CD4026A and CD4033A decade counter/dividers are applied to the segments of the digital display devices through display-driving circuits. The following paragraphs briefly describe a pair of COS/MOS integrated circuits and a pair of bipolar integrated circuits suitable for use as display drivers.

COS/MOS Drivers (CD4009A and CD4010A)

Figs. 523 and 524 show the circuit diagrams for the CD4009A (Inverting Hex Buffer) and CD4010A (Non-Inverting Hex Buffer), respectively. Six buffers are provided per package. Figs. 525 and 526 show V_{OL} output characteristics (n-channel) and V_{OH} output characteristics (combined p-and-n-channel devices) for both types.

Bipolar IC Drivers (CA3081 and CA3082)

Fig. 527(a) shows the schematic diagram of the CA3081 (common-emitter array). Fig. 527(b) shows V_{CE}(sat) as a function of collector current for one of 7 identical transistors.

Fig. 528(a) shows the schematic diagram of CA3082 (common-collector array). Fig. 528(b) shows h_{FE} as a function of collector current at $V_{CE} = 3V$.

INTERFACING DECADE COUNTER/DIVIDERS WITH DISPLAY DEVICES

The CD4026A and CD4033A decade counter/dividers can be used with most popular types of digital display devices. The following paragraphs describe the interfacing methods for various types of display devices.

Light-Emitting Diodes

The MAN 3 (Monsanto or equivalent) is a low-power monolithic 7-segment diffused planar GaAsP

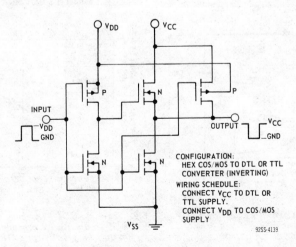

Fig. 523—Schematic diagram (1 of 6 identical stages) of CD4009A.

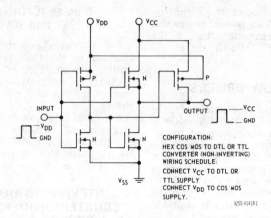

Fig. 524—Schematic diagram (1 of 6 identical stages) of CD4010A.

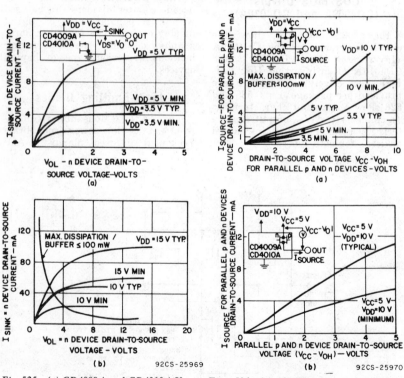

Fig. 525—(a) CD4009A and CD4010A V_{OL} output drive capability, $V_{DD} = V_{CC} = 3.5$, 5V; (b) CD4009A and CD4010A V_{OL} output drive capability, $V_{DD} = V_{CC} = 10$, 15V.

Fig. 526—(a) CD4009A and CD4010A $V_{CC} - V_{OL}$ output drive capability, $V_{DD} = V_{CC} = 10.5$, and 3.5V; (b) CD4009A and CD4010A $V_{CC} - V_{OL}$ output drive capability, $V_{DD} = 10V$ and $V_{CC} = 5V$.

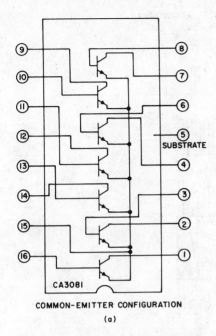

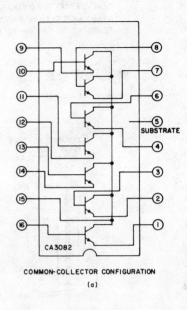

COMMON-COLLECTOR CONFIGURATION

(a)

COMMON-EMITTER CONFIGURATION

(a)

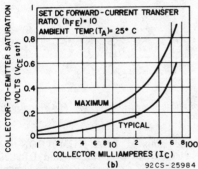

92CS-25984

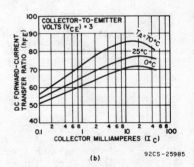

92CS-25985

(b)

Fig. 527—(a) Functional diagram CA3081; (b) $V_{CE}(sat)$ as a function of I_C at $T_A = 25°C$.

Fig. 528—(a) Functional diagram of CA3082; (b) h_{FE} as a function of I_C.

light-emitting-diode display. Figs. 529(a) and 529(b) show the equivalent schematic and physical dimensions of the MAN 3. A fairly bright display (200 foot-lamberts) is achieved at a typical power dissipation of 8.5 milliwatts (1.7 volts x 5 milliamperes) per segment for a 100-per-cent duty-cycle drive mode.

Greater display intensities can be realized by the use of higher-forward-current drives for a duty cycle of less than 100 per cent.

Fig. 530 shows techiques for the interface of the CD4033A or CD4026A to the MAN 3 display at various supply-voltage conditions. Fig. 530(a) shows a direct-drive

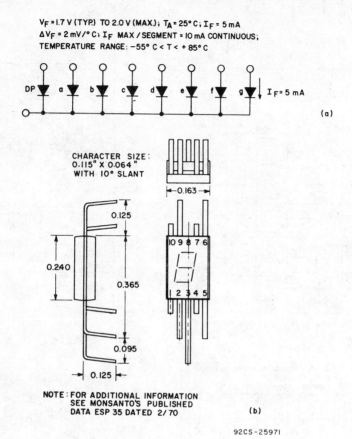

$V_F = 1.7\,V\,(TYP.)\,TO\,2.0\,V\,(MAX.);\,T_A = 25°C;\,I_F = 5\,mA$
$\Delta V_F = 2\,mV/°C;\,I_F\,MAX\,/\,SEGMENT = 10\,mA\,CONTINUOUS;$
TEMPERATURE RANGE: $-55°\,C < T < +85°\,C$

DP a b c d e f g $I_F = 5\,mA$

(a)

CHARACTER SIZE:
0.115" X 0.064"
WITH 10° SLANT

←0.163→

0.125

0.240   0.365

10 9 8 7 6

1 2 3 4 5

0.095

0.125

NOTE: FOR ADDITIONAL INFORMATION
SEE MONSANTO'S PUBLISHED
DATA ESP 35 DATED 2/70

(b)

92CS-25971

Fig. 529—(a) Schematic diagram and (b) physical dimensions of the MAN-3 display device.

condition at a duty cycle of 100 per cent for V_{DD} between 9 and 15 volts. A typical CD4033A can supply a forward current of 5 milliamperes per segment with a supply voltage as low as 9 volts and yields a fairly bright display. The power supply dissipation is approximately 45 milliwatts per segment (9 volts x 5 milliamperes).

Incandescent Readouts

Fig. 531 shows the physical di-

mensions of RCA NUMITRON devices DR2000, DR2100, and DR2200 low-power miniature incandescent readouts. Brightness and segment current as functions of segment voltage are shown in Fig. 532.

Fig. 533 shows the CD4033A or CD4026A being interfaced with DR2000-series displays. In the arrangement shown in Fig. 533(a), if V_{DD} is less than 8 volts, a CD4010A buffer must be used between the CD4033A or CD4026A segments and the CA3081. In the arrangement shown in Fig. 533(b), care should be taken not to exceed

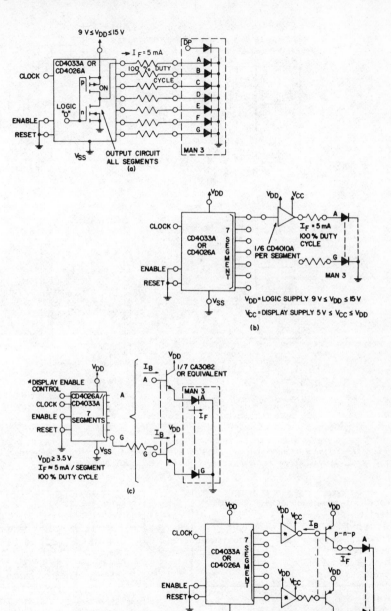

Fig. 530—CD4033A or CD4026A being interfaced with MAN-3 at various supply voltages.

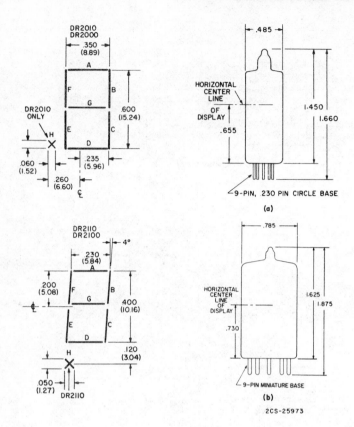

Fig. 531—Segment arrangement and physical dimensions of RCA NUMITRON devices: (a) DR2000-series devices; (b) DR2100- and DR2200-series devices.

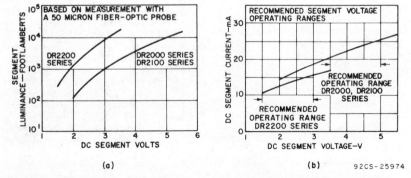

Fig. 532—(a) Brightness as a function of segment voltage and (b) segment current as a function of segment voltage for RCA NUMITRON devices.

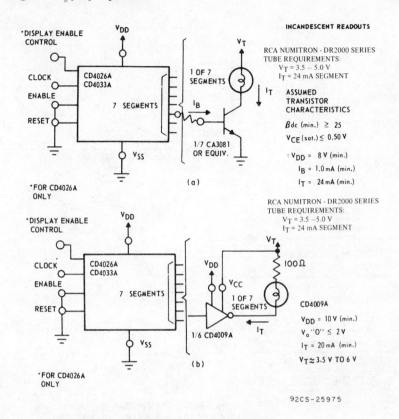

Figure 533—CD4033A or CD4026A driving RCA DR2000-Series NUMITRON.

the maximum power dissipation of the CD4009A (100 milliwatts per buffer unit, 200 milliwatts per package).

The Pinlite* Series "O" and "R" low-power miniaturized incandescent readouts can be used in much the same way as the RCA NUMITRON devices.

Low-Voltage Vacuum Fluorescent Readouts

TungSol Digivac S/G** Types DT1704B and DT1705D, Nippon Electric (NEC) Type DG12E/LD915, and Sylvania Type 8894 are

---

* Trademark of Pinlite, Inc.
** Trademark of Tung-Sol Division Wagner Electric Corp.

low-voltage and low-power vacuum fluorescent 7-segment readouts. Fig. 534 shows the physical dimensions and brightness characteristics of the TungSol Digivac DT1704B and DT1705D. A brightness level of 150 foot-lamberts (typical) is indicated at a plate voltage of 25 volts. Fig. 535 illustrates a method of driving these devices with the CD4033A. The requirements are 100 to 300 microamperes per segment at tube voltages of 12 to 25 volts, depending on the required brightness level. The filaments require 45 milliamperes at 1.6 volts (ac or dc). With an E_b of 18 volts, medium brightness in low ambient light background will result. The point of no noticeable glow occurs at $E_b = 4.5$ volts.

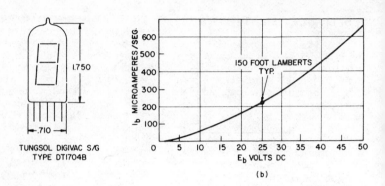

TUNGSOL DIGIVAC S/G
TYPE DT1704B

(b)

NOTE: CHARACTER SIZE APPROXIMATELY:
.360" x .570" SLANTING TO THE
RIGHT 8°.
SEE PUBLISHED DATA OF TUNGSOL
DIVISION WAGNER ELECTRIC CORP.
T438 DATED 8/69.

92CS-25976

(a)

Fig. 534—(a) Physical dimensions and (b) plate current as a function of plate voltage for
Tung-Sol S/G type DT1704B and DT1705D display devices.

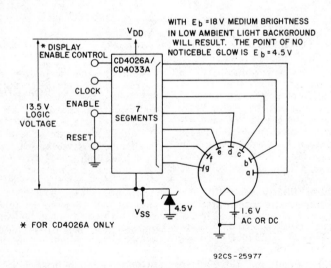

92CS-25977

Fig. 535—Interfacing CD4033A or CD4026A with Tung-Sol S/G type DT1704B or
DT1705D.

Liquid-Crystal Displays

The RCA developmental types TA8054 and TA8055, shown in Figs. 536 and 537, are representative 4-digit 7-segment liquid-crystal cells. They are available in both reflective and transmissive models. The reflective versions, designated by suffix R, are intended for use with available front lighting; they utilize a mirrored area on the inner surface of the back plate to enhance contrast. The transmissive versions,

The TA8054R, TA8054T, TA8055R, and TA8055T are designed for ac operation over a wide range of frequencies from 30 to 400 Hz; typically 32 Hz is used in many applications (dc is not recommended). Operation at approximately 15 volts is recommended for optimum response time and compatibility with COS/MOS integrated circuits. The TA8054 and TA8055 can be easily addressed and driven by the RCA COS/MOS circuit shown in Fig. 538. Both this circuit and the clock display shown

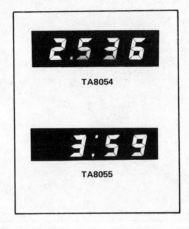

92CS-25978

Fig. 536—Typical seven-segment liquid-crystal digital displays.

designated by suffix T, are intended for use in applications where backlighted or edge-lighted readouts are desired; in these types, both plates are transparent. The TA8054R and TA8054T are numeric readout devices having decimal points before each digit, and are intended for a wide variety of display applications in test equipment, electronic meters, etc. In such applications only the desired decimal points will be visible. The TA8055R and TA8055T are numeric devices having a colon between the second and third digit, and are intended for a wide variety of applications in clock and timing equipment.

in Fig. 539 use the RCA CD4026A or CD4033A COS/MOS counter/divider.

DIGITAL-METER APPLICATIONS

The CD4033A and CD4026A are unique in that they have both counting and decoding on a single chip. The block diagram of Fig. 540 demonstrates the use of the CD4033A and CD4026A in digital-meter applications where time multiplexing or sampling-and-display functions are employed (to eliminate the need for holding circuits). An analog input is converted to a

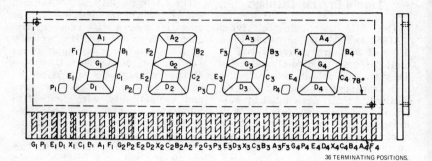

36 TERMINATING POSITIONS.

NOTES:
1. REFLECTIVE CELL IS MIRRORED INTERNALLY
   IN AREAS OF DIGITS AND DECIMALS
2. X_1, X_2, X_3, AND X_4 TERMINATIONS ARE GROUND
   ON BACK PLATE FOR DIGITS 1, 2, 3, AND 4
   RESPECTIVELY

92CS-25981

(a)

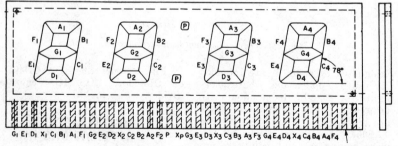

36 TERMINATING POSITIONS.

NOTES:
1. REFLECTIVE CELL IS MIRRORED INTERNALLY
   IN AREAS OF DIGITS AND COLON
2. X_1, X_2, X_p, X_3, AND X_4 TERMINATIONS ARE GROUND
   ON BACK PLATE FOR DIGITS AND COLON 1, 2, P, 3
   AND 4 RESPECTIVELY

92CS-25982

(b)

Fig. 537—(a) TA8054 and (b) TA8055 liquid-crystal cells.

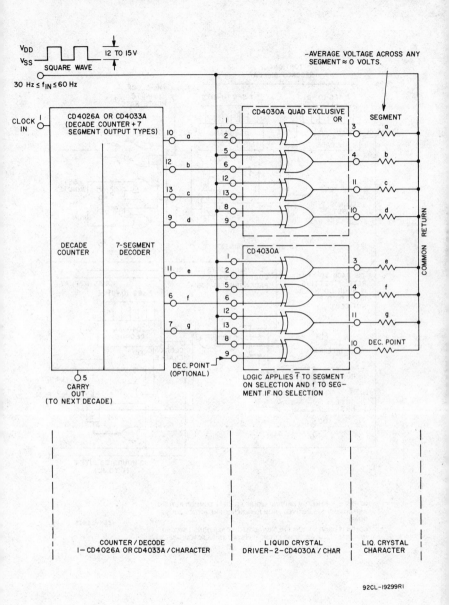

Fig. 538—COS/MOS driver circuit for TA8054R, TA8054T, TA8055R, TA8055T.

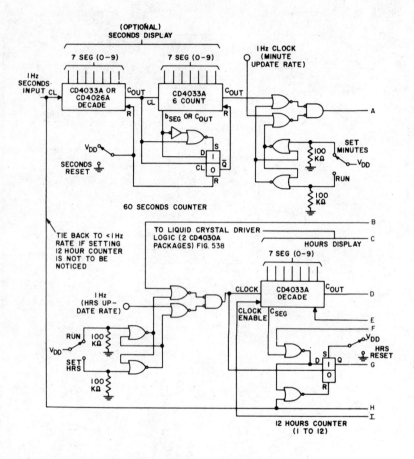

30 Hz ≤ f$_{in}$ ≤ 60 Hz FOR DRIVING LIQUID-CRYSTAL COMMON RETURN
AND CD4030A'S IS DERIVED FROM REFERENCE FREQUENCY COUNT-
DOWN.

COLON IS DRIVEN FROM 1-Hz SIGNAL TO GIVE FLASHING 1-SECOND
INDICATIONS (USE CD4034A LIQUID-CRYSTAL DRIVE SCHEME—SEE
FIG. 538.

92CS-25983

*Fig. 539—COS/MOS logic and driving circuitry for liquid-crystal clock displays TA8055R
and TA8055T (continued on page 447).*

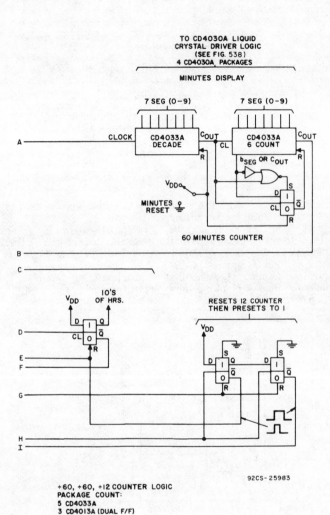

Fig. 539—COS/MOS logic and driving circuitry for liquid-crystal clock displays TA8055R and TA8055T (continued from page 446).

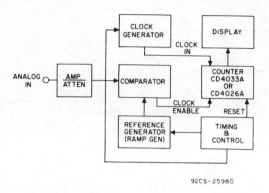

92CS-25980

Fig. 540—Block diagram of dc meter using CD4033A or CD4026A.

sequence of pulses in which the pulse count is proportional to the analog input level. Sampling rates may be fast enough to result in an apparent fixed display, or they may be slower and cause noticeable changes in display during sampling. Longer display times or display blanking dur-ing sampling may be used to prevent flicker.

Fig. 541 shows a general block diagram using the CD4033A or CD4026A in Digital Counter/Timers. As in the dc meter applications, time multiplexing of sampling and display functions is utilized.

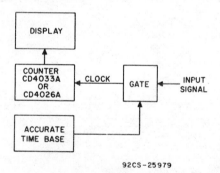

92CS-25979

Fig. 541—Block diagram of digital counter/timer using CD4033A or CD4026A.

Ignition Systems

THE increasing use of transistor and thyristor ignition systems in the automotive industry has stimulated demands for improved performance, reduction in emissions that results because more accurate spark timing is achieved with magnetic-pickup distributors, and greater reliability. The following discussion covers the requirements for automobile ignition systems and compares the relative merits of capacitive and inductive types. Both systems are described in terms of operation, performance, and limitations; practical circuits are shown. The application of the capacitive-discharge principle in small-engine ignition crcuits and in gas-system igniters is also described.

BASIC CONSIDERATIONS FOR AUTOMOTIVE SYSTEMS

Under worst-case conditions, about 22 kilovolts are required to ignite the combustible mixture in the cylinder of an automobile engine. In addition, a minimum energy of about 20 millijoules must be available in the spark to assure propagation of a stable flame front originating at the spark. The exact values of voltage and energy required under all operating conditions depend on many factors including those described in the following paragraphs.

Condition of Spark Plugs

Fouled plugs reduce both the volt-age and the energy available for ignition. The plug gap also affects both the voltage and the energy required. As the plug gap is increased, the required voltage increases, but the required energy decreases.

Cylinder Pressure

The cylinder pressure depends on both the compression at the point of ignition and the air-fuel mixture. The minimum breakover voltage in any gas is a function of the product of gas pressure and electrode spacing (Paschen's Law). In automobile engines, the minimum voltage increases as this product increases. Therefore, higher pressures also require higher voltages. However, the energy required decreases as the pressure increases and increases as the fuel-air mixture moves away from the optimum ratio. Worst-case conditions occur when the engine is started, at idle speeds, and during acceleration from a low speed because carburetion is poor and the fuel-air mixture is lean. The combination of a lower cylinder pressure and a dilute fuel-air mixture results in a high energy requirement under these conditions.

Plug Polarity

The center electrode is hotter than the outside electrode because of the thermal resistance of the ceramic sleeve that supports it. If the center electrode is made negative, the effect of thermionic emission from this electrode can reduce the required ignition voltage by 20 to 40 per cent.

Spark-Plug Voltage Waveshape

The spark-plug voltage waveshape is shown qualitatively in Fig. 542. The voltage starts to rise at

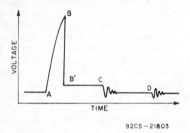

92CS—21803

Fig. 542—Ignition-voltage waveshape.

point A, and reaches ignition at point B. The region from B′ to C represents the sustaining voltage for ionization across the spark plug. When there is insufficient energy left to maintain the discharge (at point C), current flow ceases and the remaining energy is dissipated by ringing. The final small spike at point D occurs when the ignition coil again starts to pass current.

The two most important characteristics of the voltage waveshape are its rise time (from A to B) and the spark duration (from B′ to C). A rise time that is too long results in excessive energy dissipation with fouled plugs; a rise time that is too short can lead to radiation losses of the high-frequency voltage components through the ignition harness. The minimum rise time should be about 10 microseconds; a 50-microsecond rise time is acceptable. Conventional systems have a typical rise time of about 100 microseconds. It should be noted that, at an engine speed of 5000 revolutions per minute, one revolution takes 12 milliseconds. Engine timing accuracy is usually no better than 2 degrees, which corresponds to 67 microseconds. The

error caused by the rise time is therefore comparable to normal timing errors. At normal cruising speeds (about 2000 revolutions per minute), the 2-degree timing error corresponds to about 165 microseconds and rise-time effects are negligible.

Energy Storage

The energy delivered to the spark plug can be stored in either an inductor or a capacitor. Although the inductive storage method is the more common approach, both are used; both are discussed. One requirement common to both methods is that, after the storage element is discharged by ignition, it must be recharged before the next spark plus is fired. For an eight-cylinder engine that has a dwell angle of 30 degrees, the time τ between ignition pulses (in milliseconds) is equal to 15000 divided by the engine rpm, and the time τ_{ON} during which the points are closed is equal to 10000/rpm. When the engine rpm is 5000, τ_{ON} is 2 milliseconds. For either an inductive or a capacitive storage system, therefore, the charging-time constant should be small compared to 2 milliseconds.

INDUCTIVE-DISCHARGE AUTOMOTIVE SYSTEM

Fig. 543 shows the basic circuit for an inductive-discharge system. The total primary circuit resistance (ballast plus coil) is represented by R_p; the coil primary inductance is represented by L_p.

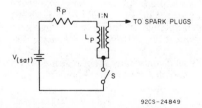

92CS—24849

Fig. 543—Basic inductive-discharge ignition circuit (Kettering system).

Switch S represents the points in a conventional system. The transformer step-up turns ratio is N. When the points close, current increases with a time constant τ_L equal to L_p/R_p. When the points open, a voltage V_P is generated across the primary terminals; the secondary voltage V_S, which is delivered to the spark plugs through the distributor, is equal to NV_p.

The maximum current is limited to about 4 amperes by possible burn-out of the points. The total energy stored in the coil must be about 50 millijoules to provide for energy losses by radiation, fouled plugs, and the like. For a battery voltage of 12 volts and a primary circuit resistance of 3 ohms, L_p must have a value of about 6 millihenries. The time constant τ_L is then about 2 milliseconds; the coil current does not reach its maximum value at high engine speeds. Fig. 544 shows primary current and secondary voltage as a function of engine speed for a typical nontransistorized ignition circuit. The degradation in secondary voltage follows the primary current. The available energy decreases even more rapidly because it is proportional to the square of the current.

This problem can be even more severe than indicated because some conventional ignition coils have inductances as high as 12 millihenries, and the time constant is correspondingly longer.

Basic Considerations

In an inductive-discharge transistor ignition system, a transistor switch is used instead of the points to control the coil current. The points control the base current of the transistor, which may be only a few hundred milliamperes. As a result, burnout of points is eliminated. In addition, because a transistor can switch higher currents, a low-inductance coil can be used for the same coil energy. For example, a 10-ampere coil current(fairly typical for some systems) requires a coil inductance of only 1 millihenry. The corresponding primary resistance is then only about 1.2 ohms, and the value of τ_L is about 0.8 millisecond. This value is sufficiently short for good high-speed performance. Typical results are given later.

Limitations—One disadvantage of this type of transistorized ignition system is the higher current drain from the battery or alternator. However, this disadvantage can be more than offset by easier starting, better high-speed performance, and reduced maintenance. Another important consideration is temperature capability. Ambient temperatures under the hood of modern automobiles may reach 125°C, and the engine itself may reach 150°C. The junction temperature of a transistor may rise 10 or 15°C above ambient or case temperature. Available low-cost silicon transistors can operate at junction temperatures up to 200°C. In addition to increased reliability, these transistors also make feasible the use of active-mode switching. This factor in itself offers some important advantages over the saturated-mode switching techniques

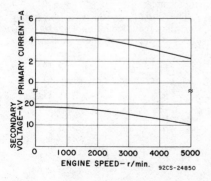

Fig. 544—Performance of conventional inductive-discharge ignition circuit.

currently used in transistorized auto ignition systems.

Ignition-Coil and Transistor Characteristics—Because the transistor acts as a simple switch in series with the ignition coil, its required characteristics are largely determined by the characteristics of the coil itself. The most important coil parameters are shown in the equivalent circuit of Fig. 545. In this circuit, L_p is the primary inductance, and C_{pt} is the total primary capacitance (including all capacitance in the secondary circuit, referred to the primary circuit). Switch S represents the transistor.

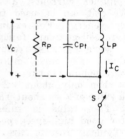

92CS-25956

Fig. 545—Simplified equivalent circuit of ignition coil primary.

With switch S closed, energy is stored in the inductance. When the switch is opened, this energy is transferred to the capacitor, and the capacitor voltage V_C appears directly across the transistor terminals. The $V_{CE}(sus)$ of the transistor (the sustaining value of the common-emitter avalanche breakdown voltage for the particular operating conditions used) must therefore be greater than V_C. If $V_{CE}(sus)$ is less than V_C, the transistor voltage is clamped to the $V_{CE}(sus)$ value until the secondary capacitance discharges sufficiently for the voltage to decrease below $V_{CE}(sus)$. The output voltage is limited to $NV_{CE}(sus)$.

Even more important is the possible destruction of the transistor. Although operation in the sustain-

ing breakdown region is not necessarily destructive, the transistor may enter a second-breakdown region in which it can be destroyed. Operation in a particular circuit may bring even high-voltage transistors into the sustained breakdown region for a brief period. Therefore, the ability of a transistor to withstand second-breakdown destruction becomes an important device characteristic. Because the second-breakdown phenomenon is energy-dependent, the transistor can be characterized as safe for operation in the second-breakdown region in terms of energy, current, and circuit inductance, or in terms of power and pulse duration. Second breakdown can occur under conditions of either forward or reverse bias. In ignition systems, only the reverse-bias case is of interest. For reliable operation, the transistor must be able to withstand the total energy stored by the coil safely.

The current-voltage capability can be represented by the transistor load-line characteristic, as shown in Fig. 546. The transistor is driven into saturation (point A) and held there until an ignition pulse is required. When the transistor is turned off, a very fast voltage spike occurs initially (point B) because of the small amount of energy stored in the coil-primary leakage inductance.

The initial pulse, which usually contains insufficient energy to damage the transistor, does not appear

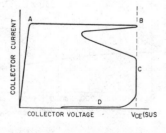

92CS-25957

Fig. 546—Typical transistor load line characteristic.

at the spark plug. However, the energy stored in the coil magnetizing inductance results in a second voltage pulse (point C) which is coupled to the ignition-coil secondary. The time required to switch from A to C may be in the order of microseconds. After the transistor current decreases to its leakage value (essentially zero, compared with the on current of several amperes), the collector voltage continues to oscillate (point D). The voltage may even exceed $V_{CE}(sus)$ and be limited instead by the transistor breakdown voltage $V_{(BR)CE}$. Most of the stored energy is transferred to the spark plug. Ignition takes place during this time because the build-up in coil voltage is determined by the resonant frequency of the parallel-tuned circuit of Fig. 545 and not by the transistor turn-off time. When ignition coils designed for transistorized operation are used, voltage rise-times in the order of 50 to 100 microseconds can be achieved. The much shorter transistor turn-off time (1 or 2 microseconds) is then negligible.

In addition to the transistor characteristics discussed thus far (current, voltage, and reverse-bias second-breakdown capabilities; reverse breakdown voltage; and switching speed), there are two other characteristics of importance in ignition systems. The first is the saturation voltage $V_{CE}(sat)$ which, together with the primary-coil current I_p, governs the transistor power dissipation. The other is the dc common-emitter forward-current transfer ratio h_{FE}, which determines the required base-drive current and thus the current through the points in a one-transistor circuit.

Typical Inductive-Discharge Circuits

Fig. 547 shows a **two-transistor ignition circuit** for negative-ground automobiles. Current flows from the 12-volt dc supply through the coil

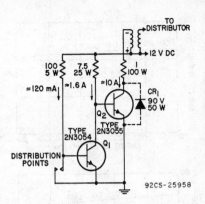

Fig. 547—Transistor ignition circuit.

and 1-ohm resistor, through Q_2 and CR_1, and then to ground. Transistor Q_2 is turned off when Q_1 is turned on, and vice versa. When the points close, Q_1 is turned off ($V_{BE-1} = 0$), and Q_2 is turned full on (is in saturation). The base-drive current for Q_2 is set by the 7.5-ohm resistor. When the points open, base current through the 100-ohm resistor turns Q_1 on, drives it into saturation, and turns Q_2 off. The $V_{CE}(sat)$ of Q_1 is less than 0.25 volt, which is much less than the V_{BE} turn-on threshold for Q_2 (about 0.6 volt). This relationship is necessary to assure that Q_2 is turned off. The zener diode CR_1 should be used to protect Q_2 from second breakdown.

The primary current and secondary voltage for the circuit of Fig. 547 are shown as functions of engine speed in Fig. 548. Two points should be noted. First, coil current and secondary voltage are nearly constant with engine speed, especially when compared with the curves for a nontransistorized system shown in Fig. 544. This improvement is a result of the much shorter inductive time constant τ_L for the transistorized system. Second, the output voltage decreases significantly under simulated fouled-plug conditions, although stored energy or coil current is independent of the shunt resistance. The decrease

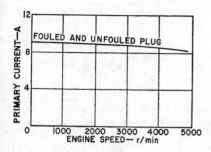

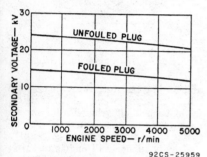

Fig. 548—*Performance of a transistorized ignition circuit.*

of experimental RCA transistors under varied conditions.

The operation of the circuit shown in Fig. 549 differs from that of most transistor ignition circuits in one important respect. For normal battery voltages and engine speeds, the transistor switches from the active region rather than from saturation. Current regulation is accomplished by means of current feedback through the emitter resistor. The two diodes in the base circuit provide a reference voltage that remains constant at 1.5 volts ± 5 per cent when the battery voltage varies from 8 to 16 volts. When the points close, current builds up in the coil and the transistor along the saturation line. The negative feedback voltage developed across the emitter resistor reduces the net base-to-emitter drive voltage. Thus, the transistor comes out of saturation and reaches a steady-state current in the active region. Fig. 550 shows waveforms for this type of high-current operation.

is caused by the energy lost in the 1-megohm shunt resistor used to simulate the fouled plugs.

Fig. 549 shows an **active-mode, current-regulated transistor ignition circuit** in its simplest form. The circuit shown can be applied directly to a positive-ground system, or it can be modified for use with a negative-ground system. Because of its simplicity, this circuit has been used to test performance

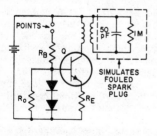

Fig. 549—*Active-mode, current-regulated circuit.*

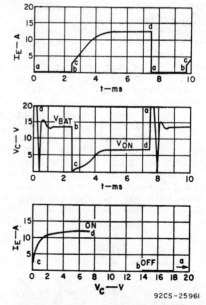

Fig. 550—*Typical waveforms for active-mode switching.*

The upper curve in Fig. 550 shows the collector-current waveform. The center curve shows the collector-voltage waveform. When the points close, V_C drops from the upper flat portion (equal to the battery voltage) to zero. The voltage then rises to a constant value during the on condition. When the points open, a large oscillating voltage is generated. The lower curve shows the operating load line on an expanded scale. For the particular test conditions used, approximately two milliseconds are required for the transistor to come out of saturation. This time corresponds to an engine speed of 5000 revolutions per minute in an eight-cylinder engine with a 30-degree dwell angle. For lower speeds, the coil current and, therefore, the output voltage remain constant, as discussed later.

Advantages of this type of operation include excellent regulation with engine speed, reduction of transistor storage time during switching to the off condition, and great simplicity and design flexibility. Performance of the basic circuit shown in Fig. 549 is good for battery voltages down to 3.5 volts. Further improvement can be obtained if the base resistor R_B is replaced by a second transistor which acts as a base-drive voltage regulator to compensate for battery-voltage variations.

The major disadvantage of this type of operation is the higher average power dissipation in the transistor. With some additional circuit modifications, however, it is possible to make the transistor on time a controlled function of engine speed. The advantages of active-mode switching can then be obtained without the chief disadvantage, the high average power dissipation.

The basic circuit of Fig. 549 has been used for both high-current and low-current operation. The choice is determined by the selection of the ignition coil and by the values used for R_E and R_B, as follows:

Component	High-Current	Low-Current	
R_E	0.02	0.1	ohm
R_B	5	10	ohms
R_D	220	220	ohms
Coil: L_p	0.9	4	mH
N	0.3	0.9	
R_p	240	113	ohm

Fig. 551 shows characteristic curves for the output voltage of the high-current circuit of Fig. 549 into a 50-picofarad, 1-megohm load as a function of engine speed for several different battery voltages. For normal cruising speeds and a battery voltage of 13.5 volts, the peak voltage is about 27 kilovolts. Even at 5000 revolutions per minute, the voltage is still about 25 kilovolts, or

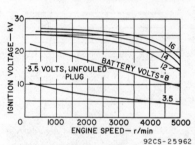

Fig. 551—Ignition voltage as a function of engine speed for the high-current system.

substantially greater than the normal 12 to 14 kilovolts required at this speed. Even at a battery voltage of 8 volts, the output voltage exceeds 15 kilovolts and satisfies normal requirements at this speed. Under extreme conditions (such as cranking the engine on a very cold morning), the voltage available from a normal 12-volt battery may be as low as 3.5 volts. At this low voltage level, the high-current circuit provides almost 12 kilovolts for a fouled plug (1-megohm load) and 17.5 kilovolts for an unfouled plug.

Fig. 552 shows the output-voltage waveform at two different time scales for a battery voltage of 16 volts and an engine speed of 5000 revolutions per minute. The output voltage is about 28 kilovolts and the rise time (measured between the 10-

and 90-percent points) is only 60 microseconds. This short rise time minimizes timing errors and permits the firing of fouled plugs. Fouled plugs require short rise times to reduce the amount of energy that is dissipated by the shunt resistance and thus made unavailable for ignition.

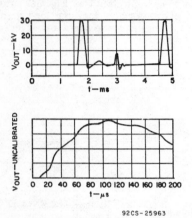

92CS-25963

Fig. 552—Output voltage for high-current operation.

Fig. 553 shows the collector current (top), collector voltage (middle), and load-line characteristic (bottom) for the same conditions. The peak voltage-current product is 4300 volt-amperes (250 volts and 18 amperes), as compared with a typical product of 1000 volt-amperes for most other auto-ignition transistors. At idling speeds, the average current drain and transistor power dissipation are excessively high in the simple test circuit of Fig. 549. However, controlled timing can be used to limit the transistor on time, for all engine speeds, to the minimum value required to just reach the desired current level. Such timing reduces both current drain and power dissipation to acceptable levels.

Fig. 554 shows the performance characteristics of the low-current circuit of Fig. 549 under fouled-plug conditions. At normal cruising

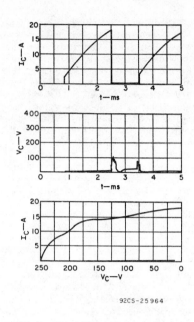

92CS-25964

Fig. 553—Collector current and voltage for high-current operation.

speeds and a battery voltage of 13.5 volts, the peak voltage is about 25 kilovolts. At 5000 revolutions per minute, the voltage is about 16 kilovolts, which is still adequate for ignition. For the extreme case of the 3.5-volt supply, the output voltage is greater than 10 kilovolts for a fouled plug and is about 16.5 kilovolts for an unfouled plug.

The somewhat poorer high-speed performance of the low-current circuit results directly from the use of

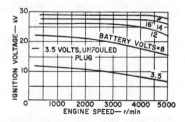

92CS-25965

Fig. 554—Ignition voltage as a function of engine speed for the low-current system.

an ignition coil that has higher primary inductance, as discussed previously. However, these performance characteristics are still superior to those of a conventional ignition system, which uses a coil of even higher inductance. A contributing factor to the improved performance is the use of current regulation in this transistorized system. Further improvements can be achieved by use of ignition coils with improved characteristics.

Fig. 555 shows the output-voltage waveform at two different time scales for a battery supply of 14 volts and an engine speed of 1000 revolutions per minute. The output voltage is 26 kilovolts, and the rise time is 66 microseconds. The somewhat longer rise time, which results from the higher coil inductance, is still much shorter than the 100-to-150-microsecond rise time typical of conventional ignition systems.

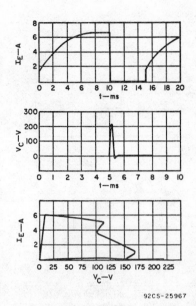

92CS-25967

Fig. 556—Emitter current and collector voltage for low-current operation.

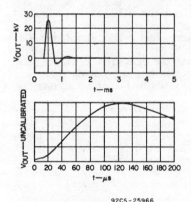

92CS-25966

Fig. 555—Output voltage for low-current operation.

Fig. 556 shows the emitter current (top), collector voltage (middle), and load-line characteristics (bottom) for the same conditions. The peak voltage-current product is only 1400 volt-amperes (220 volts and 6.6 amperes). This value, which is determined by the circuit operation, is well within the 4000-volt-ampere capability of the transistor.

CAPACITIVE-DISCHARGE SYSTEMS

The basic capacitive-discharge system is illustrated in Fig. 557. It is important to note that the transformer serves simply as a pulse transformer. Therefore, performance at high engine speeds is not affected by the transformer primary inductance but, instead, is governed by the time required to charge capacitor C to the desired voltage level.

Basic Circuit Operation

The trigger-control circuit (which can be a transistor switch) is controlled by the distributor points. More sophisticated distributor control, such as that available from distributors in which the voltage

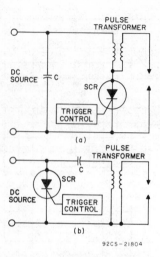

92CS-21804

Fig. 557—Basic configuration for capacitive-discharge ignition circuits: (a) storage capacitor connected across input voltage source; (b) storage capacitor connected in series with input voltage source and pulse transformer.

pulses are derived magnetically or photo-optically, can also be used to control the trigger circuit. (Such distributors are also useful in the inductive-discharge transistor ignition circuits prevously discussed.) The capacitor is charged to the dc voltage; the stored energy ε is equal to $C(V_c)^2/2$, where V_c is the capacitor voltage. At the appropriate time, the trigger-control circuit fires the silicon controlled rectifier (SCR). The capacitor discharges through the transformer, which steps up the voltage to a value V_s equal to KNV_c, where N is the transformer turns ratio and K is a constant that is dependent mainly on the value of the capacitance and of the transformer leakage inductance and generally ranges between 1 and 1.5. The stored energy is thus delivered to the spark plug in the form of a high-voltage pulse. Typical values for V_c and C are about 350 volts and 1 microfarad, respectively. Thus the energy ε is about 60 millijoules.

Because the energy dissipated in the spark gap is equal to the energy stored in the capacitor minus the losses in the transformer and SCR, the energy available in the system is relatively easy to calculate. Examination of the basic circuits shows that the energy is transferred only when the SCR is forward conducting with the gate biased on. However, part of the energy is not available in the basic circuit because the capacitor and inductor form a tuned circuit when the SCR is on, and the energy that would normally flow back from the inductor to the capacitor is stopped by the high reverse impedance of the SCR. This energy is, therefore, lost as available spark energy. The duration of the spark is limited, then, to approximately one-half cycle of the natural LC frequency of oscillation. Some of the energy lost can be regained and used to increase the spark duration by installing a diode in the basic circuits of Fig. 557 as shown in Fig. 558. The diode not only bypasses the reverse impedance of the SCR but eliminates the possibility that the SCR might conduct in the reverse direction should the gate of the SCR be biased on at this time.

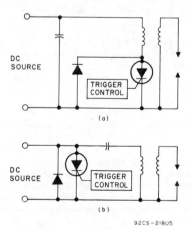

92CS-21805

Fig. 558—Basic circuit configurations shown in Fig. 557 modified by the addition of a diode in shunt with the SCR switching device.

Thus, in addition to improving system low-temperature performance by increasing spark duration, the diode reduces the possibility of excessive heating and damage to the SCR that could accompany reverse conduction and thereby reduces the over-all cost of the system by reducing the reverse blocking requirement of the SCR. The ratio of spark duration to charging time decreases with increasing RPM so that in some applications an RPM limit may be reached below the desired maximum because of the charging-time requirements.

Economic Considerations

The capacitive-discharge system is generally acknowledged to be technically superior to the inductive-discharge system. Its chief drawback has been (and continues to be) economic. Although the transformer may be less expensive than an ignition coil, the capacitor must be of fairly high quality. The SCR and its associated trigger circuit are generally more expensive than comparable transistors and trigger circuits for inductive-discharge systems. Finally, the capacitor charging circuit in an automotive ignition system is a dc-to-dc converter, which represents an additional cost element. Such converter circuits typically require a transformer, two or four diodes for rectification, and one or two transistors. With the increasing use of the ac alternators in modern automobiles, it may eventually be possible to tap off the ac voltage, step it up to the desired voltage by use of a simple transformer, and rectify it. However, it is not clear at this time that this approach is desirable or less costly. Despite these considerations, there are a number of capacitive-discharge systems available on the replacement market.

Technical Considerations

One significant feature of the

capacitive-discharge system is that the input power increases directly with the increased spark-plug power required as engine speed increases. In the inductive-discharge system, on the other hand, the opposite is true, as shown in Fig. 559. The required power is the product of the energy required per igniton pulse and the number of ignition pulses per second. The upper curve of Fig. 559 was determined experimentally for the circuit of Fig. 543 for a battery supply of 14 volts. In the capacitive-discharge system, input power can be made proportional to the required power because the capacitor is charged once per ignition pulse and holds this charge until needed.

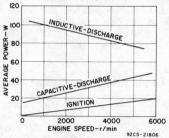

Fig. 559—Ignition power requirements.

In addition, feedback can be used to turn the converter off after the capacitor is charged and thus cut off the input power to the system. The input power can therefore be made proportional to engine speed; and higher efficiency can be achieved at all speeds. The curve shown in Fig. 559 applies for a commercially available capacitive-discharge system. The higher efficiency of this system is apparent.

A second important advantage of the capacitive-discharge system is that faster rise times are more readily obtained because the transformer acts only as a pulse transformer and not as an energy-storage element. Therefore, its high-frequency response characteristic is governed by its leakage inductance, which is much smaller than the primary magnetizing inductance. This advantage is obtained even when a conventional ignition coil is used as the pulse transformer. Secondary-voltage rise times of about 15 to 30 microseconds are readily obtained. As discussed previously, the shorter rise time greatly enhances the ability of this system to fire fouled plugs.

A major operating point that must be considered in the capacitive-discharge system is when to charge the capacitor. In some systems the capacitor is charged soon after discharge, in others, just before discharge. The second method is the better in that it minimizes the losses resulting from leakage, but this advantage is somewhat negated because of the precise timing required to institute the charge just prior to discharge. This requirement can result in complex mechanical or electrical arrangements.

Component Requirements

In a capacitive-discharge ignition system, the forward blocking voltage of the SCR must be greater than about 400 volts, and its current-handling capability must be about 5 amperes. The SCR firing characteristics as a function of temperature are important, and must be taken into account in the design of the trigger circuit. Because the specifications of the spark coil are usually known, the design of an ignition circuit is usually begun there; capacitor size and voltage are determined from these known quantities. The capacitor chosen determines the spark-gap duration and the charging time required; the reliability of the system is almost directly related to the reliability of this component. The high peak current which flows through the SCR also flows through this capacitor, necessitating that the capacitor be as carefully chosen as the SCR.

The limit on charging time is the dv/dt rating of the SCR. By using the capacitor size and the dv/dt rating of the SCR, the charg-

ing current required is determined by applying the formula $I = C dv/dt$. This formula can also be used to determine the peak current and its duration, and the rate of rise of current. The rate of rise of the current should be checked against the limiting circuit value, the di/dt rating of the SCR.

Table L—SCR Parameter Values of Importance in Ignition Circuits

I_{DROM}	when $V_D = V_{Capacitor} + 20\%$ at a case temperature T_C of 100°C.
I_{RRDM}	when $V_R = 25$ volts when using a diode as in the circuit of Fig. 558 or when $V_R = V_{peak\ reverse}$ due to flywheel effect in the flywheel charged system. In both cases, T_C is 100°C.
v_T	when $I = I_{peak}$, the value of v_T is approximately 2.5 to 4 volts depending on current pulse amplitude, repetion rate, and case temperature.
V_{gate}	at 12 volts with $R_A = 30$ ohms. V_{gate} and I_{gate} will be maximum or minimum limits depending on trigger-circuit requirements. Higher limits help prevent spurious firing as a result of noise.

The SCR parameter values that must be specified to assure reliable device operation in ignition circuit applications are shown in Table L.

The ignition coil used in the capacitive-discharge circuit can be a specially wound low-inductance unit or the existing coil. The existing coil has an advantage in that it can be used with a breaker-point distributor to provide the ignition function in the event of an electronic failure. The major disadvantage of the use of the existing coil is that the benefits of ignition pulses with sharp rise times, the type of pulses needed to fire fouled spark plugs, are reduced because of the inductance of the coil. Because of the ease of obtaining a trigger pulse and the circuit simplicity, the SCR capacitor-discharge system is used almost exclusively on small engines.

Types of Capacitor-Discharge Systems

There are three systems in which capacitor-discharge ignition circuits can be used to good advantage; the flywheel-charged small-engine system, the line-charged ignitor used with gas-operated appliances such as dryers and furnaces, and the inverter-charged system such as that used in automotive and stationary engine systems. All of these systems are operated similarly: energy stored in a capacitor is transferred to a spark gap through a transformer and SCR; the SCR assures a short-duration spark.

The circuit of Fig. 560 is typical

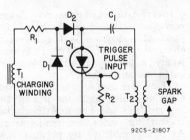

92CS-21807

Fig. 560—Typical circuit configuration for a capacitor-discharge ignition system.

of that used in the three systems. The ac potential across transformer T_1 is rectified by diode D_2, and charges capacitor C_1 to the required voltage. Resistor R_1 limits the current and prevents the SCR from firing as a result of the imposition of a dv/dt value in excess of the capability of the SCR. The combination of diodes D_1 and D_2 prevents the charging winding of transformer T_1

from impressing a high reverse voltage across the SCR. Resistor R_2 damps variations in the input impedance of the SCR. The SCR is triggered at the appropriate time, and the energy stored in the capacitor is transferred into the primary of T_2, thus causing a spark at the spark gap. The voltage required to break down the spark gap is a function of the spacing of the electrodes and pressure in the cylinder in the vicinity of the gap. The spark in the gap lasts until the value of current passing through the SCR is below its holding current. When the SCR stops conducting, D_1 and D_2 start conducting in the reverse direction and lengthen spark duration. After the SCR turns off, C_1 is discharged, and the circuit is ready to repeat the cycle.

Flywheel-Charged Systems—Some of the simplest ignition systems are constructed using the flywheel-charging method, because this method affords a reliable circuit with a minimum of active components. The system makes use of a rotating magnetic field to charge the capacitor and to trigger the SCR; mechanical position determines timing. The designer has several options in the determination of when the charging of the capacitor takes place in the flywheel system. The most advantageous time occurs just before the capacitor is to be discharged. However, some voltage regulation problems must be considered. Because $V = Nd\phi/dt$ where $d\phi$ and N are constant, the voltage produced across the charging winding varies with RPM. At low flywheel speeds, there may not be enough voltage available to produce the energy required; at high flywheel speeds, it is possible to have too high a voltage and therefore to exceed the voltage breakdown rating of the SCR and cause premature triggering. If the breakdown voltage rating of the capacitor is also exceeded, the capacitor will be damaged. Therefore, some means of accommodating or regulating the voltage must be considered.

The design of the trigger coil is also important. It must be capable of providing voltages and currents high enough to gate the SCR into conduction at all temperatures. In addition, consideration should be given to the fact that the gate pulse should end before the current through the SCR ceases to flow so that the device is not gated during the period of reverse voltage.

As is evident in the above discussion, a major factor in the performance of the flywheel ignition circuit is the design of the magnetic components used for the triggering and charging functions. A typical example of a flywheel-charged ignition circuit is shown in Fig. 561.

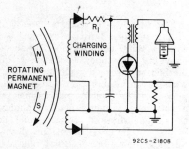

92CS-21808

Fig. 561—Flywheel-charged igniter circuit.

Line-Charged Igniter—The line-charged ignition circuit finds greatest use in heating systems and large gas-operated appliances. The line-charged igniter has small regulation or charging problems as it uses the power lines as its charging source. Normally, circuits of this type charge at the line voltage that produces the lowest-cost circuit. The gate trigger pulse is derived by using a diac and an RC phase-shift network such as that shown in Fig. 562.

On each positive half-cycle, the circuit operates once until the flame is burning; the sensor then turns on SCR_2 which prevents SCR_1

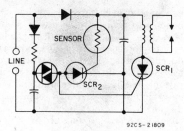

92CS-21809

Fig. 562—Line-charged ignition circuit.

from being triggered and causing further sparks. This arrangement will cause the igniter to operate whenever the flame goes out and thus provides a valuable safety feature. If a bimetallic element were used in series with the line to sense the flame, SCR_2 could be eliminated. The use of the bimetallic element would also contribute to a low standby power drain; however, at higher energy levels, the bimetallic strip method of gating is probably only marginally reliable, because the gate current produced is not sufficient for the anode-current pulse-amplitude required.

Inverter-Charged Systems—A system that eliminates the need for flywheel magnetics is the inverter-charged system. This system is used where a battery is available, such as in an automobile.

There are some practical considerations which limit the use of the inverter system. The first limitation is starting under low temperature. At an ambient temperature of −40°C, the available battery voltage in a "12-volt" automotive system (12-volts nominal at 25°C) may be as low as 6 volts dc because of the starter current required at this temperature and the reduced battery capability. In addition, at this temperature, the fuel-air mixture is wet, particularly in a two-cycle engine. For reliable starting, the full spark energy must be available immediately. This means that the inverter must be capable of producing the full energy at low supply voltages.

The voltage step-up ratio of the system transformer is constant and therefore cannot be increased as the temperature decreases; such an action would assure sufficient voltage at low temperatures, but would subject the capacitor and SCR to voltages in excess of their ratings under normal conditions and after starting. The problem of starting at low temperature may be circumvented by regulating the voltage on the capacitor or by using a transformer with a higher step-up ratio than required and then shutting down or removing the inverter, with its transformer, from the circuit at a time that will prevent any voltages from becoming a problem.

As the maximum RPM of the engine increases, the demands on the inverter also increase; this variation in demand can be alleviated by ballasting. When ballasting of the ignition is accomplished by means of a regulator circuit, external ballasts are not needed. A typical example of an inverter-type ignition system with regulator ballasting is shown in Fig. 563. The trigger cir-

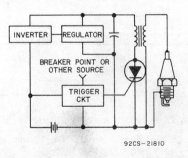

92CS–21810

Fig. 563—Block diagram of an inverter-charged ignition system.

cuit shown in the figure is subject to the same variations in potential as the inverter circuit in addition to others arising from the need to gate the SCR with a high-current at low temperature when the available voltage is low. This gating problem can only be solved by a compromise between overdriving of

the gate at high temperature and maintaining only an adequate drive at low temperatures; there are many circuits that can be used to achieve this compromise.

The inverter must be capable of handling a power level, typically between 20 and 50 watts, representing an energy level of 80 millijoules per pulse, for a 4-cycle, 8-cylinder engine. The inverter circuit should operate at a frequency high enough to make use of smaller transformer-core sizes and yet be able to incorporate low-cost power devices. The RCA line of hometaxial power transistors is generally very reliable and economical in inverter ignition systems.

Capacitive-Discharge Automotive Ignition System

Fig. 564 shows the circuit diagram for a low-cost transistor/SCR capacitor-discharge ignition system for passenger automobiles. This system offers the advantages of reduced maintenance, smaller current drain on the automobile battery, full output voltage at low battery voltage (down to 4 volts), and a high-voltage output pulse that has a rapid rate of rise. As pointed out previously, this latter factor provides greater assurance of the firing of fouled spark plugs.

The SCR ignition system is essentially a combination of eight basic circuit units, as follows: (1) a single-ended, self-oscillating swinging-choke inverter is used to provide the dc-to-ac inversion and the step-up of the battery voltage. (2) An output circuit that includes an SCR, a storage capacitor, an ignition coil (a standard automotive ignition coil is used), and a commutating diode develops the fast-rising high-voltage pulse for the spark plugs. (3) The commutating diode and a single rectifying diode form a capacitor-charging circuit to transfer energy from the inverter to the output-circuit storage capacitor. (4) A regu-

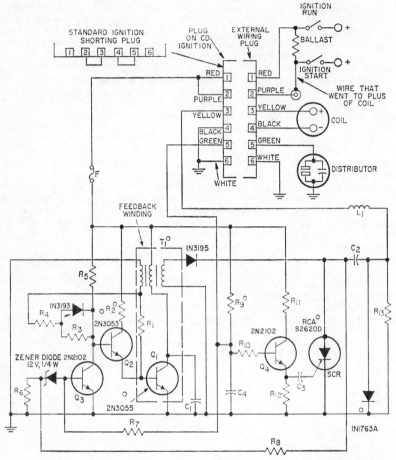

$C_1 = 0.25 \ \mu F, \ 200 \ V$
$C_2 = 1 \ \mu F, \ 400 \ V$
$C_3 = 1 \mu F, \ 25 \ V$
$C_4 = 0.25 \ \mu F, \ 25 \ V$
$F = 5 \ A$
$L_1 = 10 \ \mu H$; 100 turns of No. 28 wire wound on a 2-W resistor (100 ohms or more)
$R_1 = 1000$ ohms
$R_2 = 50$ ohms, 5 W
$R_3 = 22000$ ohms
$R_4 = 1000$ ohms
$R_5 = 10000$ ohms
$R_6 = 15000$ ohms
$R_7 = 8200$ ohms
$R_8 = 0.39$ megohm
$R_9 = 220$ ohms, 1 W
$R_{10} = 1000$ ohms

$R_{11} = 68$ ohms
$R_{12} = 4700$ ohms
$R_{13} = 27000$ ohms
T_1 = Transformer, wound as follows: A ½ in. bobbin and an El stack of grain-oriented silicon steel are used; first, 150 turns of No. 28 wire are wound and labeled start 1 and finish 1 on the winding; second, 50 turns of No. 24 and No. 30 wires are wound bifilar and labeled start 2 and finish 2; third, 150 turns of No. 28 wire are wound and labeled start 3 and finish 3. All windings are

wound in the same direction. A total air gap of 70 mils (35 mil spacer) is used. Connections are made as shown in Fig. 565.

All the resistors are ½ W unless otherwise indicated.

° These components are subject to excessive temperature rise unless provision is made to transfer the heat to the ambient air by means of an appropriate heat sink.

Fig. 564—An SCR capacitor-discharge automobile ignition circuit.

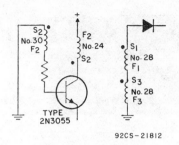

Fig. 565—Details of inverter transformer (T₁) shown in Fig. 564.

lator stage controls the frequency of the inverter stage to provide efficient regulation of the voltage across the storage capacitor. (5) A protection circuit (limiting inductance and resistance) prevents damage to the system by transients that may be developed in the case of an open or shorted ignition coil or because of high-voltage arcing to either primary terminal of the ignition coil. (6) A shut-down circuit holds the inverter inoperative when the ignition breaker points are open. (7) A trigger circuit suppresses the normal bounce of the breaker points and also prevents SCR triggering by the residual voltage across the closed points. (8) A method of SCR commutation is used that involves the interplay of several parts of the over-all system.

Inverter, Regulator, and Capacitor-Charging Circuit—The inverter uses a 2N3055 transistor (Q_1) in a single-ended output stage, a 2N3053 transistor (Q_2) in an emitter-follower driver stage, and a 2N2102 transistor (Q_3) in a control stage that is part of the shutdown circuit which holds the inverter inoperative when the ignition breaker points are open. Regenerative feedback is coupled from the feedback winding of the inverter output transformer T_1 back to the bases of the driver and output transistors. The high gain provided by the combination of transistors Q_1 and Q_2 assures oscillation and low drive-power requirements for the inverter. The

starting resistor R_5 provides a forward bias that drives transistors Q_1 and Q_2 into conduction to initiate oscillation in the inverter. The regenerative action of the circuit very quickly drives the output transistor Q_1 into conduction, and essentially the full battery voltage is then applied across the primary of the inverter transformer T_1. The resultant current increase in the transformer primary winding induces a voltage across the feedback winding that supplies sufficient current through resistors R_1 and R_4 and diode D_1 to maintain the output transistor Q_1 in saturation. During this part of the operating cycle (i.e., during the conduction of transistor Q_1), the voltage across the secondary winding of transformer T_1 reverse-biases the rectifying diode D_2 in the capacitor-charging circuit, and no energy is transferred to the output circuit of the ignition system.

With transistor Q_1 operating with fixed base current (in saturation), its collector current rises to a value beyond which it cannot increase. As a result, the feedback voltage is decreased, and no longer maintains base drive to transistor Q_1, and the transistor starts to turn off. The regenerative action of the inverter circuit causes a rapid reversal of the base drive for transistors Q_1 and Q_2. These transistors, therefore, are quickly cut off, and a "flyback" voltage pulse is generated at the collector of the output transistor Q_1. Diode D_3 blocks the reverse voltage and limits the reverse base drive. The reverse-bias current that turns off transistors Q_1 and Q_2 is applied through resistors R_3 and R_1 respectively. The flyback-pulse voltage is stepped up across the secondary of transformer T_1. The polarity of this pulse, however, is such that the rectifying diode D_2 becomes forward-biased. As a result, the energy previously stored in the primary winding of transformer T_1 is transferred through the secondary winding, rectifying diode D_2, and commutating diode D_3 to charge the output-circuit storage capacitor C_2. The capacitor C_1 connected across transistor Q_1 reduces the amplitude of the leakage-inductance pulse and restricts the rate of rise of the collector voltage of transistor Q_1. The charging current for capacitor C_2 is shunted around the ignition coil by the commutating diode D_3 so that no energy is transferred into the ignition coil and from there to the spark plugs.

When the collector voltage of transistor Q_1 decreases to a value less than the battery voltage, it again begins to conduct, and the cycle is repeated to charge the storage capacitor C_2 to a higher voltage. Until the voltage across the storage capacitor rises above a predetermined value, the voltage applied to the zener diode D_4 from the voltage divider formed by resistors R_6 and R_8 is insufficient to cause the zener diode to conduct. If the ignition breaker points are closed during this time, resistor R_7 is returned to ground and transistor Q_3 cannot conduct. When the capacitor voltage rises to a level high enough to cause zener diode D_4 to conduct, transistor Q_3 turns on and shunts base drive current from transistor Q_2. This effect reduces the base drive of transistor Q_1 and causes this transistor to pull out of saturation at a lower collector-current level which, in turn, increases the frequency of oscillation. The cutback in peak primary current reduces the charging rate of the storage capacitor C_2 to the level required to replenish circuit losses and prevents further rise in the output voltage.

Transistor Q_3 also holds the inverter inoperative when the ignition breaker points are open. When these points open, the current fed from the voltage at the breaker points through resistor R_7 causes transistor Q_3 to conduct heavily. This effect shorts the base of transistor Q_2 and stops the oscillation.

Fig. 566 shows that the collector voltage of transistor Q_1 swings alternately between the saturation

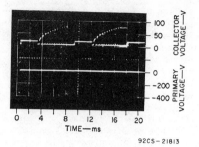

92CS–21813

Fig. 566—Collector voltage of the inverter output transistor Q_1 (top) and ignition-coil primary voltage (bottom) as functions of time (2000 rpm; $V_{CC} = 12$ V).

level and the peaks of flyback pulses of increasing amplitude. The change in frequency that results from regulator action is apparent in the voltage waveform. The collector voltage then decreases to the supply voltage when the ignition breaker points open and shut down the inverter. Fig. 567 shows an expanded view

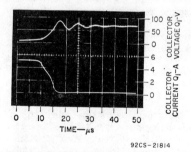

92CS–21814

Fig. 567—Expanded collector voltage (top) and current (bottom) of the inverter output transistor (Q_1) as functions of time during turn off when the storage capacitor (C_2) is being charged ($V_{CC} = 12$ V).

of the turn-off and flyback characteristics of transistor Q_1 at a point when the storage capacitor is being charged.

Output Circuit—When a high-voltage pulse is required, the RCA-S2620D SCR in the output circuit is gated on. As a result, the anode voltage of the SCR decreases to approximately zero, and the voltage across the charged storage capacitor is applied to the primary of the ignition coil. (The value of the inductance L_1 is negligible in comparison to the inductance of the ignition coil and is not considered in this analysis.) The ungrounded (+) side of the ignition-coil primary (terminal 3 on the connecting plug) is driven negative with respect to the capacitor potential. Diode D_3, in parallel with the coil, is reverse-biased at this time. The discharge of the capacitor into the primary of the ignition coil generates a high-voltage pulse across the secondary.

The capacitor discharges into the primary inductance of the ignition coil and builds up the primary current in the coil. When the voltage across the capacitor (and coil primary) decreases to zero and starts to reverse, the commutating diode D_3 becomes forward-biased and begins to conduct. The current through the primary of the ignition coil is at a peak at the time the diode begins to conduct. The current then suddenly switches out of the SCR and into the diode. The primary-coil voltage remains clamped at zero, and the primary current decays at a rate determined by the L/R ratio of the coil. Because of the clamping action of the commutating diode D_3, the duration of the spark in the spark plug is lengthened.

When the SCR is on, it effectively places a short across the secondary of the inverter transformer. However, the inverter is off when the SCR is on (because of the shutdown circuitry); the inverter, therefore, does not operate into the short.

Figs. 568 and 569 show the SCR voltage and current as a function of time. The starting point of the waveform shown in Fig. 568 occurs at the instant the ignition points open. The anode voltage of the SCR decreases to zero and the anode current builds up to the peak value in a quarter cycle. The current is then switched out of the SCR, and SCR

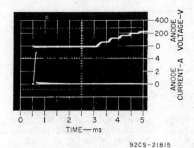

92CS-21815

Fig. 568—SCR voltage (top) and current (bottom) as a function of time (2000 rpm; $V_{CC} = 12$ V).

current decreases suddenly almost to zero. The small residual current is a result of the energy stored in the inverter transformer during the period that the inverter is inoperative. This stored energy causes a current to circulate from the secondary of the transformer through the SCR. When the ignition points close and the capacitor recharges, the SCR blocks the voltage on the capacitor.

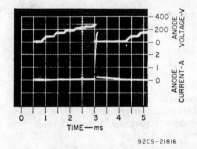

92CS-21816

Fig. 569—SCR voltage (top) and current (bottom) as a function of time (oscilloscope sweep triggered at instant ignition breaker points are closing; 4000 rpm; $V_{CC} = 12$ V).

The starting point for the waveform shown in Fig. 569 occurs at the instant that the ignition points close. The period between the instant at which the points close and that at which the voltage first begins to rise is the time during which the collector current of transistor Q_1 builds up to the switching level. The significance of this time is explained subsequently during the discussion on commutation of the SCR.

Fig. 570 shows the waveforms for voltage and current in the primary of the ignition coil that result when a 7-millihenry inductor is used to simulate the primary of the ignition coil. The primary voltage increases rapidly to a peak negative value, then decays sinusoidally to zero as the current builds to a peak. The primary voltage is then clamped at zero, and the current decays exponentially.

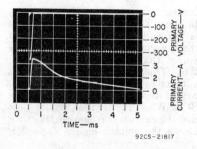

92CS-21817

Fig. 570—Primary voltage (top) and current (bottom) as a function of time (7-mH coil used in place of standard ignition coil; 2000 rpm; $V_{CC} = 12$ V).

It should be noted that an actual ignition coil, operated with the secondary open, will reflect a tuned circuit into the primary. This operation causes a ringing on top of the waveforms. The anode voltage of the SCR may actually reverse for a short time because of this ringing. The SCR essentially blocks this reverse voltage except for a small current that flows because of the presence of positive gate signal. As a result, some instantaneous dissipation occurs in the reverse-blocking function of the SCR. The SCR can safely withstand this dissipation for the short period of time required. The gate signal is kept positive through the ringing cycle so that the SCR continues to conduct when the anode voltage rings back positive. This ringing does not occur

when the secondary voltage fires a plug because the ionized plug shorts the secondary winding.

Protection Circuit—Inductance L_1 is used to protect the system against a shorted primary in the ignition coil. The limiting inductance controls the rate of change of current (di/dt) and peak current that occurs when the SCR is turned on with a short across the primary of the ignition coil. Resistance R_{13} is used to assure that the voltage across the primary of the ignition coil is not negative when the coil is open. If this voltage were not clamped, the regulator would not operate properly, and the peak collector voltage of transistor Q_1 would exceed the limits specified for the device.

Trigger Circuit—The triggering circuit performs the following functions: (1) triggers and holds the SCR on when the ignition points open (at battery voltage down to 4 volts), (2) applies a signal back into the inverter shutdown circuitry when the ignition points are open, (3) suppresses the inverter signal that rides on the power supply so that it does not trigger the SCR, (4) prevents the residual voltage across the closed points from triggering the SCR, (5) prevents normal point bounce that occurs when the points close, and (6) maintains proper operation whether or not the capacitor is present across the breaker points. The 2N2102 transistor Q_4 is used to perform these functions.

The trigger current for the gate of the SCR is initiated when the base voltage of transistor Q_4 reaches approximately 0.6 volt above the emitter voltage. The trigger current flows from the supply through resistor R_{11}, transistor Q_4, and capacitor C_3 to the gate of the SCR.

When the ignition points open, capacitor C_4 (and the capacitor across the points) charges because of the current through resistor R_8.

If the points are open long enough (without bouncing), the voltage across capacitor C_4 becomes high enough to turn on transistor Q_4. The voltage required to turn on transistor Q_4 is the sum of the gate-cathode voltage of transistor Q_5, the voltage across capacitor C_3, the emitter-base voltage of transistor Q_4, and the voltage drop across resistor R_{10}. (Resistor R_{10} ensures that the voltage across the open ignition points rises to a value high enough to supply sufficient current through resistor R_7 to shut down the inverter.) At normal engine speeds, the average voltage level across capacitor C_3 keeps both the gate-cathode junction of the SCR and the emitter-base junction of the transistor reverse-biased until capacitor C_4 charges high enough to turn on transistor Q_4. Because transistor Q_4 is off and the gate of the SCR is reverse-biased when the points are closed, the desired suppression of inverter signal and residual point voltage is achieved.

If the points bounce during normal operation, they discharge C_4 (and the distributor capacitor) almost instantly each time they close. Thus, each time they bounce open, these capacitors must recharge from zero toward the triggering level. With normal bouncing, the points do not stay open long enough for the triggering level to be reached, so the SCR is not triggered. If severe bouncing occurs at very high speeds, the points can stay open long enough to cause triggering of the SCR.

Better filtering is achieved when the automobile distributor capacitor is retained, but satisfactory operation is achieved without this capacitor. With the capacitor left in the distributor, it is possible to switch back to standard ignition by switching the plug shown in Fig. 564.

Commutating the SCR—All the parts of the system work together in such a manner as to cause the SCR current to go to zero for a sufficient length of time to cause commutation (turn off). As explained earlier, the current through the primary of the ignition coil is switched from the SCR and into the commutating diode when the primary voltage decreases to zero. From that time on, the diode keeps the coil current clamped out of the SCR. The SCR then conducts only the small current that results from the energy stored in the inverter transformer. When the points close again, the inverter restarts and, as explained previously, the rectifying diode D_2 is reverse-biased during the time the collector current of transistor Q_1 builds back up to the switching level. No current then flows in the SCR, and the SCR is allowed to turn off. Fig. 568 shows that the current is zero for about 2.5 milliseconds before anode voltage is re-applied at a rate of 0.15 volt per microsecond. A worst-case SCR commutates in less than 100 microseconds at a temperature of 100°C under these operating conditions.

Performance—Fig. 571 shows several performance curves for the SCR capacitor-discharge ignition system. Fig. 572 shows the open-circuit output voltage as a function of time, and Fig. 573 shows the output voltage when the load on the secondary consists of a 1-megohm resistor in parallel with a 50-picofarad capacitor.

Figs. 574 and 575 show the secondary voltage under sparking conditions. Arc duration, as shown in Fig. 574, is 600 microseconds (single polarity) under wide-gap conditions. The narrower-gap conditions shown in Fig. 575 result in an arc duration of 200 microseconds.

Mounting Considerations—The SCR ignition circuit must be protected from moisture. Heat-generating components, however, should not be enclosed in non-circulating atmosphere. Still air has a thermal-

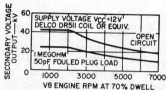

(a) Output voltage as a function of engine rpm at 12 volts for both an open secondary and a fouled-plug load.

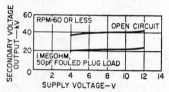

(b) Output voltage as a function of battery voltage at cranking speeds for both an open secondary and a fouled-plug load.

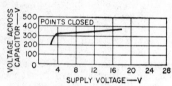

(c) Regulation curve showing peak capacitor (and SCR) voltage as a function of battery voltage.

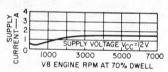

(d) Battery drain as a function of engine rpm at a battery voltage of 12 volts.

92CS–21818

Fig. 571—Performance of the capacitor-discharge ignition circuit.

resistance 12,000 times that of copper, and generation of heat in this high thermal resistance could cause the inside ambient temperature to rise about the specified limits. All components subject to high temperature rise (marked in Fig. 564 with a small circle) should be thermally connected to a low-

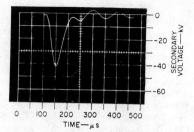

92CS–21819

Fig. 572—Open-circuit output voltage (standard ignition coil, Delco D511 or equivalent; 2000 rpm; $V_{CC} = 12$ V).

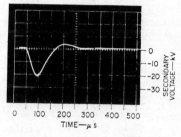

92CS–21820

Fig. 573—Output voltage with fouled spark plugs (standard ignition coil, Delco D511 or equivalent; 50-pF load in parallel with 1-megohm resistance; 2000 rpm; $V_{CC} = 12$ V).

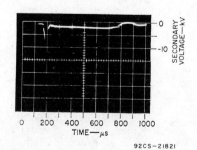

92CS–21821

Fig. 574—Output voltage showing duration of spark arc (standard ignition coil, Delco D511 or equivalent; 2000 rpm; $V_{CC} = 12$ V).

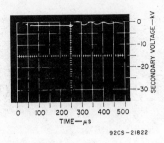

Fig. 575—Output voltage with spark gap shortened (standard ignition coil, Delco D511 or equivalent; 2000 rpm; $V_{CC} = 12$ V).

thermal-resistance path to the outside environment. For example, the SCR may be mounted to an aluminum plate on a mica insulating washer. This plate should then be fastened to the inside of the chassis wall that provides the thermal path to the outside environment. The resistors and diodes which require a heat sink should be attached to the chassis with a thermally conductive epoxy.

Thyristor Power Controls

THE use of thyristors is becoming increasingly important for power-control applications ranging from low voltages to more than 1000 volts at current levels from less than half an ampere to more than 1000 amperes. When power control involves conversion of ac voltages and/or currents to dc and control of their magnitude, SCR's are used because of their inherent rectifying properties. SCR's are also used in dc switching applications, such as pulse modulators and inverters, because the currents in the switching device are unidirectional. In addition, SCR's are generally used whenever the desired function can be accomplished adequately by this type of device because of the economics involved.

A triac provides symmetrical bidirectional electrical characteristics. Triacs have been developed specifically for control of ac power, and are used primarily for control of power to a load from ac power lines.

FUNDAMENTALS OF THYRISTOR POWER CONTROL

Thyristors are excellent devices for use in the control of ac power. In general, thyristors initially assume a blocking, or high-impedance state, and remain in that state until triggered to the on or low-impedance state. Once triggered, the thyristor remains on until the current is reduced to zero. The thyristor then returns to its blocking state. Because the current decreases to zero during every half-cycle in an ac supply,

turn-off is guaranteed every half-cycle. All that is necessary for ac power control, therefore, is a trigger circuit to control thyristor turn-on so that whole or partial cycles may be switched to the load.

In many power-control applications of thyristors, partial cycles of the applied ac voltage are switched to the load. Because the power delivered to the load is controlled by variation of the phase angle at which the thyristor switching initiates current flow, this type of operation is usually referred to as **phase control**.

Power to an ac load may be controlled by switching of complete half-cycles or integral numbers of whole cycles of the ac power to the load. This type of control is usually referred to as **integral-cycle** or **zero-voltage-switching control**.

Phase Control

The electrical angle of the applied ac voltage waveform at which thyristor current is initiated is termed the **firing angle** (θ_F). It is usually more important, however, to know and to refer to the **conduction angle** (θ_C), which is the number of electrical degrees of the applied ac voltage waveform during which the thyristor is in conduction. The conduction angle is equal to $180° - \theta_F$ for a half-wave circuit and $2(180° - \theta_F)$ for a full-wave circuit. The voltage waveforms across the thyristor and the load for each type of circuit are illustrated in Fig. 576.

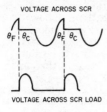

VOLTAGE ACROSS SCR

VOLTAGE ACROSS SCR LOAD

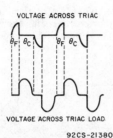

VOLTAGE ACROSS TRIAC

VOLTAGE ACROSS TRIAC LOAD

92CS-21380

Fig. 576—Voltage waveforms showing conduction angle for half-wave operation (SCR) and full-wave operation (triac) of thyristor phase-control circuits.

Basic Current Relationships—In the design of thyristor power-control circuits, it is often necessary to determine the specific values of peak, average, and rms current that flow through the thyristors. For conventional rectifiers, these values are readily determined by use of the current ratios shown in Table VII, given in the section on **Silicon Rectifiers**. For thyristors, however, the calculations are more difficult because the current ratios become functions of the conduction angle and the firing angle of the device.

The curves in Figs. 577, 578, and 579 show several current ratios as functions of conduction or firing angles for three basic SCR circuits. These curves can be used in a number of ways to calculate desired current values. For example, they can be used to determine the peak or rms current in an SCR when a cer-

tain average current is to be delivered to a load during a specific part of the conduction period. It is also possible to work backwards and determine the necessary period of conduction if, for example, a specified peak-to-average current ratio must be maintained in a particular application. Another use of the curves in Figs. 577, 578, and 579 is in the calculation of the rms current at various conduction angles when it is necessary to determine the power delivered to a load, or power losses in transformers, motors, leads, or bus bars. Although the curves are presented in terms of device current, they are equally useful for the calculation of load current and voltage ratios.

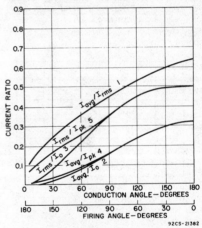

92CS-21382

Fig. 577—Ratio of SCR current as a function of conduction and firing angles for single-phase half-wave conduction into a resistive load.

The curves provide ratios that relate average current I_{avg}, rms current I_{rms}, peak current I_{pk}, and reference current I_o. The reference current is a circuit constant and is equal to the peak source voltage V_{pk} divided by the load resistance R_L. The term I_{pk} refers to the peak current that flows through the SCR during its period of forward conduction.

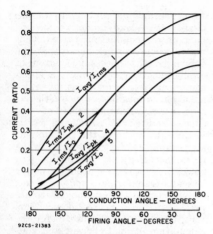

Fig. 578—Ratio of thyristor current as a function of conduction and firing angles for single-phase full-wave conduction into a resistive load.

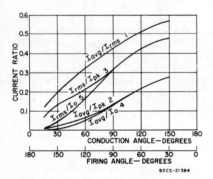

Fig. 579—Ratio of SCR current as a function of conduction and firing angles for three-phase half-wave circuit having a resistive load.

I_o is the maximum possible peak-current value during the peak of the sine wave. For conduction angles greater than 90 degrees, I_{pk} is equal to I_o; for conduction angles smaller than 90 degrees, I_{pk} is smaller than I_o.

The general procedure for the use of the curves is as follows:
(1) Identify the unknown or desired parameter.
(2) Determine the values of the parameters fixed by the circuit specifications.
(3) Use the appropriate curve to find the unknown quantity as a function of two of the fixed parameters.

Example No. 1: In the single-phase half-wave circuit shown in Fig. 580, a 2N685 SCR is used to

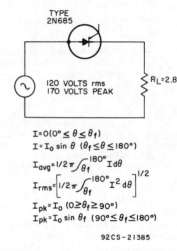

$$I = 0 \ (0° \leq \theta \leq \theta_f)$$
$$I = I_o \sin \theta \ (\theta_f \leq \theta \leq 180°)$$
$$I_{avg} = 1/2\pi \int_{\theta_f}^{180°} I \, d\theta$$
$$I_{rms} = \left[1/2\pi \int_{\theta_f}^{180°} I^2 \, d\theta \right]^{1/2}$$
$$I_{pk} = I_o \ (0 \geq \theta_f \geq 90°)$$
$$I_{pk} = I_o \sin \theta_f \ (90° \leq \theta_f \leq 180°)$$

Fig. 580—Single-phase half-wave circuit that operates into a resistive load and the respective equation for SCR current.

control power from a sinusoidal 120-volt-rms (170-volt-peak) ac source into a 2.8-ohm load. This application requires a load current that can be varied from 2 to 25 amperes rms. It is necessary to determine the range of conduction angles required to obtain this range of load current.

First, the reference current I_o is calculated as follows:

$$I_o = \frac{V_{pk}}{R_L} = \frac{170}{2.8} = 61 \text{ amperes}$$

The ratio of I_{rms} to I_o for the

minimum and maximum load-current values is then calculated as follows:

$$(I_{rms}/I_o)_{min} = 2/61 = 0.033$$

$$(I_{rms}/I_o)_{max} = 25/61 = 0.41$$

These current-ratio values are then applied to curve 3 of Fig. 579 to determine the corresponding conduction angles:

$$\Theta_{C(min)} = 15 \text{ degrees}$$
$$\Theta_{C(max)} = 106 \text{ degrees}$$

Example No. 2: In the single-phase full-wave bridge circuit (two legs controlled) shown in Fig. 581,

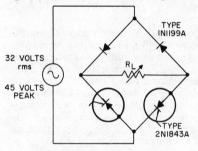

$$I = 0 \quad (0° \leq \theta \leq \theta_f)$$
$$I = I_o \sin \theta \quad (\theta_f \leq \theta \leq 180°)$$
$$I_{avg} = 1/\pi \int_{\theta_f}^{180°} I \, d\theta$$
$$I_{rms} = \left[1/\pi \int_{\theta_f}^{180°} I^2 \, d\theta \right]^{1/2}$$
$$I_{pk} = I_o \quad (0° \leq \theta_f \leq 90°)$$
$$I_{pk} = I_o \sin \theta_f \quad (90° \leq \theta_f \leq 180°)$$

92CS-21392

Fig. 581—Single-phase full-wave bridge circuit that operates into a resistive load and the respective equations for SCR current.

a constant average load current of 7 amperes is to be maintained while the load resistance varies from 0.2 to 4 ohms. In this case, it is necessary to determine the variation required in the conduction angle. The average current through the SCR is one-half the load current, or 3.5 amperes. The applicable current ratios for this circuit are shown in Fig. 577 (the individual device currents are half-wave although the load current is full-wave).

Again, the first quantity to be calculated is the reference current I_o. Because the reference current varies with the load resistance, the maximum and minimum values are determined as follows:

$$I_{o(max)} = V_{pk}/R_{L(min)} = 45/0.2$$
$$= 225 \text{ amperes}$$

$$I_{o(min)} = V_{pk}/R_{L(max)} = 45/4$$
$$= 11.2 \text{ amperes}$$

The corresponding ratios of I_{avg} to I_o are then calculated, as follows:

$$(I_{avg}/I_o)_{min} = 3.5/225 = 0.015$$

$$(I_{avg}/I_o)_{max} = 3.5/11.2 = 0.312$$

Finally, these ratios are applied to curve 2 of Fig. 577 to determine the desired conduction values:

$$\Theta_{C(min)} = 25 \text{ degrees}$$
$$\Theta_{C(max)} = 165 \text{ degrees}$$

Example No. 3: In the three-phase half-wave circuit shown in Fig. 582, the firing angle is varied continuously from 30 to 155 degrees. In this case, it is necessary to determine the resultant variation in the attainable load power. Reference current for this circuit is determined as follows:

$$I_o = V_{pk}/R_L = 85/3 = 28 \text{ amperes}$$

Rectifier current ratios are determined from Fig. 579 for the extremes of the firing range, as follows:

$$\Theta_F = 30 \text{ degrees}; \quad I_{rms}/I_o = 0.49$$

$$\Theta_F = 155 \text{ degrees}; \quad I_{rms}/I_o = 0.06$$

These ratios, together with the reference current, are then used to determine the range of rms current in the rectifiers, as follows:

$$I_{rms(max)} = 0.49 \times 28 = 13.7 \text{ amperes}$$

$$I_{rms(min)} = 0.06 \times 28 = 1.7 \text{ amperes}$$

In this circuit, the rms current in the load is equal to the rms rectifier current multiplied by the square

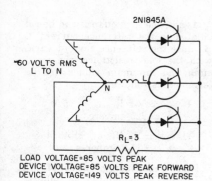

$$I = I_0 \sin \theta \quad (30° \leq \theta \leq 180°)$$

$$I_{avg} = 1/2\pi \int_{\theta_f}^{\theta_f + 120} I \, d\theta \quad (30° \leq \theta_f \leq 60°)$$

$$I_{avg} = 1/2\pi \int_{\theta_f}^{180°} I \, d\theta \quad (60° \leq \theta_f \leq 180°)$$

$$I_{rms} = \left[1/2\pi \int_{\theta_f}^{\theta_f + 120} I^2 \, d\theta \right]^{1/2} \quad (30° \leq \theta_f \leq 60°)$$

$$I_{rms} = \left[1/2\pi \int_{\theta_f}^{180°} I^2 \, d\theta \right]^{1/2} \quad (60° \leq \theta_f \leq 180°)$$

$$I_{pk} = I_0 \quad (30° \leq \theta_f \leq 90°)$$

$$I_{pk} = I_0 \sin \theta_f \quad (90° \leq \theta_f \leq 180°)$$

92CS-21393

Fig. 582—Three-phase half-wave circuit that uses a resistive load and the respective equation for SCR current.

root of three; as a result, the desired power range of the load is as follows:

$$P_{max} = [I_{rms(max)} \sqrt{3}]^2 R_L$$
$$= 1700 \text{ watts}$$

$$P_{min} = [I_{rms(min)} \sqrt{3}]^2 R_L$$
$$= 26 \text{ watts}$$

Basic Circuit Configurations—The simplest form of phase-controlled thyristor circuit is the half-wave power control as shown in Fig. 583.

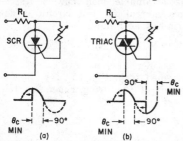

92CS-26289

Fig. 583—Degree of control over conduction angles when ac resistive network is used to trigger (a) SCR's and (b) triacs.

This circuit provides a simple, non-regulating half-wave power control that begins at the 90-degree conduction (peak-voltage) point and may be adjusted to within a few degrees of full conduction (180-degree half-cycle).

The half-wave proportional control shown in Fig. 584 is a non-regulating circuit whose function depends upon an RC delay network for gate phase-lag control. This circuit is better than simple resistance firing circuits because the phase-shifting characteristics of the RC network permit the firing of the SCR beyond the peak of the impressed voltage, resulting in small conduction angles. On the positive half-cycle of the applied voltage, capacitor C is charged through the network R_a and R_b. When the voltage across capacitor C exceeds the gate-firing voltage of the SCR, the SCR is turned on; during the remaining portion of the half-cycle, ac power is applied to the load.

The delay in firing the SCR depends upon the time-constant network (R_a, R_b, C) which produces a gate-firing voltage that is shifted in phase with respect to the supply voltage. The amount of phase shift is adjusted by R_b. With maximum

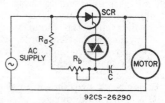

92CS-26290

Fig. 584—SCR half-wave proportional power control circuit.

resistance in the circuit, the RC time constant is longest. This condition results in a large phase shift with a correspondingly small conduction angle. With minimum resistance, the phase shift is small, and essentially the full line voltage is applied to the load.

The control circuit uses the breakdown voltage of a diac as a threshold setting for firing the SCR. The diac is specifically designed for handling the high-current pulses required to trigger SCR's. When the voltage across capacitor C reaches the breakdown voltage of the diac, it fires and C discharges through the diac to its maintaining voltage. At this point, the diac again reverts to its high-impedance state. The discharge of the capacitor from breakdown to maintaining voltage of the diac provides a current pulse of sufficient magnitude to fire the SCR. Once the SCR has fired, the voltage across the phase-shift network reduces to the forward voltage drop of the SCR for the remainder of the half-cycle.

Two SCR's are usually required to provide full-wave power control. Because of the bidirectional switching characteristics of triacs, however, only one of these devices is needed to provide the same type of control. Fig. 585 shows three full-wave power controls that employ thyristors.

In circuits of this type, a rapidly rising off-state voltage can occur across the thyristor when the device changes from a conducting state to a blocking state (commutates). The influence of this dv/dt stress on the operation of the power-control element is described below. Consideration is given only to those circuit applications that utilize a triac as the main power-control element.

The dv/dt stress in a circuit with a resistive load (such as those just described) can be illustrated by consideration of a circuit with a 6-ampere load that has a power factor close to unity. The load resistance in this circuit is 20 ohms for a source voltage of 120 volts. If the total circuit inductance is assumed to be 500 microhenries and the total triac and stray capacitance is 500 picofarads, the circuit factor for the conducting state is 0.99996, lagging. Thus, the load current lags the line voltage by the small phase delay of approximately 25 microseconds. At the time that the triac commutates current, the line voltage is 1.6 volts. At this time, a transient damped oscillation occurs as a result of the interaction of the triac junction capacitance and the circuit inductance. For the circuit parameter values given (R = 20 ohms, L = 500 microhenries, and C = 500 picofarads), the frequency of oscillation is 3.2×10^6 Hz. Calculation of the

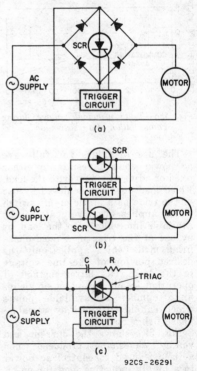

92CS-26291

Fig. 585—Full-wave thyristor motor control circuits using (a) bridge rectifier and a single SCR; (b) inverse parallel SCR's; (c) a triac.

maximum dv/dt stress across the triac yields a value of 1.97 volts per microsecond. The voltage at the time of commutation is then 1.6 volts, and the maximum commutating dv/dt becomes 3.15 volts per microsecond.

Thus, it can be seen that a definite dv/dt stress is imposed on the triac even when the load is primarily resistive. Because all resistive circuit configurations have some small inductance associated with them, a commutating dv/dt stress is produced in all resistive circuits. Fig. 586 shows a commutating dv/dt waveshape for a resistive load of 6 amperes in a 120-volt triac control circuit.

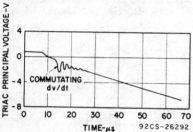

Fig. 586—*Triac principal voltage during commutation of a resistive load.*

The use of triacs for full-wave ac power control results in either fixed or adjustable power to the load. Fixed load power is achieved by use of the triac as a static on-off switch which applies effectively all of the available line voltage to the load, or by use of the triac in a fixed-phase firing mode which applies only the desired portion of the line voltage to the load. The latter method of operation is but one point of an infinite number of available points which can be attained by variable-phase firing operation.

Fig. 587 shows the current and voltage waveshapes produced when a triac is used to control ac power to a highly inductive load for on-off triac operation; Fig. 588 illustrates the waveshapes for phase-control operation. Because the load is highly

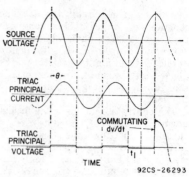

Fig. 587—*Principal voltage and current for static-switch triac operation with an inductive load.*

inductive ($\omega L \gg R$), the load current lags the line voltage by some phase angle θ. When the current through the triac (i.e., the load current) goes to zero (commutates), the triac turns off. In static control operation, the triac is immediately turned on by continuous application, or re-application, of the gate triggering signal; thus, this signal causes the triac to continue conducting for

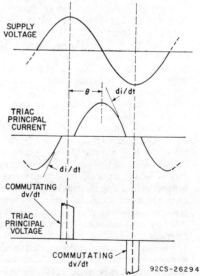

Fig. 588—*Principal voltage and current for phase-control triac operation with an inductive load.*

the desired number of successive half-cycles.

As shown in Fig. 587, at time t_1, the gate is opened and the triac continues to conduct for the remainder of that half-cycle of load current. At the end of the half-cycle, commutation occurs and the triac is subjected to an off-state blocking voltage which has a polarity opposite to the conducted current and a magnitude equal to the value of line voltage at that instant. Because the triac goes from a conducting state to a blocking state in a very short period of time, the rate of rise of off-state voltage is very rapid. This rapidly rising off-state voltage produces a dv/dt across the main power terminals of the triac and can result in the triac going into conduction if the triac is incapable of withstanding the dv/dt.

Fig. 588 shows the waveshapes produced for phase-control operation with an inductive load. The oscillations which are present on the peaks of the voltage waveform are the result of interaction of the circuit inductance and capacitance. For this type of operation, the stress caused by the commutating dv/dt is produced each time the current crosses the zero-axis and, therefore, occurs at a frequency equal to twice the line-voltage frequency. If the triac is incapable of sustaining the dv/dt which is produced, it goes into a conducting state and remains in continuous conduction, supplying current to the load. This malfunction is illustrated in Fig. 589.

Fig. 590(a) shows the circuit diagram of a series connection of voltage source, triac, and load. An equivalent circuit for this series connection is shown in Fig. 590(b). When the triac is in conduction, the triac junction capacitance is shunted by a low-value, nonlinear resistance which minimizes the effect of triac capacitance. However, when the triac goes out of conduction, the resistive component becomes very large and the equivalent triac shunting capa-

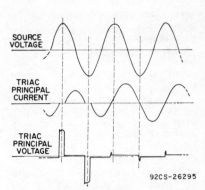

Fig. 589—Principal voltage and current showing malfunction of triac as a result of commutating dv/dt produced by an inductive load.

citance becomes significant. Because the circuit is basically a series RLC circuit, the voltage waveshape and the rate of rise of voltage across the triac at commutation are determined by the magnitude of source voltage and the circuit inductance, capacitance, and resistance. Thus the rising off-state voltage across the triac can be an overdamped, critically damped, or underdamped oscillation.

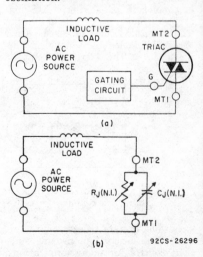

Fig. 590—(a) Series-circuit connection of triac, inductive load, and ac power source; and (b) equivalent circuit.

The increased complexity of aircraft control systems, and the need for greater reliability than electromechanical switching can offer, has led to the use of solid-state power switching in aircraft. Because 400-Hz power is used almost universally in aircraft systems, triacs employed for power switching and control in such systems must have a substantially higher commutating dv/dt capability than are those employed similarly in 60-Hz systems. (The increase in commutating dv/dt stresses on triacs with increases in frequency was explained previously in the section on **Thyristors**.) RCA offers an extensive line of triacs rated for 400-Hz applications.

Areas of application for 400-Hz triacs on aircraft include:

1. Heater controls for food-warming ovens and for windshield defrosters.

2. Lighting controls for instrument panels and cabin illumination.

3. Motor controls.

4. Solenoid controls.

5. Power supply switches

Fig. 591 shows a low-current triac in use in a simple, common, proportional-control application; the circuit consists of a single RC time constant and a threshold device. The trigger diac is used as a threshold device to remove the dependence of the trigger circuit on

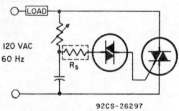

92CS-26297

Fig. 591—Simple control circuit using a single time constant.

variations in gate trigger characteristics. The circuit can provide sufficient control for many applications, such as heaters and motor-speed and switching controls. Because of its simplicity, the circuit can be packaged in confined areas where space is at a premium. Electrically, it displays a hysteresis effect and initially turns on for resistive loads with a conduction angle which may be too large; however, it provides maximum power output at the full "on" position of the control potentiometer.

The hysteresis effect produced by a single-time-constant circuit can be reduced by addition of a resistor (R_s) in series with the trigger diac, as shown by the dotted lines in Fig. 591. The series resistor reduces the capacitor discharge time and thus provides reduced time lag because of the diac turn-on-characteristics.

The circuit shown in Fig. 592 uses a double-time-constant control to improve on the performance of the single-time-constant control circuit.

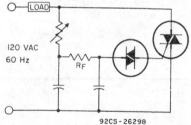

92CS-26298

Fig. 592—Control circuit using a double time constant.

This circuit minimizes the hysteresis effect and allows the triac to turn on at small conduction angles. The circuit has the advantages of low hysteresis, bidirectional operation at small conduction angles, and continuous control up to the maximum conduction angle. In addition, the fixed resistor R_f can be replaced by a trimmer potentiometer for minimum control at low conduction angles.

The circuit shown in Fig. 593 uses a neon bulb as a threshold device rather than the solid-state diac. This circuit has the advantages of low hysteresis, bidirectional operation at small conduction angles, and continuous control up to the maximum conduction angle. Because the neon-bulb threshold voltage is higher than that of a solid-stage diac, however, full 360-degree control may not be achieved.

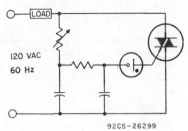

120 VAC
60 Hz

92CS-26299

Fig. 593—Control circuit using a neon-bulb threshold device.

General Application Considerations—Phase control of thyristor-and-diode combinations may be employed to provide many different ac and dc output waveforms to a load circuit. Some basic combinations, together with the corresponding voltage waveforms at the load for two complete cycles of operation, are shown in Fig. 594. In general, triac circuits are more economical for full-wave power control than are circuits that use two SCR's. For partial range control when the load is not sensitive to a nonsymmetrical waveform such as resistive loads, a control circuit that uses a diode and an SCR is acceptable.

Phase-control techniques can be used very effectively and efficiently to control ac input power in lamp-dimming, motor-speed-control, low-power electric-heating, and many other similar types of applications. Phase-control systems generate radio-frequency noise as a result of the random thyristor switching and often must include suppression circuits to minimize radio-frequency interference (RFI) in other electrical systems. In higher-power applications, the RFI is of such magnitude that the suppression circuits become excessively bulky and expensive.

Zero-Voltage Switching

Fig. 595 shows the relationship between line and load voltages for both SCR (half-wave) and triac (full-wave) power-control circuits that employ this control technique. With this type of control, the RFI associated with phase-control circuits is substantially decreased, or even eliminated, because thyristor switching occurs at or near the 0- or 180-degree (zero-voltage) points on the ac line voltage.

Fig. 596 shows a circuit in which an SCR controls the triggering and operation of a triac in an integral-cycle control circuit which is radio-frequency-interference free. A basic SCR gate-trigger or gate-control circuit can be represented by a voltage source and a series resistance, as shown in Fig. 597. The series resistance should include both the external circuit resistance and the internal generator resistance. With this type of equivalent circuit, the conventional load-line approach to gate trigger-circuit design can be used. With pulse-type triggering, it is assumed initially that the time required to trigger all SCR's of the same type is known, and that the maximum allowable gate trigger pulse widths for specific peak gate power inputs are to be determined. The magnitude of gate trigger current required to turn on an SCR of a given type can be determined from the turn-on characteristics shown in the section on **Thyristors**.

The triac in Fig. 596 is not triggered as long as the SCR is on. When the SCR is turned off by removal of the gate signal and application of a negative anode potential, the triac is triggered on at the beginning of the next half-cycle. When the triac conducts, the capacitor charges up to the peak supply voltage and retains its charge to trigger the triac on in the next half-cycle. When the triac conducts in the reverse direction, the negative charge on the capacitor is held to a low value so that it does not trigger the triac when the supply voltage reverses. If the SCR is still off, the triac repeats its conduction angle. If the SCR is con-

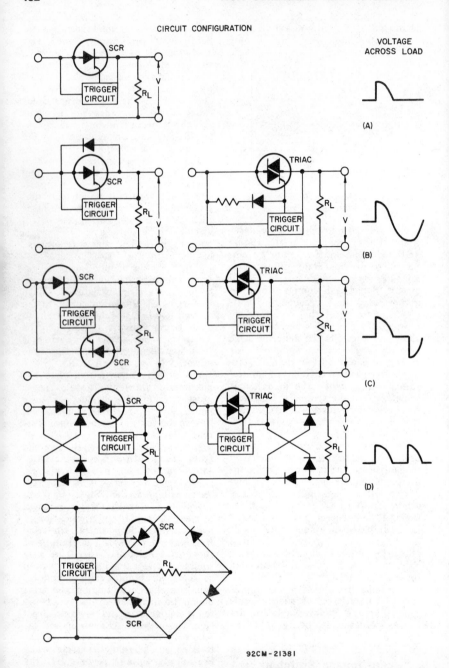

92CM-21381

Fig. 594—Basic circuit configurations for thyristor power controls and voltage wave-form across the load for two complete cycles of operation.

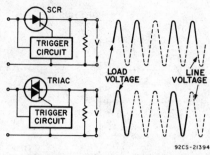

Fig. 595—Integral-cycle thyristor power control circuits.

ducting, the triac does not trigger on, but remains off until the SCR is again turned off. This circuit provides the unique function of integral-cycle switching, i.e., once the triac is triggered on, it completes one full cycle before turning off. This type of switching eliminates dc components present with half-wave control. The circuit also provides synchronous switching, i.e., the triac turns on at the beginning of the cycle and does not generate RFI.

In zero-voltage-switching controls, only two levels of input power are delivered to the load. The load receives the full amount of power for a period of time and zero power for a period of time; the average power delivered to the load, therefore, depends upon the ratio of the

power-on interval to the power-off interval.

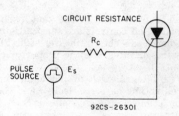

Fig. 597—Equivalent gate trigger circuit.

In solid-state power-control systems that employ zero-voltage-switching techniques, two modes of operation are possible. The controlled variable (for example, temperature in a heat-control system) may be sensed and used to turn the power on or off. Because the power-control element is a solid-state device and, therefore, is free of wear-out mechanisms, the differential in the control variable that causes the switching can be made very small, and accurate control is achieved. Fig. 598 shows the response characteristics for a heating system that uses this type of control.

In control systems that have large time constants, such as a home heating system, on-off controls of the type described above may produce relatively large overshoots and undershoots. In this type of system, better regulation may be achieved by use of a control method referred to as proportional integral-cycle control.

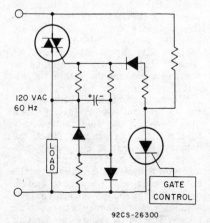

Fig. 596—Integral-cycle control circuit.

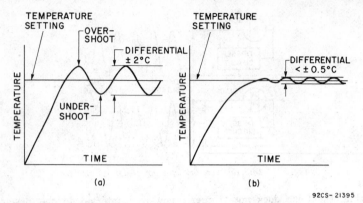

92CS-21395

Fig. 598—Transfer characteristics of (a) on-off and (b) proportional control systems.

The power excursions that result with on-off controls are substantially reduced, as shown in Fig. 598(b), by use of integral-cycle proportional control with synchronous switching. With integral-cycle proportional control, a time base is selected, and the "on" time of the thyristor is varied within the time base. The ratio of on-to-off times of the thyristor during this interval depends upon the amount of power to the load required to maintain a predetermined average level for the system. As this level is approached (as determined by a sensing element), less power is delivered to the load (i.e., the duty cycle is reduced). This type

of control is usually selected for heating systems.

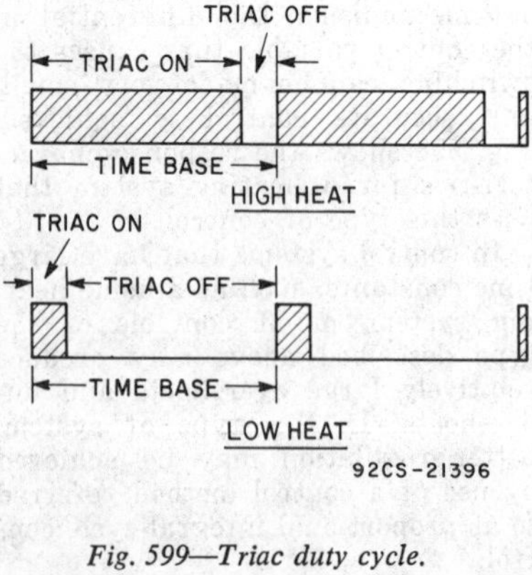

Fig. 599 shows the on-off-ratio of the triac. Within the time period, the on-time varies by an integral number of cycles from full on to a single cycle of input voltage.

One method of achieving integral-cycle proportional control is to use a fixed-frequency sawtooth generator signal which is summed with a dc control signal. The sawtooth generator establishes the period or time base of the system. The dc control is obtained from the output of the temperature-sensing network. The principle is illustrated in Fig. 600. As the sawtooth voltage increases, a level is reached which turns on power to the heating elements. As the temperature at the sensor changes, the dc level shifts accordingly and changes the length of time that the power is applied to the heating elements within the established time.

When the demand for heat is high, the dc control signal is high and high power is supplied continuously to the heating elements. When the demand for heat is completely satisfied, the dc control signal is low and low power is supplied to the heating elements. Usually a system using this principle operates continuously somewhere between full on and full off to satisfy the demand for heat.

TRIAC OFF

◄── TRIAC ON ──► │◄─►│

◄── TIME BASE ──►

HIGH HEAT

TRIAC ON

►│ │◄── TRIAC OFF ──►│

◄── TIME BASE ──►

LOW HEAT

92CS-21396

Fig. 599—Triac duty cycle.

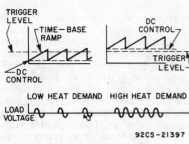

*Fig. 600—Proportional-controller wave
shapes.*

LAMP DIMMERS

A simple, inexpensive light-dimmer circuit can be constructed with a diac, a triac, and an RC charge-control network. It is important to remember that a triac in this type of circuit dissipates power at the rate of about one watt per ampere. Therefore, some means of removing heat must be provided to keep the device within its safe operating-temperature range. On a small light-control circuit such as one built into a lamp socket, the lead-in wire serves as an effective heat sink. Attachment of the triac case directly to one of the lead-in wires provides sufficient heat dissipation for operating currents up to 2 amperes (rms). On wall-mounted controls operating up to 6 amperes, the combination of faceplate and wallbox serves as an effective heat sink. For higher-power controls, however, the ordinary faceplate and wallbox do not provide sufficient heat-sink area. In this case, additional area may be obtained by use of a finned face plate that has a cover plate which stands out from the wall so air can circulate freely over the fins.

On wall-mounted controls, it is also important that the triac be electrically isolated from the face plate, but at the same time be in good thermal contact with it. Although the thermal conductivity of most electrical insulators is relatively low when compared with metals, a low-thermal-resistance, electrically

isolated bond of triac to faceplate can be obtained if the thickness of the insulator is minimized and the area for heat transfer through the insulator is maximized. Suitable insulating materials are fiberglass tape, ceramic sheet, mica, and polyimide film. Fig. 601 shows an example of isolated mounting for triacs in a plastic package. Electrical insulating tape is first placed over the inside of the faceplate. The triac is then mounted to the insulated faceplate by use of epoxy-resin cement.

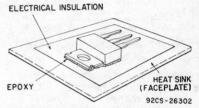

*Fig. 601—Example of isolated mounting
of a triac in a plastic package.*

Because the light output of an incandescent lamp depends upon the voltage impressed upon the lamp filament, changes in the lamp voltage vary the brightness of the lamp. When ac source voltages are used, a triac can be used in series with an incandescent lamp to vary the voltage to the lamp by changing its conduction angle; i.e., the portion of each half-cycle of ac line voltage in which the triac conducts to provide voltage to the lamp filament. The triac, therefore, is very attractive as a switching element in light-dimming applications.

To switch incandescent-lamp loads reliably, a triac must be able to withstand the inrush current of the lamp load. The inrush current is a result of the difference between the cold and hot resistance of the tungsten filament. The cold resistance of the tungsten filament is much lower than the hot resistance. The resulting inrush current is approximately 12 times the normal operating current of the lamp.

Single-Time-Constant Circuit

The simplest circuit that can be used for light-dimming applications is shown in Fig. 602. This circuit uses a diac in series with the gate of a triac to minimize the variations in

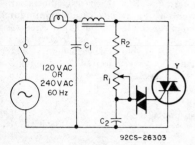

Fig. 602—Single-time-constant light-dimmer circuit.

gate trigger characteristics. Changes in the resistance in series with the capacitor change the conduction angle of the triac. Because of its simplicity, this circuit can be packaged in confined areas where space is at a premium.

The capacitor C_2 in the circuit of Fig. 602 is charged through the control potentiometer R_1 and the series resistance R_2. The series resistance is used to protect the potentiometer by limiting the capacitor charging current when the control potentiometer is at its minimum resistance setting. This resistor may be eliminated if the potentiometer can withstand the peak charging current until the triac turns on. The diac conducts when the voltage on the capacitor reaches its breakover voltage. The capacitor then discharges through the diac to produce a current pulse of sufficient amplitude and width to trigger the triac. Because the triac can be triggered with either polarity of gate signal, the same operation occurs on the opposite half-cycle of the applied voltage. The triac, therefore, is triggered and conducts on each half-cycle of the input supply voltage.

The interaction of the RC network and the trigger diode results in a

hysteresis effect when the triac is initially triggered at small conduction angles. The hysteresis effect is characterized by a difference in the control potentiometer setting when the triac is first triggered and when the circuit turns off. Fig. 603 shows the interaction between the RC network and the diac to produce the hysteresis effect. The capacitor voltage and the ac line voltage are shown as solid lines. As the resistance in the circuit is decreased from its maximum value, the capacitor voltage reaches a value which fires the diac. This point is designated A on the capacitor-voltage waveshape. When the diac fires, the capacitor discharges and triggers the triac at an initial conduction angle θ_1. During the forming of the gate trigger pulse, the capacitor voltage drops suddenly. The charge on the capacitor is smaller than when the diac did not conduct. As a result of the different voltage conditions on the capacitor, the breakover voltage of the diac is reached earlier in the next half-cycle. This point is labeled B on the capacitor-voltage waveform. The conduction angle θ_2 corresponding to point B is greater than θ_1. All succeeding conduction angles are equal to θ_2 in magnitude. When the circuit resistance is increased by a change

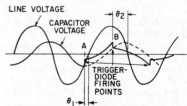

Fig. 603—Waveforms showing interaction of control network and trigger diode.

in the potentiometer setting, the triac is still triggered, but at a smaller conduction angle. Eventually, the resistance in series with the capacitance becomes so great that the voltage on the capacitor does not reach the breakover voltage of the diac. The circuit then turns off and does not turn on until the circuit resistance is again reduced to allow the diac to be fired. The hysteresis effect makes the voltage load appear much greater than would normally be expected when the circuit is initially turned on.

Double-Time-Constant Circuit

The double-time-constant circuit in Fig. 604 improves on the perform-

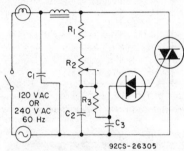

92CS-26305

Fig. 604—Double-time-constant light-dimmer circuit.

ance of the single-time-constant control circuit. This circuit uses an additional RC network to extend the phase angle so that the triac can be triggered at small conduction angles. The additional RC network also minimizes the hysteresis effect. Fig. 605 shows the voltage waveforms for the ac supply and the trigger capacitor of the circuit of Fig. 604. Because of the voltage drop across R_3, the input capacitor C_2 charges to a higher voltage than the trigger capacitor C_3. When the voltage on C_3 reaches the breakover voltage

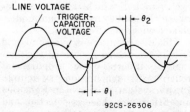

92CS-26306

Fig. 605—Voltage waveforms of double-time-constant control circuit.

of the diac, it conducts and causes the capacitor to discharge and produce the gate-current pulse to trigger the triac. After the diac turns off, the charge on C_3 is partially restored by the charge from the input capacitor C_2. The partial restoration of charge on C_3 results in better circuit performance with a minimum of hysteresis.

Zero-Voltage-Switched Circuit

Fig. 606 shows a lamp-dimmer circuit in which the use of an RCA-CA3059 integrated-circuit zero-voltage switch in conjunction with a 400-Hz triac results in minimum RFI. (The CA3059 is described briefly in the subsequent paragraph on **Heater Controls.** A detailed description of this integrated circuit is given in the manual on **RCA Linear Integrated Circuits,** Technical Series IC-42, in RCA Application Note ICAN-6182, or in the Technical Bulletin on the RCA CA3058, CA3059, and CA3079 Integrated-Circuit Zero-Voltage Switches. File No. 490.

Lamp dimming is a simple triac application that demonstrates an advantage of 400-Hz power over 60 Hz. Fig. 607 shows the adjustment of lamp intensity by phase control of the 60-Hz line voltage. Because RFI is generated by the step functions of power each half cycle, extensive filtering is required. Fig. 608 shows a means of controlling power to the lamp by the zero-voltage-switching technique. Use of 400-Hz power makes possible the elimination of complete or half cycles within a period (typically 17.5 milliseconds) without noticeable flicker. Fourteen different levels of lamp intensity can be obtained in this manner. In the circuit shown in Fig. 606, a line-synced ramp is set up with the desired period and applied to terminal No. 9 of the differential amplifier within the CA3059. The other side of the differential amplifier (terminal No. 13) uses a variable reference level, set by the potentiometer R_2.

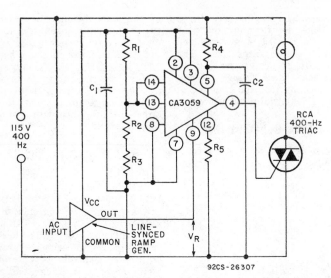

Fig. 606—Circuit diagram for 400-Hz zero-voltage-switched lamp dimmer.

A change of the potentiometer setting changes the lamp intensity.

In 400-Hz applications, it may be necessary to widen and shift the CA3059 output pulse (which is typically 12 microseconds wide and centered on zero voltage crossing) to assure that sufficient latching current is available. The resistor R_5 (terminal No. 12 to common) and the capacitor C_2 (terminal No. 5 to common) are used for this adjustment.

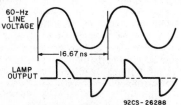

60-Hz PHASE-CONTROL LAMP DIMMER

Fig. 607—Waveforms for 60-Hz phase-controlled lamp dimmer.

HEAT CONTROLS

There are three general categories of solid-state control circuits for electric heating elements: on-off con-

trol, phase control, and proportional control using integral-cycle (zero-voltage) synchronous switching. Phase-control circuits such as those used for light dimming are very effective and efficient for electric heat control except for the problem of radio-frequency interference (RFI). In higher-power applications, the RFI is of such magnitude that suppression circuits to minimize the interference become quite bulky and expensive.

On-off controls have only two levels of power input to the load. The heating coils are either energized to full power or are at zero power. Because of thermal time constants, on-off controls produce a cyclic action which alternates between thermal overshoots and undershoots with poor resolution.

This disadvantage is overcome and RFI is minimized by use of the concept of integral-cycle proportional control with synchronous switching. In this system, as shown in Figs. 609 and 610, a time base is selected, and the on-time of the triac is varied within the time base. The ratio of

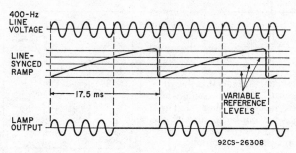

Fig. 608—Waveforms for 400-Hz zero-voltage-switched lamp dimmer.

the on-to-off time of the triac within this time interval depends upon the power required to the heating elements to maintain the desired temperature.

The RCA-CA3059 integrated-circuit zero-voltage switch is intended primarily as a trigger circuit for the control of thyristors and is particularly suited for use in thyristor temperature-control applications. This multistage circuit employs a diode limiter, a threshold detector, a differential amplifier, and a Darlington output driver to provide the basic switching action. The dc supply voltage for these stages is supplied by an internal zener-diode-regulated power supply that has sufficient current capability to drive external circuit elements, such as transistors and other integrated circuits. The

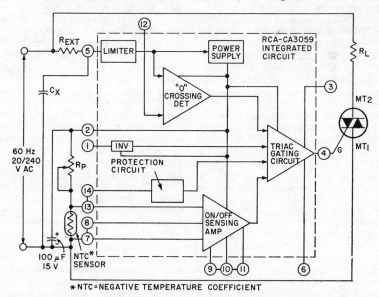

Note: Detailed descriptive information and the complete circuit diagram for the CA3059 are given in the RCA Linear Integrated Circuits Manual, Technical Series IC-42, or in RCA Application Note ICAN-6182 and the RCA Technical Bulletin on the CA3058, CA3059, and CA3079 Integrated-Circuit Zero-Voltage Switches, File No. 490.

Fig. 609—Functional block diagram of the integrated-circuit zero-voltage switch.

trigger pulse developed by this circuit can be applied directly to the gate of an SCR or a triac. A built-in protection circuit inhibits the application of these pulses to the thyristor gate circuit in the event that the external sensor for the integrated-circuit switch should be inadvertently opened or shorted. The CA3059 may be employed as either an on-off type of controller or a proportional controller, depending upon the degree of temperature regulation required.

Fig. 609 shows a functional block diagram of the CA3059 integrated-circuit zero-voltage switch. Any triac that is driven directly from the output terminal of this circuit should be characterized for operation in the I(+) or III(+) triggering modes, i.e., with positive gate current (current flows into the gate for both polarities of the applied ac voltage). The circuit operates directly from a 60-Hz ac line voltage of 120 or 240 volts.

The limiter stage of the CA3059 clips the incoming ac line voltage to approximately plus and minus 8 volts. This signal is then applied to the zero-voltage-crossing detector, which generates an output pulse during each passage of the line voltage through zero. The limiter output is also applied to a rectifying diode and an external capacitor that comprise the dc power supply. The power supply provides approximately 6 volts as the dc supply to the other stages of the CA3059. The on/off sensing amplifier is basically a differential comparator. The triac gating circuit contains a driver for direct triac triggering. The gating circuit is enabled when all the inputs are at a high voltage, i.e., the line voltage must be approximately zero volts, the sensing-amplifier output must be "high," the external voltage to terminal 1 must be a logical "1," and the output of the fail-safe circuit must be "high."

Fig. 610 shows the position and width of the pulses supplied to the gate of a thyristor with respect to

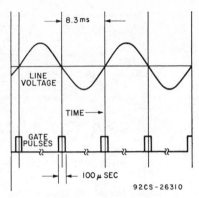

Fig. 610—Timing relationship between the output pulses of the CA3059 and the ac line voltage.

the incoming ac line voltage. The CA3059 can supply sufficient gate voltage and current to trigger most RCA thyristors at ambient temperatures of 25°C. However, under worst-case conditions (i.e., at ambient-temperature extremes and maximum trigger requirements), selection of the higher-current thyristors may be necessary for particular applications.

As shown in Fig. 609, when terminal 13 is connected to terminal 14, the fail-safe circuit of the CA3059 is operable. If the sensor should then be accidentally opened or shorted, power is removed from the load (i.e., the triac is turned off). The internal fail-safe circuit functions properly, however, only when the ratio of the sensor impedance at 25°C, if a thermistor is the sensor, to the impedance of the potentiometer R_p is less than 4 to 1.

On-Off Temperature Controller— Fig. 611 shows a triac and a CA3059 used in an on-off temperature-controller configuration. The triac is turned on at zero voltage whenever the voltage V_s exceeds the reference voltage V_r. The transfer characteristic of this system, shown in Fig. 612, indicates significant thermal

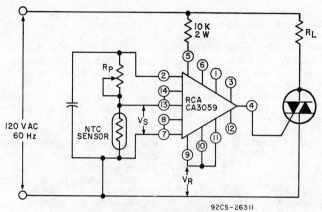

Fig. 611—On-off temperature controller.

overshoots and undershoots, a well-known characteristic of such a system. The differential or hysteresis of this system, however, can be further increased, if desired, by the addition of positive feedback.

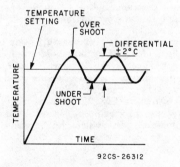

Fig. 612—Transfer characteristics of an on-off temperature-control system.

Proportional Temperature Controller—For precise temperature-control applications, the proportional-control technique with synchronous switching is employed. The transfer curve for this type of controller is shown in Fig. 613. In this case, the duty cycle of the power supplied to the load is varied with the demand for heat required and the thermal time constant (inertia)

of the system. For example, when the temperature setting is increased in an "on-off" type of controller, full power (100 per cent duty cycle) is supplied to the system. This effect results in significant temperature excursions **because there is no anticipatory circuit** to reduce the power gradually before the actual set temperature is achieved. However, in a proportional control technique, less power is supplied to the load (reduced duty cycle) as the error signal is reduced (sensed temperature approaches the set temperature).

Before such a system is implemented, a time base is chosen so that the on-time of the triac is varied within this time base. The

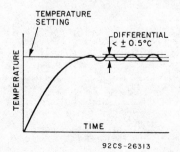

Fig. 613—Transfer characteristics of a proportional temperature-control system.

ratio of the on-to-off time of the triac within this time interval depends on the thermal time constant of the system and the selected temperature setting. Fig. 614 illustrates the principle of proportional control. For this operation, power is supplied to the load until the ramp voltage reaches a value greater than

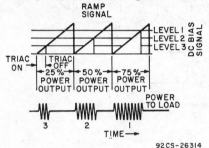

Fig. 614—Principles of proportional control.

the dc control signal supplied to the opposite side of the differential amplifier. The triac then remains off for the remainder of the time-base period. As a result, power is "proportioned" to the load in a direct relation to the heat demanded by the system.

For this application, a simple ramp generator can be realized with a minimum number of active and passive components. Exceptional ramp linearity is not necessary for

proportional operation because of the nonlinearity of the thermal system and the closed-loop type of control. In the circuit shown in Fig. 615, ramp voltage is generated when the capacitor C_2 charges through resistors R_4 and R_5. The time base of the ramp is determined by resistors R_1 and R_2, capacitor C_1, and the breakover voltage of the D2202U diac. When the voltage across C_1 reaches approximately 32 volts, the diac switches and turns on the 2N697S transistor. The capacitor C_2 then discharges through the collector-to-emitter junction of the transistor. This discharge time is the retrace or flyback time of the ramp. The circuit shown can generate ramp times ranging from 0.3 to 2.0 seconds through adjustment of R_2. For precise temperature regulations, the time base of the ramp should be shorter than the thermal time constant of the system, but long with respect to the period of the 60-Hz line voltage. Fig. 616 shows a triac and a CA3059 connected for the proportional mode.

Integral-Cycle Temperature Controllers (No half-cycling)—If a temperature controller which is completely devoid of half-cycling and hysteresis is required, then the circuit shown in Fig. 617 may be used. This type of circuit is essential for applications in which half-cycling

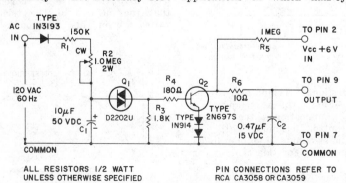

Fig. 615—Ramp generator.

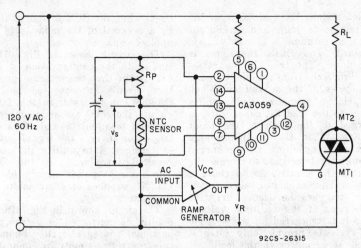

Fig. 616—*Proportional temperature controller.*

and the resultant dc component could cause overheating of a power transformer on the utility lines.

In the circuit shown in Fig. 616, the sensor is connected between terminals 7 and 9 of the CA3059. This arrangement is required because of the phase reversal introduced by the SCR. With this configuration, terminal 12 is connected to terminal 7 for operation of the CA3059 in the dc mode (however, the load is switched at zero voltage). Because the position of the sensor has been changed for this configuration, the internal fail-safe circuit cannot be used (terminals 13 and 14 are not connected).

In the integral-cycle controller, when the temperature being con-

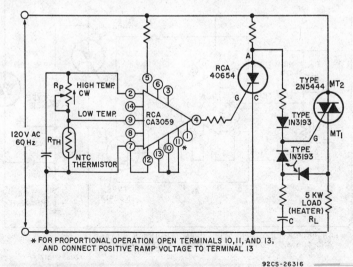

* FOR PROPORTIONAL OPERATION OPEN TERMINALS 10,11, AND 13, AND CONNECT POSITIVE RAMP VOLTAGE TO TERMINAL 13

Fig. 617—*Integral-cycle temperature controller in which half-cycling effect is eliminated.*

trolled is low, the resistance of the thermistor is high and an output signal at terminal 4 of zero volts is obtained. The SCR, therefore, is turned off. The triac is then triggered directly from the line on positive cycles of the ac voltage. When the triac is triggered and supplies power to the load R_L, capacitor C is charged to the peak of the input voltage. When the ac line swings negative, capacitor C discharges through the triac gate to trigger the triac on the negative half-cycle. The diode-resistor-capacitor "slaving network" triggers the triac on negative half-cycles of the ac input voltage after it is triggered on the positive half-cycle to provide only **integral cycles** of ac power to the load.

When the temperature being controlled reaches the desired value, as determined by the thermistor, then a positive voltage level appears at terminal 4 of the CA3059. The SCR then starts to conduct at the beginning of the positive input cycle to shunt the trigger current away from the gate of the triac. The triac is then turned off. The cycle re-

peats when the SCR is again turned by a reversal of the polarity of the applied voltage.

The circuit shown in Fig. 618 is similar to the configuration in Fig. 617 except that the internal protection circuit incorporated in the CA3059 can be used. In this latter circuit, the NTC sensor is connected between terminals 7 and 13, and a transistor inverts the signal output at terminal 4 to nullify the phase reversal introduced by the SCR. The internal power supply of the CA3059 supplies bias current to the transistor.

The circuit shown in Fig. 618 can readily be converted to a true proportional integral-cycle temperature controller simply by connection of a positive-going ramp voltage to terminal 9 (with terminals 10 and 11 open).

For applications that require complete elimination of half-cycling without the addition of hysteresis, the circuit shown in Fig. 619 may be employed. This circuit uses a CA3099E integrated-circuit programmable comparator with a zero-

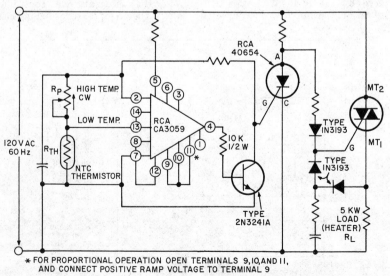

★ FOR PROPORTIONAL OPERATION OPEN TERMINALS 9, 10, AND 11 AND CONNECT POSITIVE RAMP VOLTAGE TO TERMINAL 9

92CS-26317

Fig. 618—CA3059 integral-cycle temperature controller that features a protection circuit and no half-cycling effect.

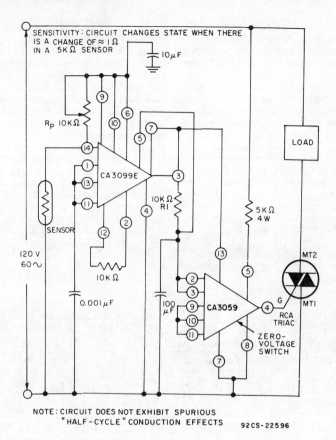

Fig. 619—Sensitive temperature control.

voltage switch. The CA3099E was described earlier in the section **Integrated Circuits for Linear Applications**. Because the CA3099E contains an integral flip-flop, its output will be in either a "0" or "1" state. Consequently the zero-voltage switch cannot operate in the linear mode, and spurious half-cycling operation is prevented. When the signal-input voltage at terminal 14 of the CA3099E is equal to or less than the "low" reference voltage (LR), current flows from the power supply through resistor R_1, and a logic "0" is applied to terminal 13 of the zero-voltage switch. This condition turns off the triac. The triac remains off until the signal-input voltage rises to or exceeds the "high" reference voltage (HR), thereby effecting a change in the state of the flip-flop so that a logic "1" is applied to terminal 13 of the zero-voltage switch, and triggers the triac on.

LIGHT-ACTIVATED CONTROLS

The RCA CA3062 integrated-circuit photosensitive detector and power amplifier can provide a light-activated switch which will drive a variety of practical loads; such as solenoids, relays, triacs, and SCR's.

"Normally on" and "normally off" outputs are available simultaneously.

The circuit diagram and terminal connections for the CA3062 are shown in Fig. 620. The circuit consists of a photo-Darlington pair and a differential amplifier which is emitter-follower coupled to a pair of high-current output transistors, Q_6 and Q_7. The CA3062 is designed for operation from power supply voltages of 5 to 15 volts between terminal Nos. 4 and 8, and voltages as high as 30 volts V+ on the output transistors.

as 10:1 can be achieved with this circuit.

Fig. 622 shows the use of the CA3062 and two triacs in an automatic shut-off and alarm circuit. In this system, ac is supplied to the load as long as the light source is "on". If the light path to the CA3062 is broken, then the ac to the load and light source is opened, thereby activating the alarm circuit. The system can be reset with the push-button shown.

Fig. 623 shows an intrusion-alarm system that uses a triac and the

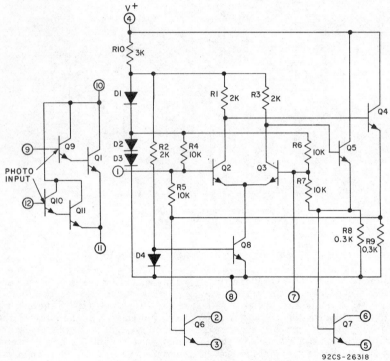

Fig. 620—Schematic diagram of CA3062.

Fig. 621 illustrates the use of the CA3062 and a triac in an optically coupled isolator circuit that is used to transfer signals that are at substantially different voltage levels. Both polarity outputs are available. Current transfer ratios of as high

CA3062. If the light path is broken or the ac is interrupted, the alarm system will be activated, provided the battery is adequately charged.

The V+ acts as a charging circuit for the battery while the circuit is operating from the ac supply.

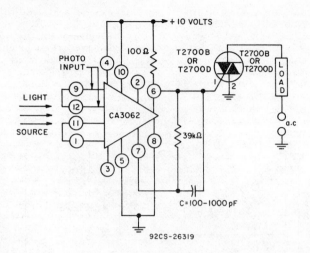

Fig. 621—Light activated triac control.

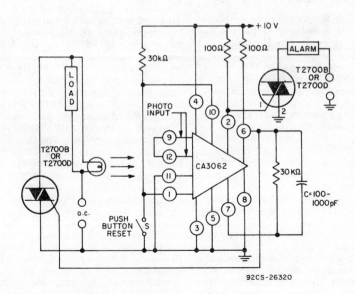

Fig. 622—Triac automatic shunt off and alarm.

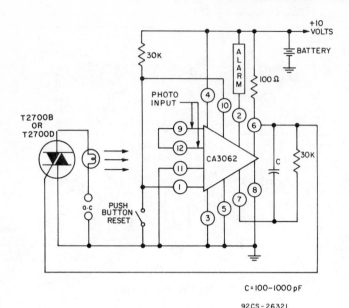

C = 100–1000 pF

92CS-26321

Fig. 623—Triac intrusion alarm system.

TRAFFIC-LIGHT FLASHER

Another application which illustrates the versatility of the CA3059 zero-voltage switch, when used with RCA thyristors, involves switching traffic-control lamps. In this type of application, it is essential that a triac withstand a current surge of the lamp load on a continuous basis. This surge results from the difference between the cold and hot resistance of the tungsten filament. If it is assumed that triac turn-on is at 90 degrees from the zero-voltage crossing, the first current-surge peak is approximately ten times the peak steady-state value or fifteen times the steady-state rms value. The second current-surge peak is approximately four times the steady-state rms value.

When the triac randomly switches the lamp, the rate of current rise di/dt is limited only by the source inductance. The triac di/dt rating may be exceeded in some power systems. In many cases, exceeding the

rating results in excessive current concentrations in a small area of the device which may produce a hot spot and lead to device failure. Critical applications of this nature require adequate drive to the triac gate for fast turn-on. In this case, some inductance may be required in the load circuit to reduce the initial magnitude of the load current when the triac is passing through the active region. Another method may be used which involves the switching of the triac at zero line voltage. This method involves the supply of pulses to the triac gate only during the presence of zero voltage on the ac line.

Fig. 624 shows a circuit in which the lamp loads are switched at zero line voltage. This approach reduces the initial di/dt, decreases the required triac surge-current ratings, increases the operating lamp life, and eliminates RFI problems. This circuit consists of two triacs, a flip-flop (FF-1), the zero-voltage switch, and a diac pulse generator. The flashing rate in this circuit is con-

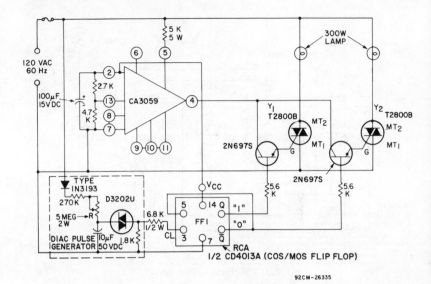

Fig. 624—Synchronous-switching traffic flasher.

trolled by potentiometer R, which provides between 10 and 120 flashes per minute. The state of FF-1 determines the triggering of triacs Y_1 or Y_2 by the output pulses at terminal 4 generated by the zero-crossing circuit. Transistors Q_1 and Q_2 inhibit these pulses to the gates of the triacs until the triacs turn on by the logical "1" (V_{cc} high) state of the flip-flop.

The arrangement described can also be used for a synchronous, sequential traffic-controller system by addition of one triac, one gating transistor, a "divide-by-three" logic circuit, and modification in the design of the diac pulse generator. Such a system can control the familiar red, amber, and green traffic signals that are found at many intersections.

Fig. 625 shows a simplified version of the synchronous-switching traffic light flasher shown in Fig. 624. Flash rate is set by use of the curve shown in Fig. 626. If a more precise

flash rate is required, the ramp generator described previously may be used. In this circuit, ZVS_1 is the master control unit and ZVS_2 is slaved to the output of ZVS_1 through its inhibit terminal (terminal 1). When power is applied to lamp No. 1, the voltage of terminal 6 on ZVS_1 is high and ZVS_2 is inhibited by the current in R_x. When lamp No. 1 is off, ZVS_2 is not inhibited, and triac Y_2 can fire. The power supplies operate in parallel. The on-off sensing amplifier in ZVS_2 is not used.

MOTOR CONTROLS

Triacs and SCR's can be used very effectively to apply power to motors and perform switching, or any other desired operating condition that can be obtained by a switching action. Because most motors are line-operated, the triac can be used as a direct replacement for electromechanical switches.

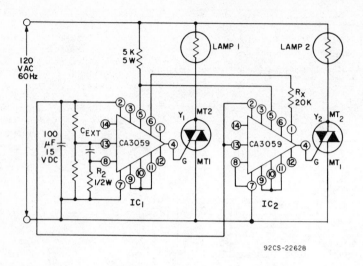

Fig. 625—Synchronous light flasher.

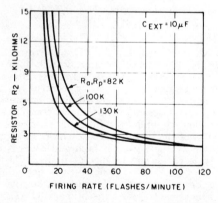

Fig. 626—Effect of variations in time-constant elements on period.

Motor Switch Circuits

A very simple triac static switch for control of ac motors is shown in Fig. 627. The low-current switch controlling the gate trigger current can be any type of transducer, such as a pressure switch, a thermal switch, a photocell, or a magnetic reed relay. This simple type of circuit allows the motor to be switched directly from the transducer switch without any intermediate power switch or relay.

Triacs can also be used to change the operating characteristics of motors to obtain many different speed and torque curves.

For dc control, the circuit of Fig. 628 can be used. By use of the dc triggering modes, the triac can be directly triggered from transistor circuits by either a pulse or continuous signal.

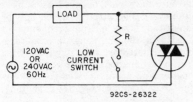

Fig. 627—Simple triac static switch.

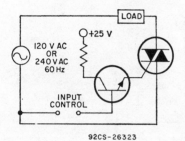

Fig. 628—AC triac switch control from dc input.

Induction-Motor Controls

Fig. 629 shows a single-time-constant circuit which can be used as a satisfactory proportional speed control for some applications and with certain types of induction motors, such as shaded pole or permanent split-capacitor motors, when the load is fixed. This type of circuit is best suited to applications which require speed control in the medium to full-power range. It is specifically useful in applications such as fans or blower-motor controls, where a small change in motor speed produces a large change in air velocity. Caution must be exercised if this type of circuit is used with induction motors because the motor may stall suddenly if the speed of the motor is reduced below the drop-out speed for the specific operating condition determined by the conduction angle of the triac. Because the single-time-constant circuit cannot provide speed control of an induction motor load from maximum power to full off, but only down to some fraction of the full-power speed, the effects of hysteresis described previously are not present. Speed ratios as high as 3:1 can be obtained from the single-time-constant circuit used with certain types of induction motors.

Because motors are basically inductive loads and because the triac turns off when the current reduces to zero, the phase difference between the applied voltage and the device current causes the triac to turn off when the source voltage is at a value other than zero. When the triac turns off, the instantaneous

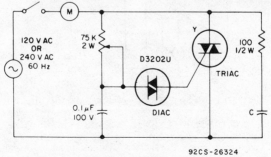

	120VAC, 60Hz	240VAC, 60Hz
C	0.22μF, 200V	0.22μF, 400V
Y	T2700B	T2700D

Fig. 629—Induction motor control.

value of input voltage is applied directly to the main terminals of the triac. This commutating voltage may have a rate of rise whch can re-trigger the triac. The commutating dv/dt can be limited to the capability of the triac by use of an RC net-work across the device, as shown in Fig. 629.

Fig. 630 illustrates the use of a CA3059 integrated-circuit zero-volt-age switch, together with two CA3086 integrated-circuit transistor arrays, in a phase-control circuit that is specifically designed for speed control of ac induction motors. (This circuit may also be used as a light dimmer.) The circuit, which can be operated from a line frequency of 50-Hz to 400-Hz, consists of a zero-voltage detector, a line-synchronized ramp generator, a zero-current de-tector, and a line-derived control cir-cuit (i.e., the zero-voltage switch).

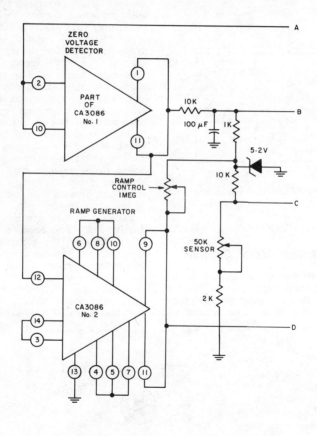

92CM-26325

Fig. 630—Phase-control using a CA3059 and two CA3086 integrated-circuits (continued on page 503).

The zero-voltage detector (part of CA3086 No. 1) and the ramp generator (CA3086 No. 2) provide a line-synchronized ramp-voltage output to terminal 13 of the zero-voltage switch. The ramp voltage, which has a starting voltage of 1.8 volts, starts to rise after the line voltage passes the zero point. The ramp generator has an oscillation frequency of twice the incoming line frequency. The slope of the ramp voltage can be adjusted by variation of the resistance of the 1-megohm ramp-control potentiometer. The output phase can be controlled easily to provide 180° firing of the triac by programming the voltage at terminal 9 of the zero-voltage switch. The basic operation of the zero-voltage switch driving a thyristor was explained previously in the discussion on **Heat Controls**.

Reversing Motor Control—In many

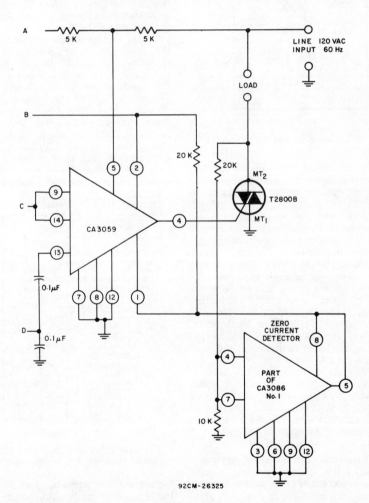

92CM-26325

Fig. 630—Phase-control using a CA3059 and two CA3086 integrated-circuits (continued from page 502).

industrial applications, it is necessary to reverse the direction of a motor, either manually or by means of an auxiliary circuit. Fig. 631

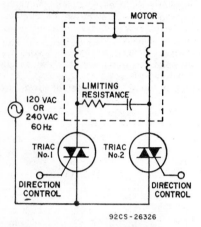

Fig. 631—Reversing motor control.

shows a circuit which uses two triacs to provide this type of reversing motor control for a split-phase capacitance motor. The reversing switch can be either a manual switch or an electronic switch used with some type of sensor to reverse the direction of the motor. A resistance is added in series with the capacitor to limit capacitor discharge current to a safe value whenever both triacs are conducting simultaneously. If triac No.1 is turned on while triac No.2 is on, a loop current resulting from capacitor discharge will occur and may damage the triacs.

The circuit operates as follows: when triac No.1 is in the off state, motor direction is controlled by triac No.2; when triac No.2 reverts to the off state and triac No.1 turns on, the motor direction is reversed.

Electronic Garage-Door System— The triac motor-reversing circuit can be extended to electronic garage-door systems which use the principle for garage-door direction control. The system contains a

transmitter and a receiver and provides remote control of door opening and closing. The block diagram in Fig. 632 shows the functions required for a complete solid-state system. When the garage door is closed, the gate drive to the DOWN triac is disabled by the lower-limit closure and the gate drive to the UP triac is inactive because of the state of the flip-flop. If the transmitter is momentarily keyed, the receiver activates the time-delay monostable multivibrator so that it then changes the flip-flop state and provides continuous gate drive to the UP triac. The door then continues to travel in the UP direction until the upper-limit switch closure disables gate drive to the UP triac. A second keying of the transmitter provides the DOWN triac with gate drive and causes the door to travel in the DOWN direction until the gate drive is disabled by the lower limit closure. The time in which the monostable multivibrator is active should override normal transmitter

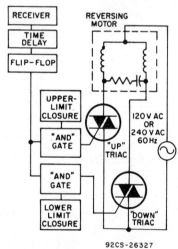

Fig. 632—Block diagram for remote-control solid-state garage-door system.

keying for the purpose of eliminating erroneous firing. A feature of this system is that, during travel, transmitter keying provides motor reversing independent of the upper- or lower-limit closures. Additional features, such as obstacle clearance, manual control, or time delay for overhead garage lights can be included very economically.

Speed Controller System for a DC Motor

Fig. 633 illustrates the use of the CA3094 in a motor-speed controller system. Circuitry associated with rectifiers D_1 and D_2 comprises a full-wave rectifier which develops a train of half-sinusoid voltage pulses to power the dc motor. The motor speed depends on the peak value of the half-sinusoids and the period of time (during each half-cycle) the SCR is conductive. The SCR conduction, in turn, is controlled by the time duration of the positive signal supplied to the SCR by the phase comparator.

that of a fixed-amplitude ramp wave generated synchronously with the ac-line-voltage frequency. The comparator output at terminal 6 is "high" (to trigger the SCR into conduction) during the period when the ramp potential is less than that of the error voltage on terminal 3. The motor-current conduction period is increased as the error voltage at terminal 3 is increased in the positive direction. Motor-speed accuracy of ± 1 per cent is easily obtained with this system.

Motor-Speed Error Detector—Fig. 634(a) shows a motor-speed error detector suitable for use with the circuit of Fig. 633. A CA3080 operational transconductance amplifier is used as a voltage comparator. The reference for the comparator is established by setting the potentiometer R so that the voltage at terminal 3 is more positive than that at termi-

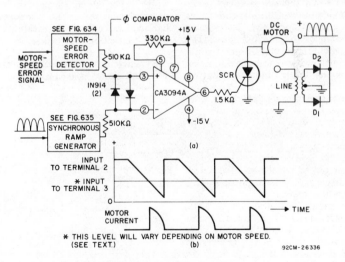

Fig. 633—Motor-speed controller system.

The magnitude of the positive dc voltage supplied to terminal 3 of the phase comparator depends on motor-speed error as detected by a circuit such as that shown in Fig. 634. This dc voltage is compared to

nal 2 when the motor speed is too low. An error voltage E_1 is derived from a tachometer driven by the motor. When the motor speed is too low, the voltage at terminal 2 of the voltage comparator is less positive

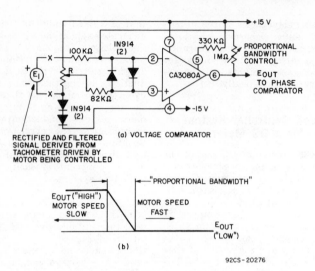

Fig. 634—Motor-speed error detector.

than that at terminal 3, and the output voltage at terminal 6 goes "high". When the motor speed is too high, the opposite input conditions exist, and the output voltage at terminal 6 goes "low". Fig. 634(b) also shows these conditions graphically, with a linear transition region between the "high" and "low" output levels. This linear transition region is known as "proportional bandwidth". The slope of this region is determined by the proportional bandwidth control to establish the error-correction response time.

Synchronous Ramp Generator— Fig. 635 shows a schematic diagram

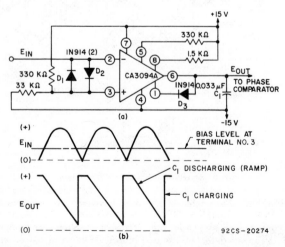

Fig. 635—Synchronous ramp generator with input and output waveforms.

and signal waveforms for a synchronous ramp generator suitable for use with the motor-controller circuit shown in Fig. 633. Terminal 3 is biased at approximately +2.7 volts (above the negative supply voltage). The input signal E_{IN} at terminal 2 is a sample of the half-sinusoids (at line frequency) used to power the motor in Fig. 633. A synchronous ramp signal is produced by using the CA3094 to charge and discharge capacitor C_1 in response to the synchronous toggling of E_{IN}. The charging current for C_1 is supplied by terminal 6. When terminal 2 swings more positive than terminal 3, capacitor C_1 discharges linearly through the external diode D_3 and the CA3094 to produce the ramp wave. The E_{out} signal is supplied to the phase comparator shown in Fig. 633.

Speed Controls for Universal Motors

Many fractional-horsepower motors are series-wound "universal" motors, so named because of their ability to operate directly from either ac or dc power sources. Fig. 636 is a schematic of this type of motor operated from an ac supply. Because most domestic applications today require 60-Hz power, universal motors are usually designed to have optimum performance characteristics at this frequency. Most universal motors run faster at a given dc voltage than at the same 60-Hz ac voltage.

The field winding of a universal motor, whether distributed or lumped (salient pole), is in series with the armature and external circuit, as shown in Fig. 636. The current through the field winding produces a magnetic field which cuts across the armature conductors. The action of this field in opposition to the field set up by the armature current subjects the individual conductors to a lateral thrust which results in armature rotation.

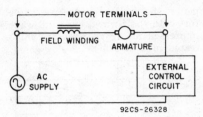

Fig. 636—Series-wound universal motor.

AC operation of a universal motor is possible because of the nature of its electrical connections. As the ac source voltage reverses every half-cycle, the magnetic field produced by the field winding reverses its direction simultaneously. Because the armature windings are in series with the field windings through the brushes and commutating segments, the current through the armature winding also reverses. Because both the magnetic field and armature current are reversed, the direction of the lateral thrust on the armature windings remains constant. Typical performance characteristic curves for a universal motor are shown in Fig. 637.

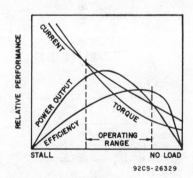

Fig. 637—Typical performance curves for a universal motor.

One of the simplest and most efficient means of varying the impressed voltage to a load on an ac power system is by control of the conduction angle of a thyristor placed in series with the load. Typical curves showing the variation of motor speed with conduction angle for both half-wave and full-wave impressed motor voltages are illustrated in Fig. 638.

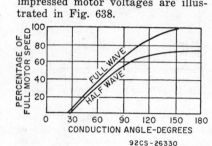

92CS-26330

Fig. 638—Typical performance curves for a universal motor with phase-angle control.

Half-Wave Control—There are many good circuits available for half-wave control of universal motors. The circuits are divided into two classes: regulating and non-regulating. Regulation in this instance implies load sensing and compensation of the system to prevent changes in motor speed.

The half-wave proportional control circuit shown in Fig. 639 is a

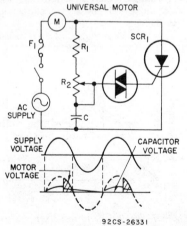

92CS-26331

Fig. 639—Half-wave motor control with no regulation.

non-regulating circuit that depends upon an RC delay network for gate phase-lag control. This circuit is better than simple resistance firing circuits because the phase-shifting characteristics of the RC network

permit the firing of the SCR beyond the peak of the impressed voltage, resulting in small conduction angles and very slow speed.

Fig. 640 shows a fundamental circuit of direct-coupled SCR control with voltage feedback. This circuit is highly effective for speed control of universal motors. The circuit makes use of the counter emf induced in the rotating armature because of the residual magnetism in the motor on the half-cycle when the SCR is blocking.

The counter emf is a function of speed and, therefore, can be used as an indication of speed changes as mechanical load varies. The gate-firing circuit is a resistance network consisting of R_1 and R_2. During the positive half-cycle of the source voltage, a fraction of the voltage is developed at the center-tap of the potentiometer and is compared with the counter emf developed in the rotating armature of the motor. When the bias developed at the gate of the SCR from the potentiometer exceeds the counter emf of the motor, the SCR fires. AC power is then applied to the motor for the remaining portion of the positive half-cycle. Speed control is accomplished by adjustment of potentiometer R_1. If the SCR is fired early in the cycle, the motor operates at high speed because essentially the full rated line voltage is applied to the motor. If the SCR is fired later in the cycle, the average value of voltage applied to the motor is reduced, and a corresponding reduction in motor speed occurs. On the negative half-cycle, the SCR blocks voltage to the motor. The voltage applied to the gate of the SCR is a sine wave because it is derived from the sine-wave line voltage. The minimum conduction angle occurs at the peak of the sine wave and is restricted to 90 degrees. Increasing conduction angles occur when the gate bias to the SCR is increased to allow firing at voltage values which are less than the peak value.

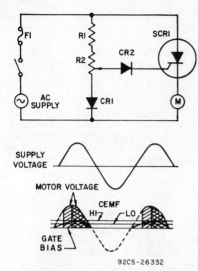

92CS-26332

Fig. 640—Half-wave motor control with regulation.

At no load and low speed, skip cycling operation occurs. This type of operation results in erratic motor speeds. Because no counter EMF is induced in the armature when the motor is standing still, the SCR will fire at low bias-potentiometer settings and causes the motor to accelerate to a point at which the counter emf induced in the rotating armature exceeds the gate firing bias of the SCR and prevents the SCR from firing. The SCR is not able to fire again until the speed of the motor has reduced, as a result of friction losses, to a value at which the induced voltage in the rotating armature is less than the gate bias. At this time the SCR fires again. Because the motor deceleration occurs over a number of cycles, there is no voltage applied to the motor; hence, the term skip-cycling.

When a load is applied to the motor, the motor speed decreases and thus reduces the counter emf induced in the rotating armature. With a reduced counter emf, the SCR fires earlier in the cycle and provides increased motor torque to the load. Fig. 640 also shows variations of

conduction angle with changes in counter emf. The counter emf appears as a constant voltage at the motor terminals when the SCR is blocking.

If a universal motor is operated at low speed under a heavy mechanical load, it may stall and cause heavy current flow through the SCR. For this reason, low-speed heavy-load conditions should be allowed to exist for only a few seconds to prevent possible circuit damage. In any case, fuse ratings should be carefully determined and observed.

Nameplate data for some universal motors are given in developed horsepower to the load. This mechanical designation can be converted into its electrical current equivalent through the following procedure. Internal motor losses are taken into consideration by assigning a figure of merit. This figure, 0.5, represents motor operation at 50-percent efficiency, and indicates that the power input to the motor is twice the power delivered to the load. With this figure of merit and the input voltage V_{ac}, the rms input current to the motor can be calculated as follows:

$$\text{rms current} = \frac{\text{mechanical horsepower} \times 746}{0.5 \ V_{ac}}$$

For an input voltage of 120 volts, the rms input current becomes

$$\text{rms current} = \text{horsepower} \times 12.4$$

For an input voltage of 240 volts, the rms input current becomes

$$\text{rms current} = \text{horsepower} \times 6.2$$

The motor-control circuits described above should not be used with universal motors that have calculated rms current exceeding the values given. The circuits will accommodate universal motors with ratings up to ¾ horsepower at 120 volts input and up to 1½ horsepower at 240 volts input.

Full-Wave Control—In applications in which the hysteresis effect can be tolerated or which require speed control primarily in the medium to full-power range, a single-time-constant circuit such as that shown in Fig. 629 for induction motors can also be used for universal motors. However, it is usually desirable to extend the range of speed control from full-power on to very low conduction angles. The double-time-constant circuit shown in Fig. 641 pro-

In the power-control circuits described, the integrated-circuit zero-voltage switch is used as the trigger circuit for the power triacs. The following conditions are also imposed in the design of the triac control circuits:

1. The load should be connected in a three-wire configuration with the triacs placed external to the load; either delta or wye arrangements may be used. Four-wire loads in wye configurations can be handled as

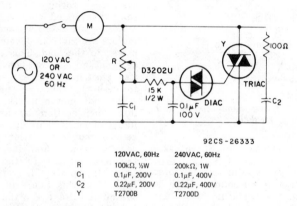

92CS-26333

	120VAC, 60Hz	240VAC, 60Hz
R	100kΩ, ½W	200kΩ, 1W
C_1	0.1µF, 200V	0.1µF, 400V
C_2	0.22µF, 200V	0.22µF, 400V
Y	T2700B	T2700D

Fig. 641—Universal motor speed control.

vides the delay necessary to trigger the triac at very low conduction angles with a minimum of hysteresis, and also provides practically full power to the load at the minimum-resistance position of the control potentiometer. When this type of control circuit is used, an infinite range of motor speeds can be obtained from very low to full-power speeds.

TRIAC POWER CONTROLS FOR THREE-PHASE SYSTEMS

This section describes recommended configurations for power-control circuits intended for use with both inductive and resistive balanced three-phase loads. The specific design requirements for each type of loading condition are discussed.

three independent single-phase systems. Delta configurations in which a triac is connected within each phase rather than in the incoming lines can also be handled as three independent single-phase systems.

2. Only one logic command signal is available for the control circuits. This signal must be electrically isolated from the three-phase power system.

3. Three separate triac gating signals are required.

4. For operation with resistive loads, the zero-voltage switching technique should be used to minimize any radio-frequency interference (RFI) that may be generated.

Isolation of DC Logic Circuitry

As explained earlier, isolation of

the dc logic circuitry * from the ac line, the triac, and the load circuit is often desirable even in many single-phase power-control applications. In control circuits for polyphase power systems, however, this type of isola-

* The dc logic circuitry provides the low-level electrical signal that dictates the state of the load. For temperature controls, the dc logic circuitry includes a temperature sensor for feedback. The RCA integrated-circuit zero-voltage switch, when operated in the dc mode with some additional circuitry, can replace the dc logic circuitry for temperature controls.

tion is essential, because the common point of the dc logic circuitry cannot be referenced to a common line in all phases.

In the three-phase circuits described in this section, photo-optic techniques (i.e., photo-coupled isolators) are used to provide the electrical isolation of the dc logic command signal from the ac circuits and the load. The photo-coupled isolators consist of an infrared light-emitting diode aimed at a silicon photo transistor, coupled in a common package. The light-emitting diode is the input section, and the

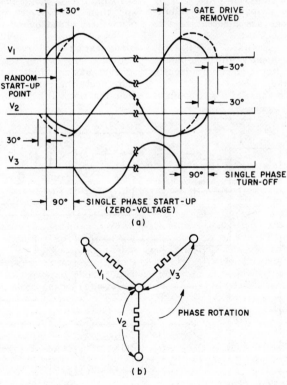

92CS-22636

Fig. 642—Voltage phase relationship for a three-phase resistive load when the application of load power is controlled by zero-voltage switching: (a) voltage waveforms, (b) load-circuit orientation of voltages. (The dashed lines indicate the normal relationship of the phases under steady-state conditions. The deviation at start-up and turn-off should be noted.)

photo transistor is the output section. The two components provide a voltage isolation typically of 1500 volts. Other isolation techniques, such as pulse transformers, magnetoresistors, or reed relays, can also be used with some circuit modifications.

Resistive Loads

Fig. 642 (shown on page 511) illustrates the basic phase relationships of a balanced three-phase resistive load, such as may be used in heater applications, in which the application of load power is controlled by zero-voltage switching. The following conditions are inherent in this type of application:

1. The phases are 120 degrees apart; consequently, all three phases cannot be switched on simultaneously at zero voltage.

2. A single phase of a wye configuration type of three-wire system cannot be turned on.

3. Two phases must be turned on for initial starting of the system. These two phases form a single-phase circuit which is out of phase with both of its component phases. The single-phase circuit leads one phase by 30 degrees and lags the other phase by 30 degrees.

These conditions indicate that in order to maintain a system in which no appreciable RFI is generated by the switching action from initial starting through the steady-state operating condition, the system must first be turned on, by zero-voltage switching, as a single-phase circuit and then must revert to synchronous three-phase operation.

Fig. 643 shows a simplified circuit configuration of a three-phase heater control that employs zero-voltage synchronous switching in the steady-state operating condition, with random starting. In this system, the

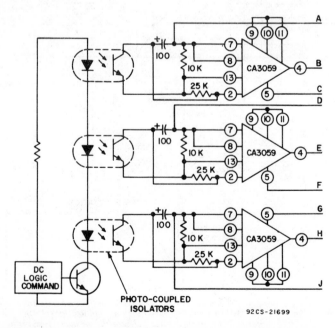

Fig. 643—Simplified diagram of a three-phase heater control that employs zero-voltage synchronous switching in the steady-state operating conditions (continued on page 513).

logic command to turn on the system is given when heat is required, and the command to turn off the system is given when heat is not required. Time proportioning heat control is also possible through the use of logic commands.

The three photo-coupled inputs to the three zero-voltage switches change state simultaneously in response to a "logic command". The zero-voltage switches then provide a positive pulse, approximately 100 microseconds in duration, only at a zero-voltage crossing relative to their particular phase. A balanced three-phase sensing circuit is set up with the three zero-voltage switches each connected to a particular phase on their common side (terminal 7) and referenced at their high side (terminal 5), through the current-limiting resistors R_4, R_5, and R_6, to an established artificial neutral point. This artificial neutral point is elec-

trically equivalent to the inaccessible neutral point of the wye type of three-wire load and, therefore, is used to establish the desired phase relationships. The same artificial neutral point is also used to establish the proper phase relationships for a delta type of three-wire load. Because only one triac is pulsed on at a time, the diodes (D_1, D_2, and D_3) are necessary to trigger the opposite-polarity triac and, in this way, to assure initial latching-on of the system. The three resistors (R_1, R_2, and R_3) are used for current limiting of the gate drive when the opposite-polarity triac is triggered on by the line voltage.

In critical applications that require suppression of all generated RFI, the circuit shown in Fig. 644 may be used. In addition to synchronous steady-state operating conditions, this circuit also incorporates a zero-voltage starting circuit. The

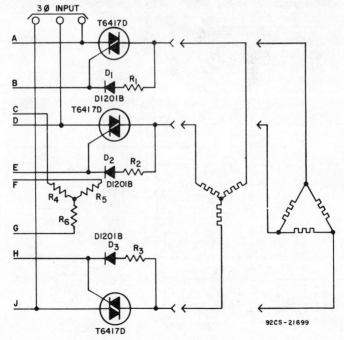

92CS-21699

Fig. 643—Simplified diagram of a three-phase heater control that employs zero-voltage synchronous switching in the steady-state operating conditions (continued from page 512).

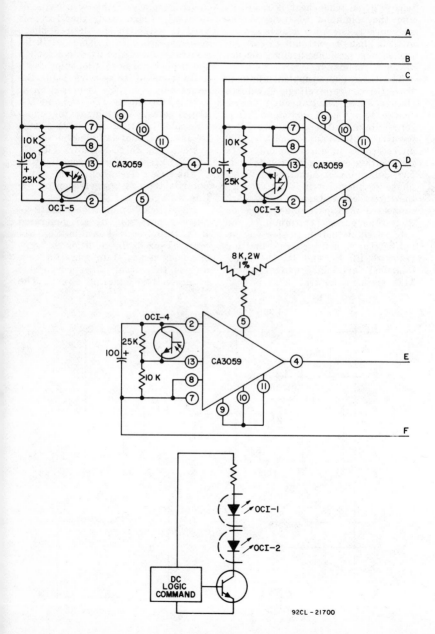

Fig. 644—Three-phase power control that employs zero-voltage synchronous switching both for steady-state operation and for starting (continued on page 515).

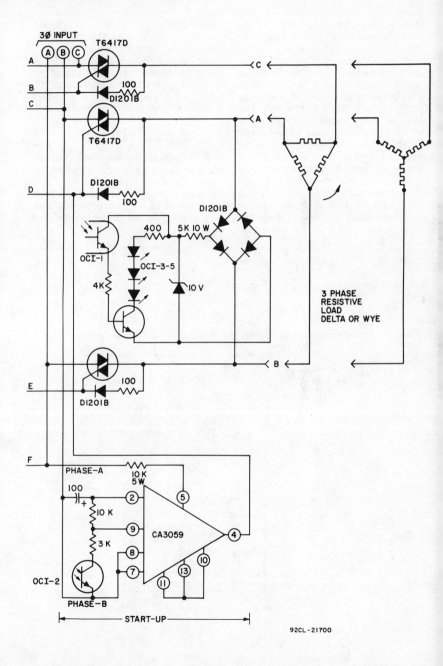

Fig. 644—*Three-phase power control that employs zero-voltage synchronous switching both for steady-state operation and for starting (continued from page 514).*

start-up condition is zero-voltage synchronized to a single-phase, 2-wire, line-to-line circuit, comprised of phases A and B. The logic command engages the single-phase start-up zero-voltage switch and three-phase photo-coupled isolators OCI-3, OCI-4, OCI-5 through the photo-coupled isolators OCI-1 and OCI-2. The single-phase zero-voltage switch, which is synchronized to phases A and B, starts the system at zero voltage. As soon as start-up is accomplished, the three photo-coupled isolators OCI-3, OCI-4, and OCI-5 take control, and three-phase synchronization begins. When the

"logic command" is turned off, all control is ended, and the triacs automatically turn off when the sine-wave current decreases to zero. Once the first phase turns off, the other two will turn off simultaneously, 90° later, as a single-phase line-to-line circuit, as is apparent from Fig. 642.

Inductive Loads

For inductive loads, zero-voltage turn-on is not generally required because the inductive current cannot increase instantaneously; therefore, the amount of RFI generated is usually negligible. Also, because of the

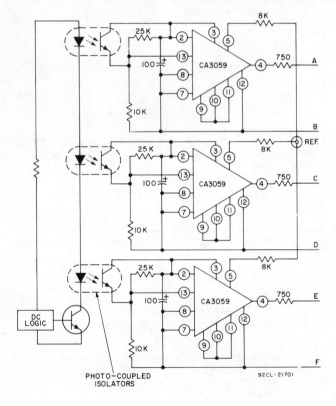

Fig. 645—Triac three-phase control circuit for an inductive load, i.e., three-phase motor
(continued on page 517).

lagging nature of the inductive current, the triacs cannot be pulse-fired at zero voltage. There are several ways in which the zero-voltage switch may be interfaced to a triac for inductive-load applications. The most direct approach is to use the zero-voltage switch in the dc mode, i.e., to provide a continuous dc output instead of pulses at points of zero-voltage crossing. This mode of operation is accomplished by connection of terminal 12 to terminal 7, as shown in Fig. 645. The output of

the zero-voltage switch should also be limited to approximately 5 milli-amperes in the dc mode by the 750-ohm series resistor. Use of a triac such as the T2301D is recommended for this application. Terminal 3 is connected to terminal 2 to limit the steady-state power dissipation within the zero-voltage switch. For most three-phase inductive load applications, the current-handling capability of the T2301D triac (2.5 amperes) is not sufficient. Therefore, the T2301D is used as a trigger triac

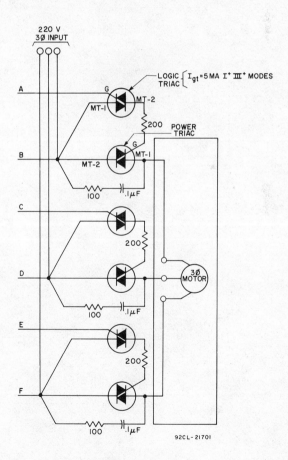

Fig. 645—Triac three-phase control circuit for an inductive load, i.e., three-phase motor (continued from page 516).

to turn on any other currently available power triac that may be used. The trigger triac is used only to provide trigger pulses to the gate of the power triac (one pulse per half cycle); the power dissipation in this device, therefore, will be minimal.

Simplified circuits using pulse transformers and reed relays will also work quite satifactorily in this type of application. The RC networks across the three power triacs are used for suppression of the commutating dv/dt when the circuit operates into inductive loads.

TV Deflection Systems

FOR reproduction of a transmitted picture in a television receiver, the face of a cathode-ray tube is scanned with an electron beam while the intensity of the beam is varied to control the emitted light at the phosphor screen. The scanning is synchronized with a scanned image at the TV transmitter, and the black-through-white picture areas of the scanned image are converted into an electrical signal that controls the intensity of the electron beam in the picture tube at the receiver.

SCANNING FUNDAMENTALS

The scanning procedure used in the United States employs horizontal linear scanning in an odd-line interlaced pattern. The standard scanning pattern for television systems includes a total of 525 horizontal scanning lines in a rectangular frame having an aspect ratio of 4 to 3. The frames are repeated at a rate of 30 per second, with two fields interlaced in each frame. The first field in each frame consists of all odd-number scanning lines, and the second field in each frame consists of all even-

number scanning lines. The field repetition rate is thus 60 per second, and the vertical scanning rate is 60 Hz. (For color systems, the vertical scanning rate is 59.94 Hz.)

The geometry of the standard odd-line interlaced scanning pattern is illustrated in Fig. 646. The scanning beam starts at the upper left corner of the frame at point A, and sweeps across the frame with uniform velocity to cover all the picture elements in one horizontal line. At the end of each trace, the beam is rapidly returned to the left side of the frame, as shown by the dashed line, to begin the next horizontal line. The horizontal lines slope downward in the direction of scanning because the vertical deflecting signal simultaneously produces a vertical scanning motion, which is very slow compared with the horizontal scanning speed. The slope of the horizontal line trace from left to right is greater than the slope of the retrace from right to left because the shorter time of the retrace does not allow as much time for vertical deflection of the beam. Thus, the beam is continuously and slowly deflected downward as it scans

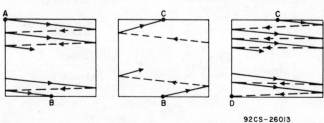

92CS-26013

Fig. 646—The odd-line interlaced scanning procedure.

the horizontal lines, and its position is successively lower as the horizontal scanning proceeds.

At the bottom of the field, the vertical retrace begins, and the beam is brought back to the top of the frame to begin the second or even-number field. The vertical "flyback" time is very fast compared to the trace, but is slow compared to the horizontal scanning speed; therefore, some horizontal lines are produced during the vertical flyback.

All odd-number fields begin at point A in Fig. 646 and are the same All even-number fields begin at point C and are the same. Because the beginning of the even-field scanning at C is on the same horizontal level as A, with a separation of one-half line, and the slope of all lines is the same, the even-number lines in the even fields fall exactly between the odd-number lines in the odd field.

Sync Pulses

In addition to picture information, the composite video signal from the video detector of a television receiver contains timing pulses to assure that the picture is produced on the faceplate of the picture tube at the right instant and in the right location. These pulses, which are called sync pulses, control the horizontal and vertical scanning generators of the receiver.

Fig. 647 shows a portion of the detected video signal. When the picture is bright, the amplitude of the signal is low. Successively deeper grays are represented by higher amplitudes until, at the "blanking level" shown in the diagram, the amplitude represents a complete absence of light. This "black level" is held constant at a value equal to 75 per cent of the maximum amplitude of the signal during transmission. The remaining 25 per cent of the signal amplitude is used for synchronization information. Portions of the signal in this region (above the black level) cannot produce light.

In the transmission of a television picture, the camera becomes inactive at the conclusion of each horizontal line and no picture information is transmitted while the scanning beam is retracing to the beginning of the next line. The scanning beam of the receiver is maintained at the black level during this retrace interval by means of the blanking pulse shown in Fig. 647. Immediately after the beginning of the blanking period, the signal amplitude rises further above the black level to provide a horizontal-synchronization pulse that initiates the action of the horizontal scanning generator. When the bottom line of the picture is reached, a similar vertical-synchronization pulse initiates the action of the vertical scanning generator to move the scanning spot back to the top of the pattern.

Sync Separation

The sync pulses in the composite video signal are separated from the picture information in a **sync-sepa-**

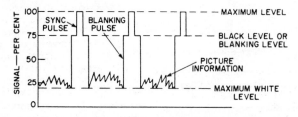

92CS-26014

Fig. 647—Detected video signal.

rator stage, as shown in Figs. 648 and 649. This stage is biased sufficiently beyond cutoff so that current flows and an output signal is produced only at the peak positive swing of the input signal. In the diode circuit of Fig. 648, negative bias for the diode is developed by R and C as a result of the flow of diode current on the positive extreme of signal input. The bias automatically adjusts itself so

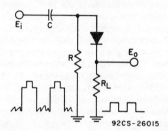

Fig. 648—Diode sync-separator circuit.

that the peak positive swing of the input signal drives the anode of the diode positive and allows the flow of current only for the sync pulse. In the circuit shown in Fig. 649, the base-emitter junction of the transistor functions in the same manner as the diode in Fig. 648, but in addition the pulses are amplified.

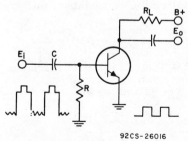

Fig. 649—Transistor sync-separator circuit.

After the synchronizing signals are separated from the composite video signal, it is necessary to filter out the horizontal and vertical sync signals so that each can be applied to its respective deflection generator. This filtering is accomplished by RC

circuits designed to filter out all but the desired synchronizing signals. Although the horizontal, vertical, and equalizing pulses are all rectangular pulses of the same amplitude, they differ in frequency and pulse width, as shown in Fig. 650. The horizontal sync pulses have a repetition rate of 15,750 per second (one for each horizontal line) and a pulse width of 5.1 microseconds. (For color system, the repetition rate of the horizontal sync pulses is 15,734 per second.) The equalizing pulses have a width approximately half the horizontal pulse width, and a repetition rate of 31,500 per second; they occur at half-line intervals, with six pulses immediately preceding and six following the vertical synchronizing pulse. The vertical pulse is repeated at a rate of 60 per second (one for each field), and has a width of approximately 190 microseconds. The serrations in the vertical pulse occur at half-line intervals, dividing the complete pulse into six individual pulses that provide horizontal synchronization during the vertical retrace. (Although the picture is blanked out during the vertical retrace time, it is necessary to keep the horizontal scanning generator synchronized.)

All the pulses described above are produced at the transmitter by the synchronizing-pulse generator; their waveshapes and spacings are held within very close tolerances to provide the required synchronization of receiver and transmitter scanning.

The horizontal sync signals are separated from the total sync in a differentiating circuit that has a short time constant compared to the width of the horizontal pulses. When the total sync signal is applied to the differentiating circuit shown in Fig. 651, the capacitor charges completely very soon after the leading edge of each pulse, and remains charged for a period of time equal to practically the entire pulse width. When the applied voltage is removed at the time corresponding to the trailing edge of each pulse, the capa-

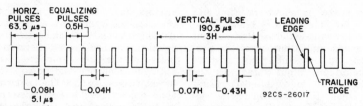

Fig. 650—Waveform of TV synchronizing pulses (H = horizontal line period of 1/15,750 seconds, or 63.5 μs).

citor discharges completely within a very short time. As a result, a positive peak of voltage is obtained for each leading edge and a negative peak for the trailing edge of every pulse. One polarity is produced by the charging current for the leading edge of the applied pulse, and the opposite polarity is obtained from the discharge current corresponding to the trailing edge of the pulse.

As mentioned above, the serrations in the vertical pulse are inserted to provide the differentiated output needed to synchronize the horizontal scanning generator during the time of vertical synchronization. During the vertical blanking period, many more voltage peaks are available than are necessary for horizontal synchronization (only one pulse is used for each horizontal line period). The check marks above the differentiated output in Fig. 651 indicate the voltage peaks used to synchronize

the horizontal deflection generator for one field. Because the sync system is made sensitive only to positive pulses occurring at approximately the right horizontal timing, the negative sync pulses and alternate differentiated positive pulses produced by the equalizing pulses and the serrated vertical information have no effect on horizontal timing. It can be seen that although the total sync signal (including vertical synchronizing information) is applied to the circuit of Fig. 651, only horizontal synchronization information appears at the output.

The vertical sync signal is separated from the total sync in an integrating circuit which has a time constant that is long compared with the duration of the 5-microsecond horizontal pulses, but short compared with the 190-microsecond vertical pulse width. Fig. 652 shows the general circuit configuration used, to-

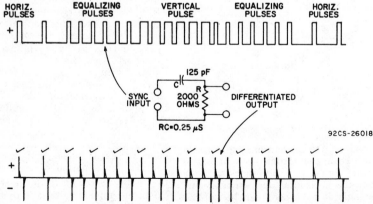

Fig. 651—Separation of the horizontal sync signals from the total sync by a differentiating circuit.

gether with the input and output signals for both odd and even fields. The period between horizontal pulses, when no voltage is applied to the RC circuit, is so much longer than the horizontal pulse width that the capacitor has time to discharge almost down to zero. When the vertical pulse is applied, however, the integrated voltage across the capacitor builds

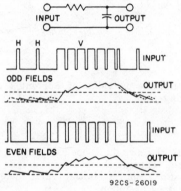

Fig. 652—Separation of vertical sync signals from the total sync for odd and even fields with no equalizing pulses. (Dashed line indicates triggering level for vertical scanning generator.)

up to the value required for triggering the vertical scanning generator. This integrated voltage across the capacitor reaches its maximum amplitude at the end of the vertical pulse, and then declines practically to zero, producing a pulse of the triangular wave shape shown for the complete vertical synchronizing pulse. Although the total sync signal (including horizontal information) is applied to the circuit of Fig. 649, therefore, only vertical synchronization information appears at the output.

The vertical synchronizing pulses are repeated in the total sync signal at the field frequency of 60 per second (59.94 per second in color systems). Therefore, the integrated output voltage across the capacitor of the RC circuit of Fig. 652 can be coupled to the vertical scanning generator to provide vertical syn-

chronization. The six equalizing pulses immediately preceding and following the vertical pulse improve the accuracy of the vertical synchronization for better interlacing. The equalizing pulses that precede the vertical pulses make the average value of applied voltage more nearly the same for even and odd fields, so that the integrated voltage across the capacitor adjusts to practically equal values for the two fields before the vertical pulse begins. The equalizing pulses that follow the vertical pulse minimize any difference in the trailing edge of the vertical synchronizing signal for even and odd fields.

HORIZONTAL DEFLECTION

The main functions of the horizontal-deflection system in a television receiver are to deflect the electron beam linearly (from left to right) across the picture-tube screen, return the beam rapidly to the left side of the screen, and then repeat the process. Fig. 653 shows an idealized waveform of the current that passes through the horizontal-deflection-yoke windings during one complete scanning cycle. As pointed out previously, 525 such "scanning lines" are required to produce each picture in a United States television system.

In addition to beam deflection, the over-all horizontal system performs a number of auxiliary functions. These functions may include:

1. Generation of the high voltage for the picture tube.
2. High-voltage regulation.
3. Scan-linearity correction.
4. Retrace blanking.
5. Gating signal for automatic gain control (agc).
6. Timing reference for automatic frequency control (afc).
7. Bias voltage for grid 2 of the picture tube.
8. Focus voltage for the picture tube.
9. Convergence waveforms.
10. Low-voltage supplies.

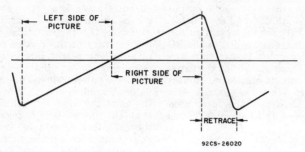

92CS-26020

Fig. 653—Current waveform applied to the horizontal-yoke windings during one complete scanning cycle.

Basic Analysis of Horizontal-System Switching

In the horizontal-deflection system of a television receiver, a current I that varies linearly with time and has a sufficient peak-to-peak amplitude must be passed through the horizontal-deflection-yoke winding to develop a magnetic field adequate to deflect the electron beam of the television picture tube. After the beam is deflected completely across the face of the picture tube, it must be returned very quickly to its starting point. (As explained previously, the beam is extinguished during this retrace by the blanking pulse incorporated in the composite video signal, or in some cases by additional external blanking derived from the horizontal-deflection system.)

The simplest form of a deflection circuit is shown in Fig. 654(a). In this circuit, the yoke impedance L is assumed to be a perfect inductor. When the switch S is closed, the yoke current starts from zero and increases linearly. The rate of increase in current (di/dt) is determined as follows:

$$\frac{di}{dt} = \frac{E}{L}$$

Integration of this equation yields the following expression for the in-

stantaneous value of current i at any time t:

$$i = \frac{Et}{L}$$

If the switch is opened at $t = t_1$, the current I instantly drops to zero from an initial value determined as follows:

$$I = \frac{Et_1}{L}$$

Although the simple circuit shown in Fig. 654(a) satisfies the basic requirement for horizontal beam deflection, it presents some serious problems and limitations.

The voltage across the switch is given by the following equation:

$$e = L \frac{di}{dt}$$

Because the rate of change in current di/dt is infinite, the voltage across the switch also is infinite.

In addition, if very little of the total time were spent at zero current, the average supply current would be I/2. This current would require a tremendous amount of dc power because the voltage-current product ($E \times I$) for standard deflection systems ($E = 18$ kilovolts for 110-degree U.S. black-and-white systems) is in the order of 300 watts. (The product of the peak inverse voltage and the peak-to-peak current, which is often used to describe a system, is closer to 2500 watts, as

(a) SIMPLE DEFLECTION CIRCUIT

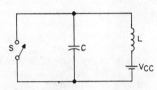

(b) ADDITION OF CAPACITOR

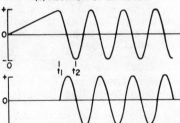

(c) YOKE CURRENT (top) AND SWITCH VOLTAGE (bottom) FOR CIRCUIT (b)

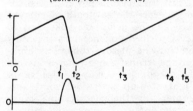

(d) YOKE CURRENT (top) AND SWITCH VOLTAGE (bottom) FOR SWITCH CLOSED AT t_2

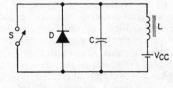

(e) ADDITION OF DAMPER DIODE

92CM-26021

Fig. 654—Development of horizontal-deflection circuit.

explained later.) In addition, the operation of the switch would be rather critical with regard to both its opening and its closing.

A final limitation would be the fact that the deflection field would be poled or phased in one direction only, so that the beam would have to be centered at the extreme left of the screen for zero yoke current.

If a capacitor is placed across the switch, the yoke current still increases linearly when the switch is closed at time $t = 0$. When the switch is opened, however, at time $t = t_1$, a parallel resonant circuit is formed by the parallel combination of L and C, as shown in Fig. 654(b). The intial conditions of this simple transient network are as follows:

$$i_y = \frac{Et_1}{L}$$

$$e_c = 0$$

where i_y is the yoke current and e_c is the capacitor or switch voltage. The resulting yoke currents and switch voltages are shown in Fig. 654(c). The current is maximum when the voltage equals zero, and the voltage is maximum when the current equals zero. The ringing frequency (if zero losses are assumed) is given by

$$f_{osc} = \frac{1}{2\pi\sqrt{LC}}$$

If the switch is closed again at any time the capacitor voltage is not equal to zero, an infinite switch current flows as a result of the capacitive discharge. However, if the switch is closed at the precise moment that the capacitor voltage equals zero, the capacitor current effortlessly transfers to the switch and a new transient condition results, as described below.

At the time of the proposed switching, time $t = t_2$ and the yoke current $i_y = -I$. The yoke current increases at the rate dictated by the ratio E/L, but it starts from

—I instead of from zero. Fig. 654(d) shows the yoke current and switch voltage waveforms.

If the switch is again opened at t_4, closed at t_5, and so on, the desired sweep will result, the peak switch voltage will be finite, and the average supply current will be zero. The deflection system is then lossless and efficient. Because the average yoke current is also zero, beam decentering is avoided.

The only fault in the final circuit is the critical timing of the switch, particularly at time $t = t_2$. If the switch is shunted by a properly poled diode (the damper diode), as shown in Fig. 654(e), the diode acts as a closed switch as shown as the capacitor voltage reverses slightly. The switch may then be closed any time from $t = t_2$ to $t = t_3$.

The horizontal scanning rate for television systems in the United States is 15,750 scans per second for black-and-white types and 15,734 scans per second for color types. Obviously mechanical switches cannot operate at such high rates. In practice, the switch S will be an active device, such as an electron tube, a transistor, or a thyristor.

Fig. 655 shows a basic horizontal deflection circuit that uses a transistor in place of switch S. High voltage is generated by use of the

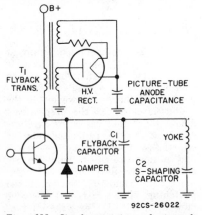

Fig. 655—Simple transistor horizontal-deflection circuit.

step-up transformer T_1 in parallel with the yoke. This step-up transformer is designed so that its leakage inductance, distributed capacitance, and output stray capacitance complement the yoke inductance and retrace tuning capacitance in such a manner that the peak voltage across the primary winding is reduced and the peak voltage across the secondary winding is increased, as compared to the values that would be obtained in a perfect transformer. This technique, which is referred to as **third-harmonic tuning**, yields a voltage ratio of secondary-to-primary peak voltage of approximately 1.7 times the value expected in a perfect transformer.

The following paragraphs describe the use of power transistors and thyristors (SCR's) as the main switching element in practical horizontal deflection systems.

Transistor Horizontal-Deflection Circuits

Fig. 656 shows the functional relationship among the various circuit elements of a horizontal-deflection circuit that uses a power transistor to generate the sawtooth of current through the deflection yoke and to develop the beam accelerating voltage for the picture tube. The high-voltage transformer shown across the output stage may be used as a slight step-up or step-down transformer for the picture-tube high-voltage supply, the yoke, the damper diode, the capacitor, or any combination of these elements.

In the following paragraphs, the design factors and technical considerations used in the development of a typical horizontal-deflection-system circuit are explained. This system is assumed to provide the deflection energy and high voltage required for a 19-inch, 20-kilovolt, 114-degree monochrome receiver from a power supply having a 12-microsecond retrace time. Basic circuit configurations for practical

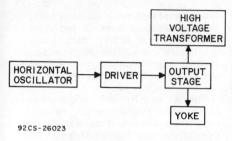

92CS-26023

Fig. 656—Block diagram of a transistor horizontal-deflection system.

horizontal-deflection systems for both monochrome and color television receivers are then shown and analyzed.

Voltage Considerations—For an idealized horizontal-deflection circuit, the peak voltage E_{max} across the transistor is given by

$$E_{max} = \left(1.79 + 1.57\, \frac{T_t}{T_R} \right) E_{dc}$$

where T_t is the scanning or trace time, T_R is the retrace time, and E_{dc} is the supply voltage. If third-harmonic tuning is employed, the peak voltage is reduced by approximately 20 per cent.

The highest anticipated value of E_{max} is determined by use of the value of E_{dc} obtained at high ac line voltage and at the lowest horizontal-oscillator frequency, i.e., the longest trace time. (For these conditions, of course, the receiver is out of sync.) The tolerances on the inductors and capacitors alter the trace time only slightly and usually may be ignored if a 10-per-cent tolerance is used for the tuning capacitor.

When a capacitor is used in series with the yoke for linearity correction, the peak-to-peak yoke current and the flyback voltage are both increased by about 10 per cent. In a first-order approximation, this effect may be ignored if the system is designed without S-shaping. If shaping is employed, however, the supply voltage must be reduced by

5 to 10 per cent to restore the scan conditions originally observed.

An abnormality that must be considered is high-voltage arcing. Fig. 657 shows the normal transistor load for third-harmonic tuning of the flyback transformer in which the leakage inductance, secondary-winding capacitances, and anode stray capacitances are reflected to the primary. In a properly designed system, the leakage inductance is about one-half the shunt inductance (yoke plus flyback primary inductance).

When a high-voltage arc occurs, the secondary is momentarily shorted, placing the leakage inductance in parallel with the shunt inductance. As a result, the peak collector current is increased by a factor of about three, and the retrace time is decreased by a factor of about two (if the transistor is still operating as an ideal switch).

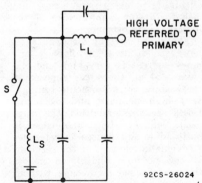

HIGH VOLTAGE REFERRED TO PRIMARY

92CS-26024

Fig. 657—Equivalent output circuit for third-harmonic tuning (referred to primary side.)

Because the flyback voltage would then be increased by a factor of 2.5, avalanche breakdown occurs at a high current level, second breakdown is initiated, and the transistor is destroyed. Since occasional high-voltage arcing is unavoidable in the picture-tube gun, the output transistor must be protected.

If a diode and capacitor are connected in series and placed across the transistor, the flyback pulse is clamped at a level equal to the normal peak value when a high-voltage arc occurs. When the arcing is sustained long enough for an appreciable increase in the capacitor voltage, the increased drain (caused by the very high peak collector current) opens a fuse in the B-supply, and the transistor is adequately protected. A bleeder resistor is placed across the capacitor to protect against intermittent arcs. This circuit also protects against several other types of high-voltage short circuits.

Another method used to reduce the effect of high-voltage arcing is to make the leakage inductance of the secondary very high compared to the shunt inductance by designing the secondary to resonate at the fundamental frequency (15 kHz). In addition to protection during high-voltage arcs, this method reduces peak collector current (caused by higher primary inductance) and also facilitates manufacture of the flyback transformer.

There are several disadvantages, however. Because the flyback primary current is very high and circulates at all times (as opposed to the case of third-harmonic tuning), very high primary and secondary losses occur. In addition, the magnetic field of the transformer is quite high and causes interference problems in the rest of the receiver. It also becomes difficult to enclose the transformer in a cage without causing an excessive shorted-turn problem because the cage is magnetically coupled.

A third and rather significant disadvantage is the high peak-to-peak secondary voltage developed for a given value of dc high voltage. In third-harmonic tuning, the secondary voltage waveform exhibits a narrow spike for approximately 10 per cent of the cycle and a low constant voltage for the remainder of the cycle. As a result, the peak-inverse rating on the high-voltage rectifier is approximately 1.1 times the dc high voltage developed. In the

"fundamental-tuned" arrangement, the secondary voltage is nearly a sine wave and results in a peak-inverse rating on the high-voltage rectifier approximately twice that of the third-harmonic-tuned system.

Choice of Retrace Time—The choice of a slightly longer retrace time offers the following significant advantages for circuit design:

1. As retrace time is lengthened, the product of peak voltage and peak current is reduced directly.

2. The peak stored energy, as well as the voltage-current product, is reduced because more primary inductance can be used in the flyback transformer.

3. The retrace losses are reduced with the square of the retrace time.

4. Losses in the yoke and flyback that result from skin effect are reduced.

5. The core losses in the flyback transformer are reduced because of the greater inductance.

6. The supply voltage may be increased because of the lower flyback pulse.

7. The flyback transformer secondary becomes easier to wind.

High-Voltage Power—High voltage is obtained by means of a tertiary winding on the flyback transformer which through auto-transformer action steps up the yoke pulse to a high value. In monochrome receivers, the energy typically extracted is seldom greater than 0.3 millijoule for 5 watts of beam power and results in a typical circuit Q of approximately 60, if other degenerative losses are neglected. When an LC network is damped to a Q of 60, the voltage and current waveshapes for the first π radians show very little change (except for phase relationship) over the infinite-Q condition. Therefore, the losses, which are determined by the voltage and current waveshapes, do not increase when beam current flows; i.e., 5 watts of beam power reflects only an added demand of 5 watts in the power supply.

A further point of interest in transistor deflection circuits is the excellent high-voltage regulation encountered. This improvement is the result of the high efficiency of these circuits, which keeps the extracted energy to a minimum and results in a fairly high circuit Q. As noted, the anode-voltage amplitude does not change much as energy is extracted and thus accounts for good high-voltage regulation.

Scan Linearity—For accurate reproduction of pictures on the picture-tube screen, the electron beam must move at a linear rate across the faceplate of the picture tube. If the faceplate were a section of a sphere with its center at the center of deflection, a linear sawtooth of current through the deflection yoke would be required for linear deflection. Although the faceplate is a section of a sphere, its radius is much greater than the distance to the center of deflection. For all practical purposes, the faceplate can be considered as a flat plane. The distance from the center of deflection to the faceplate, therefore, is greater at the edges than at the center of the picture tube. Consequently, a given amount of deflection of the beam at the deflection center produces a greater movement of the electron beam on the face of the picture tube. For this reason, the required current waveform through the deflection yoke should be somewhat "S"-shaped rather than an absolutely linear sawtooth. Much of this S-shaping is accomplished by the capacitor connected in series with the yoke. At the start of trace (left side of the picture), maximum energy is stored in the yoke. Current flows from the yoke into the S-shaping capacitor and causes the capacitor to charge at an exponential rate. Yoke current rises linearly during the first half of trace and gradually decreases near

the right-hand side of the screen as the S-shaping capacitor accumulates charge. However, because the first part of trace is linear, the left side of the picture will be stretched in comparison to the right side, and additional linearity correction is required.

Methods for eliminating this non-linearity include the use of a saturable reactor in series with the yoke, the use of a permanent magnet near the yoke to distort the field, and the use of a damped series-resonant circuit connected in parallel with the S-shaping capacitor and in series with the yoke.

Deflection Energy Requirement— The peak deflection energy required by the yoke for complete scanning of picture tubes varies directly with the high voltage, the $5/2$ power of the deflection angle (approximately), and the neck diameter (where all geometries of the yoke are adjusted in direct proportion). The peak energy required for minimum scanning of a 114-degree picture tube having a 1-1/8-inch neck diameter and an anode voltage of 20 kilovolts is 2.4 millijoules (scan from center to either side).

When full scan is obtained at low line voltage and at an anode voltage that corresponds to low line, the peak stored energy equals ε. When the line voltage is increased, the peak energy increases in proportion to the square of the voltage. If the low line voltage is 105 volts and high line is 135 volts, the increase in energy is a factor of 1.65.

If the receiver is adjusted out of sync by 2 microseconds (or a 480-cycle pullout range), the energy, which is proportional to the square of the trace time in a fixed circuit, increases by a factor of 1.08. If the yoke is shunted by a practical flyback transformer, the inductance is reduced by a factor of approximately 1.3 and, therefore, the peak stored energy in the system is increased by a factor of 1.3. When all three items are considered, the transistor must handle 2.3 times the peak stored energy normally expected.

Transistor Drive Considerations— Transformer drive is usually employed for the output transistor. When this type of drive is used, the collector load may be placed in series with either the collector or the emitter because, in either case, the transformer secondary appears from base to emitter. If the load is in series with the emitter (emitter loading), the collector is directly at the supply-voltage potential. If a positive-supply is used, the transistor case it at chassis potential. The damper diode is constructed with its anode at case potential so that it is also at chassis potential.

This method has a disadvantage in that a high potential is placed between the primary and secondary windings of the driver transformer. Because the driver transformer is very tightly coupled, insulation breakdown must be carefully considered.

While the output stage is cut off, the driver stage should be conducting; the transformer secondary can then provide any current demanded. (The current, however, is limited by the leakage inductance.) When the driver stage is cut off, the energy stored in the transformer flows in the secondary in the form of a constant current. If this mode of drive is employed, and if the base-to-emitter voltage of one transistor varies from that of another, the turn-on current still starts at the same value but decays at a different rate. If all charge is removed from the base of the output transistor during turn-off, no more transformer current is required and the transistor stays at a reverse-bias mode.

No impedance should be placed in the base. Transistor interchangeability is thus improved because the voltage level remains low enough to

prevent breakdown of the base-emitter junction during the turn-off period. The primary and secondary windings of the driver transformer must be very tightly coupled to obtain a large spike of current during the turn-off period (for a fast turn-off time).

The circuit shown in Fig. 658 is

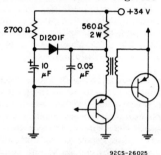

92CS-26025

Fig. 658—Waveforming circuit.

used to develop the all-important waveshaping. The 560-ohm resistor in combination with the 0.05-microfarad capacitor increases the amplitude and rise time of the turn-off base current for the first few microseconds. The D1201F diode, together with the 2700-ohm resistor and 10-microfarad capacitor, serves as a clamp circuit which assures that the ouput transistor is always reverse-biased during the entire turn-off period, even in the presence of high I_{CBO} at several hundred volts and elevated temperatures. With this circuit, the 560-ohm, 0.05-microfarad combination can be optimized for the best turn-off time without regard for the remainder of the off signal.

The turn-off pulse developed by this circuit is 3 amperes for approximately 2 microseconds, followed by a constant voltage of approximately 1/2 volt for 18 microseconds. The on-pulse then initiates at 650 milliamperes and, 45 microseconds later, decreases to 500 milliamperes.

Deflection Circuit for Monochrome Receiver—The following paragraphs describe a practical horizontal-deflection system for a 19-inch black-and-white (monochrome) television receiver. The deflection system operates from a regulated dc supply of 100 volts.

The power-supply voltage of 100 volts is decoupled to 85 volts for raster regulation with brightness. A retrace time of 14 microseconds is selected to present the maximum usable picture, although a value of 17 microseconds could have been used with no sacrifice in performance as compared to present-day receivers.

The picture tube used, the 19DQP4, has minimum usable screen dimensions of 15⅛ inches horizontally and 12 inches vertically. These dimensions establish the front mask size for the cabinet and fix the aspect ratio at 1.26. The diagonal deflection angle of the 19DQP4 is 114 degrees, and the neck diameter is a nominal 1⅛ inches. The zero-beam accelerating potential is 20 kilovolts. The horizontal circuit should be capable of providing an average beam current of 400 microamperes with virtually no change in raster height or width at any brightness setting between zero and full current. An over-scan of 4 per cent is desired.

Energy requirements for horizontal deflection show that the peak stored energy in the yoke must be 2.4 millijoules to fulfill the requirements for the 19DQP4. If a trace time of 49.5 microseconds and a power-supply voltage of 85 volts are used, and if it is assumed that the use of "S" shaping (by use of a capacitor in series with the yoke) has the effect of increasing the scan by 5 to 10 per cent, the yoke inductance must be 1 millihenry and the peak-to-peak yoke current must be 4.4 amperes.

Driver and output circuit: The horizontal-drive-and-output circuit for the receiver is shown in Fig. 659. The output circuit (Q_5) is basically a self-oscillator which requires the 27,000-ohm resistor to initiate oscillation. Drive current is obtained through the 33-ohm resistor in parallel with the picture-tube heater and through diode D_5, and is applied from the feedback winding of the transformer T_1 through the 50-microfarad capacitor in parallel with the 10-ohm resistor to the base of the output transistor Q_5. If this drive circuit is correctly designed, transistor Q_5 does not come out of saturation during normal operation.

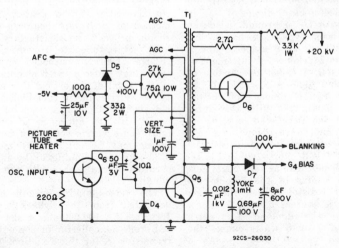

Fig. 659—Horizontal driver and output circuit for a black-and-white television receiver.

When retrace is to be initiated, the driver transistor Q_6 is driven heavily into saturation. The drive current is then shunted to ground, and the base of transistor Q_5 is simultaneously reverse-biased by means of the charge stored on the 50-microfarad capacitor. If the resistor shunting this capacitor is large compared to 2 ohms, most of the drive current flows through the capacitor. When transistor Q_6 is saturated, therefore, a capacitor current of opposite polarity results through Q_6 and D_4. As a result, the turn-off drive to transistor Q_5 is always a reverse-bias voltage equal to the forward drop across diode D_4. If the value of the parallel resistor is made extremely large, the circuit may not start.

When the output transistor Q_5 is turned off, the collector voltage comes out of saturation and goes through the normal flyback pulse. During this time, the feedback drive (already shunted to ground) decreases and, as the collector voltage passes the supply voltage, a heavy reverse drive current results. However, the driver diode D_5 blocks this reverse current flow. D_5 also permits the starting current (through the 27,000-ohm resistor) to flow through the base of Q_5.

The damper current flows through the collector-base diode of Q_5 in series with D_4 to ground. Diode D_4 is a silicon type that has a low forward drop at 2 amperes and a minimum breakdown requirement of only 1 volt. It must be capable of dissipating 300 milliwatts. D_4 is called a damper diode, even though it only partially fulfills this function. The 50-microfarad coupling capacitor must have a low series resistance to obtain proper turn-off.

The 100-volt supply voltage is reduced 15 per cent at zero beam current by means of the 75-ohm decoupling resistor. When the beam current is increased to 400 microamperes, the demand for extra power of 7 or 8 watts causes the decoupled voltage to drop. As a result, the high voltage and the scanning current decrease linearly with the decoupled voltage. The high voltage also decreases because of the lack of perfect high-voltage regulation. If the circuit is designed correctly, the high voltage decreases with the square of decoupled voltage so that the scanning-energy requirement approximately tracks the scanning energy provided. This decoupled voltage is also fed back to the vertical circuit in the size-determining portion of the circuit so that the vertical scan energy also tracks the high voltage as a function of picture-tube average brightness setting.

A separate winding on the flyback transformer T_1 provides gating for the agc circuit. A signal taken from the driver diode D_5 provides a timing reference for the horizontal phase circuit (afc). A positive voltage of approximately 500 volts is available from the clamp circuit provided by diode D_7 to supply bias to grid No. 2 or grid No. 4 of the picture tube. (The current drain should be kept below 1 milliampere.)

Picture-tube heater power is also derived from the horizontal-driver circuit. When the receiver is first turned on, the base drive current to transistor Q_5 is larger than normal because of the thermally nonlinear characteristic of the heater. This method of providing heater power should prove to be satisfactory for long picture-tube life. However, excessive heater-to-cathode capacitance may cause a video modulation in the form of a vertical line similar to a drive line in tube deflection. No such problem has been experienced with the approach shown. A more conservative control of heater power may be obtained by means of a separate winding or by incorporation of the heater with the agc winding.

The video-blanking circuit must be gated from the flyback transformer T_1. A 100,000-ohm resistor is fed from the collector of the output transistor Q_5 for this function. This resistor provides blanking whenever the collector voltage is more positive than 25 volts.

The picture-tube heater has a dc voltage across it, together with a large ac voltage. After adequate decoupling, this dc voltage provides a convenient source of negative potential to power the agc and sync-separator circuits.

Various forms of arcing protection are provided in the horizontal output circuit.

A voltage-clamp circuit is provided by the clamp diode D_7 in conjunction with the 8-microfarad capacitor. Sufficient curent drain must be provided across the capacitor to discharge it between arcs. The capacitor must be large enough to absorb most of the energy stored by the picture-tube capacitance. The purpose of this clamp circuit is to assure that the transistor does not go into voltage breakdown during high-voltage arcs.

The 75-ohm decoupling resistor provides raster regulation, as mentioned previously, and also limits the maximum power that may be delivered to the entire horizontal-scanning circuit to 30 watts.

If the output transistor Q_5 is pulled out of saturation at a high collector-current level as a result of high-voltage arcing, the feedback drive circuit turns off Q_5 and thus controls the transistor load line. The transistor turns off fairly fast under this condition because it is in an unsaturated state. If the driver transistor Q_6 is turned on or off when Q_5 is reverse-biased, no change in state occurs because the drive is basically self-oscillating and transistor Q_6 functions merely as a gate. If drive is available, Q_6 may exclude it. If drive is being applied, Q_6 may turn it off. However, Q_6 may not provide drive if Q_5 is not saturated.

Although drive is available at the beginning of trace time, it should be excluded by Q_6 until about 5 microseconds prior to normal need. As a result, Q_6 should receive a drive pulse that saturates it for approxi-mately 30 microseconds, and turns it off for the remaining 34 microseconds.

The clamp diode D_7 must not fail. If it does, destruction of Q_6 is almost assured.

Horizontal oscillator: Fig. 660 shows a simplified diagram of a multivibrator type of horizontal-oscillator circuit. It should be noted

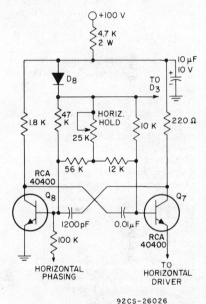

92CS-26026

Fig. 660—Horizontal oscillator circuit.

that a gated dc feedback signal is provided from the power supply. If the 100-volt supply becomes excessively high as a result of a fault, the horizontal-oscillator frequency is raised to such a point that the flyback voltage remains within specifications.

Horizontal phasing (afc): The horizontal phasing used is novel. Gating is obtained from a 1-milliampere sync pulse that is only 2 microseconds wide (and can be much narrower if desired). Nearly any desired average control current up to several milliamperes can be provided. When the picture is correctly phased, the circuit is open to receive a sync pulse for only 4 microseconds, and thus is relatively immune to noise. Because the circuit functions on the leading edge of the sync pulse, rather than on the entire area of the differentiated pulse, the effects of the vertical equalizing pulses and the serrations in the vertical pulse are greatly minimized. As a result, the top of the picture exhibits proper synchronization at essentially all settings of the hold control (as is not the case with normal afc).

The horizontal-phasing circuit is shown in Fig. 661. A control cur-

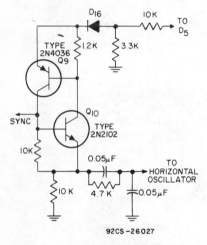

Fig. 661—Horizontal-phasing (afc) circuit.

rent of only 20 microamperes is required for the low-level oscillator. Transistors Q_9 and Q_{10} are connected in a latching configuration that resembles a thyristor. (A thyristor could be used if it were fast enough and sensitive enough.) A replica of the flyback pulse is applied to the emitter of transistor Q_9 from diode D_5 through the voltage divider consisting of the 10,000-ohm and 3300-ohm resistors. During trace time, this voltage is slightly negative, and any signal appearing on

the base of transistor Q_{10} is ineffective. When the flyback voltage appears across transistor Q_9, however, the gate is open to receive a sync pulse. When the pulse fires transistor Q_{10}, the combination of transistors Q_9 and Q_{10} becomes regenerative; the transistors then become heavily saturated and pass any amount of current the voltage divider will permit. Transistors Q_9 and Q_{10} remain in saturation until the voltage reverses and resets the latch. Control current of the opposite polarity may be obtained from the emitter of transistor Q_9 if desired. Diode D_{16} serves to block reverse current.

Deflection Circuit for Color Receiver—Fig. 662 shows a schematic of a transistorized horizontal-deflection circuit for a color TV receiver. The horizontal output transistor, Q_4, is a high-voltage silicon transistor. The normal collector-to-emitter pulse voltage across Q_4 includes an ample safety factor that allows for any increased pulse that may result from out-of-sync operation, line surges, and other abnormal conditions.

A unique feature of the horizontal-deflection circuit is the low-voltage supply of approximately 23 volts that is derived from it. This feature makes it possible to eliminate the power transformer in the power supply. The low-voltage power is used to operate all but the high-voltage receiver stages, such as the video-output stage, the audio-output stage, and the horizontal oscillator and driver. The vertical oscillator is supplied from the same point which supplies the horizontal output in such a way that the actual voltage is a function of beam current; this connection compensates for the tendency for picture height to change with brightness settings.

The transistorized deflection circuit achieves commercially acceptable high-voltage regulation without the use of the high-voltage shunt regulator used with tube-type deflection circuits. With a flyback

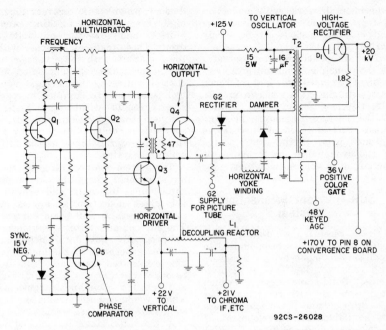

Fig. 662—Horizontal-deflection system for a color television receiver.

transformer of normal design and a low-voltage power supply with about 3-per-cent regulation, high-voltage regulation from zero beam to full load of 750 microamperes is about 3 kilovolts, and is accompanied by a considerable increase in picture width. Improvement of this behavior with brightness changes is achieved by utilizing the accompanying changes of direct current to the deflection circuit in two ways. First, the air gap of the transformer is reduced to permit core saturation to decrease the system inductance as the high-voltage load is increased. When this method is used, regulation is improved to about half that of the normal transformers with no circuit instabilities, but picture-width change is still greater than desired. Second, series resistance is added to the B supply to decrease power input at full load and thereby reduce the change in picture width (at some sacrifice in high-voltage

regulation). The net result of both changes is a regulation of about 2.8 kilovolts for the high voltage, with very little variation in picture size.

A secondary benefit of the inherently good regulation of the transistor deflection system is a reduction in the size of the flyback transformer. The size reduction is accomplished by a reduction in the area of the "window" in the flyback core. A reduction in the size of the high-voltage cage required to maintain adequate isolation of the high-voltage winding from ground is possible because of the smaller flyback transformer.

The transformer-coupled driver stage takes advantage of the high-voltage capability and switching speed of the horizontal driver transistor which is designed primarily for video-output use. A sine-wave stabilized multivibrator type of horizontal oscillator is used. This type of oscillator is especially useful in experimental work with deflection systems because it permits on-time and off-time periods to be easily varied.

The afc phase detector operates on the principle of pulse-width variation of combined sync and reference pulses. In the circuit shown in Fig. 662, timing information is related to the leading edges of the sync pulses, and the retrace process is initiated prior to the leading edge of the sync pulse; performance of the circuit is very satisfactory.

SCR Horizontal-Deflection System

Until recently, solid-state horizontal deflection has been limited to small-screen monochrome receivers with relatively low energy requirements. Solid-state devices which would compete with receiving tubes for both cost and performance were not available. The development of silicon controlled rectifiers (SCR's) and fast-recovery diodes capable of operating at the horizontal scanning rates used in television has made possible the design of a horizontal-deflection system that is economically competitive and, at the same time, provides greater reliability than any other known deflection system.

In this system, the switching action required to generate the scan current in the horizontal-yoke windings and the high-voltage pulse used to derive the dc operating voltages for the picture tube is controlled by two SCR's that are used in conjunction with associated fast-recovery diodes to form bipolar switches.

The SCR's used to control the trace current and to provide the commutating action to initiate trace-retrace switching exhibit high voltage- and current-handling capabilities together with the excellent switching characteristics required for reliable operation in deflection-system applications. The switching diodes, (trace and commutating

diodes), provide fast recovery times, high reverse-voltage blocking capabilities, and low turn-on voltage drops. These features, together with the fact that, with the exception of one noncritical triggering pulse, all control voltages, timing, and control polarities are supplied by passive elements within the system (rather than by external drive sources), contribute substantially to the excellent reliability of the SCR deflection system.

The system operates directly from a conventional, unregulated dc power supply of $+155$ volts, and provides full-screen deflection at angles up to 90 degrees at full beam current. The current and voltage waveforms required for horizontal deflection and for generation of the high voltage are derived essentially from LC resonant circuits. As a result, fast and abrupt switching transients which would impose strains on the solid-state device are avoided.

A regulator stage is included in the SCR horizontal-deflection circuit to maintain the scan and the high voltage within acceptable limits with variations in the ac line voltage or picture-tube beam current. The system also contains circuits that provide full protection against the effects of arcs in the picture tube or the high-voltage rectifier, and linearity and pincushion correction circuits.

Basic Deflection Circuit—The essential components in the SCR horizontal-deflection system required to develop the scan current in the yoke windings are shown in Fig. 663. Essentially the trace-switch diode D_T and the trace-switch controlled rectifier SCR_T provide the switching action which controls the current in the horizontal yoke windings L_y during the picture-tube beam-trace interval. The commutating-switch diode D_C and the commutating-switch controlled rectifier SCR_C initiate retrace and control the yoke current during the retrace interval.

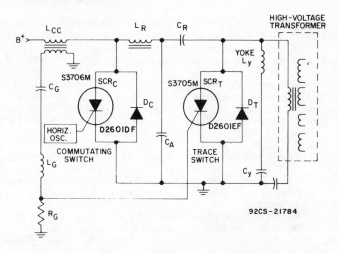

Fig. 663—Basic circuit for generation of the deflection-current waveform in the horizontal yoke winding.

Inductor L_R and capacitors C_R, C_A, and C_y provide the necessary energy storage and timing cycles. Inductor L_{CC} supplies a charge path for capacitor C_R from the dc supply voltage (B+) so that the system can be recharged from the receiver power supply. The secondary of inductor L_{CC} provides the gate trigger voltage for the trace-switch SCR. Capacitor C_R establishes the optimum retrace time by virtue of its resonant action with inductor L_R.

The complete horizontal-deflection cycle may best be described as a sequence of discrete intervals, each terminated by a change in the conduction state of a switching device. In the following discussion, the action of the auxiliary capacitor C_A and the flyback high-voltage transformer are initially neglected to simplify the explanation.

First half of the trace interval: Fig. 664 shows the circuit elements involved and the voltage and current relationships during the first half of the trace deflection-current interval, the period from T_0 to T_2. At time T_0, the magnetic field has

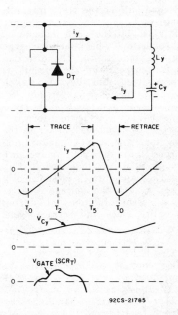

Fig. 664—Effective configuration of the deflection circuit during the first half of the trace interval, time T_0 to T_2, and operating voltage and current waveforms for the complete trace-retrace cycle.

been established about the horizontal yoke windings L_y by the circuit action during the retrace period of the preceding cycle (explained in the subsequent discussion of retrace intervals). This magnetic field generates a decaying yoke current i_y that decreases to zero when the energy in the yoke winding is depleted (at time T_2). This current charges capacitor C_y to a positive voltage V_{Cy} through the trace-switch diode D_T.

During the first half of the trace interval (just prior to time T_2) the trace controlled rectifier SCR_T is made ready to conduct by application of an appropriate gate voltage pulse V_{GATE}. SCR_T does not conduct, however, until a forward bias is also applied between its anode and cathode. This voltage is applied during the second half of the trace interval.

Second half of the trace interval: At time T_2, current is no longer maintained by the yoke inductance, and capacitor C_y begins to discharge into this inductance. The direction of the current in the circuit is then reversed, and the trace-switch diode D_T becomes reverse-biased. The trace-switch controlled rectifier SCR_T, however, is then forward-biased by the voltage V_{Cy} across the capacitor, and the capacitor discharges into the yoke inductance through SCR_T, as indicated in Fig. 665. The capacitor C_y is sufficiently large so that the voltage V_{Cy} remains essentially constant during the entire trace and retrace cycle.

This constant voltage results in a linear rise in current through the yoke inductance L_y over the entire scan interval from T_0 to T_5.

Start of the retrace interval: The circuit action to initiate retrace starts before the trace interval is completed. Fig. 666 shows the circuit elements and the voltage and current waveforms required for this action. At time T_3, prior to the end of the trace period, the commutating-switch controlled rectifier SCR_C is turned on by application of a pulse from the horizontal oscillator to its gate. Capacitor C_R is then allowed to discharge through SCR_C and inductor L_R. The current in this loop, referred to as the commutating circuit, builds up in the form of a half-sine-wave pulse. At time T_4, when the magnitude of this current pulse exceeds the yoke current, the trace-switch diode D_T again becomes forward-biased. The excess current in the commutating pulse is then bypassed around the yoke winding by the shunting action of diode D_T. During the time from T_4 to T_5, the trace-switch controlled rectifier SCR_T is reverse-biased by the amount of the voltage drop across diode D_T. The trace-switch controlled rectifier, therefore, is turned off during this interval and is allowed to recover its ability to block the forward voltage that is subsequently applied.

First half of the retrace interval: At time T_5, the commutating pulse is no longer greater than the yoke

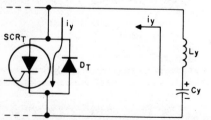
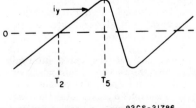

92CS-21786

Fig. 665—Effective configuration of the deflection circuit during the second half of the trace interval, time T_2 to T_5, and the complete scan-current waveform.

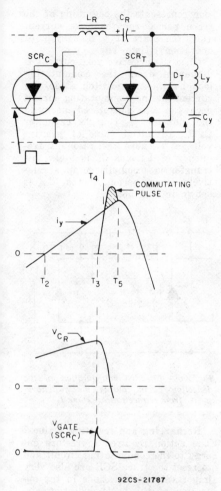

Fig. 666—Effective configuration of the deflection circuit and significant voltage and current waveforms for initiation of retrace, time T_3 to T_5.

supplied by this current charges capacitor C_R with an opposite-polarity voltage in a resonant oscillation. At time T_6, the yoke current is zero, and capacitor C_R is charged to its maximum negative-voltage value. This action completes the first half of retrace.

Second half of the retrace interval: At time T_6, the energy in the yoke inductance is depleted, and the stored energy on the retrace capacitor C_R is then returned to the yoke inductance. This action reverses the direction of current flow in the yoke.

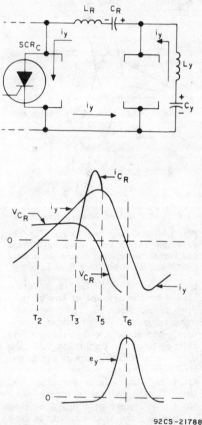

Fig. 667—Effective configuration of the deflection circuit and operating voltage and current waveforms during the first half of retrace time, T_5 to T_6.

current, as shown in Fig. 667; trace-switch diode D_T then ceases to conduct. The yoke inductance maintains the yoke current but, with SCR_T in the off state, this current now flows in the commutating loop formed by L_R, C_R, and SCR_C. Time T_5 is the beginning of retrace.

As the current in the yoke windings decreases to zero, the energy

During the reversal of yoke current, the commutating-switch diode D_C provides the return path for the loop current, as indicated in Fig. 668. The commutating-switch con-

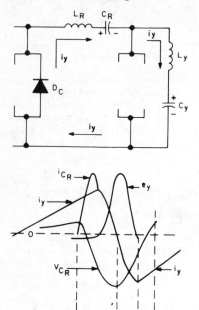

Fig. 668—Effective configuration of the deflection circuit and operating voltage and current waveforms during the second half of retrace, time T_6 to T_0.

trolled rectifier SCR_C is reverse-biased by the amount of the voltage drop across diode D_C. The commutating-switch controlled rectifier, therefore, turns off and recovers its voltage-blocking capability. As the yoke current builds up in the negative direction, the voltage on the retrace capacitor C_R is decreased. At time T_0, the voltage across capacitor C_R no longer provides a driving voltage for the yoke current to flow in the loop formed by L_R, C_R, and L_y. The yoke current finds an easier path up through trace-switch diode D_T, as shown in Fig. 669. This ac-

tion represents the beginning of the trace period for the yoke current (i.e., the start of a new cycle of operation), time T_0.

Once the negative yoke current is decoupled from the commutating loop by the trace-switch diode, the current in the commutating circuit decays to zero. The stored energy in the inductor L_R charges capacitor C_R to an initial value of positive voltage. Because the resonant frequency of L_R and C_R is high, this transfer is accomplished in a relatively short period, T_0 to T_1, as shown in Fig. 668.

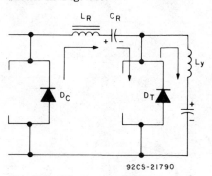

Fig. 669—Effective configuration of the deflection circuit during the switchover from retrace to trace, time T_0.

Recharging and resetting actions:
The action required to restore energy to the commutating circuit and to reset the trace SCR are also very important considerations in the operation of the basic deflection circuit. Both actions involve the inductor L_{CC}.

During the retrace period, inductor L_{CC} is connected between the dc supply voltage (B+) and ground by the conduction of either the commutating-switch SCR or diode (SCR_C or D_C), as indicated in Fig. 670. When the diode and the SCR cease to conduct, however, the path from L_{CC} to ground is opened. The energy stored in inductor L_{CC} during the retrace interval then charges capacitor C_R through the B+ supply, as shown in Fig. 671. This

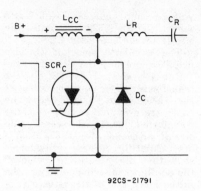

92CS-21791

Fig. 670—Circuit elements and current path used to supply energy to the charging choke L_{CC} during period from the start of retrace switching action to the end of the first half of the retrace interval, time T_3 to T_1.

charging process continues through the trace period until retrace is again initiated. The resultant charge on capacitor C_R is used to resupply energy to the yoke circuit during the retrace interval.

The voltage developed across inductor L_{CC} during the charging of capacitor C_R is used to forward-bias the gate electrode of the trace SCR properly so that this device is made ready to conduct. This voltage is inductively coupled from L_{CC} and applied to the gate of SCR_T through a wave-shaping network formed by inductor L_G, capacitor C_G and resistor R_G. The resulting voltage signal applied to the gate of SCR_T has the desired shape and amplitude so that SCR_T conducts when a forward bias is applied from anode to cathode, approximately midway through the trace interval.

Effect of auxiliary capacitor C_A: In the preceding discussions of the operation of the deflection circuit, the effect of capacitor C_A was neglected. Inclusion of this capacitor affects some of the circuit waveforms, as shown in Fig. 672, aids in the turn-off of the trace SCR, reduces the retrace time, and provides additional energy-storage capability for the circuit.

During most of the trace interval (from T_0 to T_4), including the interval (T_3 to T_4) during which the commutating pulse occurs, the trace switch is closed, and capacitor C_A is in parallel with the retrace capacitor C_R. From the start of retrace at time T_4 to the beginning of the next trace interval at time T_0, the trace switch is open. For this condition, capacitor C_A is in series with the yoke L_y and the retrace

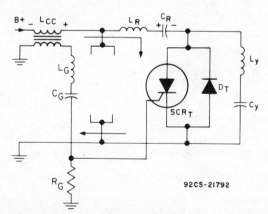

92CS-21792

Fig. 671—Effective configuration of the deflection circuit for resetting (application of forward bias to) the trace SCR and recharging the retrace capacitor C_R, during time interval from T_1 to T_3.

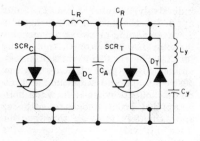

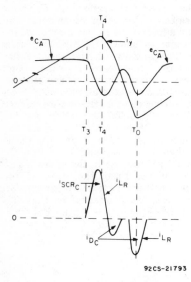

92CS-21793

Fig. 672—Circuit configuration showing the addition of auxiliary capacitor C_A and current and voltage waveforms showing the effect of this capacitor.

capacitor C_R so that the capacitance in the retrace circuit is effectively decreased. As a result, the resonant frequency of the retrace is increased, and the retrace time is reduced.

The auxiliary capacitor C_A is also in parallel with the retrace inductor L_R. The waveshapes in the deflection circuit are also affected by the resultant higher-frequency resonant discharge around this loop. The

voltage and current waveforms shown in Fig. 672 illustrate the effects of the capacitor C_A.

The auxiliary capacitor C_A also helps to prevent the fast-rise-time voltages developed by the flyback transformer from appearing across the trace switch. Fig. 673 shows the basic trace-switch circuit and the waveforms developed across this circuit with and without the auxiliary capacitor. These waveforms show that the integrating action of the auxiliary capacitor eliminate the initial steep rise of the retrace voltage pulse.

High-Voltage Generation—T h e SCR horizontal-deflection system generates the high voltage for the picture tube in essentially the same manner as has been used for many years in television receivers, i.e., by transformation of the horizontal retrace pulse to a high voltage with a voltage step-up transformer and subsequent rectification of this stepped-up voltage. In common with other

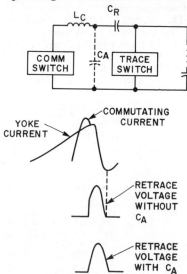

92CS-21794

Fig. 673—Simplified schematic of trace-switch circuit and waveforms showing effect of auxiliary capacitor on the rise time of the retrace voltage pulse.

solid-state receiver designs, a solid-state voltage multiplier is used as the high-voltage rectifier. A high-voltage rectifier tube such as the 3CZ3 could also be used although the increased source impedance in the high-voltage transformer would result in slightly poorer high-voltage regulation.

High-Voltage Regulation—The use of a silicon voltage multiplier for the high-voltage rectifier, together with very tight coupling between the primary and secondary of the high-voltage transformer, results in a high-voltage system that has very low internal impedance. As a result, it is necessary to regulate the high voltage only against changes in line voltage. The regulator, shown in Fig. 674, is reactive (non-dissipating) and provides good reliability at low cost.

A supplementary winding on the high-voltage transformer provides a pulse voltage proportional to the supply voltage. Because the supply voltage varies directly with line voltage, the voltage pulse provides an excellent reference for sensing variations in the line voltage. This voltage pulse is rectified and used as the collector voltage source for the regulator transistor. By means of a resistive voltage divider and the zener diode, it also provides base bias to the regulator transistor. When the voltage, as determined by the resistive divider, exceeds the zener voltage, the transistor conducts and current flows through the control winding of the saturable reactor. This current saturates the core of the saturable reactor (to a degree dependent upon the base voltage applied to the regulator transistor) and the inductance of the

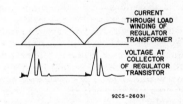

CURRENT THROUGH LOAD WINDING OF REGULATOR TRANSFORMER

VOLTAGE AT COLLECTOR OF REGULATOR TRANSISTOR

92CS-26031

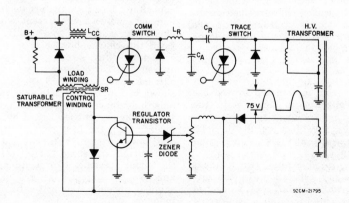

Fig. 674—High voltage regulator circuit and operating voltage and current waveforms.

load winding drops sharply. Because the load winding is in parallel with the input reactor, L_{CC}, it limits the amount of energy that can be stored in L_{CC} and, therefore, the amount of energy that can be stored in L_R, C_R, and C_A. Consequently, as line voltage increases, the amount of energy stored in these components is limited, and the increase in high-voltage is limited accordingly. If line voltage decreases, the pulse voltage applied to the regulator circuit is reduced, less current is drawn by the regulator transistor, and the degree of core saturation of the saturable reactor is reduced. Consequently, more "relative" energy can be stored in L_{CC}, even though the input voltage to the system is reduced, and the high voltage remains constant.

The high-voltage regulator system, as mentioned previously, dissipates very little energy and keeps the high-voltage constant with variations in horizontal-oscillator frequency or with component values. The response time of the system is very short so that essentially every horizontal line is regulated. Although the major function of the system is to maintain a constant high voltage (and scan) with variations in line voltage, it does provide some supplemental regulation of high voltage with picture-tube beam current, because there is some variation in supply voltage with the power drawn from the high-voltage supply.

Linearity Correction—Some S-shaping can be obtained by the use of a capacitor in series with the yoke. The nonlinearity caused by the yoke resistance results in left-hand stretch and right-hand compression.

Fig. 675 illustrates two methods of linearity correction. In the circuit shown in Fig. 675(a), a damped series-resonant circuit is connected between an auxiliary winding on the high-voltage transformer and the

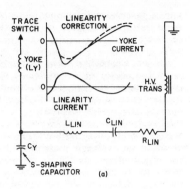

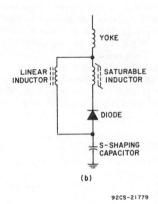

92CS-21779

Fig. 675—Two methods of linearity correction.

ungrounded side of the S-shaping capacitor. This circuit produces a damped sine-wave of current which effectively adds to and subtracts from the charge on the S-shaping capacitor, thus altering the yoke current to correct for any trace-current nonlinearity. The circuit shown in Fig. 675(b) acts as a variable inductance in series with the yoke. Yoke current is blocked by the diode during the first part of trace and flows through the linear inductor. During the second part of trace, the diode becomes forward-biased and yoke current is gradually shunted through the self-saturable inductor. With the proper values for the two inductors, the equivalent inductance in series with the yoke varies just the right amount to produce the proper degree of linearity correction.

Raster Correction—The distance from the center of deflection to the outside edge of the raster on the picture tube is greatest at the corners of the raster, decreasing to a minimum at the center. Because the electron beam must travel a greater distance to reach the corners of the raster, a given deflection of the beam produces a greater movement on the faceplate of the picture tube, and a type of distortion known as "pincushion" is produced. This effect is shown in Fig. 676. The degree

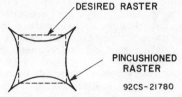

Fig. 676—Effect of pincushion distortion.

of pincushion distortion increases with deflection angle. Correction of this type of raster distortion, therefore, is of greater importance with wide-angle picture tubes.

Correction of pincushion distortion can be accomplished by decreasing the deflection (yoke current) at the corners of the raster or by increasing the deflection at the center

of the raster. The usual method is to reduce the yoke current as the beam approaches the corners of the raster. One method of pincushion correction is shown in Fig. 677. In this circuit, the collector supply for transistor Q_2 is the voltage across the 0.68-microfarad capacitor C_1 in the primary of the high-voltage transformer. Loading of this capacitor by transistor Q_2 increases the energy being drawn from the high-voltage transformer and thus reduces scan. Transistor Q_2 is driven by a vertical sawtooth. During the second half of vertical trace, the 10-microfarad capacitor C_2 is discharged by the collector current of Q_2, thereby loading the capacitor C_1 with an increasing current from the middle to the end of vertical scan. When transistor Q_2 is turned off at the end of vertical scan, capacitor C_2 is again charged by the energy stored in the capacitor C_1. The loading of capacitor C_1 decreases toward zero from the top toward the middle of vertical scan as soon as the capacitor C_2 charges.

Another method of correcting pincushion distortion is shown in Fig. 678. In this circuit, the "control winding" of the saturable transformer is supplied with a vertical sawtooth (preferably somewhat parabolic in shape) which determines the degree of core saturation of the transformer and thus the im-

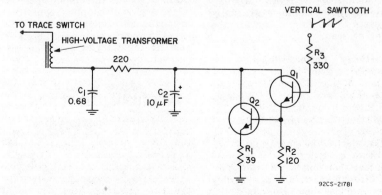

Fig. 677—Active side-pincushion correction circuit.

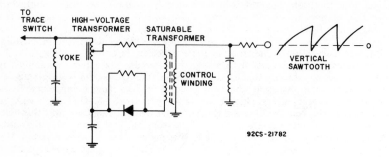

Fig. 678—Saturable reactor side-pincushion correction circuit.

pedance of the secondary windings, which are shunted across a portion of the primary of the high-voltage transformer. During the second half of vertical scan, the current through the control winding gradually increases from zero at the center of scan to maximum at the end of scan. The degree of core saturation also increases, and the inductance of the windings in shunt with the high-voltage transformer (and yoke) decreases accordingly. The loading of the high-voltage transformer thus gradually increases from a minimum at the center of vertical scan to a maximum at the end of scan. At the beginning of scan, the current through the control winding is at a maximum, and the degree of loading of the yoke is maximum. The current in the control winding (and the yoke loading) gradually decreases to a minimum toward the center of scan.

Auxiliary Power Supplies—An important area of potential cost reduction in the SCR deflection system is the power supply. SCR's have much greater current-carrying capability than that required for deflection. This extra capability can be used to derive the operating power for other portions of the receiver from the horizontal-deflection system. Fig 679 shows several possible methods of deriving power from the deflection system.

An advantage of the circuits shown in (c) through (e) of Fig. 679 is that they are regulated against changes in line voltage by the same regulator circuit used for regulating scan against changes in line voltage.

Rectifiers used to obtain dc power from the horizontal-deflection system operate at 15.75 kHz. Therefore, fast-recovery types must be used.

Over-all SCR Deflection System—Fig. 680 shows the circuit diagram of a complete SCR horizontal deflection system. This system is designed to operate directly from the rectified line voltage. A 250-volt unregulated supply for the video amplifier is obtained by rectifying the voltage obtained from the input reactor (L_{CC}). A 40-volt supply (regulated) for the small-signal stages is obtained by rectifying a pulse obtained from an auxiliary winding on the high-voltage transformer.

VERTICAL DEFLECTION

The vertical-deflection circuit in a television receiver is essentially a class A audio amplifier with a complex load line, severe low-frequency requirements (much lower than 60 Hz), and a need for controlled linearity. The equivalent low-frequency response for a 10-per cent

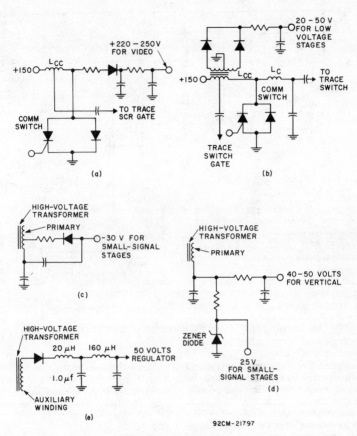

92CM-21797

Fig. 679—Various circuit arrangements for developing an auxiliary low voltage from the SCR deflection system.

deviation from linearity is 1 Hz. A simple circuit configuration is shown in Fig. 681.

The required performance can be obtained in a vertical-deflection circuit in any of three ways. The amplifier may be designed to provide a flat response down to 1 Hz. This design, however, requires an extremely large output transformer and immense capacitors. Another arrangement is to design the amplifier for fairly good low-frequency response and predistort the generated signal.

The third method is to provide

extra gain so that feedback techniques can be used to provide linearity. If loop feedback of 20 or 30 dB is used, transistor gain variations and nonlinearities become fairly insignificant. The feedback automatically provides the necessary "predistortion" to correct low-frequency limitations. In addition, the coupling of miscellaneous signals (such as power-supply hum or horizontal-deflection signals) in the amplifying loop is suppressed.

The inductance of the output transformer must be fairly low for maximum efficiency. When a circuit

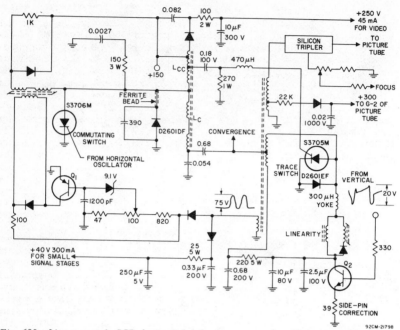

Fig. 680—Line-operated, SCR horizontal-deflection system for 90-degree color picture tube.

is designed for maximum efficiency, the transistor dissipation must be at least three times the yoke power. When interchangeability, line-voltage variations, and bias instability are considered, the dissipation may reach high levels (e.g., 14 watts in a 25-inch color receiver); as a result, expensive bias techniques and extruded-aluminum heat sinks must be used.

Use of a toroid yoke having an L/R time constant of 3.2 milliseconds

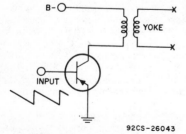

Fig. 681—Simple vertical-deflection circuit.

reduces the maximum dissipation to 3 or 4 watts and allows the plated steel chassis to be used as the heat sink for the transistor. The output transformer may also be reduced in size.

The higher Q of the toroid yoke normally results in a long retrace time or a very high flyback voltage.

Basic Design Approach

In recent commercial television receivers, the Miller-integrator concept is employed in the generation of the linear ramp of current required in the vertical-deflection yoke. Fig. 682 shows the basic configuration of a Miller-integrator type of vertical-deflection circuit. In this circuit, a high-gain amplification system is used to develop the drive current for the yoke winding, and the integrating capacitor is connected in shunt with the yoke and the amplifier system. In effect, the

Miller circuit multiplies the capacitor charging current by a factor equal to the gain of the amplifier without feedback. This technique results in an extremely linear output current waveform. In addition, variations in supply voltage, amplifier gain, and other factors that drastically affect the output of conventional vertical-deflection circuits have but slight adverse effects in the Miller circuit because of the large degenerative feedback.

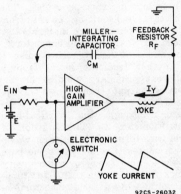

Fig. 682—Basic Miller Sweep Circuit.

At the beginning of the vertical-trace interval, the integrating capacitor C_M is charged from a voltage source E. The resulting voltage across the capacitor causes the amplifier to supply current to the yoke

winding and to the feedback resistor R_F, which is directly coupled to the integrating capacitor. The feedback action of the integrating capacitor tends to maintain a constant input to the amplifier so that the voltage across the capacitor builds up (integrates) at a constant rate. Because the voltage across the feedback resistor, which is essentially the same as the voltage across the integrating capacitor, is directly proportional to the yoke current, the yoke current increases at a constant rate, and a linear scan results. The sweep rate is determined by an electronic switch which discharges the integrating capacitor at the end of each scan period.

The amplifiers used in the vertical-deflection system are similar to those used in any high-gain audio-amplifier system. Either conventional transformer-coupled types or transformerless true-complementary-symmetry or quasi-complementary-symmetry types may be used. The following paragraphs describe the use of different types of output amplifiers and their associated circuitry in vertical-system applications.

Vertical Circuit that Uses a Conventional Output Stage

Fig. 683 shows the basic functional relationship among the vari-

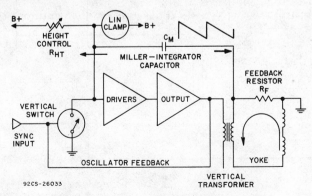

Fig. 683—Basic configuration of a Miller-integrator vertical-deflection system that uses a conventional transformer-coupled output stage.

ous stages of a Miller-integrator vertical-deflection circuit used in a recent commercial color-television receiver. The vertical-switch circuit controls the trace and retrace times and, therefore, the over-all operating frequency of the circuit. The switching action of the vertical switch is made self-sustaining by use of positive feedback from the output stage. Vertical synchronizing pulses applied to the switch from the sync separator determine the exact instant at which the switch is triggered on and, in this way, synchronize the switching action with the transmitted scanning interval. The Miller high-gain amplification system includes predriver and driver stages in addition to a conventional transformer-coupled output power-amplifier stage. The Miller-integrator capacitor is connected between the yoke winding and the input to the predriver so that it shunts the gain stages. The linearity-clamp circuit provides the initial charging current for this capacitor.

Vertical Switch—The vertical switch discharges the Miller-integrator capacitor at the end of the vertical scanning interval and, in this way, causes beam retrace and

prepares the circuit for a subsequent scanning interval. Fig. 684 shows the schematic diagram and operating waveforms for the vertical-switch circuit. The operation of the circuit is made self-sustaining by two feedback signals.

One feedback signal is applied to the base of the vertical-switch transistor from a secondary winding on the vertical-output transformer through resistors R_3 and R_4. This feedback signal is referred to as the triggering or turn-on pulse. The vertical synchronizing pulses from the sync separator are integrated by resistors R_1 and R_2 and capacitor C_2 and added to the triggering pulse.

Another feedback signal from a different secondary winding on the vertical-output transformer is applied to the base of the switch transistor through the vertical-hold potentiometer R_H. The addition of this waveform to the turn-on waveform causes the voltage at the base of the switch transistor to pass very quickly through the transistor turn-on voltage. As a result, the turn-on action of the vertical switch is very stable and relatively immune to noise voltages. The vertical-hold potentiometer provides some control over the shape of the latter feedback waveform and, therefore, offers

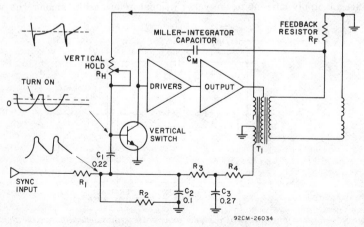

Fig. 684—Vertical-switch circuit.

limited control over the exact point at which the switch turns on.

Driver Stages—Two common-emitter stages (predriver and driver) provide the amplification required to increase the amplitude of the vertical-switch output sufficiently to drive the vertical-output stage. Fig. 685 shows a simplified

the Miller-integrator capacitor C_1, which is charged through the height-control potentiometer R_{HT}. The height-control supply voltage is made relatively immune to temperature-caused variations by the thermistor R_T. This supply also receives some dynamic regulation from a voltage supplied from the horizontal-deflection system. The addition of

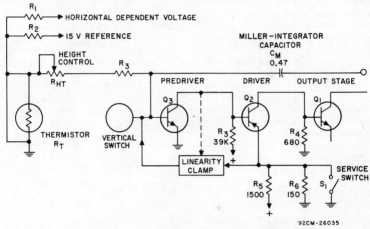

Fig. 685—Vertical predriver and driver stages.

circuit diagram of the driver stage.

The vertical predriver employs an n-p-n transistor Q_3 that is directly coupled to the p-n-p transistor Q_2 used in the driver stage. The emitter supply voltage for the driver is obtained from the voltage-divider network formed by resistors R_5 and R_6. The collector load of the driver consists of the parallel combination of the 680-ohm resistor R_4 and the base-emitter junction of the output-stage transistor Q_1. The service switch S_1 included in the emitter circuit of the driver can be used to cut off the vertical scanning during set-up adjustments of the picture tube if desired. When this switch is closed, the emitter of the driver is shorted to ground, and no vertical-deflection signals are developed.

The predriver input waveform is supplied by the charging action of

this regulating voltage helps to maintain a constant vertical height with respect to horizontal-scan and high-voltage variations.

Vertical Output Stage—Fig. 686 shows the circuit details for the output stage of the vertical system. This stage, which is directly driven by the driver circuit, uses a transistor operated in a common-emitter amplifier configuration to develop the power necessary to produce the required vertical deflection of the picture-tube beams. The collector load circuit consists of the vertical-output transformer T_1 and the vertical convergence circuitry. The secondary of the vertical-output transformer is loaded by the vertical yoke windings, two feedback paths, and the pincushion-correction circuitry. The Miller-integrator capa-

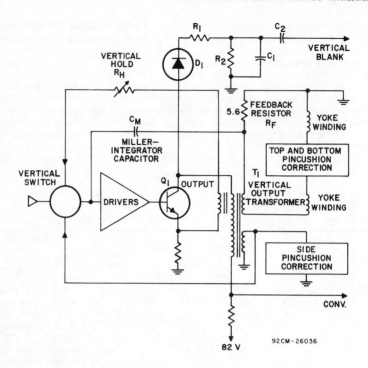

Fig. 686—Transformer-coupled vertical output circuit.

citor C_M is coupled to the 5.6-ohm feedback resistor R_F, which is connected in series with the output-transformer secondary and the windings of the vertical-deflection yoke. Two feedback waveforms are provided from the output stage (from separate secondary windings on the output transformer) to the vertical switch to assure stable, self-sustaining switch operation.

The diode D_1 and the filter network formed by resistor R_2 and caacitor C_1 form a protective clamp circuit for the output transistor. Positive-going retrace pulses cause the diode D_1 to conduct and capacitor C_1 charges rapidly through the short-time-constant path provided by diode D_1 and resistor R_2. After the retrace pulse is removed, the capacitor attempts to discharge through the resistor R_2. Because of the long-time-constant path provided by this

resistor, the capacitor is only allowed to discharge an amount sufficient to assure a voltage differential across the diode when the retrace pulses occur. This action effectively clamps the collector output of transistor Q_1 to the voltage across capacitor C_1. The pulses that appear across this capacitor during the conduction of the diode are coupled by capacitor C_2 to the television-receiver video-amplifier circuit for use in vertical-retrace blanking.

Linearity Clamp—A circuit referred to as the linearity clamp is included in the vertical-deflection system to assure that sufficient initial-scan charging current is provided for the Miller-integrator capacitor. Fig. 687 illustrates the action of this circuit.

When the Miller-integrator capacitor C_M is discharged by the vertical switch at the end of a vertical-scan interval, the capacitor discharges into the base circuit of the predriver stage, and the predriver transistor is cut off. The positive voltage that then appears at the collector of the predriver transistor forward-biases the p-n-p linearity-clamp transistor, and current flows through this transistor, resistor R_2, and the vertical switch.

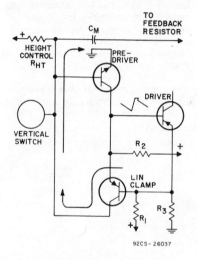

Fig. 687—Linearity clamp.

After approximately 700 microseconds, the vertical switch turns off, and the current through the linearity clamp is used to provide rapid initial charging of the Miller-integrator capacitor C_M. As the initial charge quickly builds up on the capacitor, the predriver and driver stages start to conduct, and the base-emitter junction of the linearity-clamp transistor is reverse-biased by the voltage drop across the base-emitter junction of the driver transistor. This action cuts off the linearity-clamp circuit and initiates another vertical-scan interval. The Miller-integrator capacitor continues to charge through the height-control potentiometer R_{HT} for the duration of the scan interval.

Vertical Circuit that Uses Complementary-Symmetry Output Stage

The introduction of complementary pairs of power transistors has led to the development of class B transformerless output stages that are both economical and efficient. In vertical-output applications, such circuits may be capacitively coupled to the yoke, and the output transformer, together with the problems of nonlinearity, low-frequency phase shift, and excessive retrace pulse amplitudes associated with it, can be eliminated. Regardless of the type of output stage used, the generation of a linear sawtooth by use of the Miller-integrator circuit has become widespread.

Fig. 688 shows a block diagram of a vertical-deflection system of this type that uses a true-complementary-symmetry output stage. The vertical switch controls the free-running frequency of the vertical system. The high-gain amplifier consists of a direct-coupled predriver and driver, in addition to the

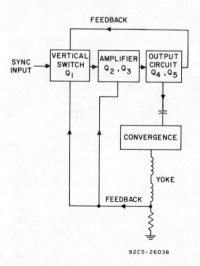

Fig. 688—Block diagram of a vertical-deflection system that uses a true-complementary-symmetry output stage.

true-complementary-symmetry output stage. The output stage is capacitively coupled to the convergence circuitry and the vertical-deflection yoke.

Vertical-Switch and Predriver Circuit—Fig. 689 shows the circuit configuration of the predriver circuit and its interconnection with the vertical-switch circuit. An increase in the positive voltage at the junction of resistors R_{12} and R_{13} will increase the raster height. Because the major source of the voltage to R_{13} is obtained from the +15-volt regulated supply through resistors R_{11} and R_{12}, as the resistance of R_{12} is decreased, raster height increases. If the setup switch S_1 is closed to

the service position, the supply to R_{13} is diminished practically to zero, and the raster collapses.

Because the output-stage transistor Q_5 is cut off while the top half of the raster is scanned, the voltage at the collector of this transistor is zero until vertical scan reaches the center. During the bottom half of scan, the collector current of output transistor Q_5 increases linearly, so that the voltage fed back to R_{13} tends to "stretch" the lower part of the raster, to overcome some tendency towards bottom compression. Feedback to the vertical-switch transistor also is derived from resistor R_1.

The remaining input to resistor R_{13} is obtained from the horizontal

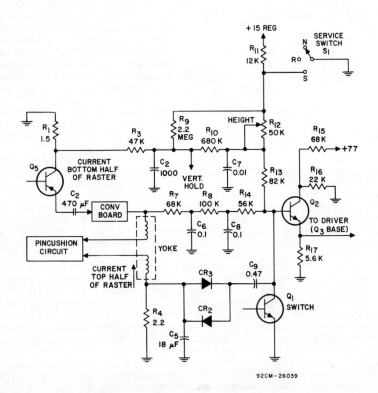

92CM-26039

Fig. 689—The vertical predriver with its inputs.

system. As high-voltage return current to the brightness limiter increases or decreases, the voltage at the junction of R_5 and R_6 also varies. An increase in picture-tube current, therefore, reduces slightly the voltage to the height control and causes a slight decrease in vertical deflection. This action causes scanning height to track scanning width.

A feedback signal is fed to capacitor C_9 from the junction of the system feedback resistor R_4 and the yoke. If the effect of capacitor C_5 is ignored, the voltage at this point reaches its maximum positive value at the beginning of scan, passes through zero, and reaches maximum negative just before vertical retrace. Therefore, the feedback to the predriver transistor Q_2 is degenerative, because voltage at the base of Q_2 tends to rise throughout the scanning interval. Capacitor C_5 is used to filter out any horizontal-deflection voltage which may be present.

The transistor vertical-switch circuit shown in Fig. 690 performs three functions. It controls the free-running frequency of the vertical-deflection system, allows synchronization with the received signal, and determines the duration of vertical retrace. The overall vertical system may be considered as a free-running oscillator. The base of switch transistor Q_1 is returned to the supply voltage (height-control B+) through resistor R_7, the hold control, and a 680-kilohm resistor R_8. If no sync pulses are present at the moment after the end of retrace, capacitor C_6 begins to charge, and the base of Q_1 begins to swing positive. When Q_1 begins conducting (about 17 milliseconds later), predriver and driver transistors Q_2 and Q_3, shown in Figs. 689 and 691 respectively, conduct less, and the output transistor Q_4, which was cut off during the lower half of vertical scan, resumes conduction. Because the voltage across the yoke inductance leads the current through it, a sharp positive

pulse appears at the input to resistor R_1, and this pulse, coupled to the base of Q_1, drives Q_1 into saturation. This transistion of Q_1 from cutoff to saturation is very rapid.

Capacitor C_5 and inductor L_1, connected from the junction of resistor R_1 and capacitor C_4 to ground, are series resonant at the horizontal-scan frequency, and shunt to ground any 15.734-kHz energy which may be present. The presence of horizontal ripple at the vertical switch tends to synchronize the vertical scan with the horizontal scan and causes a degradation of interlace. Resistor R_6 and capacitor C_1 shape the feedback pulse so that the transition of Q_1 from cutoff to saturation is as rapid as possible.

When Q_1 saturates, Q_4 reaches maximum conduction, and the yoke current rises to maximum in the direction which produces maximum upward deflection. During retrace, the base current of transistor Q_1 charges capacitor C_3 negatively. The duration of the scanning is determined by the length of time required for the base of Q_1 to become forward-biased once more.

A second feedback circuit improves the frequency stability of the oscillator circuit. During the top half of scan, output transistor Q_5 is cut off, and the voltage at the junction of resistors R_9 and R_{10} is essentially zero. Therefore, the voltage rise at the base of switch transistor Q_1 is exponential. But, as scan nears the bottom of the raster, transistor Q_4 conducts, and causes a positive voltage to be developed across resistor R_9. This voltage sharpens the voltage rise at the base of Q_1, so that its transistion from cutoff to saturation is more rapid. Similarly, the sharp drop in voltage across R_9 (from maximum to zero during the first half of retrace) enhances the cut-off characteristics of the Q_1 circuit.

The composite sync signal is introduced into the vertical system at terminal 12. Resistor R_2 and capacitor C_2 integrate the input so that the horizontal sync pulses are reduced in

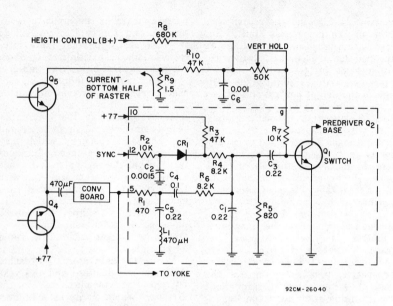

Fig. 690—Vertical-switch circuit.

amplitude to about 8 volts and the vertical pulses about twice this amplitude. Since diode CR_1 has about 12 volts of positive bias on its cathode, only the vertical sync pulse can pass to the switch transistor. If the free-running frequency of the vertical system is slightly less than the vertical-sync rate, Q_1 is at the threshold of conduction when each sync pulse arrives, so that the vertical system is synchronized at the vertical-sync pulse rate.

Vertical Driver and Output Stage —Fig. 691 shows the schematic diagram of the vertical-system driver and of the output stage with the yoke circuit simplified. The circuit configuration is very similar to that of a high-quality audio power amplifier. The yoke itself is analogous to the speaker voice coil, C_C is the coupling capacitor, and R_y is the equivalent of the total resistance of the yoke and convergence circuit. The value of capacitor C_C is selected to provide maximum energy transfer at the vertical scanning frequency. Feed-

back to the Miller capacitor is developed across resistor R_{13}, and capacitor C_{10} is a filter.

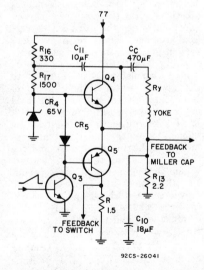

Fig. 691—Vertical driver, output, and simplified yoke circuit.

During retrace, transistor Q_3 is cut off, and its collector voltage rises towards the supply voltage; however, the 65-volt zener diode CR_4 limits the maximum base bias of transistor Q_4 and, in this way, limits the yoke retrace current. During the scanning interval, the bases of transistors Q_4 and Q_5 are driven progressively less positive at a linear rate. Conduction is through Q_4 during most of the retrace time and as scan passes from the top of the raster to center. The voltage across capacitor C_C at vertical scan center has

reached maximum (90° out of phase with the current), and during the lower half of scan, capacitor C_C discharges back through the yoke and transistor Q_4. This current increases at a linear rate, because the forward bias on the base of transistor Q_4 is increasing at a linear rate.

The diode connected between the bases of transistors Q_4 and Q_5 improves the switching characteristics of the transistors at mid-scan. Q_5 has zero bias as long as Q_4 is conducting. Therefore, only slight voltage swings are necessary to cut off

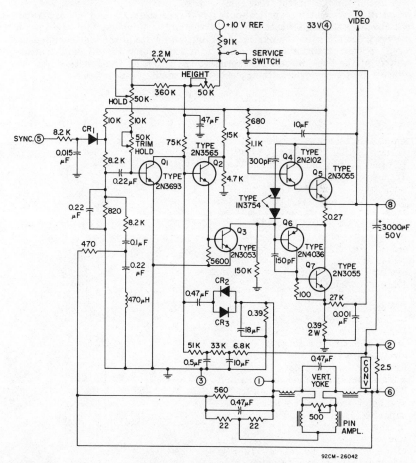

92CM - 26042

Fig. 692—Complete transistor vertical-deflection system that uses a quasi-complementary-symmetry output stage.

Q_4 and turn on Q_5 at the center of the raster. If the diode were shorted or bypassed, reverse bias would exist between base and emitter of Q_5 while Q_4 was conducting, and consequently there would be appreciably more disturbance in the circuit during transition time.

Vertical Circuit that Uses a Quasi-Complementary-Symmetry Output Stage

A disadvantage of the true-complementary-symmetry vertical-output circuit is the higher cost of p-n-p power transistors in comparison to n-p-n power transistors. Because control of the base diffusion is more difficult in p-n-p devices, their cost is generally 25 per cent more than comparable n-p-n devices.

Fig. 692 shows the complete circuit diagram for a vertical-deflection system that uses a quasi-complementary-symmetry output stage to drive a low-impedance toroidal yoke (L = 950 microhenries, R = 1.5 ohms). This system is basically the same as described earlier in which a true-complementary symmetry output stage is used, with the exception of some minor modifications necessary to supply the higher deflection current required for the toroidal yoke.

Transistors Q_3 and Q_5 are functionally equivalent to the n-p-n output device in the true-complementary-symmetry circuit and transistors Q_4 and Q_6 function as an equivalent p-n-p device.

TV Chroma Circuits

I N the transmission of picture signals for color-television receivers, all the color information is contained in three signals, a luminance (black-and-white) or monochrome signal and two chrominance signals. The luminance signal, which is called the Y signal, contains brightness information only. The voltage response of the Y signal is made similar to the brightness response of the human eye by use of a composite signal that contains definite proportions of the red, green, and blue signals from the color-television camera (30 per cent red, 59 per cent green, and 11 per cent blue). This Y signal, which includes sync and blanking pulses, provides a correct monochrome picture in a conventional black-and-white television receiver.

BASIC SYSTEM REQUIREMENTS

For the generation of color-television signals, the Y signal is subtracted from the red, green, and blue signals to provide a new set of color-difference signals, which are designated as R-Y, B-Y, and G-Y. All of the original picture information is contained in the Y signal, the R-Y signal, and the B-Y signal. Therefore, the G-Y signal is not contained in the transmitted signal, but is synthesized in the receiver by proper combinations of the R-Y and B-Y signals.

(Color signals transmitted under present color-television standards are not R-Y and B-Y, but a similar pair of signals designated as I and Q. In the color-television receiver, R-Y and B-Y sginals are demodulated directly from the I and Q signals with negligible loss of color quality. For purposes of simplicity, only R-Y and B-Y signals are considered in this explanation. In addition, a 90-degree phase-shift network is shown; the phase-shift angle could be, and often is, some other value.)

Because the luminance signal and the two color-difference signals must be transmitted with a standard 6-MHz channel, the two color signals are combined into one signal at the transmitter and are independently recovered at the receiver by proper detection techniques. A color subcarrier of approximately 3.58 MHz is used for transmitting the color information within the 6-MHz spectrum of the television station. As shown in Fig. 693, the 3.58-MHz subcarrier and one of the color-difference signals are applied directly to a balanced AM modulator. The other color-difference signal is applied directly to a second balanced AM modulator, and the 3.58-MHz subcarrier is applied to this second modulator through a 90-degree phase-shifting network. The balanced modulators effectively cancel both the individual color-difference signals and the subcarrier signal, and the output contains only

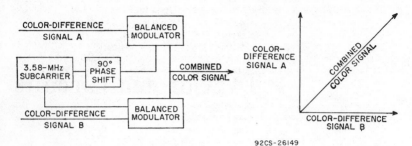

92CS-26149

Fig. 693—Formation of combined color signal for transmission.

the sidebands of the combined chrominance signal.

Recovery of the color information at the receiver involves a process called **synchronous detection.** In this process, two separate detectors are used to recover the separate color information, just as two separate modulators were used to combine the information at the transmitter. The 3.58-MHz subcarrier, which was suppressed during transmission, must be reinserted at the receiver for recovery of the color information. The basis of synchronous detection is the phase relationship of this reinserted 3.58-MHz subcarrier.

For example, the original color information is represented in Fig. 693 by the color-difference signals A and B. At the receiver, the combined color signal is fed to two demodulators A and B, as shown in Fig. 694. At the same time, a 3.58-MHz subcarrier is also fed to the two demodulators, with the same phase relationship that was used in the

modulators at the transmitter. This locally generated subcarrier essentially duplicates or replaces the original subcarrier, which was removed at the transmitter.

The local 3.58-MHz oscillator in the color-television receiver is made to function at the proper frequency and phase by means of a synchronizing signal sent out by the transmitter. The synchronizing signal consists of a short **burst** of 3.58-MHz signals transmitted during the horizontal blanking interval, immediately after the horizontal sync pulse, as shown in Fig. 695.

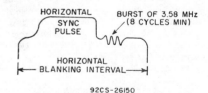

92CS-26150

Fig. 695—Waveform for synchronizing signal.

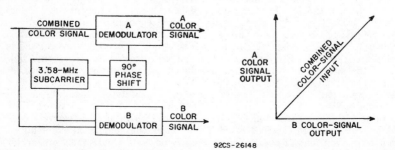

92CS-26148

Fig. 694—Separation of combined color signal into two signals at the receiver.

INTEGRATED-CIRCUIT CHROMA SYSTEMS

New color-TV receiver designs indicate a rapidly changing trend toward all solid-state circuitry with special emphasis on monolithic integrated circuits. The solid-state color-demodulator circuits in these receivers provide excellent demodulation gain, linearity, and uniformity. These advantages are easily achieved by the integrated-circuit design of the balanced demodulator. At present, the integrated circuit is economically adaptable to the current design requirements of the color-picture-tube circuits. Two complete integrated-circuit chroma systems are described in the following sections. One system uses a CA3126 for signal processing, and a CA3067 for demodulation; the other system uses a CA3070 signal processor and a CA3121 demodulator.

CA3126/CA3067 System

Fig. 696 shows a functional diagram of the complete TV-receiver chroma system that uses RCA-CA3126 and RCA-CA3067 monolithic integrated circuits. It is a phase-locked-loop system, using sample-and-hold circuit techniques.

CA3126 Signal Processor—The signal flow and organization of the CA3126 are shown in block form in Fig. 697. The composite chroma signal is applied to the first chroma amplifier. The output from this stage proceeds along three paths. The first path leads to the doubly-balanced wide-band AFPC detector. Here the burst signal is compared with the reference carrier to produce the required error signal for synchronization. Two sample-and-hold circuits serve to achieve high detection efficiency and bias stability. One sample-and-hold circuit samples the detected signal during the horizontal keying interval and stores the peak error signal in a filter capacitor. A second similar circuit provides an accurate reference potential as described later. The bias stability of this system is sufficient to eliminate the need for the adjustments required in conventional circuit design.

The detected and filtered burst signal controls the frequency and phase of a voltage-controlled oscillator (VCO) by operating on an electronic phase-shifter. The VCO consists of an amplifier-limiter followed by the electronic phase-shifter. A crystal filter located between the output of the phase-shifter and the

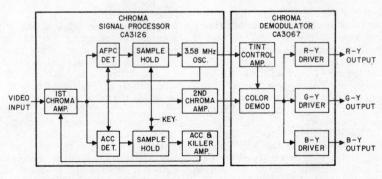

92CS-26077

Fig. 696—Simplified functional diagram of a two-package TV chroma system utilizing the CA3126 and CA3067.

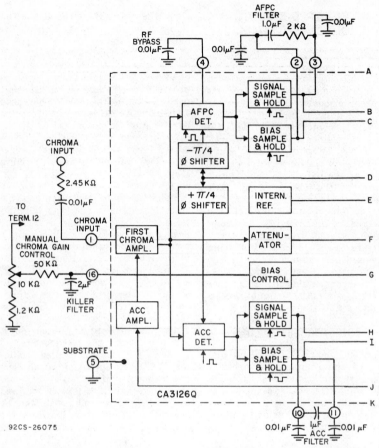

Fig. 697—Functional diagram of the RCA-CA3126 (continued on page 563).

input of the amplifier-limiter closes the loop of the VCO. The filtered oscillator signal is amplified to produce the required reference carriers for the AFPC and ACC synchronous detectors. The required quadrature relationship is obtained by $+ \pi/4$ and $-\pi/4$ radian integrated phase-shift networks.

The ACC-killer detector is similar in structure to the AFPC detector, and is also driven from the first chroma amplifier stage. It detects synchronously the in-phase component of the burst signal and produces a pulse signal proportional in amplitude to the level of the burst signal. The resulting control signal passes through a sampling circuit, as described above, and is applied to the killer and ACC amplifiers. The action of both amplifiers is delayed so that the unkill action takes place prior to ACC and the latter is fully activated upon reaching the predetermined burst level. The ACC amplifier controls the gain of the first chroma amplifier so as to maintain the burst signal constant while the killer amplifier enables the output stage in the presence of the burst signal.

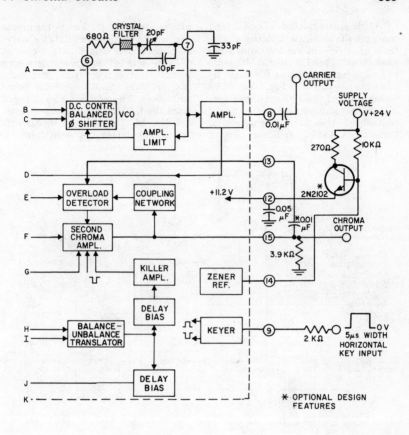

Fig. 697—Functional diagram of the RCA-CA3126 (continued from page 562).

92CS-26075

The signal level to the second chroma amplifier is reduced to one fourth of the available signal level to allow for the extremes of the chroma signal excursions. A horizontal rate keyer operating on this stage removes the burst signal so that the output stage is activated only during the horizontal scanning interval. A saturation control, available for front panel control, allows a continuous gain adjustment of this amplifier. A desirable feature of this control is the linear correspondence between the control bias and the chroma output signal. The chroma maximum level corresponds to the maximum bias potential without a dead spot at the extreme of the control range. A threshold type overload detector monitors the output signal and maintains the output from the second chroma amplifier below an arbitrary set level. This prevents the overload of the picture tube usually experienced on noisy or excessively large chroma signals. The required keying signals for the various functions are generated by two cascaded keyer stages in which either polarity pulses can be generated.

CA3067 Demodulator—The CA3067 contains the separate functional systems of a dc tint control and a demodulator. The functional diagram and external connection for this circuit are shown in Fig. 698. The phase shift of the tint amplifier system is accomplished by functional control of the fixed phase signal from the CA3126 oscillator output.

proximately 76°). These terminals are inputs to the demodulator drive amplifiers. The demodulators consist of two sets of balanced detectors which receive their reference subcarrier from the demodulator drive amplifiers. The chroma signal input from the CA3126 is applied to terminal No. 14. The chroma signal differentially drives the demodulators.

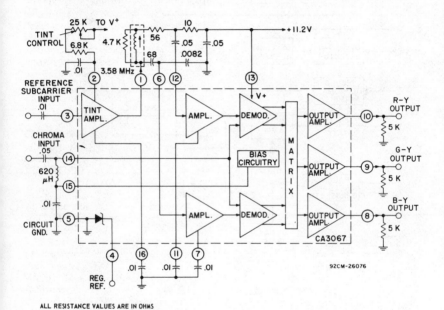

Fig. 698—Functional diagram of the RCA-CA3067.

This regenerated reference subcarrier is applied to terminal No. 3 and driven differentially into phase shift circuits. The tint adjustment controls the vector addition of phase shifted signals after which a limiting amplifier removes any remaining amplitude modulation. The output of the tint amplifier at terminal No. 1 is phase separated for the required reference subcarrier phase at terminals No. 6 and No. 12 (terminal No. 12 lags terminal No. 6 by approximately 76°).

The demodulation components are matrixed and dc-shifted in voltage to give R-Y, G-Y, and B-Y color difference components with close dc balance and proper amplitude ratios. The output amplifiers of the CA3067 are specially designed to meet the low-impedance driving source requirements of the high-level color output amplifiers. A special feature of the CA3067 is RC filtering of high-frequency demodulation components. Terminal No. 4 is a zener diode for use as a regulated voltage reference at 11.9 volts. When the zener reference element is not used, the power supply should be maintained at +11.2 ± 0.5 volts.

CA3070/CA3121 System

Fig. 699 shows a functional diagram of a TV receiver chroma system that uses RCA-CA3070 and RCA-CA3121 monolithic integrated circuits. This system, which is also a phase-locked-loop type, provides a high-level output.

CA3070 Signal Processor—The CA3070, shown in Fig. 700, is a complete subcarrier regeneration system with automatic phase control applied to the oscillator. As shown in Fig. 699, an amplified chroma signal from the CA3121 is applied to terminals No. 13 and No. 14, which are the automatic phase control (APC) and the automatic chroma control (ACC) inputs. APC and ACC detection is keyed by the horizontal pulse which also inhibits the oscillator output amplifier during the burst interval.

The ACC system uses a synchronous detector to develop a correction voltage at the differential output terminal Nos. 15 and 16. This control signal is applied to the input terminal Nos. 1 and 16 of the CA3121. The APC system also uses a synchronous detector. The APC error voltage is internally coupled to the 3.58-MHz oscillator at balance; the phase of the signal at terminal No. 13 is in quadrature with the oscillator.

To accomplish phasing requirements, an RC phase shift network is used between the chroma input and terminal Nos. 13 and 14. The feedback loop of the oscillator is from terminal Nos. 7 and 8 back to No. 6. The same oscillator signal is available at terminal Nos. 7 and 8, but the dc output of the APC detector controls the relative signal levels at terminal No. 7 or No. 8. Because

the output at terminal No. 8 is shifted in phase compared to the output at terminal No. 7, which is applied directly to the crystal circuit, control of the relative amplitudes at terminal Nos. 7 and 8 alters the phase in the feedback loop, thereby changing the frequency of the crystal oscillator. Balance adjustments of dc offsets are provided to establish an initial no-signal offset control in the ACC output, and a no-signal, on-frequency adjustment through the APC detector-amplifier circuit which controls the oscillator frequency. The oscillator output stage is differentially controlled at terminal Nos. 2 and 3 by the hue control input to terminal No. 1. The hue phase shift is accomplished by the external R, L, and C components that couple the oscillator output to the demodulator input terminals. The CA3070 includes a shunt regulator to establish a 12-volt dc supply.

CA3121 Demodulator — The CA3121, shown in Fig. 701, consists of three basic circuit sections: (1) amplifier No. 1, (2) amplifier No. 2, and (3) demodulator. Amplifier No. 1 contains the circuitry for automatic chroma control (ACC) and color-killer sensing. The output of amplifier No. 1 (Terminal 3) is coupled to the chroma signal processor (CA3070) for ACC and automatic phase control (APC) operation and to the input of amplifier No. 2 (Terminal 4) containing the chroma gain control circuitry. The signal from the color-killer circuit in amplifier No. 1 acts upon amplifier No. 2 to greatly reduce its gain.

The output from amplifier No. 2 (Terminal 14) is applied, through a filtering network, to the demodulator input (Terminal 13). The demodulator also receives the R-Y and B-Y demodulation subcarrier signals (Terminals 7 and 8) from the oscillator output of the chroma signal processor. The R-Y and B-Y demodulators and the matrix network contained in the demodulator section of the CA3121 reconstruct

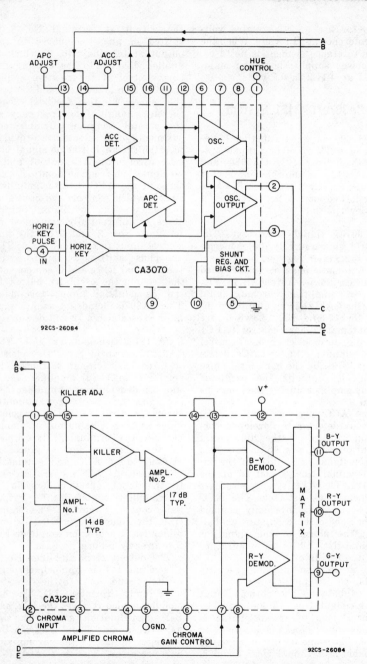

92CS-26084

92CS-26084

Fig. 699—Simplified functional diagram of a two-package TV chroma system utilizing the CA3070 and CA3121.

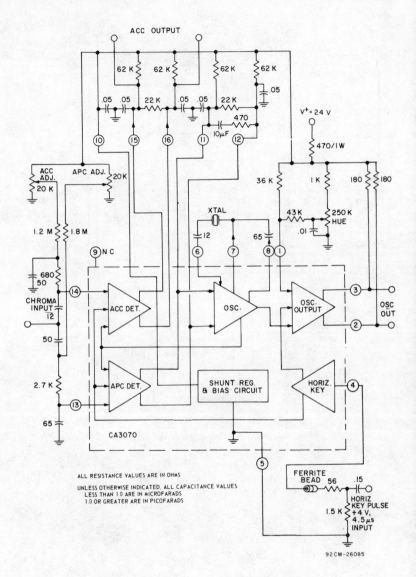

Fig. 700—Functional diagram of the RCA-CA3070.

the G-Y signal to achieve the R-Y, G-Y, and B-Y color difference signals. These high-level outputs signals with low impedance outputs are suitable for driving high-level R, G, B output amplifiers. Internal capacitors are included on each output to filter out unwanted harmonics.

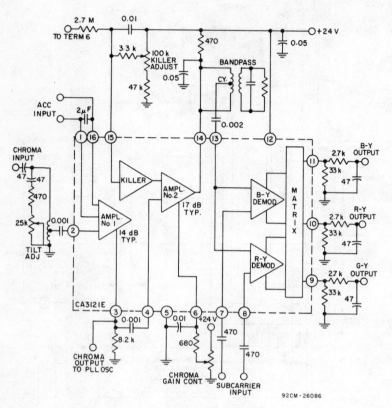

Fig. 701—Functional diagram of the RCA-CA3121.

Guide To RCA
Solid-State Products

THIS section is intended as a guide to the use of RCA transistors, thyristors, rectifiers, and integrated circuits. Package considerations are discussed, and techniques for mounting and handling are described. Tables present all of the types in the line of standard RCA solid-state devices, including military-specification (JAN, JANTX, and JANTXV) types and RCA high-reliability types and also all the RCA "top-of-the-line" SK-series replacement solid-state devices. Dimensional outlines are shown for all types, both standard and SK.

PACKAGES FOR SOLID-STATE DEVICES

RCA solid-state devices are supplied in both hermetic packages (metal and/or ceramic) and plastic packages. The photographs in Fig. 702 show the packages that are used for RCA transistors, thyristors, and rectifiers.

Fig. 703 shows the packages that are used for RCA integrated circuits. The different package designs offer variety in package parasitics, in mounting and connection considerations, in cooling techniques, in resistance to environmental effects, and in cost.

The volume and area of the package are important in determining the power dissipation capability of a solid-state device; chip mounting and encapsulation are also factors. The maximum allowable power dissipation in the device is limited by its junction temperature, which depends upon the ability of the thermal circuit to conduct heat away from the chip. The predominant mode of heat transfer is conduction through the silicon chip and through the case; the effects of internal free convection and radiation and lead conduction are small and may be neglected. The thermal resistance from pellet to case depends upon the pellet dimensions and the package configuration.

When the device is operated in free air, without a heat sink, the steady-state thermal circuit is defined by the junction-to-free-air thermal resistance. Thermal considerations require that there be a free flow of air around the device and that the power dissipation be maintained below that which would cause the junction temperature to rise above the maximum rating. When the device is mounted on a heat sink, however, care must be taken to assure that all portions of the thermal circuit are considered.

Fig. 704 shows the thermal circuit for a heat-sink-mounted device. This figure shows that the junction-to-ambient thermal circuit includes three series thermal-resistance components, i.e., junction-to-case, $R\theta_{JC}$; case-to-heat-sink, $R\theta_{CS}$; and heat-sink-to-ambient, $R\theta_{SA}$. The junction-to-case thermal resistance of the various device types is given in the in-

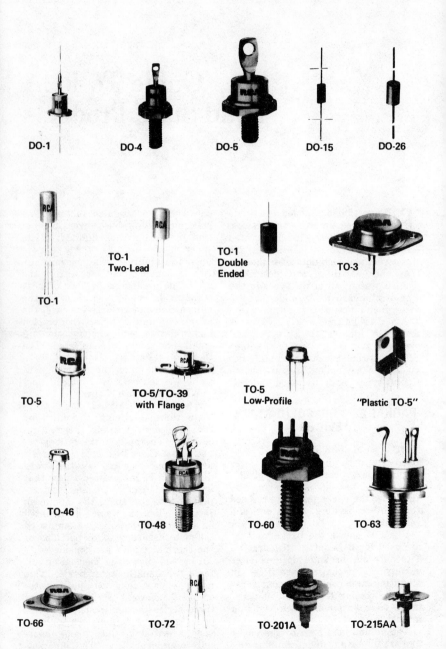

Fig. 702—*Hermetic and plastic packages used for RCA transistors, thyristors, and rectifiers (Continued on the next page).*

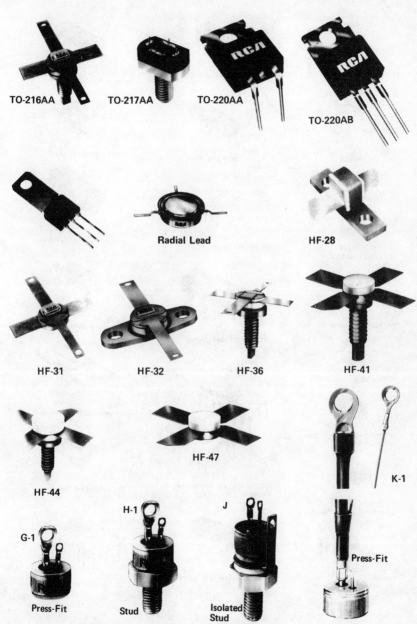

Fig. 702—Hermetic and plastic packages used for RCA transistors, thyristors, and rectifiers (Continued from page 570).

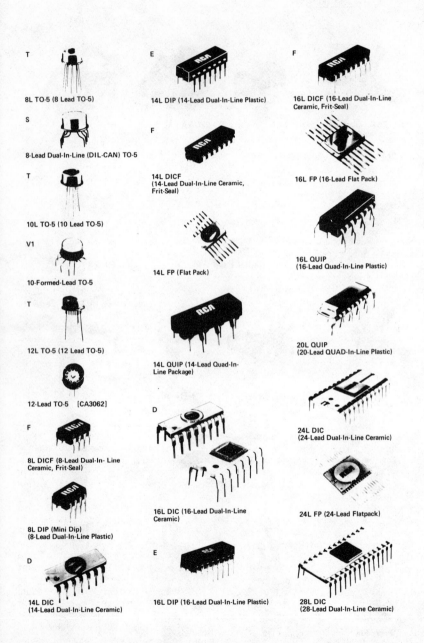

T
8L TO-5 (8 Lead TO-5)

S
8-Lead Dual-In-Line (DIL-CAN) TO-5

T
10L TO-5 (10 Lead TO-5)

V1
10-Formed-Lead TO-5

T
12L TO-5 (12 Lead TO-5)

12-Lead TO-5   [CA3062]

F
8L DICF (8-Lead Dual-In- Line Ceramic, Frit-Seal)

8L DIP (Mini Dip)
(8-Lead Dual-In-Line Plastic)

D
14L DIC
(14-Lead Dual-In-Line Ceramic)

E
14L DIP (14-Lead Dual-In-Line Plastic)

F
14L DICF
(14-Lead Dual-In-Line Ceramic, Frit-Seal)

14L FP (Flat Pack)

14L QUIP (14-Lead Quad-In-Line Package)

D
16L DIC (16-Lead Dual-In-Line Ceramic)

E
16L DIP (16-Lead Dual-In-Line Plastic)

F
16L DICF (16-Lead Dual-In-Line Ceramic, Frit-Seal)

16L FP (16-Lead Flat Pack)

16L QUIP
(16-Lead Quad-In-Line Plastic)

20L QUIP
(20-Lead QUAD-In-Line Plastic)

24L DIC
(24-Lead Dual-In-Line Ceramic)

24L FP (24-Lead Flatpack)

28L DIC
(28-Lead Dual-In-Line Ceramic)

Fig. 703—Hermetic and plastic packages used for RCA integrated circuits.

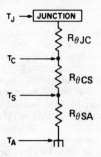

T_J =junction temperature
T_C =case temperature
T_S =heat-sink temperature
T_A =ambient temperature
$R_{\theta JC}$=junction-to-case thermal resistance
$R_{\theta CS}$=case-to-heat-sink thermal resistance
$R_{\theta SA}$=heat-sink-to-ambient thermal resistance

92CS-26386

Fig. 704—Thermal equivalent cicuit for a solid-state device mounted on a heat sink.

dividual technical bulletins on specific types. The heat-sink-to-ambient thermal resistance can be determined from the technical data provided by the heat-sink manufacturer, or from published heat-sink nomographs. The case-to-heat-sink thermal resistance depends on several factors, which include the condition of the heat-sink surface, the type of material and thickness of the insulator, the type of thermal compound, the mounting torque, and the diameter of the mounting hole in the heat-sink.

The dimensional outlines for all RCA solid-state device packages are shown at the end of this section.

MOUNTING AND CONNECTION TECHNIQUES

The selection of a particular method for mounting and connection of solid-state devices in equipment depends on the type of package involved; on the equipment available for mounting and interconnection; on the connection method used (soldered, welded, crimped, etc.); on the size, shape, and weight of the equipment package; on the degree of reliability and maintainability (ease of replacement) required; and, of course, on cost considerations.

In the following discussion, the information given applies to the package rather than the device unless otherwise specified. In other words, the discussion of handling and mounting of the TO-5 package is understood to cover mounting of transistors, silicon rectifiers, and thyristors in TO-5 packages.

Packages with Flexible Leads

Some solid-state device packages have flexible leads; these leads are usually soldered to the circuit elements. In all soldering operations, some slack or an expansion elbow should be provided in each lead to prevent excessive tension on the leads. Excessive heat should be avoided during the soldering operation to prevent possible damage to the devices. Some of the heat can be absorbed if the flexible lead of the device is grasped between the case and the soldering point with a pair of long-nosed pliers.

Although flexible leads can be bent into almost any configuration to fit any mounting requirement, they are not intended to take repeated bending. In particular, repeated bending at the point at which the lead enters the case should be avoided. The leads are not especially brittle at this point, but the sharp edge of the case produces an excessively small radius of curvature in a bend made at the case. Repeated bending with a small radius of curvature at a fixed point will cause fatigue and breakage in almost any material. For this reason, right-angle bends should be made at least 0.020 inch from the case. This practice will avoid sharp bends and maintain sufficient electrical isolation between lead connections and header. A safe bend can be assured if the lead is gripped with pliers close to the case and then bent the requisite amount with the fingers, as shown in Fig. 705. When the leads of

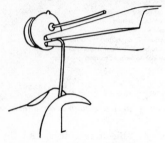

92CS-26387

Fig. 705—Method of bending leads on a flexible-lead package.

a number of devices are to be bent into a particular configuration, it may be advantageous to use a lead-bending fixture to assure that all leads are bent to the same shape and in the correct place the first time, so that there is no need for the repeated bending.

Transistors, thyristors, and rectifiers should be mounted on heat sinks when they are operated at high power levels. An efficient heat-sink method for transistors in JEDEC TO-5 and modified TO-5 packages is to provide intimate contact between the heat sink and at least one-half of the base of the device opposite the leads. TO-5 packages can be mounted to the heat sink mechanically, with glue or an epoxy adhesive; soldering, however, is preferable for thyristors and rectifiers. Not only is the solder bond both permanent and most efficient, but the thermal resistance $R_{\theta CS}$ from the case to the heat sink is easily kept below 1°C per watt under normal soldering conditions. Oven or hot-plate batch-soldering techniques are recommended because of their low cost. Transistors should not be soldered to the heat sink.

Packages with Mounting Flanges

The mounting flanges of packages such as the JEDEC-type TO-3 or

TO-66 often serve as the collector or anode terminal. In such cases, it is essential that the mounting flange be securely fastened to the heat sink, which may be the equipment chassis. Under no circumstances, however, should the mounting flange be soldered directly to the heat sink or chassis because the heat of the soldering operation could permanently damage the device.

Such devices can be installed in commercially available sockets. Electrical connections may also be made by soldering directly to the terminal pins. Such connections may be soldered to the pins close to the pin seats provided care is taken to conduct excessive heat away from the seals; otherwise, the heat of the soldering operation could crack the pin seals and damage the device.

During operation, the mounting-flange temperature is higher than the ambient temperature by an amount which depends on the heat sink used. The heat sink must provide sufficient thermal conduction to the ambient environment to assure that the temperature of the device mounting flange does not rise above the rated value. The heat sink or chassis may be connected to either the positive or negative supply.

Fig. 706 shows methods of mounting flanged packages. Zinc-oxide-filled silicone grease should be used between the device and the heat sink to eliminate surface voids and to help conduct heat across the interface. Although glue or epoxy adhesive provides good bonding, a significant amount of thermal resistance may exist at the interface. To minimize this interface resistance, an adhesive material with low thermal resistance, such as Hysol* Epoxy Patch Material No. 6C or Wakefield* Delta Bond No. 152, or their equivalent, should be used.

---

* Products of Hysol Corporation. Olean, New York, and Wakefield Engineering, Inc., Wakefield, Massachusetts, respectively.

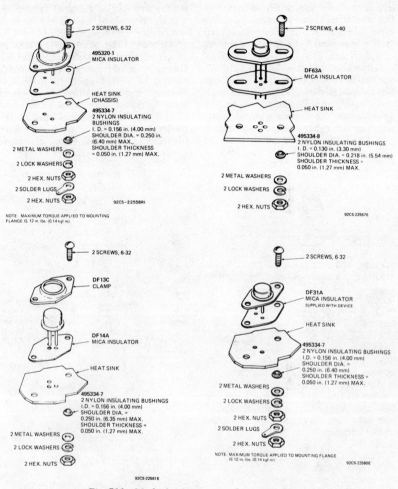

Fig. 706—Methods of mounting flanged packages.

Stud Packages

Some high-power solid-state devices are housed in stud packages like the TO-48 or DO-5 shown in Fig. 702. Connection of these packages to the chassis or heat sink should be made at the flat surface of the device perpendicular to the threaded stud. A large mating surface should be provided to avoid hot spots and high thermal drop. The hole for the stud should be only as large as necessary for clearance and should contain no burrs or ridges on its perimeter. As mentioned in the discussion of flanged packages, the use of a zinc-oxide-filled silicone grease between the device and the heat sink eliminates surface voids, prevents insulation buildup due to oxidation, and helps conduct heat across the interface. The package can be screwed directly into the heat sink or can be fastened by means of a nut. In either case, care must be taken to avoid the application of too much torque lest the semiconductor junction be damaged. Maximum limitations are given in the technical data for the particular devices. (CAUTION: Flexible, stranded wire should be used for all connections to the terminals that extend through the glass seals in both stud and press-fit packages. Excessive torque on these terminals may damage the seals and cause a loss in package hermeticity, which leads to premature device failure. These terminals, therefore, should not be bent under any circumstances.)

Although the studs are made of relatively soft copper to provide high thermal conductivity, the threads cannot be relied upon to provide a mating surface. The actual heat transfer must take place on the underside of the hexagonal part of the package. Fig. 707 shows suggested mounting arrangements of some

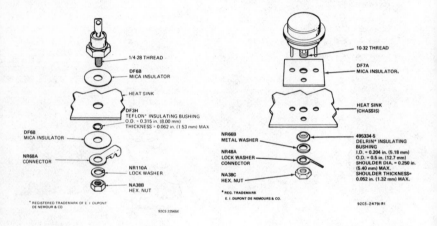

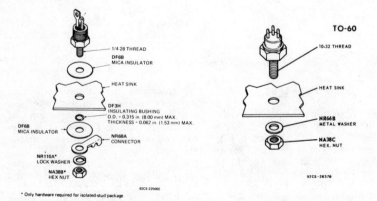

Fig. 707—Mounting arrangements for higher-current-type stud packages.

higher-current-type stud packages. Mounting components are shown with each package. With these mounting components, the increase in thermal resistance Rθ_{CS} from the case to the heat-sink surface can be maintained as low as 0.1°C per watt.

Press-fit Packages

Press-fit packages are used for some thyristors. Press-fit mounting depends upon an interference fit between the thyristor case and the heat sink. As the thyristor is forced into the heat-sink hole, metal from the heat sink flows into the knurl voids of the thyristor case. The resulting close contact between the heat sink and thyristor case assures low-thermal resistances.

The recommended mounting method shown in Fig. 708 shows press-fit

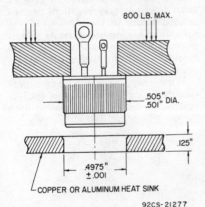

800 LB. MAX.

.505"
.501" DIA.

.125"

.4975"
±.001

COPPER OR ALUMINUM HEAT SINK

92CS-21277

Fig. 708—Recommended mounting method for press-fit packages.

knurl and heat-sink hole dimensions. If these dimensions are maintained, a "worst-case" condition of 0.0085-inch interference fit will allow press-fit insertion below the maximum allowable insertion force of 800 pounds. A slight chamfer in the heat-sink hole will help center and guide the press-fit package properly into the heat sink. The insertion tool should be a hollow shaft having an inner diameter of 0.380 ± 0.010 inch and an

outer diameter of 0.500 inch. These dimensions provide sufficient clearance for the leads and assure that no direct force is applied to the glass seal of the thyristor. (Refer to CAUTION note shown in section on **Stud Packages.**)

The press-fit package is not restricted to a single mounting arrangement; direct soldering and the use of epoxy adhesives have been successfully employed. The press-fit case is tin-plated to facilitate direct soldering to the heat sink. A 60-40 solder should be used, and heat should be applied only long enough to allow the solder to flow freely.

Molded-Plastic Packages

RCA power transistors and thyristors (SCR's and triacs) in molded-silicone-plastic packages are available in a wide range of power-dissipation ratings and a variety of package configurations.

The most popular molded-plastic packages are the VERSAWATT packages, which are designed for ease of use in many applications. The JEDEC TO-220AB in-line-lead version, shown in Fig. 702, represents the basic style. This configuration features leads that can be formed to meet a variety of specific mounting requirements. The JEDEC TO-220AA version of the VERSAWATT package can replace the JEDEC TO-66 transistor package in a commercial socket or printed-circuit board without retooling. The pin-connection arrangement of thyristors supplied in TO-220AA packages, however, differs from that of thyristors supplied in conventional TO-66 packages so that some hardware changes are required to effect a replacement. The TO-220AA VERSAWATT package can also be obtained with an integral heat sink.

RCA VERSAWATT plastic packages are both rugged and versatile within the confines of commonly accepted standards for such devices. Although these versatile packages lend themselves to numerous arrangements, provision of a wide variety of lead configuration to conform to the specific requirements of many different mounting arrangements is highly impractical. However, the leads of the VERSAWATT in-line package can be formed to a custom shape, provided that they are not indiscriminately twisted or bent. Although these leads can be formed, they are not flexible in the general sense, nor are they sufficiently rigid for unrestrained wire wrapping.

Before an attempt is made to form the leads of an in-line package to meet the requirements of a specific application, the desired lead configuration should be determined, and a lead-bending fixture should be designed and constructed. The use of a properly designed fixture for this operation eliminates the need for repeated lead bending. When the use of a special bending fixture is not practical, a pair of long-nosed pliers may be used. The pliers should hold the lead firmly between the bending point and the case, but should not touch the case. Fig. 709 illustrates the use of long-nosed pliers for lead bending. Fig. 709(a) shows techniques that should be avoided; Fig. 709(b) shows the correct method.

When the leads of an in-line plastic package are to be formed, whether by use of long-nosed pliers or a special bending fixture, the following precautions must be observed to avoid internal damage to the device:

1. Restrain the lead between the bending point and the plastic case to prevent relative movement between the lead and the case.

2. When the bend is made in the plane of the lead (spreading), bend only the narrow part of the lead.

3. When the bend is made in the plane perpendicular to that of the leads, make the bend at least ⅛ inch from the plastic case.

4. Do not use a lead-bend radius of less than 1/16 inch.

5. Avoid repeated bending of leads.

The leads of the TO-220AB VERSAWATT in-line package are not designed to withstand excessive axial pull. Force in this direction greater than 4 pounds may result in permanent damage to the device. If the mounting arrangement tends to impose axial stress on the leads, some method of strain relief should be devised.

Wire wrapping of the leads is permissible, provided that the lead is restrained between the plastic case

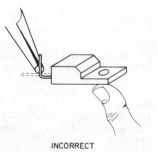

INCORRECT
LEAD IS NOT RESTRAINED BETWEEN
BENDING POINT AND PLASTIC CASE.

(a)

CORRECT

(b)

92CS-26385

Fig. 709—Use of long-nosed pliers for lead bending: (a) incorrect method; (b) correct method.

and the point of the wrapping. Soldering to the leads is also allowed; the maximum soldering temperature, however, must not exceed 275°C and must be applied for not more than 5 seconds at a distance greater than ⅛ inch from the plastic case. When wires are used for connections, care should be exercised to assure that movement of the wire does not cause movement of the lead at the lead-to-plastic junctions.

Fig. 710 shows recommended mounting arrangements and suggested hardware for VERSAWATT devices. The rectangular washer

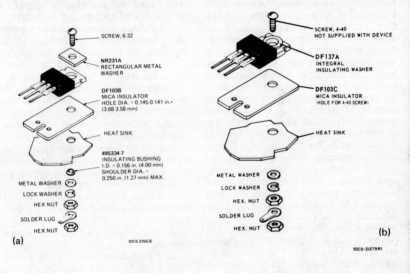

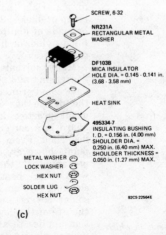

Fig. 710—Mounting arrangements for VERSAWATT transistors: (a) and (b) methods of mounting in-line-lead types; (c) chassis mounting.

(NR231A) shown in Fig. 710(a) is designed to minimize distortion of the mounting flange when the device is fastened to a heat sink. Excessive distortion of the flange could cause damage to the device. The washer is particularly important when the size of the mounting hole exceeds 0.140 inch (6-32 clearance). Larger holes are needed to accommodate insulating bushings; however, the holes should not be larger than necessary to provide hardware clearance and, in any case, should not exceed a diameter of 0.250 inch. Flange distortion is also possible if excessive torque is used during mounting. A maximum torque of 8 inch-pounds is recommended. The tool used to drive the mounting screw should never come in contact with the plastic body during driving operation. Such contact can result in damage to the plastic body and internal device connections. An excellent method of avoiding this problem is to use a spacer or combination spacer-isolating bushing which raises the screw head or nut above the top surface of the plastic body, as shown in Fig. 711. Suggested materials for these bushings are diallphthalate, fiberglass-filled nylon, or fiberglass-filled polycarbo-

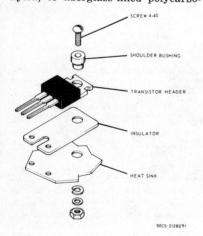

SCREW 4-40

SHOULDER BUSHING

TRANSISTOR HEADER

INSULATOR

HEAT SINK

92CS-21282R1

Fig. 711—Mounting arrangements in which an isolating bushing is used to raise the head of the mounting screw above the plastic body of the VERSAWATT device.

nate. Unfilled nylon should be avoided.

Modification of the flange can also result in flange distortion and should not be attempted. The flange should not be soldered to the heat sink by use of lead-tin solder because the heat required with this type of solder will cause the junction temperature of the device to become excessive.

TO-220AA devices can be mounted in commercially available TO-66 sockets, such as UID Electronics Corp. Socket No. PTS-4 or equivalent. For testing purposes, the TO-220AB in-line package can be mounted in a Jetron Socket No. CD74-104 or equivalent. Regardless of the mounting method, the following precautions should be taken:

1. Use appropriate hardware.

2. Always fasten the devices to the heat sink before the leads are soldered to fixed terminals.

3. Never allow the mounting tool to come in contact with the plastic case.

4. Never exceed a torque of 8 inch-pounds.

5. Avoid oversize mounting holes.

6. Provide strain relief if there is any probability that axial stress will be applied to the leads.

7. Use insulating bushings made of materials that do not have hot-creep problems. Such bushings should be made of diallphthalate, fiberglass-filled nylon, or fiberglass-filled polycarbonate.

Many solvents are available for degreasing and removal of flux from device and printed-circuit board after the device has been mounted. The usual practice is to submerge the board in a solvent bath for a specified time. From a reliability standpoint, however, it is extremely important that the solvent, together with other chemicals in the solder-cleaning system (such as flux and solder covers), not adversely affect the life of the device. This consideration applies to all non-hermetic and molded-plastic devices.

It is, of course, impractical to evaluate the effect on long-term device life of all cleaning solvents, which are marketed under a variety of brand names with numerous additives. Chlorinated solvents, gasoline, and other hydrocarbons cause the inner encapsulant to swell and damage the transistor. Alcohols are acceptable solvents and are recommended for flux removal whenever possible. Several examples of suitable alcohols are listed below:

1. methanol
2. ethanol
3. isopropanol
4. blends of the above

When considerations such as solvent flammability are of concern, selected freon-alcohol blends are usable when exposure is limited. Solvents such as those listed below should be safe when used for normal flux removal operations, but care should be taken to assure their suitability in the cleaning procedure:

1. Freon TE
2. Freon TE-35
3. Freon TP-35 (Freon PC)

These solvents may be used for a maximum of 4 hours at 25°C or for a maximum of 1 hour at 50°C.

Care must also be used in the selection of fluxes in the soldering of leads. Rosin or activated-rosin fluxes are recommended; organic fluxes are not.

In addition to the VERSAWATT packages, the "plastic TO-5" package shown in Fig. 712 is also used for some power transistors. The leads of plastic TO-5 package are similar in every way to those on the standard TO-5 package, so that the plastic can replace the metallic TO-5 in most sockets.

Integrated-Circuit Packages

RCA integrated circuits are supplied in a wide variety of packages. The three basic styles are: (1) metal hermetic packages of the TO-5 type; (2) ceramic flat-packs; (3) plastic or ceramic dual-in-line and quad-in-line packages. The TO-5 styles have 8,

Fig. 712—Photograph of the "plastic TO-5" package.

10, or 12 leads, which may be straight or may be formed into a dual-in-line configuration (8-lead only) or into a circular configuration with a diameter that is larger than the diameter of the TO-5 package (radial off-set). The ceramic flat-packs have 14, 16, 24, or 28 leads. Welded-seal ceramic dual-in-line packages are supplied with 14, 16, 24, or 28 leads, and frit-seal ceramic dual-in-line packages have 8, 14 or 16 leads. Plastic dual-in-lines have either 14 or 16 leads, and the plastic quad-in-line packages have 14, 16, or 20 leads.

Sockets are available for most of these package configurations, and are useful as temporary mountings in design and experimental work with the integrated circuit. For permanent connection, however, the integrated circuits are usually mounted directly on a printed circuit board by soldering the leads to the printed wiring. Sometimes the leads are welded to a post or to a tab.

For RCA integrated circuits that use the TO-5-style package, the most direct method of mounting is the straight-through technique shown in Fig. 713. The leads of the device are simply inserted in plated-through holes in the printed-circuit board, and the connection is completed by dip-soldering or wave-soldering.

Several methods for connecting

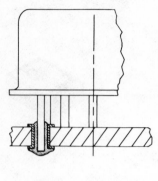

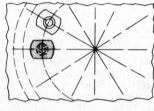

Mounting and connection techniques for dual-in-line and quad-in-line packages are similar to those used in the in-line methods already discussed for the flat packages. The dual-in-line leads are larger than those of the flat-packs, however, so that they are easier to insert in the printed-circuit board; also, the lead spacing is greater, so that soldering is simplified. One other feature of the dual-in-line leads is their shoulder that holds the package up off the board.

The quad-in-line packages can also be provided with staggered leads for even greater ease in mounting.

RCA STANDARD SOLID-STATE DEVICES

Table LI provides a detailed listing of RCA solid-state devices currently available as standard or high-reliability products. Because detailed data for these devices are available in RCA data bulletins and in the RCA SSD-200 DATABOOK series, only brief descriptive data are included here.

The File Numbers listed in the table indicate the data bulletins for specific device types.

92CS-26383

Fig. 713—Straight-through method for mounting TO-5 packages.

flat-pack packages are shown in Fig. 714. Lead and terminal arrangements for through-the-board mounting and for surface mounting of these packages are shown in Fig. 715.

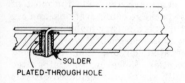
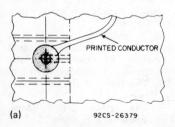
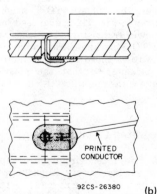

(a)   92CS-26379

Fig. 714—Methods of Connecting Flat Packages: (a) straight-through method; (b) clinched-lead full-pad method

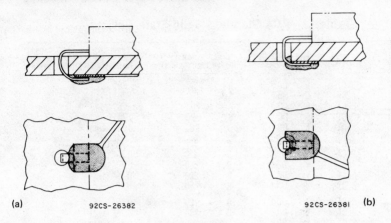

Fig. 715—Methods of Connecting Flat Packages: (a) clinched-lead offset-pad method;
(b) clinched-lead half-pad method

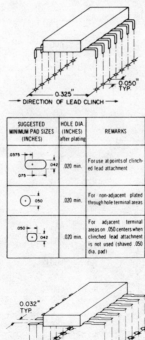

SUGGESTED MINIMUM PAD SIZES (INCHES)	HOLE DIA. (INCHES) after plating	REMARKS
0.375 × .075, .042	.020 min.	For use at points of clinched lead attachment
⊙ .050	.020 min.	For non-adjacent plated through hole terminal areas
.050 × .042	.020 min.	For adjacent terminal areas on .050 centers when clinched lead attachment is not used (shaved .050 dia. pad)

92CS-26388

Fig. 716—"In-Line Lead and Terminal Arrangements for RCA Flat Packages.

Table LI—RCA Standard Solid-State Devices

Type No.	Prod. Line	File No.	Package▲	Lead Code■	Voltage (V) (Note 1)	Power (W), Current* (A) or Noise Figure (dB)	DC Current Transfer Ratio or Frequency
1N248C	RECT	6	DO-5	A#	50	20* (avg.)	—
1N249C	RECT	6	DO-5	A#	100	20* (avg.)	—
1N250C	RECT	6	DO-5	A#	200	20* (avg.)	—
1N440B	RECT	5	DO-1	B	100	0.75* (avg.)	—
1N441B	RECT	5	DO-1	B	200	0.75* (avg.)	—
1N442B	RECT	5	DO-1	B	300	0.75* (avg.)	—
1N443B	RECT	5	DO-1	B	400	0.75* (avg.)	—
1N444B	RECT	5	DO-1	B	500	0.75* (avg.)	—
1N445B	RECT	5	DO-1	B	600	0.75* (avg.)	—
1N536	RECT	3	DO-1	B	50	0.75* (avg.)	—
1N537	RECT	3	DO-1	B	100	0.75* (avg.)	—
1N538	RECT	3	DO-1	B	200	0.75* (avg.)	—
1N539	RECT	3	DO-1	B	300	0.75* (avg.)	—
1N540	RECT	3	DO-1	B	400	0.75* (avg.)	—
1N547	RECT	3	DO-1	B	600	0.75* (avg.)	—
1N1095	RECT	3	DO-1	B	500	0.75* (avg.)	—
1N1183A	RECT	38	DO-5	A#	50	40* (avg.)	—
1N1184A	RECT	38	DO-5	A#	100	40* (avg.)	—
1N1186A	RECT	38	DO-5	A#	200	40* (avg.)	—
1N1187A	RECT	38	DO-5	A#	300	40* (avg.)	—
1N1188A	RECT	38	DO-5	A#	400	40* (avg.)	—
1N1189A	RECT	38	DO-5	A#	500	40* (avg.)	—
1N1190A	RECT	38	DO-5	A#	600	40* (avg.)	—
1N1195A	RECT	6	DO-5	A#	300	20* (avg.)	—
1N1196A	RECT	6	DO-5	A#	400	20* (avg.)	—
1N1197A	RECT	6	DO-5	A#	500	20* (avg.)	—
1N1198A	RECT	6	DO-5	A#	600	20* (avg.)	—
1N1199A	RECT	20	DO-4	A#	50	12* (avg.)	—
1N1200A	RECT	20	DO-4	A#	100	12* (avg.)	—
1N1202A	RECT	20	DO-4	A#	200	12* (avg.)	—
1N1203A	RECT	20	DO-4	A#	300	12* (avg.)	—
1N1204A	RECT	20	DO-4	A#	400	12* (avg.)	—
1N1205A	RECT	20	DO-4	A#	500	12* (avg.)	—
1N1206A	RECT	20	DO-4	A#	600	12* (avg.)	—
1N1341B	RECT	58	DO-4	A#	50	6* (avg.)	—
1N1342B	RECT	58	DO-4	A#	100	6* (avg.)	—
1N1344B	RECT	58	DO-4	A#	200	6* (avg.)	—
1N1345B	RECT	58	DO-4	A#	300	6* (avg.)	—
1N1346B	RECT	58	DO-4	A#	400	6* (avg.)	—
1N1347B	RECT	58	DO-4	A#	500	6* (avg.)	—
1N1348B	RECT	58	DO-4	A#	600	6* (avg.)	—
1N1763A	RECT	89	DO-1	B	400	1* (avg.)	—
1N1764A	RECT	89	DO-1	B	500	1* (avg.)	—
1N2858A	RECT	91	DO-1	B	50	1* (avg.)	—
1N2859A	RECT	91	DO-1	B	100	1* (avg.)	—
1N2860A	RECT	91	DO-1	B	200	1* (avg.)	—
1N2861A	RECT	91	DO-1	B	300	1* (avg.)	—
1N2862A	RECT	91	DO-1	B	400	1* (avg.)	—
1N2863A	RECT	91	DO-1	B	500	1* (avg.)	—
1N2864A	RECT	91	DO-1	B	600	1* (avg.)	—

▲ Dimensional outlines are shown following data charts.   # Reverse-polarity type available.
■ Lead codes are explained in Table LII.
See Notes on page 619.

Table LI—RCA Standard Solid-State Devices (cont'd)

Type No.	Prod. Line	File No.	Package▲	Lead Code■	Voltage (V) (Note 1)	Power (W), Current* (A) or Noise Figure (dB)	DC Current Transfer Ratio or Frequency
1N3193	RECT	41	DO-26	B	200	0.75* (avg.)	—
1N3194	RECT	41	DO-26	B	400	0.75* (avg.)	—
1N3195	RECT	41	DO-26	B	600	0.75* (avg.)	—
1N3196	RECT	41	DO-26	B	800	0.75* (avg.)	—
1N3253	RECT	41	DO-26	B	200	0.75* (avg.)	—
1N3254	RECT	41	DO-26	B	400	0.75* (avg.)	—
1N3255	RECT	41	DO-26	B	600	0.75* (avg.)	—
1N3256	RECT	41	DO-26	B	800	0.5* (avg.)	—
1N3563	RECT	41	DO-26	B	1000	0.4* (avg.)	—
1N3879	RECT	726	DO-4	A#	50	6* (av.g)	—
1N3880	RECT	726	DO-4	A#	100	6* (av.g)	—
1N3881	RECT	726	DO-4	A#	200	6* (av.g)	—
1N3882	RECT	726	DO-4	A#	300	6* (av.g)	—
1N3883	RECT	726	DO-4	A#	400	6* (av.g)	—
1N3889	RECT	727	DO-4	A#	50	12* (avg.)	—
1N3890	RECT	727	DO-4	A#	100	12* (avg.)	—
1N3891	RECT	727	DO-4	A#	200	12* (avg.)	—
1N3892	RECT	727	DO-4	A#	300	12* (avg.)	—
1N3893	RECT	727	DO-4	A#	400	12* (avg.)	—
1N3899	RECT	728	DO-5	A#	50	20* (avg.)	—
1N3900	RECT	728	DO-5	A#	100	20* (avg.)	—
1N3901	RECT	728	DO-5	A#	200	20* (avg.)	—
1N3902	RECT	728	DO-5	A#	300	20* (avg.)	—
1N3903	RECT	728	DO-5	A#	400	20* (avg.)	—
1N3909	RECT	729	DO-5	A#	50	30* (avg.)	—
1N3910	RECT	729	DO-5	A#	100	30* (avg.)	—
1N3911	RECT	729	DO-5	A#	200	30* (avg.)	—
1N3912	RECT	729	DO-5	A#	300	30* (avg.)	—
1N3913	RECT	729	DO-5	A#	400	30* (avg.)	—
1N5211	RECT	245	DO-26	B	200	1* (avg.)	—
1N5212	RECT	245	DO-26	B	400	1* (avg.)	—
1N5213	RECT	245	DO-26	B	600	1* (avg.)	—
1N5214	RECT	245	DO-26	B	800	0.75* (avg.)	—
1N5215	RECT	245	DO-26	B	200	1* (avg.)	—
1N5216	RECT	245	DO-26	B	400	1* (avg.)	—
1N5217	RECT	245	DO-26	B	600	1* (avg.)	—
1N5218	RECT	245	DO-26	B	800	0.75* (avg.)	—
1N5391	RECT	478	DO-15	B	50	1.5* (avg.)	—
1N5392	RECT	478	DO-15	B	100	1.5* (avg.)	—
1N5393	RECT	478	DO-15	B	200	1.5* (avg.)	—
1N5394	RECT	478	DO-15	B	300	1.5* (avg.)	—
1N5395	RECT	478	DO-15	B	400	1.5* (avg.)	—
1N5396	RECT	478	DO-15	B	500	1.5* (avg.)	—
1N5397	RECT	478	DO-15	B	600	1.5* (avg.)	—
1N5398	RECT	478	DO-15	B	800	1.5* (avg.)	—
1N5399	RECT	478	DO-15	B	1000	1.5* (avg.)	—
2N681	SCR	96	TO-48	I	25	25* (rms)	—
2N682	SCR	96	TO-48	I	50	25* (rms)	—
2N683	SCR	96	TO-48	I	100	25* (rms)	—
2N684	SCR	96	TO-48	I	150	25* (rms)	—

▲ Dimensional outlines are shown following data charts.
■ Lead codes are explained in Table LII.
See **Notes** on page 619.

\# Reverse-polarity type available.

Table LI—RCA Standard Solid-State Devices (cont'd)

Type No.	Prod. Line	File No.	Package▲	Lead Code■	Voltage (V) (Note 1)	Power (W), Current* (A) or Noise Figure (dB)	DC Current Transfer Ratio or Frequency
2N685	SCR	96	TO-48	I	200	25* (rms)	—
2N686	SCR	96	TO-48	I	250	25* (rms)	—
2N687	SCR	96	TO-48	I	300	25* (rms)	—
2N688	SCR	96	TO-48	I	400	25* (rms)	—
2N689	SCR	96	TO-48	I	500	25* (rms)	—
2N690	SCR	96	TO-48	I	600	25* (rms)	—
2N697	PWR	16	TO-39, TO-5	F	60	2	40-120
2N699	PWR	22	TO-39, TO-5	F	120	2	40-120
2N918	RF	83	TO-72	S	6-15 (V_{CE})	6(NF)	60 MHz
2N1479	PWR	135	TO-39, TO-5	F	60	5	20-60
2N1480	PWR	135	TO-39, TO-5	F	100	5	20-60
2N1481	PWR	135	TO-39, TO-5	F	60	5	35-100
2N1482	PWR	135	TO-39, TO-5	F	100	5	35-100
2N1483	PWR	137	TO-8	F	60	25	20-60
2N1484	PWR	137	TO-8	F	100	25	20-60
2N1485	PWR	137	TO-8	F	60	25	35-100
2N1486	PWR	137	TO-8	F	100	25	35-100
2N1487	PWR	139	TO-3	D	60	75	15-45
2N1488	PWR	139	TO-3	D	100	75	15-45
2N1489	PWR	139	TO-3	D	60	75	25-75
2N1490	PWR	139	TO-3	D	100	75	25-75
2N1491	RF	10	TO-39	F	20	0.01	70 MHz
2N1492	RF	10	TO-39	F	30	0.1	70 MHz
2N1493	RF	10	TO-39	F	50	0.5	70 MHz
2N1613	PWR	106	TC-39, TO-5	F	75	3	20 min.
2N1711	PWR	26	TO-39, TO-5	F	75	3	35 min.
2N1842A	SCR	28	TO-48	I	25	16* (rms)	—
2N1483A	SCR	28	TO-48	I	50	16* (rms)	—
2N1844A	SCR	28	TO-48	I	100	16* (rms)	—
2N1845A	SCR	28	TO-48	I	150	16* (rms)	—
2N1846A	SCR	28	TO-48	I	200	16* (rms)	—
2N1847A	SCR	28	TO-48	I	250	16* (rms)	—
2N1848A	SCR	28	TO-48	I	300	16* (rms)	—
2N1849A	SCR	28	TO-48	I	400	16* (rms)	—
2N1850A	SCR	28	TO-48	I	500	16* (rms)	—
2N1893	PWR	34	TO-39, TO-5	F	120	3	40-120
2N2015	PWR	12	TO-36	F	100	150	15-50
2N2016	PWR	12	TO-36	F	130	150	15-50
2N2102	PWR	106	TO-39, TO-5	F	120	5	20 min.
2N2270	PWR	24	TO-39, TO-5	F	60	5	50-200
2N2405	PWR	34	TO-39, TO-5	F	120	5	60-200
2N2631	RF	32	TO-39	F	28	7.5	50 MHz
2N2857	RF	61	TO-72	S	6-15 (V_{CE})	4.5(NF)	450 MHz
2N2876	RF	32	TO-60	E	10	28	50 MHz
2N2895	PWR	143	TO-18	F	120	1.8	40-120
2N2896	PWR	143	TO-18	F	140	1.8	60-200
2N2897	PWR	143	TO-18	F	60	1.8	50-200
2N3053	PWR	432	TO-39, TO-5	F	60	5	50-250
2N3054	PWR	527	TO-66	D	90	25	25-150
2N3055	PWR	524	TO-3	D	100	115	20-70

▲ Dimensional outlines are shown following data charts.
■ Lead codes are explained in Table LII.
See **Notes** on page 619.

Table LI—RCA Standard Solid-State Devices (cont'd)

Type No.	Prod. Line	File No.	Package▲	Lead Code■	Voltage (V) (Note 1)	Power (W), Current* (A) or Noise Figure (dB)	DC Current Transfer Ratio or Frequency
2N3118	RF	42	TO-5	F	28	1	50 MHz
2N3119	RF	44	TO-5	F	28	1	50 MHz
2N3228	SCR	114	TO-66	I	200	5* (rms)	—
2N3229	RF	50	TO-60	E	50	15	50 MHz
2N3262	RF	56	TO-39	F	High-speed switching		
2N3263	PWR	54	Radial	C	150	20	25-75
2N3264	PWR	54	Radial	C	120	30	20-80
2N3265	PWR	54	TO-63	F	150	24	25-75
2N3266	PWR	54	TO-63	F	120	28	20-80
2N3375	RF	386	TO-60	E	28	3	400 MHz
2N3439	PWR	64	TO-39, TO-5	F	450	10	40-160
2N3440	PWR	64	TO-39, TO-5	F	300	10	40-160
2N3441	PWR	529	TO-66	D	160	25	25-100
2N3442	PWR	528	TO-3	D	160	117	20-70
2N3478	RF	77	TO-72	S	6-15 (V_{CE})	4.5(NF)	200 MHz
2N3525	SCR	114	TO-66	I	400	5* (rms)	—
2N3528	SCR	114	TO-8	K	200	2* (rms)	—
2N3529	SCR	114	TO-8	K	400	2* (rms)	
2N3553	RF	386	TO-39	F	28	2.5	175 MHz
2N3583	PWR	138	TO-66	D	250	35	40 min.
2N3584	PWR	138	TO-66	D	375	35	25-100
2N3585	PWR	138	TO-66	D	500	35	25-100
2N3600	RF	83	TO-72	S	6-15 (V_{CE})	4.5(NF)	200 MHz
2N3632	RF	386	TO-60	E	28	13.5	175 MHz
2N3650	SCR	408	TO-48	I	100	35* (rms)	—
2N3651	SCR	408	TO-48	I	200	35* (rms)	—
2N3652	SCR	408	TO-48	I	300	35* (rms)	—
2N3653	SCR	408	TO-48	I	400	35* (rms)	—
2N3654	SCR	724	TO-48	I	50	35* (rms)	—
2N3655	SCR	724	TO-48	I	100	35* (rms)	—
2N3656	SCR	724	TO-48	I	200	35* (rms)	—
2N3657	SCR	724	TO-48	I	300	35* (rms)	—
2N3658	SCR	724	TO-48	I	400	35* (rms)	—
2N3668	SCR	116	TO-3	I	100	12.5	—
2N3669	SCR	116	TO-3	I	200	12.5	—
2N3670	SCR	116	TO-3	I	400	12.5	—
2N3733	RF	72	TO-60	E	28	10	400 MHz
2N3771	PWR	525	TO-3	D	50	150	15-60
2N3772	PWR	525	TO-3	D	100	150	15-60
2N3773	PWR	526	TO-3	D	160	150	15-60
2N3839	RF	229	TO-72	S	6-15 (V_{CE})	3.9(NF)	450 MHz
2N3866	RF	80	TO-39	F	28	1	400 MHz
2N3870	SCR	578	Press-fit	I	100	35* (rms)	—
2N3871	SCR	578	Press-fit	I	200	35* (rms)	—
2N3872	SCR	578	Press-fit	I	400	35* (rms)	—
2N3873	SCR	578	Press-fit	I	600	35* (rms)	—
2N3878	PWR	766	TO-66	D	120	35	50-200
2N3879	PWR	766	TO-66	D	120	35	20-80
2N3896	SCR	578	Stud	I	100	35* (rms)	—
2N3897	SCR	578	Stud	I	200	35* (rms)	—

▲ Dimensional outlines are shown following data charts.
■ Lead codes are explained in Table LII.
See **Notes** on page 619.

Table LI—RCA Standard Solid-State Devices (cont'd)

Type No.	Prod. Line	File No.	Package▲	Lead Code■	Voltage (V) (Note 1)	Power (W), Current* (A) or Noise Figure (dB)	DC Current Transfer Ratio or Frequency
2N3898	SCR	578	Stud	I	400	35* (rms)	—
2N3899	SCR	578	Stud	I	600	35* (rms)	—
2N4012	RF	90	TO-60	E	28	2.5	1 GHz
2N4036	PWR	216	TO-39, TO-5	F	—90	7	20-200
2N4037	PWR	216	TO-39, TO-5	F	—60	7	50-250
2N4063	PWR	64	Flange	F	450	—	40-160
2N4064	PWR	64	Flange	F	300	10	40-160
2N4101	SCR	114	TO-66	I	600	5* (rms)	—
2N4102	SCR	114	TO-8	K	600	2* (rms)	—
2N4103	SCR	116	TO-3	I	600	12.5	—
2N4240	PWR	135	TO-66	D	500	2.5	30-150
2N4314	PWR	216	TO-39, TO-5	F	—90	7	50-250
2N4347	PWR	528	TO-3	D	140	100	15-60
2N4348	PWR	526	TO-3	D	140	120	15-60
2N4427	RF	228	TO-39	F	12	1	175 MHz
2N4440	RF	217	TO-60	E	28	5	400 MHz
2N4932	RF	249	TO-60	U	13.5	12	88 MHz
2N4933	RF	249	TO-60	U	24	20	88 MHz
2N5016	RF	255	TO-60	G	28	15	400 MHz
2N5034	PWR	244	TO-219AA	D	55	83	20-80
2N5035	PWR	244	TO-219AB	D	55	83	20-80
2N5036	PWR	244	TO-219AA	D	70	83	20-80
2N5037	PWR	244	TO-219AB	D	70	83	20-80
2N5038	PWR	698	TO-3	D	150	140	50-200
2N5039	PWR	698	TO-3	D	120	140	30-150
2N5070	RF	268	TO-60	E	28	25(PEP)	30 MHz
2N5071	RF	269	TO-60	U	24	24	76 MHz
2N5090	RF	270	TO-60	E	28	1.2	400 MHz
2N5102	RF	279	TO-60	G	24	15	136 MHz
2N5109	RF	281	TO-39	F	15	3(NF)	200 MHz
2N5179	RF	288	TO-72	S	6(V_{CE})	4.5(NF)	200 MHz
2N5180	RF	289	TO-72	S	8(V_{CE})	2.5(NF)	200 MHz
2N5189	PWR	296	TO-39 (Mod.)	F	60	5	30 min.
2N5202	PWR	766	TO-66	D	120	35	10-100
2N5239	PWR	321	TO-3	D	300	100	20-80
2N5240	PWR	321	TO-3	D	375	100	20-80
2N5262	PWR	313	TO-39 (Mod.)	F	75	5	35 min.
2N5293	PWR	322	TO-220AA	M	80	36	30-120
2N5294	PWR	322	TO-220AB	N	80	36	30-120
2N5295	PWR	322	TO-220AA	M	60	36	30-120
2N5296	PWR	322	TO-220AB	N	60	36	30-120
2N5297	PWR	322	TO-220AA	M	80	36	20-80
2N5298	PWR	322	TO-220AB	N	80	36	20-80
2N5320	PWR	325	TO-39, TO-5	F	100	10	30-130
2N5321	PWR	325	TO-39, TO-5	F	75	10	40-250
2N5322	PWR	325	TO-39, TO-5	F	—100	10	30-130
2N5223	PWR	325	TO-39, TO-5	F	—75	10	40-250
2N5415	PWR	336	TO-39, TO-5	F	—200	10	30-150
2N5416	PWR	336	TO-39, TO-5	F	—350	10	30-120
2N5441	TRI	593	Press-fit	J	200	40* (rms)	—

▲ Dimensional outlines are shown following data charts.
■ Lead codes are explained in Table LII.
See **Notes** on page 619.

Table LI—RCA Standard Solid-State Devices (cont'd)

Type No.	Prod. Line	File No.	Package▲	Lead Code■	Voltage (V) (Note 1)	Power (W), Current* (A) or Noise Figure (dB)	DC Current Transfer Ratio or Frequency
2N5442	TRI	593	Press-fit	J	400	40* (rms)	—
2N5443	TRI	593	Press-fit	J	600	40* (rms)	—
2N5444	TRI	593	Stud	J	200	40* (rms)	—
2N5445	TRI	593	Stud	J	400	40* (rms)	—
2N5446	TRI	593	Stud	J	600	40* (rms)	—
2N5470	RF	350	TO-215AA	E	28	1	2 GHz
2N5490	PWR	353	TO-220AB	M	60	50	20-100
2N5491	PWR	353	TO-220AA	N	60	50	20-100
2N5492	PWR	353	TO-220AB	M	75	50	20-100
2N5493	PWR	353	TO-220AA	N	75	50	20-100
2N5494	PWR	353	TO-220AB	M	60	50	20-100
2N5495	PWR	353	TO-220AA	N	60	50	20-100
2N5496	PWR	353	TO-220AB	M	90	50	20-100
2N5497	PWR	353	TO-220AA	N	90	50	20-100
2N5567	TRI	457	Press-fit	J	200	10* (rms)	—
2N5568	TRI	457	Press-fit	J	400	10* (rms)	—
2N5569	TRI	457	Stud	J	200	10* (rms)	—
2N5570	TRI	457	Stud	J	400	10* (rms)	—
2N5571	TRI	458	Press-fit	J	200	15* (rms)	—
2N5572	TRI	458	Press-fit	J	400	15* (rms)	—
2N5573	TRI	458	Stud	J	200	15* (rms)	—
2N5574	TRI	458	Stud	J	400	15* (rms)	—
2N5575	PWR	359	TO-3 (Mod.)	D	70	300	10-40
2N5578	PWR	359	TO-3 (Mod.)	D	90	300	10-40
2N5671	PWR	383	TO-3	D	120	140	20-100
2N5672	PWR	383	TO-3	D	150	140	20-100
2N5754	TRI	414	TO-5 (Mod.)	L	100	2.5* (rms)	—
2N5755	TRI	414	TO-5 (Mod.)	L	200	2.5* (rms)	—
2N5756	TRI	414	TO-5 (Mod.)	L	400	2.5* (rms)	—
2N5757	TRI	414	TO-5 (Mod.)	L	600	2.5* (rms)	—
2N5781	PWR	413	TO-39, TO-5	F	—80	10	20-100
2N5782	PWR	413	TO-39, TO-5	F	—65	10	20-100
2N5783	PWR	413	TO-39, TO-5	F	—45	10	20-100
2N5784	PWR	413	TO-39, TO-5	F	80	10	20-100
2N5785	PWR	413	TO-39, TO-5	F	65	10	20-100
2N5786	PWR	413	TO-39, TO-5	F	45	10	20-100
2N5804	PWR	407	TO-3	D	300	110	10-100
2N5805	PWR	407	TO-3	D	375	110	10-100
2N5838	PWR	410	TO-3	D	275	100	8-40
2N5839	PWR	410	TO-3	D	300	100	10-50
2N5840	PWR	410	TO-3	D	375	100	10-50
2N5913	RF	423	TO-39	D	12	1	470 MHz
2N5914	RF	424	TO-216AA	W	12	2	470 MHz
2N5915	RF	424	TO-216AA	W	12	6	470 MHz
2N5916	RF	425	TO-216AA	W	28	2	400 MHz
2N5917	RF	425	RCA HF-31	W	28	2	400 MHz
2N5918	RF	448	TO-216AA	W	28	10	400 MHz
2N5919A	RF	505	TO-216AA	W	28	16	400 MHz
2N5920	RF	440	TO-215AA	E	28	2	2 GHz
2N5921	RF	427	TO-201AA	E	28	5	2 GHz

▲ Dimensional outlines are shown following data charts.
■ Lead codes are explained in Table LII.
See Notes on page 619.

Table LI—RCA Standard Solid-State Devices (cont'd)

Type No.	Prod. Line	File No.	Package▲	Lead Code■	Voltage (V) (Note 1)	Power (W), Current* (A) or Noise Figure (dB)	DC Current Transfer Ratio or Frequency
2N5954	PWR	675	TO-66	D	—85	40	20-100
2N5955	PWR	675	TO-66	D	—70	40	20-100
2N5956	PWR	675	TO-66	D	—50	40	20-100
2N5995	RF	454	TO-216AA	W	12.5	7	175 MHz
2N6032	PWR	462	TO-3 (Mod.)	D	120	140	10-50
2N6033	PWR	462	TO-3 (Mod.)	D	150	140	10-50
2N6055	PWR	563	TO-3	D	60	100	750-18000
2N6056	PWR	563	TO-3	D	80	100	750-18000
2N6077	PWR	492	TO-66	D	300	45	12-70
2N6078	PWR	492	TO-66	D	275	45	12-70
2N6079	PWR	492	TO-66	D	375	45	12-50
2N6093	RF	484	TO-217AA	Y	28	75 (PEP)	30 MHz
2N6098	PWR	485	TO-220AA	M	70	75	20-80
2N6099	PWR	485	TO-220AB	N	70	75	20-80
2N6100	PWR	485	TO-220AA	M	80	75	20-80
2N6101	PWR	485	TO-220AB	N	80	75	20-80
2N6102	PWR	485	TO-220AA	M	45	75	15-60
2N6103	PWR	485	TO-220AB	N	45	75	15-60
2N6104	RF	504	RCA HF-32	W	28	30	400 MHz
2N6105	RF	504	TO-216AA	W	28	30	400 MHz
2N6106	PWR	676	TO-220AA	M	—80	40	30-150
2N6107	PWR	676	TO-220AB	N	—80	40	30-150
2N6108	PWR	676	TO-220AA	M	—60	40	30-150
2N6109	PWR	676	TO-220AB	N	—60	40	30-150
2N6110	PWR	676	TO-220AA	M	—40	40	30-150
2N6111	PWR	676	TO-220AB	N	—40	40	30-150
2N6175	PWR	508	Plastic TO-5	E	300	20	30-190
2N6176	PWR	508	Plastic TO-5	E	350	20	30-150
2N6177	PWR	508	Plastic TO-5	E	450	20	30-150
2N6178	PWR	562	Plastic TO-5	F	100	25	30-130
2N6179	PWR	562	Plastic TO-5	F	75	25	40-250
2N6180	PWR	562	Plastic TO-5	F	—100	25	30-130
2N6181	PWR	562	Plastic TO-5	F	—75	25	40-250
2N6211	PWR	507	TO-66	D	—275	35	10-100
2N6212	PWR	507	TO-66	D	—350	35	10-100
2N6213	PWR	507	TO-66	D	—400	35	10-100
2N6214	PWR	507	TO-66	D	—450	35	10-100
2N6246	PWR	677	TO-3	D	—70	125	20-100
2N6247	PWR	677	TO-3	D	—90	125	20-100
2N6248	PWR	677	TO-3	D	—110	125	20-100
2N6249	PWR	523	TO-3	D	300	175	10-50
2N6250	PWR	523	TO-3	D	375	175	8-50
2N6251	PWR	523	TO-3	D	450	175	6-50
2N6253	PWR	524	TO-3	D	55	115	20-70
2N6254	PWR	524	TO-3	D	100	150	20-70
2N6257	PWR	525	TO-3	D	50	150	15-75
2N6258	PWR	525	TO-3	D	100	250	20-60
2N6259	PWR	526	TO-3	D	170	250	15-60
2N6260	PWR	527	TO-66	D	50	29	20-100
2N6261	PWR	527	TO-66	D	90	50	25-100

▲ Dimensional outlines are shown following data charts.
■ Lead codes are explained in Table LII.
See **Notes** on page 619.

Table LI—RCA Standard Solid-State Devices (cont'd)

Type No.	Prod. Line	File No.	Package▲	Lead Code■	Voltage (V) (Note 1)	Power (W), Current* (A) or Noise Figure (dB)	DC Current Transfer Ratio or Frequency
2N6262	PWR	528	TO-3	D	170	150	20-70
2N6263	PWR	529	TO-66	D	140	20	20-100
2N6264	PWR	529	TO-66	D	170	50	20-60
2N6265	RF	543	RCA HF-28	T	28	2	2 GHz
2N6266	RF	544	RCA HF-28	T	28	5	2 GHz
2N6267	RF	545	RCA HF-28	T	28	10	2 GHz
2N6268	RF	546	RCA HF-28	T	22	2	2.3 GHz
2N6269	RF	546	RCA HF-28	T	22	6.5	2.3 GHz
2N6288	PWR	676	TO-220AB	N	40	40	30-150
2N6289	PWR	676	TO-220AA	M	40	40	30-150
2N6290	PWR	676	TO-220AB	N	60	40	30-150
2N6291	PWR	676	TO-220AA	M	60	40	30-150
2N6292	PWR	676	TO-220AB	N	80	40	30-150
2N6293	PWR	676	TO-220AA	M	80	40	30-150
2N6354	PWR	582	TO-3	D	150	140	20-150
2N6371	PWR	607	TO-3	D	50	117	15-60
2N6372	PWR	675	TO-66	D	50	40	20-100
2N6373	PWR	675	TO-66	D	70	40	20-100
2N6374	PWR	675	TO-66	D	90	40	20-100
2N6383	PWR	609	TO-3	D	40	100	1000-20000
2N6384	PWR	609	TO-3	D	60	100	1000-20000
2N6385	PWR	609	TO-3	D	80	100	1000-20000
2N6386	PWR	610	TO-220AB	N	40	40	1000-20000
2N6387	PWR	610	TO-220AB	N	60	40	1000-20000
2N6388	PWR	610	TO-220AB	N	80	40	1000-20000
2N6389	RF	617	TO-72	F	12(V_{CE})	¾(NF)	450/890 MHz
2N6467	PWR	675	TO-66	D	—110	40	15-150
2N6468	PWR	675	TO-66	D	—130	40	15-150
2N6469	PWR	677	TO-3	D	—50	125	20-150
2N6470	PWR	677	TO-3	D	50	125	20-150
2N6471	PWR	677	TO-3	D	70	125	20-150
2N6472	PWR	677	TO-3	D	90	125	20-150
2N6473	PWR	676	TO-220AB	N	110	40	15-150
2N6474	PWR	676	TO-220AB	N	130	40	15-150
2N6475	PWR	676	TO-220AB	N	—110	40	15-150
2N6476	PWR	676	TO-220AB	N	—130	40	15-150
2N6477	PWR	680	TO-220AB	N	140	20	25-100
2N6478	PWR	680	TO-220AB	N	160	25	25-100
2N6479	PWR	702	Radial	C	100	87	20-300
2N6480	PWR	702	Radial	C	100	87	20-300
2N6481	PWR	702	Radial	C	100	117	20-300
2N6482	PWR	702	Radial	C	100	117	20-300
2N6486	PWR	678	TO-220AB	N	50	75	20-150
2N6487	PWR	678	TO-220AB	N	70	75	20-150
2N6488	PWR	678	TO-220AB	N	90	75	20-150
2N6489	PWR	678	TO-220AB	N	—50	75	20-150
2N6490	PWR	678	TO-220AB	N	—70	75	20-150
2N6491	PWR	678	TO-220AB	N	—90	75	20-150
2N6496	PWR	698	TO-3	D	150	140	12-100
2N6500	PWR	766	TO-66	D	120	35	15-60

▲ Dimensional outlines are shown following data charts.
■ Lead codes are explained in Table LII.
See **Notes** on page 619.

Table LI—RCA Standard Solid-State Devices (cont'd)

Type No.	Prod. Line	File No.	Package▲	Lead Code■	Voltage (V) (Note 1)	Power (W), Current* (A) or Noise Figure (dB)	DC Current Transfer Ratio or Frequency
2N6510	PWR	848	TO-3	D	250#	125	10 min.
2N6511	PWR	848	TO-3	D	300#	125	10 min.
2N6512	PWR	848	TO-3	D	350#	125	10 min.
2N6513	PWR	848	TO-3	D	400#	125	10 min.
2N6514	PWR	848	TO-3	D	350#	125	10 min.
2N6530	PWR	873	TO-220AB	N	80	65	1000-10,000
2N6531	PWR	873	TO-220AB	N	100	65	500-10,000
2N6532	PWR	873	TO-220AB	N	100	65	1000-10,000
2N6533	PWR	873	TO-220AB	N	120	65	1000-10,000
2N6534	PWR	874	TO-66	D	80	36	1000-10,000
2N6535	PWR	874	TO-66	D	100	36	500-10,000
2N6536	PWR	874	TO-66	D	100	36	1000-10,000
2N6537	PWR	874	TO-66	D	120	36	1000-10,000
3N128	MOS/FET	309	TO-72	R	20	single-gate amplifier	
3N138	MOS/FET	283	TO-72	R	35	single-gate chopper, multiplexer	
3N139	MOS/FET	284	TO-72	R	35	single-gate af/rf amplifier	
3N140	MOS/FET	285	TO-72	P	20	dual-gate rf amplifier	
3N141	MOS/FET	285	TO-72	P	20	dual-gate mixer	
3N142	MOS/FET	286	TO-72	R	20	single-gate rf amplifier	
3N143	MOS/FET	309	TO-72	R	20	single-gate vhf mixer/oscillator	
3N152	MOS/FET	314	TO-72	R	20	single-gate vhf amplifier	
3N153	MOS/FET	320	TO-72	R	20	single-gate chopper/multiplexer	
3N154	MOS/FET	335	TO-72	R	20	single-gate vhf amplifier	
3N159	MOS/FET	326	TO-72	P	20	dual-gate rf amplifier	
3N187	MOS/FET	436	TO-72	P	20	dual-gate rf amplifier	
3N200	MOS/FET	437	TO-72	P	20	dual-gate rf amplifier	
40080	RF	301	TO-5	F	12	0.1	27 MHz
40081	RF	301	TO-5	F	12	0.4	27 MHz
40082	RF	301	TO-39	F	12	3	27 MHz
40250	PWR	112	TO-66	D	50	29	25-100
40250V1	PWR	112	Radiator	D	50	5.8	25-100
40251	PWR	112	TO-3	D	50	117	25-100
40279	RF	46	TO-60	E	premium high-reliability version of		2N3375
40280	RF	68	TO-39	F	13.5	1	175 MHz
40281	RF	68	TO-60	G	13.5	4	175 MHz
40282	RF	68	TO-60	G	13.5	12	175 MHz
40290	RF	70	TO-39	F	12.5	2	135 MHz
40291	RF	70	TO-60	E	12.5	2	135 MHz
40292	RF	70	TO-60	E	12.5	6	135 MHz
40294	RF	202	TO-72	S	premium high-reliability version of		2N2857
40296	RF	603	TO-72	F	premium high-reliability version of		2N3839
40305	RF	144	TO-39	F	premium high-reliability version of		2N3553
40306	RF	144	TO-60	E	premium high-reliability version of		2N3375
40307	RF	144	TO-60	E	premium high-reliability version of		2N2632
40309	PWR	78	TO-39, TO-5	F	18(V_{CEO})	5	70-350
40310	PWR	78	TO-66	D	35(V_{CEO})	29	20-120
40311	PWR	78	TO-39, TO-5	F	30(V_{CEO})	5	70-350
40312	PWR	78	TO-66	D	60(V_{CER})	29	20-120
40313	PWR	78	TO-66	D	300(V_{CER})	35	40-250
40314	PWR	78	TO-39, TO-5	F	40(V_{CEO})	5	70-350

▲ Dimensional outlines are shown following data charts.
■ Lead codes are explained in Table LII.
See **Notes** on page 619.

Table LI—RCA Standard Solid-State Devices (cont'd)

Type No.	Prod. Line	File No.	Package▲	Lead Code■	Voltage (V) (Note 1)	Power (W), Current* (A) or Noise Figure (dB)	DC Current Transfer Ratio or Frequency
40315	PWR	78	TO-39, TO-5	F	35(V_{CEO})	5	70-350
40316	PWR	78	TO-66	D	40(V_{CER})	29	20-120
40317	PWR	78	TO-39, TO-5	F	40(V_{CEO})	5	40-200
40318	PWR	78	TO-66	D	300(V_{CER})	35	50 min.
40319	PWR	78	TO-39, TO-5	F	—40(V_{CEO})	5	35-200
40320	PWR	78	TO-39, TO-5	F	40(V_{CEO})	5	40-200
40321	PWR	78	TO-39, TO-5	F	300(V_{CER})	5	25-200
40322	PWR	78	TO-66	D	300(V_{CER})	35	75 min.
40323	PWR	78	TO-39, TO-5	F	10(V_{CEO})	5	70-350
40324	PWR	78	TO-66	D	35(V_{CEO})	29	20-120
40325	PWR	78	TO-3	D	35(V_{CEO})	117	12-60
40326	PWR	78	TO-39, TO-5	F	40(V_{CEO})	5	40-200
40327	PWR	78	TO-39, TO-5	F	300(V_{CER})	5	40-250
40328	PWR	78	TO-66	D	300(V_{CER})	35	40 min.
40340	RF	74	TO-60	U	13.5	25	50 MHz
40341	RF	74	TO-60	U	24	30	50 MHz
40346	PWR	211	TO-39, TO-5	F	175(V_{CER})	10	25 min.
40346V1	PWR	211	Radiator	F	175(V_{CER})	10	25 min.
40346V2	PWR	211	Flange	F	175(V_{CER})	4	25 min.
40347	PWR	88	TO-39, TO-5	F	60	8.75	25-100
40347V1	PWR	88	Radiator	F	60	4.4	25-100
40347V2	PWR	88	Flange	F	60	11.7	25-100
40348	PWR	88	TO-39, TO-5	F	90	8.75	30-125
40348V1	PWR	88	Radiator	F	90	4.4	30-125
40348V2	PWR	88	Flange	F	90	11.7	30-125
40349	PWR	88	TO-39, TO-5	F	160	8.75	30-125
40349V1	PWR	88	Radiator	F	160	4.4	30-125
40349V2	PWR	88	Flange	F	160	11.7	30-125
40360	PWR	78	TO-39, TO-5	F	70(V_{CEO})	5	40-200
40361	PWR	78	TO-39, TO-5	F	70(V_{CER})	5	70-350
40362	PWR	78	TO-39, TO-5	F	70(V_{CER})	5	35-200
40363	PWR	78	TO-3	D	70(V_{CER})	115	20-70
40364	PWR	78	TO-66	D	60(V_{CER})	35	35-175
40366	PWR	215	TO-39, TO-5	F	120	5	40-120
40367	PWR	215	TO-39, TO-5	F	100	5	35-100
40368	PWR	215	TO-8	F	100	25	35-100
40369	PWR	215	TO-3	D	100	75	25-75
40372	PWR	527	Radiator	D	90	25	25-150
40373	PWR	529	Radiator	D	160	25	25-100
40374	PWR	128	Radiator	D	250	5.8	40 min.
40375	PWR	766	Radiator	D	120	5.8	50-200
40385	PWR	215	TO-39, TO-5	F	450	5	40-160
40389	PWR	432	Radiator	F	60	3.5	50-250
40390	PWR	64	Radiator	F	300	3.5	40-160
40391	PWR	216	Radiator	F	—60	3.5	50-250
40392	PWR	432	Flange	F	60	7	50-250
40394	PWR	216	Flange	F	—60	7	50-250
40406	PWR	219	TO-39, TO-5	F	—50(V_{CEO})	1	30-200
40407	PWR	219	TO-39, TO-5	F	50(V_{CEO})	1	40-200
40408	PWR	219	TO-39, TO-5	F	90(V_{CEO})	1	40-200

▲ Dimensional outlines are shown following data charts
■ Lead codes are explained in Table LII.
See Notes on page 619.

Table LI—RCA Standard Solid-State Devices (cont'd)

Type No.	Prod. Line	File No.	Package▲	Lead Code■	Voltage (V) (Note 1)	Power (W), Current* (A) or Noise Figure (dB)	DC Current Transfer Ratio or Frequency
40409	PWR	219	Radiator	F	$90(V_{CER})$	3	50-250
40410	PWR	219	Radiator	F	$-90(V_{CER})$	3	50-250
40411	PWR	219	TO-3	D	$90(V_{CER})$	150	35-100
40412	PWR	211	TO-39, TO-5	F	$250(V_{CER})$	10	40 min.
40412V1	PWR	211	Radiator	F	$250(V_{CER})$	4	40 min.
40412V2	PWR	211	Flange	F	$250(V_{CER})$	10	40 min.
40414	RF	259	TO-72	S	premium high-reliability version of 2N2857		
40446	RF	301	Flange	F	12	3	27 MHz
40467A	MOS/FET	324	TO-72	R	20	single-gate vhf amplifier	
40468A	MOS/FET	323	TO-72	R	20	single-gate rf amplifier	
40513	PWR	244	TO-219AB	D	$45(V_{CER})$	83	25-100
40514	PWR	244	TO-219AA	D	$45(V_{CER})$	83	25-100
40537	PWR	320	TO-39, TO-5	F	$-55(V_{CER})$	5	50-300
40538	PWR	320	TO-39, TO-5	F	$-55(V_{CER})$	5	15-90
40539	PWR	303	TO-39, TO-5	F	$55(V_{CER})$	5	15-90
40542	PWR	304	Plastic TO-3	D	$50(V_{CER})$	83	20-70
40543	PWR	304	Plastic TO-3	D	$60(V_{CER})$	83	20-70
40544	PWR	303	Flange	F	$50(V_{CER})$	7	35-200
40559A	MOS/FET	323	TO-72	R	20	single-gate mixer	
40577	RF	297	TO-5	F	premium high-reliability version of 2N3118		
40578	RF	298	TO-39	F	premium high-reliability version of 2N3866		
40581	RF	301	TO-39	F	12	3.5	27 MHz
40582	RF	301	Flange	F	12	3.5	27 MHz
40594	PWR	358	TO-39, TO-5	F	$95(V_{CER})$	5	70-350
40595	PWR	358	TO-39, TO-5	F	$-95(V_{CER})$	10	70-350
40600	MOS/FET	333	TO-72	P	20	dual-gate rf amplifier	
40601	MOS/FET	333	TO-72	P	20	dual-gate if amplifier	
40602	MOS/FET	333	TO-72	P	20	dual-gate mixer	
40603	MOS/FET	334	TO-72	P	20	dual-gate rf amplifier	
40604	MOS/FET	334	TO-72	P	20	dual-gate mixer	
40605	RF	389	TO-39	F	premium high-reliability version of 2N3553		
40606	RF	600	TO-60	U	premium high-reliability version of 2N3632		
40608	RF	356	TO-39	F	15	3(NF)	200 MHz
40611	PWR	358	TO-39, TO-5	F	$25(V_{CEO})$	5	70-500
40613	PWR	358	TO-220AA	M	$25(V_{CEO})$	36	30-120
40616	PWR	358	TO-39, TO-5	F	$32(V_{CEO})$	5	70-500
40618	PWR	358	TO-220AA	M	$30(V_{CEO})$	36	30-120
40621	PWR	358	TO-220AA	M	$32(V_{CEO})$	36	25-100
40622	PWR	358	TO-220AA	M	$40(V_{CEO})$	36	25-100
40624	PWR	358	TO-220AA	M	$45(V_{CEO})$	50	20-100
40625	PWR	358	Radiator	F	$45(V_{CEO})$	3.5	100-300
40627	PWR	358	TO-220AA	M	$55(V_{CEO})$	50	20-100
40628	PWR	358	Radiator	F	$55(V_{CEO})$	3.5	100-300
40629	PWR	358	TO-220AA	M	$35(V_{CER})$	36	20-70
40630	PWR	358	TO-220AA	M	$40(V_{CER})$	36	20-70
40631	PWR	358	TO-220AA	M	$45(V_{CER})$	36	20-70
40632	PWR	358	TO-220AA	M	$60(V_{CER})$	50	20-70
40634	PWR	358	TO-39, TO-5	F	$-75(V_{CER})$	5	50-250
40635	PWR	358	TO-39, TO-5	F	$75(V_{CER})$	5	50-250
40636	PWR	358	TO-3	D	$95(V_{CER})$	115	20-70

▲ Dimensional outlines are shown following data charts.
■ Lead codes are explained in Table LII.
See **Notes** on page 619.

Table LI—RCA Standard Solid-State Devices (cont'd)

Type No.	Prod. Line	File No.	Package▲	Lead Code■	Voltage (V) (Note 1)	Power (W), Current* (A) or Noise Figure (dB)	DC Current Transfer Ratio or Frequency
40665	RF	386	TO-60	U	28	13.5	175 MHz
40666	RF	386	TO-60	U	28	3	400 MHz
40673	MOS/FET	381	TO-72	P	20	dual-gate rf amplifier	
40819	MOS/FET	463	TO-72	P	25	dual-gate rf amplifier	
40820	MOS/FET	464	TO-72	P	20	dual-gate rf amplifier	
40821	MOS/FET	464	TO-72	P	20	dual-gate mixer	
40822	MOS/FET	465	TO-72	P	18	dual-gate rf amplifier	
40823	MOS/FET	465	TO-72	P	18	dual-gate mixer	
40829	PWR	675	Radiator	D	—90	40	20-100
40830	PWR	675	Radiator	D	—70	40	20-100
40831	PWR	675	Radiator	D	—50	40	20-100
40836	RF	497	TO-215AA	D	21	0.5	2 GHz
40837	RF	497	TO-215AA	D	28	1.5	2 GHz
40841	MOS/FET	489	TO-72	P	18	dual-gate general-purpose type	
40850	PWR	498	TO-66	D	450	35	25 min.
40851	PWR	498	TO-66	D	450	45	12 min.
40852	PWR	498	TO-3	D	450	100	12 min.
40853	PWR	498	TO-3	D	450	100	10 min.
40854	PWR	498	TO-3	D	450	110	10 min.
40871	PWR	699	TO-220AB	N	100(V_{CEO})	40	50-250
40872	PWR	699	TO-220AB	N	—100(V_{CEO})	40	50-250
40873	PWR	699	TO-220AB	N	70(V_{CEO})	40	30-150
40874	PWR	699	TO-220AB	N	—70(V_{CEO})	40	30-150
40875	PWR	699	TO-220AB	N	50(V_{CEO})	40	20-120
40876	PWR	699	TO-220AB	N	—50(V_{CEO})	40	20-120
40885	PWR	508	Heat Clip	E	300	20	30-190
40886	PWR	508	Heat Clip	E	350	20	30-190
40887	PWR	508	Heat Clip	E	450	20	30-190
40893	RF	514	RCA HF-36	W	12.5	15	470 MHz
40894	RF	548	TO-72	S	12	15(G_{pe})	200 MHz
40895	RF	548	TO-72	S	12	15(G_{pe})	200 MHz
40896	RF	548	TO-72	S	12	15(G_{pe})	200 MHz
40897	RF	548	TO-72	S	12	18(G_{pe})	200 MHz
40898	RF	538	TO-215AA	E	22	2	2.3 GHz
40899	RF	538	TO-201AA	E	22	6	2.3 GHz
40909	RF	547	TO-201AA	D	25	2	2 GHz
40910	PWR	527	Radiator	D	50	29	20-100
40911	PWR	527	Radiator	D	90	50	25-100
40912	PWR	529	Radiator	D	140	20	20-100
40913	PWR	529	Radiator	D	170	50	20-60
40915	RF	574	TO-72	S	10	2.5(NF)	450 MHz
40934	RF	550	RCA HF-31	W	12.5	2	470 MHz
40936	RF	551	TO-60	U	21	0.5	2 GHz
40940	RF	553	TO-216AA	W	28	5	400 MHz
40941	RF	554	RCA HF-31	W	28	1	400 MHz
40953	RF	579	TO-39	F	12.5	1.75	156 MHz
40964	RF	581	TO-39	F	12	0.4	470 MHz
40965	RF	581	TO-39	F	12	0.5	470 MHz
40968	RF	596	RCA HF-44	W	12.5	6	470 MHz
40970	RF	656	RCA HF-40	W	12.5	30	470 MHz

▲ Dimensional outlines are shown following data charts.
■ Lead codes are explained in Table LII.
See **Notes** on page 619.

Table LI—RCA Standard Solid-State Devices (cont'd)

Type No.	Prod. Line	File No.	Package▲	Lead Code■	Voltage (V) (Note 1)	Power (W), Current* (A) or Noise Figure (dB)	DC Current Transfer Ratio or Frequency
40971	RF	656	RCA HF-40	W	12.5	45	470 MHz
40972	RF	597	TO-39	F	12.5	1.75	175 MHz
40975	RF	606	TO-39	F	12.5	0.05	118 MHz
40976	RF	606	TO-39	F	12.5	0.5	118 MHz
41024	RF	658	TO-39	F	28	1	1 GHz
41038	RF	679	TO-46	F	20	0.75	1.68 GHz
41039	RF	764	TO-39	F	$15(V_{CE})$	$15(G_{pe})$	50-250 MHz
41042	RF	758	RCA HF-44	W	13	20	136 MHz
41044	RF	783	RCA HF-56	N	$-20(V_{EE})$	0.4	4.36 GHz
41500	PWR	772	TO-220AB	N	35	40	25 min.
41501	PWR	770	TO-220AB	N	-35	40	25 min.
41502	PWR	773	TO-39, TO-5	F	$30(V_{CEO})$	3	20 min.
41503	PWR	774	TO-39, TO-5	F	$-30(V_{CEO})$	7	20 min.
41504	PWR	775	TO-220AB	N	$35(V_{CER})$	36	25 min.
41505	PWR	771	Plastic TO-5	F	$200(V_{CEO})$	20	20 min.
41506	PWR	776	TO-3	D	200	100	8 min.
43104	PWR	622	TO-3	D	160	150	15-60
CA101AT	LIC	786	Note 3	Note 6	general-purpose operational amplifier		
CA101T	LIC	786	Note 3	Note 6	general-purpose operational amplifier		
CA107S	LIC	785	Note 3	Note 6	general-purpose operational amplifier		
CA107T	LIC	785	Note 3	Note 6	general-purpose operational amplifier		
CA108AS	LIC	621	Note 3	Note 6	precision operational amplifier		
CA108AT	LIC	621	Note 3	Note 6	precision operational amplifier		
CA108S	LIC	621	Note 3	Note 6	precision operational amplifier		
CA108T	LIC	621	Note 3	Note 6	precision operational amplifier		
CA111S	LIC	797	Note 3	Note 6	voltage comparator		
CA111T	LIC	797	Note 3	Note 6	voltage comparator		
CA201AT	LIC	786	Note 3	Note 6	general-purpose operational amplifier		
CA201T	LIC	786	Note 3	Note 6	general-purpose operational amplifier		
CA207S	LIC	785	Note 3	Note 6	general-purpose operational amplifier		
CA207T	LIC	785	Note 3	Note 6	general-purpose operational amplifier		
CA208AS	LIC	621	Note 3	Note 6	precision operational amplifier		
CA208AT	LIC	621	Note 3	Note 6	precision operational amplifier		
CA208S	LIC	621	Note 3	Note 6	precision operational amplifier		
CA208T	LIC	621	Note 3	Note 6	precision operational amplifier		
CA211S	LIC	797	Note 3	Note 6	voltage comparator		
CA211T	LIC	797	Note 3	Note 6	voltage comparator		
CA301AE	LIC	786	Note 3	Note 6	general-purpose operational amplifier		
CA301AT	LIC	786	Note 3	Note 6	general-purpose operational amplifier		
CA307E	LIC	785	Note 3	Note 6	general-purpose operational amplifier		
CA307S	LIC	785	Note 3	Note 6	general-purpose operational amplifier		
CA307T	LIC	785	Note 3	Note 6	general-purpose operational amplifier		
CA308AS	LIC	621	Note 3	Note 6	precision operational amplifier		
CA308AT	LIC	621	Note 3	Note 6	precision operational amplifier		
CA308H	LIC	516	—	Note 6	precision operational amplifier		
CA308S	LIC	621	Note 3	Note 6	precision operational amplifier		
CA308T	LIC	621	Note 3	Note 6	precision operational amplifier		
CA311S	LIC	797	Note 3	Note 6	voltage comparator		
CA311T	LIC	797	Note 3	Note 6	voltage comparator		
CA324E	LIC	796	Note 3	Note 6	quad operational amplifier		

▲ Dimensional outlines are shown following data charts.
■ Lead codes are explained in Table LII.
See **Notes** on page 619.

Table LI—RCA Standard Solid-State Devices (cont'd)

Type No.	Prod. Line	File No.	Package▲	Lead Code■	Voltage (V) (Note 1)	Power (W), Current* (A) or Noise Figure (dB)	DC Current Transfer Ratio or Frequency
CA339AE	LIC	795	Note 3	Note 6	quad operational amplifier		
CA339E	LIC	795	Note 3	Note 6	quad operational amplifier		
CA723CE	LIC	788	Note 3	Note 6	voltage regulator		
CA723CT	LIC	788	Note 3	Note 6	voltage regulator		
CA723E	LIC	788	Note 3	Note 6	voltage regulator		
CA723T	LIC	788	Note 3	Note 6	voltage regulator		
CA741/	LIC	718	Note 4	Note 6	high-reliability operational amplifier		
CA741CE	LIC	531	Note 3	Note 6	operational amplifier		
CA741CH	LIC	516	—	Note 6	operational amplifier chip		
CA741CS	LIC	531	Note 3	Note 6	operational amplifier		
CA741CT	LIC	531	Note 3	Note 6	operational amplifier		
CA741L	LIC	515	—	Note 6	beam-lead operational amplifier		
CA741S	LIC	531	Note 3	Note 6	operational amplifier		
CA741T	LIC	531	Note 3	Note 6	operational amplifier		
CA747/	LIC	718	Note 4	Note 6	high-reliability operational amplifier		
CA747CE	LIC	531	Note 3	Note 6	operational amplifier		
CA747CF	LIC	531	Note 3	Note 6	operational amplifier		
CA747CH	LIC	516	—	Note 6	operational amplifier chip		
CA747CT	LIC	531	Note 3	Note 6	operational amplifier		
CA747E	LIC	531	Note 3	Note 6	operational amplifier		
CA747F	LIC	531	Note 3	Note 6	operational amplifier		
CA747L	LIC	531	—	Note 6	beam-lead operational amplifier		
CA747T	LIC	531	Note 3	Note 6	operational amplifier		
CA748/	LIC	718	Note 4	Note 6	high-reliability operational amplifier		
CA748CE	LIC	531	Note 3	Note 6	operational amplifier		
CA748CH	LIC	516	—	Note 6	operational amplifier chip		
CA748CS	LIC	531	Note 3	Note 6	operational amplifier		
CA748CT	LIC	531	Note 3	Note 6	operational amplifier		
CA748S	LIC	531	Note 3	Note 6	operational amplifier		
CA748T	LIC	531	Note 3	Note 6	operational amplifier		
CA758E	LIC	760	Note 3	Note 6	rc stereo decoder		
CA1310E	LIC	761	Note 3	Note 6	rc stereo decoder		
CA1398E	LIC	686	Note 3	Note 6	TV chroma processor		
CA1458E	LIC	531	Note 3	Note 6	operational amplifier		
CA1458S	LIC	531	Note 3	Note 6	operational amplifier		
CA1458T	LIC	531	Note 3	Note 6	operational amplifier		
CA1541D	LIC	536	Note 3	Note 6	memory sense amplifier		
CA1541H	LIC	516	—	Note 6	memory-sense-amplifier chip		
CA1558/	LIC	718	Note 4	Note 6	high-reliability operational amplifier		
CA1558S	LIC	531	Note 3	Note 6	operational amplifier		
CA1558T	LIC	531	Note 3	Note 6	operational amplifier		
CA2111AE	LIC	612	Note 3	Note 6	FM if subsystem		
CA2111AQ	LIC	612	Note 3	Note 6	FM if subsystem		
CA3000	LIC	121	Note 3	Note 6	dc amplifier		
CA3000/	LIC	705	Note 3	Note 6	high-reliability dc amplifier		
CA3000H	LIC	516	—	Note 6	dc-amplifier chip		
CA3001	LIC	122	Note 3	Note 6	video amplifier		
CA3001/	LIC	714	Note 4	Note 6	high-reliability video amplifier		
CA3001H	LIC	516	—	Note 6	video-amplifier chip		
CA3002	LIC	123	Note 3	Note 6	if amplifier		

▲ Dimensional outlines are shown following data charts.
■ Lead codes are explained in Table LII.
See **Notes** on page 619.

Table LI—RCA Standard Solid-State Devices (cont'd)

Type No.	Prod. Line	File No.	Package▲	Lead Code■	Voltage (V) (Note 1)	Power (W), Current* (A) or Noise Figure (dB)	DC Current Transfer Ratio or Frequency
CA3002/	LIC	713	Note 4	Note 6	high-reliability if amplifier		
CA3002H	LIC	516	—	Note 6	If-amplifier chip		
CA3004	LIC	124	Note 3	Note 6	rf amplifier		
CA3004/	LIC	712	Note 4	Note 6	high-reliability rf amplifier		
CA3005	LIC	125	TO-5	Note 6	rf amplifier		
CA3005H	LIC	516	—	Note 6	rf-amplifier chip		
CA3006	LIC	125	TO-5	Note 6	rf amplifier		
CA3006/	LIC	763	Note 4	Note 6	high-reliability rf amplifier		
CA3007	LIC	126	TO-5	Note 6	af amplifier		
CA3008	LIC	316	Flat Pack	Note 6	operational amplifier		
CA3008A	LIC	310	Flat Pack	Note 6	operational amplifier		
CA3010	LIC	316	TO-5	Note 6	operational amplifier		
CA3010A	LIC	310	TO-5	Note 6	operational amplifier		
CA3011	LIC	128	TO-5	Note 6	wide-band amplifier		
CA3012	LIC	128	TO-5	Note 6	wide-band amplifier		
CA3012H	LIC	516	—	Note 6	wide-band-amplifier chip		
CA3013	LIC	471	TO-5	Note 6	wide-band amplifier-discriminator		
CA3014	LIC	129	TO-5	Note 6	wide-band amplifier-discriminator		
CA3015	LIC	316	TO-5	Note 6	operational amplifier		
CA3015A	LIC	310	TO-5	Note 6	operational amplifier		
CA3015A/	LIC	715	Note 4	Note 6	high-reliability operational amplifier		
CA3015H	LIC	516	—	Note 6	operational-amplifier chip		
CA3015L	LIC	515	—	Note 6	beam-lead operational amplifier		
CA3016	LIC	316	Flat Pack	Note 6	operational amplifier		
CA3016A	LIC	310	Flat Pack	Note 6	operational amplifier		
CA3018	LIC	338	TO-5	Note 6	transistor array		
CA3018A	LIC	338	TO-5	Note 6	transistor array		
CA3018A/	LIC	762	Note 4	Note 6	high-reliability transistor array		
CA3018H	LIC	516	—	Note 6	transistor-array chip		
CA3018L	LIC	515	—	Note 6	beam-lead transistor array		
CA3019	LIC	236	TO-5	Note 6	diode array		
CA3019/	LIC	722	Note 4	Note 6	high-reliability diode array		
CA3019H	LIC	516	—	Note 6	diode-array chip		
CA3020	LIC	339	TO-5	Note 6	wide-band power amplifier		
CA3020A	LIC	339	TO-5	Note 6	wide-band power amplifier		
CA3020A/	LIC	767	Note 4	Note 6	high-reliability wide-band power amplifier		
CA3020H	LIC	516	—	Note 6	wide-band-power-amplifier chip		
CA3021	LIC	243	TO-5	Note 6	low-power video amplifier		
CA3022	LIC	243	TO-5	Note 6	low-power video amplifier		
CA3023	LIC	243	TO-5	Note 6	low-power video amplifier		
CA3023H	LIC	516	—	Note 6	low-power-video-amplifier chip		
CA3026	LIC	388	TO-5	Note 6	dual differential amplifier		
CA3026/	LIC	706	Note 4	Note 6	high-reliability dual differential amplifier		
CA3026H	LIC	516	—	Note 6	dual-differential-amplifier chip		
CA3028A	LIC	382	TO-5	Note 6	differential/cascode amplifier		
CA30228AF	LIC	382	Note 3	Note 6	differential/cascode amplifier		
CA3028AH	LIC	516	—	Note 6	differential/cascode-amplifier chip		
CA3028AL	LIC	515	—	Note 6	beam-lead differential/cascode amplifier		
CA3028AS	LIC	382	Note 3	Note 6	differential/cascode amplifier		
CA3028B	LIC	382	TO-5	Note 6	differential/cascode amplifier		

▲ Dimensional outlines are shown following data charts.
■ Lead codes are explained in Table LII.
See **Notes** on page 619.

Table LI—RCA Standard Solid-State Devices (cont'd)

Type No.	Prod. Line	File No.	Package▲	Lead Code■	Voltage (V) (Note 1)	Power (W), Current* (A) or Noise Figure (dB)	DC Current Transfer Ratio or Frequency
CA3028B/	LIC	711	Note 4	Note 6	high-reliability differential/cascode amplifier		
CA3028BF	LIC	382	Note 3	Note 6	differential/cascode amplifier		
CA3028BS	LIC	382	Note 3	Note 6	differential/cascode amplifier		
CA3029	LIC	316	DIP	Note 6	operational amplifier		
CA3029A	LIC	310	DIP	Note 6	operational amplifier		
CA3030	LIC	316	DIP	Note 6	operational amplifier		
CA3030A	LIC	310	DIC	Note 6	operational amplifier		
CA3033	LIC	360	DIC	Note 6	operational amplifier		
CA3033A	LIC	360	DIC	Note 6	operational amplifier		
CA3033H	LIC	516	—	Note 6	operational-amplifier chip		
CA3035	LIC	274	TO-5	Note 6	wide-band amplifier array		
CA3035H	LIC	516	—	Note 6	wide-band amplifier-array chip		
CA3035V1	LIC	274	Note 3	Note 6	wide-band amplifier array		
CA3036	LIC	275	TO-5	Note 6	dual Darlington array		
CA3037	LIC	316	DIC	Note 6	operational amplifier		
CA3037A	LIC	310	DIC	Note 6	operational amplifier		
CA3038	LIC	316	DIC	Note 6	operational amplifier		
CA3038A	LIC	310	DIC	Note 6	operational amplifier		
CA3039	LIC	310	TO-5	Note 6	diode array		
CA3039/	LIC	704	Note 4	Note 6	high-reliability diode array		
CA3039H	LIC	516	—	Note 6	diode-array chip		
CA3039L	LIC	515	—	Note 6	beam-lead diode array		
CA3040	LIC	363	TO-5	Note 6	wide-band amplifier		
CA3041	LIC	318	DIP	Note 6	TV if sound subsystem		
CA3042	LIC	319	DIP	Note 6	TV if sound subsystem		
CA3043	LIC	331	TO-5	Note 6	FM if subsystem		
CA3043H	LIC	516	—	Note 6	FM-if-subsystem chip		
CA3044	LIC	340	TO-5	Note 6	TV automatic-fine-tuning subsystem		
CA3044V1	LIC	340	Note 3	Note 6	TV automatic-fine-tuning subsystem		
CA3045	LIC	341	DIC	Note 6	transistor array		
CA3045/	LIC	710	Note 4	Note 6	high-reliability transistor array		
CA3045F	LIC	341	Note 3	Note 6	transistor array		
CA3045H	LIC	516	—	Note 6	transistor-array chip		
CA3045L	LIC	515	—	Note 6	beam-lead transistor array		
CA3046	LIC	341	DIP	Note 6	transistor array		
CA3047	LIC	360	DIP	Note 6	operational amplifier		
CA3047A	LIC	360	DIP	Note 6	operational amplifier		
CA3048	LIC	377	DIP	Note 6	amplifier array		
CA3048H	LIC	516	—	Note 6	amplifier-array chip		
CA3049/	LIC	707	Note 4	Note 6	high-reliability dual differential amplifier		
CA3049H	LIC	516	—	Note 6	dual-differential-amplifier chip		
CA3049L	LIC	515	—	Note 6	beam-lead dual differential amplifier		
CA3049T	LIC	611	TO-5	Note 6	dual differential amplifier		
CA3050	LIC	361	DIC	Note 6	dual differential amplifier		
CA3051	LIC	361	DIP	Note 6	dual differential amplifier		
CA3052	LIC	387	DIP	Note 6	stereo preamplifier		
CA3053	LIC	382	TO-5	Note 6	differential/cascode amplifier		
CA3053F	LIC	382	Note 3	Note 6	differential/cascode amplifier		
CA3053S	LIC	382	Note 3	Note 6	differential/cascode amplifier		
CA3054	LIC	388	DIP	Note 6	dual differential amplifier		

▲ Dimensional outlines are shown following data charts.
■ Lead codes are explained in Table LII.
See **Notes** on page 619.

Table LI—RCA Standard Solid-State Devices (cont'd)

Type No.	Prod. Line	File No.	Package▲	Lead Code■	Voltage (V) (Note 1)	Power (W), Current* (A) or Noise Figure (dB)	DC Current Transfer Ratio or Frequency
CA3054H	LIC	516	—	Note 6	dual-differential-amplifier chip		
CA3054L	LIC	515	—	Note 6	beam-lead dual differential amplifier		
CA3058	LIC	490	DIC	Note 6	zero-voltage switch		
CA3058/	LIC	703	Note 4	Note 6	high-reliability zero-voltage switch		
CA3059	LIC	490	DIP	Note 6	zero-voltage switch		
CA3059H	LIC	516	—	Note 6	zero-voltage-switch chip		
CA3060AD	LIC	537	Note 3	Note 6	OTA array		
CA3060BD	LIC	537	Note 3	Note 6	OTA array		
CA3060D	LIC	537	Note 3	Note 6	OTA array		
CA3060E	LIC	537	Note 3	Note 6	OTA array		
CA3060H	LIC	516	—	Note 6	OTA-array chip		
CA3062	LIC	421	TO-5	Note 6	photo detector and power amplifier		
CA3064	LIC	396	TO-5	Note 6	TV automatic-fine-tuning subsystem		
CA3064E	LIC	396	Note 3	Note 6	TV automatic-fine-tuning subsystem		
CA3065	LIC	412	DIP	Note 6	TV if sound system		
CA3066	LIC	466	DIP	Note 6	TV chroma signal processor		
CA3067	LIC	466	DIP	Note 6	TV chroma demodulator		
CA3068	LIC	467	QUIP	Note 6	TV video if system		
CA3070	LIC	468	DIP	Note 6	TV chroma signal processor		
CA3071	LIC	468	DIP	Note 6	TV chroma amplifier		
CA3072	LIC	468	DIP	Note 6	TV chroma demodulator		
CA3075	LIC	429	QUIP	Note 6	FM if subsystem		
CA3075H	LIC	516	—	Note 6	FM-if-subsystem chip		
CA3076	LIC	430	TO-5	Note 6	FM if gain block		
CA3076H	LIC	516	—	Note 6	FM-if-gain-block chip		
CA3078AS	LIC	535	Note 3	Note 6	micropower operational amplifier		
CA3078AT	LIC	535	Note 3	Note 6	micropower operational amplifier		
CA3078H	LIC	516	—	Note 6	micropower-operational-amplifier chip		
CA3078S	LIC	535	Note 3	Note 6	micropower operational amplifier		
CA3078T	LIC	535	Note 3	Note 6	micropower operational amplifier		
CA3079	LIC	490	DIP	Note 6	zero-voltage switch		
CA3080	LIC	475	TO-5	Note 6	operational transconductance amplifier		
CA3080/	LIC	709	Note 4	Note 6	high-reliability OTA		
CA3080A	LIC	475	TO-5	Note 6	operational transconductance amplifier		
CA3080A/	LIC	709	Note 4	Note 6	high-reliability OTA		
CA3080AS	LIC	475	Note 3	Note 6	operational transconductance amplifier		
CA3080E	LIC	475	Note 3	Note 6	operational transconductance amplifier		
CA3080H	LIC	516		Note 6	OTA chip		
CA3080S	LIC	475	Note 3	Note 6	operational transconductance amplifier		
CA3081	LIC	480	DIP	Note 6	transistor array (n-p-n)		
CA3081F	LIC	480	Note 3	Note 6	transistor array (n-p-n)		
CA3081H	LIC	516	—	Note 6	transistor-array chip (n-p-n)		
CA3082	LIC	480	DIP	Note 6	transistor array (n-p-n)		
CA3082F	LIC	480	Note 3	Note 6	transistor array (n-p-n)		
CA3082H	LIC	516	—	Note 6	transistor-array chip (n-p-n)		
CA3083	LIC	481	DIP	Note 6	beam-lead transistor array (n-p-n)		
CA3083F	LIC	481	Note 3	Note 6	transistor array (n-p-n)		
CA3083H	LIC	516	—	Note 6	transistor array chip (n-p-n)		
CA3083L	LIC	515	—	Note 6	beam-lead transistor array (n-p-n)		
CA3084	LIC	482	DIP	Note 6	transistor-array chip (p-n-p)		

▲ Dimensional outlines are shown following data charts.
■ Lead codes are explained in Table LII.
See **Notes** on page 619.

Table LI—RCA Standard Solid-State Devices (cont'd)

Type No.	Prod. Line	File No.	Package▲	Lead Code■	Voltage (V) (Note 1) Power (W), Current* (A) or Noise Figure (dB) DC Current Transfer Ratio or Frequency
CA3084H	LIC	516	—	Note 6	transistor-array chip (p-n-p)
CA3084L	LIC	515	—	Note 6	beam-lead transistor array (p-n-p)
CA3085	LIC	491	TO-5	Note 6	positive voltage regulator
CA3085/	LIC	708	Note 4	Note 6	high-reliability positive voltage regulator
CA3085A	LIC	491	TO-5	Note 6	positive voltage regulator
CA3085A/	LIC	708	Note 4	Note 6	high-reliability positive voltage regulator
CA3085AE	LIC	491	Note 3	Note 6	positive voltage regulator
CA3085AF	LIC	491	Note 3	Note 6	positive voltage regulator
CA3085AS	LIC	491	Note 3	Note 6	positive voltage regulator
CA3085B	LIC	491	TO-5	Note 6	positive voltage regulator
CA3085B/	LIC	708	Note 4	Note 6	positive voltage regulator
CA3085BF	LIC	491	Note 3	Note 6	positive voltage regulator
CA3085BS	LIC	491	Note 3	Note 6	positive voltage regulator
CA3085E	LIC	491	Note 3	Note 6	positive voltage regulator
CA3085F	LIC	491	Note 3	Note 6	positive voltage regulator
CA3085H	LIC	516	—	Note 6	positive-voltage-regulator chip
CA3085L	LIC	515	—	Note 6	beam-lead positive voltage regulator
CA3085S	LIC	491	Note 3	Note 6	positive voltage regulator
CA3086	LIC	483	DIP	Note 6	transistor array (n-p-n)
CA3086F	LIC	483	Note 3	Note 6	transistor array (n-p-n)
CA3088E	LIC	560	Note 3	Note 6	AM receiver subsystem
CA3089E	LIC	561	Note 3	Note 6	FM if system
CA3090AQ	LIC	684	Note 3	Note 6	stereo multiplex decoder
CA3091D	LIC	534	Note 3	Note 6	four-quadrant multiplier
CA3091H	LIC	516	—	Note 6	four-quadrant multiplier chip
CA3093E	LIC	533	Note 3	Note 6	transistor-diode array
CA3093H	LIC	516	—	Note 6	transistor-diode-array chip
CA3094/	LIC	692	Note 4	Note 6	high-rel. programmable power switch/amplifier
CA3094A/	LIC	692	Note 4	Note 6	high-rel. programmable power switch/amplifier
CA3094AE	LIC	598	Note 3	Note 6	programmable power switch/amplifier
CA3094AT	LIC	598	Note 3	Note 6	programmable power switch/amplifier
CA3094B/	LIC	692	Note 4	Note 6	high-rel. programmable power switch/amplifier
CA3094BT	LIC	598	Note 3	Note 6	programmable power switch/amplifier
CA3094E	LIC	598	Note 3	Note 6	programmable power switch/amplifier
CA3094H	LIC	516	—	Note 6	programmable power switch/amplifier chip
CA3094T	LIC	598	Note 3	Note 6	programmable power switch/amplifier
CA3095E	LIC	591	Note 3	Note 6	super-beta transistor array
CA3096AE	LIC	595	Note 3	Note 6	n-p-n/p-n-p transistor array
CA3096E	LIC	595	Note 3	Note 6	n-p-n/p-n-p transistor array
CA3096H	LIC	516	—	Note 6	n-p-n/p-n-ptransistor-array chip
CA3097E	LIC	633	Note 3	Note 6	thyristor/transistor array
CA3097H	LIC	516	—	Note 6	thyristor/transistor-array chip
CA3099E	LIC	620	Note 3	Note 6	programmable comparator
CA3099H	LIC	516	—	Note 6	programmable-comparator chip
CA3100H	LIC	516	—	Note 6	wide-band operational-amplifier chip
CA3100S	LIC	625	Note 3	Note 6	wide-band operational amplifier
CA3100T	LIC	625	Note 3	Note 6	wide-band operational amplifier
CA3102E	LIC	611	Note 3	Note 6	dual differential amplifier
CA3102H	LIC	516	—	Note 6	dual differential-amplifier chip
CA3118AT	LIC	532	Note 3	Note 6	high-voltage transistor array n-p-n

▲ Dimensional outlines are shown following data charts.
■ Lead codes are explained in Table LII.
See **Notes** on page 619.

Table LI—RCA Standard Solid-State Devices (cont'd)

Type No.	Prod. Line	File No.	Package▲	Lead Code■	Voltage (V) (Note 1)	Power (W), Current* (A) or Noise Figure (dB) / DC Current Transfer Ratio or Frequency
CA3118H	LIC	532	—	Note 6	high-voltage transistor array	
CA3118T	LIC	532	Note 3	Note 6	high-voltage transistor array	
CA3120E	LIC	691	Note 3	Note 6	TV signal processor	
CA3121E	LIC	688	Note 3	Note 6	TV chroma amplifier/demodulator	
CA3123E	LIC	631	Note 3	Note 6	AM radio receiver subsystem	
CA3125E	LIC	685	Note 3	Note 6	TV chroma processor	
CA3126Q	LIC		Note 3	Note 6	TV chroma processor	
CA3127E	LIC	662	Note 3	Note 6	HF array	
CA3127H	LIC	516	—	Note 6	HF-array chip	
CA3130AS	LIC	817	Note 3	Note 6	COS/MOS operational amplifier	
CA3130AT	LIC	817	Note 3	Note 6	COS/MOS operational amplifier	
CA3130BS	LIC	817	Note 3	Note 6	COS/MOS operational amplifier	
CA3130BT	LIC	817	Note 3	Note 6	COS/MOS operational amplifier	
CA3130S	LIC	817	Note 3	Note 6	COS/MOS operational amplifier	
CA3130T	LIC	817	Note 3	Note 6	COS/MOS operational amplifier	
CA3140E	LIC	630	Note 3	Note 6	quad operational amplifier	
CA3140H	LIC	516	—	Note 6	quad-operational-amplifier chip	
CA3146AE	LIC	532	Note 3	Note 6	high-voltage transistor array (n-p-n)	
CA3146E	LIC	532	Note 3	Note 6	high-voltage transistor array (n-p-n)	
CA3146H	LIC	516	—	Note 6	high-voltage-transistor-array chip (n-p-n)	
CA3183AE	LIC	532	Note 3	Note 6	high-voltage transistor array (n-p-n)	
CA3183E	LIC	532	Note 3	Note 6	high-voltage transistor array (n-p-n)	
CA3183H	LIC	516	—	Note 6	high-voltage-transistor-array cihp (n-p-n)	
CA3401E	LIC	630	Note 3	Note 6	dual operational amplifier	
CA3600E	LIC	619	Note 3	Note 6	COS/MOS transistor array	
CA6078AS	LIC	592	Note 3	Note 6	low-noise operational amplifier	
CA6078AT	LIC	592	Note 3	Note 6	low-noise operational amplifier	
CA6741S	LIC	592	Note 3	Note 6	low-noise operational amplifier	
CA6741T	LIC	592	Note 3	Note 6	low-noise operational amplifier	
CD2150	LIC	308	Flat Pack	Note 6	ultra-high-speed ECCSL gate	
CD2151	LIC	308	Flat Pack	Note 6	ultra-high-speed ECCSL gate	
CD2152	LIC	308	Flat Pack	Note 6	ultra-high-speed ECCSL gate	
CD2153	LIC	308	Flat Pack	Note 6	ultra-high-speed ECCSL gate	
CD2154	LIC	402	Flat Pack	Note 6	ultra-high-speed ECCSL gate	
CD2500E	LIC	392	DIP	Note 6	BCD-to-7-segment decoder/driver	
CD2501E	LIC	392	DIP	Note 6	BCD-to-7-segment decoder/driver	
CD2502E	LIC	392	DIP	Note 6	BCD-to-7-segment decoder/driver	
CD2503E	LIC	392	DIP	Note 6	BCD-to-7-segment decoder/driver	
CD4000A/	COS/MOS	687	Note 5	Note 6	dual 3-input NOR gate plus inverter	
CD4000AD	COS/MOS	479	Note 3	Note 6	dual 3-input NOR gate plus inverter	
CD4000AE	COS/MOS	479	Note 3	Note 6	dual 3-input NOR gate plus inverter	
CD4000AF	COS/MOS	479	Note 3	Note 6	dual 3-input NOR gate plus inverter	
CD4000AH	COS/MOS	517	—	Note 6	dual-3-input-NOR-gate-plus-inverter chip	
CD4000AK	COS/MOS	479	Note 3	Note 6	dual 3-input NOR gate plus inverter	
CD4001A/	COS/MOS	687	Note 5	Note 6	high-reliability quad 2-input NOR gate	
CD4001AD	COS/MOS	479	Note 3	Note 6	quad 2-input NOR gate	
CD4001AE	COS/MOS	479	Note 3	Note 6	quad 2-input NOR gate	
CD4001AF	COS/MOS	479	Note 3	Note 6	quad 2-input NOR gate	
CD4001AH	COS/MOS	517	—	Note 6	quad-2-input-NOR-gate chip	
CD4001AK	COS/MOS	479	Note 3	Note 6	quad 2-input NOR gate	

▲ Dimensional outlines are shown following data charts.
■ Lead codes are explained in Table LII.
See **Notes** on page 619.

Table LI—RCA Standard Solid-State Devices (cont'd)

Type No.	Prod. Line	File No.	Package▲	Lead Code■	Voltage (V) (Note 1)	Power (W), Current* (A) or Noise Figure (dB)	DC Current Transfer Ratio or Frequency
CD4002A/	COS/MOS	687	Note 5	Note 6	high-reliability dual 4-input NOR gate		
CD4002AD	COS/MOS	479	Note 3	Note 6	dual 4-input NOR gate		
CD4002AE	COS/MOS	479	Note 3	Note 6	dual 4-input NOR gate		
CD4002AF	COS/MOS	479	Note 3	Note 6	dual 4-input NOR gate		
CD4002AH	COS/MOS	517	—	Note 6	dual-4-in-put-NOR-gate chip		
CD4002AK	COS/MOS	479	Note 3	Note 6	dual 4-input NOR gate		
CD4006A/	COS/MOS	689	Note 5	Note 6	high-reliability 18-stage static shift register		
CD4006AD	COS/MOS	479	Note 3	Note 6	18-stage static shift register		
CD4006AE	COS/MOS	479	Note 3	Note 6	18-stage static shift register		
CD4006AF	COS/MOS	479	Note 3	Note 6	18-stage static shift register		
CD4006AH	COS/MOS	517	—	Note 6	18-stage-static-shift-register chip		
CD4006AK	COS/MOS	479	Note 3	Note 6	18-stage static shift register		
CD4007A/	COS/MOS	695	Note 5	Note 6	high-reliability dual complementary pair plus inverter		
CD4007AD	COS/MOS	479	Note 3	Note 6	dual complementary pair plus inverter		
CD4007AE	COS/MOS	479	Note 3	Note 6	dual complementary pair plus inverter		
CD4007AF	COS/MOS	479	Note 3	Note 6	dual complementary pair plus inverter		
CD4007AH	COS/MOS	517	—	Note 6	dual-complementary-pair-plus-inverter chip		
CD4007AK	COS/MOS	479	Note 3	Note 6	dual complementary pair plus inverter		
CD4008A/	COS/MOS	696	Note 5	Note 6	high-reliability 4-bit full adder with parallel carry		
CD4008AD	COS/MOS	479	Note 3	Note 6	4-bit full adder with parallel carry		
CD4008AE	COS/MOS	479	Note 3	Note 6	4-bit full adder with parallel carry		
CD4008AF	COS/MOS	479	Note 3	Note 6	4-bit full adder with parallel carry		
CD4008AH	COS/MOS	517	—	Note 6	4-bit-full-adder-with-parrallel-carry chip		
CD4008AK	COS/MOS	479	Note 3	Note 6	4-bit full adder with parallel carry		
CD4009A/	COS/MOS	719	Note 5	Note 6	high-reliability hex buffer/converter (inverting)		
CD4009AD	COS/MOS	479	Note 3	Note 6	hex buffer/converter (inverting)		
CD4009AE	COS/MOS	479	Note 3	Note 6	hex buffer/converter (inverting)		
CD4009AH	COS/MOS	517	—	Note 6	hex-buffer/converter chip (inverting)		
CD4009AK	COS/MOS	479	Note 3	Note 6	hex buffer/converter (inverting)		
CD4010A/	COS/MOS	719	Note 5	Note 6	high-reliability hex buffer/converter (non-inverting)		
CD4010AD	COS/MOS	479	Note 3	Note 6	hex buffer/converter (non-inverting)		
CD4010AE	COS/MOS	479	Note 3	Note 6	hex buffer/converter (non-inverting)		
CD4010AH	COS/MOS	517	—	Note 6	hex-buffer/converter chip (non-inverting)		
CD4010AK	COS/MOS	479	Note 3	Note 6	hex buffer/converter (non-inverting)		
CD4011A/	COS/MOS	717	Note 5	Note 6	high-reliability quad 2-input NAND gate		
CD4011AD	COS/MOS	479	Note 3	Note 6	quad 2-input NAND gate		
CD4011AE	COS/MOS	479	Note 3	Note 6	quad 2-input NAND gate		
CD4011AF	COS/MOS	479	Note 3	Note 6	quad 2-input NAND gate		
CD4011AH	COS/MOS	517	—	Note 6	quad-2-input-NAND-gate chip		
CD4011AK	COS/MOS	479	Note 3	Note 6	quad 2-input NAND gate		
CD4012A/	COS/MOS	717	Note 5	Note 6	high-reliability dual 4-input NAND gate		
CD4012AD	COS/MOS	479	Note 3	Note 6	dual 4-input NAND gate		
CD4012AE	COS/MOS	479	Note 3	Note 6	dual 4-input NAND gate		
CD4012AF	COS/MOS	479	Note 3	Note 6	dual 4-input NAND gate		
CD4012AH	COS/MOS	517	—	Note 6	dual-4-input-NAND-gate chip		

▲ Dimensional outlines are shown following data charts.
■ Lead codes are explained in Table LII.
See **Notes** on page 619.

Table LI—RCA Standard Solid-State Devices (cont'd)

Type No.	Prod. Line	File No.	Package▲	Lead Code■	Voltage (V) (Note 1)	Power (W), Current* (A) or Noise Figure (dB)	DC Current Transfer Ratio or Frequency
CD4012AK	COS/MOS	479	Note 3	Note 6	dual 4-input NAND gate		
CD4013A/	COS/MOS	697	Note 5	Note 6	high-reliability dual "D" flip-flop with set/reset		
CD4013AD	COS/MOS	479	Note 3	Note 6	dual "D" flip-flop with set/reset		
CD4013AE	COS/MOS	479	Note 3	Note 6	dual "D" flip-flop with set/reset		
CD4013AF	COS/MOS	479	Note 3	Note 6	dual "D" flip-flop with set/reset		
CD4013AH	COS/MOS	517	—	Note 6	dual-"D"-flip-flop-with-set/reset chip		
CD4013AK	COS/MOS	479	Note 3	Note 6	dual "D" flip-flop with set/reset		
CD4014A/	COS/MOS	720	Note 5	Note 6	high-reliability 8-stage static shift register		
CD4014AD	COS/MOS	479	Note 3	Note 6	8-stage static shift register		
CD4014AE	COS/MOS	479	Note 3	Note 6	8-stage static shift register		
CD4014AF	COS/MOS	479	Note 3	Note 6	8-stage static shift register		
CD4014AH	COS/MOS	517	—	Note 6	8-stage-static-shift-register chip		
CD4014AK	COS/MOS	479	Note 3	Note 6	8-stage static shift register		
CD4015A/	COS/MOS	721	Note 5	Note 6	high-reliability dual 4-stage static shift register		
CD4015AD	COS/MOS	479	Note 3	Note 6	dual 4-stage static shift register		
CD4015AE	COS/MOS	479	Note 3	Note 6	dual 4-stage static shift register		
CD4015AF	COS/MOS	479	Note 3	Note 6	dual 4-stage static shift register		
CD4015AH	COS/MOS	517	—	Note 6	dual-4-stage-static-shift-register chip		
CD4015AK	COS/MOS	479	Note 3	Note 6	dual 4-stage static shift register		
CD4016A/	COS/MOS	744	Note 5	Note 6	high-reliability quad bilateral switch		
CD4016AD	COS/MOS	479	Note 3	Note 6	quad bilateral switch		
CD4016AE	COS/MOS	479	Note 3	Note 6	quad bilateral switch		
CD4016AF	COS/MOS	479	Note 3	Note 6	quad bilateral switch		
CD4016AH	COS/MOS	517	—	Note 6	quad-bilateral-switch chip		
CD4016AK	COS/MOS	479	Note 3	Note 6	quad bilateral switch		
CD4017A/	COS/MOS	741	Note 5	Note 6	high-reliability decade counter/divider		
CD4017AD	COS/MOS	479	Note 3	Note 6	decade counter/divider		
CD4017AE	COS/MOS	479	Note 3	Note 6	decade counter/divider		
CD4017AF	COS/MOS	479	Note 3	Note 6	decade counter/divider		
CD4017AH	COS/MOS	517	—	Note 6	decade-counter/divider chip		
CD4017AK	COS/MOS	479	Note 3	Note 6	decade counter/divider		
CD4018A/	COS/MOS	742	Note 5	Note 6	high-reliability presettable divide-by-"N" counter		
CD4018AD	COS/MOS	479	Note 3	Note 6	presettable divide-by-"N" counter		
CD4018AE	COS/MOS	479	Note 3	Note 6	presettable divide-by-"N" counter		
CD4018AF	COS/MOS	479	Note 3	Note 6	presettable divide-by-"N" counter		
CD4018AH	COS/MOS	517	—	Note 6	presettable-divide-by-"N"-counter chip		
CD4018AK	COS/MOS	479	Note 3	Note 6	presettable divide-by-"N" counter		
CD4019A/	COS/MOS	743	Note 5	Note 6	high-reliability quad AND-OR select gate		
CD4019AD	COS/MOS	479	Note 3	Note 6	quad AND-OR select gate		
CD4019AE	COS/MOS	479	Note 3	Note 6	quad AND-OR select gate		
CD4019AF	COS/MOS	479	Note 3	Note 6	quad AND-OR select gate		
CD4019AH	COS/MOS	517	—	Note 6	quad-AND-OR-select-gate chip		
CD4019AK	COS/MOS	479	Note 3	Note 6	quad AND-OR select gate		
CD4020A/	COS/MOS	750	Note 5	Note 6	high-reliability 14-stage binary/ripple counter		
CD4020AD	COS/MOS	479	Note 3	Note 6	14-stage binary/ripple counter		
CD4020AE	COS/MOS	479	Note 3	Note 6	14-stage binary/ripple counter		
CD4020AF	COS/MOS	479	Note 3	Note 6	14-stage binary/ripple counter		
CD4020AH	COS/MOS	517	—	Note 6	14-stage-binary/ripple-counter chip		
CD4020AK	COS/MOS	479	Note 3	Note 6	14-stage binary/ripple counter		
CD4021A/	COS/MOS	730	Note 5	Note 6	high-reliability 8-stage static shift register		

▲ Dimensional outlines are shown following data charts.
■ Lead codes are explained in Table LII.
See **Notes** on page 619.

Table LI—RCA Standard Solid-State Devices (cont'd)

Type No.	Prod. Line	File No.	Package▲	Lead Code■	Voltage (V) (Note 1)	Power (W), Current* (A) or Noise Figure (dB)	DC Current Transfer Ratio or Frequency
CD4021AD	COS/MOS	479	Note 3	Note 6	8-stage static shift register		
CD4021AE	COS/MOS	479	Note 3	Note 6	8-stage static shift register		
CD4021AF	COS/MOS	479	Note 3	Note 6	8-stage static shift register		
CD4021AH	COS/MOS	517	—	Note 6	8-stage-static-shift-register chip		
CD4021AK	COS/MOS	479	Note 3	Note 6	8-stage static shift register		
CD4022A/	COS/MOS	731	Note 5	Note 6	high-reliability divide-by-8 counter/divider		
CD4022AD	COS/MOS	479	Note 3	Note 6	divide-by-8 counter/divider		
CD4022AE	COS/MOS	479	Note 3	Note 6	divide-by-8 counter/divider		
CD4022AF	COS/MOS	479	Note 3	Note 6	divide-by-8 counter/divider		
CD4022AH	COS/MOS	517	—	Note 6	divide-by-8-counter/divider chip		
CD4022AK	COS/MOS	479	Note 3	Note 6	divide-by-8 counter/divider		
CD4023A/	COS/MOS	717	Note 5	Note 6	high-reliability triple-3-input NAND gate		
CD4023AD	COS/MOS	479	Note 3	Note 6	triple 3-input NAND gate		
CD4023AE	COS/MOS	479	Note 3	Note 6	triple 3-input NAND gate		
CD4023AF	COS/MOS	479	Note 3	Note 6	triple 3-input NAND gate		
CD4023AH	COS/MOS	517	—	Note 6	triple-3-input-NAND-gate chip		
CD4023AK	COS/MOS	479	Note 3	Note 6	triple 3-input NAND gate		
CD4024A/	COS/MOS	732	Note 5	Note 6	high-reliability 7-stage binary counter		
CD4024AD	COS/MOS	503	Note 3	Note 6	7-stage binary counter		
CD4024AE	COS/MOS	503	Note 3	Note 6	7-stage binary counter		
CD4024AF	COS/MOS	503	Note 3	Note 6	7-stage binary counter		
CD4024AH	COS/MOS	517	—	Note 6	7-stage-binary-counter chip		
CD4025A/	COS/MOS	687	Note 5	Note 6	high-reliability triple 3-input NOR gate		
CD4025AD	COS/MOS	479	Note 3	Note 6	triple 3-input NOR gate		
CD4025AE	COS/MOS	479	Note 3	Note 6	triple 3-input NOR gate		
CD4025AF	COS/MOS	479	Note 3	Note 6	triple 3-input NOR gate		
CD4025AH	COS/MOS	517	—	Note 6	triple-3-input-NOR-gate chip		
CD4025AK	COS/MOS	479	Note 3	Note 6	triple 3-input NOR gate		
CD4026A/	COS/MOS	733	Note 5	Note 6	high-reliability decade counter/divider		
CD4026AD	COS/MOS	503	Note 3	Note 6	decade counter/divider		
CD4026AE	COS/MOS	503	Note 3	Note 6	decade counter/divider		
CD4026AF	COS/MOS	503	Note 3	Note 6	decade counter/divider		
CD4026AH	COS/MOS	517	—	Note 6	decade-counter/divider chip		
CD4026AK	COS/MOS	503	Note 3	Note 6	decade counter/divider		
CD4027A/	COS/MOS	734	Note 5	Note 6	high-reliability dual J-K master-slave flip-flop		
CD4027AD	COS/MOS	503	Note 3	Note 6	dual J-K master-slave flip-flop		
CD4027AE	COS/MOS	503	Note 3	Note 6	dual J-K master-slave flip-flop		
CD4027AH	COS/MOS	517	—	Note 6	dual-J-K-master-slave-flip-flop chip		
CD4027AK	COS/MOS	503	Note 3	Note 6	dual J-K master-slave flip-flop		
CD4028A/	COS/MOS	735	Note 5	Note 6	high-reliability BCD-to-decimal decoder		
CD4028AD	COS/MOS	503	Note 3	Note 6	BCD-to-decimal decoder		
CD4028AE	COS/MOS	503	Note 3	Note 6	BCD-to-decimal decoder		
CD4028AF	COS/MOS	503	Note 3	Note 6	BCD-to-decimal decoder		
CD4028AH	COS/MOS	517	—	Note 6	BCD-to-decimal-decoder chip		
CD4028AK	COS/MOS	503	Note 3	Note 6	BCD-to-decimal decoder		
CD4029A/	COS/MOS	736	Note 5	Note 6	high-reliability presettable up/down counter		
CD4029AD	COS/MOS	503	Note 3	Note 6	presettable up/down counter		
CD4029AE	COS/MOS	503	Note 3	Note 6	presettable up/down counter		
CD4029AH	COS/MOS	517	—	Note 6	presettable-up/down-counter chip		
CD4029AK	COS/MOS	503	Note 3	Note 6	presettable up/down counter		

▲ Dimensional outlines are shown following data charts.
■ Lead codes are explained in Table LII.
See **Notes** on page 619.

Table LI—RCA Standard Solid-State Devices (cont'd)

Type No.	Prod. Line	File No.	Package▲	Lead Code■	Voltage (V) (Note 1)	Power (W), Current* (A) or Noise Figure (dB)	DC Current Transfer Ratio or Frequency
CD4030A/	COS/MOS	737	Note 5	Note 6	high-reliability quad exclusive-OR-gate		
CD4030AD	COS/MOS	503	Note 3	Note 6	quad exclusive-OR gate		
CD4030AE	COS/MOS	503	Note 3	Note 6	quad exclusive-OR gate		
CD4030AF	COS/MOS	503	Note 3	Note 6	quad exclusive-OR gate		
CD4030AH	COS/MOS	503	—	Note 6	quad-exclusive-OR-gate chip		
CD4030AK	COS/MOS	503	Note 3	Note 6	quad exclusive-OR gate		
CD4031A/	COS/MOS	738	Note 5	Note 6	high-reliability 64-stage static shift register		
CD4031AD	COS/MOS	569	Note 3	Note 6	64-stage static shift register		
CD4031AE	COS/MOS	569	Note 3	Note 6	64-stage static shift register		
CD4031AH	COS/MOS	517	—	Note 6	64-stage-static-shift-register chip		
CD4031AK	COS/MOS	569	Note 3	Note 6	64-stage static shift register		
CD4032A/	COS/MOS	739	Note 5	Note 6	high-reliability triple serial adder (positive logic)		
CD4032AD	COS/MOS	503	Note 3	Note 6	triple serial adder (positive logic)		
CD4032AE	COS/MOS	503	Note 3	Note 6	triple serial adder (positive logic)		
CD4032AH	COS/MOS	517	—	Note 6	triple-serial-adder chip (positive logic)		
CD4032AK	COS/MOS	503	Note 3	Note 6	triple serial adder (positive logic)		
CD4033A/	COS/MOS	733	Note 5	Note 6	high-reliability decade counter/divider		
CD4033AD	COS/MOS	503	Note 3	Note 6	decade counter/divider		
CD4033AE	COS/MOS	503	Note 3	Note 6	decade counter/divider		
CD4033AF	COS/MOS	503	Note 3	Note 6	decade counter/divider		
CD4033AH	COS/MOS	517	—	Note 6	decade-counter/divider chip		
CD4033AK	COS/MOS	503	Note 3	Note 6	decade counter/divider		
CD4034A/	COS/MOS	740	Note 5	Note 6	high-reliability MSI 8-stage static shift register		
CD4034AD	COS/MOS	575	Note 3	Note 6	MIS 8-stage static shift register		
CD4034AE	COS/MOS	575	Note 3	Note 6	MIS 8-stage static shift register		
CD4034AH	COS/MOS	517	—	Note 6	MSI-8-stage-static-shift-register chip		
CD4034AK	COS/MOS	575	Note 3	Note 6	MIS 8-stage static shift register		
CD4035A/	COS/MOS	751	Note 5	Note 6	high-reliability 4-stage parallel in/out shift register		
CD4035AD	COS/MOS	568	Note 3	Note 6	4-stage parallel in/out shift register		
CD4035AE	COS/MOS	568	Note 3	Note 6	4-stage parallel in/out shift register		
CD4035AH	COS/MOS	517	—	Note 6	4-stage-parallel-in/out-shift-register chip		
CD4035AK	COS/MOS	568	Note 3	Note 6	4-stage parallel in/out shift register		
CD4036A/	COS/MOS	517	Note 5	Note 6	high-reliability 4-word x 8-bit RAM (binary addressing)		
CD4036AD	COS/MOS	613	Note 3	Note 6	4-word x 8-bit RAM (binary addressing)		
CD4036AE	COS/MOS	613	Note 3	Note 6	4-word x 8-bit RAM (binary addressing)		
CD4036AH	COS/MOS	517	—	Note 6	4-word x 8-bit-RAM chip (binary addressing)		
CD4036AK	COS/MOS	613	Note 3	Note 6	4-word x 8-bit RAM (binary addressing)		
CD4037AD	COS/MOS	576	Note 3	Note 6	triple AND-OR bi-phase pairs		
CD4037AE	COS/MOS	576	Note 3	Note 6	triple AND-OR bi-phase pairs		
CD4037AF	COS/MOS	576	Note 3	Note 6	triple-AND-OR-bi-phase-pairs		
CD4037AH	COS/MOS	517	—	Note 6	triple-AND-OR-bi-phase-pairs chip		
CD4037AK	COS/MOS	576	Note 3	Note 6	triple AND-OR bi-phase pairs		
CD4038A/	COS/MOS	739	Note 5	Note 6	high-reliability triple serial adder (negative logic)		
CD4038AD	COS/MOS	503	Note 3	Note 6	triple serial adder (negative logic)		
CD4038AE	COS/MOS	503	Note 3	Note 6	triple serial adder (negative logic)		

▲ Dimensional outlines are shown following data charts.
■ Lead codes are explained in Table LII.
See **Notes** on page 619.

Table LI—RCA Standard Solid-State Devices (cont'd)

Type No.	Prod. Line	File No.	Package▲	Lead Code■	Voltage (V) (Note 1)	Power (W), Current* (A) or Noise Figure (dB)	DC Current Transfer Ratio or Frequency
CD4038AH	COS/MOS	517	—	Note 6	triple-serial-adder chip (negative logic)		
CD4038AK	COS/MOS	503	Note 3	Note 6	triple serial adder (negative logic)		
CD4039A/	COS/MOS	749	Note 5	Note 6	high-reliability 4-word x 8-bit RAM (word addressing)		
CD4039AD	COS/MOS	613	Note 3	Note 6	4-word x 8-bit RAM (word addressing)		
CD4039AH	COS/MOS	517	—	Note 6	4-word x 8-bit-RAM chip (word addressing)		
CD4039AK	COS/MOS	613	Note 3	Note 6	4-word x 8-bit RAM (word addressing)		
CD4040A/	COS/MOS	748	Note 5	Note 6	high-reliability 12-stage binary/ripple counter		
CD4040AD	COS/MOS	624	Note 3	Note 6	12-stage binary/ripple counter		
CD4040AE	COS/MOS	624	Note 3	Note 6	12-stage binary/ripple counter		
CD4040AF	COS/MOS	624	Note 3	Note 6	12-stage binary/ripple counter		
CD4040AH	COS/MOS	517	—	Note 6	12-stage-binary/ripple-counter chip		
CD4040AK	COS/MOS	624	Note 3	Note 6	12-stage binary/ripple counter		
CD4041A/	COS/MOS	753	Note 5	Note 6	high-reliability quad true/complement buffer		
CD4041AD	COS/MOS	572	Note 3	Note 6	quad true/complement buffer		
CD4041AE	COS/MOS	572	Note 3	Note 6	quad true/complement buffer		
CD4041AH	COS/MOS	517	—	Note 6	quad-true/complement-buffer chip		
CD4041AK	COS/MOS	572	Note 3	Note 6	quad true/complement buffer		
CD4042A/	COS/MOS	756	Note 5	Note 6	high-reliability quad clocked "D" latch		
CD4042AD	COS/MOS	589	Note 3	Note 6	quad clocked "D" latch		
CD4042AE	COS/MOS	589	Note 3	Note 6	quad clocked "D" latch		
CD4042AF	COS/MOS	589	Note 3	Note 6	quad clocked "D" latch		
CD4042AH	COS/MOS	517	—	Note 6	quad-clocked-"D"-latch chip		
CD4042AK	COS/MOS	589	Note 3	Note 6	quad clocked "D" latch		
CD4043A/	COS/MOS	754	Note 5	Note 6	high-reliability quad 3-state NOR R/S latch		
CD4043AD	COS/MOS	590	Note 3	Note 6	quad 3-state NOR R/S latch		
CD4043AE	COS/MOS	590	Note 3	Note 6	quad 3-state NOR R/S latch		
CD4043AH	COS/MOS	517	—	Note 6	quad-3-state-NOR-R/S latch chip		
CD4043AK	COS/MOS	590	Note 3	Note 6	quad 3-state NOR R/S latch		
CD4044A/	COS/MOS	754	Note 5	Note 6	high-reliability quad 3-state NAND R/S latch		
CD4044AD	COS/MOS	590	Note 3	Note 6	quad 3-state NAND R/S latch		
CD4044AE	COS/MOS	590	Note 3	Note 6	quad 3-state NAND R/S latch		
CD4044AH	COS/MOS	517	—	Note 6	quad-3-state-NAND-R/S-latch chip		
CD4044AK	COS/MOS	590	Note 3	Note 6	quad 3-state NAND R/S latch		
CD4045A/	COS/MOS	755	Note 5	Note 6	high-reliability 21-stage counter		
CD4045AD	COS/MOS	614	Note 3	Note 6	21-stage counter		
CD4045AE	COS/MOS	614	Note 3	Note 6	21-stage counter		
CD4045AH	COS/MOS	517	—	Note 6	21-stage-counter chip		
CD4045AK	COS/MOS	614	Note 3	Note 6	21-stage counter		
CD4046A/	COS/MOS	752	Note 5	Note 6	high-reliability micropower phase-locked loop		
CD4046AD	COS/MOS	637	Note 3	Note 6	micropower phase-locked loop		
CD4046AE	COS/MOS	637	Note 3	Note 6	micropower phase-locked loop		
CD4046AH	COS/MOS	517	—	Note 6	micropower-phase-locked-loop chip		
CD4046AK	COS/MOS	637	Note 3	Note 6	micropower phase-locked loop		
CD4047A/	COS/MOS	745	Note 5	Note 6	high-reliability monostable/astable multivibrator		
CD4047AD	COS/MOS	623	Note 3	Note 6	monostable/astable multivibrator		

▲ Dimensional outlines are shown following data charts.
■ Lead codes are explained in Table LII.
See **Notes** on page 619.

Table LI—RCA Standard Solid-State Devices (cont'd)

Type No.	Prod. Line	File No.	Package▲	Lead Code■	Voltage (V) (Note 1)	Power (W), Current* (A) or Noise Figure (dB)	DC Current Transfer Ratio or Frequency
CD4047AE	COS/MOS	623	Note 3	Note 6	monostable/astable multivibrator		
CD4047AH	COS/MOS	517	—	Note 6	monostable/astable-multivibrator chip		
CD4047AK	COS/MOS	623	Note 3	Note 6	monostable/astable multivibrator		
CD4048A/	COS/MOS	747	Note 5	Note 6	high-reliability expandable 8-input gate		
CD4048AD	COS/MOS	636	Note 3	Note 6	expandable 8-input gate		
CD4048AE	COS/MOS	636	Note 3	Note 6	expandable 8-input gate		
CD4048AH	COS/MOS	517	—	Note 6	expandable-8input-gate chip		
CD4048AK	COS/MOS	636	Note 3	Note 6	expandable 8-input gate		
CD4049A/	COS/MOS	746	Note 5	Note 6	high-reliability hex buffer/converter (inverting)		
CD4049AD	COS/MOS	599	Note 3	Note 6	hex buffer/converter (inverting)		
CD4049AE	COS/MOS	599	Note 3	Note 6	hex buffer/converter (inverting)		
CD4049AF	COS/MOS	599	Note 3	Note 6	hex buffer/converter (inverting)		
CD4049AH	COS/MOS	517	—	Note 6	hex-buffer/converter chip (inverting)		
CD4049AK	COS/MOS	599	Note 3	Note 6	hex buffer/converter (inverting)		
CD4050A/	COS/MOS	746	Note 5	Note 6	high-reliability hex buffer/converter (non-inverting)		
CD4050AD	COS/MOS	599	Note 3	Note 6	hex buffer/converter (non-inverting)		
CD4050AE	COS/MOS	599	Note 3	Note 6	hex buffer/converter (non-inverting)		
CD4050AF	COS/MOS	599	Note 3	Note 6	hex buffer/converter (non-inverting)		
CD4050AH	COS/MOS	517	—	Note 6	hex-buffer/converter chip (non-inverting)		
CD4050AK	COS/MOS	599	Note 3	Note 6	hex buffer/converter (non-inverting)		
CD4051AD	COS/MOS	Prel.	Note 3	Note 6	single 8-channel multiplexer		
CD4051AE	COS/MOS	Prel.	Note 3	Note 6	single 8-channel multiplexer		
CD4051AK	COS/MOS	Prel.	Note 3	Note 6	single 8-channel multiplexer		
CD4052AD	COS/MOS	Prel.	Note 3	Note 6	differential 4-channel multiplexer		
CD4052AE	COS/MOS	Prel.	Note 3	Note 6	differential 4-channel multiplexer		
CD4052AK	COS/MOS	Prel.	Note 3	Note 6	differential 4-channel multiplexer		
CD4053AD	COS/MOS	Prel.	Note 3	Note 6	triple 2-channel multiplexer		
CD4053AE	COS/MOS	Prel.	Note 3	Note 6	triple 2-channel multiplexer		
CD4053AK	COS/MOS	Prel.	Note 3	Note 6	triple 2-channel multiplexer		
CD4054AD	COS/MOS	634	Note 3	Note 6	4-line liquid-crystal display driver		
CD4054AE	COS/MOS	634	Note 3	Note 6	4-line liquid-crystal display driver		
CD4054AH	COS/MOS	517	—	Note 6	4-line-liquid-crystal-display-driver chip		
CD4054AK	COS/MOS	634	Note 3	Note 6	4-line liquid-crystal display driver		
CD4055AD	COS/MOS	634	Note 3	Note 6	BCD-7-segment decoder/driver		
CD4055AE	COS/MOS	634	Note 3	Note 6	BCD-7-segment decoder/driver		
CD4055AK	COS/MOS	634	Note 3	Note 6	BCD-7-segment decoder/driver		
CD4056AD	COS/MOS	634	Note 3	Note 6	BCD-7-segment decoder/driver		
CD4056AE	COS/MOS	634	Note 3	Note 6	BCD-7-segment decoder/driver		
CD4056AH	COS/MOS	517	—	Note 6	BCD-7-segment-decoder/driver chip		
CD4056AK	COS/MOS	634	Note 3	Note 6	BCD-7-segment decoder/driver		
CD4057AD	COS/MOS	635	Note 3	Note 6	LSI 4-bit arithmetic logic unit		
CD4057AH	COS/MOS	517	—	Note 6	LSI-4-bit-arithmetic-logic-unit chip		
CD4059AD	COS/MOS	Prel.	Note 3	Note 6	programmable divide-by-"N"counter		
CD4060AE	COS/MOS	813	Note 3	Note 6	14-stage counter and oscillator		
CD4061AD	COS/MOS	768	Note 3	Note 6	256-word x 1-bit static RAM		
CD4061AH	COS/MOS	768	—	Note 6	256-word x 1-bit-static-RAM chip		
CD4062AK	COS/MOS	816	Note 3	Note 6	200-stage dynamic shift register		
CD4062AT	COS/MOS	816	Note 3	Note 6	200-stage dynamic shift register		
CD4063BE	COS/MOS	805	Note 3	Note 6	4-bit magnitude comparator		
CD4066AD	COS/MOS	769	Note 3	Note 6	quad bilateral switch		

▲ Dimensional outlines are shown following data charts.
■ Lead codes are explained in Table LII.
See **Notes** on page 619.

Table LI—RCA Standard Solid-State Devices (cont'd)

Type No.	Prod. Line	File No.	Package▲	Lead Code■	Voltage (V) (Note 1)	Power (W), Current* (A) or Noise Figure (dB)	DC Current Transfer Ratio or Frequency
CD4066AE	COS/MOS	769	Note 3	Note 6	quad bilateral switch		
CD4066AH	COS/MOS	769	—	Note 6	quad-bilateral-switch chip		
CD4066AK	COS/MOS	769	Note 3	Note 6	quad bilateral switch		
CD4068AE	COS/MOS	Prel.	Note 3	Note 6	8-input NAND gate		
CD4069AE	COS/MOS	Prel.	Note 3	Note 6	hex inverter		
CD4071AE	COS/MOS	Prel.	Note 3	Note 6	quad 2-, dual 4-, and triple 3-input OR gates		
CD4072AE	COS/MOS	Prel.	Note 3	Note 6	quad 2-, dual 4-, and triple 3-input OR gates		
CD4073AE	COS/MOS	Prel.	Note 3	Note 6	triple 3-, quad 2-, and dual-4-input AND gates		
CD4075AE	COS/MOS	Prel.	Note 3	Note 6	quad 2-, dual 4-, and triple 3-input OR gates		
CD4078AE	COS/MOS	Prel.	Note 3	Note 6	8-input NOR gate		
CD4081AE	COS/MOS	Prel.	Note 3	Note 6	triple 3-, quad 2-, and dual-4-input AND gates		
CD4082AE	COS/MOS	Prel.	Note 3	Note 6	triple 3-, quad 2-, and dual-4-input AND gates		
CD4085AE	COS/MOS	Prel.	Note 3	Note 6	dual 2-wide, 2-input AND-OR invert gate		
CD4086AE	COS/MOS	Prel.	Note 3	Note 6	expandable 4-wide 2-input AND-OR invert		
CD4093AE	COS/MOS	Prel.	Note 3	Note 6	Schmitt trigger		
CD4514BD	COS/MOS	Prel.	Note 3	Note 6	4-bit latch/4-to-16 line decoder		
CD4515BD	COS/MOS	Prel.	Note 3	Note 6	4-bit latch/4-to-16 line decoder		
CD4518BE	COS/MOS	Prel.	Note 3	Note 6	dual BCD up counter		
CD4520BE	COS/MOS	Prel.	Note 3	Note 6	dual binary up counter		
CH2102	PWR	632	—	—	60 (V_{CEO})	—	50 min.
CH2270	PWR	632	—	—	45 (V_{CEO})	—	50 min.
CH2405	PWR	632	—	—	90 (V_{CEO})	—	50 min.
CH3053	PWR	632	—	—	30 (V_{CEO})	—	50 min.
CH3439	PWR	632	—	—	325 (V_{CEO})	—	30 min.
CH3440	PWR	632	—	—	250 (V_{CEO})	—	30 min.
CH4036	PWR	632	—	—	—65 (V_{CEO})	—	35 min.
CH4037	PWR	632	—	—	—40 (V_{CEO})	—	35 min.
CH5320	PWR	632	—	—	80 (V_{CEO})	—	30 min.
CH5321	PWR	632	—	—	55 (V_{CEO})	—	30 min.
CH5322	PWR	632	—	—	—80 (V_{CEO})	—	30 min.
CH5323	PWR	632	—	—	—55 (V_{CEO})	—	30 min.
CH5262	PWR	632	—	—	35 (V_{CEO})	—	30 min.
CH6479	PWR	632	—	—	40 (V_{CEO})	—	40 min.
D1201A	RECT	495	DO-15	A	100	1* (avg.)	—
D1201B	RECT	495	DO-15	A	200	1* (avg.)	—
D1201D	RECT	495	DO-15	A	400	1* (avg.)	—
D1201F	RECT	495	DO-15	A	50	1* (avg.)	—
D1201M	RECT	495	DO-15	A	600	1* (avg.)	—
D1201N	RECT	495	DO-15	A	800	1* (avg.)	—
D1201P	RECT	495	DO-15	A	1000	1* (avg.)	—
D1300A	RECT	784	TO-1 (Mod.)	B	100	0.25* (avg.)	—
D1300B	RECT	784	TO-1 (Mod.)	B	200	0.25* (avg.)	—
D1300D	RECT	784	TO-1 (Mod.)	B	400	0.25* (avg.)	—
D2101S	RECT	522	DO-1	A	700	1* (avg.)	—
D2103S	RECT	522	DO-1	A	700	1* (avg.)	—
D2103SF	RECT	522	DO-1	A	750	1* (avg.)	—
D2201A	RECT	629	DO-15	A	100	1* (avg.)	—
D2201B	RECT	629	DO-15	A	200	1* (avg.)	—
D2201D	RECT	629	DO-15	A	400	1* (avg.)	—
D2201F	RECT	629	DO-15	A	50	1* (avg.)	—

▲ Dimensional outlines are shown following data charts.
■ Lead codes are explained in Table LII.
See **Notes** on page 619.

Table LI—RCA Standard Solid-State Devices (cont'd)

Type No.	Prod. Line	File No.	Package▲	Lead Code■	Voltage (V) (Note 1)	Power (W), Current* (A) or Noise Figure (dB)	DC Current Transfer Ratio or Frequency
D2201M	RECT	629	DO-15	A	600	1* (avg.)	—
D2201N	RECT	629	DO-15	A	800	1* (avg.)	—
D2406A	RECT	663	DO-4	A#	100	6* (avg.)	—
D2406B	RECT	663	DO-4	A#	200	6* (avg.)	—
D2406C	RECT	663	DO-4	A#	300	6* (avg.)	—
D2406D	RECT	663	DO-4	A#	400	6* (avg.)	—
D2406F	RECT	663	DO-4	A#	50	6* (avg.)	—
D2406M	RECT	663	DO-4	A#	600	6* (avg.)	—
D2412A	RECT	664	DO-4	A#	100	12* (avg.)	—
D2412B	RECT	664	DO-4	A#	200	12* (avg.)	—
D2412C	RECT	664	DO-4	A#	300	12* (avg.)	—
D2412D	RECT	664	DO-4	A#	400	12* (avg.)	—
D2412F	RECT	664	DO-4	A#	50	12* (avg.)	—
D2412M	RECT	664	DO-4	A#	600	12* (avg.)	—
D2520A	RECT	665	DO-5	A#	100	20* (avg.)	—
D2520B	RECT	665	DO-5	A#	200	20* (avg.)	—
D2520C	RECT	665	DO-5	A#	300	20* (avg.)	—
D2520D	RECT	665	DO-5	A#	400	20* (avg.)	—
D2520F	RECT	665	DO-5	A#	50	20* (avg.)	—
D2520M	RECT	665	DO-5	A#	600	20* (avg.)	—
D2540A	RECT	580	DO-5	A#	100	40* (avg.)	—
D2540B	RECT	580	DO-5	A#	200	40* (avg.)	—
D2540D	RECT	580	DO-5	A#	400	40* (avg.)	—
D2540F	RECT	580	DO-5	A#	50	40* (avg.)	—
D2540M	RECT	580	DO-5	A#	600	40* (avg.)	—
D2600EF	RECT	354	DO-26	B	550	1* (avg.)	—
D2601A	RECT	723	DO-26	B	100	1* (avg.)	—
D2601B	RECT	723	DO-26	B	200	1* (avg.)	—
D2601D	RECT	723	DO-26	B	400	1* (avg.)	—
D2601DF	RECT	354	DO-26	B	450	1* (avg.)	—
D2601EF	RECT	354	DO-26	B	550	1* (avg.)	—
D2601F	RECT	723	DO-26	B	50	1* (avg.)	—
D2601M	RECT	723	DO-26	B	600	1* (avg.)	—
D2601N	RECT	723	DO-26	B	800	1* (avg.)	—
D3202U	DIAC	577	DO-15	A/B	25-40	2* (pk.)	—
D3202Y	DIAC	577	DO-15	A/B	29-35	2* (pk.)	—
HC2000H	HYB	566	RCA	Note 6	75	35	—
HC2500	HYB	681	RCA	Note 6	75	100	—
JAN2N918	RF	—	TO-72	S	6-15 (V_{CE})	6 (NF)	60 MHz
JAN2N1482	PWR	—	TO-39, TO-5	F	100	5	35-100
JAN2N1486	PWR	—	TO-8	F	100	25	35-100
JAN2N1490	PWR	—	TO-3	D	100	75	25-75
JAN2N1493	RF	—	TO-39	F	50	0.5	70 MHz
JAN2N2016	PWR	—	TO-36	X	130	150	15-50
JAN2N2857	RF	—	TO-72	S	6-15 (V_{CE})	4.5 (NF)	450 MHz
JAN2N3055	PWR	—	TO-3	D	100	115	20-70
JAN2N3375	RF	—	TO-60	E	28	3	400 MHz
JAN2N3439	PWR	—	TO-5	F	450	10	40-160
JAN2N3441	PWR	—	TO-66	D	160	25	25-100
JAN2N3442	PWR	—	TO-3	D	160	117	20-70

▲ Dimensional outlines are shown following data charts.
■ Lead codes are explained in Table LII.
See **Notes** on page 619.

Table LI—RCA Standard Solid-State Devices (cont'd)

Type No.	Prod. Line	File No.	Package▲	Lead Code■	Voltage (V) (Note 1)	Power (W), Current* (A) or Noise Figure (dB)	DC Current Transfer Ratio or Frequency
JAN2N3553	RF	—	TO-39	F	28	2.5	175 MHz
JAN2N3585	PWR	—	TO-66	D	500	2.5	25-100
JAN2N3772	PWR	—	TO-3	D	100	150	15-60
JAN2N3866	RF	—	TO-39	F	28	1	400 MHz
JAN2N4440	RF	—	TO-60	E	28	5	400 MHz
JAN2N5038	PWR	—	TO-3	D	150	140	50-200
JAN2N5071	RF	—	TO-60	U	24	24	76 MHz
JAN2N5109	RF	—	TO-39	F	15	3 (NF)	200 MHz
JAN2N5416	PWR	—	TO-5	F	−350	10	30-120
JAN2N5672	PWR	—	TO-3	D	150	140	20-100
JAN2N5840	PWR	—	TO-3	D	375	100	10-50
JAN2N5918	RF	—	TO-216AA	W	28	10	400 MHz
JAN2N6213	PWR	—	TO-66	D	−400	35	30-150
JANTX2N1486	PWR	—	TO-8	F	100	25	35-100
JANTX2N2857	RF	—	TO-72	S	6-15 (V_{CE})	4.5 (NF)	450 MHz
JANTX2N3055	PWR	—	TO-3	D	100	115	20-70
JANTX2N3375	RF	—	TO-60	E	28	3	400 MHz
JANTX2N3439	PWR	—	TO-5	F	450	10	40-160
JANTX2N3441	PWR	—	TO-66	D	160	25	25-100
JANTX2N3553	RF	—	TO-39	F	28	2.5	175 MHz
JANTX2N3585	PWR	—	TO-66	D	500	2.5	25-100
JANTX2N4440	RF	—	TO-60	E	28	5	400 MHz
JANTX2N5038	PWR	—	TO-3	D	150	140	50-200
JANTX2N5071	RF	—	TO-60	U	24	24	76 MHz
JANTX2N5109	RF	—	TO-39	F	15	3 (NF)	200 MHz
JANTX2N5416	PWR	—	TO-5	F	−350	10	30-120
JANTX2N5672	PWR	—	TO-3	D	150	140	20-100
JANTX2N5840	PWR	—	TO-3	D	375	100	10-50
JANTX2N6213	PWR	—	TO-66	D	−400	35	30-150
JANTXV2N3375	RF	—	TO-60	E	28	3	400 MHz
JANTXV2N3553	RF	—	TO-39	F	28	2.5	175 MHz
JANTXV2N4440	RF	—	TO-60	E	28	5	400 MHz
RCA1A01	PWR	651	TO-39	F	70 (V_{CEO})	5	40-200
RCA1A02	PWR	651	TO-39	F	−50 (V_{CEO})	7	30-200
RCA1A03	PWR	651	TO-39	F	95	10	70-300
RCA1A04	PWR	651	TO-39	F	−95	10	70-300
RCA1A05	PWR	651	TO-39	F	−75	5	50-250
RCA1A06	PWR	651	TO-39	F	75	5	50-250
RCA1A07	PWR	651	TO-39	F	50	5	50-250
RCA1A08	PWR	651	TO-39	F	−50	7	70-250
RCA1A09	PWR	651	TO-39	F	175 (V_{CEO})	10	20-100
RCA1A10	PWR	651	TO-39	F	−175 (V_{CEO})	10	10-250
RCA1A11	PWR	651	TO-39	F	175 (V_{CEO})	10	40-250
RCA1A15	PWR	651	TO-39	F	100 (V_{CEO})	10	20-100
RCA1A16	PWR	651	TO-39	F	−100 (V_{CEO})	10	40-250
RCA1A17	PWR	651	TO-39	F	90 (V_{CEO})	5	40-200
RCA1A18	PWR	651	TO-39	F	10 (V_{CEO})	7	40-250

▲ Dimensional outlines are shown following data charts.
■ Lead codes are explained in Table LII.
See **Notes** on page 619.

Table LI—RCA Standard Solid-State Devices (cont'd)

Type No.	Prod. Line	File No.	Package▲	Lead Code■	Voltage (V) (Note 1)	Power (W), Current* (A) or Noise Figure (dB)	DC Current Transfer Ratio or Frequency
RCA1A19	PWR	651	TO-39	F	-10 (V_{CEO})	7	40-250
RCA1B01	PWR	647	TO-3	D	95	115	20-70
RCA1B04	PWR	649	TO-3	D	225	150	15-75
RCA1B05	PWR	650	TO-3	D	275	150	15-75
RCA1B06	PWR	648	TO-3	D	120	150	10-50
RCA1B07	PWR	791	TO-3	D	80	100	1000-15,000
RCA1B08	PWR	791	TO-3	D	-80	100	1000-15,000
RCA1C03	PWR	652	TO-220AB	N	120	40	50-250
RCA1C04	PWR	652	TO-220AB	N	-120	40	50-250
RCA1C05	PWR	644	TO-220AB	N	60	40	20-120
RCA1C06	PWR	644	TO-220AB	N	-60	40	20-120
RCA1C07	PWR	646	TO-220AB	N	75	75	20-120
RCA1C08	PWR	646	TO-220AB	N	-75	75	20-120
RCA1C09	PWR	645	TO-220AB	N	75	75	20-120
RCA1C10	PWR	642	TO-220AB	N	40	40	50-250
RCA1C11	PWR	642	TO-220AB	N	-40	40	50-250
RCA1C12	PWR	652	TO-220AB	N	140	40	40-250
RCA1C13	PWR	652	TO-220AB	N	-140	40	40-250
RCA1C14	PWR	643	TO-220AB	N	60	50	20-70
RCA1E02	PWR	653	TO-66	D	200	35	30-150
RCA1E03	PWR	653	TO-66	D	-200	35	30-150
RCA29	PWR	583	TO-220AB	N	40	30	15-75
RCA29A	PWR	583	TO-220AB	N	60	30	15-75
RCA29B	PWR	583	TO-220AB	N	80	30	15-75
RCA29C	PWR	583	TO-220AB	N	100	30	15-75
RCA29/SDH	PWR	792	TO-220AB	N	40	36	40 min.
RCA29A/SDH	PWR	792	TO-220AB	N	60	36	40 min.
RCA29B/SDH	PWR	792	TO-220AB	N	80	36	40 min.
RCA29C/SDH	PWR	792	TO-220AB	N	100	50	40 min.
RCA30	PWR	584	TO-220AB	N	-40	30	15-75
RCA30A	PWR	584	TO-220AB	N	-60	30	15-75
RCA30B	PWR	584	TO-220AB	N	-80	30	15-75
RCA30C	PWR	584	TO-220AB	N	-100	30	15-75
RCA31	PWR	585	TO-220AB	N	40	40	10-50
RCA31A	PWR	585	TO-220AB	N	60	40	10-50
RCA31B	PWR	585	TO-220AB	N	80	40	10-50
RCA31C	PWR	585	TO-220AB	N	100	40	10-50
RCA31/SDH	PWR	793	TO-220AB	N	40	36	25 min.
RCA31A/SDH	PWR	793	TO-220AB	N	60	36	25 min.
RCA31B/SDH	PWR	793	TO-220AB	N	80	36	25 min.
RCA31C/SDH	PWR	793	TO-220AB	N	100	50	25 min.
RCA32	PWR	586	TO-220AB	N	-40	40	10-50
RCA32A	PWR	586	TO-220AB	N	-60	40	10-50
RCA32B	PWR	586	TO-220AB	N	-80	40	10-50
RCA32C	PWR	586	TO-220AB	N	-100	40	10-50
RCA41	PWR	587	TO-220AB	N	40	65	15-75
RCA41A	PWR	587	TO-220AB	N	60	65	15-75
RCA41B	PWR	587	TO-220AB	N	80	65	15-75
RCA41C	PWR	587	TO-220AB	N	100	65	15-75
RCA41/SDH	PWR	794	TO-220AB	N	40	75	30 min.

▲ Dimensional outlines are shown following data charts.
■ Lead codes are explained in Table LII.
See **Notes** on page 619.

Table LI—RCA Standard Solid-State Devices (cont'd)

Type No.	Prod. Line	File No.	Package▲	Lead Code■	Voltage (V) (Note 1)	Power (W), Current* (A) or Noise Figure (dB)	DC Current Transfer Ratio or Frequency
RCA41A/SDH	PWR	794	TO-220AB	N	60	75	30 min.
RCA1B/SDH	PWR	794	TO-220AB	N	80	75	30 min.
RCA42	PWR	588	TO-220AB	N	—40	65	15-75
RCA42A	PWR	588	TO-220AB	N	—60	65	15-75
RCA42B	PWR	588	TO-220AB	N	—80	65	15-75
RCA42C	PWR	588	TO-220AB	N	—100	65	15-75
RCA120	PWR	840	TO-220AB	N	60	60	1000 min.
RCA121	PWR	840	TO-220AB	N	80	60	1000 min.
RCA122	PWR	840	TO-220AB	N	100	60	1000 min.
RCA125	PWR	841	TO-220AB	N	—60	60	1000 min.
RCA126	PWR	841	TO-220AB	N	—80	60	1000 min.
RCA410	PWR	509	TO-3	D	200	125	30-90
RCA411	PWR	510	TO-3	D	300	125	30-90
RCA413	PWR	511	TO-3	D	400	125	20-80
RCA423	PWR	512	TO-3	D	400	125	30-90
RCA431	PWR	513	TO-3	D	400	125	15-35
RCA1000	PWR	594	TO-3	D	60	90	1000 min.
RCA1001	PWR	594	TO-3	D	80	90	1000 min.
RCA2001	RF	759	RCA HF-46	T	28	1	2 GHz
RCA2310	RF	765	RCA HF-46	T	24	10	2.3 GHz
RCA3054	PWR	618	TO-220AB	N	90	36	25-100
RCA3055	PWR	618	TO-220AB	N	100	75	20-70
RCA3441	PWR	666	TO-220AB	N	160	36	20-150
RCA6263	PWR	666	TO-220AB	N	140	36	20-150
RCA8203	PWR	835	TO-220AB	N	—40	60	1000-20,000
RCA8203A	PWR	835	TO-220AB	N	—60	60	1000-20,000
RCA8203B	PWR	835	TO-220AB	N	—80	60	1000-20,000
RCA8350	PWR	861	TO-3	D	—40	70	1000-20,000
RCA8350A	PWR	861	TO-3	D	—60	70	1000-20,000
RCA8350B	PWR	861	TO-3	D	—80	70	1000-20,000
RCP111A	PWR	822	Plastic	F	200	6.25	50-300
RCP111B	PWR	822	Plastic	F	250	6.25	50-300
RCP111C	PWR	822	Plastic	F	300	6.25	50-300
RCP111D	PWR	822	Plastic	F	350	6.25	50-300
RCP113A	PWR	822	Plastic	F	200	6.25	30-150
RCP113B	PWR	822	Plastic	F	250	6.25	30-150
RCP113C	PWR	822	Plastic	F	300	6.25	30-150
RCP113D	PWR	822	Plastic	F	350	6.25	30-150
RCP115	PWR	822	Plastic	F	100	6.25	50 min.
RCP115B	PWR	822	Plastic	F	250	6.25	50 min.
RCP117	PWR	822	Plastic	F	100	6.25	20 min.
RCP117B	PWR	822	Plastic	F	250	6.25	20 min.
RCP700A	PWR	821	Plastic	F	—55	10	50-250
RCP700B	PWR	821	Plastic	F	—85	10	50-250
RCP700C	PWR	821	Plastic	F	—105	10	50-250
RCP700D	PWR	821	Plastic	F	—125	10	50-250
RCP701A	PWR	820	Plastic	F	55	10	50-250
RCP701B	PWR	820	Plastic	F	85	10	50-250
RCP701C	PWR	820	Plastic	F	105	10	50-250
RCP701D	PWR	820	Plastic	F	125	10	50-250

▲ Dimensional outlines are shown following data charts.
■ Lead codes are explained in Table LII.
See **Notes** on page 619.

Table LI—RCA Standard Solid-State Devices (cont'd)

Type No.	Prod. Line	File No.	Package▲	Lead Code■	Voltage (V) (Note 1)	Power (W), Current* (A) or Noise Figure (dB)	DC Current Transfer Ratio or Frequency
RCP702A	PWR	821	Plastic	F	—55	10	50-150
RCP702B	PWR	821	Plastic	F	—80	10	30-150
RCP702C	PWR	821	Plastic	F	—105	10	30-150
RCP702D	PWR	821	Plastic	F	—125	10	30-150
RCP703A	PWR	820	Plastic	F	55	10	30-150
RCP703B	PWR	820	Plastic	F	85	10	30-150
RCP703C	PWR	820	Plastic	F	105	10	30-150
RCP703D	PWR	820	Plastic	F	125	10	30-150
RCP704	PWR	821	Plastic	F	—45	10	30-150
RCP704B	PWR	821	Plastic	F	—85	10	50 min.
RCP705	PWR	820	Plastic	F	45	10	50 min.
RCP705B	PWR	820	Plastic	F	85	10	50 min.
RCP706	PWR	821	Plastic	F	—45	10	20 min.
RCP706B	PWR	821	Plastic	F	—85	10	20 min.
RCP707	PWR	820	Plastic	F	45	10	20 min.
RCP707B	PWR	820	Plastic	F	85	10	20 min.
RCS242	PWR	778	TO-3	D	50	115	20 min.
RCS559	PWR	782	TO-66	D	—275	35	10-100
RCS560	PWR	782	TO-66	D	—250	35	7.5 min.
RCS564	PWR	779	TO-3	D	300	175	5 min.
RCS880	PWR	777	TO-39	D	—150*	0.75	20-150
RCS881	PWR	780	TO-39	D	—250	0.75	20 min.
RCS882	PWR	781	TO-39	D	—350	7.5	20 min.
S2060A	SCR	654	TO-220AB	H	100	4* (rms)	—
S2060B	SCR	654	TO-220AB	H	200	4* (rms)	—
S2060C	SCR	654	TO-220AB	H	300	4* (rms)	—
S2060D	SCR	654	TO-220AB	H	400	4* (rms)	—
S2060E	SCR	654	TO-220AB	H	500	4* (rms)	—
S2060F	SCR	654	TO-220AB	H	50	4* (rms)	—
S2060M	SCR	654	TO-220AB	H	600	4* (rms)	—
S2060Q	SCR	654	TO-220AB	H	15	4* (rms)	—
S2060Y	SCR	654	TO-220AB	H	30	4* (rms)	—
S2061A	SCR	654	TO-220AB	H	100	4* (rms)	—
S2061B	SCR	654	TO-220AB	H	200	4* (rms)	—
S2061C	SCR	654	TO-220AB	H	300	4* (rms)	—
S2061D	SCR	654	TO-220AB	H	400	4* (rms)	—
S2061E	SCR	654	TO-220AB	H	500	4* (rms)	—
S2061F	SCR	654	TO-220AB	H	50	4* (rms)	—
S2061M	SCR	654	TO-220AB	H	600	4* (rms)	—
S2061Q	SCR	654	TO-220AB	H	15	4* (rms)	—
S2061Y	SCR	654	TO-220AB	H	30	4* (rms)	—
S2062A	SCR	654	TO-220AB	H	100	4* (rms)	—
S2062B	SCR	654	TO-220AB	H	200	4* (rms)	—
S2062C	SCR	654	TO-220AB	H	300	4* (rms)	—
S2062D	SCR	654	TO-220AB	H	400	4* (rms)	—
S2062E	SCR	654	TO-220AB	H	500	4* (rms)	—
S2062F	SCR	654	TO-220AB	H	60	4* (rms)	—
S2062M	SCR	654	TO-220AB	H	600	4* (rms)	—
S2062Q	SCR	654	TO-220AB	H	15	4* (rms)	—
S2062Y	SCR	654	TO-220AB	H	30	4* (rms)	—

▲ Dimensional outlines are shown following data charts.
■ Lead codes are explained in Table LII.
See **Notes** on page 619.

Table LI—RCA Standard Solid-State Devices (cont'd)

Type No.	Prod. Line	File No.	Package▲	Lead Code■	Voltage (V) (Note 1)	Power (W), Current* (A) or Noise Figure (dB)	DC Current Transfer Ratio or Frequency
S2400A	SCR	567	TO-8	K	100	4.5* (rms)	—
S2400B	SCR	567	TO-8	K	200	4.5* (rms)	—
S2400D	SCR	567	TO-8	K	400	4.5* (rms)	—
S2400M	SCR	567	TO-8	K	600	4.5* (rms)	—
S2600B	SCR	496	TO-5 (Mod.)	K	200	7* (rms)	—
S2600D	SCR	496	TO-5 (Mod.)	K	400	7* (rms)	—
S2600M	SCR	496	TO-5 (Mod.)	K	600	7* (rms)	—
S2610B	SCR	496	Radiator	K	200	3.3* (rms)	—
S2610D	SCR	496	Radiator	K	400	3.3* (rms)	—
S2610M	SCR	496	Radiator	K	600	3.3* (rms)	—
S2620B	SCR	496	Heat Spdr.	K	200	7* (rms)	—
S2620D	SCR	496	Heat Spdr.	K	400	7* (rms)	—
S2620M	SCR	496	Heat Spdr.	K	600	7* (rms)	—
S2710B	SCR	266	Radiator	I	200	1.7* (rms)	—
S2710D	SCR	266	Radiator	I	400	1.7* (rms)	—
S2710M	SCR	266	Radiator	I	600	1.7* (rms)	—
S2800A	SCR	501	TO-220AB	Q	100	8* (rms)	—
S2800B	SCR	501	TO-220AB	Q	200	8* (rms)	—
S2800D	SCR	501	TO-220AB	Q	400	8* (rms)	—
S3700B	SCR	306	TO-66	I	200	5* (rms)	—
S3700D	SCR	306	TO-66	I	400	5* (rms)	—
S3700M	SCR	306	TO-66	I	600	5* (rms)	—
S3701M	SCR	476	TO-66	I	600	5* (rms)	—
S3702SF	SCR	522	TO-66	I	750	5* (rms)	—
S3703SF	SCR	522	TO-66	I	750	5* (rms)	—
S3704A	SCR	690	TO-66	I	100	5* (rms)	—
S3704B	SCR	690	TO-66	I	200	5* (rms)	—
S3704D	SCR	690	TO-66	I	400	5* (rms)	—
S3704M	SCR	690	TO-66	I	600	5* (rms)	—
S3704S	SCR	690	TO-66	I	700	5* (rms)	—
S3705M	SCR	354	TO-66	I	600	5* (rms)	—
S3706M	SCR	354	TO-66	I	600	5* (rms)	—
S3714A	SCR	690	Radiator	I	100	5* (rms)	—
S3714B	SCR	690	Radiator	I	200	5* (rms)	—
S3714D	SCR	690	Radiator	I	400	5* (rms)	—
S3714M	SCR	690	Radiator	I	600	5* (rms)	—
S3714S	SCR	690	Radiator	I	700	5* (rms)	—
S3800D	ITR	639	TO-66	I	400	5* (rms)	—
S3800E	ITR	639	TO-66	I	500	5* (rms)	—
S3800EF	ITR	639	TO-66	I	550	5* (rms)	—
S3800M	ITR	639	TO-66	I	600	5* (rms)	—
S3800MF	ITR	639	TO-66	I	650	5* (rms)	—
S3800S	ITR	639	TO-66	I	700	5* (rms)	—
S3800SF	ITR	639	TO-66	I	750	5* (rms)	—
S5210B	SCR	757	Stud	I	200	10* (rms)	—
S5210D	SCR	757	Stud	I	400	10* (rms)	—
S5210M	SCR	757	Stud	I	600	10* (rms)	—
S6200A	SCR	418	Press-fit	I	100	20* (rms)	—
S6200B	SCR	418	Press-fit	I	200	20* (rms)	—
S6200D	SCR	418	Press-fit	I	400	20* (rms)	—

▲ Dimensional outlines are shown following data charts.
■ Lead codes are explained in Table LII.
See **Notes** on page 619.

Table LI—RCA Standard Solid-State Devices (cont'd)

Type No.	Prod. Line	File No.	Package▲	Lead Code■	Voltage (V) (Note 1)	Power (W), Current* (A) or Noise Figure (dB)	DC Current Transfer Ratio or Frequency
S6200M	SCR	418	Press-fit	I	600	20* (rms)	—
S6210A	SCR	418	Stud	I	100	20* (rms)	—
S6210B	SCR	418	Stud	I	200	20* (rms)	—
S6210D	SCR	418	Stud	I	400	20* (rms)	—
S6210M	SCR	418	Stud	I	600	20* (rms)	—
S6220A	SCR	418	Iso. Stud	I	100	20* (rms)	—
S6220B	SCR	418	Iso. Stud	I	200	20* (rms)	—
S6220D	SCR	418	Iso. Stud	I	400	20* (rms)	—
S6220M	SCR	418	Iso. Stud	I	600	20* (rms)	—
S6400N	SCR	578	Press-fit	I	800	35* (rms)	—
S6410N	SCR	578	Stud	I	800	35* (rms)	—
S6420A	SCR	578	Iso. Stud	I	100	35* (rms)	—
S6420B	SCR	578	Iso. Stud	I	200	35* (rms)	—
S6420D	SCR	578	Iso. Stud	I	400	35* (rms)	—
S6420M	SCR	578	Iso. Stud	I	600	35* (rms)	—
S6420N	SCR	578	Iso. Stud	I	800	35* (rms)	—
S6431M	SCR	247	TO-48	I	600	35* (rms)	—
S7430M	SCR	408	TO-48	I	600	35* (rms)	—
S7432M	SCR	724	TO-48	I	600	35* (rms)	—
T2300A	TRI	470	TO-5 (Mod.)	L	100	2.5* (rms)	—
T2300B	TRI	470	TO-5 (Mod.)	L	200	2.5* (rms)	—
T2300D	TRI	470	TO-5 (Mod.)	L	400	2.5* (rms)	—
T2301A	TRI	431	TO-5 (Mod.)	L	100	2.5* (rms)	—
T2301B	TRI	431	TO-5 (Mod.)	L	200	2.5* (rms)	—
T2301D	TRI	431	TO-5 (Mod.)	L	400	2.5* (rms)	—
T2302A	TRI	470	TO-5 (Mod.)	L	100	2.5* (rms)	—
T2302B	TRI	470	TO-5 (Mod.)	L	200	2.5* (rms)	—
T2302D	TRI	470	TO-5 (Mod.)	L	400	2.5* (rms)	—
T2304B	TRI	441	TO-5 (Mod.)	L	200	0.5* (rms)	—
T2304D	TRI	441	TO-5 (Mod.)	L	400	0.5* (rms)	—
T2305B	TRI	441	TO-5 (Mod.)	L	200	0.5* (rms)	—
T2305D	TRI	441	TO-5 (Mod.)	L	400	0.5* (rms)	—
T2306A	TRI	406	Radiator	L	100	2.5* (rms)	—
T2306B	TRI	406	TO-5 (Mod.)	L	200	2.5* (rms)	—
T2306D	TRI	406	TO-5 (Mod.)	L	400	2.5* (rms)	—
T2310A	TRI	470	Radiator	L	100	1.6* (rms)	—
T2310B	TRI	470	Radiator	L	200	1.6* (rms)	—
T2310D	TRI	470	Radiator	L	400	1.6* (rms)	—
T2311A	TRI	431	Radiator	L	100	1.6* (rms)	—
T2311B	TRI	431	Radiator	L	200	1.6* (rms)	—
T2311D	TRI	431	Radiator	L	400	1.6* (rms)	—
T2312A	TRI	470	Radiator	L	100	1.9* (rms)	—
T2312B	TRI	470	Radiator	L	200	1.9* (rms)	—
T2312D	TRI	470	Radiator	L	400	1.9* (rms)	—
T2313A	TRI	414	Radiator	L	100	1.9* (rms)	—
T2313B	TRI	414	Radiator	L	200	1.9* (rms)	—
T2313D	TRI	414	Radiator	L	400	1.9* (rms)	—
T2313M	TRI	414	Radiator	L	600	1.9* (rms)	—
T2316A	TRI	406	Radiator	L	100	2.5* (rms)	—
T2316B	TRI	406	Radiator	L	200	2.5* (rms)	—

▲ Dimensional outlines are shown following data charts.
■ Lead codes are explained in Table LII.
See Notes on page 619.

Table LI—RCA Standard Solid-State Devices (cont'd)

Type No.	Prod. Line	File No.	Package▲	Lead Code■	Voltage (V) (Note 1)	Power (W), Current* (A) or Noise Figure (dB)	DC Current Transfer Ratio or Frequency
T2316D	TRI	406	Radiator	L	400	2.5* (rms)	—
T2500B	TRI	615	TO-220AB	Z	200	6* (rms)	—
T2500D	TRI	615	TO-220AB	Z	400	6* (rms)	—
T2700B	TRI	351	TO-66	J	200	6* (rms)	—
T2700D	TRI	351	TO-66	J	400	6* (rms)	—
T2706B	TRI	406	TO-66	J	200	6* (rms)	—
T2706D	TRI	406	TO-66	J	400	6* (rms)	—
T2710B	TRI	351	Radiator	J	200	3.3* (rms)	—
T2710D	TRI	351	Radiator	J	400	3.3* (rms)	—
T2716B	TRI	406	Radiator	J	200	3.3* (rms)	—
T2716D	TRI	406	Radiator	J	400	3.3* (rms)	—
T2800B	TRI	364	TO-220AB	Z	200	8* (rms)	—
T2800D	TRI	364	TO-220AB	Z	400	8* (rms)	—
T2800M	TRI	364	TO-220AB	Z	600	8* (rms)	—
T2801DF	TRI	493	TO-220AB	Z	450	6* (rms)	—
T2806B	TRI	406	TO-220AB	Z	200	8* (rms)	—
T2806D	TRI	406	TO-220AB	Z	400	8* (rms)	—
T2850A	TRI	540	TO-220AB	Za	100	8* (rms)	—
T2850B	TRI	540	TO-220AB	Za	200	8* (rms)	—
T2850D	TRI	540	TO-220AB	Za	400	8* (rms)	—
T4100M	TRI	458	Press-fit	J	600	15* (rms)	—
T4101M	TRI	457	Press-fit	J	600	10* (rms)	—
T4103B	TRI	443	Press-fit	J	200	15* (rms)	—
T4103D	TRI	443	Press-fit	J	400	15* (rms)	—
T4104B	TRI	443	Press-fit	J	200	10* (rms)	—
T4104D	TRI	443	Press-fit	J	400	10* (rms)	—
T4105B	TRI	443	Press-fit	J	200	6* (rms)	—
T4105D	TRI	443	Press-fit	J	400	6* (rms)	—
T4106B	TRI	406	Press-fit	J	200	15* (rms)	—
T4106D	TRI	406	Press-fit	J	400	15* (rms)	—
T4107B	TRI	406	Press-fit	J	200	10* (rms)	—
T4107D	TRI	406	Press-fit	J	400	10* (rms)	—
T4110M	TRI	458	Stud	J	600	15* (rms)	—
T4111M	TRI	457	Stud	J	600	10* (rms)	—
T4113B	TRI	443	Stud	J	200	15* (rms)	—
T4113D	TRI	443	Stud	J	400	15* (rms)	—
T4114B	TRI	443	Stud	J	200	10* (rms)	—
T4114D	TRI	443	Stud	J	400	10* (rms)	—
T4115B	TRI	443	Stud	J	200	6* (rms)	—
T4115D	TRI	443	Stud	J	400	6* (rms)	—
T4116B	TRI	406	Stud	J	200	15* (rms)	—
T4116D	TRI	406	Stud	J	400	15* (rms)	—
T4117B	TRI	406	Stud	J	200	10* (rms)	—
T4117D	TRI	406	Stud	J	400	10* (rms)	—
T4120B	TRI	458	Iso. Stud	J	200	15* (rms)	—
T4120D	TRI	458	Iso. Stud	J	400	15* (rms)	—
T4120M	TRI	458	Iso. Stud	J	600	15* (rms)	—
T4121B	TRI	457	Iso. Stud	J	200	10* (rms)	—
T4121D	TRI	457	Iso. Stud	J	400	10* (rms)	—
T4121M	TRI	457	Iso. Stud	J	600	10* (rms)	—

▲ Dimensional outlines are shown following data charts.
■ Lead codes are explained in Table LII.
See **Notes** on page 619.

Table LI—RCA Standard Solid-State Devices (cont'd)

Type No.	Prod. Line	File No.	Package▲	Lead Code■	Voltage (V) (Note 1)	Power (W), Current* (A) or Noise Figure (dB)	DC Current Transfer Ratio or Frequency
T4700B	TRI	300	TO-66	J	200	15* (rms)	—
T4700D	TRI	300	TO-66	J	400	15* (rms)	—
T4706B	TRI	406	TO-66	J	200	15* (rms)	—
T4706D	TRI	406	TO-66	J	400	15* (rms)	—
T6400N	TRI	593	Press-fit	J	800	40* (rms)	—
T6401B	TRI	459	Press-fit	J	200	30* (rms)	—
T6401D	TRI	459	Press-fit	J	400	30* (rms)	—
T6401M	TRI	459	Press-fit	J	600	30* (rms)	—
T6404B	TRI	487	Press-fit	J	200	40* (rms)	—
T6404D	TRI	487	Press-fit	J	400	40* (rms)	—
T6405B	TRI	487	Press-fit	J	200	25* (rms)	—
T6405D	TRI	487	Press-fit	J	400	25* (rms)	—
T6406B	TRI	406	Press-fit	J	200	40* (rms)	—
T6406D	TRI	406	Press-fit	J	400	40* (rms)	—
T6406M	TRI	406	Press-fit	J	600	40* (rms)	—
T6407B	TRI	406	Press-fit	J	200	30* (rms)	—
T6407D	TRI	406	Press-fit	J	400	30* (rms)	—
T6407M	TRI	406	Press-fit	J	600	30* (rms)	—
T6410N	TRI	593	Stud	J	800	40* (rms)	—
T6411B	TRI	459	Stud	J	200	30* (rms)	—
T6411D	TRI	459	Stud	J	400	30* (rms)	—
T6411M	TRI	459	Stud	J	600	30* (rms)	—
T6414B	TRI	487	Stud	J	200	40* (rms)	—
T6414D	TRI	487	Stud	J	400	40* (rms)	—
T6415B	TRI	487	Stud	J	200	25* (rms)	—
T6415D	TRI	487	Stud	J	400	25* (rms)	—
T6416B	TRI	406	Stud	J	200	40* (rms)	—
T6416D	TRI	406	Stud	J	400	40* (rms)	—
T6416M	TRI	406	Stud	J	600	40* (rms)	—
T6417B	TRI	406	Stud	J	200	30* (rms)	—
T6417D	TRI	406	Stud	J	400	30* (rms)	—
T6417M	TRI	406	Stud	J	600	30* (rms)	—
T6420B	TRI	593	Iso. Stud	J	200	40* (rms)	—
T6420D	TRI	593	Iso. Stud	J	400	40* (rms)	—
T6420M	TRI	593	Iso. Stud	J	600	40* (rms)	—
T6420N	TRI	593	Iso. Stud	J	800	40* (rms)	—
T6421B	TRI	459	Iso. Stud	J	200	30* (rms)	—
T6421D	TRI	459	Iso. Stud	J	400	30* (rms)	—
T6421M	TRI	459	Iso. Stud	J	600	30* (rms)	—
T8401B	TRI	725	Press-fit +	J	200	60* (rms)	—
T8401D	TRI	725	Press-fit +	J	400	60* (rms)	—
T8401M	TRI	725	Press-fit +	J	600	60* (rms)	—
T8411B	TRI	725	Stud +	J	200	60* (rms)	—
T8411D	TRI	725	Stud +	J	400	60* (rms)	—
T8411M	TRI	725	Stud +	J	600	60* (rms)	—
T8421B	TRI	725	Iso. Stud +	J	200	60* (rms)	—
T8421D	TRI	725	Iso. Stud +	J	400	60* (rms)	—
T8421M	TRI	725	Iso. Stud +	J	600	60* (rms)	—
T8430B	TRI	549	Press-fit	J	200	80* (rms)	—
T8430D	TRI	549	Press-fit	J	400	80* (rms)	—

+ Flying leads
▲ Dimensional outlines are shown following data charts.
■ Lead codes are explained in Table LII.
See Notes on page 619.

Table LI—RCA Standard Solid-State Devices (cont'd)

Type No.	Prod. Line	File No.	Package▲	Lead Code■	Voltage (V) (Note 1)	Power (W), Current* (A) or Noise Figure (dB)	DC Current Transfer Ratio or Frequency
T8430M	TRI	549	Press-fit	J	600	80* (rms)	—
T8440B	TRI	549	Stud	J	200	80* (rms)	—
T8440D	TRI	549	Stud	J	400	80* (rms)	—
T8440M	TRI	549	Stud	J	600	80* (rms)	—
T8450B	TRI	549	Iso. Stud	J	200	80* (rms)	—
T8450D	TRI	549	Iso. Stud	J	400	80* (rms)	—
T8450M	TRI	549	Iso. Stud	J	600	80* (rms)	—

▲ Dimensional outlines are shown following data charts.
■ Lead codes are explained in Table LII.
..See NOTES below.

NOTES for Table LI:

(1) Voltage shown is V_{CBO} for power transistors, V_{CC} for rf transistors, V_{DS} for MOS/FET's, V_{DROM} for thyristors, or V_{RRM} for diodes or rectifiers. Any exceptions are noted in the table.

(2) Power shown is power dissipation (P_T) for power transistors, or output power (P_{OB} or P_{OE}) for rf transistors.

(3) Integrated-circuit packages are designated by the last letter of the type number. The package type corresponding to each letter is shown in Fig. 703.

(4) High-reliability linear integrated circuits are supplied in hermetic packages in conformance with MIL-STD-883. Refer to the data bulletin for package details.

(5) All RCA high-reliability COS/MOS integrated circuits are available in dual-in-line and flat-pack ceramic packages, except the CD4061/ . . . , which are supplied only in the dual-in-line package. Package details are shown on the data bulletins.

(6) Terminal connections, together with package details, are shown on the data bulletins.

Table LII—Lead Codes

Lead Code	Lead Configuration			
	1	2	3	4
A	A	K		
B	K	A		
C	B	C	E	
D	B	E	C*	
E	E	B	C	
F	E	B	C*	
G	E*	B	C	

Table LII—Lead Codes (cont'd)

H	G	A*	K	
I	G	K	A*	
J	G	T1	T2*	
Q	K	A*	G	
K	K	G	A*	
L	T1	G	T2*	
M	B	—	E	C
N	B	C	E	C
P	D	G2	G1	SX
R	D	S	G	X
S	E	B	C	Case
T	E	B	C	B
U	E*	B	C	E
W	E	B	E	C
X	E	C	B	C
Y	EK	C	B	A
Z	T1	T2	G	T2
Za	T1	T2	G	Tab Isol.

* Element connected to case, mounting flange, or heat spreader or radiator

Abbrev.	Terminal
A	Anode
B	Base
C	Collector
D	Drain
E	Emitter
G	Gate
G1, G2	Gate 1, Gate 2
K	Cathode
S	Source
T1, T2	Main Terminal 1, 2
X	Substrate, Case

RCA SK-SERIES SOLID-STATE REPLACEMENT DEVICES

The RCA "top-of-the-line" SK-series entertainment and industrial solid-state devices are a group of high-quality types specifically intended for replacement purposes in line-operated and battery-operated electronic equipment. Each transistor, rectifier, or integrated circuit included in the SK-series is designed to provide outstanding performance in a specific application or type of service and can be used to replace a broad variety of solid-state devices used in that application or type of service in original equipment.

SK devices are precisely engineered, manufactured, and tested specifically for use as replacements. Each device has electrical characteristics comparable with, or superior to, those of the devices that it replaces. In some instances, the case of an SK device may be slightly taller or thicker than that of the original device or may have a slightly different shape. These slight mechanical differences will not affect the performance of the equipment in which the replacement is made and normally will not prevent or complicate the installation of the SK device. In most cases, therefore, the recommended SK replacement device can be installed without changes in mechanical mounting arrangements, circuit wiring, or operating conditions. Dimensional outlines for the SK devices are shown in the **Outlines** Section of this Manual.

Because the SK-series transistors, integrated circuits, rectifiers, and thyristors are intended specifically for replacement purposes, RCA does not publish technical data sheets for these types. However, for the benefit of users who may require some information on the performance capability of these devices, data are given in Tables G-I through G-XIII for safe operation in the intended applications. Operation outside of the limit conditions may result in damage to the device.

For more detailed information on the use and capabilities of RCA SK solid-state devices, the reader should refer to the **SK-Series Top of the Line Replacement Guide**, RCA Publication No. SPG-202R. This Guide lists in numerical-alphabetical sequence more than 103,000 solid-state devices widely used in electronic equipment and the recommended SK replacement device for each type. The Guide also provides detailed instruction and precautionary measures that should be followed to assure successful use of SK types for replacement of original-equipment devices.

Table LIII—SK-Series Bipolar Transistors

RCA Type	Polarity and Material	LIMIT CONDITIONS					CHARACTERISTICS				PACKAGE	
		P_T W	I_C A	V_{CBO} V	V_{CEO} V	V_{EBO} V	h_{FE}	V_{CE} V	I_C A	f_T MHz	Case Style	Dim. Outline *
AF Small-Signal Applications												
SK3003	PNP Ge	0.15	−0.5	−20	−18	−2.5	90	−1	−0.05	1	TO-1	1
SK3004	PNP Ge	0.6	−1	−32	−25	−12	90	−1	−0.15	1	TO-1	1
SK3010	NPN Ge	0.15	0.1	30	30	12	90	1	0.05	2	To-1	1
SK3038	NPN Si	0.3	0.3	30	25	7	175	10	0.1	150	TO-5	4A
SK3114	PNP Si	0.4	−0.5	−60	−50	−5	175	−10	−0.15	200	TO-92	6E
SK3122	NPN Si	0.4	0.5	50	50	4	150	3	0.01	200	—	6F
SK3124	NPN Si	0.3	0.3	50	30	5	175	5	0.02	175	TO-92	6E
SK3156	NPN Si	0.5	0.3	40	30	10	10K	5	0.01	100	—	6F

* Terminal connections are indicated on the outline drawing at end of this section.

Table LIII—SK-Series Bipolar Transistors (cont'd)

RCA Type	Polarity and Material	LIMIT CONDITIONS					CHARACTERISTICS				PACKAGE	
		P_T W	I_C A	V_{CBO} V	V_{CEO} V	V_{EBO} V	h_{FE}	V_{CE} V	I_C A	f_T MHz	Case Style	Dim. Outline *
AF Medium Power Applications												
SK3020	NPN Si	1	0.8	30	25	7	125	10	0.1	150	TO-5	4A
SK3024	NPN Si	5	1	100	90	7	100	10	0.15	150	TO-5	4A/10B
SK3025	PNP Si	7	—1	—90	—90	—7	100	—10	—0.15	100	TO-5	4A/10B
SK3045	NPN Si	10	1	300	300	6	80	10	0.05	30	TO-5	8
SK3052	PNP Ge	6	—2	—60	—60	—12	110	—1	—0.5	0.45	TO-66	12A
SK3082	PNP Ge	12	—2	—35	—35	—6	110	—1.5	—0.2	0.45	—	11D
SK3086	PNP Ge	*Matched Pair of SK3082 Transistors, for Data See SK3082*										11D
SK3123	PNP Ge	9	—5	—60	—40	—5	80	—2	—0.4	1	TO-8	9D
SK3137	NPN Si	0.6	1	60	50	5	150	10	0.15	200	TO-92	7
SK3138	PNP Si	0.6	—1	—60	—50	—5	150	—10	—0.15	200	TO-92	7
SK3512	NPN Si	10	2	100	90	7	90	4	0.4	70	TO-5	4A
SK3513	PNP Si	10	—2	—100	—90	—7	70	—4	—0.4	70	TO-5	4A
SK3536	NPN Si	4.4	3	160	140	7	100	4	0.15	1	—	31
AF High Power Applications												
SK3009	PNP Ge	30	—10	—60	—50	—10	90	—1.5	—0.2	0.45	TO-3	11A
SK3012	PNP Ge	150	—15	—50	—30	—20	50	—2	—5	0.1	TO-36	13
SK3013	PNP Ge	*Matched Pair of SK3009 Transistors, for Data See SK3009*									TO-3	11A
SK3014	PNP Ge	40	—11	—75	—50	—5	150	—1.5	—8	2.5	TO-3	11A
SK3015	PNP Ge	*Matched Pair of SK3014 Transistors, for Data See SK3014*									TO-3	11A
SK3026	NPN Si	29	4	90	60	7	70	4	0.1	1	TO-66	12A
SK3027	NPN Si	115	15	100	80	7	70	4	0.3	2	TO-3	11A
SK3028	NPN Si	*Matched Pair of SK3026 Transistors, for Data See SK3026*									TO-66	12A
SK3029	NPN Si	*Matched Pair of SK3027 Transistors, for Data See SK3027*									TO-3	11A
SK3036	NPN Si	150	30	90	80	5	100	4	1.5	1.5	TO-3	11A
SK3037	NPN Si	*Matched Pair of SK3036 Transistors, for Data See SK3036*									TO-3	11A
SK3041	NPN Si	36	4	35	35	5	100	4	0.5	2	TO-220	10A
SK3054	NPN Si	50	7	90	70	5	70	4	1.	2	TO-220	10A
SK3083	PNP Si	40	—7	—80	—70	—5	70	—2	—1.5	20	TO-220	10A
SK3084	PNP Si	40	—7	—40	—30	—5	100	—2	—3	20	TO-220	10A
SK3173	PNP Si	125	—15	—90	—80	—5	70	—4	—0.3	15	TO-3	11A
SK3510	NPN Si	115	15	100	60	7	45	4	2	1.5	TO-3	11A
SK3511	NPN Si	150	30	90	80	5	40	4	8	1.5	TO-3	11A
SK3530	NPN Si	25	3	100	55	12	30	4	1	1	TO-8	9D
SK3534	NPN Si	75	10	80	70	8	55	4	1	2	TO-220	10A
SK3535	NPN Si	250	30	170	150	7	40	4	4	2	TO-3	11A
SK3538	NPN Si	40	4	160	140	7	30	4	0.5	1	TO-66	12A
AF High Voltage Applications												
SK3021	NPN Si	6	0.15	300	300	4	105	10	0.05	25	TO-66	12A
SK3045	NPN Si	10	1	300	300	6	80	10	0.05	30	TO-5	8
SK3053	PNP Si	10	—1	—200	—200	—4	90	—10	—0.05	50	TO-5	4A
SK3103	NPN Si	20	1	450	350	6	100	10	0.05	30	TO-5	10B
SK3104	NPN Si	20	1	300	250	6	75	10	0.02	30	TO-5	10B
SK3131	NPN Si	12.5	2	200	140	6	70	10	0.2	1	TO-66	12A
SK3528	PNP Si	10	—1	—350	—250	—6	50	—10	—0.05	50	TO-5	4A
SK3537	NPN Si	4	1	200	200	3	75	10	0.05	10	—	31
AM Broadcast Band Receiver Applications												
SK3005	PNP Ge	0.08	—0.01	—40	—40	—0.5	90	—9	—0.001	10	TO-1	1
SK3007	PNP Ge	0.08	—0.01	—24	—15	—0.5	120	—12	—0.001	130	TO-7	2B
SK3008	PNP Ge	0.08	—0.01	—34	—15	—0.5	150	—12	—0.001	45	TO-1	1
SK3011	NPN Ge	0.15	0.1	25	25	12	70	6	0.001	8	TO-1	1
SK3018	NPN Si	0.18	0.02	30	30	3	80	10	0.002	550	TO-72	6A
SK3118	PNP Si	0.5	—0.75	—50	—40	—5	100	—2	—0.002	200	TO-92	6C
SK3122	NPN Si	0.4	0.5	50	50	4	150	3	0.01	200	—	6F

* Terminal connections are indicated on the outline drawings at end of this section.

Table LIII—SK-Series Bipolar Transistors (cont'd)

RCA Type	Polarity and Material	P_T W	I_C A	V_{CBO} V	V_{CEO} V	V_{EBO} V	h_{FE}	V_{CE} V	I_C A	f_T MHz	Case Style	Dim. Outline *
		LIMIT CONDITIONS						**CHARACTERISTICS**			**PACKAGE**	
Short Wave Band Receiver Applications												
SK3007	PNP Ge	0.08	—0.01	—24	—15	—0.5	120	—12	—0.001	130	TO-7	2B
SK3018	NPN Si	0.18	0.02	30	30	3	80	10	0.002	550	TO-72	6A
SK3118	PNP Si	0.5	0.75	—50	—40	—5	100	—2	—0.002	200	TO-92	6C
SK3122	NPN Si	0.4	0.5	50	50	4	150	3	0.01	200	—	6F
FM Broadcast Band Receiver Applications												
SK3006	PNP Ge	0.08	—0.01	—30	—15	—0.5	100	—12	—0.001	400	TO-45	2A
SK3018	NPN Si	0.18	0.02	30	30	3	80	10	0.002	550	TO-72	6A
SK3118	PNP Si	0.5	—0.75	—50	—40	—5	100	—2	—0.002	200	TO-92	6C
SK3122	NPN Si	0.4	0.5	50	50	4	150	3	0.01	200	—	6F
TV VHF Tuner Applications												
SK3018	NPN Si	0.18	0.02	30	30	3	80	10	0.002	550	TO-72	6A
SK3039	NPN Si	0.15	0.02	20	20	3	60	10	0.002	1000	TO-72	6A
TV UHF Tuner Applications												
SK3019	NPN Si	0.15	0.02	20	20	3	130	10	0.002	1000	TO-72	6A
SK3039	NPN Si	0.15	0.02	20	20	3	60	10	0.002	1000	TO-72	6A
TV Sound IF Amplifier Applications												
SK3006	PNP Ge	0.08	—0.01	—30	—15	—0.5	100	—12	—0.001	400	TO-45	2A
SK3018	NPN Si	0.18	0.02	30	30	3	80	10	0.002	550	TO-72	6A
SK3118	PNP Si	0.5	—0.75	—50	—40	—5	100	—2	—0.002	200	TO-92	6C
TV Video IF Amplifier Applications												
SK3018	NPN Si	0.18	0.02	30	30	3	80	10	0.002	550	TO-72	6A
SK3117	NPN Si	0.18	0.02	20	20	3	60	10	0.002	550	TO-72	6A
SK3132	NPN Si	0.3	0.05	40	40	3	100	10	0.01	800	TO-72	6A
Citizens Band Power Applications												
SK3046	NPN Si	0.5	0.25	60	30	2	50	12	0.015	300	TO-5	4A
SK3047	NPN Si	2	0.25	60	30	2	50	12	0.05	300	TO-5	4A
SK3048	NPN Si	5	1.5	60	30	2.5	60	12	0.3	200	TO-5	4A
SK3049	NPN Si	10	1.5	60	30	2.5	60	12	0.3	200	TO-5	8
TV Horizontal Driver Deflection Applications												
SK3034	PNP Ge	32	—10	—200	—	—1.5	35	—1.5	—4	2.5	TO-3	11A
SK3104	NPN Si	20	1	300	250	6	75	10	0.02	30	TO-5	10B
TV Horizontal Output Deflection Applications												
SK3035	PNP Ge	32	—10	—220	—	—1.5	25	—1.5	—4	2.5	TO-3	11A
SK3104	NPN Si	20	1	300	250	6	75	10	0.02	30	TO-5	10B
SK3111	NPN Si	22	10	1200	500	6	25	2	1.5	10	TO-3	11A
SK3115	NPN Si	65	7	1500	1500	5	5	10	2.5	10	TO-3	11A
TV Vertical Output Deflection Applications												
SK3034	PNP Ge	32	—10	—200	—	—1.5	35	—1.5	—4	2.5	TO-3	11A
SK3054	NPN Si	50	7	90	70	5	70	4	1	2	TO-220	10A
SK3079	NPN Si	117	15	160	140	7	80	4	0.5	1	TO-3	11A
SK3083	PNP Si	40	—7	—80	—70	—5	70	—2	—1.5	20	TO-220	10A
SK3085	PNP Si	40	—6	—90	—80	—5	60	—2	—1.5	12	TO-66	12A
SK3131	NPN Si	12.5	2	200	140	6	70	10	0.2	1	TO-66	12A
SK3133	NPN Si	22	1	1500	550	5	70	10	0.1	1	TO-3	11A

* Terminal connections are indicated on the outline drawings at end of this section.

Table LIII—SK-Series Bipolar Transistors (cont'd)

RCA Type	Polarity and Material	LIMIT CONDITIONS P_T W	I_C A	V_{CBO} V	V_{CEO} V	V_{EBO} V	CHARACTERISTICS h_{FE}	V_{CE} V	I_C A	f_T MHz	PACKAGE Case Style	Dim. Outline *
TV High Voltage Regulator Applications												
SK3044	NPN Si	1	1	300	300	7	80	10	0.05	30	TO-5	4A
SK3045	NPN Si	10	1	300	300	6	80	10	0.05	30	TO-5	8
SK3104	NPN Si	20	1	300	250	6	75	10	0.02	30	TO-5	10B
TV Video Amplifier Applications												
SK3040	NPN Si	1	0.1	200	200	6	55	10	0.05	150	TO-5	4A
SK3045	NPN Si	10	1	300	300	6	80	10	0.05	30	TO-5	8
SK3103	NPN Si	20	1	450	350	6	60	10	0.05	30	TO-5	10B
TV Chroma Amplifier Applications												
SK3044	NPN Si	1	1	300	300	7	80	10	0.05	30	TO-5	4A
SK3104	NPN Si	20	1	300	250	6	75	10	0.02	30	TO-5	10B
TV AGC Amplifier Applications												
SK3044	NPN Si	1	1	300	300	7	80	10	0.05	30	TO-5	4A
High Voltage/Power Switching Applications												
SK3510	NPN Si	115	15	100	80	7	30	4	6	1.5	TO-3	11A
SK3511	NPN Si	150	30	90	80	5	40	4	8	1.5	TO-3	11A
SK3512	NPN Si	10	2	100	90	7	90	4	0.4	70	TO-5	4A
SK3513	PNP Si	10	−2	−100	−90	−7	90	−4	−0.4	70	TO-5	4A
SK3528	PNP Si	10	−1	−350	−250	−6	50	−10	−0.05	50	TO-5	4A
SK3529	NPN Si	5	3	75	50	5	50	1	0.5	350	TO-5	4A
SK3530	NPN Si	25	3	100	55	12	30	4	1	1	TO-8	9D
SK3534	NPN Si	75	10	80	70	8	55	4	1	2	TO-220	10A
SK3535	NPN Si	250	35	170	150	7	40	2	4	2	TO-3	11A
SK3536	NPN Si	4.4	3	160	140	7	100	4	0.15	1	—	31
SK3537	NPN Si	4	1	200	200	3	75	10	0.05	10	—	31
SK3538	NPN Si	40	4	160	140	7	30	4	0.5	1	TO-66	12A

Table LIV—SK-Series Field Effect Transistors, N-Channel Depletion Mode

RCA Type	Device Class	LIMIT CONDITIONS P_T W	I_D or (I_G) mA	V_{DS} V	V_{G1S} V	V_{G2S} V	CHARACTERISTICS g_{fs} μmho	I_{DS} mA	PACKAGE Case Style	Dim. Outline *
AF Small-Signal Applications										
SK3112	Junction	0.1	(10)	−50	−50	—	2,000	1.5	—	6D
FM Broadcast Band Receiver Applications										
SK3116	Junction	0.2	(10)	−18	−18	—	7,500	8	TO-72	5
TV VHF Tuner Applications										
SK3050	Dual Gate	0.33	50	−0.2 to +20	−6 to +6	−6 to +6	12,000		TO-72	6B

* Terminal connections are indicated on the outline drawings at end of this section.

Table LIV—SK-Series Field Effect Transistors, N-Channel Depletion Mode (cont'd)

RCA Type	Device Class	P_T W	I_D or (I_G) mA	V_{DS} V	V_{G1S} V	V_{G2S} V	g_{fs} μmho	I_{DS} mA	Case Style	Dim. Outline *
				LIMIT CONDITIONS			CHARACTERISTICS		PACKAGE	
VHF Communications Receiver Applications										
SK3065	Dual Gate	0.33	50	—0.2 to +20	—6 to +6	—6 to +6	13,000	15	TO-72	6B
Chopper/Multiplex Applications										
SK3531	Single Gate	0.33	50	35	—10 to +10	—	6,000	15	TO-72	5

Table LV—SK-Series Silicon Controlled Rectifiers

RCA Type	V_{DROM} V	RMS A	SURGE A	di/dt A/μs	P_G W	Max. V_{GT} V	Max. I_{GT} mA	Max. I_{HO} mA	dv/dt V/μs	I_{DROM} μA	Max. t_{off} μs	Case Style	Dim. Outline *
		FORWARD CURRENT											
TV Horizontal Deflection Applications													
SK3042	600	5	80	200	25	4	30	—	—	500	2.5	TO-66	12C
Power Control/Switching Applications													
SK3502	600	5	60	200	13	2	15	20	10	500	50	TO-66	12C
SK3503	600	7.5	100	200	40	1.5	15	20	200	500	50	TO-5	3
SK3504	600	20	200	200	40	2	15	20	75	200	40	TO-48	20A
SK3505	600	35	300	200	40	2	40	70	100	350	40	TO-48	20A
SK3527	600	12.5	200	200	40	2	40	50	100	350	50	TO-3	11C
SK3557	500	4	35	100	0.5	0.8	0.002	3	8	10	100	TO-220	10D
SK3558	400	8	200	100	16	1.5	15	20	200	2000	35	TO-220	10D

Table LVI—SK-Series Triacs

RCA Type	V_{DROM} V	RMS A	SURGE A	di/dt A/μs	P_G W	Max. V_{GT} V	Max. I_{GT} mA	Max. I_{HO} mA	dv/dt V/μs	I_{DROM} μA	Max. t_{off} μs	Case Style	Dim. Outline *
		FORWARD CURRENT											
Power Control/Switching Applications, 240-Volt Line Operation													
SK3506	400	2.5	25	—	10	2.2	10	15	10	750	—	TO-5	4B
SK3507	400	15	100	—	16	2.5	80	60	10	200	2.5	TO-66	12B
SK3508	400	15	100	150	16	2.5	80	75	10	200	2.5	TO-48	20B

* Terminal connections are indicated on the outline drawings at end of this section.

Table LVI—SK-Series Triacs (cont'd)

RCA Type	V$_{DROM}$ V	RMS A	SURGE A	di/dt A/μs	P$_G$ W	Max. V$_{GT}$ V	Max. I$_{GT}$ mA	Max. I$_{HO}$ μA	dv/dt V/μs	I$_{DROM}$ μA	Max. t$_{off}$ μs	Case Style	Dim. Outline *
	LIMIT CONDITIONS						**CHARACTERISTICS**					**PACKAGE**	
	FORWARD CURRENT												
Power Control/Switching Applications, 240-Volt Line Operation													
SK3509	400	40	300	100	40	2.5	80	60	30	200	3	TO-48	20B+
SK3532	450	6	100	—	16	4	50	—	10	200	—	TO-5	4C
SK3533	400	8	100	—	16	2.5	60	30	10	100	2.5	TO-220	10C
High-Voltage Power Control/Switching Applications, V$_{DROM}$ = 600 V													
SK3519	600	2.5	25	—	10	2.2	25	35	10	200	—	TO-5	4B
SK3520	600	15	100	150	16	2.5	80	75	10	200	2.5	TO-48	20B
SK3521	600	30	300	100	40	2.5	80	60	20	200	3	TO-48	20B
SK3522	600	40	300	100	40	2.5	80	60	30	200	3	TO-48	20B

Table LVII—SK-Series Low-Voltage Rectifiers/Diodes

RCA Type	Material	V$_{RRM}$ V	V$_{RSM}$ V	RMS I$_F$ A	SURGE I$_{FM}$ A	Max. I$_R$ μA	Max. V$_F$ V	Typ. t$_\pi$ μs	Case Style	Dim. Outline *
		LIMIT CONDITIONS					**CHARACTERISTICS**		**PACKAGE**	
		PEAK REVERSE VOLTAGE		**FORWARD CURRENT**						
Single Unit Power Supply Applications										
SK3016	Si	500	—	1	35	5	1.2	—	DO-1	15A
SK3017A	Si	600	—	1	50	5	1.2	—	DO-26	15A
SK3030	Si	200	—	1	50	5	1.2	—	DO-26	15A
SK3031	Si	400	—	1	50	5	1.2	—	DO-26	15A
SK3032	Si	800	—	1	50	5	1.2	—	DO-26	1A
SK3033	Si	1000	—	1	35	5	1.2	—	DO-1	14
SK3051	Si	1000	—	3	200	100	1	—	—	15B
SK3080	Si	1000	1200	1	30	10	1.1	1.5	DO-15	15D
SK3081	Si	1000	1200	2	50	10	1.4	1.5	DO-15	15D
SK3174	Si	400	525	0.25	30	10	1	30	TO-1	1B
SK3500	Si	600	700	12	240	5	0.55	—	DO-4	18A
SK3501	Si	600	700	40	850	5	0.65	—	DO-5	19
SK3517	Si	Reverse Polarity Version of SK3500, for Data See SK3500							DO-4	18C
SK3518	Si	Reverse Polarity Version of SK3501, For Data See SK3501							DO-5	19A
Full-Wave Bridge Power Applications										
SK3105	Si	100	—	2	75	—	—	—	—	27
SK3106	Si	400	—	2	75	—	—	—	—	27
SK3107	Si	1000	—	2	75	—	—	—	—	27

* Terminal connections are indicated on the outline drawings at end of this section.

Table LVII—SK-Series Low-Voltage Rectifiers/Diodes
(cont'd)

RCA Type	Material	LIMIT CONDITIONS				CHARACTERISTICS			PACKAGE	
		PEAK REVERSE VOLTAGE		FORWARD CURRENT						Dim. Outline *
		Repetitive PRV V	Non-Repetitive V_{RM} V	RMS I_F A	SURGE I_{FM} A	Max. I_R μA	Max. V_F V	Typ. t_{rr} μs	Case Style	
TV Horizontal Deflection Applications: Trace and Commutating Diodes										
SK3043	Si	850	900	1	70	10	1.1	1	DO-26	15A
SK3127	Si	850	950	1	70	10	1.1	1	DO-1	18D
SK3128	Si	850	950	1	70	10	1.1	1	DO-1	14A
TV Switching Applications: Sync, Detector, and Clamp Diode										
SK3100	Si	80	—	0.2	—	0.2	1.1	0.004	DO-7	16
SK3175	Si	800	1000	1.5	50	15	1.9	0.5	DO-15	15D
TV Damper Diode Applications										
SK3113	Ge	—320	—	10	25	—	—	—	TO-3	11B
SK3125	Si	1500	—	1.5	100	1υ	1.2	1	DO-1	14A
SK3130	Si	1200	—	0.5	35	10	1.3	0.5	DO-1	14
Convergence Rectifier Applications Diode Array										
SK3110	Se	35	—	0.065	—	300	—	—	—	28
TV Dural Diode Discriminator/Phase Detector Applications Common Cathode Connected										
SK3119	Se	20	—	0.0011	—	4	—	--	—	29
TV Dual Diode Discriminator/Phase Detector Applications Series Connected										
SK3120	Se	20	—	0.0011	—	4	—	—	—	29
TV Dual Diode Discriminator/Phase Detector Applications Common Anode Connected										
SK3121	Se	20	—	0.0011	—	4	—	—	—	29
AF Rectifier/RF Detector Signal Diode Applications										
SK3087	Ge	60	—	0.05	—	—	—	—	DO-7	16
SK3090	Ge	80	—	0.09	—	—	—	—	DO-7	16
TV Video Detector Applications										
SK3088	Ge	25	—	0.05	—	—	—	—	DO-7	16
SK3091	Ge	40	—	0.35	—	—	—	—	DO-7	16
TV UHF Mixer Diode Applications										
SK3089	Si	5	—	—	—	—	—	—	DO-7	16
Fast Recovery Rectifiers										
SK3175	Si	800	1000	1.5	35	15	1.9	0.5	DO-26	15A
SK3515	Si	800	1000	4.5	75	10	1.9	0.5	DO-4	18B
SK3516	Si	800	1000	1.5	50	15	1.9	0.5	DO-15	15D

* Terminal connections are indicated on the outline drawings at end of this section.

Table LVIII—SK-Series High-Voltage Dectifiers

		LIMIT CONDITIONS		OPERATING CONDITIONS					PACKAGE	
		PEAK REVERSE VOLTAGE								
RCA Type	Material	DC (No Load) V	Pulse V	DC Output Voltage V	Input Pulse V	I_F mA	I_R μA	T_A °C	Case Style	Dim. Out-line *
TV Focus Rectifier Applications										
SK3066	Se	8,000	9,000	6,500	7,800	2	30	50	—	17D
TV Picture Tube Anode Rectifier Applications										
SK3067	Se	12,400	14,500	11,000	13,000	0.3	20	50	—	17A
SK3068	Se	14,400	17,300	13,000	15,600	0.3	20	50	—	17A
SK3108	Se	21,400	25,400	20,000	24,000	0.3	20	50	—	17B
TV Voltage Tripler Module										
SK3069	Se	27,500	9,400	25,000	8,600	1	—	60	—	21
TV Boost Rectifier/Grid Clamp Diode Applications										
SK3109	Se	560	827	—	550	2	250	85	—	17C

Table LIX—SK-Series Zener Diodes

		LIMIT CONDITIONS			CHARACTERISTICS			PACKAGE	
RCA Type	Material	P_D W	V_Z V	I_Z mA	Z_Z Ω	f Hz	I_Z mA	Case Style	Dim. Out-line *
Voltage Regulator Applications									
SK3055	Si	1	3.6	69	10	60	6.9	DO-27	15C
SK3056	Si	1	5.1	49	7	60	4.9	DO-27	15C
SK3057	Si	1	5.6	45	5	60	4.5	DO-27	15C
SK3058	Si	1	6.2	41	2	60	4.1	DO-27	15C
SK3059	Si	1	7.5	34	6	60	3.4	DO-27	15C
SK3060	Si	1	9.1	28	6	60	2.8	DO-27	15C
SK3061	Si	1	10	30	7	60	30	DO-7	16
SK3062	Si	1	12	20	9	60	20	DO-7	16
SK3063	Si	1	15	20	14	60	20	DO-7	16
SK3064	Si	1	27	9.5	35	60	0.95	DO-27	15C
SK3092	Si	1	11.5	21	9	60	2.1	DO-27	15C
SK3093	Si	1	12.8	19	11	60	1.9	DO-27	15C
SK3094	Si	1	14	18	12	60	1.8	DO-27	15C
SK3095	Si	1	33	7.5	45	60	0.75	DO-27	15C
SK3096	Si	1	55	4.5	110	60	0.45	DO-27	15C

* Terminal connections are indicated on the outline drawings at end of this section.

Table LIX—SK-Series Zener Diodes (cont'd)

RCA Type	Material	LIMIT CONDITIONS			CHARACTERISTICS			PACKAGE	
		P_D W	V_Z V	I_Z mA	Z_Z Ω	f Hz	I_Z mA	Case Style	Dim. Outline *
Voltage Regulator Applications									
SK3097	Si	1	62	4	125	60	0.4	DO-27	15C
SK3098	Si	1	82	3	200	60	0.3	DO-27	15C
SK3099	Si	1	110	2.3	450	60	0.23	DO-27	15C
SK3136	Si	1	8.2	31	4.5	60	3.1	DO-27	15C
SK3139	Si	1	11	20	8	60	20	DO-7	16
SK3142	Si	1	16.2	20	16	60	20	DO-7	16
SK3145	Si	1	17.7	20	20	60	20	DO-7	16
SK3148	Si	1	26	10	28	60	10	DO-7	16
SK3150	Si	1	29	10	33	60	10	DO-7	16
SK3151	Si	1	24	10	25	60	10	DO-7	16

Table LX—SK-Series Variable-Capacitance Diode

RCA Type	Material	LIMIT CONDITIONS				CHARACTERISTICS		Test Conditions		PACKAGE	
		PEAK REVERSE VOLTAGE									
		V_{RRM} V	V_{RSM} V	I_F mA	I_R μA	C pF	pF/V	V_R V	f MHz	Case Style	Dim. Outline *
FM Receiver AFC Applications											
SK3126	Si	—14	—20	50	5	22	2.6	—10	1	DO-7	16

Table LXI—SK-Series Bidirectional Diode

RCA Type	Material	LIMIT CONDITIONS		CHARACTERISTICS					PACKAGE	
		P_D W	Ipeak (pulsed) A	V_{BO} V Min.	V Max.	$\|+V_{BO}\|-\|-V_{BO}\|$ V	I_{BO} μA	Peak Output Current mA	Case Style	Dim. Outline *
Thyristor Triggering Diode										
SK3523	Si	1	2	29	35	± 3	25	190	DO-15	15D

* Terminal connections are indicated on the outline drawings at end of this section.

Table LXII—SK-Series Monolithic Linear Integrated-Circuit Subsystems

RCA Type	Applications	PACKAGE	
		Case Style	Dim. Outline *
SK3022	Sound IF-Amplifier Stages for TV-Receivers. Supply Voltage, $V_{CC} = 7.5$ volts.	TO-5	9A
SK3023	Sound IF-Amplifier Stages for TV-Receivers. Supply Voltage, $V_{CC} = 7.5$ volts.	TO-5	9B
SK3070	Automatic Frequency Control, AFC Stages for TV-Receivers. Typical Input Voltage Sensitivity, $V_{IN} = 75mV$. Supply Voltage, $V_{CC} = 30$ volts	TO-5	9A
SK3071	Full Function Stereo Preamplifier Stages for Hi-Fi Equipment Supply Voltage, $V_{CC} = 14$ volts.	MO-001-AC	22
SK3072	Sound IF-Amplifier Stages for TV-Receivers. Supply Voltage, $V_{CC} = 140$ volts.	MO-001-AB	23
SK3073	Chroma Signal Processor Stages for Color TV-Receivers. Supply Voltage, $V_{CC} = 12$ volts.	MO-001-AC	22
SK3074	Chroma Signal Processor Stages for Color TV-Receivers. Supply Voltage, $V_{CC} = 12$ volts.	MO-001-AC	22
SK3075	Chroma Signal Processor Stages for Color TV-Receivers. Supply Voltage, $V_{CC} = 24$ volts.	MO-001-AC	22
SK3076	Chroma Amplifier Stages for Color TV-Receivers. Supply Voltage, $V_{CC} = 24$ volts.	MO-001-AB	23
SK3077	Chroma Demodulator Stages for Color TV-Receivers. Supply Voltage, $V_{CC} = 24$ volts.	MO-001-AB	23
SK3078	Stereo Decoder System in FM-Multiplex Stages for Hi-Fi-Equipment. Supply Voltage, $V_{CC} = 12$ volts.	MO-001-AC	22
SK3101	Sound IF Amplifier, Limiter FM-Detection, AF-Preamplifier, and AF-Driver Stages for TV Receivers using Tube Type AF-Output Amplifiers. Supply Voltage, $V_{CC} = 140$ volts.	MO-001-AB	23
SK3102	Sound IF-Amplifier, Limiter, FM-Detection, AF-Preamplifier, and AF-Driver Stages for TV Receivers using Transistor Type AF-Output Amplifiers. Supply Voltage, $V_{CC} = 140$ volts.	MO-001-AB	23
SK3129	IF-Amplifier Stages for FM-Brodcast Band Receivers. Supply Voltage, $V_{CC} = 9$ volts.	TO-5	9A
SK3134	Chroma Demodulator Stages for Color TV-Receivers. Supply Voltage, $V_{CC} = 28$ volts.		30
SK3135	Sound IF Amplifier-Limiter and Quadrature Detector for TV and FM Broadcast Band Receivers. Supply Voltage, $V_{CC} = 12$ volts.	MO-001-AB	23
SK3140	IF Amplifier, Limiter, Detector, AF Preamplifier and Driver Stages in FM Broadcast Band and Communications Receivers. Supply Voltage, $V_{CC} = 30$ volts.	TO-5	9E
SK3141	Automatic Frequency Control AFC Stages in TV Receivers. Typical Input Voltage Sensitivity, $V_{IN} = 18mV$. Supply Voltage, $V_{CC} = 140$ volts.	TO-5	9B
SK3143	Video IF Amplifier System for Color and Black/White TV Receivers. Supply Voltage, $V_{CC} = 11$ volts.	—	26
SK3144	IF Amplifier, Limiter, Detector, and AF Preamplifier Stages in FM Broadcast Band and Communications Receivers. Supply Voltage, $V_{CC} = 11.2$ volts.	MO-001-AB	23

* Terminal connections are indicated on the outline drawings at end of this section.

Table LXII—SK-Series Monolithic Linear Integrated-Circuit Subsystems (cont'd)

RCA Type	Applications	Case Style (PACKAGE)	Dim. Outline * (PACKAGE)
SK3146	AM Broadcast Band System. Includes RF Converter, IF Amplifier, Detector, and Audio Preamplifier Stages. Supply Voltage, V_{cc} = 9 volts.	MO-001-AC	22
SK3147	FM Broadcast Band System. Includes IF Amplifier, Quadrature Detector, and AF Preamplifier Stages. Supply Voltage, V_{cc} = 12 volts.	MO-001-AC	22
SK3149	Chroma Amplifier and Demodulator System Stages for Color TV Receivers. Supply Voltage, V_{cc} = 24 volts.	MO-001-AC	22
SK3157	RF/IF Amplifier for FM and Communications Receivers. Supply Voltage, V_{cc} = 12 volts.	TO-5	9F
SK3158	Chroma Signal Processor Stages for Color TV-Receivers. Supply Voltage, V_{cc} = 24 volts.	MO-001-AC	22
SK3159	Stereo Decoder System in FM-Multiplex Stages for Hi-Fi-Equipment. Supply Voltage, V_{cc} = 15 volts.	MO-001-AB	23
SK3160	Stereo Decoder (RC Phase-Lock-Loop) System in FM-Multiplex Stages for Hi-Fi-Equipment. Supply Voltage, V_{cc} = 14 volts.	MO-001-AB	23
SK3161	Stereo Decoder System in FM-Multiplex Stages for Hi-Fi-Equipment. Supply Voltage, V_{cc} = 15 volts.	MO-001-AB	23
SK3162	Full Function Stereo Preamplifier Stages for Hi-Fi-Equipment. Supply Voltage, V_{cc} = 33 volts.	MO-001-AB	23
SK3163	Three Stage IF- Amplifier System for FM Receivers. Supply Voltage, V_{cc} = 18 volts.	—	23A
SK3164	Voltage Regulator Stages for Color and Black/White TV Receivers. Supply Voltage, V_{cc} = 40 volts.	TO-5	24
SK3165	Voltage Regulator Stages for Color and Black/White TV Receivers. Supply Voltage, V_{cc} = 40 volts.	MO-001-AB	23
SK3166	Full Function Stereo Preamplifier Stages for Hi-Fi Equipment Supply Voltage V_{cc} = 18 volts.	TO-5	24
SK3167	Chroma Signal Processor Stages for Color TV-Receivers Supply Voltage, V_{cc} = 20 volts.	MO-001-AB	23
SK3168	Video IF Amplifier System for Color and Black/White TV Receivers. Supply Voltage, V_{cc} = 12 volts.	MO-001-AB	23
SK3169	IF Amplifier and Limiter for FM Recivers. Supply Voltage, V_{cc} = 15 volts.	TO-5	24
SK3170	Chroma Signal Processor Stages for Color TV-Receivers. Supply Voltage, V_{cc} = 24 Volts.	MO-001-AC	22
SK3171	AM Broadcast Band System. Includes RF/IF Amplifier, Mixer, Oscillator, AGC Detector, and Voltage Regulator. Supply Voltage, V_{cc} = 12 volts.	MO-001-AB	23
SK3172	Stereo Decoder System in FM-Multiplex Stages for Hi-Fi Equipment. Supply Voltage, V_{cc} = 16 volts.	MO-001-AC	22

* Terminal connections are indicated on the outline drawings at end of this section.

Table LXIII—SK-Series Monolithic Linear Integrated-Circuit Subsystems (cont'd)

RCA Type	Description	OPERATING CONDITIONS								PACKAGE	
		Max. P_D mW	V_{CC} V	Max. V_{IN} mV	Max. Input Off-Set mV	nA	Max. Input Bias nA	Volt. Gain dB	MUF MHz	Case Style	Dim. Outline *
SK3514	Single Operational Amplifier	500	36	—	6	200	500	95	—	TO-5	9C
SK3524	AF Power/Wide Band Amplifier	1000	12	100	Power Gain = 75dB Typ.				8	TO-5	9E
SK3525	VHF Differential Cascade Amplifier	450	12	400	5	.006	80 mA	40	100	TO-5	9C
SK3526	Dual Operational Amplifier	800	36	—	6	200	500	95	—	TO-5	9A
SK3539	Single Operational Amplifier	600	12	1000	2	1.6 mA	6 mA	70	—	MO-001-AD	25
SK3540	Single Operational Amplifier	600	12	1000	2	1.6 mA	6 mA	70	—	TO-5	9E
SK3541	Zero Voltage Switch	700	14	—	—	—	1000	—	—	MO-001-AD	25
SK3542	Transistor Array	450	*Unit Ratings:* V_{CBO} = 30V, V_{CEO} = 15V, IC = 50 mA						120	TO-5	9E
SK3543	Transistor Array	750	*Unit Ratings:* V_{CBO} = 20V, V_{CEO} = 15V, IC = 50 mA						120	MO-001-AD	25
SK3544	Dual Independent Differential Ampl.	750	*Unit Ratings:* V_{CBO} = 20V, V_{CEO} = 15V, IC = 50 mA						120	MO-001-AB	23
SK3545	Diode Array	600	*Unit Ratings:* PRV = 20V, IF = 25mA, IF (surge) = 100 mA							TO-5	9E
SK3546	Diode Array	120	*Unit Ratings:* PRV = 25V, IF = 1mA, IR = 10 µA, CD = 1.8 pF							TO-5	9A
SK3547	DC Differential Amplifier	450	6	± 4	5	10 µA	36 mA	38	30	TO-5	9A
SK3548	VHF Dual Differential Amplifier	600	*Unit Ratings:* V_{CBO} = 20V, V_{CEO} = 15V, I_C = 50 mA, f_T = 1.35 GHz typ.						500	TO-5	9E
SK3549	Video/Wide Band Amplifier	450	6	± 4	1.5	10	36 mA	14	50	TO-5	9A
SK3550	Transistor Array	750	*Unit Ratings:* V_{CBO} = 20V, V_{CEO} = 16V, I_C = 100 mA, hfe = 68 typ.							MO-001-AC	22
SK3551	Dual Operational Amplifier	680	36	—	6	200	800	100	—	TO-5	9G
SK3552	Single Operational Amplifier	500	44	—	5	200	500	100	—	MO-001-AB	23
SK3553	Single Operational Amplifier	500	44	—	5	200	500	100	—	TO-5	9B
SK3554	Memory Sense Amplifier	750	10	—	6	2µA	25 µA	—	—	MO-001-AD	25
SK3555	Dual Operational Amplifier	680	36	—	6	200	500	100	—	TO-5	9C
SK3556	Dual Operational Amplifier	300	36	—	5	200	500	100	—	MO-001-AB	23

* Terminal connections are indicated on the outline drawings at end of this section.

Table LXIV—Power Hybrid Integrated Circuits

RCA Type	Material	LIMIT CONDITIONS						OPERATING CHARACTERISTICS			PACKAGE
		V_{CC} V	I_{CC} A	V_{CC} V	I_{CC} A	mA	A dB	P_{OUT} at $R_L=4\Omega$ W	Sensitivity at Rated Output V	R_{IN} Ω	Dim. Outline •
AF Power Amplifier Applications											
SK3152	Si	+ 35	0.9	+ 25	0.5	20	35	10	0.15	40,000	60
SK3153	Si	+ 43	1.1	+ 32	0.6	20	33	15	0.25	40,000	60
SK3154	Si	+ 50	1.3	+ 38	0.7	25	33	20	0.25	40,000	61
SK3155	Si	± 22	1.1	± 16	0.7	30	33	20	0.25	27,000	62

Table LXV—SK-Series COS/MOS Digital Integrated Circuits

RCA Type	LIMITS		TYPICAL CHARACTERISTICS										PACKAGE	
			STATIC					DYNAMIC						
				Output Voltage				Prop. Delay		Transition Time				
	P_D μW	V_{DD}-V_{SS} V	DC Input Current I_1 pA	"0" V	"1" I	Noise Immunity V_{NL} V	Input Cap. C_1 pF	"0" ns	"1" ns	"0" ns	"1" ns	Case Style JEDEC MO-001-	Dim. Outline •
Gates Dual 3-Input NOR Plus Inverter													
SK4000	0.01	3 to 15	10	0	10	4.5	5	25	25	35	35	AD	25
Dual 4-Input NOR													
SK4002	0.01	3 to 15	10	0	10	4.5	5	25	25	35	35	AD	25
Triple 3-Input NOR													
SK4025	0.01	3 to 15	10	0	10	4.5	5	25	25	35	35	AD	25
Quad 2-Input NOR													
SK4001	0.01	3 to 15	10	0	10	4.5	5	25	25	35	35	AD	25
Dual Input NAND													
SK4012	0.01	3 to 15	10	0	10	4.5	5	50	25	125	40	AD	25
Triple 3-Input NAND													
SK4023	0.01	3 to 15	10	0	10	4.5	5	25	25	50	40	AD	25
Quad 2-Input NAND													
SK4011	0.01	3 to 15	10	0	10	4.5	5	25	25	50	40	AD	25
Quad Exclusive OR													
SK4030	0.1	3 to 15	10	0	10	4.5	5	100	100	25	30	AD	25
Quad AND-OR Select													
SK4019	0.5	3 to 15	10	0	10	4.5	12	50	50	40	40	AE	24

Table LXV—SK-Series COS/MOS Digital Integrated Circuits (con

RCA Type	P_D μW	V_{DD}-V_{SS} V	DC Input Current I_I pA	Output Voltage "0" V	"1"	Noise Immunity V_{NL} V	Input Cap. C_I pF	Prop. Delay "0" ns	"1" ns	Transition Time "0" ns	"1" ns	Case Style JEDEC MO-001-	Dim. Out-line *
Dual Complementary Pair Plus Inverter													
SK4007	0.01	3 to 15	10	0	10	4.5	5	20	20	30	30	AD	25
Flip-Flops **Dual "D", Typical Clock Toggle Rate = 10 MHz**													
SK4013	0.05	3 to 15	10	0	10	4.5	5	75	75	50	50	AD	25
Dual J-K Master-Slave, Typical Clock Toggle Rate = 8 MHz													
SK4027	0.05	3 to 15	10	0	10	4.5	5	75	75	50	50	AE	24
HEX Buffers/Converters **Inverting**													
SK4009	0.1	3 to 15	10	0	10	4.5	5	10	25	16	50	AE	24
Non Inverting													
SK4010	0.1	3 to 15	10	0	10	4.5	5	10	25	16	50	AE	24
Multiplexer **Quad Bilateral Switch, Typical Frequency Response = 40 MHz**													
SK4016	0.1	3 to 15	± 10	—	—	—	4	Propogation Delay = 20 ns				AD	25
Counters **Decade Counter/Divider, plus 10 decoded decimal outputs**													
SK4017	5	3 to 15	10	0	10	4.5	5	125	200	50	125	AE	24
7-Stage Binary Counter, with buffered reset													
SK4024	5	3 to 15	10	0	10	4.5	5	80	80	80	80	AD	25
Register **Dual 4-Stage Static Shift Register, with serial input/parallel output**													
SK4015	10	3 to 15	10	0	10	4.5	5	100		75	75	AE	24

* Terminal connections are indicated on the outline drawings at end of this section.

PACKAGE OUTLINES

This section shows the dimensional outlines for all RCA solid-state de- vice packages. Separate outlines are shown for standard-product RCA solid-state devices, and for RCA "top-of-the-line" SK-series solid-state replacement devices.

Dimensional Outlines for Thyristors and Rectifiers

DO-1

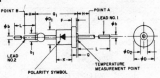

92CS-17423RI

SYMBOL	INCHES		MILLIMETERS		NOTES
	MIN.	MAX.	MIN.	MAX.	
ϕb	0.027	0.035	0.69	0.89	2
b_1	–	0.125	–	3.18	1
ϕD	0.360	0.400	9.14	10.16	
ϕD_1	0.245	0.280	6.22	7.11	
ϕD_2	–	0.200	–	5.08	
F	–	0.075	–	1.91	
G_1	–	0.725	–	18.42	
H	0.5	–	12.7	–	
K	0.220	0.260	5.59	6.60	
L	1.000	1.625	25.40	41.28	
Q	–	0.025	–	0.64	

NOTES:
1. Dimensions to allow for pinch or seal deformation anywhere along tubulation (optional).

2. Diameter to be controlled from free end of lead to within 0.188 inch (4.78 mm) from the point of attachment to the body. Within the 0.188 inch (4.78 mm) dimension, the diameter may vary to allow for lead finishes and irregularities.

DO-4

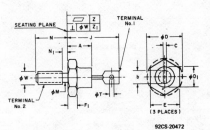

92CS-20472

SYMBOL	INCHES		MILLIMETERS		NOTES
	MIN.	MAX.	MIN.	MAX.	
A	–	0.405	–	10.28	
b	–	0.250	–	6.35	2
c	0.020	0.065	0.51	1.65	
ϕD	–	0.505	–	12.82	
ϕD_1	0.265	0.424	6.74	10.76	
E	0.423	0.438	10.75	11.12	
F_1	0.075	0.175	1.91	4.44	1
J	0.600	0.800	15.24	20.32	
ϕM	0.163	0.189	4.15	4.80	
N	0.422	0.453	10.72	11.50	
N_1	–	0.078	–	1.98	
ϕT	0.060	0.095	1.53	2.41	
ϕW	10-32 UNF-2A		10-32 UNF-2A		3
Z	–	0.002	–	0.050	
Z_1	–	0.006	–	0.152	

NOTES:
1: Chamfer or undercut on one or both sides of hexagonal base is optional.
2: Angular orientation and contour of Terminal No. 1 is optional.

3: ϕW is pitch diameter of coated threads. REF: Screw Thread Standards for Federal Services, Handbook H 28 Part I. Recommended torque: 15 inch-pounds.

DO-5

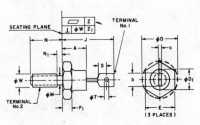

92CS-20473RI

SYMBOL	INCHES		MILLIMETERS		NOTES
	MIN.	MAX.	MIN.	MAX.	
A	–	0.450	–	11.43	
b	–	0.375	–	9.52	
c	0.030	0.080	0.77	2.03	
ϕD	–	0.794	–	20.16	
ϕD_1	–	0.667	–	16.94	
E	0.669	0.688	17.00	17.47	
F_1	0.115	0.200	2.93	5.08	
J	0.750	1.000	19.05	25.40	
ϕM	0.220	0.249	5.59	6.32	
N	0.422	0.453	10.72	11.50	
N_1	–	0.090	–	2.28	
S	0.156	–	3.97	–	
ϕT	0.140	0.175	.3.56	4.44	
ϕW	1/4-28 UNF 2A		1/4-28 UNF 2A		1
Z	–	0.002	–	0.050	
Z_1	–	0.006	–	0.152	

NOTE
1: ϕW is pitch diameter of coated threads. REF: Screw-Thread Standards for Federal Services, Handbook H 28 Part I. Recommended torque: 30 inch-pounds.

DO-15

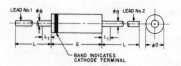

BAND INDICATES
CATHODE TERMINAL

92CS-17313RI

SYMBOL	INCHES		MILLIMETERS	
	MIN.	MAX.	MIN.	MAX.
ϕb	0.027	0.035	0.686	0.889
ϕD	0.104	0.140	2.64	3.56
G	0.230	0.300	5.84	7.62
L	1.000	–	25.40	–
L_1*	–	0.050	–	1.27

*Within this zone the diameter may vary to allow for lead finishes and irregularities.

DO-26

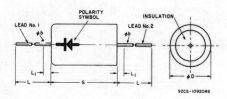

92CS-10920R6

SYMBOL	INCHES		MILLIMETERS		NOTES
	MIN.	MAX.	MIN.	MAX.	
ϕb	0.027	0.039	0.69	0.99	
ϕD	0.220	0.260	5.59	6.60	1
G	0.344	0.410	8.74	10.41	1
L	1.400	–	35.56	–	
L_1	–	0.080	–	2.03	2

NOTES:
1. Package contour optional within cylinder of diameter, ϕD, and length, G. Slugs, if any, shall be included within this cylinder but shall not be subject to the minimum limit of ϕD.
2. Lead diameter not controlled in this zone to allow for flash, lead-finish build up, and minor irregularities other than slugs.

DO-26 With Insulating Sleeve

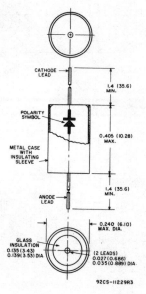

92CS-11229R3

Mod. TO-1 2-Lead

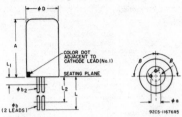

92CS-11676R5

SYMBOL	INCHES		MILLIMETERS		NOTES
	MIN.	MAX.	MIN.	MAX.	
ϕa	0.061	0.081	1.55	2.06	
A	–	0.410	–	10.41	
ϕb	–	0.021	–	0.533	1
ϕb2	0.016	0.019	0.406	0.483	1
ϕD	–	0.240	–	6.10	
L	1.500	–	38.10	–	1
L1	–	0.05	–	1.27	
L2	0.25	–	6.35	–	1
β	90° NOMINAL				

NOTE:

1. ϕb_2 applies between L_1 and L_2. ϕb applies between L_2 and

1.5 in. (38.10 mm) from seating plane. Diameter is uncontrolled in L_1 and beyond 1.5 in. (38.10 mm) from seating plane.

TO-3

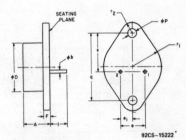

92CS-15222

SYMBOL	INCHES		MILLIMETERS		NOTES
	MIN.	MAX.	MIN.	MAX.	
A	0.250	0.450	6.35	11.43	
ϕb	0.038	0.043	0.97	1.09	2
ϕD		0.875		22.23	
e	0.420	0.440	10.67	11.18	
e1	0.205	0.225	5.21	5.72	
F		0.135		3.43	
1	0.312		7.92		2
ϕP	0.151	0.161	3.84	4.09	
q	1.177	1.197	29.90	30.40	
r1		0.525		13.34	
r2		0.188		4.78	
s	0.655	0.675	16.64	17.15	1

NOTES:

1. These dimensions should be measured at points 0.050 in. (1.27 mm) to 0.055 in. (1.40 mm) below seating plane.

When gage is not used, measurement will be made at seating plane.

2. Two pins.

MODIFIED TO-3 (2N5575, 2N5578)

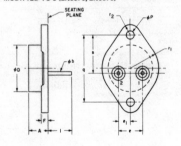

SYMBOL	INCHES		MILLIMETERS		NOTES
	MIN.	MAX.	MIN.	,MAX.	
A	0.300	0.350	7.62	8.89	
ϕb	0.059	0.061	1.50	1.55	2
ϕD		0.800		20.32	
e	0.420	0.440	10.67	11.18	
e1	0.205	0.225	5.21	5.72	
F		0.114		2.90	
I	0.440	0.470	11.18	11.94	2
ϕp	0.151	0.161	3.84	4.09	
q	1.177	1.197	29.90	30.40	
r1		0.525		13.34	
r2		0.188		4.78	
s	0.655	0.675	16.64	17.15	1

NOTES:

1. THESE DIMENSIONS SHOULD BE MEASURED AT POINTS 0.050" (1.27 mm) TO 0.055" (1.40 mm) BELOW SEATING PLANE. WHEN GAGE IS NOT USED, MEASUREMENT WILL BE MADE AT SEATING PLANE.

2. TWO LEADS.

92CS-17432

JEDEC TO-8

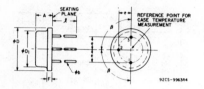

92CS-9963R4

SYMBOL	INCHES		MILLIMETERS		NOTES
	MIN.	MAX.	MIN.	MAX.	
A	0.270	0.330	6.86	8.38	--
ϕb	0.027	0.033	0.686	0.838	1
ϕD	0.550	0.650	13.97	16.51	--
ϕD_1	0.444	0.524	11.28	13.31	--
e	0.136	0.146	3.45	3.71	--
F	--	0.115	--	2.92	--
ℓ	0.360	0.440	9.14	11.18	1
$\jmath$	90 NOMINAL				

NOTE:

1. Three leads.

JEDEC TO-18

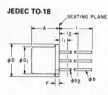

92CS-20223

JEDEC TO-36

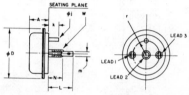

92CS-24690

SYMBOL	INCHES		MILLIMETERS		NOTES
	MIN.	MAX.	MIN.	MAX.	
A	0.170	0.210	4.32	5.33	
ϕb	0.016	0.021	0.406	0.533	1
ϕb_2	0.016	0.019	0.406	0.483	1
ϕD	0.209	0.230	5.31	5.84	
ϕD_1	0.178	0.195	4.52	4.95	
e	0.100 T.P.		2.54 T.P.		2, 4
e_1	0.050 T.P.		1.27 T.P.		2, 4
F		0.030		0.762	
j	0.036	0.046	0.914	1.17	4
k	0.028	0.048	0.711	1.22	3
l	0.500		12.70		1
l_1		0.050		1.27	1
l_2	0.250		6.35		1
α	45° T.P.				5

SYMBOL	INCHES		MILLIMETERS		NOTES
	MIN.	MAX.	MIN.	MAX.	
A	--	0.520	--	13.21	
ϕD	--	1.250	--	31.75	
ϕj	--	0.140	--	3.56	
k	--	0.312	--	7.92	1
L	0.610	0.710	15.49	18.03	
m	--	0.190	--	4.83	
N	0.375	0.500	9.53	12.70	
r	0.345 NOMINAL		8.76 NOMINAL		
W					2

NOTES:

1. INSULATED LOCATOR PIN.

2. 10-32 UNF-2A. MAXIMUM PITCH DIAMETER OF PLATED THREADS SHALL BE BASIC PITCH DIAMETER 0.1697 in. (4.31 mm) REFERENCE (SCREW THREAD STANDARDS FOR FEDERAL SERVICES 1957) HANDBOOK H28 1957 P1.

3. CONTROLLING DIMENSION: INCH.

NOTES:

1. (Three leads) ϕb_2 applies between l_1 and l_2. ϕb applies between l_2 and 0.5 in. (12.70 mm) from seating plane. Diameter is uncontrolled in l_1 and beyond 0.5 in. (12.70 mm) from seating plane.

2. Leads having maximum diameter 0.019 in. (0.483 mm) measured in gaging plane 0.054 in. (1.37 mm) + 0.001 in. (0.025 mm) − 0.00 in. (0.00 mm) below the seating plane of the device shall be within 0.007 in. (0.178 mm) of their true positions relative to a maximum-width tab.

3. Measured from maximum diameter of the actual device.

4. The device may be measured by direct methods or by the gage and gaging procedure described on gage drawing GS-2.

5. Tab centerline.

JEDEC TO-39/TO-5

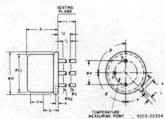

SYMBOL	INCHES		MILLIMETERS		NOTES
	MIN.	MAX.	MIN.	MAX.	
ϕa	0.190	0.210	4.83	5.33	
A	0.240	0.260	6.10	6.60	
ϕb	0.016	0.021	0.406	0.533	2
$\phi b2$	0.016	0.019	0.406	0.483	2
ϕD	0.350	0.370	8.89	9.40	
ϕD_1	0.305	0.335	8.00	8.51	
h	0.009	0.041	0.229	1.04	
j	0.028	0.034	0.711	0.864	
k	0.029	0.040	0.737	1.02	3
L long lead	1.500		38.10		2
L short lead	0.500		12.70		2
l_1		0.050		1.27	2
l_2	0.250		6.35		2
P	0.100		2.54		1
Q					4
a	45° NOMINAL				
β	90° NOMINAL				

Note 1: This zone is controlled for automatic handling. The variation in actual diameter within this zone shall not exceed 0.010 in. (0.254 mm).

Note 2: (Three leads) $\phi b2$ applies between l_1 and l_2. ϕb applies between l_2 and l. Diameter is uncontrolled in l_1.

Note 3: Measured from maximum diameter of the actual device.

Note 4: Details of outline in this zone optional.

MODIFIED TO-39

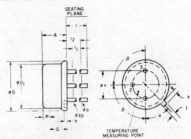

SYMBOL	INCHES		MILLIMETERS		NOTES
	MIN.	MAX.	MIN.	MAX.	
ϕa	0.190	0.210	4.83	5.33	
A	0.160	0.180	4.07	4.57	
ϕb	0.016	0.021	0.406	0.533	2
$\phi b2$	0.016	0.019	0.406	0.483	2
ϕD	0.350	0.370	8.89	9.40	
ϕD_1	0.315	0.335	8.00	8.51	
h	0.009	0.125	0.229	3.18	
j	0.028	0.034	0.711	0.864	
k	0.029	0.040	0.737	1.02	3
l	0.500		12.70		2
l_1		0.050		1.27	2
l_2	0.250		6.35		2
P					1
Q					4
a	45° NOMINAL				
β	90° NOMINAL				

Note 1: This zone is controlled for automatic handling. The variation in actual diameter within this zone shall not exceed 0.010 in. (0.254 mm).

Note 2: (Three leads) $\phi b2$ applies between l_1 and l_2 and 0.5 in. (12.70 mm) from seating plane. Diameter is controlled in l_1 and beyond 0.5 in. (12.70 mm) from seating plane.

Note 3: Measured from maximum diameter of the actual device.

Note 4: Details of outline in this zone optional.

92CS-20893

TO-39/TO-5 WITH HEAT RADIATOR

TO-5 PACKAGE WELDED TO HEAT-RADIATOR

MOUNTING TAB (LEAD NO. 2 BEHIND MOUNTING TAB)

4 DIMPLED STANDOFFS

HEAT RADIATOR (NOTE 1)

2 MOUNTING TABS (NOTE 2)

SYMBOL	INCHES		MILLIMETERS		NOTES
	MIN.	MAX.	MIN.	MAX.	
A	–	0.630	–	16.00	
D	1.205	1.235	30.61	31.37	
D_1	0.775	0.785	19.69	19.93	
E	0.875	0.905	22.22	22.99	
F	0.040	0.055	1.02	1.40	
F_1	0.160	0.195	4.06	4.95	
L long lead	1.410	–	35.81	–	
L short lead	0.410	–	10.41	–	
ϕP	0.295	0.305	7.493	7.747	
ϕP_1	0.093	0.095	2.362	2.413	
N	0.048	0.062	1.21	1.57	
N_1	0.998	1.002	25.349	25.450	3
W	0.048	0.052	1.219	1.320	

NOTES:

1. 0.035 C.R.S., finish—electroless nickel plate.

2. Recommended hole size for printed-circuit board is 0.070 in. (1.78 mm) dia.

3. Measured at bottom of heat-radiator.

92CS-22335

TO-39/TO-5 WITH FLANGE

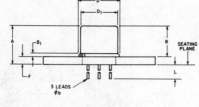

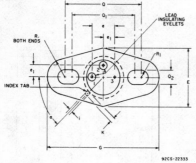

92CS-22333

SYMBOL	INCHES		MILLIMETERS		NOTES
	MIN.	MAX.	MIN.	MAX.	
A	–	0.328	–	8.33	
B	0.240	0.260	6.10	6.60	
B_1	0.009	0.125	0.229	3.18	
ϕ_b	0.016	0.019	0.406	0.483	
D	0.335	0.370	8.51	9.40	
D_1	0.305	0.335	7.75	8.51	
E	0.495	0.505	12.57	12.83	
e	0.200 T.P.		5.08 T.P.		1
e_1	0.100 T.P.		2.54 T.P.		1
F	0.062	0.068	1.57	1.74	
G	0.995	1.005	25.27	25.53	
i	0.028	0.034	0.711	0.864	
k	0.029	0.045	0.737	1.14	
L long lead	1.430		36.32		
L short lead	0.430		10.92		
Q	0.685	0.691	17.40	17.55	
Q_1	0.559	0.565	14.20	14.35	
Q_2	0.128	0.132	3.25	3.35	
R	0.156 T.P.		3.96 T.P.		1
R_1	0.064	0.066	1.63	1.67	
a	45° T.P.				1, 2

NOTES:
1. True position.
2. Tab centerline.

TO-48

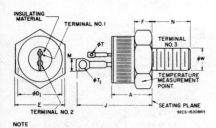

92CS-1520BR3

NOTE
1. ϕW is pitch diameter of coated threads.

SYMBOL	INCHES		MILLIMETERS		NOTES
	MIN.	MAX.	MIN.	MAX.	
A	0.330	0.505	8.4	12.8	–
ϕD_1	–	0.544	–	13.81	–
E	0.544	0.562	13.82	14.28	–
F	0.113	0.200	2.87	5.08	–
J	0.950	1.100	24.13	27.94	–
M	0.215	0.225	5.46	5.71	–
N	0.422	0.453	10.72	11.50	–
ϕT	0.058	0.068	1.47	1.73	–
ϕT_1	0.138	0.148	3.51	3.75	–
ϕW	1/4-28 UNF-2A		1/4-28 UNF-2A		1

REF: Screw-Thread Standards for Federal Services, Handbook H28, Part I. Recommended Torque: 25 inch-pounds.

JEDEC TO-63

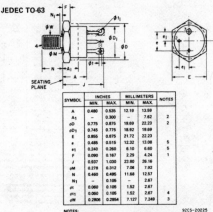

SYMBOL	INCHES		MILLIMETERS		NOTES
	MIN.	MAX.	MIN.	MAX.	
A	0.480	0.535	12.19	13.59	
A₁	–	0.300	–	7.62	2
øD	0.775	0.875	19.69	22.23	2
øD₁	0.745	0.775	18.92	19.69	
E	0.855	0.875	21.72	22.23	
e	0.485	0.515	12.32	13.08	5
e₁	0.240	0.260	6.10	6.60	5
F	0.090	0.167	2.29	4.24	1
J	0.937	1.030	23.80	26.16	
øM	0.278	0.312	7.06	7.92	
N	0.460	0.495	11.68	12.57	
N₁	–	0.105	–	2.67	
øt	0.060	0.105	1.52	2.67	
øt₁	0.060	0.105	1.52	2.67	4
øW	0.2806	0.2854	7.127	7.249	3

NOTES: 92CS-20225
1. DIMENSION DOES NOT INCLUDE SEALING FLANGES.
2. PACKAGE CONTOUR OPTIONAL WITHIN DIMENSIONS
   SPECIFIED.
3. PITCH DIAMETER - THREAD 5/16-24 UNF-2A (COATED).
   REFERENCE (SCREW THREAD STANDARDS FOR FED-
   ERAL SERVICES - HANDBOOK H-28).
4. THIS TERMINAL CAN BE FLATTENED AND PIERCED
   OR HOOK TYPE.
5. POSITION OF LEADS IN RELATION TO THE HEXAGON
   IS NOT CONTROLLED.

TO-60

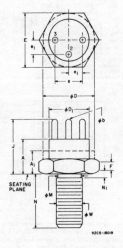

92CS-18019

SYMBOL	INCHES		MILLIMETERS		NOTES
	MIN.	MAX.	MIN.	MAX.	
A	0.215	0.320	5.46	8.13	
A₁	–	0.165	–	4.19	2
øb	0.030	0.046	0.762	1.17	4
øD	0.360	0.437	9.14	11.10	2
øD₁	0.320	0.360	8.13	9.14	
E	0.424	0.437	10.77	11.10	
e	0.185	0.215	4.70	5.46	
e₁	0.090	0.110	2.29	2.79	
F	0.090	0.135	2.29	3.43	1
J	0.355	0.480	9.02	12.19	
øM	0.163	0.189	4.14	4.80	
N	0.375	0.455	9.53	11.56	
N₁	–	0.078	–	1.98	
øW	0.1658	0.1697	4.212	4.310	3, 5

MILLIMETER DIMENSIONS ARE DERIVED
FROM ORIGINAL INCH DIMENSIONS

NOTES:

1. Dimension does not include sealing flanges
2. Package contour optional within dimensions specified
3. Pitch diameter – 10-32 UNF 2A thread (coated)
4. Pin spacing perimts insertion in any socket having a
   pin-circle diameter of 0.200 in. (5.08 mm) and contacts which will accommodate pins with a diameter
   of 0.030 in. (0.762 mm) min., 0.046 in. (1.17mm) max.
5. The torque applied to a 10-32 hex nut assembled on the
   thread during installation should not exceed 12 inch-
   pounds.

JEDEC TO-66

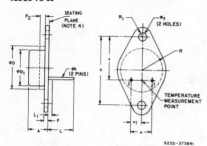

92SS-3738RI

SYMBOL	INCHES		MILLIMETERS		NOTES
	MIN.	MAX.	MIN.	MAX.	
A	0.250	0.340	6.35	8.64	
φb	0.028	0.034	0.711	0.863	
φD	–	0.620	–	15.75	1
φD₁	0.470	0.500	11.94	12.70	
e	0.190	0.210	4.83	5.33	2
e₁	0.093	0.107	2.36	2.72	2
F₁	0.050	0.075	1.27	1.91	
F₂	–	0.050	–	1.27	1
L	0.360	–	9.14	–	
L₁	–	0.050	–	1.27	3
φp	0.142	0.152	3.61	3.86	
q	0.958	0.962	24.33	24.43	
R	–	0.350	–	8.89	
R₁	–	0.145	–	3.68	
s	0.570	0.590	14.48	14.99	

NOTES:
1. Body contour is optional within zone defined by φD and F₂.

2. These dimensions should be measured at points 0.050 in. (1.27 mm) to 0.055 in. (1.40 mm) below seating plane. When gage is not used, measurement will be made at seating plane.

3. φb applies between L₁ and L. Diameter is uncontrolled in L₁.

4. The seating plane of header shall be flat within 0.001 in. (0.025 mm) concave to 0.004 in. (0.10 mm) convex inside a 0.520 in. (13.21 mm) diameter circle on the center of the header and flat within 0.001 in. (0.025 mm) concave to 0.006 in. (0.15 mm) convex overall.

TO-66 WITH HEAT RADIATOR

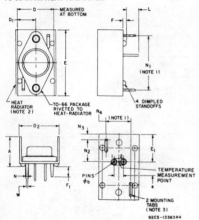

92CS-13383R4

SYMBOL	INCHES		MILLIMETERS		NOTES
	MIN.	MAX.	MIN.	MAX.	
A	–	0.620	–	15.75	
φb	0.028	0.034	0.711	0.864	
D	0.750	0.760	19.05	19.30	
D₁	0.370	0.385	9.40	9.78	
D₂	0.820	0.920	20.83	23.37	
E	1.297	1.327	32.94	33.70	
E₁	0.546	0.566	13.87	14.37	
e	0.190	0.210	4.83	5.33	
F	0.30	0.55	7.62	13.97	
F₁	0.175	0.210	4.44	5.33	
L	0.270	–	6.86	–	
N	0.052	0.065	1.32	1.65	
N₁	1.098	1.102	27.89	27.99	1
N₂	0.448	0.452	11.38	11.47	
N₃	0.099	0.113	0.25	0.29	
N₄	0.498	0.502	12.66	12.75	
W	0.048	0.060	1.22	1.52	

NOTES:
1. Measured at bottom of heat radiator.
2. 0.035 in. (0.889) C.R.S. tin plated.
3. Recommended hole size for printed-circuit board is 0.070 in. (1.778) dia.

JEDEC TO-72

92CS-17444RI

SYMBOL	INCHES		MILLIMETERS		NOTES
	MIN.	MAX.	MIN.	MAX.	
A	0.170	0.210	4.32	5.33	2
φb	0.016	0.021	0.406	0.533	2
φb₂	0.016	0.019	0.406	0.483	2
φD	0.209	0.230	5.31	5.84	
φD₁	0.178	0.195	4.52	4.95	
e	0.100 T.P.		2.54 T.P.		4
e1	0.050 T.P.		1.27 T.P.		4
h	–	0.030	–	0.762	
j	0.036	0.046	0.914	1.17	
k	0.028	0.048	0.711	1.22	3
l	0.500	–	12.70	–	2
l₁	–	0.050	–	1.27	2
l₂	0.250	–	6.35	–	2
α	45° T.P.		45° T.P.		4, 6

Note 1: (Four leads). Maximum number leads omitted in this outline, "none" (0). The number and position of leads actually present are indicated in the product registration. Outline designation determined by the location and minimum angular or linear spacing of any two adjacent leads.

Note 2: (All leads) φb₂ applies between l₁ and l₂. φb applies between l₂ and 0.50 in. (12.70 mm) from seating plane, Diameter is uncontrolled in l₁ and beyond 0.50 in. (12.70 mm) from seating plane.

Note 3: Measured from maximum diameter of the product.

Note 4: Leads having maximum diameter 0.019 in. (0.484 mm) measured in gaging plane 0.054 in. (1.37 mm) +0.001 in. (0.025 mm) − 0.000 (0.000 mm) below the seating plane of the product shall be within 0.007 in. (0.178 mm) of their true position relative to a maximum width tab.

Note 5: The product may be measured by direct methods or by gage.

Note 6: Tab centerline.

RADIAL PACKAGE

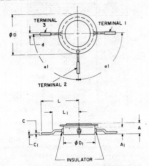

SYMBOL	INCHES		MILLIMETERS		NOTES
	MIN.	MAX.	MIN.	MAX.	
A	–	0.200	–	5.08	
A_1	–	0.125	–	3.17	1
C	0.015	0.019	0.38	0.48	
C_1	–	0.015	–	0.38	
ϕD	–	0.710	–	18.03	
ϕD_1	0.615	0.690	15.62	17.52	1
d	0.042	0.046	1.06	1.16	
L	–	0.705	–	17.90	
L_1	–	0.510	–	12.95	
a_1	$90^o \pm 2^o$		$90^o \pm 2^o$		

92CS- 20224

NOTE:
1. CONTROLLED AREA OF THE DIAMETER DOES NOT INCLUDE THE BRAZED AREA AROUND THE CERAMIC AND TERMINAL 2.

PLASTIC PACKAGES
PLASTIC TO-5 AND PLASTIC TO-5 WITH HEAT CLIP

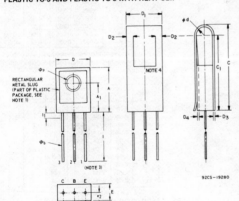

92CS-19280

SYMBOL	INCHES		MILLIMETERS		NOTES
	MIN.	MAX.	MIN.	MAX.	
A	0.385	0.395	9.78	10.03	
A_1	0.251	0.261	6.37	6.63	
ϕb	0.016	0.019	0.41	0.48	2
C	0.858		21.79		
C_1	0.750		19.05		
D	0.305	0.315	7.75	8.00	
D_1	0.300		7.62		
D_2	0.070		1.77		
D_3	0.0329		0.813		
D_4	0.021	0.041	0.533	1.04	
ϕd	0.073	0.077	1.85	1.95	
E	0.145	0.155	3.68	3.94	
e	0.195	0.205	4.95	5.21	
e_1	0.095	0.105	2.41	2.67	
e_2	0.070	0.080	1.78	2.03	
ℓ	0.725	0.745	18.41	18.91	
ℓ_1	0.125	0.250	3.17	6.35	
ϕp	0.112	0.118	2.84	2.99	

NOTE 1: To attach to heat-sink, use a 4-40 binding-head screw and a No. 4 flat washer. The recommended screw torque (for even distribution of mounting pressure and optimum thermal contact) is 6 in.-lb.
NOTE 2: Three leads. Leads are pretinned to the ℓ_1 dimension.
NOTE 3: Lead numbering from right to left with rectangular metal slug facing observer.
NOTE 4: Tab to be sheared through and set inward as shown.

TO-72

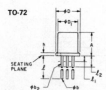

92CS-17444RI

SYMBOL	INCHES		MILLIMETERS		NOTES
	MIN.	MAX.	MIN.	MAX.	
A	0.170	0.210	4.32	5.33	
φb	0.016	0.021	0.406	0.533	2
φb₂	0.016	0.019	0.406	0.483	2
φD	0.209	0.230	5.31	5.84	
φD₁	0.178	0.195	4.52	4.95	
e	0.100 T.P.		2.54 T.P.		4
e₁	0.050 T.P.		1.27 T.P.		4
h		0.030		0.762	
j	0.036	0.046	0.914	1.17	
k	0.028	0.048	0.711	1.22	3
l	0.500		12.70		2
l₁		0.050		1.27	2
l₂	0.250		6.35		2
α	45° T.P.		45° T.P.		4, 6

MILLIMETER DIMENSIONS ARE DERIVED
FROM ORIGINAL INCH DIMENSIONS

Note 1: (Four leads). Maximum number leads omitted in this outline, "none" (0). The number and position of leads actually present are indicated in the product registration. Outline designation determined by the location and minimum angular or linear spacing of any two adjacent leads.

Note 2: (All leads) φb₂ applies between l₁ and l₂. φb applies between l₂ and .500" (12.70 mm) from seating plane. Diameter is uncontrolled in l₁ and beyond .500" (12.70 mm) from seating plane.

Note 3: Measured from maximum diameter of the product.

Note 4: Leads having maximum diameter .019" (.483 mm) measured in gaging plane .054" (1.37 mm) + .001" (.025 mm) − .000" (.000 mm)

below the seating plane of the product shall be within .007" (.178 mm) of their true position relative to a maximum width tab.

Note 5: The product may be measured by direct methods or by gage.

Note 6: Tab centerline.

TO-201AA

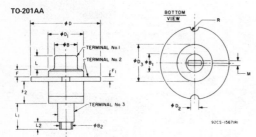

92CS-1567IRI

SYMBOL	INCHES		MILLIMETERS	
	MIN.	MAX.	MIN.	MAX.
φB	0.165	0.175	4.19	4.44
φB₁	0.115	0.125	2.92	3.17
φB₂	0.090	0.110	2.29	2.79
φD	0.495	0.505	12.57	12.83
φD₁	0.245	0.255	6.22	6.48
φD₂	0.055	0.065	1.39	1.65
φD₃	0.245	0.255	6.22	6.48
F	0.045	0.060	1.14	1.52
F₁	0.025	0.035	0.63	0.88
F₂	0.145	0.175	3.68	4.44
L	0.095	0.115	2.41	2.92
L₁	0.165	0.195	4.19	4.95
L₂	0.040	0.060	1.02	1.52
M	0.045	0.055	1.14	1.39
R	0.027	0.033	0.68	0.83

MILLIMETER DIMENSIONS ARE DERIVED
FROM ORIGINAL INCH DIMENSIONS

TO-215AA

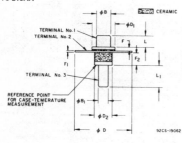

92CS-19062

SYMBOL	INCHES		MILLIMETERS		NOTES
	MIN.	MAX.	MIN.	MAX.	
φB	0.118	0.122	2.997	3.098	1
φB₁	0.090	0.094	2.286	2.387	2
φD	0.497	0.503	12.624	12.776	3
φD₁	0.180	NOM.	4.57	NOM.	
φD₂	0.162	NOM.	4.11	NOM.	
F	0.028	0.039	0.71	0.99	
F₁	0.009	0.011	0.229	0.279	
F₂	0.114	0.126	2.90	3.20	
L	0.098	0.104	2.49	2.64	
L₁	0.179	0.191	4.55	4.85	

MILLIMETER DIMENSIONS ARE DERIVED
FROM ORIGINAL INCH DIMENSIONS

NOTES:

1. Silver or KOVAR*
2. Solid silver
3. Gold-plated KOVAR

*Trademark, Westinghouse Electric Corp.

TO-216AA

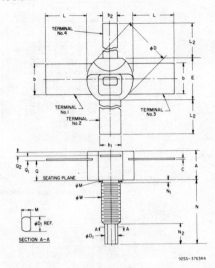

92SS-3763R4

SYMBOL	INCHES		MILLIMETERS		NOTES
	MIN.	MAX.	MIN.	MAX.	
A	0.150	0.230	3.81	5.84	–
b	0.195	0.205	4.953	5.207	–
b_1	0.135	0.145	3.429	3.683	–
b_2	0.095	0.105	2.413	2.667	–
C	0.004	0.010	0.102	0.254	3
ϕD	0.305	0.320	7.75	8.12	5
ϕD_1	0.110	0.130	2.80	3.30	1
E	0.275	0.300	6.99	7.62	5
L	0.265	0.290	6.74	7.36	–
L_2	0.455	0.510	11.56	12.95	–
M	0.053	0.064	1.35	1.62	–
ϕM	0.120	0.163	3.05	4.14	–
N	0.425	0.470	10.80	11.93	–
N_1	–	0.078	–	1.98	4
N_2	0.110	0.150	2.80	3.81	–
Q	0.120	0.170	3.05	4.31	–
Q_1	0.025	0.045	0.64	1.14	–
Q_2	–	–	–	–	5
ϕW	–	–	–	–	2

MILLIMETER DIMENSIONS ARE DERIVED
FROM ORIGINAL INCH DIMENSIONS

NOTES:

1. 0.053 - 0.064 INCH (1.35 - 1.62 mm) WRENCH FLAT.
2. PITCH DIA. OF 8-32 UNC-2A COATED THREADS (REF: UNITED SCREW THREADS ANS B1.1 - 1960). THE APPLIED TORQUE SHOULD NOT EXCEED 5 IN. LBS. CLAMPING FORCES MUST BE APPLIED ONLY TO THE FLAT SUR-FACES OF THE STUD.
3. TYPICAL FOR ALL LEADS.
4. LENGTH OF INCOMPLETE OR UNDERCUT THREADS OF ϕW.
5. BODY CONTOUR OPTIONAL WITHIN Q_2, ϕD, AND E.

TO-217AA

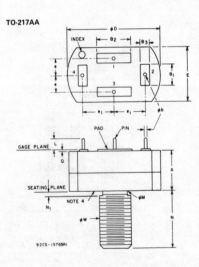

92CS-15765R1

SYMBOL	INCHES		MILLIMETERS		NOTES
	MIN.	MAX.	MIN.	MAX.	
A	0.295	0.325	7.50	8.25	–
B_1	0.135	0.150	3.43	3.81	–
B_2	0.235	0.250	5.97	6.35	–
B_3	0.055	0.065	1.40	1.65	5
ϕb	0.020	0.025	0.508	0.635	4 Pins
ϕD	0.650	0.680	16.51	17.27	–
E	0.360	0.380	9.15	9.65	–
e	0.111	0.131	2.82	3.32	1
e_1	0.213	0.233	5.42	5.91	1
L	0.114	0.133	2.90	3.37	–
ϕM	0.220	0.249	5.59	6.23	–
N	0.420	0.460	10.67	11.68	–
N_1	–	0.090	–	2.28	–
Q	–	0.015	–	0.038	–
ϕW	–	–	–	–	2

MILLIMETER DIMENSIONS ARE DERIVED
FROM ORIGINAL INCH DIMENSIONS

NOTES:

1. The pin center-to-center dimensions are measured at the gage plane.
2. ¼ in. 28 UNF 2A (Mod). Applied torque not to exceed 12 inch-pounds.
3. This device may be operated in any position.
4. Seating plate to be flat within 0.003 inches.
5. Typical 4 places.

JEDEC TO-219AA

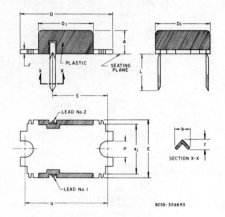

SYMBOL	INCHES		MILLIMETERS		NOTES
	MIN.	MAX.	MIN.	MAX.	
A	0.160	0.200	4.07	5.08	
b	0.045	0.060	1.15	1.52	
c	0.025	0.045	0.64	1.14	
D	0.890	0.910	22.61	23.11	
D_1	0.480	0.515	12.20	13.08	
E	0.480	0.520	12.20	13.20	
e_1	0.460	0.505	11.69	12.82	1
F	0.055	0.070	1.40	1.77	
L	0.370	0.450	9.40	11.43	2
P	0.128	0.150	3.26	3.81	
q	0.740	0.760	18.80	19.30	
s	0.500	0.520	12.70	13.20	

NOTES:
1. e_1 is measured at seating plane.
2. Terminal end configurations are optional.

9255-3598R3

JEDEC TO-219AB

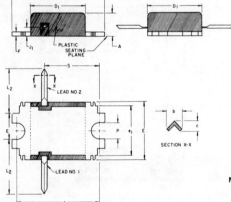

SYMBOL	INCHES		MILLIMETERS	
	MIN.	MAX.	MIN.	MAX.
A	0.160	0.200	4.07	5.08
b	0.045	0.060	1.15	1.52
c	0.025	0.045	0.64	1.14
D	0.890	0.910	22.61	23.11
D_1	0.480	0.515	12.20	13.03
E	0.480	0.520	12.20	13.20
F	0.055	0.070	1.40	1.77
J_1	0.100	0.120	2.54	3.04
L_2	0.415	0.560	10.54	14.22
P	0.128	0.150	3.26	3.81
q	0.740	0.760	18.80	19.30
s	0.500	0.520	12.70	13.20

NOTE: Terminal end configurations are optional.

9255-3599R2

JEDEC TO-220AA

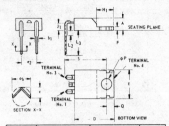

SYMBOL	INCHES		MILLIMETERS		NOTES
	MIN.	MAX.	MIN.	MAX.	
A	0.140	0.190	3.56	4.82	–
φb	0.02	0.045	0.51	1.14	–
b₁	0.045	0.070	1.15	1.77	–
c	0.015	0.030	0.38	0.762	–
D	0.560	0.625	14.23	15.87	–
E	0.380	0.420	9.66	10.66	1
e₂	0.190	0.210	4.83	5.33	2
F	0.045	0.055	1.15	1.39	–
H₁	0.230	0.270	5.85	6.85	1
J₁	0.080	0.115	2.04	2.92	–
L₂	–	0.050	–	1.27	–
L₃	0.360	0.422	9.15	10.71	–
φP	0.139	0.147	3.531	3.733	–
Q	0.100	0.120	2.54	3.04	–
S	0.580	0.610	14.74	15.49	–

92CS-17990R1

NOTES:

1. Tab contour optional within H_1 and E.
2. Position of lead to be measured 0.050 – 0.055 in. (1.27 – 1.40 mm) below seating plane.

JEDEC TO-220AB

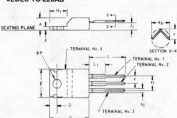

SYMBOL	INCHES		MILLIMETERS		NOTES
	MIN.	MAX.	MIN.	MAX.	
A	0.140	0.190	3.56	4.82	–
φb	0.020	0.045	0.51	1.14	–
b₁	0.045	0.070	1.15	1.77	–
c	0.015	0.030	0.38	0.762	–
D	0.560	0.625	14.23	15.87	–
E	0.380	0.420	9.66	10.66	1
e	0.090	0.110	2.29	2.79	2
e₁	0.190	0.210	4.83	5.33	2
F	0.045	0.055	1.15	1.39	–
H₁	0.230	0.270	5.85	6.85	1
J₁	0.080	0.115	2.04	2.92	–
L	0.500	0.562	12.70	14.27	–
L₁	–	0.250	–	6.35	–
φP	0.139	0.147	3.531	3.733	–
Q	0.100	0.120	2.54	3.04	–

92CS-17991RI

NOTES:

1. Tab contour optional within H_1 and E.
2. Position of lead to be measured 0.250 – 0.255 in. (6.35 – 6.48 mm) from case.

RCP PLASTIC PACKAGE

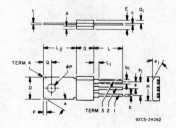

92CS-24062

SYMBOL	INCHES		MILLIMETERS		NOTES
	MIN.	MAX.	MIN.	MAX.	
A	–	0.05	–	1.270	1
b	0.023	0.029	0.584	0.736	
b₁	0.045	0.055	1.143	1.397	1
c	0.018	0.026	0.457	0.660	
D	0.305	0.325	7.747	8.255	
E	0.130	0.150	3.302	3.810	
e	0.095	0.105	2.413	2.667	
e₁	0.190	0.210	4.826	5.334	
F	–	0.08	–	2.032	1
G	0.230	0.250	5.842	6.350	
H	0.330	0.370	8.382	9.398	
L	0.400	0.450	10.16	11.43	
L₁	–	0.100	–	2.54	1,2
L₂	0.540	0.580	13.71	14.73	
φP	0.123	0.127	3.124	3.225	
Q	0.120	0.130	3.048	3.302	
Q₁	0.039	0.050	0.990	1.270	
α	–	35⁰	–	35⁰	1
α₁	–	50⁰	–	50⁰	1

NOTES:

1. Package contour optional within dimensions specified.

2. Lead dimensions uncontrolled in this zone.

3. Chamfer on tab optional.

4. Controlling dimensions: inch.

Press-Fit
6-, 10-, and 15-A Triacs; 20- and 35-A SCR's

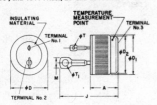

SYMBOL	INCHES		MILLIMETERS		NOTES
	MIN.	MAX.	MIN.	MAX.	
A	–	0.380	–	9.65	
ϕ D	0.501	0.510	12.73	12.95	
ϕ D$_1$	–	0.505	–	12.83	1
ϕ D$_2$	0.465	0.475	11.81	12.07	
J	–	0.750	–	19.05	
M	–	0.155	–	3.94	
ϕ T	0.058	0.068	1.47	1.73	
ϕ T$_1$	0.080	0.090	2.03	2.29	

NOTE 1: Outer diameter of knurled surface.

92CS-23I34

Press-Fit
25-, 30-, and 40-A Triacs

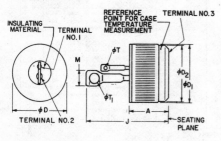

SYMBOL	INCHES		MILLIMETERS		NOTES
	MIN.	MAX.	MIN.	MAX.	
A	–	0.380	–	9.65	–
ϕD	0.501	0.510	12.73	12.95	–
ϕD$_1$	–	0.505	–	12.83	1
ϕD$_2$	0.465	0.475	11.81	12.07	–
J	0.825	1.000	20.95	25.40	–
M	0.215	0.225	5.46	5.71	–
ϕT	0.058	0.068	1.47	1.73	–
ϕT$_1$	0.138	0.148	3.51	3.75	–

NOTE:
1. Outer diameter of knurled surface.

92CS-I5207R3

Press-Fit
60- and 80-A Triacs

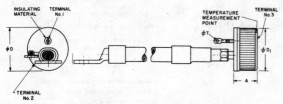

SYMBOL	INCHES		MILLIMETERS		NOTES
	MIN.	MAX.	MIN.	MAX.	
A	–	0.466	–	11.84	
ϕD	0.751	0.760	19.08	19.30	
ϕD$_1$	–	0.7535	–	19.139	
J	6.8 NOM.		172.72 NOM.		1
J$_1$	6.3 NOM.		160.02 NOM.		1
ϕT	0.060	0.065	1.52	1.65	
ϕT$_1$	0.266	–	6.75	–	
ϕT$_2$	0.144	–	3.70	–	

NOTE:
1: Leads J and J$_1$ available at various lengths. For information, contact the RCA Sales Office in your locale.

92CM-22833

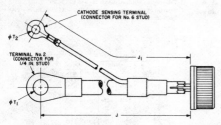

Stud 6-, 10, and 15-A Triacs; 20- and 35-A SCR's

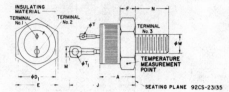

SEATING PLANE 92CS-23135

SYMBOL	INCHES		MILLIMETERS		NOTES
	MIN.	MAX.	MIN.	MAX.	
A	0.330	0.505	8.40	12.80	
φD₁	–	0.544	–	13.81	
E	0.544	0.562	13.82	14.28	
F	0.113	0.200	2.87	5.08	
J	–	0.950	–	24.13	
M	–	0.155	–	3.94	
N	0.422	0.453	10.72	11.50	
φT	0.058	0.068	1.47	1.73	
φT₁	0.080	0.090	2.03	2.29	
φW	1/4-28 UNF-2A		1/4-28 UNF-2A		1

NOTE 1: φW is pitch diameter of coated threads.
REF. Screw-Thread Standard for Federal Services
Handbook H28, Part I.
Recommended torque: 35 inch-pounds.

Stud 25-, 30-, and 40-A Triacs

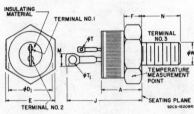

SEATING PLANE
92CS-15208R3

TERMINAL NO. 2

NOTE
1. φW is pitch diameter of coated threads.

SYMBOL	INCHES		MILLIMETERS		NOTES
	MIN.	MAX.	MIN.	MAX.	
A	0.330	0.505	8.4	12.8	–
φD₁	–	0.544	–	13.81	–
E	0.544	0.562	13.82	14.28	–
F	0.113	0.200	2.87	5.08	–
J	0.950	1.100	24.13	27.94	–
M	0.215	0.225	5.46	5.71	–
N	0.422	0.453	10.72	11.50	–
φT	0.058	0.068	1.47	1.73	–
φT₁	0.138	0.148	3.51	3.75	–
φW	1/4-28 UNF-2A		1/4-28 UNF-2A		1

REF: Screw-Thread Standards for Federal Services, Handbook H28,
Part I. Recommended Torque: 25 inch-pounds.

Stud 60- and 80-A Triacs

NOTES:
1: Leads J and J₁ available at various lengths. For information,
   contact the RCA Sales Office in your locale.

2: φW is pitch diameter of coated threads. REF: Screw Thread
   Standard for Federal Services, Handbook H 28 Part I.
   Recommended torque: 125 inch-pounds.

SYMBOL	INCHES		MILLIMETERS		NOTES
	MIN.	MAX.	MIN.	MAX.	
A	–	0.620	–	15.75	
φD	0.751	0.760	19.08	19.30	
E	0.866	0.872	21.99	22.14	
F	0.182	0.192	4.62	4.87	
J	6.8 NOM.		172.72 NOM.		1
J₁	6.3 NOM.		160.02 NOM.		1
N	0.740	0.760	18.79	19.30	
φT	0.060	0.065	1.52	1.65	
φT₁	0.266	–	6.75	–	
φT₂	0.144	–	3.70	–	
φW	1/2-20	NF-2A	1/2-20	NF-2A	2

CATHODE SENSING TERMINAL
(CONNECTOR FOR No. 6 STUD)

TERMINAL No. 2
(CONNECTOR FOR
1/4 IN. STUD)

TERMINAL No. 3

SEATING
PLANE

TEMPERATURE
MEASUREMENT
POINT   92CM-22834

Isolated-Stud 6-, 10-, and 15-A Triacs; 20- and 35-A SCR's

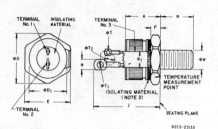

92CS-23133

SYMBOL	INCHES		MILLIMETERS		NOTES
	MIN.	MAX.	MIN.	MAX.	
A	–	0.673	–	17.09	
$\diamond$D	0.604	0.614	15.34	15.59	
$\diamond$D$_1$	0.501	0.505	12.72	12.82	
E	0.551	0.557	13.99	14.14	
F	0.175	0.185	4.44	4.69	
J	–	1.055	–	26.79	
M	–	0.155	–	3.94	
M$_1$	0.200	0.210	5.08	5.33	
N	0.422	0.452	10.72	11.48	
$\diamond$T	0.058	0.068	1.47	1.73	
$\diamond$T$_1$	0.080	0.090	2.03	2.29	
$\diamond$T$_2$	0.138	0.148	3.50	3.75	
$\diamond$W	1/4-28 UNF-2A	1/4-28 UNF-2A			1

NOTE 1: $\diamond$W is pitch diameter of coated threads.
REF. Screw-Thread Standard for Federal Services Handbook H28, Part I.
Recommended torque: 35 inch-pounds.
NOTE 2:
Isolating material (ceramic) between hex (stud) and terminal No. 3 is beryllium oxide. Minimum isolation breakdown voltage is 2100 V rms for 1 minute duration.

Isolated-Stud 25-, 30-, and 40-A Triacs

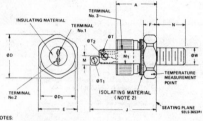

92LS-3653(R)

SYMBOL	INCHES		MILLIMETERS		NOTES
	MIN.	MAX.	MIN.	MAX.	
A	–	0.673	–	17.09	–
$\diamond$D	0.604	0.614	15.34	15.59	–
$\diamond$D$_1$	0.501	0.505	12.72	12.82	–
E	0.551	0.557	13.99	14.14	–
F	0.175	0.185	4.44	4.69	–
J	–	1.298	–	32.96	–
M	0.210	0.230	5.33	5.84	–
M$_1$	0.200	0.210	5.08	5.33	–
N	0.422	0.452	10.72	11.48	–
$\diamond$T	0.058	0.068	1.47	1.73	–
$\diamond$T$_1$	0.125	0.165	3.18	4.19	–
$\diamond$T$_2$	0.138	0.148	3.50	3.75	–
$\diamond$W	1/4-28 UNF-2 A	1/4-28 UNF-2 A			1

NOTES:

1. $\diamond$W is pitch diameter of coated threads. REF: Screw-Thread Standards for Federal Services, Handbook H28, Part I. Recommended Torque: 25 inch-pounds.

2. Isolating material (ceramic) between hex (stud) and terminal No. 3 is beryllium oxide. Minimum isolation breakdown voltage is 2100 V rms for 1 minute duration.

Isolated-Stud 60- and 80-A Triacs

NOTES:

1: Leads J and J$_1$ available at various lengths. For information, contact the RCA Sales Office in your locale.

2: $\diamond$W is pitch diameter of coated threads. REF: Screw Thread Standards for Federal Services, Handbook H 28 Part I.

Recommended torque: 125 inch-pounds.

3: Isolating material (ceramic) between hex (stud) and terminal No. 3 is beryllium oxide. Minimum isolation breakdown voltage is 2100 V rms for 1 minute duration.

SYMBOL	INCHES		MILLIMETERS		NOTES
	MIN.	MAX.	MIN.	MAX.	
A	–	0.710	–	18.03	
$\diamond$D$_1$	0.751	0.760	19.08	19.30	
E	0.866	0.872	21.99	22.14	
F	0.182	0.192	4.62	4.87	
J	6.8 NOM.		172.72 NOM.		1
J$_1$	6.3 NOM.		160.02 NOM.		1
M	0.375	0.385	9.52	9.78	
N	0.740	0.760	18.79	19.30	
$\diamond$T	0.060	0.065	1.52	1.65	
$\diamond$T$_1$	0.266	–	6.75	–	
$\diamond$T$_2$	0.144	–	3.70	–	
$\diamond$T$_3$	0.195	0.205	4.95	5.20	
$\diamond$W	1/2-20 NF 2A	1/2-20 NF 2A			2

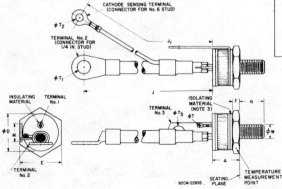

92CM-22835

HF-28

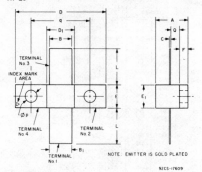

NOTE: EMITTER IS GOLD PLATED

92CS-17609

SYMBOL	INCHES		MILLIMETERS	
	MIN.	MAX.	MIN.	MAX.
A	0.225	0.250	5.72	6.35
B	0.145	0.160	3.69	4.06
B_1	0.165	0.180	4.20	4.57
C	0.004	0.010	0.102	0.254
D	0.657	0.667	16.69	16.94
D_1	0.190	0.210	4.83	5.33
E	0.155	0.165	3.94	4.19
E_1	0.140	0.165	3.56	4.19
F	0.058	0.063	1.48	1.72
L	0.235	0.265	5.97	6.73
ϕp	0.090	0.096	2.286	2.438
Q	0.062	0.077	1.58	1.95
q	0.420	0.440	10.67	11.17

MILLIMETER DIMENSIONS ARE DERIVED
FROM ORIGINAL INCH DIMENSIONS

HF-31

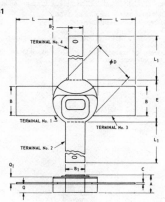

SYMBOL	INCHES		MILLIMETERS		NOTES
	MIN.	MAX.	MIN.	MAX.	
A	0.090	0.135	2.29	3.42	—
B	0.195	0.205	4.96	5.20	—
B_1	0.135	0.145	3.43	3.68	—
B_2	0.095	0.105	2.42	2.66	—
C	0.004	0.010	0.11	0.25	1
ϕD	0.305	0.320	7.48	8.12	—
E	0.275	0.300	6.99	7.62	—
L	0.265	0.290	6.74	7.36	—
L_1	0.455	0.510	11.56	12.95	—
Q	0.055	0.070	1.40	1.77	—
Q_1	0.025	0.045	0.64	1.14	—

MILLIMETER DIMENSIONS ARE DERIVED
FROM ORIGINAL INCH DIMENSIONS
NOTE: 1. TYPICAL FOR ALL LEADS

92CS-4462 R1

HF-32

92CS-19827

SYMBOL	INCHES		MILIMETERS		NOTES
	MIN.	MAX.	MIN.	MAX.	
A	0.160	0.210	4.07	5.33	
b	0.135	0.145	3.429	3.683	
b_1	0.095	0.105	2.413	2.667	
c	0.004	0.010	0.102	0.254	1
ϕD	0.305	0.320	7.75	8.12	
E	0.275	0.300	6.99	7.62	
F_1	0.057	0.067	1.448	1.701	
L	0.455	0.510	11.56	12.95	
ϕP	0.115	0.125	2.921	3.175	
Q	0.085	0.105	2.16	2.66	
Q_1	—	—	—	—	2
q	0.590	0.610	14.99	15.49	
R	0.115	0.125	2.921	3.175	

MILLIMETER DIMENSIONS ARE DERIVED
FROM ORIGINAL INCH DIMENSIONS

NOTES: 1. TYPICAL TWO LEADS.
2. BODY CONTOUR OPTIONAL WITHIN Q_1, ϕD,
AND E.

HF-36

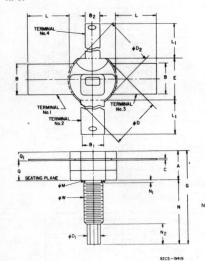

SYMBOL	INCHES		MILLIMETERS		NOTES
	MIN.	MAX.	MIN.	MAX.	
A	0.185	0.240	4.70	6.11	–
B	0.195	0.205	4.96	5.20	–
B_1	0.135	0.145	3.43	3.68	–
B_2	0.095	0.105	2.42	2.66	–
C	0.004	0.010	0.11	0.25	3
ϕD	0.319	0.335	8.12	8.52	–
ϕD_1	0.033	0.065	0.84	1.65	1
ϕD_2	0.305	0.320	7.48	8.12	–
E	0.275	0.300	6.99	7.62	–
G	0.635	0.730	16.11	18.51	–
L	0.265	0.290	6.74	7.36	–
L_1	0.455	0.510	11.56	12.95	–
ϕM	0.120	0.163	3.05	4.14	–
N	0.450	0.490	11.41	12.45	–
N_1	–	0.078	–	1.98	4
N_2	0.095	0.135	2.42	3.43	–
Q	0.145	0.170	3.68	4.31	–
Q_1	0.025	0.045	0.64	1.14	–
ϕW	0.1399	0.1437	3.531	3.632	2

MILLIMETER DIMENSIONS ARE DERIVED
FROM ORIGINAL INCH DIMENSIONS

NOTES: 1.   0.053–0.064 INCH (1.35 – 1.62 mm) WRENCH FLAT.
           2.   PITCH DIA. OF 8-32 UNC-2A COATED THREAD. (ASA B1. 1-1960).
           3.   TYPICAL FOR ALL LEADS.
           4.   LENGTH OF INCOMPLETE OR UNDERCUT THREADS OF ϕW
           5.   RECOMMENDED TORQUE: 5 INCH-POUNDS

92CS–19419

HF-40

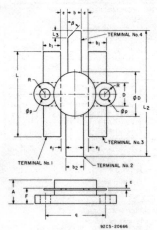

SYMBOL	INCHES		MILLIMETERS	
	MIN.	MAX.	MIN.	MAX.
A	0.260	0.280	6.604	7.112
b	0.153	0.157	3.866	3.987
b_1	0.210	0.220	5.334	5.588
b_2	0.203	0.207	5.156	5.257
c	0.006	0.007	0.153	0.178
D	0.240	0.250	6.096	6.350
ϕD	0.490	0.510	12.446	12.954
e	0.070	0.080	1.778	2.032
e_1	0.045	0.055	1.143	1.397
F	0.165	0.185	4.191	4.699
L	0.970	0.990	24.638	25.146
L_2	1.430	1.470	36.322	37.338
L_3	0.070	0.080	1.778	2.032
ϕP	0.115	0.125	2.921	3.175
q	0.723	0.728	18.364	18.491
R	0.120	0.130	3.048	3.302
B	45°		45°	

MILLIMETER DIMENSIONS ARE DERIVED
FROM ORIGINAL INCH DIMENSIONS

92CS–20666

HF-41

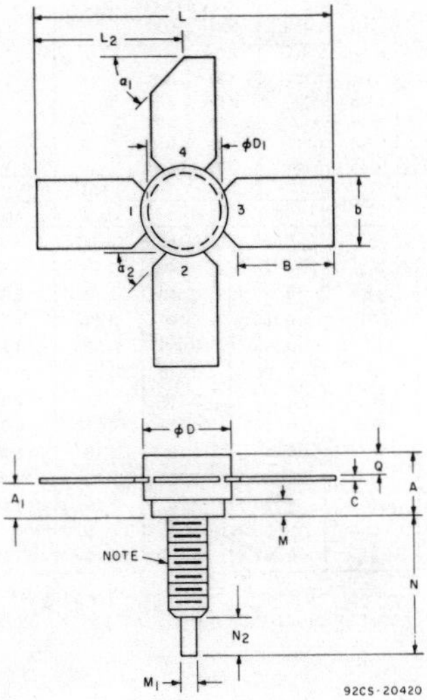

92CS-20420

SYMBOL	INCHES		MILLIMETERS	
	MIN.	MAX.	MIN.	MAX.
A	0.185	0.215	4.70	5.46
A₁	0.114	0.122	2.90	3.10
B	0.380	0.390	9.66	9.90
b	0.220	0.230	5.58	5.84
C	0.002	0.008	0.05	0.20
φD	0.270	0.290	6.86	7.38
φD₁	0.245	0.255	6.22	6.48
L	1.040	1.060	26.42	26.93
L₂	0.520	0.530	13.20	13.45
M	0.058	0.062	1.47	1.57
M₁	C.056	0.064	1.42	1.62
N	0.445	0.455	11.29	11.55
N₂	0.125	0.135	3.18	3.43
Q	0.070	0.090	1.78	2.28
α₁	45° NOM.		45° NOM.	
α₂	45° NOM.		45° NOM.	

MILLIMETER DIMENSIONS ARE DERIVED
FROM ORIGINAL INCH DIMENSIONS

NOTE: PITCH DIA. OF 8-32 UNF-2A COATED
THREAD (ASA B1. 1-1960)

HF-44

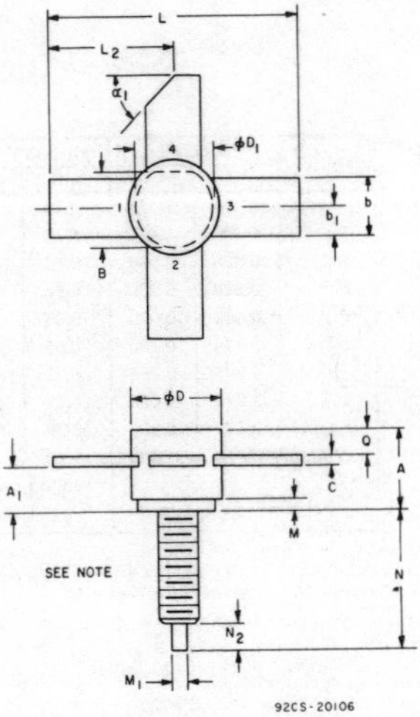

92CS-20106

SYMBOL	INCHES		MILLIMETERS	
	MIN.	MAX.	MIN.	MAX.
A	0.250	0.275	6.35	6.98
A₁	0.163	0.173	4.141	4.394
B	0.299	0.307	7.595	7.797
b	0.221	0.229	5.614	5.816
b₁	0.110	C.115	2.794	2.921
C	0.0045	0.006	0.113	0.152
φD	0.370	0.390	9.40	9.90
φD₁	0.320	0.330	8.128	8.382
L	1.040	1.055	26.42	26.79
L₂	0.520	0.530	13.208	13.462
M	0.070	0.080	1.778	2.032
M₁	0.055	0.065	1.397	1.651
N	0.455	0.475	11.56	12.06
N₂	0.100	0.130	2.54	3.30
Q	0.085	0.095	2.159	2.413
α₁	45° NOM.		45° NOM.	

MILLIMETER DIMENSIONS ARE DERIVED
FROM ORIGINAL INCH DIMENSIONS

NOTE: PITCH DIA. OF 8-32 UNC-2A COATED THREAD
(ASA B1. 1-1960)

HF-46

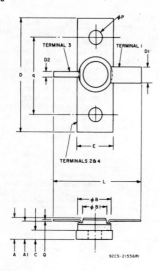

SYMBOL	INCHES		MILLIMETERS	
	Min.	Max.	Min.	Max.
A	0.155	0.165	3.937	4.191
A1	0.120	0.140	3.05	3.55
ϕB	0.225	0.240	5.72	6.00
ϕB1	0.160	0.180	4.07	4.57
C	0.055	0.065	1.397	1.651
D	0.790	0.810	20.07	20.57
D1	0.113	0.117	2.871	2.971
D2	0.028	0.032	0.712	0.812
E	0.240	0.260	6.10	6.60
L	0.740	0.760	18.80	19.30
ϕP	0.120	0.132	3.26	3.35
Q	0.005 Nom.		0.127 Nom.	
q	0.557	0.567	14.15	14.40

MILLIMETER DIMENSIONS ARE DERIVED
FROM ORIGINAL INCH DIMENSIONS

92CS-21556RI

HF-47

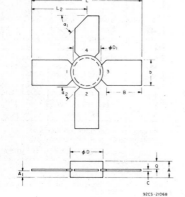

SYMBOL	INCHES		MILLIMETERS	
	MIN.	MAX.	MIN.	MAX.
A	0.127	0.153	3.23	3.89
A1	0.056	0.060	1.43	1.53
B	0.380	0.390	9.66	9.90
b	0.220	0.230	5.58	5.84
C	0.002	0.008	0.05	0.20
ϕD	0.270	0.290	6.86	7.38
ϕD1	0.245	0.255	6.22	6.48
L	1.040	1.060	26.42	26.93
L2	0.520	0.530	13.20	13.45
Q	0.070	0.090	1.78	2.28
α1	45° NOM.		45° NOM.	
α2	45° NOM.		45° NOM.	

MILLIMETER DIMENSIONS ARE DERIVED
FROM ORIGINAL INCH DIMENSIONS

92CS-21068

HF-50

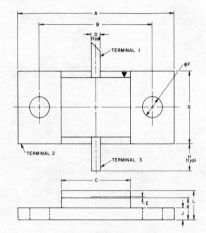

SYMBOL	INCHES		MILLIMETERS	
	MIN.	MAX.	MIN.	MAX.
A	0.890	0.910	22.61	23.11
B	0.645	0.655	16.39	16.63
C	0.390	0.410	9.91	10.41
D	0.045	0.055	1.14	1.35
E	0.004	0.010	0.10	0.25
ϕF	0.117	0.125	2.97	3.17
G	0.390	0.410	9.91	10.41
H	0.115	0.150	2.92	3.81
J	0.057	0.067	1.45	1.70
K	0.110	0.130	2.79	3.30
L	0.150	0.230	3.81	5.84

MILLIMETER DIMENSIONS ARE DERIVED
FROM ORIGINAL INCH DIMENSIONS

92CS-23888RI

HF-55

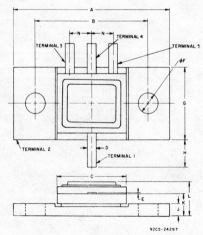

SYMBOL	INCHES		MILLIMETERS	
	MIN.	MAX.	MIN.	MAX.
A	0.890	0.910	22.61	23.11
B	0.645	0.655	16.39	16.63
C	0.390	0.405	9.91	10.29
D	0.045	0.055	1.14	1.35
E	0.004	0.010	0.10	0.25
ϕF	0.117	0.125	2.97	3.17
G	0.390	0.410	9.91	10.41
H	0.115	0.150	2.92	3.81
J	0.057	0.067	1.45	1.70
K	0.110	0.130	2.79	3.30
L	0.150	0.230	3.81	5.84
N	0.135	0.145	3.23	3.68

MILLIMETER DIMENSIONS ARE DERIVED
FROM ORIGINAL INCH DIMENSIONS

92CS-24297

HF-56

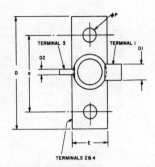

SYMBOL	INCHES		MILLIMETERS	
	Min.	Max.	Min.	Max.
A	0.155	0.165	3.937	4.191
A1	0.120	0.140	3.05	3.55
ϕB	0.225	0.240	5.72	6.00
ϕB1	0.160	0.180	4.07	4.57
C	0.055	0.065	1.397	1.651
D	0.790	0.810	20.07	20.57
D1	0.113	0.117	2.871	2.971
D2	0.028	0.032	0.712	0.812
E	0.240	0.260	6.10	6.60
L	0.440	0.460	11.18	11.68
ϕP	0.120	0.132	3.26	3.35
Q	0.005 Nom.		0.127 Nom.	
q	0.557	0.567	14.15	14.40

MILLIMETER DIMENSIONS ARE DERIVED
FROM ORIGINAL INCH DIMENSIONS

92CM-23960RI

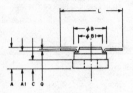

HYBRID-CIRCUIT PACKAGE

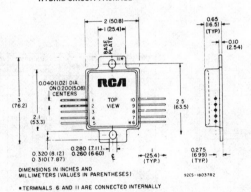

DIMENSIONS IN INCHES AND
MILLIMETERS (VALUES IN PARENTHESES)

92CS-18037R2

*TERMINALS 6 AND II ARE CONNECTED INTERNALLY

Ceramic Flat Packs

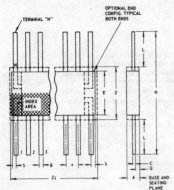

TERMINAL "N"

OPTIONAL END
CONFIG. TYPICAL
BOTH ENDS

INDEX AREA

BASE AND SEATING PLANE

NOTES:

1. Refer to Rules for Dimensioning Peripheral Lead Outlines.
2. Leads within .005'' (.12 mm) radius of True Position (TP) at maximum material condition.
3. N is the maximum quantity of lead positions.
4. Z and Z_1 determine a zone within which all body and lead irregularities lie.

JEDEC MO-004-AF 14-LEAD

SYMBOL	INCHES		NOTE	MILLIMETERS	
	MIN.	MAX.		MIN.	MAX.
A	0.008	0.100		0.21	2.54
B	0.015	0.019	1	0.381	0.482
C	0.003	0.006	1	0.077	0.152
e	0.050 TP		2	1.27 TP	
E	0.200	0.300		5.1	7.6
H	0.600	1.000		15.3	25.4
L	0.150	0.350		3.9	8.8
N	14		3	14	
Q	0.005	0.050		0.13	1.27
S	0.000	0.050		0.00	1.27
Z	0.300		4	7.62	
Z_1	0.400		4	10.16	

92SS-4300RI

JEDEC MO-004-AG 16-LEAD

SYMBOL	INCHES		NOTE	MILLIMETERS	
	MIN.	MAX.		MIN.	MAX.
A	0.008	0.100		0.21	2.54
B	0.015	0.019	1	0.381	0.482
C	0.003	0.006	1	0.077	0.152
e	0.050 TP		2	1.27 TP	
E	0.200	0.300		5.1	7.6
H	0.600	1.000		15.3	25.4
L	0.150	0.350		3.9	8.8
N	16		3	16	
Q	0.005	0.050		0.13	1.27
S	0.000	0.025		0.00	0.63
Z	0.300		4	7.62	
Z_1	0.400		4	10.16	

92CS-17271RI

24-LEAD

SYMBOL	INCHES		NOTE	MILLIMETERS	
	MIN.	MAX.		MIN.	MAX.
A	0.075	0.120		1.91	3.04
B	0.018	0.022	1	0.458	0.558
C	0.004	0.007	1	0.102	0.177
e	0.050 TP		2	1.27 TP	
E	0.600	0.700		15.24	17.78
H	1.150	1.350		29.21	34.29
L	0.225	0.325		5.72	8.25
N	24		3	24	
Q	0.035	0.070		0.89	1.77
S	0.060	0.110	1	1.53	2.79
Z	0.700		4	17.78	
Z_1	0.750		4	19.05	

92CS-19949

28-LEAD

SYMBOL	INCHES		NOTE	MILLIMETERS	
	MIN.	MAX.		MIN.	MAX.
A	0.075	0.120		1.91	3.04
B	0.018	0.022	1	0.458	0.558
C	0.004	0.007	1	0.102	0.177
e	0.050 TP		2	1.27 TP	
E	0.600	0.700		15.24	17.78
H	1.150	1.350		29.21	34.29
L	0.225	0.325		5.72	8.25
N	28		3	28	
Q	0.035	0.070		0.89	1.77
S	0	0.060	1	0	1.53
Z	0.700		4	17.78	
Z_1	0.750		4	19.05	

92CS-20972

When these devices are supplied solder-dipped, the maximum lead thickness (narrow portion) will not exceed 0.013''.

Ceramic Dual-in-Line Packages

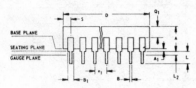

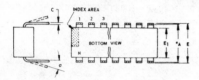

NOTES:

1. Refer to Rules for Dimensioning (JEDEC Publication No. 13) for Axial Lead Product Outlines.
2. Leads within 0.005" (0.12 mm) radius of True Position (TP) at gauge plane with maximum material condition and unit installed.
3. e_A applies in zone L_2 when unit installed.
4. a applies to spread leads prior to installation.
5. N is the maximum quantity of lead positions.
6. N_1 is the quantity of allowable missing leads.

JEDEC MO-001-AD
14-Lead Welded-Seal

SYMBOL	INCHES		NOTE	MILLIMETERS	
	MIN.	MAX.		MIN.	MAX.
A	0.120	0.160		3.05	4.06
A_1	0.020	0.065		0.51	1.65
B	0.014	0.020		0.356	0.508
B_1	0.050	0.065		1.27	1.65
C	0.008	0.012		0.204	0.304
D	0.745	0.770		18.93	19.55
E	0.300	0.325		7.62	8.25
E_1	0.240	0.260		6.10	6.60
e_1	0.100 TP		2	2.54 TP	
e_A	0.300 TP		2, 3	7.62 TP	
L	0.125	0.150		3.18	3.81
L_2	0.000	0.030		0.000	0.76
a	0º	15º	4	0º	15º
N	14		5	14	
N_1	0		6	0	
Q_1	0.050	0.085		1.27	2.15
S	0.065	0.090		1.66	2.28

92SS-441(R)

JEDEC MO-001-AE
16-Lead Welded-Seal

SYMBOL	INCHES		NOTE	MILLIMETERS	
	MIN.	MAX.		MIN.	MAX.
A	0.120	0.160		3.05	4.06
A_1	0.020	0.065		0.51	1.65
B	0.014	0.020		0.356	0.508
B_1	0.035	0.065		0.89	1.65
C	0.008	0.012		0.204	0.304
D	0.745	0.785		18.93	19.93
E	0.300	0.325		7.62	8.25
E_1	0.240	0.260		6.10	6.60
e_1	0.100 TP		2	2.54 TP	
e_A	0.300 TP		2, 3	7.62 TP	
L	0.125	0.150		3.18	3.81
L_2	0.000	0.300		0.000	0.76
a	0º	15º	4	0º	15º
N	16		5	16	
N_1	0		6	0	
Q_1	0.050	0.085		1.27	2.15
S	0.015	0.060		0.39	1.52

92SS-4286R2

When these devices are supplied solder-dipped, the maximum lead thickness (narrow portion) will not exceed 0.013".

Ceramic Dual-in-Line Packages (Cont'd)

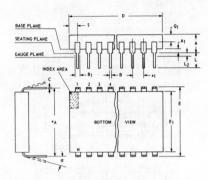

NOTES:

1. REFER TO RULES FOR DIMENSIONING (JEDEC PUBLICATION No. 13) AXIAL LEAD PRODUCT OUTLINES.

2. WHEN BASE OF BODY IS TO BE ATTACHED TO HEAT SINK, TERMINAL LEAD STAND-OFFS ARE NOT REQUIRED AND $A_1 = 0$. WHEN $A_1 = 0$, THE LEADS EMERGE FROM THE BODY WITH THE B_1 DIMENSION AND REDUCE TO THE B DIMENSION ABOVE THE SEATING PLANE.

3. e_1 AND e_A APPLY IN ZONE L_2 WHEN UNIT INSTALLED. LEADS WITHIN .005" RADIUS OF TRUE POSITION (TP) AT GAUGE PLANE WITH MAXIMUM MATERIAL CONDITION.

4. APPLIES TO SPREAD LEADS PRIOR TO INSTALLATION.

5. N IS THE MAXIMUM QUANTITY OF LEAD POSITIONS.

6. N_1 IS THE QUANTITY OF ALLOWABLE MISSING LEADS.

24-Lead Welded-Seal

SYMBOL	INCHES		NOTE	MILLIMETERS	
	MIN.	MAX.		MIN.	MAX.
A	0.090	0.150		2.29	3.81
A_1	0.020	0.065	2	0.51	1.65
B	0.015	0.020		0.381	0.508
B_1	0.045	0.055		1.143	1.397
C	0.008	0.012		0.204	0.304
D	1.15	1.22		29.21	30.98
E	0.600	0.625		15.24	15.87
E_1	0.480	0.520		12.20	13.20
e_1	0.100 TP		3	2.54 TP	
e_A	0.600 TP		3	15.24 TP	
L	0.100	0.180		2.54	4.57
L_2	0.000	0.030	3	0.00	0.76
a	0°	15°	4	0°	15°
N	24		5	24	
N_1	0		6	0	
Q_1	0.020	0.080		0.51	2.03
S	0.020	0.060		0.51	1.52

92CS-19948

JEDEC MO-015-AH
28-Lead Welded-Seal

SYMBOL	INCHES		MILLIMETERS		NOTES
	MIN.	MAX.	MIN.	MAX.	
A	.100	.200	2.6	5.0	2
A_1	.000	.070	0	1.77	
B	.015	.020	.381	.508	
B_1	.015	.055	.39	1.39	
C	.008	.012	.204	.304	
D	1.380	1.420	35.06	36.06	
E	.600	.625	15.24	15.87	
E_1	.485	.515	12.32	13.08	
e_1	.100 TP		2.54 TP		3
e_A	.600 TP		15.24 TP		3
L	.100	.200	2.6	5.0	
L_2	.000	.030	0	.76	
a	0	15	0°	15°	4
N	28		28		5
N_1	0		0		6
Q_1	.020	.070	.51	1.77	
S	.040	.070	1.02	1.77	
See Note	1				

92CM-20250

When these devices are supplied solder-dipped, the maximum lead thickness (narrow portion) will not exceed 0.013".

Ceramic Dual-in-Line Packages (Cont'd)

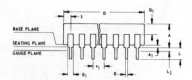

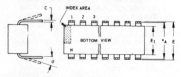

NOTES

1. Refer to Rules for Dimensioning (JEDEC Publication No. 13) for Axial Lead Product Outlines.

2. Leads within 0.005" (0.12 mm) radius of True Position (TP) at gauge plane with maximum material condition and unit installed.

3. e_A applies in zone L_2 when unit installed.

4. a applies to spread leads prior to installation.

5. N is the maximum quantity of lead positions.

6. N_1 is the quantity of allowable missing leads.

7. B_1 applies to all leads except the four end leads which have one-half the normal width (B_1 min. = 0.025 in.)

JEDEC MO-001-AC
16-Lead
(except types CD4026AF, CD4029AF, CD4031AF, CD4033AF)

SYMBOL	INCHES		NOTE	MILLIMETERS	
	MIN.	MAX.		MIN.	MAX.
A	0.155	0.200		3.94	5.08
A$_1$	0.020	0.050		0.51	1.27
B	0.014	0.020		0.356	0.508
B$_1$	0.035	0.065		0.89	1.65
C	0.008	•0.012		0.204	0.304
D	0.745	0.785		18.93	19.93
E	0.300	0.325		7.62	8.25
E$_1$	0.240	0.260		6.10	6.60
e$_1$	0.100 TP		2	2.54 TP	
e$_A$	0.300 TP		2, 3	7.62 TP	
L	0.125	0.150		3.18	3.81
L$_2$	0.000	0.030		0.000	0.76
a	0⁰	15⁰	4	0⁰	15⁰
N	16		5	16	
N$_1$	0		6	0	
Q$_1$	0.040	0.075		1.02	1.90
S	0.015	0.060		0.39	1.52

92CM-15967RI

JEDEC MO-001-AG
16-Lead
(Types CD4026AF, CD4029AF, CD4031AF, CD4033AF)

SYMBOL	INCHES		NOTE	MILLIMETERS	
	MIN.	MAX.		MIN.	MAX.
A	*0.165*	*0.210*		*4.20*	*5.33*
A$_1$	*0.015*	*0.045*		*0.381*	*1.14*
B	*0.015*	0.020		*0.381*	0.508
B$_1$	*0.045*	*0.070*	7	*1.15*	*1.77*
C	*0.009*	•*0.011*		*0.229*	*0.279*
D	*0.750*	*0.795*		*19.05*	*20.19*
E	*0.295*	0.325		*7.50*	8.25
E$_1$	*0.245*	*0.300*		*6.23*	*7.62*
e$_1$	0.100 TP		2	2.54 TP	
e$_A$	0.300 TP		2, 3	7.62 TP	
L	*0.120*	*0.160*		*3.05*	*4.06*
L$_2$	0.000	0.030		0.000	0.76
a	*2⁰*	15⁰	4	*2⁰*	15⁰
N	16		5	16	
N$_1$	0		6	0	
Q$_1$	*0.050*	*0.080*		*1.27*	*2.03*
S	*0.010*	0.060		*0.254*	1.52

92CM-22284

This outline differs from the standard 16-Lead frit-seal ceramic package MO-001-AC as indicated by the values in italics shown in the chart above.

JEDEC MO-001-AB
14-Lead

SYMBOL	INCHES		NOTE	MILLIMETERS	
	MIN.	MAX.		MIN.	MAX.
A	0.155	0.200		3.94	5.08
A$_1$	0.020	0.050		0.51	1.27
B	0.014	0.020		0.356	0.508
B$_1$	0.050	0.065		1.27	1.65
C	0.008	•0.012		0.204	0.304
D	0.745	0.770		18.93	19.55
E	0.300	0.325		7.62	8.25
E$_1$	0.240	0.260		6.10	6.60
e$_1$	0.100 TP		2	2.54 TP	
e$_A$	0.300 TP		2, 3	7.62 TP	
L	0.125	0.150		3.18	3.81
L$_2$	0.000	0.030		0.000	0.76
a	0⁰	15⁰	4	0⁰	15⁰
N	14		5	14	
N$_1$	0		6	0	
Q$_1$	0.040	0.075		1.02	1.90
S	0.065	0.090		1.66	2.28

92SS-4296RI

When these devices are supplied solder-dipped, the maximum lead thickness (narrow portion) will not exceed 0.013".

Ceramic Dual-in-Line Welded Seal

16-Lead Side-Brazed

Plastic Dual-in-Line Packages

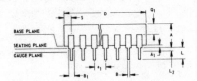

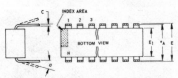

*WHEN THIS DEVICE IS SUPPLIED SOLDER-DIPPED, THE MAX. LEAD THICKNESS (NARROW PORTION) WILL NOT EXCEED 00I3 (O.33mm)

NOTE: DIMENSIONS IN PARENTHESES ARE IN MILLIMETERS AND ARE DERIVED FROM THE BASIC INCH DIMENSIONS

92CS-2I2I9

NOTES:

1. Refer to Rules for Dimensioning (JEDEC Publication No. 13) for Axial Lead Product Outlines.

2. Leads within 0.005'' (0.12 mm) radius of True Position (TP) at guage plane with maximum material condition and unit installed.

3. e_A applies in zone L_2 when unit installed.

4. a applies to spread leads prior to installation.

5. N is the maximum quantity of lead positions.

6. N_1 is the quantity of allowable missing leads.

JEDEC MO-001-AB
14-Lead

SYMBOL	INCHES		NOTE	MILLIMETERS	
	MIN.	MAX.		MIN.	MAX.
A	0.155	0.200		3.94	5.08
A_1	0.020	0.050		0.51	1.27
B	0.014	0.020		0.356	0.508
B_1	0.050	0.065		1.27	1.65
C	0.008	0.012		0.204	0.304
D	0.745	0.770		18.93	19.55
E	0.300	0.325		7.62	8.25
E_1	0.240	0.260		6.10	6.60
e_1	0.100 TP		2	2.54 TP	
e_A	0.300 TP		2, 3	7.62 TP	
L	0.125	0.150		3.18	3.81
L_2	0.000	0.030		0.000	0.76
a	0⁰	15⁰	4	0⁰	15⁰
N	14		5	14	
N_1	0		6	0	
Q_1	0.040	0.075		1.02	1.90
S	0.065	0.090		1.66	2.28

92SS-4296RI

JEDEC MO-001-AC
16-Lead

SYMBOL	INCHES		NOTE	MILLIMETERS	
	MIN.	MAX.		MIN.	MAX.
A	0.155	0.200		3.94	5.08
A_1	0.020	0.050		0.51	1.27
B	0.014	0.020		0.356	0.508
B_1	0.035	0.065		0.89	1.65
C	0.008	0.012		0.204	0.304
D	0.745	0.785		18.93	19.93
E	0.300	0.325		7.62	8.25
E_1	0.240	0.260		6.10	6.60
e_1	0.100 TP		2	2.54 TP	
e_A	0.300 TP		2, 3	7.62 TP	
L	0.125	0.150		3.18	3.81
L_2	0.000	0.030		0.000	0.76
a	0⁰	15⁰	4	0⁰	15⁰
N	16		5	16	
N_1	0		6	0	
Q_1	0.040	0.075		1.02	1.90
S	0.015	0.060		0.39	1.52

92CM-I5967RI

When these devices are supplied solder-dipped, the maximum lead thickness (narrow portion) will not exceed 0.013''.

MODIFIED 16-LEAD QUAD-IN-LINE PLASTIC PACKAGE
WITH INTEGRAL BENT DOWN WING-TAB HEAT SINK

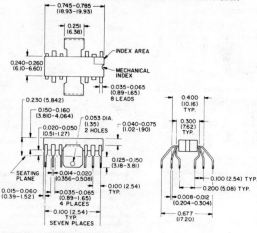

DIMENSIONS IN PARENTHESES ARE MILLIMETER
EQUIVALENTS OF THE BASIC INCH DIMENSIONS

92CM-25044

MODIFIED 16-LEAD QUAD-IN-LINE PLASTIC PACKAGE
WITH INTEGRAL FLAT WING—TAB HEAT SINK

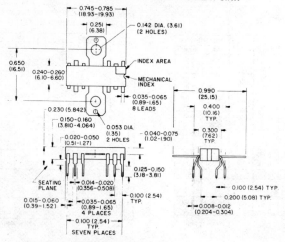

DIMENSIONS IN PARENTHESES ARE MILLIMETER
EQUIVALENTS OF THE BASIC INCH DIMENSIONS

92CM-25045

JEDEC MO-002-AL 8-LEAD TO-5 STYLE

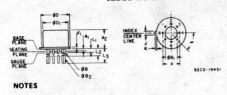

SYMBOL	INCHES MIN.	INCHES MAX.	NOTE	MILLIMETERS MIN.	MILLIMETERS MAX.
a	0.200 TP		2	5.88 TP	
A_1	0.010	0.050		0.26	1.27
A_2	0.165	0.185		4.20	4.69
øB	0.016	0.019	3	0.407	0.482
$øB_1$	0.125	0.160		3.18	4.06
$øB_2$	0.016	0.021	3	0.407	0.533
øD	0.335	0.370		8.51	9.39
øD	0.305	0.335		7.75	8.50
F_1	0.020	0.040		0.51	1.01
j	0.028	0.034		0.712	0.863
k	0.029	0.045	4	0.74	1.14
L_1	0.000	0.050	3	0.00	1.27
L_2	0.250	0.500	3	6.4	12.7
L_3	0.500	0.562	3	12.7	14.27
α	45° TP			45° TP	
N	8		6	8	
N_1	3		5	3	

92CS-19431

NOTES

1. Refer to Rules for Dimensioning Axial Lead Product Outlines.

2. Leads at gauge plane within 0.007'' (0.178 mm) radius of True Position (TP) at maximum material condition.

3. øB applies between L_1 and L_2. $øB_2$ applies between L_2 and 0.500'' (12.70 mm) from seating plane. Diameter is uncontrolled in L_1 and beyond 0.500'' (12.70 mm).

4. Measure from Max. øD.

5. N_1 is the quantity of allowable missing leads.

6. N is the maximum quantity of lead positions.

8-LEAD TO-5 STYLE WITH DUAL-IN-LINE FORMED LEADS (DIL-CAN)

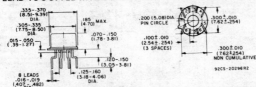

92CS-20296R2

10-LEAD TO-5 STYLE PACKAGE JEDEC MO-006-AF

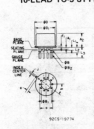

92CS-19774

SYMBOL	INCHES MIN.	INCHES MAX.	NOTE	MILLIMETERS MIN.	MILLIMETERS MAX.
a	0.230 TP		2	5.84 TP	
A_1	0	0		0	0
A_2	0.165	0.185		4.19	4.70
øB	0.016	0.019	3	0.407	0.482
$øB_1$	0	0		0	0
$øB_2$	0.016	0.021	3	0.407	0.533
øD	0.335	0.370		8.51	9.39
$øD_1$	0.305	0.335		7.75	8.50
F_1	0.020	0.040		0.51	1.01
j	0.028	0.034		0.712	0.863
k	0.029	0.045	4	0.74	1.14
L_1	0.000	0.050	3	0.00	1.27
L_2	0.250	0.500	3	6.4	12.7
L_3	0.500	0.562	3	12.7	14.27
α	360° TP			360° TP	
N	10		6	10	
N_1	1		5	1	

NOTES:

1. Refer to Rules for Dimensioning Axial Lead Product Outlines.

2. Leads at gauge plane within 0.007'' (0.178 mm) radius of True Position (TP) at maximum material condition.

3. øB applies between L_1 and L_2. $øB_2$ applies between L_2 and 0.500'' (12.70 mm) from seating plane. Diamerer is uncontrolled in L_1 and beyond 0.500'' (12.70 mm).

4. Measure from Max. øD.

5. N_1 is the quantity of allowable missing leads.

6. N is the maximum quantity of lead positions.

10 FORMED LEADS RADIALLY ARRANGED TO-5 TYPE

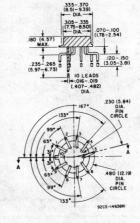

92CS-14638RI

12-LEAD TO-5 STYLE PACKAGE JEDEC MO-006-AG

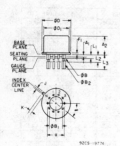

SYMBOL	INCHES MIN.	INCHES MAX.	NOTE	MILLIMETERS MIN.	MILLIMETERS MAX.
a	0.230		2	5.84 TP	
A_1	0	0		0	0
A_2	0.165	0.185		4.19	4.70
øB	0.016	0.019	3	0.407	0.482
$øB_1$	0	0		0	0
$øB_2$	0.016	0.021	3	0.407	0.533
øD	0.335	0.370		8.51	9.39
$øD_1$	0.305	0.335		7.75	8.50
F_1	0.020	0.040		0.51	1.01
j	0.028	0.034		0.712	0.863
k	0.029	0.045	4	0.74	1.14
L_1	0.000	0.050	3	0.00	1.27
L_2	0.250	0.500	3	6.4	12.7
L_3	0.500	0.562	3	12.7	14.27
.	30 TP			30 TP	
N	12		6	12	
N_1	1		5	1	

NOTES:

1. Refer to Rules for Dimensioning Axial Lead Product Outlines.

2. Leads at gauge plane within 0.007" (0.178 mm) radius of True Position (TP) at maximum material condition.

3. øB applies between L_1 and L_2. $øB_2$ applies between L_2 and 0.500" (12.70 mm) from seating plane. Diameter is uncontrolled in L_1 and beyond 0.500" (12.70 mm).

4. Measure from Max. øD.

5. N_1 is the quantity of allowable missing leads.

6. N is the maximum quantity of lead positions.

92CS-19774

8-LEAD DUAL-IN-LINE PLASTIC PACKAGE (MINI-DIP)

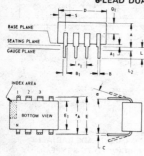

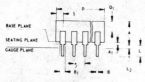

SYMBOL	INCHES MIN.	INCHES MAX.	NOTE	MILLIMETERS MIN.	MILLIMETERS MAX.
A	0.155	0.200		3.94	5.08
A_1	0.020	0.050		0.508	1.27
B	0.014	0.020		0.356	0.508
B_1	0.035	0.066		0.889	1.65
C	0.008	0.012*		0.203	0.304
D	0.370	0.400		9.40	10.16
E	0.300	0.325		7.62	8.25
E_1	0.240	0.260		6.10	6.60
e_1	0.100 TP		2	2.54 TP	
e_A	0.300 TP		2, 3	7.62 TP	
L	0.125	0.150		3.18	3.81
L_2	0.000	0.030		0.000	0.762
α	0°	15°	4	0°	15°
N	8		5	8	
N_1	0		6		
Q_1	0.040	0.075		1.02	1.90
S	0.015	0.060		0.381	1.52

NOTES:

1. Refer to Rules for Dimensioning (JEDEC Publication No. 13) for Axial Lead Product Outlines.

2. Leads within 0.005" (0.12 mm) radius of True Position (TP) at gauge plane with maximum material condition and unit installed.

3. e_A applies in zone L_2 when unit installed.

4. α applies to spread leads prior to installation.

5. N is the maximum quantity of lead positions.

6. N_1 is the quantity of allowable missing leads.

* When device is supplied solder dipped, the maximum lead thickness (narrow portion) will not exceed 0.013".

92CS-24026

8-LEAD DUAL-IN-LINE FRIT-SEAL PACKAGE

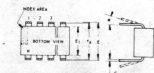

SYMBOL	INCHES MIN.	INCHES MAX.	NOTE	MILLIMETERS MIN.	MILLIMETERS MAX.
A	0.155	0.200		3.94	5.08
A_1	0.020	0.050		0.508	1.27
B	0.014	0.020		0.356	0.508
B_1	0.050	0.065		1.27	1.65
C	0.008	0.012		0.203	0.304
D	0.376	0.396		9.55	10.05
E	0.315	0.345*		8.00	8.76
E_1	0.240	0.260		6.10	6.60
e_1	0.100 TP		2	2.54 TP	
e_A	0.300 TP		2, 3	7.62 TP	
L	0.100	0.150		2.54	3.81
L_2	0.000	0.030		0.000	0.762
α	0°	15°	4	0°	15°
N	8		5	8	
N_1	0		6		
Q_1	0.040	0.075		1.02	1.90
S	0.020	0.060		0.508	1.52

92CM-20827

NOTES:

1. Refer to Rules for Dimensioning (JEDEC Publication No. 13) for Axial Lead Product Outlines.

2. Leads within 0.005" (0.12 mm) radius of True Position (TP) at gauge plane with maximum material condition and unit installed.

3. e_A applies in zone L_2 when unit installed.

4. α applies to spread leads prior to installation.

5. N is the maximum quantity of lead positions.

6. N_1 is the quantity of allowable missing leads.

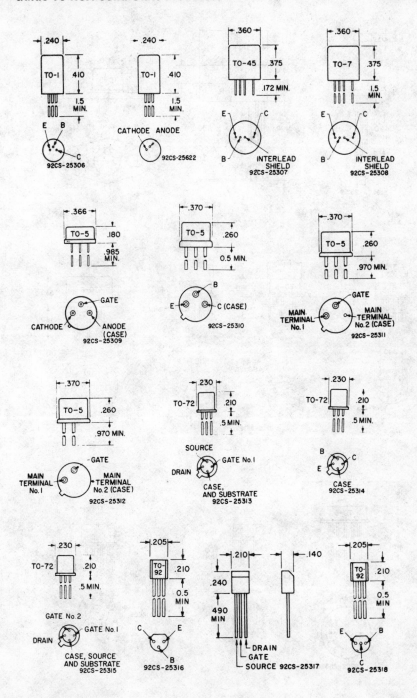

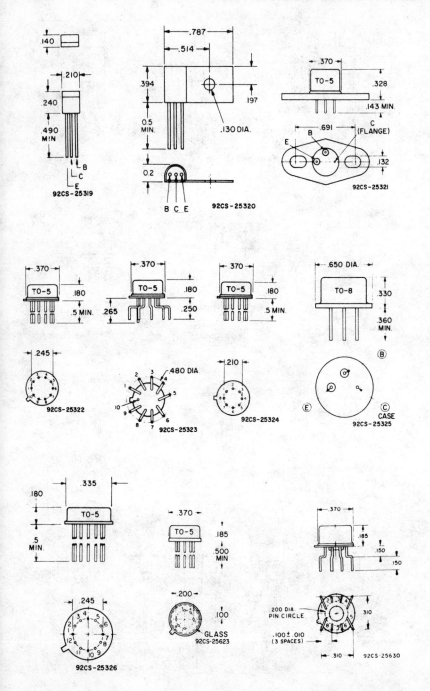

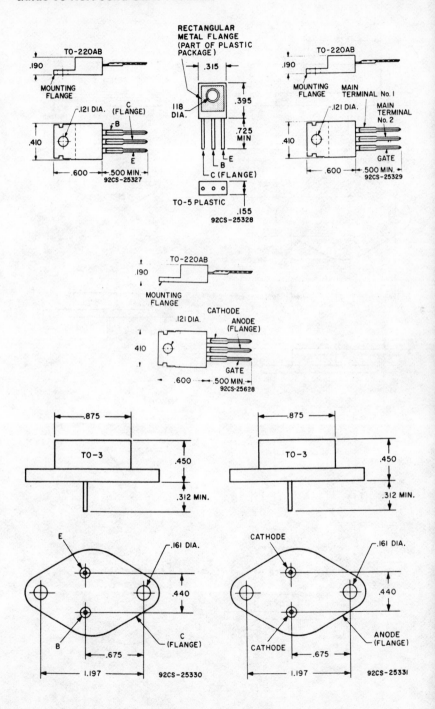

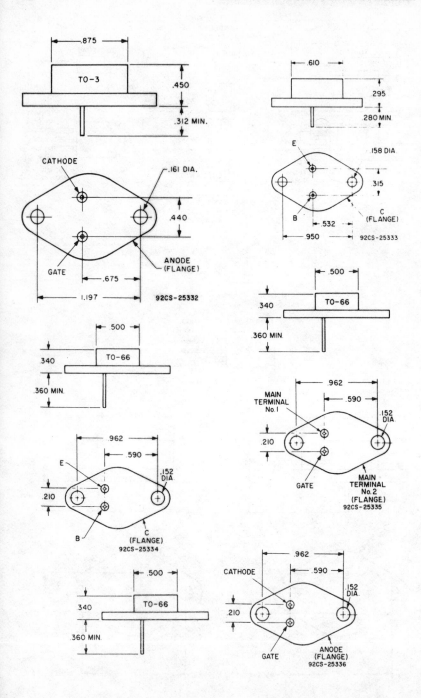

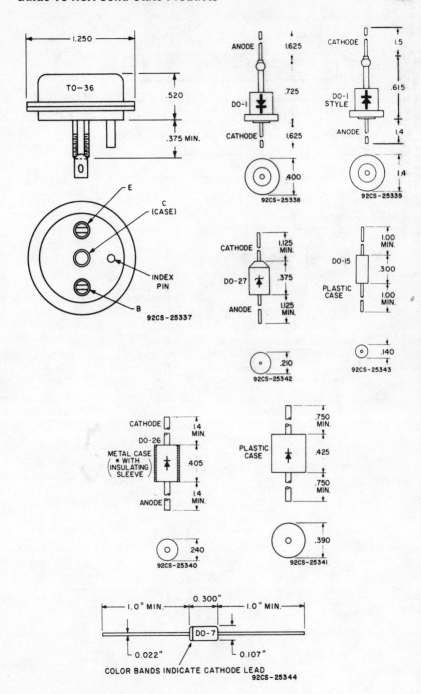

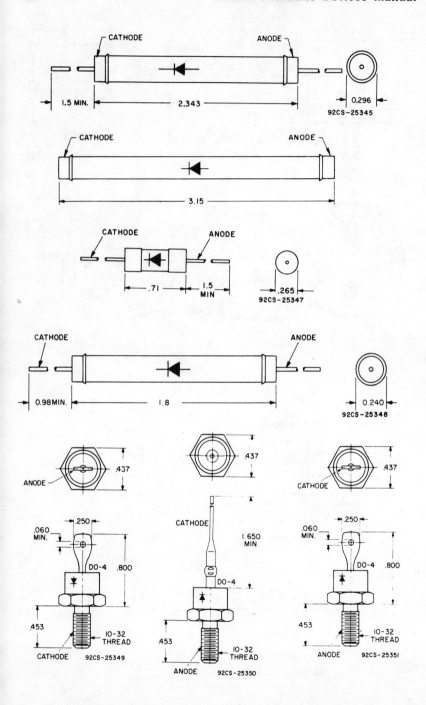

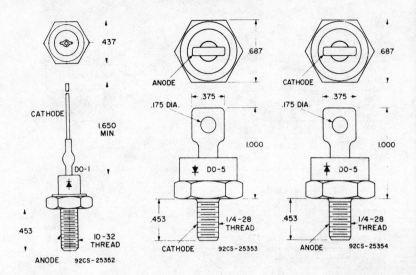

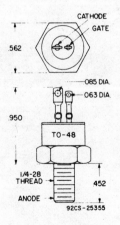

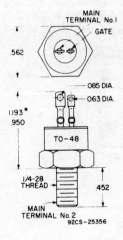

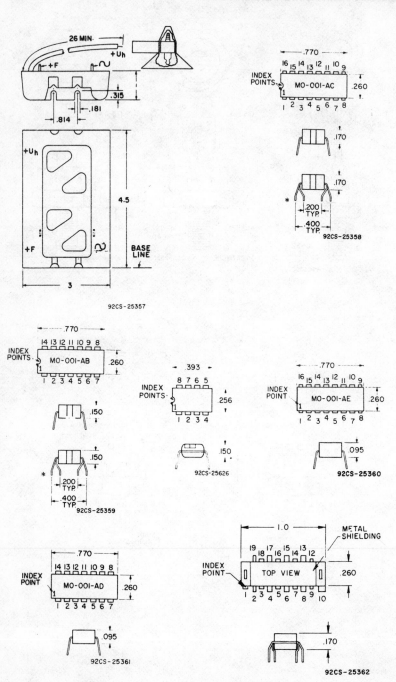

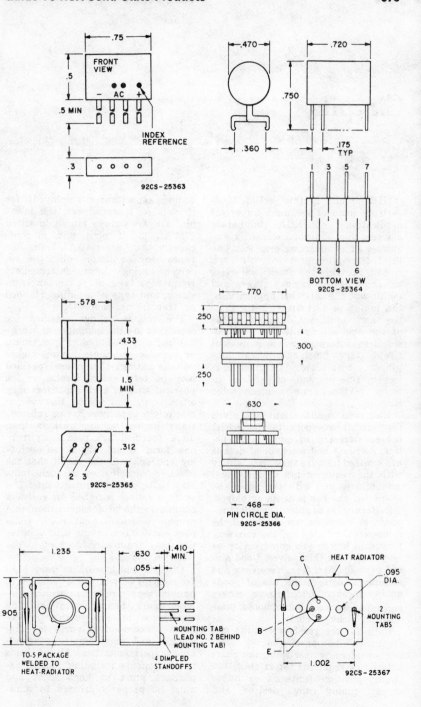

92CS-25363

92CS-25364

92CS-25365

92CS-25366

92CS-25367

Circuits

THE circuits in this section illustrate some of the more important applications of RCA solid-state devices; they are not necessarily examples of commercial practice. The brief description provided with each circuit explains the functional relationships of the various stages and points out the intended applications, the major performance characteristics, and significant design features of the over-all circuit. Detailed descriptive information on individual circuit stages (such as detectors, amplifiers, or oscillators) is given earlier in this Manual, as well as in many textbooks on semiconductor circuits.

Electrical specifications are given for circuit components to assist those interested in home construction. Layouts and mechanical details are omitted because they vary widely with the requirements of individual set builders and with the sizes and shapes of the components employed.

Performance of these circuits depends as much on the quality of the components selected and the care employed in layout and construction as on the circuits themselves. Good signal reproduction from receivers and amplifiers requires the use of good-quality speakers, transformers, chokes and input sources (microphones, phonograph pickups, etc.).

Coils for the receiver circuits can frequently be purchased at local parts dealers by specifying the characteristics required: for rf coils, the circuit position (antenna or interstage), tuning range desired, and

tuning capacitances employed; for if coils or transformers, the intermediate frequency, circuit position (1st if, 2nd if, etc.), and, in some cases, the associated transistor types; for oscillator coils, the receiver tuning range, intermediate frequency, type of converter transistor, and type of winding (tapped or transformer-coupled).

The voltage ratings specified for capacitors are the minimum dc working voltages required. Paper, mica, or ceramic capacitors having higher voltage ratings than those specified may be used except insofar as the physical sizes of such capacitors may affect equipment layout. However, if electrolytic capacitors having substantially higher voltage ratings than those specified are used, they may not "form" completely at the operating voltage, with the result that the effective capacitances of such units may be below their rated value. The wattage ratings specified for resistors assume methods of construction that provide adequate ventilation; compact installations having poor ventilation may require resistors of higher wattage ratings.

Circuits which work at very high frequencies or which are required to handle very wide bandwidths demand more than ordinary skill and experience in construction. Placement of component parts is quite critical and may require considerable experimentation. All rf leads to components including bypass capacitors must be kept short and must be properly dressed to minimize undesirable coupling and capacitance effects. Correct circuit alignment and oscillator tracking may require the use of a cathode-ray oscilloscope, a high-impedance vacuum-tube voltmeter, and a signal generator capable of supplying a properly modulated signal at the appropriate frequencies. Unless the builder has had considerable experience with broad-band, high-frequency circuits, he should not undertake the construction of such circuits.

List of Circuits

MANUFACTURERS OF SPECIAL COMPONENTS AND MATERIALS
REFERRED TO IN PARTS LISTS

AirDux, trade name of
Icore Electro-Plastics, Inc.
Subsidiary of Icore Industries
1050 Kifer Road
Sunnyvale, Calif.

Allen-Bradley Co.
1201 S. 2nd Street
Milwaukee, Wis.

Alpha Wire Corporation
180 Varick Street
New York, N. Y.

Amphenol Connector Division
Amphenol-Borg Electronics Corp.
1830 South 54th Street
Chicago, Ill.

Arco Electronics, Inc.
Community Drive
Great Neck, N. Y.

Arnold Magnetics
11520 W. Jefferson Blvd.
Culver City, Calif.

B and W, Inc.
Canal and Beaver Dam Road
Bristol, Pa.

Bud Radio, Inc.
4605 E. 355th Street
Willoughby, Ohio

Cambion, trade name of Cambridge
Thermionic Corp.

Cambridge Thermionic Corp. (CTC)
445 Concord Avenue
Cambridge, Mass.

Centralab
Division of Globe Union, Inc.
P.O. Box 591
Milwaukee, Wisc.

Dow Corning Corp.
S. Saginaw Road
Midland, Mich.

Elmwood Sensors, Inc.
1563 Elmwood Avenue
Cranston, R.I.

Freed Transformer Co.
1718 Weirfield Street
Brooklyn, N. Y.

General Ceramics Corp.
Crows Mill Road
Keasby, N. J.

Hammarlund Manufacturing Co.
Hammarlund Drive
Mars Hill, N. C.

International Resistor Corp.
401 N. Broad Street
Philadelphia, Pa.

E. F. Johanson Mfg. Corp.
P.O. Box 329
Boonton, N. J.

Litz, trade name of
Alpha Wire Corp.
180 Varick Street
New York, N. Y.

Magnetic Metals Corp.
Hayes Avenue at 21st Street
Camden, N. J.

MANUFACTURERS (cont'd)

P. R. Mallory and Co. Inc.
3029 E. Washington Street
Indianapolis, Ind.

J. W. Miller Co.
5917 South Main Street
Los Angeles, Calif.

Signal Transformer Co.
1 Junius Street
Brooklyn, N. Y.

Simpson Electric Co.
5200 West Kinzie Street
Chicago, Ill.

Sprague Electric Co.
481 Marshall St.
North Adams, Mass.

Stancor (Chicago-Stancor)
3501 West Addison Street
Chicago, Ill.

Thordarson-Meissner
7th and Bellmont
Mt. Carmel, Ill.

Triad
305 North Briant Street
Huntington, Ind.

United Transformer Corp.
Div. Thompson-Ramo-Wooldridge,
150 Varick Street
New York, N. Y.

Vitramon, Inc.
Box 544
Bridgeport, Conn.

Wakefield Engineering, Inc.
139 Foundry Street
Wakefield, Mass.

NOTES: Components and materials identified by RCA stock numbers may be obtained through authorized RCA distributors. In general, all components specified in the circuit parts lists can be purchased from local radio and electronic supply stores or mail-order houses. If the parts are not available from these sources, they may be obtained from the pertinent manufacturers listed above.

16-1   VOLTAGE REGULATOR, SHUNT TYPE
Regulation 0.5%

Circuit Description

This simple two-transistor shunt-type voltage regulator can provide a constant (within 0.5 per cent) dc output of 28 volts for load currents up to 0.5 ampere and dc inputs from 45 to 55 volts. The two transistors operate as variable resistors to provide the output regulation. A 27-volt zener reference diode is used as the control, or sensing, element.

With a 28-volt output, the reverse-bias-connected reference diode, CR, operates in the breakdown-voltage region. In this region, the voltage drop across the diode remains constant (at the reference potential of 27 volts) over a wide range of reverse currents through the diode.

The output voltage tends to rise with an increase in either the applied voltage or the load-circuit impedance. The current through resistor R_2 and reference diode CR then increases. However, the voltage drop across CR remains constant at 27 volts, and the full increase in the output voltage is developed across R_2. This increased voltage across R_2 is directly coupled to the base of the 2N1481 transistor and increases the forward bias so that the 2N1481 conducts more heavily.

The rise in the emitter current of the 2N1481 increases the forward bias on the 2N3054, and the current through this transistor also increases.

As the increased currents of the transistors flow through resistor R_1, which is in series with the load impedance, the voltage drop across R_1 becomes a larger proportion of the total applied voltage. In this way, any tendency for an increase in the output voltage is immediately reflected as an increased voltage drop across R_1 so that the output voltage delivered to the load circuit remains constant.

If the output voltage tends to decrease slightly, the voltage drop across reference diode CR still remains constant, and the full decrease occurs across R_2. As a result, the forward bias of both transistors decreases so that less current flows through R_1. The resultant decrease in the proportional amount of the applied voltage dropped across this resistor immediately cancels any tendency for a decrease in the output voltage, and the voltage applied to the load circuit again remains constant.

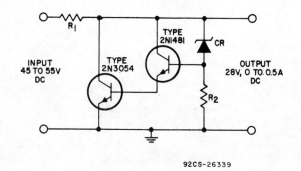

92CS-26339

Parts List

CR = reference diode, 27 V, 0.5 watt
R_1 = 28 ohms, 50 watts (includes source resistance of transformers, rectifiers, etc.)
R_2 = 1000 ohms, 0.5 watt

16-2   **VOLTAGE REGULATOR, SERIES TYPE**
With Adjustable Output
Line Regulation within 1.0%   **Load Regulation within 0.5%**

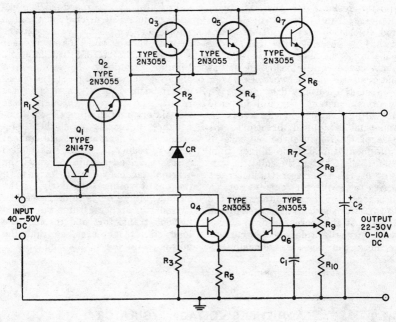

92CS-26338

Parts List

$C_1 = 1\ \mu F$, paper, 25 V	$R_1 = 1200$ ohms, 0.5 watt	$R_7 = 270$ ohms, 0.5 watt
$C_2 = 100\ \mu F$, electrolytic, 50 V	$R_2, R_4, R_6 = 0.1$ ohm, 5 watts	$R_8\ R_{10} = 1000$ ohms, 0.5 watt
$D_1 =$ zener diode, 12 V, 1 watt	$R_3 = 2000$ ohms, 0.5 watt	$R_9 =$ potentiometer, 1000 ohms, 0.5 watt
	$R_5 = 570$ ohms, 0.5 watt	

Circuit Description

In this series-type voltage regulator, regulation is accomplished by varying the current through three paralleled 2N3055 transistors connected in series with the load circuit. A reverse-bias-connected zener diode provides the reference voltage for the circuit. The voltage drop across this diode remains constant at the reference potential of 12 volts over a wide range of current through the diode.

If the output voltage tends to rise for any reason, the total increase in voltage is distributed across bleeder resistors R_8, R_9, and R_{10}. If potentiometer R_9, the output-voltage adjustment, is set to the mid-point of its range, one-half the increase in output voltage is applied to the base of the 2N3053 transistor Q_6. This increased voltage is coupled (through the emitter-to-base junction of transistor Q_6) to the base of the 2N3053

16-2   **VOLTAGE REGULATOR, SERIES TYPE (cont'd)**

Circuit Description (cont'd)

transistor Q_4 by R_5, the common emitter resistor for the two transistors. The zener diode D_1 and its series resistor R_3 are connected in parallel with the bleeder resistors, and the increase in output voltage is also reflected across the diode-resistor network. However, because the voltage drop across D_1 remains constant, the full increase in voltage is developed across R_3 and thus is applied directly to the base of Q_4. Because the increase in voltage at the base is higher than that at the emitter, the collector current of the transistor Q_4 increases.

As the collector current of Q_4 increases, the base voltage of the 2N1479 transistor Q_1 decreases by the amount of the increased drop across R_1. The resultant decrease in current through the 2N1479 transistor Q_1 causes a decrease in the emitter voltage of this transistor. The resultant decrease in current through transistor Q_1 causes a decrease in the emitter voltage and thus in the base voltage of the 2N3055 transistor Q_2. Similar action by Q_2 results in a negative-going voltage at the bases of transistors Q_3, Q_5, and Q_7. As a result, the current through these transistors, and through the load impedance in series with them, decreases. The decrease in load current tends to reduce the voltage developed across the load circuit to cancel the original tendency for an increase in the ouput voltage. Similarly, if the output voltage tends to decrease, the current through the three paralleled 2N3055 transistors and through the load circuit increases, so that the output voltage remains constant.

16-3   **ADJUSTABLE VOLTAGE REGULATOR**
With Current Limiting
0.1 V to 50 V at 1 A

Circuit Description

This voltage regulator employs a CA3130 integrated-circuit comparator as an error amplifier, a CA3086 integrated-circuit transistor array, and several discrete transistors to provide a regulated dc output voltage that is continuously adjustable from 0.1 to 50 volts at currents up to 1 ampere.

The CA3130 integrated circuit is an ideal choice for an error amplifier in regulated power supplies in which the regulated output voltage is required to approach zero. Transistors Q_3 and Q_4 in the CA3086 array function as zener diodes to provide the supply voltage for the CA3130 comparator. Transistors Q_1 (connected as a diode), Q_2 (operated as a zener), and Q_5 are connected to operate as a low-impedance, temperature-compensated source of adjustable reference voltage for the error amplifier.

The output of the CA3130 error amplifier is boosted by a 2N2102 discrete transistor (Q_4) to provide adequate base drive for the Darlington-connected series pass transistors Q_1 (2N2102) and Q_2 (2N3055). The 2N5294 transistor Q_3 functions as a current-limiting device by diverting base drive from the series pass transistors in accordance with the setting of the Current Limit Adjust potentiometer R_9

16-3   **ADJUSTABLE VOLTAGE REGULATOR (cont'd)**

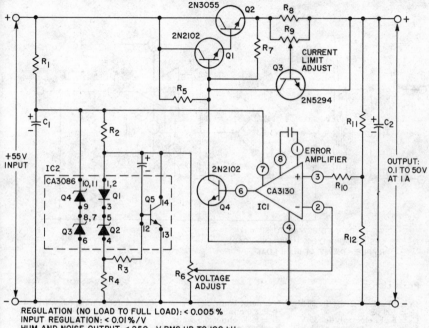

REGULATION (NO LOAD TO FULL LOAD): < 0.005 %
INPUT REGULATION: < 0.01%/V
HUM AND NOISE OUTPUT: < 250 μV RMS UP TO 100 kHz

92CM-26100

Parts List

C_1, C_2 = 100 μF, electrolytic 75 V
R_1 = 4300 ohms, 1 watt
R_2 = 2200 ohms, 0.5 watt

R_3, R_7 = 1000 ohms, 0.5 watt
R_4 = 62000 ohms, 0.5 watt
R_5 = 3300 ohms, 1 watt
R_6 = potentiometer, 50000 ohms, 0.5 watt

R_8 = 1 ohm, 0.5 watt
R_9 = potentiometer, 10000 ohms, 0.5 watt
R_{10} = 10000 ohms, 0.5 watt

16-4   **REGULATED DC POWER SUPPLY**
              3.5 V to 20 V; 0 to 90 mA

Circuit Description

This dc power supply uses a CA3085, CA3085A, or CA3085B integrated-circuit voltage regulator, together with four 1N3193 silicon rectifiers connected in a full-wave bridge configuration, to provide a regulaled dc output voltage that is continuously adjustable from 3.5 to 20 volts at currents from 0 to 90 milliamperes.

The RCA-CA3085, CA3085A, and CA3085B silcon monolithic integrated circuits are designed specifically for service as voltage regulators as output voltages ranging from 1.7 to 46 volts at currents up to 100 milliamperes. The voltage regulators provide important features such as: frequency compensation, short-circuit protection, temperature-compensated

16-4   **REGULATED DC POWER SUPPLY (cont'd)**

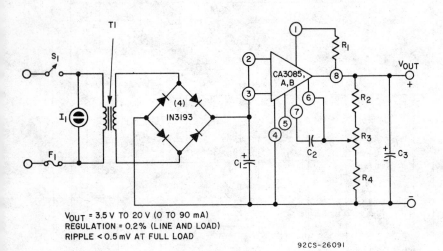

V_{OUT} = 3.5 V TO 20 V (0 TO 90 mA)
REGULATION = 0.2% (LINE AND LOAD)
RIPPLE < 0.5 mV AT FULL LOAD

92CS-26091

Parts List

C_1 = 50 μF, electrolytic, 50 V
C_2 = 100 pF
C_3 = 5 μF, electrolytic, 35 V
F_1 = fuse, 1 ampere, 120 V, slow-blow

I_1 = neon lamp, 120 V
R_1 = 5.6 ohms, 0.5 watt
R_2 = 8200 ohms, 0.5 watt
R_3 = potentiometer, 10000 ohms, 0.5 watt

R_4 = 1000 ohms, 0.5 watt
T_1 = power transformer, Stancor TP-3 or equiv.

Circuit Description (cont'd)

reference voltage, current limiting, and booster input. These devices are useful in a wide range of applications for regulating high-current, switching, shunt, and positive and negative voltages. They are also applicable for current and dual-tracking regulation.

The CA3085A and CA3085B have output current capabilities up to 100 mA and the CA3085 up to 12 mA without the use of external pass transistors. However, all the devices can provide voltage regulation at load currents greater than 100 mA with the use of suitable external pass transistors. The CA3085 Series has

an unregulated input voltage ranging from 7.5 to 30 30 V (CA3085), 7.5 to 40 V (CA3085A), and 7.5 to 50 V (CA3085B) and a minimum regulated output voltage of 26 V (CA3085), 36 V (CA3085A), and 46 V (CA-3085B).

These types are supplied in the 8-lead TO-5 style package (CA3085, CA3085A, CA3085B, and the 8-lead TO-5 with dual-in-line formed leads ("DIL-CAN", CA3085S, CA3085AS, CA3085BS). The CA 3085 is also supplied in the 8-lead dual-in-line plastic package ("MINI-DIP", CA-3085E), and in chip form (CA3085H).

16-5

FULL-RANGE VARIABLE-VOLTAGE
REGULATED DC POWER SUPPLY

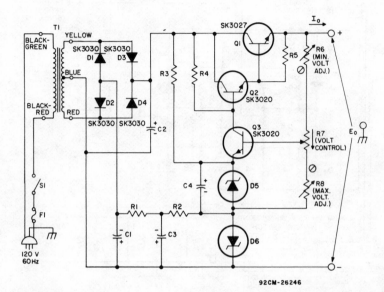

92CM-26246

Parts List

C_1, C_3 = 500 μF, electrolytic, 25 V

C_2 = 5000 μF, electrolytic, 25 V

C_4 = 100 μF, electrolytic, 10 V

D_5 = zener diode, 6.8 V, 1 watt

D_6 = zener diode, 12 V, 1 watt

F_1 = fuse, 1 ampere, 120 V, slow-blow

R_1, R_5 = 220 ohms, 0.5 watt, 10%

R_2 = 470 ohms, 0.5 watt, 10%

R_3 = 6800 ohms, 0.5 watt, 10%

R_4 = 10000 ohms, 0.5 watt, 10%

R_6, R_8 = trimmer potentiometer, 5000 ohms, Mallory MTC-1 or equiv.

R_7 = potentiometer, 5000 ohms, linear taper

S_1 = toggle switch, 120 V, 1 ampere, single-pole, single-throw

T_1 = power transformer; primary 117 V; secondary 16 V, 1.5 ampere; Stancor TP-4 or equiv.

Circuit Description

This full-range variable-voltage power supply is a continuously variable supply capable of delivering up to 12 volts at a maximum current of 1 ampere.

The regulator circuit in this supply receives full-wave rectified ac from a bridge rectifier that uses four RCA-SK3030 silicon rectifiers. This arrangement provides the regulator with a high input voltage.

The load voltage is equal to the regulator voltage minus the voltage across zener diode D_6. When the two voltages are equal, the load voltage is zero. If the regular-circuit voltage falls below 12 volts, the base-emitter junction of transistor Q_3 becomes reverse-biased and the transistor turns off. As a result, transistors Q_2 and Q_1 also turn off and prevent the load voltage from reversing polarity (becoming negative).

16-6 HIGH-VOLTAGE, HIGH-CURRENT DC POWER SUPPLY
600 Volts; 300 Volts; Total Current 330 Milliamperes
(Intermittent Duty)

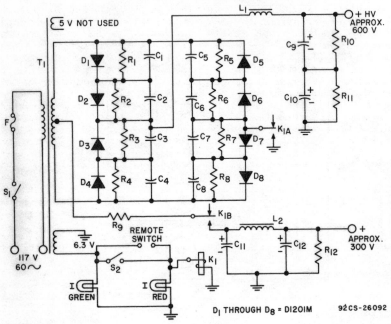

D_1 THROUGH D_8 = D1201M   92CS-26092

Parts List

C_1 C_2 C_3 C_4 C_5 C_6 C_7 C_8 =
0.001 μF, ceramic disc,
1000 V

C_9, C_{10}, C_{11}, C_{12} = 40 μF,
electrolytic, 450 V

F = fuse, 5 amperes

I = indicator lamp

K_1 = relay; Potter and
Brumfield KA11AY or
equiv.

L_1 = 2.8 henries, 300 mA;
Stancor C-2334 or equiv.

L_2 = 4 henries, 175 mA;
Stancor C-1410 or equiv.

R_1 R_2 R_3 R_4 R_5 R_6 R_7 R_8 =
0.47 megohm, 0.5 watt

R_9 = 47 ohms, 1 watt

R_{10} R_{11} = 15000 ohms, 10
watts

R_{12} = 47000 ohms, 2 watts

S_1 S_2 = toggle switch, single-
pole single-throw

T = power transformer;
Stancor P-8166 or equiv.

Circuit Description

This power supply uses eight D1201M silicon diodes in series-connected pairs in a bridge-rectifier circuit to supply a 600-volt dc output from a 117-volt ac input. One set of the diode pairs (D_5 through D_8) is also used in a conventional full-wave rectifier circuit to supply a 300-volt dc output. Series-connected pairs of diodes are used to provide the rectification in this circuit because the peak-inverse-voltage rating of such combinations is twice that of a single diode.

The operation of the power supply is controlled by two switches. When the ON-OFF switch S_1 is closed, the 117-volt 60-c/s ac input power is applied across the primary of the step-up power transformer T_1. The power supply does not become operative, however, until switch S_2 is also closed. Relay K_1 is then energized, and the closed contacts of the relay complete the ground return paths for the power-supply circuits. Switch S_2 can be used as a STANDBY switch for a

16-6 HIGH-VOLTAGE, HIGH-CURRENT DC POWER SUPPLY (cont'd)

Circuit Description (cont'd)

transmitter or other equipment in which the power supply may be used. If desired, another switch may be connected in parallel with S_2 so that the standby-to-on function can be controlled from a remote location.

During the half-cycle of ac input for which the voltage across the secondary winding of T_1 is positive at the top end and negative at the bottom end, current flows from the bottom of the secondary through diodes D_7 and D_8 (which are oriented in the proper direction), out the K_{1A} section of the relay contacts to ground, and then up through bleeder resistors R_{10} and R_{11} and the external load connected in shunt with the resistors to develop the 600-volt output. The return flow is completed through filter choke L_1, diodes D_1 and D_2 and the entire secondary winding. During the next half-cycle of the ac input, the polarity of the voltage across the secondary reverses, and the current flows

through diodes D_5 and D_6, through the bleeder resistors and the external load circuit in the same direction as before, and then through diodes D_3 and D_4. Capacitors C_9 and C_{10} and choke L_1 provide the filtering to smooth out the pulsations in the 600-volt dc output.

For the 300-volt dc output, only one-half the voltage across the secondary winding of T_1 is required. The D_5-D_6 and D_7-D_8 diode pairs are operated in a full-wave rectifier configuration to provide this output (diodes D_1 through D_4 are not included in the 300-volt circuit.) The current flow through the diode pairs is the same as described before, but the current is directed from the relay contacts up through bleeder resistor R_{12} and the external load circuit. The return flow is through choke L_2 and the transformer center tap. Capacitors C_{11} and C_{12} and choke L_2 provide the filtering for the 300-volt dc output.

16-7 MICROPHONE PREAMPLIFIER

Circuit Description

This integrated-circuit microphone preamplifier is a high-gain, low-noise, high-fidelity preamplifier that accommodates both low- and high-impedance dynamic microphones. It can be used with tape recorders and audio systems, or with radio transmitters. The maximum output of the circuit is 1.4 volts.

The integrated circuit used in the preamplifier is the RCA-CA3018 transistor array. This transistor array is connected to form a Darlington input stage in cascade with an emitter-follower output stage. The Darlington configuration provides the microphone preamplifier with a high-gain high-impedance input. The output of the Darlington stage is applied directly to the base of the emitter-

follower output stage. This stage, which has a high input impedance, isolates the output load from the input. The chart below shows voltages at the terminals of a CA3018 in a properly operating circuit.

Voltages at IC Terminals in Microphone Preamplifier Circuit

Terminal	Voltage (Volts)
1	0.5
2	1.2
6	4.5
7	3.7
8	9
9	1.8
11	4.5
12	4.5

16-7   **MICROPHONE PREAMPLIFIER (cont'd)**

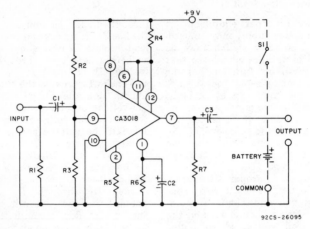

92CS-26095

Parts List

$C_1 = 10 \ \mu F$, electrolytic, 6 V
$C_2 = 25 \ \mu F$, electrolytic, 6 V
$C_3 = 25 \ \mu F$, electrolytic, 10 V
R_1, R_2, R_3 = see resistor-
   value chart.

$R_4 = 10000$ ohms, 0.5 watt,
   10%
$R_5 = 56000$ ohms, 0.5 watt,
   10%

$R_6 = 1500$ ohms, 0.5 watt,
   10%
$R_7 = 470$ ohms, 0.5 watt,
   10%

Resistor Values as a Function of Microphone Type

Resistors*	Microphone Type	
	Low Impedance	High Impedance
R_1	270 ohms	not used
R_2	220,000 ohms	1 megohm
R_3	56,000 ohms	0.27 megohm

* 0.5 watt, 10%

16-8   **LOW-NOISE PREAMPLIFIER WITH HIGH
DYNAMIC RANGE**

Circuit Description

This circuit illustrates the use of a CA3094 programmable integrated-circuit operational amplifier in a volume-controlled amplifier that has a high dynamic range. This type of amplifier is particularly useful, for example, as a microphone (or musical instrument) input amplifier, where a high dynamic range of input signals is likely.

16-8   LOW-NOISE PREAMPLIFIER (cont'd)

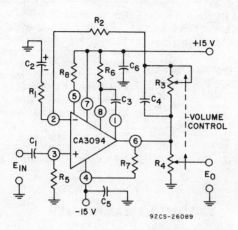

92CS-26089

$C_1 = 0.22 \ \mu F$, mylar or ceramic, 25 V
$C_2 = 2 \ \mu F$, electrolytic, 3 V to 6 V
$C_3 = 100$ pF, disc ceramic, $\pm 20\%$
$C_4 = 30$ pF, disc ceramic, $\pm 20\%$

C_5, $C_6 = 0.5 \ \mu F$, 25-Volts (GMV)
R_1, $R_2 = 10000$ ohm, 0.5 watt, composition, $\pm 10\%$
R_3, $R_4 = $ dual ganged Potentiometers, 0.25 megohm, 0.5 watt, linear taper

$R_5 = 51000$ ohms, 0.5 watt, composition
$R_6 = 100$ ohm, 0.5 watt, composition, $\pm 10\%$
$R_7 = 2700$ ohms, 0.5 watt, composition, $\pm 10\%$

Circuit Description (cont'd)

The circuit design is unique in that the overload characteristics of the amplifier improve as the volume control is decreased. This performance is accomplished by use of dual ganged volume-control potentiometers, R_3 and R_4, that reduce the gain of the operational amplifier in accordance with the reduction in output voltage E_o desired. At the maximum-volume setting of potentiometer R_4, the maximum resistance of potentiometer R_3 is added to that of resistor R_2 in the feedback path of the amplifier. Under these conditions, the voltage gain is approximated as follows:

$$\frac{E_o}{E_{in}} \cong 1 + \frac{R_2 + R_3}{R_1}$$

At the minimum-volume setting of R_4, the incremental feedback resistance due to R_3 is also at a minimum (e.g., zero). Thus, the output signal level is decreased by the setting of potentiometer R_4, and the amplifier gain is simultaneously reduced by the value of $(R_2 + R_3)$. At low volume settings, therefore, the overload point of the amplifier is increased, and the noise output is reduced. Because the input is applied to the noninverting $(+)$ terminal of the CA3094, the input impedance of the amplifier will be relatively high. With linear potentiometers, the voltage gain is approximately given by

$$\frac{E_o}{E_{in}} \cong A^2 \, \frac{R_3}{R_1} + A \, \frac{R_2}{R_1} + A$$

16-8   LOW-NOISE PREAMPLIFIER (cont'd)

Circuit Description (cont'd)

where A is the fractional rotation of the control. This relationship results in a parabolic variation of the output signal as a function of control position.

The low-noise preamplifier has a maximum gain of nearly 30 dB and can handle input signals up to about 0.25 volts (rms) without overloading. With the volume control set for a barely audible output signal, the input signal overload level is about 6-volts (rms). Supply current consumption is about 6-milliamperes at the supply-voltages shown.

Resistor R_7 is used to set the value of amplifier-bias-current (I_{ABC}). [The operation of the programmable operational amplifier (CA3094) used in this circuit is described in the section **Integrated Circuits for Linear Applications**.] Capacitor C_3 is a compensating capacitor to assure circuit stability. Capacitor C_4 is used to increase the feedback at the higher frequencies, thereby reducing high-frequency noise-signal output.

16-9   EQUALIZED PREAMPLIFIERS

The exceptional low-noise characteristics of the super-beta transistor in the CA3095E IC-array are ideal for preamplifier service in professional-grade tape-playback and magnetic phono-cartridge amplifier systems. Sketch (a) shows the schematic diagram of the CA3095E integrated-circuit transistor array. The superbeta transistors Q_1 and Q_2 are generically similar to conventional bipolar transistors except that they have betas in the range of 1000 to 5000. Because the super-beta transistors in CA3095E array have a collector-emitter breakdown voltage $V_{(BR)CEO}$ of about 2 volts, it is necessary to limit the collector-emitter voltage accordingly. This limiting is accomplished in the CA3095E by means of the network consisting of D_1, D_2, and Q_5.

Preamplifier for Tape-Head Signals—A typical preamplifier circuit

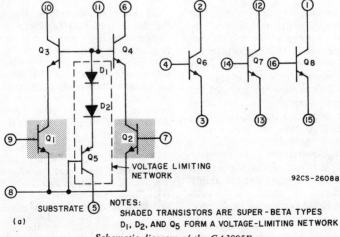

92CS-26088

NOTES:
SHADED TRANSISTORS ARE SUPER-BETA TYPES
D_1, D_2, AND Q_5 FORM A VOLTAGE-LIMITING NETWORK

Schematic diagram of the CA3095E.

EQUALIZED PREAMPLIFIERS (cont'd)

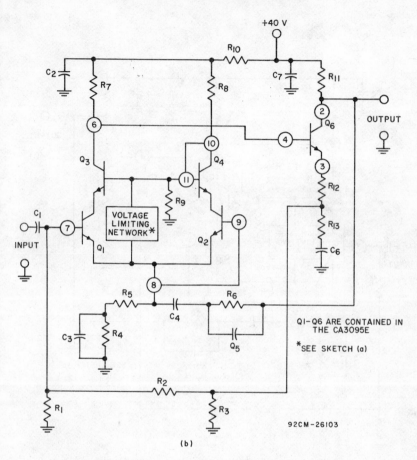

Tape play-back preamplifier equalized for NAB standards (7.5 in/s).

Parts List

Preamplifier for Tape-Head Signals

$C_1 = 0.33\ \mu F$, paper or mylar, 100 Volts
$C_2 = 10\ \mu F$, electrolytic, 50 volts
$C_3 = 10\ \mu F$, electrolytic, 5 volts
$C_4 = 0.0075\ \mu F$, paper or mylar, ±5%
$C_5 = 470$ pF, mica or ceramic, ±5%
$C_6 = 50\ \mu F$, electrolytic, 25 volts

$C_7 = 0.1\ \mu F$, 50 volts minimum
$R_1 = 47000$ ohms, 0.5 watt, composition, ±10%
$R_2 = 82000$ ohms, 0.5 watt, composition, ±10%
$R_3 = 4300$ ohms, 0.5 watt, composition, ±10%
$R_4 = 10000$ ohms, 0.5 watt, composition, ±10%
$R_5 = 330$ ohms, 0.5 watt, composition, ±5%
$R_6 = 680$ ohms, 0.5 watt, composition, ±5%

$R_7 = 0.33$ megohm, 0.5 watt, composition, ±10%
$R_8 = 0.39$ megohm, 0.5 watt, composition, ±10%
$R_9 = 24000$ ohms, 0.5 watt, composition, ±5%
$R_{10} = 10000$ ohms, 0.5 watt, composition, ±10%
$R_{11} = 18000$ ohms, 0.5 watt, composition, ±10%
$R_{12}, R_{13} = 100$ ohms, 0.5 watt, composition, ±10%

16-9   **EQUALIZED PREAMPLIFIERS (cont'd)**

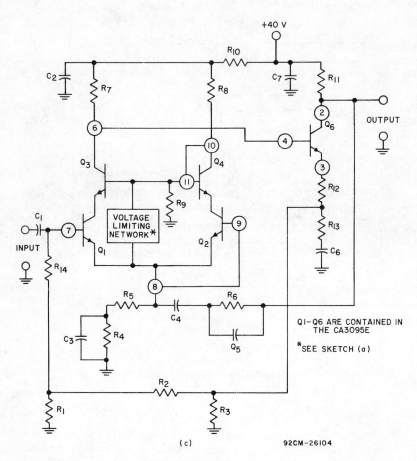

Preamplifier equalized for RIAA standards applicable to magnetic phonograph cartridges.

Parts List

Preamplifier for Magnetic Phono Pick-Up

(Parts similar to those for tape-head preamplifier with the following exceptions)

$C_1 = 0.01 \ \mu F$, paper or mylar, $\pm 5\%$

$C_5 = 0.003 \ \mu F$, paper or mylar, $\pm 5\%$

$R_8 = 33000$ ohms, 0.5 watt, composition, $\pm 5\%$

$R_{11} = 16000$ ohms, 0.5 watt, composition, $\pm 5\%$

Circuit Description (cont'd)

for tape-head signals with equalization for NAB standards [7.5 in/s] is shown in sketch (b). Transistors Q_1 and Q_3 are cascode-connected as the input stage, and transistor Q_6 is connected as a common-emitter post amplifier. Transistors Q_2 and Q_4 are nonconductive because the emitter-base junction in Q_2 and the base-collector junction in Q_4 are shunted by external wiring. Equalization for the NAB tape-playback frequency-response characteristics is provided by the R_6, C_4, C_5 network connected in the ac feedback path; dc feedback stabilization is provided by the path through resistor R_2. The amplifier has an over-all gain of about 37 dB at 1-kHz, and can deliver output voltage in the order of 25 volts, peak-to-

peak. With the gain indicated, the dynamic range of the circuit is typically about 95 dB.

Preamplifier for Signals from Magnetic Phonograph Cartridges—A typical preamplifier circuit for magnetic phono pickup with equalization for RIAA playback standards is shown in sketch (c). Circuit operation is generically similar to that described for the amplifier shown in sketch (b). Equalization is provided by the R_6, C_4, C_5 network connected in the ac feedback path. The amplifier has an over-all gain of about 40 dB at 1-kHz, and can deliver output voltages in the order of 25 volts, peak-to-peak. With the gain indicated, the dynamic range of the circuit is typically about 95 dB.

16-10 **HIGH-FIDELITY PREAMPLIFIER**
For Phono, FM, or Tape Pickup

Circuit Description

This phonograph preamplifier can be used with an audio power amplifier, such as circuits 15-12 through 15-15, to provide an excellent high-fidelity system. The circuit is designed for use with a magnetic pickup that can supply an input signal of at least 5 millivolts. Provisions are also included in the preamplifier for tape and tuner inputs. For a 5-millivolt input signal, the preamplifier delivers an output of at least 1 volt. An input of 300 millivolts from a tuner or tape recorder is required to produce an output of 1 volt. The preamplifier requires a dc supply of 20 volts at 7.5 milliamperes.

The preamplifier uses two SK3020 transistors Q_1 and Q_2 in a two-stage direct-coupled input circuit. A frequency-shaping network in the feedback circuit of transistor Q_2 provides frequency compensation when the preamplifier is used with a magnetic phonograph pickup. The output circuit of transistor Q_2 contains a level control R_{10} that feeds the loudness control R_{12} through the selector switch S_1. The loudness control, in turn, drives the tone-control circuits of the preamplifier. Tape, tuner, or phono inputs can be selected by means of the selector switch; an output connector in the arm of the selector switch permits

16-10   HIGH-FIDELITY PREAMPLIFIER (cont'd)

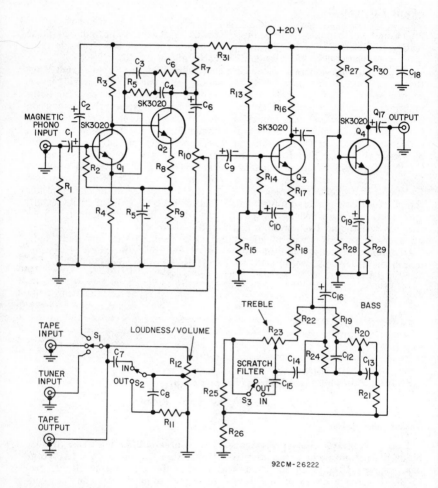

92CM-26222

Parts List

$C_1 = 2 \ \mu F$, electrolytic, 6 V

C_2, C_{17}, $C_{18} = 25 \ \mu F$, electrolytic, 25 V

$C_3 = 0.0027 \ \mu F$, paper, 200 V

$C_4 = 0.01 \ \mu F$, paper, 200 V

$C_5 = 100 \ \mu F$, electrolytic, 3 V

$C_6 = 10 \ \mu F$, electrolytic, 25 V

$C_7 = 180 \ pF$, mica, 500 V

$C_8 = 0.033 \ \mu F$, paper, 200 V

$C_9 = 1 \ \mu F$, electrolytic, 12 V

C_{10}, $C_{16} = 10 \ \mu F$, electrolytic, 10 V

$C_{11} = 10 \ \mu F$, electrolytic, 25 V

C_{12}, $C_{13} = 0.022 \ \mu F$, paper, 200 V

$C_{14} = 0.0039 \ \mu F$, paper, 200 V

$C_{15} = 0.0047 \ \mu F$, paper, 200 V

$C_{19} = 100 \ \mu F$, electrolytic, 6 V

R_1, $R_3 = 68000$ ohms, 10%, 0.5 watt

$R_2 = 0.18$ megohm, 10%, 0.5 watt

$R_4 = 470$ ohms, 10%, 0.5 watt

$R_5 = 27000$ ohms, 10%, 0.5 watt

$R_6 = 0.47$ megohm, 10%, 0.5 watt

R_7, R_{19}, R_{21}, $R_{24} = 10000$ ohms, 10%, 0.5 watt

$R_8 = 82$ ohms, 10%, 0.5 watt

$R_9 = 1800$ ohms, 10%, 0.5 watt

$R_{10} = $ potentiometer, 0.1 megohm, 0.5 watt, audio taper

$R_{11} = 8200$ ohms, 10%, 0.5 watt

$R_{12} = $ potentiometer, 0.25 megohm, 0.5 watt, audio taper with tap, Centralab F11-250K or equiv.

$R_{13} = 33000$ ohms, 10%, 0.5 watt

R_{14}, $R_{28} = 18000$ ohms, 10%, 0.5 watt

Parts List (cont'd)

R_{15}, R_{31} = 4700 ohms, 10%, 0.5 watt
R_{16} = 6800 ohms, 10%, 0.5 watt
R_{17} = 68 ohms, 10%, 0.5 watt
R_{18}, R_{22}, R_{25}, R_{29} = 1000 ohms, 10%, 0.5 watt
R_{20}, R_{23} = potentiometer, 0.1 megohm, 0.5 watt, linear taper
R_{26} = 47000 ohms, 10%, 0.5 watt
R_{27} = 56000 ohms, 10%, 0.5 watt
R_{30} = 2700 ohms, 10%, 0.5 watt
S_1 = switch, single-pole, 3-position, wafer
S_2 = switch, single-pole, double-throw, toggle
S_3 = switch, single-pole, single-throw, toggle

Circuit Description (cont'd)

tape recordings to be made without affecting volume or loudness.

The treble and bass tone controls provide boost of 10 dB and cut of 15 dB for deep bass and high treble frequencies. Each control operates independently so that precise tone shaping is possible. When both controls are in the center position, the response is flat; the bass and treble frequencies are equally mixed.

Output distortion is low at all frequencies for any setting of either the bass or the treble tone control. The collector-to-base feedback in the SK3020 transistors Q_3 and Q_4 works with the tone controls to provide the over-all tonal response of the preamplifier.

Included in the preamplifier is a loudness/volume control switch S_2. With the loudness control in, lower tones are enhanced at low output levels, and a more pleasing sound is produced. When the loudness control is switched out, the volume control attenuates all tones equally.

The scratch filter attenuates somewhat the frequencies at which scratch noise from scratched records is most prevalent.

16-11 **MULTIPLE-INPUT AUDIO MIXER**

Circuit Description

This multiple-input circuit is designed to mix inputs from up to 12 sources, (usually microphones) for input to an amplifier, tape recorder, or other audio equipment. The mixer has a gain of unity and, therefore, has no effect on the system in which it is installed.

The audio mixer can handle large signals without overloading; it is normally located in the audio system after the system gain control and ahead of the power amplifier. Each of the input resistors R_1 through R_{12} can be connected to the wiper arm of a potentiometer for independent gain control of each channel. All unused inputs should be grounded.

16-11   MULTIPLE-INPUT AUDIO MIXER (cont'd)

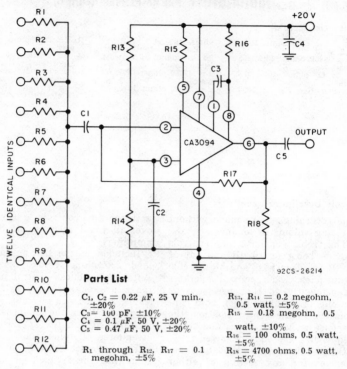

92CS-26214

Parts List

C₁, C₂ = 0.22 μF, 25 V min., ±20%
C₃ = 100 pF, ±10%
C₄ = 0.1 μF, 50 V, ±20%
C₅ = 0.47 μF, 50 V, ±20%

R₁ through R₁₂, R₁₇ = 0.1 megohm, ±5%

R₁₃, R₁₄ = 0.2 megohm, 0.5 watt, ±5%
R₁₅ = 0.18 megohm, 0.5 watt, ±10%
R₁₆ = 100 ohms, 0.5 watt, ±5%
R₁₈ = 4700 ohms, 0.5 watt, ±5%

16-12   GENERAL-PURPOSE AUDIO AMPLIFIER

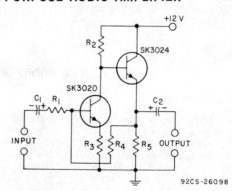

92CS-26098

Parts List

C₁ = 10 microfarads, 6 volts, electrolytic
C₂ = 50 microfarads, 25 volts, electrolytic

R₁ = 1000 ohms, 0.5 watt
R₂ = 1200 ohms, 0.5 watt
R₃ = See chart, 0.5 watt

R₄ = See chart, 0.5 watt
R₅ = 270 ohms, 0.5 watt

All resistors have a tolerance of 10 per cent.

16-12 GENERAL-PURPOSE AUDIO AMPLIFIER (cont'd)

Circuit Description

This two-stage amplifier is useful as a line driver for audio systems in which the power amplifier is located at a considerable distance from the signal source, as a driver for the line inputs of tape recorders, as an output stage for inexpensive radio receivers, and in many other general-purpose audio-amplifier applications. The amplifier has a frequency response that is flat from 20 to 20,000 Hz and can be used to drive any line that has an impedance of 250 ohms or greater. It operates from a dc supply of 12 volts and can supply a maximum undistorted output of 3-volts rms into a 250-ohm line.

The voltage gain and input impedance of the amplifier are determined by the values chosen for the emitter resistor (R_3) and feedback resistor (R_4) for the input stage. A chart shows values of these resistors for various voltage gains from unity to 166 and for input impedances from 2700 ohms to 55,000 ohms.

The amplifier employs an SK3020 transistor Q_1 in a common-emitter input stage and a SK3024 transistor

Q_2 in an emitter-follower output stage. These stages are interconnected in a self-adjusting configuration that maintains the amplifier in a stable operating state regardless of variations in dc supply voltage and ambient temperature. This stability is achieved by use of a dc feedback applied from the output (emitter) of transistor Q_2 to the input (base) of transistor Q_1 through R_4.

If the emitter current of transistor Q_1 should increase, the base voltage of transistor Q_2 would also decrease because of the rise in the voltage drop across resistor R_2. This decrease in the base voltage of transistor Q_2 results in a corresponding reduction in the emitter current of this transistor. Consequently, the amount of positive dc voltage fed back from the emitter of transistor Q_2 to the base of transistor Q_1 is reduced. This reduction in voltage at the base of Q_1 causes a decrease in current through this transistor that compensates for the original increase, and the amplifier is stabilized. Use of an emitter-follower output stage makes possible the low output impedance of the amplifier.

Resistance Data for Different Voltage Gains and Input Impedances*

Voltage Gain	Input Impedance (ohms)	R_3 (ohms)	R_4 (kilohms)
166	2700	0	680
22	7300	39	470
17	9000	68	430
10	15000	100	390
3	55000	390	360
1	100000	1200	330

* Data obtained for an output of 1 volt rms into a 250-ohm line.

16-13   SMALL-SIGNAL AUDIO AMPLIFIER
With Tone and Volume Controls

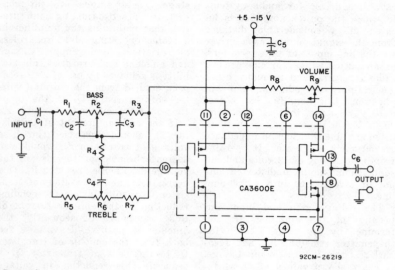

92CM-26219

Parts List

C_1, C_6 = 0.33 μF, mylar or ceramic, 25 V

C_2, C_3 = 0.0047 μF, mylar or ceramic, 25 V, $\pm10\%$

C_4 = 100 pF, disc, 100 V, $\pm10\%$

C_5 = 0.05 μF, mylar or ceramic [GMV], 25 V

R_1, R_3 = 51000 ohms, 0.5 watt, composition

R_2, R_9 = Potentiometer, 1 megohm, linear taper

R_4 = 0.47 megohm, 0.5 watt, composition, $\pm10\%$

R_5, R_7 = 4700 ohms, 0.5 watt, composition, $\pm10\%$

R_6 = Potentiometer, 0.5 megohm, linear taper

R_8 = 0.27 megohm, 0.5 watt, composition, $\pm10\%$

Circuit Description

This audio amplifier is useful in handling signals up to about 200 millivolts rms without overloading. It uses four of the six transistors in the CA3600E COS/MOS transistor array. The first stage is a Baxandall tone-control circuit which features ±15 dB bass and treble "boost-and-cut" characteristics at 100 Hz and 10 kHz, respectively. The Baxandall tone-control network is connected in the feedback loop [between terminals 12 and 10] of the first stage. The maximum volume-gain in the second stage is set by the ratio R_9/R_8; with the values shown it is about 12 dB.

The output [at terminals 8, 13] is biased at one-half the supply voltage via the feedback path through resistor R_9 to gate terminal 6. Because the gate terminal currents to the CA3600E are in the order of 5-pico-amperes, there is negligibly small dc current through the controls, thereby reducing the possibilities of their becoming noisy. Standard linear taper potentiometers are used. The circuit is operational over the supply voltage range from 5 to 15-volts. It requires about 4.5 milliamperes of supply-current for operation at 10-volts. The CA3600E integrated circuit is described in the section **Integrated Circuits for Linear Applications.**

16-14 **10-W AUTOMOBILE AUDIO AMPLIFIER**

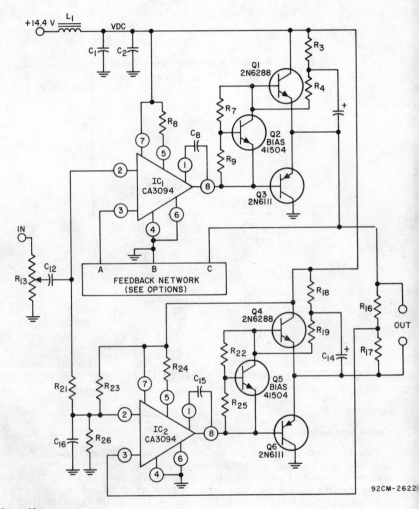

92CM-2622

Parts List

$C_1 = 1.500$ μF, electrolytic, 25 V
$C_2 = 0.02$ μF, disc, 50 V
$C_3 = 0.12$ μF, disc, 50 V
$C_4 = 0.01$ μF, disc, 50 V
C_5, $C_6 = 0.001$ μF, disc, 50 V
C_7, $C_{14} = 100$ μF, electrolytic, 25 V
$C_8 = 100$ pF, 50 V
$C_9 = 25$ μF, electrolytic, 25 V
$C_{10} = 0.2$ μF, 50 V
$C_{11} = 0.02$ μF, 50 V
$C_{12} = 0.047$ μF, 50 V
$C_{13} = 1$ μF, mylar, 50 V
$C_{15} = 250$ pF, 50 V

$C_{16} = 50$ μF, electrolytic, 25 V
$C_{17} = 1$ μF, Mylar, 50 V
$L_1 = 9$ mH, 3 A
$R_1 =$ potentiometer, 10000 ohms, 0.5 watt, linear taper
$R_2 = 820$ ohms, 0.5 watt
R_3, R_4, R_{18}, $R_{19} = 33$ ohms, 0.5 watt
R_5, R_9, $R_{25} = 68$ ohms, 0.5 watt
$R_6 = 300$ ohms, 0.5 watt
$R_7 = 51$ ohms, 0.5 watt
R_8, $R_{24} = 0.18$ megohm, 0.5 watt

$R_{10} = 3300$ ohms, 0.5 watt
$R_{11} = 910$ ohms, 0.5 watt
$R_{12} = 39000$ ohms, 0.5 watt
$R_{13} =$ potentiometer 0.25 megohm, 0.5 watt, linear taper
$R_{14} =$ potentiometer, 0.1 megohm, 0.5 watt, linear taper
$R_{15} = 10000$ ohms, 0.5 watt
R_{16}, $R_{17} = 10000$ ohms, 0.5 watt, 1% precision
$R_{20} = 1000$ ohms, 0.5 watt
$R_{21} = 0.22$ megohm, 0.5 watt
$R_{22} = 51$ ohms, 0.5 watt
R_{23}, $R_{26} = 22000$ ohms, 0.5 watt

16-14   **10 WATT AUTOMOBILE AUDIO AMPLIFIER (cont'd)**
Feedback Network Options

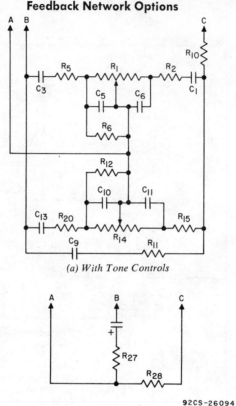

(a) With Tone Controls

(b) Fixed

92CS-26094

Circuit Description

This high-power automobile audio amplifier can supply a continuous sine-wave output of 10 watts into a 4-ohm speaker at less than 1 percent distortion (1-kHz input signal). The amplifier features a transformerless high-power output and thermal stability of the output stage. The circuit uses two CA3094T integrated circuits (IC_1 and IC_2) and six transistors [two 2N6288's (Q_1 and Q_4), two 2N6292's (Q_2 and Q_5) and two 2N6111's (Q_3 and Q_6)] connected in a full-wave bridge configuration. Tone controls may be incorporated into the feedback network, or the tone controls can be omitted, as desired. (Separate feedback networks are shown for each option.) The input signal is amplified by IC_1 which drives transistors Q_1 and Q_3 (top half of the bridge) while IC_2 is being driven from the output (junction of resistors R_{16} and R_{17}). IC_2 drives transistors Q_4 and Q_6 (the bottom half of the bridge). Transistors Q_2 and Q_5 are biasing transistors for the output stages and must be mounted on the output heatsink to assure thermal stability. Sensitivity of the amplifier is 50 millivolts for an output of 10 watts.

16-15 ## 12-W TRUE-COMPLEMENTARY-SYMMETRY AUDIO AMPLIFIER

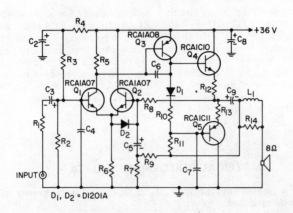

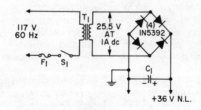

92CM-26220

Parts List

C₁ = 1000 μF, electrolytic, 50 V
C₂ = 10 μF, electrolytic, 50 V
C₃, C₅ = 5 μF, electrolytic, 25 V
C₄ = 150 pF
C₆ = 75 pF
C₇ = 470 pF
C₈ = 0.01 μF
C₉ = 1000 μF, electrolytic, 25 V

F₁ = fuse, 1 ampere, 120 V, slow-blow
L₁ = inductor, 5 μH
R₁ = 1200 ohms, 0.5 watt
R₂ = 47000 ohms, 0.5 watt
R₃ = 39000 ohms, 0.5 watt
R₄ = 10000 ohms, 0.5 watt
R₅, R₁₁ = 330 ohms, 0.5 watt
R₆ = 2700 ohms, 0.5 watt
R₇ = 47 ohms, 0.5 watt
R₈ = 18000 ohms, 0.5 watt
R₉ = 1000 ohms, 0.5 watt

R₁₀ = 7.2 ohms, 0.5 watt
R₁₂, R₁₃ = 0.47 ohms, 0.5 watt
R₁₄ = 22 ohms, 0.5 watt
S₁ = on-off switch, single-pole, single-throw
T₁ = power transformer; 120-volt primary; 25.5-secondary; Thordarson 23V118, Stancor TP-4, Triad F-93X, or equiv. (for stereo amplifiers)

NOTE: Resistors should be non-inductive.

16-15   **12-W TRUE-COMPLEMENTARY-SYMMETRY
AUDIO AMPLIFIER (cont'd)**

Circuit Description

This 12-watt audio-amplifier circuit uses RCA1C10 and RCA1C11 as output devices in conjunction with three discrete transistors, two diodes, and a single 36-volt power supply; the amplifier output is capacitively coupled to an 8-ohm speaker. The choice of a true-complementary-symmetry output stage provides excellent fidelity for a low-cost-system. RCA1C10 and RCA1C11 are n-p-n and p-n-p epitaxial-base silicon power transistors respectively, especially characterized for audio-output service. They are provided in the JEDEC TO-220AB version of the VERSA-WATT plastic package.

TYPICAL PERFORMANCE DATA

Measured at a line voltage of 120V, $T_A = 25°C$, and a frequency of 1 kHz, unless otherwise specified.

Power:
  Rated power (8-Ω load, at rated distortion) ...... 12 W
  Typical power (4-Ω load) . 12 W
  Typical power (16-Ω load). 6.5 W
  Music power (8-Ω load, at 5% THD with regulated supply) .... 15 W
  Dynamic power (8-Ω load, at 1% THD with regulated supply) ..... 13 W
Total Harmonic Distortion:
  Rated distortion 1.0%

IM Distortion:
  10 dB below continuous power output at 60 Hz and 7 kHz (4:1) 1.5%
Sensitivity:
  At continuous power-output rating .... 600 mV
Hum and Noise:
  Below continuous power output:
    Input shorted .... 90 dB
    Input open .... 70 dB
Input Resistance .... 23 kΩ

16-16   **25-W FULL-COMPLEMENTARY-SYMMETRY
AUDIO AMPLIFIER**

Circuit Description

This 25-watt audio-amplifier circuit uses RCA1C05 and RCA1C06 as output devices in conjunction with seven TO-39 discrete transistors, ten diodes, and a 52-volt split power supply. The amplifier output is directly coupled to an 8-ohm speaker. The full-complementary-symmetry output stage provides excellent high-frequency performance at moderate cost.

RCA1C05 and RCA1C06 are n-p-n and p-n-p epitaxial-base silicon power transistors, respectively. These complementary output devices for audio applications are provided in the JEDEC TO-220AB plastic package.

16-16 ## 25-W FULL-COMPLEMENTARY-SYMMETRY
AUDIO AMPLIFIER (cont'd)

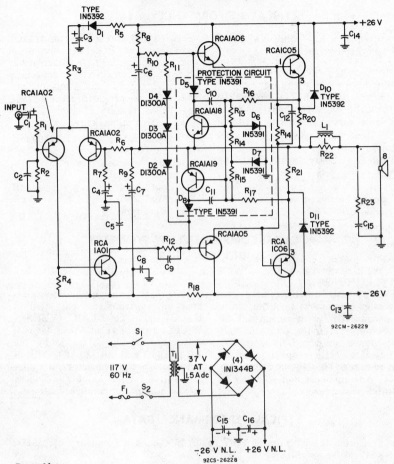

Parts List

$C_1 = 5\ \mu F$, electrolytic, 12 V
$C_2 = 180$ pF
$C_3, C_6, C_7 = 50\ \mu F$, electrolytic, 50 V
$C_4 = 50\ \mu F$, electrolytic, 12 V
$C_5 = 180$ pF
$C_8, C_9 = 0.02\ \mu F$, 50 V
$C_{10}, C_{11} = 0.01\ \mu F$
$C_{12} = 0.2\ \mu F$
$C_{13}, C_{14} = 0.05\ \mu F$, 50 V
$C_{15}, C_{16} = 2500\ \mu F$, electrolytic, 35 V
$F_1 =$ fuse, 120 V,
1.5 amperes, slow-blow

$L_1 =$ inductor, 2.4 μH, Miller 4606 or equiv.
$R_1, R_8 = 1800$ ohms, 0.5 watt
$R_2, R_6 = 18000$ ohms, 0.5 watt
$R_3 = 12000$ ohms, 0.5 watt
$R_4, R_7 = 680$ ohms, 0.5 watt
$R_5 = 180$ ohms, 0.5 watt
$R_9, R_{12} = 270$ ohms, 0.5 watt
$R_{10}, R_{11} = 2200$ ohms, 0.5 watt
$R_{11} = 75$ ohms, 0.5 watt
$R_{13}, R_{15} = 1000$ ohms, 0.5 watt

$R_{16}, R_{17} = 68$ ohms, 0.5 watt
$R_{18}, R_{19} = 100$ ohms, 0.5 watt
$R_{20}, R_{21} = 0.43$ ohm, 5 watts
$R_{22}, R_{23} = 22$ ohms, 0.5 watt
$S_1 = 80°C$ thermal cutout
$S_2 =$ on-off switch, 120 V, single-pole, single-throw
$T_1 =$ power transformer; primary 117 V; secondary 37 V, 1.5 A; Signal 36-2 or equiv.

NOTES:

1. All resistors should be non-inductive.
2. Driver transistors should be mounted on printed-circuit-board heat sink.
3. Thermal cutout S_1 should be mounted on common heat sink with output transistor.
4. Heat-sink capability of 2°C per watt should be provided for each output transistor.

16-16   FULL-COMPLEMENTARY-SYMMETRY
AUDIO AMPLIFIER (cont'd)

TYPICAL PERFORMANCE DATA

Measured at a line voltage of 120V, $T_A = 25°C$, and a frequency of 1 kHz, unless otherwise specified.

Power:

Rated power (8-Ω load, at
rated distortion) ...... 25 W
Typical power (4-Ω load) . 45 W
Typical power (16-Ω load) 16 W
Total Harmonic Distortion:
Rated distortion 1.0%
Typical at 20 W 0.05%
IM Distortion:
10 dB below continuous
power output at 60 Hz
and 7 kHz (4:1) 0.1%

IHF Power Bandwidth:
3 dB below rated
continuous power at
rated distortion .... 80 kHz
Sensitivity:
At continuous power-
output rating ....600 mV
Hum and Noise:
Below continuous power
output:
Input shorted .... 80 dB
Input open .... 75 dB
Input Resistance .... 20 kΩ

16-17   40-W QUASI-COMPLEMENTARY-SYMMETRY
AUDIO AMPLIFIER

Circuit Description

This 40-watt amplifier uses two RCA1C09 transistors as output units in conjunction with seven TO-39 transistors, 11 diodes, and a 64-volt split power supply. The amplifier output is directly coupled to an 8-ohm speaker. This 40-watt amplifier features ruggedness and economy in the mid-power range.

RCA1C09 is an n-p-n, epitaxial-base silicon transistor packaged in the JEDEC TO-220AB (VERSA-WATT) case. Two of these devices, driven in the class-B mode by the RCA1A06 and RCA1A05 silicon n-p-n and p-n-p transistors, are ideally suited for use as output devices in audio-amplifier applications.

TYPICAL PERFORMANCE DATA

Measured at a line voltage of 120V, $T_A = 25°C$, and a frequency of 1 kHz, unless otherwise specified.

Power:

Rated power (8-Ω load at
rated distortion) ...... 40 W
Typical power (4-Ω load) . 55 W
Typical power (16-Ω load) 25 W
Music power (8-Ω load, at
5% THD with regulated
supply) .... 55 W
Dynamic power (8-Ω load,
at 1% THD with
regulated supply) ..... 50 W
Total Harmonic Distortion:
Rated distortion 1.0%

IM Distortion:
10 dB below continuous
power output at 60 Hz
and 7 kHz (4:1) 0.1%
Sensitivity:
At continuous power-
output rating ....600 mV
Hum and Noise:
Below continuous power
output:
Input shorted .... 80 dB
With 2 kΩ resistance on
20-ft. cable on input ... 75 dB
Input Resistance .... 20 kΩ

16-17

40-W QUASI-COMPLEMENTARY-SYMMETRY AUDIO AMPLIFIER (cont'd)

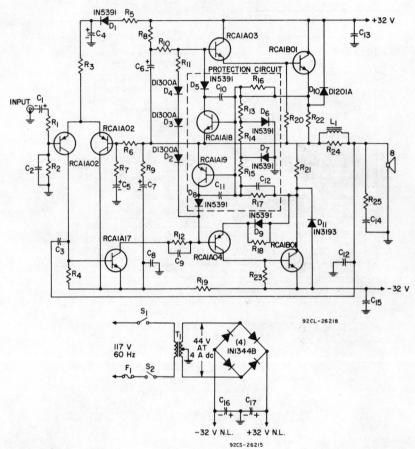

92CL-26218

92CS-26215

Parts List

$C_1 = 5$ μF, electrolytic, 12 V
$C_2 = 180$ pF
$C_3 = 39$ pF
C_4, C_5, C_6, $C_7 = 50$ μF, electrolytic, 50 V
C_8, C_9, $C_{15} = 0.02$ μF, 50 V
C_{10} $C_{11} = 0.01$ μF
C_{12}, C_{13}, $C_{14} = 0.05$ μF, 50 V
C_{16}, $C_{17} = 3500$ μF, electrolytic, 50 V
$F_1 =$ fuse, 120 V, 2 amperes, slow-blow
$L_1 = 10$ μH, Miller 4622 or equiv.

$R_1 = 1800$ ohms
R_2, $R_6 = 18000$ ohms
$R_3 = 12000$ ohms, 0.5 watt
$R_4 = 680$ ohms, 0.5 watt
$R_5 = 180$ ohms, 0.5 watt
$R_7 = 560$ ohms, 0.5 watt
$R_8 = 22000$ ohms, 0.5 watt
$R_9 = 270$ ohms, 0.5 watt
$R_{10} = 2700$ ohms, 0.5 watt
$R_{11} = 390$ ohms, 0.5 watt
$R_{12} = 47$ ohms, 0.5 watt
R_{13}, $R_{15} = 1000$ ohms, 0.5 watt
$R_{14} = 6800$ ohms, 0.5 watt

R_{16}, $R_{17} = 68$ ohms, 0.5 watt
R_{18}, R_{19}, R_{20}, $R_{23} = 100$ ohms, 0.5 watt
R_{21}, $R_{22} = 0.39$ ohm, 5 watts
R_{24}, $R_{25} = 22$ ohms, 0.5 watt
$S_1 = 90°C$ thermal cutout
$S_2 =$ on-off switch, 120 V, single-pole, single-throw
$T_1 =$ power transformer; primary 117 V; secondary 44 V, 4 A; Signal 88-2 (parallel secondary) or equiv.

NOTES:

1. All resistors should be non-inductive.
2. Driver transistors should be mounted on heat sink Wakefield No. 209AB or equiv.
3. Heat-sink capability of 1.3°C per watt should be provided for each output transistor.
4. Thermal cutout S_1 should be mounted on common heat sink with output transistors.

16-18   40-W FULL-COMPLEMENTARY-SYMMETRY
With Darlington Output Transistors

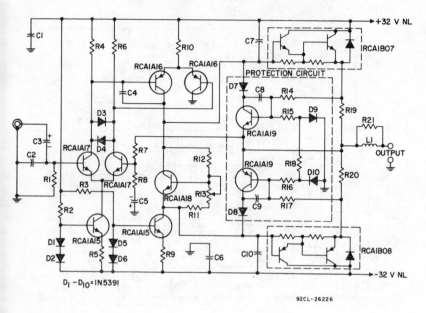

92CL-26226

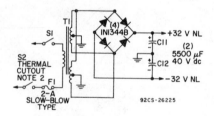

92CS-26225

Parts List

$C_1 = 0.05\ \mu F$, 50 V
$C_2, C_7, C_{10} = 200$ pF
$C_3 = 5\ \mu F$, electrolytic, 25 V
$C_4 = 100$ pF
$C_5 = 100\ \mu F$, electrolytic, 25 V
$C_6 = 0.05\ \mu F$
$C_8, C_9 = 0.01\ \mu F$
$C_{11}, C_{12} = 5500\ \mu F$, electrolytic, 40 V
$F_1 =$ fuse, 120 V, 2 amperes, slow-blow

$L_1 = 2\ \mu H$; 22 turns of No. 22 enameled wire wound around a 2-watt resistor
$R_1 = 18000$ ohms, 0.5 watt
$R_2, R_3 = 12000$ ohms, 1 watt
$R_4, R_6 = 820$ ohms, 0.5 watt
$R_5 = 200$ ohms, 0.5 watt, 5%
$R_7 = 15000$ ohms, 0.5 watt
$R_8 = 560$ ohms, 0.5 watt
$R_9 = 39$ ohms, 0.5 watt, 5%
$R_{10} = 15$ ohms, 1 watt, 5%
$R_{11} = 330$ ohms, 0.5 watt

$R_{12}, R_{15}, R_{16} = 1000$ ohms, 0.5 watt
$R_{13} =$ potentiometer, 1000 ohms, 0.5 watt
$R_{14}, R_{17} = 68$ ohms, 0.5 watt
$R_{18} = 20000$ ohms, 0.5 watt
$R_{19}, R_{20} = 0.39$ ohm, 5 watts
$S_1 =$ on-off switch, 120 V, single-pole, single-throw
$S_2 = 90°C$ thermal cutout
$T_1 =$ power transformer, Signal 88-2 (parallel secondary) or equiv.

NOTES:

1. Heat-sink capability of 1.3°C should be provided for each output transistor, based on mounting with mica washer and zinc-oxide thermal compound (Dow Corning No. 340 or equiv.) with $T_A = 45°C$ max.
2. Thermal cutout S_2 should be mounted on a common heat sink with output transistors.

16-18   40-W FULL-COMPLEMENTARY-SYMMETRY AUDIO AMPLIFIER (cont'd)

Circuit Description

This 40-watt audio amplifier uses RCA1B07 and RCA1B08 transistors as output devices in conjunction with nine TO-39 discrete transistors, and ten diodes. The amplifier uses a 64-volt split power supply with the output directly coupled to an 8-ohm speaker. This 40-watt Darlington full-complementary-symmetry amplifier combines excellent performance with economy.

RCA1B07 and RCA1B08 are n-p-n and p-n-p Darlington silicon transistors respectively. They are especially characterized for use as output devices in audio applications, and are provided in the JEDEC TO-3 package.

TYPICAL PERFORMANCE DATA

Measured at a line voltage of 120V, $T_A = 25°C$, and a frequency of 1 kHz, unless otherwise specified.

Power:
   Rated power (8-Ω load, at rated distortion) ..... 40 W
Total Harmonic Distortion:
   Rated distortion 0.5%
IM Distortion:
   10 dB below continuous power output at 60 Hz and 7 kHz (4:1) ...... <0.2%
IHF Power Bandwidth:
   3 dB below rated continuous power at rated distortion 5 Hz to 50 kHz

Bandwidth at
   1 W .... 5 Hz to 100 kHz
Sensitivity:
   At continuous power-output rating .... 700 mV
Hum and Noise:
   Below continuous power output:
      Input shorted .... 100 dB
      Input open .... 85 dB
      With 2 kΩ resistance on 20-ft. cable on input ... 97 dB
Input Resistance .... 18 kΩ

16-19   70-W QUASI-COMPLEMENTARY-SYMMETRY AUDIO AMPLIFIER

Circuit Description

This 70-watt amplifier uses the RCA1B01 in conjunction with seven TO-39 transistors, eleven diodes, and an 84-volt split power supply. The amplifier output is directly coupled to an 8-ohm speaker. This amplifier is most useful for instrumentation applications where ruggedness and raw power are essential.

The RCA1B01 is an n-p-n home-taxial-base silicon transistor in a JEDEC TO-3 package. This device is particularly suitable for audio-output use, and can be driven by either the RCA1A03 n-p-n or RCA1A04 p-n-p transistor.

16-19   **70-W QUASI-COMPLEMENTARY-SYMMETRY**
AUDIO AMPLIFIER (cont'd)

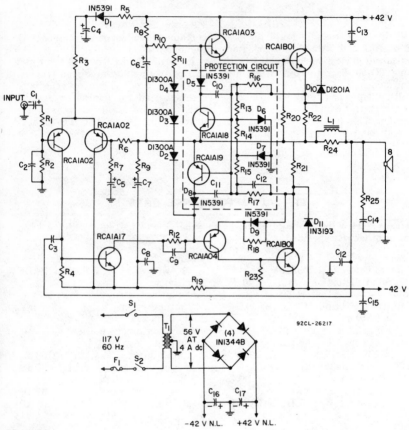

92CL-26217

92CS-26216

Parts List

$C_1 = 5\ \mu F$, electrolytic, 12 V
$C_2 = 180$ pF
$C_3 = 39$ pF
$C_4,\ C_6,\ C_7 = 50\ \mu F$, electrolytic, 50 V
$C_5 = 50\ \mu F$, electrolytic, 12 V
$C_8,\ C_9,\ C_{14} = 0.02\ \mu F$, 50 V
$C_{10},\ C_{11} = 0.01\ \mu F$
$C_{12},\ C_{13},\ C_{15} = 0.05\ \mu F$, 50 V
$C_{16},\ C_{17} = 3500\ \mu F$, electrolytic, 55 V
$F_1 = $ fuse, 120-V, 3-ampere, slow-blow
$L_1 = $ inductor, 10 μH, Miller

No. 4622 or equiv.
$R_1 = 1800$ ohms, 0.5 watt
$R_2,\ R_6 = 18000$ ohms, 0.5 watt
$R_3 = 12000$ ohms, 0.5 watt
$R_4 = 680$ ohms, 0.5 watt
$R_5 = 180$ ohms, 0.5 watt
$R_7,\ R_{12} = 470$ ohms, 0.5 watt
$R_8 = 2700$ ohms, 0.5 watt
$R_9 = 270$ ohms, 0.5 watt
$R_{10} = 33000$ ohms, 0.5 watt
$R_{11} = 47$ ohms, 0.5 watt
$R_{13},\ R_{15} = 1000$ ohms, 0.5 watt

$R_{14} = 4700$ ohms, 0.5 watt
$R_{16},\ R_{17} = 68$ ohms, 0.5 watt
$R_{18},\ R_{19},\ R_{20},\ R_{23} = 100$ ohms, 0.5 watt
$R_{21},\ R_{22} = 0.33$ ohm, 5 watts
$R_{24},\ R_{25} = 22$ ohms, 0.5 watt
$S_1 = 90°C$ thermal cutout
$S_2 = $ on-off switch, single-pole, single-throw
$T_1 = $ power transformer; primary, 117 V; secondary; 56 V, 4 amperes; Signal 56-4 or equiv.

NOTES:
1. All resistors should be non-inductive.
2. A heat-sink capability of 1°C per watt should be provided for each output transistor, based on mounting with mica washer and zinc-oxide thermal compound (Dow Corning No. 340 or equiv.) with $T_A = 45°C$ max.
3. Driver transistors should be mounted on printed circuit board heat sink.
4. Thermal cutout S_1 should be mounted on a common heat sink with RCA1B01 output transistors.

16-19 **70-W QUASI-COMPLEMENTARY-SYMMETRY AUDIO AMPLIFIER (cont'd)**

TYPICAL PERFORMANCE DATA

Measured at a line voltage of 120V, $T_A = 25°C$, and a frequency of 1 kHz, unless otherwise specified.

Power:
Rated power (8-Ω load, at rated distortion) ...... 70 W
Typical power (4-Ω load) . 100 W
Typical power (16-Ω load) 40 W
Music power (8-Ω load, at 5% THD with regulated supply) .... 100 W
Dynamic power (8-Ω load, at 1% THD with regulated supply) .... 88 W
Total Harmonic Distortion:
Rated distortion 1.0%

IM Distortion:
10 dB below continuous power output at 60 Hz and 7 kHz (4:1) 0.1%
Sensitivity:
At continuous power-output rating .... 700 mV
Hum and Noise:
Below continuous power output:
Input shorted .... 85 dB
Input open .... 80 dB
Input Resistance .... 20 kΩ

16-20 **120-W QUASI-COMPLEMENTARY-SYMMETRY AUDIO AMPLIFIER**

With Parallel Output Transistors

Circuit Description

This 120-watt amplifier circuit uses the RCA1B04 in conjunction with eleven other discrete transistors, twelve diodes, and a 130-volt split power supply. The amplifier output is directly coupled to an 8-ohm speaker. This RCA 120-watt audio amplifier is especially designed for top-of-the-line quadrasonic use in ap-plications requiring ½ kW of qua-drasonic sound with excellent tonal quality.

RCA1B04 is an n-p-n silicon pi-nu transistor in a JEDEC TO-3 package. This device is especially character-ized for audio applications, and can be driven by RCA1C12 and RCA1C13 transistors.

16-20   **120-W QUASI-COMPLEMENTARY-SYMMETRY**
AUDIO AMPLIFIER (cont'd)

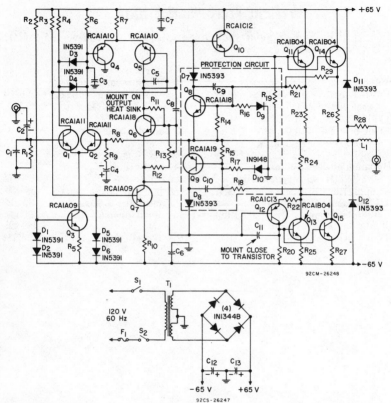

92CM-26248

92CS-26247

Parts List

$C_1 = 200$ pF
$C_2 = 5$ μF, electrolytic, 25 V
$C_3 = 680$ pF
$C_4 = 100$ μF, electrolytic, 25 V
$C_5 = 100$ pF
C_6, $C_7 = 0.05$ μF
$C_8 = 75$ pF
C_9, $C_{10} = 0.01$ μF
$C_{11} = 200$ pF
C_{12}, $C_{13} = 10000$ μF, 75 V
$F_1 =$ fuse, 120-V, 6-ampere, slow-blow
$L_1 =$ inductor, 3 μH at 10 A

R_1, $R_8 = 18000$ ohms, 0.5 watt
R_2, $R_3 = 22000$ ohms, 1 watt
R_4, R_6, $R_9 = 560$ ohms, 0.5 watt
$R_5 = 150$ ohms, 0.5 watt, 5%
$R_7 = 15$ ohms, 0.5 watt, 5%
$R_{10} = 33$ ohms, 0.5 watt, 5%
$R_{11} = 1000$ ohms, 0.5 watt
$R_{12} = 330$ ohms, 0.5 watt
$R_{13} =$ potentiometer, 1000 ohms, 0.5 watt
R_{14}, $R_{15} = 75$ ohms, 0.5 watt
R_{16}, $R_{17} = 820$ ohms, 1 watt

$R_{18} = 75$ ohms, 0.5 watt
R_{19}, $R_{20} = 47$ ohms, 1 watt
$R_{21} = 150$ ohms, 0.5 watt
$R_{22} = 10$ ohms, 0.5 watt
R_{23}, R_{25}, R_{26}, $R_{27} = 1$ ohm, 10 watts
$R_{24} = 0.5$ ohm, 10 watts
$R_{28} = 22$ ohms, 0.5 watt
$S_1 =$ on-off switch, 120-V, single-pole, single-throw
$S_2 = 100°C$ thermal cutout, Elmwood Sensor part No. 2455-88-4 or equiv.
$T_1 =$ power transformer, Signal 88-6 or equiv.

NOTES:

1. All resistors should be non-inductive.
2. Transistors Q_4, Q_5, and Q_7 should be mounted on a heat sink. Use Wakefield 209AB-series type or equiv.
3. A heat-sink capability of 1°C per watt should be provided for each output transistor, based on mounting with mica washer and zinc-oxide thermal compound (Dow Corning No. 340 or equiv.) with $T_A = 45°C$ max.
4. Thermal cutout S_2 should be mounted on common heat sink with output transistors.

16-20 **120-W QUASI-COMPLEMENTARY-SYMMETRY AUDIO AMPLIFIER (cont'd)**

TYPICAL PERFORMANCE DATA

Measured at a line voltage of 120 V, $T_A = 25°C$, and a frequency of 1 kHz, unless otherwise specified.

Power:
  Rated power (8-Ω load, at rated distortion) .... 120 W
  Typical power (4-Ω load) 180 W
  Typical power (16-Ω load) .... 80 W
Total Harmonic Distortion:
  Rated distortion 0.5%
IM Distortion:
  10 dB below continuous power output at 60 Hz and 7 kHz (4:1) 0.2%
IHF Power Bandwidth:
  3 dB below rated continuous power at rated distortion ...... 5 Hz to 50kHz
Sensitivity:
  At continuous power output rating .... 900 mV
Hum and Noise:
  Below continuous power output:
    Input shorted .... 104 dB
    Input open .... 88 dB
    With 2 kΩ resistance on 20-ft. cable on input .... 104 dB
Input Resistance .... 18 kΩ

16-21 **200-W QUASI-COMPLEMENTARY-SYMMETRY AUDIO AMPLIFIER**
With Parallel Output Transistors

Circuit Description

This 200-watt amplifier uses eight RCA 1B05 transistors, two as drivers and six as parallel units in the amplifier output stages. These devices are employed in conjunction with eleven other discrete transistors, twelve diodes, and a 160-volt split power supply. The amplifier output is directly coupled to an 8-ohm speaker. This 200-watt audio amplifier is especially designed to feature ruggedness in combination with high power output and excellent high-fidelity performance.

The RCA1B05 is a silicon n-p-n pi-nu transistor in a JEDEC TO-3 package. This device is especially suitable for applications in audio-amplifier circuits, in which it may be used as either driver or output unit.

16-21   **200-W QUASI-COMPLEMENTARY-SYMMETRY
AUDIO AMPLIFIER (cont'd)**

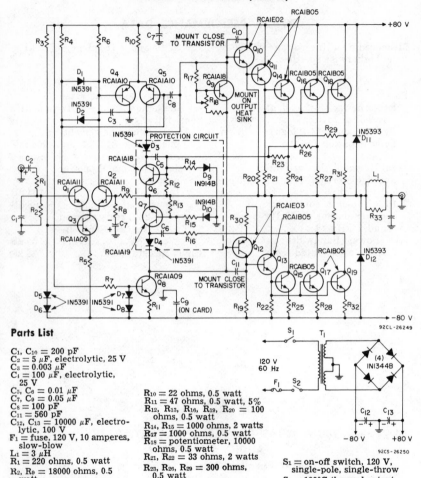

92CL-26249

Parts List

C_1, C_{10} = 200 pF
C_2 = 5 μF, electrolytic, 25 V
C_3 = 0.003 μF
C_4 = 100 μF, electrolytic, 25 V
C_5, C_6 = 0.01 μF
C_7, C_9 = 0.05 μF
C_8 = 100 pF
C_{11} = 560 pF
C_{12}, C_{13} = 10000 μF, electrolytic, 100 V
F_1 = fuse, 120 V, 10 amperes, slow-blow
L_1 = 3 μH
R_1 = 220 ohms, 0.5 watt
R_2, R_9 = 18000 ohms, 0.5 watt
R_3, R_7 = 33000 ohms, 1 watt
R_4, R_6 = 620 ohms, 0.5 watt
R_5 = 150 ohms, 0.5 watt, 5%
R_8 = 390 ohms, 0.5 watt

R_{10} = 22 ohms, 0.5 watt
R_{11} = 47 ohms, 0.5 watt, 5%
R_{12}, R_{13}, R_{16}, R_{19}, R_{20} = 100 ohms, 0.5 watt
R_{14}, R_{15} = 1000 ohms, 2 watts
R_{17} = 1000 ohms, 0.5 watt
R_{18} = potentiometer, 10000 ohms, 0.5 watt
R_{21}, R_{22} = 33 ohms, 2 watts
R_{23}, R_{26}, R_{29} = 300 ohms, 0.5 watt
R_{24}, R_{25}, R_{27}, R_{28}, R_{31}, R_{32} = 1 ohm, 10 watts
R_{30} = 0.3 ohm, 10 watts
R_{33} = 22 ohms, 2 watts

92CS-26250

S_1 = on-off switch, 120 V, single-pole, single-throw
S_2 = 100°C thermal cutout, Elmwood Sensor Part No. 2455-88-4 or equiv.
T_1 = power transformer, Signal 120-6 or equiv.

NOTES:

1. All resistors should be non-inductive.
2. Heat-sink capability of 1°C per watt should be provided for each output transistor, based on mounting with mica washer and zinc-oxide thermal compound (Dow Corning No. 340 or equiv.)
3. Thermal cutout S_2 should be mounted on a common heat sink with output transistors.
4. Transistor Q_9 should be mounted on a heat sink Wakefield No. 260-6 SH 5E or equiv.
5. Transistors Q_4, Q_5, and Q_8 should be mounted on a heat sink Wakefield No. 209AB series or equiv.

16-21   ## 200-W QUASI-COMPLEMENTARY-SYMMETRY
AUDIO AMPLIFIER (cont'd)

TYPICAL PERFORMANCE DATA

Measured at a line voltage of 120V, $T_A = 25°C$, and a frequency of 1 kHz, unless otherwise specified.

Power:

Rated power (8-Ω load, at rated distortion)	200 W
Typical power (4-Ω load)	300 W
Typical power (16-Ω load)	130 W

Total Harmonic Distortion:
Rated distortion 0.5%

IM Distortion:
10 dB below continuous power output at 60 Hz and 7 kHz (4:1) 0.2%

IHF Power Bandwidth:
3 dB below rated continuous power at rated distortion ...... 5Hz to 35 kHz

Sensitivity:
At continuous power output rating .... 900 mV

Hum and Noise:
Below continuous power output:

Input shorted	96 dB
Input open	84 dB
With 2 kΩ resistance on 20-ft. cable on input .	94 dB

Input Resistance .... 18 kΩ

16-22   ## SERVO AMPLIFIER

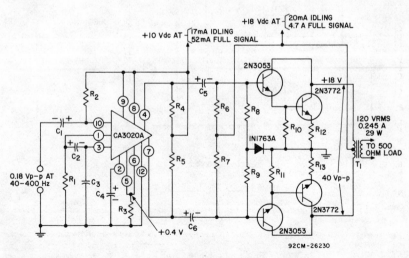

92CM-26230

Parts List

C₁, C₂, C₄, C₅, C₆ = 100 μF, electrolytic, 10 V
C₃ = 0.1 μF, paper
R₁ = 4700 ohms, 0.5 watt
R₂ = 0.47 megohm, 1 watt

R₃ = 5 ohms, 0.5 watt
R₄, R₅ = 270 ohms, 1 watt
R₆, R₇ = 22000 ohms, 1 watt
R₈, R₉, R₁₀, R₁₁ = 470 ohms, 1 watt

R₁₂, R₁₃ = 0.5 ohm, 10 watts
T₁ = output transformer, Stancor P-8358 or equiv.

16-22   SERVO AMPLIFIER (cont'd)

Circuit Description

This servo amplifier can supply up to 29 watts of power to the drive motor of a servo system with a signal input of only 180 millivolts (peak-to-peak). The driver portion of the amplifier uses a CA3020A integrated circuit. The output stage uses Darlington-connected push-pull power transistors to develop the required output power. Appropriate alternative transformers can be selected for the output transformer T_1; the impedance of the transformer

primary should be approximately 60 ohms (collector-to-collector).

The output stage in the CA3020A is a class B Amplifier. It requires a + 10-volt supply with current requirements as shown in accordance with the input-signal amplitude. The push-pull output stage also operates as a class B Amplifier from a +18-volt supply, with supply-current requirements up to 4.7 amperes at the specified output power.

16-23   FREQUENCY-SELECTIVE AUDIO AMPLIFIER

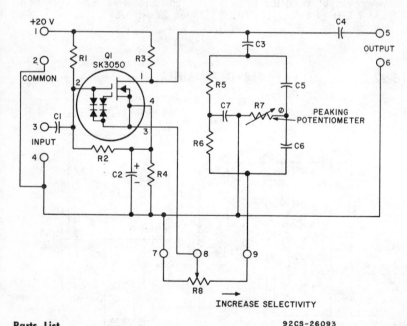

INCREASE SELECTIVITY

92CS-26093

Parts List

C_1, C_3, C_4 = 0.1 μF, 25 V or greater
C_2 = 10 μF, electrolytic, 6 V
C_5, C_6 = 680 pF, 25 V or greater
C_7 = 1500 pF, 25 V or greater

R_1 = 1 megohm, 0.5 watt, 10%
R_2 = 100000 ohms, 0.5 watt, 10%
R_3 = 6800 ohms, 0.5 watt, 10%
R_4 = 1200 ohms, 0.5 watt, 10%

R_5, R_6 = 220000 ohms, 0.5 watt, 10%
R_7 = potentiometer, 250000 ohms, linear taper, trimpot
R_8 = potentiometer, 1 megohm, linear taper

16-23 FREQUENCY-SELECTIVE AUDIO AMPLIFIER (cont'd)

Circuit Description

This frequency-selective audio-frequency amplifier amplifies signals at only one predetermined frequency. At that frequency, the voltage gain is 20 to 30; at other frequencies the voltage gain is unity or less. Circuits of this type are useful in screening out undesirable side signals when copying code or for identifying the frequency of a particular signal.

Potentiometer R8 controls the level of the feedback signal; potentiometer R7 is used to adjust or peak the twin-T bridge to the desired frequency. The SK3050 dual-gate MOS field-effect transistor Q_1 acts as a basic audio amplifier. Part of the output of the amplifier is applied to the twin-T bridge oscillator through C_3. At the predetermined frequency, where most gain occurs, the filter passes the ac at a phase angle that assures positive feedback. The feedback, adjusted by R_8, is added to the incoming signal and causes it to increase.

With the component values shown in the parts list, the frequency selected for amplification is approximately 1000 Hz; the chart below shows values of C_5, C_6, and C_7 required for other typical frequencies.

The current drain for this circuit is approximately 1.5 milliamperes. The maximum input signal is 0.1 volt rms.

Frequency-Selective AF Amplifier Bridge Capacitor Values for Various Frequencies

Approximate Frequency (Hz)	C_5, C_6 (pF)	C_7 (pF)
150	5600	12,000
300	2700	6,200
600	1300	3,000
1200	680	1,500
2400	330	750
4800	160	360
9600	82	180

16-24 AUDIO AMPLIFIER-OSCILLATOR

Circuit Description

This integrated-circuit audio amplifier-oscillator can be used as an amplifier in portable systems, such as portable phonographs, or in any application that requires a low-power, portable, light-weight unit. The amplifier requires an input signal of 40 millivolts, and provides an output power of ½ watt. Two amplifiers may be used to form a stereo system. The audio oscillator can be used with a telegraph key as a code practice oscillator or with an on-off switch to provide a continuous tone.

The CA3020 used in the amplifier-oscillator is a wideband power amplifier that includes a voltage regulator, buffer or amplifier, differential amplifier and phase splitter, driver, and power-output amplifier. The volt-

16-24   AUDIO AMPLIFIER-OSCILLATOR (cont'd)

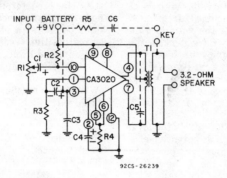

92CS-26239

Parts List

C₁, C₂ = 5 μF, electrolytic, 12 V
C₃ = 0.01 μF, 25 V or greater
C₄ = 1 μF, electrolytic, 6 V
C₅, C₆ = 0.1 μF, 25 V or greater (not used in amplifier)

R₁ = potentiometer; 100000 ohms, 0.5 watt, linear taper
R₂ = 470000 ohms, 0.5 watt, 10%
R₃ = 4700 ohms, 0.5 watt, 10%
R₄ = 1 ohm, 0.5 watt, 10%

R₅ = 2200 ohms, 0.5 watt, 10%
T₁ = transformer; primary, 200 ohms; secondary, 3.2 ohms, 500 milliwatts; United Transformer Company HCA308 or equiv.

Circuit Description (cont'd)

age regulator keeps power dissipation constant within the —55 to +125°C temperature range and supplies two voltages to the differential amplifier: a base supply voltage of about 1.4 volts and a collector supply voltage of about 2.1 volts.

The chart below shows voltages at the terminals of a CA3020 inte-

grated circuit in a properly operating circuit.

Components C₅, C₆, and R₅ are not required in the amplifier. In the oscillator circuit, the potentiometer acts as a tone control. A 3-volt power supply should be sufficient for most oscillator uses; the 9-volt supply tends to make the audio output level too high for comfort.

Voltages at IC Terminals in Audio Amplifier-Oscillator

Terminal	Voltage (Volts)
1	3.8
2	0.8
3	0.8
4	9
7	9
8	9
9	9
10	4.5

16-25   AUDIO OSCILLATOR

Circuit Description

This basic audio-oscillator circuit may be used to provide a single-tone sine-wave output at any frequency to well above 100 kHz. (A chart of capacitance values is shown for different frequencies of operation.) The circuit is excellently suited for use in the testing of high-fidelity audio equipment and amateur radio transmitters; it can also be adapted for use as a code-practice oscillator. (A keyer can be inserted between points A and B.) The oscillator operates from a dc supply of 12 volts and supplies a relatively distortion-free output waveform to any circuit that has an input impedance of 3000 ohms or more.

The SK3020 amplifier transistor Q_1, capacitors C_1, C_2, C_3, and C_4, and resistors R_1, R_2, and R_3 form a basic twin-T oscillator circuit. A portion of the signal developed at the collector of transistor Q_1 is applied to the twin-T network formed by C_1, C_2, C_3, R_1, R_2, R_3, and R_4. Potentiometer R_2 provides an adjustment of approximately ±10 per cent in the oscillator frequency. The output of this network is then coupled to the base of transistor Q_1 through capacitor C_4 to supply the positive feedback required to sustain oscillation. The oscillator-stage output from the collector of transistor Q_1 is applied to the base of the SK3020 output transistor Q_2, which is operated in an emitter-follower circuit configuration. This stage amplifies the oscillator output to provide the sine-wave output signal. Potentiometer R_7 in the emitter circuit of transistor Q_1 is adjusted to obtain the desired output waveform.

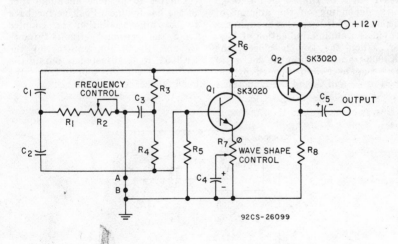

92CS-26099

Parts List

C_1, C_2 = see chart for value, mica or paper
C_3 = twice the value of C_1, mica or paper
C_4 = 1 μF, electrolytic, 12 V
C_5 = 300 μF for frequencies below 2000 Hz or 5 μF for frequencies above 2000 Hz, electrolytic, 6 V
C_6 = 20 μF, electrolytic, 6 V
R_1 = 2700 ohms, 0.5 watt
R_2 = Frequency control, potentiometer, 5000 ohms, 0.5 watt
R_3, R_4 = 51000 ohms, 0.5 watt
R_5 = 22000 ohms, 0.5 watt
R_6 = 4700 ohms, 0.5 watt
R_7 = Wave-shape control, potentiometer, 250 ohms, 0.5 watt
R_8 = 820 ohms, 0.5 watt

16-25   **AUDIO OSCILLATOR (cont'd)**

Capacitor Selection Chart for Different Operating Frequencies

Approx. Freq. (Hz)	Value of C_1 and C_2
100,000	50 pF
50,000	100 pF
10,000	500 pF
5,000	1000 pF
1,000	0.005 MF
500	0.01 μF
100	0.05 μF
50	0.1 μF
10	0.5 μF
5	1 μF

16-26   **CODE-PRACTICE OSCILLATOR**

Circuit Description

This simple audio oscillator operates from a dc supply of 1.5 to 4.5 volts, depending on the amount of output desired. Magnetic headphones provide an audible indication of keying. When the key is closed, the SK3003 transistor supplies energy to the resonant circuit formed by capacitors C_1 and C_2 and the inductance of the headphones, and this circuit resonates to produce an audio tone in the headphones. Positive feedback to sustain oscillation is coupled from the resonant circuit through C_1 and C_2 to the emitter of the SK3003. R_4 is adjusted to obtain the desired level of sound from the headphones.

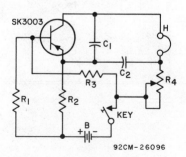

92CM-26096

Parts List

B = 1.5-4.5 V (One to three series-connected RCA VS036 dry cells may be used, depending upon the volume level desired.)

C_1, C_2 = 0.1 μF, paper, 150 V
H = Headphone, 2000-ohm, magnetic
R_1 = 2200 ohms, 0.5 watt

R_2 = 27000 ohms, 0.5 watt
R_3 = 3000 ohms, 0.5 watt
R_4 = volume control potentiometer, 50000 ohms, 0.5 watt

16-27 ### PREAMPLIFIER FOR 6-, 10-, OR 15-METER AMATEUR-BAND RECEIVER

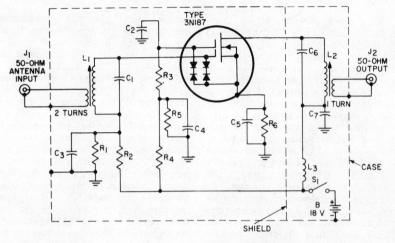

92CS-26097

Parts List

B = Two RCA type VS323 batteries for transistor service; and one case, Bud-CU2103A or equivalent.
C_1 = 8 pF, mica or ceramic tubular
C_2, C_3, C_4, C_5, C_7 = 0.01 μF, ceramic
C_6 = 10 pF, mica or ceramic tubular

J_1, J_2 = Coaxial receptacle, Amphenol BNC type UG-1094 or equiv.
L_1, L_2 = 1.6 to 3.1 μH, adjustable, Miller 4404 or equiv.
L_3 = 22 μH, Miller 74F-225A1 or equiv.
R_1 = 27,000 ohms. 0.25 watt, 10%
R_2 = 150,000 ohms. 0.25

watt, 10%, carbon
R_3 = 1,800 ohms. 0.25 watt, 10%, carbon
R_4 = 100,000 ohms. 0.25 watt, 10%, carbon
R_5 = 33,000 ohms. 0.25 watt, 10%, carbon
R_6 = 270 ohms. 0.25 watt, 10%, carbon
S_1 = toggle switch, single-pole, single-throw

Tuned-Circuit Components for 21 and 50 MHz

Component	Value	
	21 MHz	50 MHz
C_1	22 pF	8 pF
C_2, C_3, C_4, C_5 C_7	No Change	1,000 pF, ceramic
C_6	22 pF	10 pF
L_1	No Change	8 turns, No. 30 E wire on ¼-inch - diameter c o r e (Miller 4500 or equiv.) Link: 2 turns, No. 30 E wire on ground end.
L_2	No Change	Same as L_1
L_3	No Change	6.8 μH (Miller 74F686AP or equiv.)

16-27   PREAMPLIFIER FOR 6-, 10-, OR 15-METER
AMATEUR-BAND RECEIVER (cont'd)

Circuit Description

This inexpensive, easily constructed preamplifier circuit uses a 3N187 dual-gate-protected MOS transistor to provide more than 26 dB of gain ahead of a receiver operated in the 6-, 10-, or 15-meter amateur band. This additional gain, together with the low noise figure of the preamplifier (less than 2.5 dB), substantially increases both the sensitivity and signal-to-noise ratio of the receiver. The circuit as shown is intended for use in the 10-meter (28-MHz) frequency band; the 3N187 MOS transistor, however, has excellent performance characteristics at frequencies well below the 10-meter band and up to 200 MHz. The preamplifier, therefore, can be readily adapted for use in other frequency bands with only a few changes in tuned-circuit components. A chart is provided to show the changes in tuned-circuit components required for operation in the 15-meter (21-MHz) and 6-meter (50-MHz) bands. The dc operating voltage for the preamplifier may be obtained from a battery supply, as shown in the circuit diagram, or from any other reasonably well-filtered dc supply voltage of 15 to 18 volts.

The dual-gate MOS transistor in the preamplifier is operated so that essentially it is electrically equivalent to two single-gate MOS transistors connected in cascode and enclosed in the same package. The advantage of the dual gate transistor is that it provides an inexpensive cascode circuit that offers maximum resistance to cross-modulation from nearby transmitters.

The rf input is link coupled from the antenna to the input tuned circuit formed by L_1 and C_1 and applied to gate No. 1 (pin 3) of the 3N187 transistor. This gate, which is equivalent to the gate (or base) of the grounded-source (or -emitter) section of a two-transistor cascode circuit, is forward-biased by the dc voltage at the junction of the voltage-divider resistors R_1 and R_2. The source resistor R_2 is large enough to assure that gate No. 1 is always negative with respect to the source. Gate No. 2 (pin 2), in accordance with cascode-circuit requirements, is returned to ac ground through capacitor C_2. The dc bias level for this gate, established by the voltage divider R_4 and R_5, represents a compromise between optimum gain and optimum cross-modulation resistance. The amplified rf signals developed in the drain circuit of the 3N187 transistor are link coupled from the tuned-circuit drain load impedance formed by L_2 and C_6, through coaxial connector J_2 to the input of the receiver.

Tuning of the preamplifier is simplified because no special neutralization is required, even at frequencies as high as 155 MHz. Rough adjustments of coils L_1 and L_2 can be made by use of a grid-dip oscillator. The finishing adjustments are then made while listening to a weak station.

16-28   STABLE VARIABLE-FREQUENCY OSCILLATOR

Circuit Description

This VFO circuit uses a 40823 dual-gate-protected MOS transistor in a highly stable variable-frequency oscillator stage and SK3018 and SK3020 bipolar transistors in a two-stage isolation (output) amplifier to

16-28 STABLE VARIABLE-FREQUENCY OSCILLATOR (cont'd)

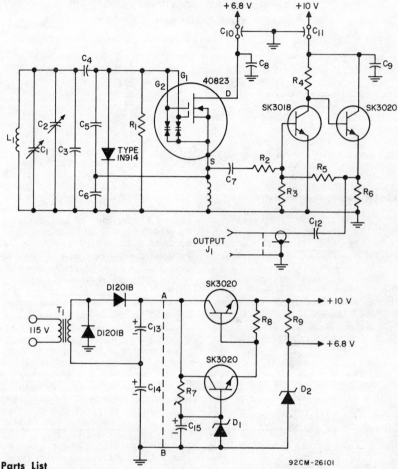

92CM-26101

Parts List

C_1 = Double-bearing variable capacitor, Millen 23100 or 23050 (or equiv.) depending upon frequency range (see Tuned-Circuit Data)

C_2 = Air-type trimmer capacitor, 25 pF maximum, Hammarlund APC-25 or equiv.

C_3, C_4, C_5, C_6 = silver-mica capacitors (see Tuned-Circuit Data for values)

C_7 = 2200 pF, silver mica

C_8 = 0.05 pF, ceramic disc, 50 V.

C_9 = 0.1 pF, ceramic disc, 50 V.

C_{10}, C_{11} = 1500 pF, feed-through

C_{12} = 0.025 µF, ceramic disc, 50 V.

C_{13} = 500 µF, electrolytic, 12 V.

C_{14} = 500 µF, electrolytic, 12 V.

C_{15} = 50 µF, electrolytic, 12 V.

D_1 = Zener diode, 12-volt, 1-watt

D_2 = Zener diode, 6.8 volt, 1-watt

J_1 = Coaxial connector

L_1 = Variable inductor (see Tuned-Circuit Data for details)

L_2 = Miniature rf choke,

R_1 = 22000 ohms, 0.5 watt

R_2 = 12000 to 47000 ohms, 0.5 watt; select value for 2-volt peak output level at input to transmitter

R_3 = 12000 ohms, 0.5 watt

R_4 = 820 ohms, 0.5 watt

R_5 = 47000 ohms, 0.5 watt

R_6 = 240 ohms, 0.5 watt

R_7 = 2200 ohms, 0.5 watt

R_8 = 220 ohms, 0.5 watt

R_9 = 180 ohms, 0.5 watt

T_1 = 6.3-volt, 1.2-ampere filament transformer

16-28 STABLE VARIABLE-FREQUENCY OSCILLATOR (cont'd)

Tuned-Circuit Data

	3.5-4.0 MHz	5.0-5.5 MHz	8.0-9.0 MHz
L_1			
No. of turns	17*	14¾*	11½**
Wire size	20	20	18
Turns/inch	16	16	8
Diam., inches	1	1	1
C_1, p.	100	50	50
C_2, pf.	25	25	25
C_3, pf.	100	None	None
C_4, pf.	390	390	270
C_5, pf.	680	680	560
C_6, pf.	680	680	560

* B & W 3015, AirDux 816T, or equiv.
** B & W 3014, AirDux 808T, or equiv.

Circuit Description (cont'd)

achieve exceptional frequency stability at low dc operating potentials. The MOS-transistor oscillator circuit is useful at any frequency up to and including the 144-MHz band. Tuned-circuit data are provided for the standard 3.5-to-4-MHz band, for the 5-to-5.5-MHz band for single-sideband transmitters, and for the 8-to-9-MHz band for 50- and 144-MHz transmitters. (See chart on page 607.)

The oscillator stage is a Colpitts type. The variable capacitor C_1 is the tuning control for the circuit. With a Millen 10037 (or equivalent) "no sting" dial coupled to the shaft of this capacitor, the oscillator tuning range encompasses essentially the full dial area. Capacitor C_2 is the trimmer adjustment for the circuit. The effect of changes in transistor-element capacitances is reduced to a minimum by use of a three-capacitor (C_4, C_5, and C_6) voltage divider. The relatively large values of the capacitors C_5 and C_6, which are connected across the gate-to-source circuit of the MOS transistor, almost completely obviate the effect of the transistor capacitances. The rf choke L_1 provides the required low voltage (IR) drop for the source current of the MOS transistor.

The 1N914 silicon rectifier in the gate circuit of the oscillator stage is used to provide the rectified gate current for the MOS transistor. This rectifier makes possible a degree of automatic bias comparable to that obtainable with an electron tube and, in this way, contributes substantially to the frequency stability of the VFO circuit. The use of silver-mica types for all fixed-value capacitors in the oscillator stage assures a stable frequency-temperature characteristic.

The output of the oscillator stage is coupled from the source of the MOS transistor, through capacitor C_7 and resistor R_1, to the base of the SK3018 bipolar transistor used in the input stage of the isolation amplifier. The output of the SK3018 transistor, in turn, drives the SK3020 emitter-follower output stage. The isolation amplifier is essentially a two-stage, direct-coupled, negative-feedback output circuit that greatly reduces the effect of a change in output conditions on oscillator performances and provides a convenient means (by a change in the value of resistor R_1) to vary the output voltage of the VFO circuit.

The dc operating potentials for the VFO circuit can be obtained directly from a 12-volt source. For operation from a 117-volt, 60-Hz ac source, a low-voltage dc supply, such as that shown in the circuit diagram, may be used to supply the re-

16-28 STABLE VARIABLE-FREQUENCY OSCILLATOR (cont'd)

Circuit Description (cont'd)

quired voltage. The 117-volt ac source voltage is stepped down to 6.3 volts ac by the power transformer T_1 and then converted to a dc voltage of 12 volts by the voltage-doubler circuit formed by the D1201B rectifier diodes and filter capacitors C_{13} and C_{14}. The two SK3020 bipolar transistors and the Zener diodes D_1 and D_2 connected between points A and B of the voltage-doubler circuit form an electronic voltage regulator that maintains constant dc output voltages with changes in the input ac voltage.

The voltage-regulator circuit is also used when the VFO is operated in a mobile system. For this type of operation, the power transformer T_1

and the voltage doubler are disconnected from the remainder of the circuit, and points A and B are connected to the positive and negative terminals, respectively, of a 12-volt battery.

The VFO circuit is characterized by its exceptional frequency stability. A unit designed to operate in the 3.5-to-4-MHz frequency range exhibits a frequency drift of less than 30 Hz in 2 hours after a 30-second warm-up. A 5-to-5.5-MHz unit has a frequency drift of less than 50 Hz for the same period, and a 8-to-9-Mz unit has a frequency drift of only slightly more than 200 Hz.

16-29 50-MHz, 40-WATT CW TRANSMITTER
With Load-Mismatch Protection

Circuit Description

This cw transmitter uses a VSWR bridge circuit to maintain a steady-state dissipation in the output stage under all conditions of antenna mismatch. This technique makes it possible to realize the full power potential of the 40341 overlay transistor used in the output stage.

The 50-MHz crystal-controlled 2N3118 oscillator stage develops the low-level excitation signal for the transmitter. The 50-MHz output signal from the collector of the oscillator transistor is coupled by L_8 to the base of a second 2N3118 used in a predriver stage (low-level amplifier). This step-down transformer matches the collector impedance of the oscillator transistor to the low-impedance base circuit of the predriver transistor. The collector circuit of the predriver is tuned to provide maximum signal output at 50 MHz. This signal is coupled from a tap on inductor L_8 to the input

(base) circuit of the driver stage, which uses a 2N3375 silicon power transistor to develop the power required to drive the output stage.

The 40341 overlay transistor used in the output stage develops 40 watts of power output at the transmitting frequency of 50 MHz. The driving power for the output stage is coupled from the collector of the driver transistor through a bandpass filter to the base of the output transistor. The filter networks in the collector circuit of the 40341 provide the required harmonic and spurious-frequency rejection. The 50-MHz output from these filter sections is coupled through a length of 50-ohm coaxial line to the antenna. Capacitors C_6, C_9, and C_{13} are adjusted to provide optimum impedance match between the transmitter and the antenna.

The output of the transmitter is sampled by a current transformer

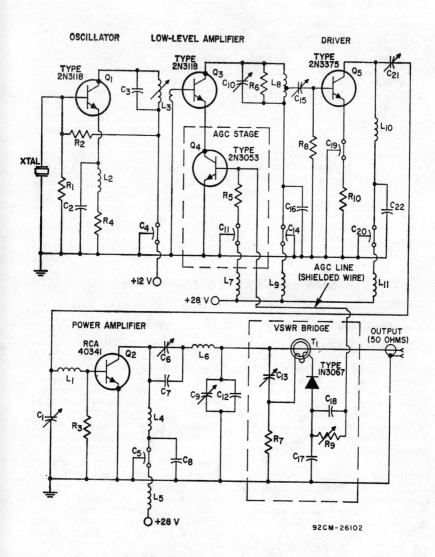

92CM-26102

16-29   50-MHz, 40-WATT CW TRANSMITTER (cont'd)

Parts List

C_1 = variable capacitor, 90 to 400 pF, Arco No. 429 or equiv.
C_2 = 51 pF, mica
C_3 = 30 pF, ceramic
C_4, C_5, C_{11}, C_{14}, C_{19}, C_{20} = feedthrough capacitor, 1000 pF
C_6 = variable capacitor, 1.5 to 20 pF, Arco No. 402 or equiv.
C_7 = 36 pF, mica
C_8, C_{16}, C_{22} = 0.02 μF, ceramic
C_9, C_{10} = variable capacitor, 8 to 60 pF, Arco No. 404 or equiv.
C_{12} = 91 pF, mica
C_{13} = variable capacitor, 0.9 to 7 pF, Vitramon No. 400 or equiv.
C_{15} = variable capacitor,

14 to 150 pF, Arco No. 426 or equiv.
C_{17} = 1000 pF, ceramic
C_{18} = 0.01 μF, ceramic
C_{21} = variable capacitor, 32 to 250 pF, Vitramon No. 464 or equiv.
L_1 = 1 turn of No. 16 wire; inner diameter, $\frac{5}{16}$ inch; length, $\frac{1}{8}$ inch
L_2 = rf choke, 1 μH
L_3 = oscillator coil; primary, 7 turns; secondary, 1-$\frac{3}{4}$ turns; wound from No. 22 wire on CTC coil form having "white dot" core
L_4 = 5 turns of No. 16 wire; inner diameter, $\frac{5}{16}$ inch; length, $\frac{1}{2}$ inch
L_5, L_7, L_9, L_{10}, L_{11} = rf choke, 7 μH

L_6 = 4 turns of B & W No. 3006 coil stock
L_8 = 6 turns of No. 16 wire; inner diameter, $\frac{3}{8}$ inch; length, $\frac{3}{4}$ inch
R_1, R_6 = 510 ohms, 0.5 watt
R_2 = 3900 ohms, 0.5 watt
R_3, R_8 = 2.2 ohms, wire-wound, 0.5 watt; International Resistor Corp. BWH type, or equiv.
R_4 = 51 ohms, 0.5 watt
R_5 = 24000 ohms, 0.5 watt
R_7 = 240 ohms, 0.5 watt
R_9 = agc control, potentiometer, 50000 ohms
R_{10} = 5.6 ohms, 1 watt
T_1 = current transformer (toroid), Arnold No. A4-437-125-SF, or equiv.
XTAL = 50-MHz transmitting crystal

Circuit Description (cont'd)

(toroid) T_1 loosely coupled about the output transmission line. This transformer is the sensor for a VSWR bridge detector used to prevent excessive dissipation in the output stage under conditions of antenna mismatch. If the antenna is disconnected or poorly matched to the transmitter, large standing waves of voltage and current occur on the output transmission line. A portion of this standing-wave energy is applied by T_1 to the 1N3067 diode in the bridge circuit. The rectified current from this diode charges capacitor C_{18} to a dc voltage proportional to the amplitude of the standing waves. This voltage, which is essentially an agc bias, is applied to the base of the 2N3053 agc amplifier stage. The output of the agc stage biases the 2N3118 predriver stage so that its gain changes in inverse proportion to the amplitude of the standing wave on the output transmission line. Therefore, as the amplitude of the standing waves increases (tending to cause higher heat dissipation in the output transistor), the input drive to the output stage is reduced. This compensating effect maintains a steady-state dissipation in the output transistor regardless of mismatch conditions between the transmitter output circuit and the antenna.

16-30   175-MHz, 35-WATT AMPLIFIER

Circuit Description

This four-stage rf power amplifier operates from a dc supply of 13.5 volts and delivers 35 watts of power output at 175 MHz for an input of 125 milliwatts. The silicon overlay transistors used in the amplifier

16-30   **175-MHz, 35-WATT AMPLIFIER (cont'd)**

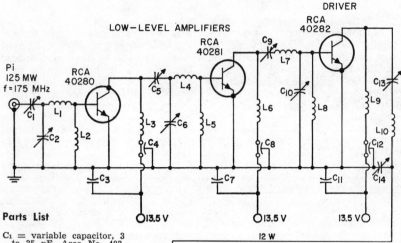

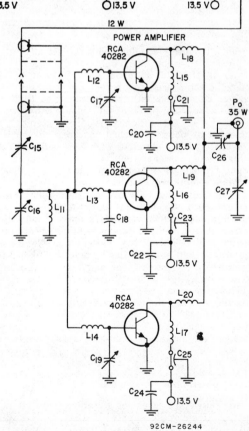

Parts List

C_1 = variable capacitor, 3 to 35 pF, Arco No. 403, or equiv.

C_2, C_6, C_{16}, C_{17}, C_{18}, C_{19}, C_{27} = variable capacitor, 8 to 60 pF, Arco No. 404, or equiv.

C_3, C_7, C_{11} = 0.1 μF, ceramic disc

C_4, C_8, C_{12}, C_{21}, C_{23}, C_{25} = feedthrough capacitor, 1500 pF

C_5, C_{10}, C_{13}, C_{14}, C_{26} = variable capacitor, 7 to 100 pF, Arco No. 423, or equiv.

C_9 = variable capacitor, 14 to 150 pF, Arco No. 424 or equiv.

C_{15} = variable capacitor, 1.5 to 20 pF, Arco No. 402 or equiv.

C_{20}, C_{22}, C_{24} = 0.2 μF, ceramic disc

L_1 = 2 turns of No. 16 wire; inner diameter, $\frac{3}{16}$ inch; length, $\frac{1}{4}$ inch

L_2, L_5, L_8 = 450-ohm ferrite rf choke

L_3, L_6, L_{11} = rf choke, 1.0 μH

L_4, L_7 = 3 turns of No. 16 wire; inner diameter, $\frac{3}{16}$ inch; length, $\frac{1}{4}$ inch

L_9 = 1-$\frac{1}{2}$ turns of No. 16 wire; inner diameter, $\frac{1}{4}$ inch; length, $\frac{3}{8}$ inch

L_{10} = 2 turns of No. 16 wire; inner diameter, $\frac{1}{4}$ inch; length, $\frac{5}{16}$ inch

L_{12}, L_{13}, L_{14} = 5 turns of No. 16 wire; inner diameter, $\frac{1}{4}$ inch; length, $\frac{1}{2}$ inch

L_{15}, L_{16}, L_{17} = 2 turns of No. 18 wire; inner diameter, $\frac{1}{8}$ inch; length, $\frac{1}{8}$ inch

L_{18}, L_{19}, L_{20} = 2 turns of No. 16 wire; inner diameter $\frac{1}{4}$ inch; length, $\frac{1}{4}$ inch

16-30   175-MHz, 35-WATT AMPLIFIER (cont'd)

Circuit Description (cont'd)

supply maximum output power at this level of dc voltage for use in mobile systems.

The low-level portion of the amplifier consists of three unneutralized, class C, common-emitter rf amplifier stages interconnected by band-pass filters tuned to provide maximum transfer of energy at 175 MHz. The 40280 input stage develops 1 watt of power output when a 125-milliwatt 175-MHz signal is applied to the amplifier input terminal. This output is increased to 4 watts by the 40281 transistor used in the second stage. The 40282 driver transistor then develops 12 watts of driving power for the output stage.

When the low-level stages and the output stage are mounted on separate chassis, the output from the driver stage is coupled to the output stage through a low-loss coaxial line. The line is terminated by variable capacitors C_{15} and C_{16} and inductor L_{11}. The capacitors are adjusted to assure a good impedance match between the output of the driver and the input of the output stage at 175 MHz. The driving signal developed across inductor L_{11} is applied to the tuned input networks of three parallel-connected 40282 transistors in the single-ended output stage. For an input of 12 watts, the three 40282 transistors deliver 35 watts of 175-MHz power to the output terminal of the amplifier. Capacitors C_{26} and C_{27} are adjusted to match the amplifier output to the load impedance at the operating frequency.

16-31   40-WATT PEAK-ENVELOPE-POWER AIRCRAFT-BAND AMPLIFIER FOR AM TRANSMITTERS

Circuit Description

This broadband rf power amplifier is intended for use in amplitude-modulated (AM) transmitters operating in the aircraft communication band (118 to 136 MHz). The circuit is simple and easy to duplicate and requires a minimum of adjustments. The amplifier uses 2N3866 and 40290 transistors in a two-stage predriver, a 40291 in the driver stage, and two 40292 transistors in a push-pull output stage. These transistors, which are epitaxial silicon planar types of the "overlay" emitter-electrode construction, are intended for low-voltage, high-power operation in amplitude-modulated class C amplifiers.

In addition to standard breakdown-voltage ratings, the 40290, 40291, and 40292 transistors have rf breakdown-voltage characteristics which assure safe operation with high rf voltage on the collector. The 40292 transistors used in the final amplifier stage are 100-per-cent tested for load mismatch at a VSWR of 3:1. During this test, the transistor is fully modulated to simulate actual operation for added reliability.

The amplifier is capable of delivering peak envelope power of 40 watts at a modulation of 95 per cent with a collector voltage of 12.5 volts dc. Unmodulated drive of 5 milliwatts is required at the input. The over-all efficiency of the amplifier is 48 to 53 per cent, and the envelope distortion is less than 5 per cent for amplitude modulation of 95 per cent.

16-31 40-WATT PEAK-ENVELOPE-POWER AIRCRAFT-BAND
AMPLIFIER FOR AM TRANSMITTERS (cont'd)

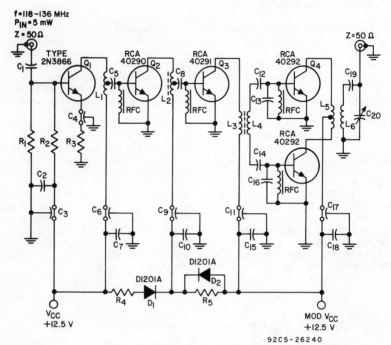

92CS-26240

Parts List

C_1 = 300 pF, silver mica
C_2 = 0.005 μF, ceramic
C_3, C_4, C_6, C_9, C_{11}, C_{17} = Feedthrough capacitor, 1000 pF
C_5 = 50 pF, silver mica
C_7, C_{10}, C_{15}, C_{18} = 0.5 μF, ceramic
C_8, C_{12}, C_{14} = 82 pF, silver mica
C_{13}, C_{16}, C_{19} = 150 pF, silver mica
C_{20} = Variable capacitor, 8 to 60 pF, Arco No. 404 or equiv.
L_1 = 7 turns of No. 22 wire,

13/64 inch in diameter, 9/16 inch long, tapped at 1.5 turns
L_2 = 5.5 turns of No. 22 wire, 13/64 inch in diameter, closely wound on Cambion IRN-9 (or equiv.) core material, tapped at 2 turns
L_3 = 6 turns of No. 22 wire, 13/64 inch in diameter, interwind with L_4 on Cambion IRN-9 (or equiv.) core material
L_4 = 4 turns of No. 22 wire, 13/64 inch in diameter, interwind with

L_3 on common core
L_5 = 5 turns of No. 22 wire, 13/64 inch in diameter, center-tapped; interwind with L_6
L_6 = Same as L_5; interwind with L_5
RFC = 1 turn of No. 28 wire, ferrite bead Ferroxcube No. 56-590-65/4B, or equiv.
R_1 = 470 ohms, 0.5 watt
R_2 = 1500 ohms, 0.5 watt
R_3 = 47 ohms, 0.5 watt
R_4 = 15 ohms, 0.5 watt
R_5 = 33 ohms, 0.5 watt

Performance Characteristics

DC Supply Voltage	12.5	V
Peak Envelope Power	40	W
Modulation	95	%
Efficiency	48-53	%
Envelope Distortion for 95% AM	≤ 5	%
Second Harmonic	> 10	dB down

16-32 16-WATT 225-TO-400-MHz POWER AMPLIFIER

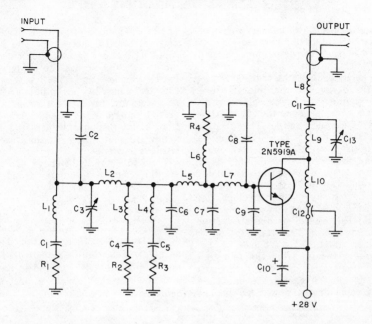

92CS-26243

Parts List

C_1 = Gimmick capacitor, 2.2 pF, Quality Components type 10% QC or equiv.
C_2 = 10 pF, silver mica
C_3 = Variable capacitor, 0.8 to 10 pF, Johanson No. 3957 or equiv.
C_4 = Gimmick capacitor, 1.0 pF, Quality Components type 10% QC or equiv.
C_5 = Gimmick capacitor, 1.5 pF, Quality Components type 10% QC or equiv.
C_6 = 36 pF, ATC-100 type or equiv.

C_7 = 51 pF, ATC-100 type or equiv.
C_8 = 68 pF, ATC-100 type or equiv.
C_9 = 47 pF, ATC-100 type or equiv.
C_{10} = 1 μF, electrolytic, 50 V
C_{11} = 12 pF, silver mica
C_{12} = Feedthrough capacitor, 1000 pF, Allen-Bradley No. FA5C or equiv.
C_{13} = Variable capacitor, 0.8 to 20 pF, Johanson No. 4802 or equiv.

L_1, L_3, L_4 = RF choke, 0.18 μH, Nytronics type P. #DD-0.18 or equiv.
L_2 = 1.5 turns*
L_5 = Copper strip, 5/8 inch long, 5/32 inch wide
L_6 = RF choke, 0.1 μH, Nytronics type P. #DD-0.10
L_7 = Transistor base lead, 0.5 inch long
L_8, L_{10} = 3 turns*
L_9 = 2 turns*
R_1 = 100 ohms, 1 watt
R_2, R_3 = 100 ohms, 0.5 watt
R_4 = 5.1 ohms, carbon, 0.5 watt

* All coils are wound from No. 18 wire with an inner diameter of 5/32 inch and a pitch of 12 turns per inch.

Circuit Description

This broadband power amplifier provides a constant power output of 16 watts with a gain variation of less than 1 dB over a bandwidth of 225 to 400 MHz for an input driving power of 3 to 4 watts. Two of these amplifiers can be connected in parallel to provide a constant power output of 25 watts over this frequency range. In a 225-to-400-MHz high-power transistor amplifier, a good transistor package is of particular

16-32 16-WATT 225-TO-400-MHz POWER AMPLIFIER (cont'd)

Circuit Description (cont'd)

importance. Low parasitic inductances are essential because the real part of the transistor input impedance is inherently low.

The RCA-2N5919A transistor used in the broadband amplifier features a stripline package specifically designed for use in the 225-to-400-MHz frequency range. This transistor is operated in the Class C mode, as is usually the case in high-power rf amplifiers. If the amplifier is to be used in an amplitude-modulated system, the linearity requirements can be met by use of envelope correction, a slight forward bias, or both. The amplifier operates from a dc supply of 28 volts.

The broad flat response of the amplifier results from the fact that the circuit is designed for the best possible match across the band and that some of the power at the low

end of the band is dissipated through dissipative RLC networks. The low input VSWR of the amplifier (maximum of 2 to 1 across the frequency band) verifies the effectiveness of this technique. A low input VSWR is necessary for protection of the driving stage in a cascade connection. A flat response reduces the dynamic range required in the output leveling system.

The collector efficiency of the amplifier has a minimum value of 63 per cent across the frequency band. The second harmonic of a 225-MHz signal is 12 dB down and that of a 400-MHz signal is 30 dB down from the fundamental. This harmonic rejection is excellent for an amplifier that is required to have a bandwidth that covers almost an octave.

16-33   DIP/WAVE METER

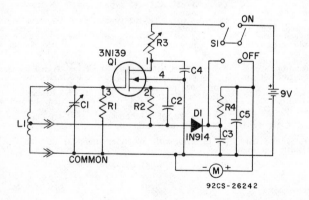

92CS-26242

Parts List

Battery = 9-volt transistor type, RCA VS323 or equiv.
C_1 = variable capacitor, 50 pF, Hammarlund HF-50 or equiv.
C_2, C_4, C_5 = 0.01 μF, ceramic, 50 V or greater

C_3 = 0.001 μF, ceramic, 50 V or greater
L_1 = see coil characteristics chart
M = microammeter, 0 to 100 μA
R_1 = 47000 ohms, 0.5 watt, 10%

R_2 = 1000 ohms, 0.5 watt, 10%
R_3 = potentiometer, 10000 ohms, linear taper
R_4 = 6800 ohms, 0.5 watt, 10%
S_1 = toggle switch, double-pole, double-throw

16-33 DIP/WAVE METER (cont'd)

Circuit Description

This MOS field-effect-transistor dip/wave meter is one of the most useful instruments available to the electronics experimenter working in rf. The meter is essentially an oscillator that can be used to measure resonant frequencies. With the power switch OFF, the meter becomes an absorption-type wavemeter that measures the resonant frequency of energized rf circuits; with the power switch ON, the meter measures the resonant frequency of unenergized rf circuits. Then, if the inductance of the circuit is known, the capacitance can be calculated; if the capacitance is known the inductance can be calculated.

In operation, the coil of the meter is placed close to the tuned circuit to be measured. Capacitor C_1 is then tuned until a movement of the meter needle is observed. If the dip/wave meter is being used to measure an energized circuit, the needle will jump upward slightly when the frequency of the meter oscillator matches that of the LC circuit being measured. If the dip/wave meter is being used to measure the frequency of an unenergized LC circuit, the needle will dip sharply at the point of resonance. If the meter gives no indication a different meter coil should be tried.

The dip/wave meter is essentially an MOS field-effect-transistor oscillator. Oscillator feedback is provided by return of the source to a tap on the coil; transistor operating bias is obtained through the by-passed source resistor R_2. Oscillator rf voltage is rectified by diode D_1 and measured with the microammeter M. Potentiometer R_3 adjusts the supply voltage to the oscillator and the intensity of oscillations or sensitivity of the dip/wave meter; R_3 also controls the meter reading. C_1 sets the frequency of the dip/wave meter.

Care should be taken when the dip/wave meter is being operated as a wave meter not to overdrive the field-effect transistor; this condition is encountered when the meter is deflected beyond full scale.

Power is supplied to the dip/wave meter by a 9-volt transistor battery. The current drain for this circuit is 2 milliamperes maximum.

Dip/Wave Meter Typical Coil Characteristics

Coil *	Inductance (μH)	Frequency (MHz) Min.	Max.	Wire Size	Turns Per Coil	Diameter of Coil (inches)	Length of Coil (inches)	Tap Location (No. of turns from common end
A	280	1.16	2.25	32+	120½	1	1½	30¼
B	99	2	4.1	30+	72½	1	1	18¼
C	25	3.9	8	28+	46½	¾	⅞	12¼
D	6.6	7.7	16.1	22+	19½	¾	9⁄16	4¾
E	1.7	15.4	32.5	20■	11⅓	¾	1	3⅛
F	0.39	32	66	20■	3¾	¾	½	⅞
G	0.16	50	110	12■	3	⅜	½	1

*Coil A to D close wound on polyethylene forms; E and F space wound on polyethylene forms; G self supporting.
+Enameled   ■Tin Plated

16-34   **BASIC ASTABLE MULTIVIBRATOR**
                         (Frequency = 7000 Hz)

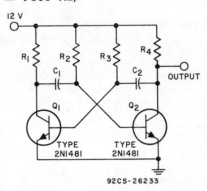

92CS-26233

$$f = \frac{1}{(0.7C_1R_2) + (0.7C_2R_3)}$$

Parts List

C_1, C_2 = 0.1 μF, paper, 25 V
R_1, R_4 = 60 ohms, 5 watts
R_2, R_3 = 1000 ohms, 0.5 watt

Circuit Description

This astable (free-running) multi-vibrator develops a square-wave output that has a peak value equal to the dc supply voltage (V_{CC} = 12 volts) and a minimum value equal to the collector saturation voltage of the transistors. The circuit is basically a two-stage nonsinusoidal oscillator in which one stage conducts at saturation while the other is cut off until a point is reached at which the stages reverse their conditions. The circuit employs two 2N1481 transistors operated in identical common-emitter amplifier stages with regenerative feedback resistance-capacitance coupled from the collector of each transistor to the base of the other transistor.

When power is initially applied to the circuit, the same amount of current tends to flow through each transistor. It is unlikely, however, that a perfect balance will be maintained, and if the current through transistor Q_1, for example, should increase slightly without an attendant increase in that through transistor Q_2, the multivibrator will oscillate to generate a square-wave output.

As the current through transistor Q_1 increases, the resultant decrease in collector voltage is immediately coupled to the base of transistor Q_2 by the discharge of capacitor C_1 through resistor R_2. This negative voltage at the base reduces the current through transistor Q_2, and its

collector voltage rises. The charge of capacitor C_2 through resistor R_3 couples the increase in voltage at the collector of transistor Q_2 to the base of transistor Q_1, and further increases the flow of current through Q_1. The collector voltage of Q_1 decreases even more, and the base of Q_2 is driven more negative. As a result of this regenerative action, transistor Q_1 is driven to saturation almost instantaneously, and, just as quickly, transistor Q_2 is cut off. This condition is maintained as long as the discharge current of C_1 develops sufficient voltage across R_2 to hold Q_2 cut off. The time constant of C_1 and R_2, therefore, determines the time that Q_2 remains cut off (i.e., the duration of the positive half-cycle of the square-wave output). During this period, the voltage at the output terminal is the dc supply voltage (12 volts).

The discharge current from C_1 decreases exponentially, as determined by the time constant of the discharge path, and eventually becomes so small that the voltage developed across R_2 is insufficient to hold Q_2 cut off. The decrease in collector voltage that results when Q_2 conducts is coupled by C_2 and R_3 to the base of Q_1. The current through Q_1 then decreases, and the collector voltage of this transistor rises. The positive swing of the voltage at the collector of Q_1 is coupled

Circuit Description (cont'd)

by C_1 and R_2 to the base of Q_2 to increase further the conduction of Q_2. The regenerative action of the multivibrator then quickly drives Q_2 to saturation and Q_1 to cutoff. The length of time that this condition is maintained is determined by the time constant of C_2 and R_3. During this period, which represents the negative half-cycle of the square-wave output, the voltage at the output terminal is the collector saturation potential of Q_2.

If desired, a square-wave output may also be obtained from the collector of transistor Q_1. This output will be equal in magnitude to that at the collector of transistor Q_2, but will be opposite in phase.

16-35   **ASTABLE POWER MULTIVIBRATOR**

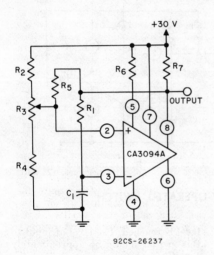

92CS-26237

Parts List

C_1 = 560 pF, 100 V
R_1 = 100 Kilohms, 0.25 watt
R_2 = 27 Kilohms, 0.25 watt
R_3 = potentiometer, 50000 ohms
R_4 = 27 Kilohms, 0.25 watt
R_5 = 100 Kilohms, 0.25 watt
R_6 = 300 Kilohms, 0.25 watt
R_7 = 1 Kilohm, 2 watts

$$f = \frac{1}{2R_1C_1 \ln \left[(\tau/R_2) + 1 \right]}$$

Circuit Description

This circuit illustrates the use of a CA3094A programmable integrated-circuit operational amplifier in an astable (free-running) power multivibrator circuit that can produce pulses of variable length and simultaneously maintain an essentially constant pulse repetition rate. Pulse-length variability is effected by ad-justment of potentiometer R_3. With the component values shown, the circuit operates at a pulse repetition rate of 20 kHz.

If the CA3094B is used in place of the CA3094A, this generic circuit can be used at supply voltages up to 44 volts.

16-36   LAMP FLASHER

Circuit Description

This lamp-flasher circuit employs a CA3094A programmable integrated-circuit operational amplifier in a single-supply astable multivibrator circuit. With the component values shown, it produces one flash per second with a 25 percent "on" time while delivering output current in excess of 100-milliamperes. During the 75 percent "off" time it idles with micropower consumption. The flashing rate can be maintained within ± 2 percent of the nominal value over a battery voltage range from 6 to 15 volts and a temperature excursion from 0 to 70°C.

If the CA3094B is used in place of the CA3094A, this generic circuit can be used at supply voltages up to 44 volts and can switch peak currents up to 300 milliamperes.

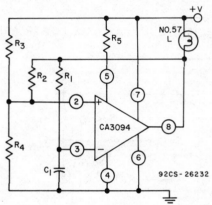

$$f = \frac{1}{2R_1C_1 \ln (\tau/R_5 + 1)}$$

Parts List

$C_1 = 0.47 \ \mu F$, 50 V
$L = $ No. 57 Lamp

$R_1 = 4.3$ megohms, 0.25 watt	$R_4 = 12$ megohms, 0.25 watt
$R_2 = 18$ megohms, 0.25 watt	$R_5 = 1.2$ megohms, 0.25 watt
$R_3 = 3$ megohms, 0.25 watt	

92CS-26232

16-37   AUDIO-FREQUENCY-OPERATED SWITCH

Circuit Description

This audio - frequency - operated-switch circuit can be used to turn on a load rated up to one kilowatt when the sound level increases above a predetermined level. The load continues to receive power until the sound level drops below the predetermined level. The circuit can be activated by an audio signal provided by a microphone preamplifier such as that described in Circuit 16-7. The circuit can be used to control electrical systems, such as alarms, transmitters and remote intercoms. It can also be used to measure noise level; in such applications, it activates some device when a predetermined noise level is reached. The level of input to this switch should be approximately 1 volt.

The audio- or radio-frequency signal applied to the input terminals is rectified by D_5 and D_6. The resulting signal is applied to the base of Q_1 through the potentiometer R_1. The amount of noise required to activate the circuit can be controlled by adjustment of the potentiometer. The signal applied to the base of Q_1 causes it to conduct provided that the emitter of Q_1 is positive. The current conducted by Q_1 charges Q_2 through diodes D_3 and D_4 and resistor R_4.

16-37   AUDIO-FREQUENCY-OPERATED SWITCH (cont'd)

Circuit Description (cont'd)

On the following half-cycle, the charge on capacitor C_2 is applied to the gate of SCR_2 and turns it on; a voltage is thus placed across the load. The load voltage is also applied to the combination of D_1, R_2, and C_1, and causes the capacitor to charge. The charge on C_1 turns on SCR_1 during the next half-cycle. This process repeats as long as there is a sufficient audio- or radio-frequency signal present at the input terminals to cause Q_1 to conduct.

When the signal is removed, Q_1 becomes nonconductive; the charging path for capacitor C_2 is thus opened. If C_2 cannot charge, SCR_2 cannot turn on. The result is an open circuit to the load. Because there is

no voltage across the load, capacitor C_1 cannot obtain the charge it needs to turn on SCR_1 during the next half-cycle. Therefore, both SCR's remain off until another signal is received at the input terminals.

The release time, or the time that it takes for the switch to turn off after the input signal ceases, can be increased so that the switch does not open during momentary interruptions (e.g., between syllables). This increase in release time is accomplished by connection of a capacitor between the emitter and the collector of Q_1. Values of capacitance up to 100 microfarads (15 volts) can be used.

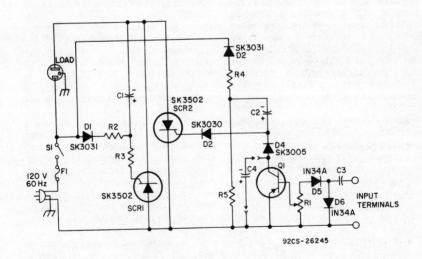

92CS-26245

Parts List

C_1, C_2 = 10 μF, electrolytic, 15 V
C_3 = 0.1 μF, 25 V or greater
C_4 = 10 to 100 μF, electrolytic, 12 V, to increase release time

F_1 = fuse, 125 V, ampere rating depends on load (10 amperes maximum)
R_1 = potentiometer, 5000 ohms, 2 watts, linear taper
R_2, R_4 = 4700 ohms, 2 watts, 10%

R_3 = 270 ohms, 0.5 watt, 10%
R_5 = 470 ohms, 0.5 watt, 10%
S_1 = toggle switch, 125 V, 15 amperes, single-pole, single-throw

Circuit Description

These triac light-dimmer circuits are designed to provide full-wave control of the light intensity of incandescent lamps. Component values and triac types are shown for operation of the circuits from a 60-Hz ac source of 120 or 240 volts. For 120-volt operation, the T2800B triac is recommended; for 240-volt operation, the higher-voltage T2800D triac should be used. A D3202U trigger diode (diac), together with associated resistance-capacitance time-constant networks, is used to develop the gate current pulses that trigger the selected triac into coduction.

In each light-dimmer circuit, the triac is connected in series with the lamp load. During the beginning of each half cycle of the input ac voltage, the triac is in the OFF state. As a result, the entire line voltage appears across the triac, and the lamp is not lighted. The entire line voltage, however, is also impressed across the resistance capacitance network connected in parallel with the triac, and this voltage charges the capacitor(s) in this network. When the voltage across the trigger capacitor, C_2 in circuit (a) or C_3 in circuit (b), rises to the breakover voltage V_{BO} of the diac, and the diac conducts. The capacitor then discharges through the diac and the triac gate to trigger the triac. At this point, the line voltage is transferred from the triac to the lamp load for the remainder of that half cycle of the input ac power. This sequence of events is repeated for each half cycle of either polarity.

The potentiometer R_2 is adjusted to control the brightness of the incandescent lamp. If the resistance of the potentiometer is decreased, the trigger capacitor charges more rapidly, and the breakover voltage of the diac is reached earlier in the cycle so that the power applied to the lamp and thus the intensity of the light is increased. Conversely,

if the resistance of the potentiometer is increased, triggering occurs later in the cycle, and the light intensity is decreased. The resistor R_1 in series with the potentiometer protects the potentiometer by limiting the current when he potentiometer is at the low-resistance end of its range.

Capacitor C_1 and inductor L_1 form an rfi suppression network. This network suppresses the high-frequency transients generated by the rapid ON-and-OFF switching of the triac so that these transients do not produce noise interference in nearby electrical equipment.

The two lamp-dimmer circuits differ in that circuit (a) employs a single-time-constant trigger network and circuit (b) uses a double-time-constant trigger circuit. As pointed out earlier in the section on **Power Switching and Control,** the use of the second time constant network reduces hysteresis effects and thereby extends the effective range of the light-control potentiometer. As applied to light dimmers, the term hysteresis refers to a difference in the control-potentiometer setting at which the lamp turns on and the setting at which the light is extinguished. The additional capacitor C_2 in circuit (b) reduces hysteresis by charging to a higher voltage than capacitor C_3. During gate triggering, C_3 discharges to form the gate current pulse. Capacitor C_2, however, has a longer discharge time constant and this capacitor restores some of the charge removed from C_3 by the gate current pulse.

It is important to realize that a triac in these circuits dissipates power at the rate of about one watt per ampere. Therefore, some means of heat removal must be provided to keep the device within its safe operating-temperature range. On a small light-control circuit such as one built into a lamp socket, the lead-in wire serves as an effective

16-38   **LIGHT DIMMERS (cont'd)**

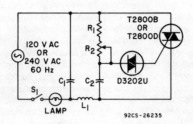

(a) Single-time-constant light-dimmer circuit.

Parts List

120-Volt, 60-Hz Operation

C_1, C_2 = 0.1 μF, 200 V
L_1 = 100 μH
R_1 = 3300 ohms, 0.5 watt
R_2 = light control, poten-

tiometer, 0.25 megohm, 0.5 watt

240-Volt, 50/60 Hz Operation

C_1 = 0.1 μF, 400 V

C_2 = 0.05 μF, 400 V
L_1 = 200 μH
R_1 = 4700 ohms, 0.5 watt
R_2 = light control, potentiometer, 0.25 megohm, 1 watt

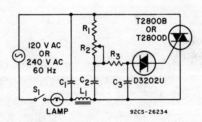

(b) Double-time-constant light-dimmer circuit.

Parts List

120-Volt, 60-Hz Operation

C_1, C_2 = 0.1 μF, 200 V
C_3 = 0.1 μF, 100 V
L_1 = 100 μH
R_1 = 1000 ohms, 0.5 watt
R_2 = light control, poten-

tiometer, 0.1 megohm, 0.5 watt

240-Volt, 60-Hz Operation

C_1 = 0.1 μF, 400 V
C_2 = 0.05 μF, 400 V

C_3 = 0.1 μF, 100 V
L_1 = 100 μH
R_1 = 7500 ohms, 2 watts
R_2 = light control, potentiometer, 0.2 megohm, 1 watt
R_3 = 7500 ohms, 2 watts

Circuit Description (cont'd)

heat sink. Attachment of the triac case directly to one of the lead-in wires provides sufficient heat dissipation for operating currents up to 2 amperes (rms). On wall mounted controls operating up to 6 amperes, the combination of face plate and wall box serves as an effective heat sink. For higher-power controls, however, the ordinary face plate and

16-38   LIGHT DIMMERS (cont'd)

Circuit Description (cont'd)

wallbox do not provide sufficient heat-sink area. In this case, additional area may be obtained by use of a finned face plate that has a cover plate which stands out from the wall so air can circulate freely over the fins.

On wall-mounted controls, it is also important that the triac be electrically isolated from the face plate, but at the same time be in good thermal contact with it. Although the termal conductivity of most electrical insulators is relatively low when compared with metals, a low-thermal-resistance, electrically isolated bond of triac to face plate can be obtained if the thickness of the insulator is minimized, and the area for heat transfer through the insulator is maximized. Suitable insulating materials are fiber-glass tape, ceramic sheet, mica, and polyimide film.

16-39   LIGHT MINDER FOR AUTOMOBILES

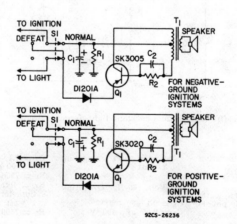

92CS-26236

Parts List

C_1 = 30 μF, electrolytic, 25 volts
C_2 = 0.22 μF, 25 volts
R_1 = 680 ohms, 0.5 watt
R_2 = 15000 ohms, 1 watt

S_1 = switch, double-pole, double-throw
Speaker = 1½-inch permanent-magnet type; voice-coil impedance, 3.2 ohms

T_1 = audio-output transformer; 400-ohm primary, 3.2-ohm secondary; Stancor No. TA-42 or equiv.

16-39   LIGHT MINDER FOR AUTOMOBILES (cont'd)

Circuit Description

This light-minder circuit sounds an alarm if the lights of a car are left on when the ignition is turned off. The alarm stops when the lights are turned off. When the lights are intentionally left on, the alarm can be "defeated" so that no warning sounds. The alarm then sounds when the ignition switch is turned on as a reminder that the system has been "defeated" and the switch should be returned to its "normal" position.

The circuit is essentially an oscillator that obtains its supply voltage from two possible sources, the ignition system or the light system of the car. In the "normal" mode of operation, the ignition system is connected to the collector circuit of the SK3005 (or SK3020) transistor, and the light system is connected through the D1201A diode to the 2N217 (or 2N647) emitter. When the ignition switch is on, the collector of the transistor is at the supply volt-age. If, at the same time, the lights are on, the emitter of the transistor is also at the supply voltage. Because both the emitter and the collector are at the same voltage, the circuit does not oscillate and no alarm sounds. When the ignition is turned off, the collector is returned to ground through R_1 and C_1, but the emitter remains at the supply voltage and provides the necessary bias for the circuit to oscillate. Turning the lights out removes the supply voltage and stops the oscillation.

In the "defeat" mode of operation, the ignition system is connected through the D1201A diode to the emitter of the transistor, and the light system is completely disconnected. The lights can then be turned on without the alarm sounding. When the ignition is turned on, it supplies the necessary voltage to the emitter of the transistor so that the circuit oscillates and causes the alarm to sound.

16-40   BATTERY CHARGERS
For 6- and 12-Volt Automobile Batteries

Circuit Description

These battery chargers can be used to recharge run-down batteries in automobiles and other vehicles without removing them from their original mounting and without the need for constant attention. When the battery is fully charged, the charger circuits automatically switch from charging current to "trickle" charge, and an indicator lamp lights to provide a visual indication of this condition.

12-Volt Battery Charger—This circuit can be used to charge 6-cell, 12-volt lead storage batteries at a maximum charging rate of 2 amperes. When switch S_1 is closed, the rectified current produced by the four 1N2860 silicon diodes in the full-wave bridge rectifier charges capacitor C_1 through resistors R_1 and R_2 and the No. 1488 indicator lamp, I_1. As C_1 charges, the anode of the 1N3754 diode is rapidly raised to a positive voltage high enough so that the diode is allowed to conduct. Gate current is then supplied to the 2N3228 SCR to trigger it into conduction. The SCR and the battery under charge then form essentially the full load on the bridge rectifier,

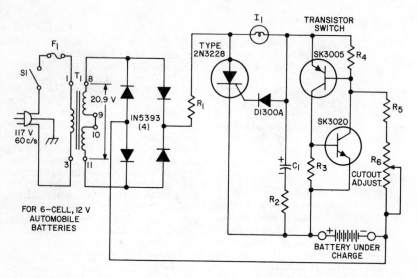

FOR 6-CELL, 12 V
AUTOMOBILE
BATTERIES

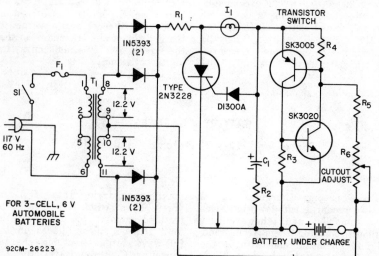

FOR 3-CELL, 6 V
AUTOMOBILE
BATTERIES

92CM-26223

NOTE: Heat sinks are required for the 1N2860 rectifiers. A simple, effective method is to mount the rectifiers in fuse clips.

Parts List

C_1 = 50 μF, electrolytic, 15 V
F_1 = fuse, 1-ampere, 3 AG
I_1 = pilot lamp, No. 1488 (14 V, 150 mA) for 12-volt system or No. 47 (6.3 V, 150 mA) for 6-volt system
R_1 = 5 ohms, 20 watts for

12-volt system or 2 ohms, 25 watts for 6-volt system
R_2 = 33 ohms, 0.5 watt
R_3 = 470 ohms, 0.5 watt
R_4 = 150 ohms, 0.5 watt
R_5 = 1800 ohms, 0.5 watt
R_6 = potentiometer, cutoff

adjustment, 10000 ohms, 2 watts
S_1 = toggle switch, single-pole, single-throw, 3-ampere, 125-volt
T_1 = power transformer, Stancor No. RT-202, or equiv.

BATTERY CHARGERS (cont'd)

Circuit Description (cont'd)

and a charging current flows through the battery that is proportional to the difference in potential between the battery voltage and the rectifier output. Resistor R_1 limits the current to a safe value to protect the 1N5393 rectifier diodes in the event that the load is a "dead" battery. The energy stored in C_1 assures that the SCR conducts and, thereby, that the charging current flows for practically the full 180 degrees of each successive half-cycle of input until the battery is fully charged. (The SCR is actually cut off near the end of each half-cycle but is re-triggered shortly after the beginning of each succeeding half-cycle by the gate current applied through the D1300A diode as a result of the steady potential on C_1.)

When the battery is fully charged, the two-transistor regenerative switch is triggered into conduction (the triggering point is preset by means of potentiometer R_0). As a result of the regenerative action, the SK3005 and SK3020 transistors in the switch are rapidly driven to saturation and thus provide a low-impedance discharge path for C_1. The capacitor then discharges through these transistors and resistor R_2 to about 1 volt (the voltage drop across the transistors). This value is too low to sustain conduction of the D1300A diode, and the 2N3228 SCR is not triggered on the succeeding half-cycle of the input. The saturated transistor switch also provides a low-resistance path for the current to the No. 1488 indicator lamp, which glows to signal the fully charged condition of the battery. The current in the lamp circuit (R_1, lamp, and transistor switch) provides a "trickle" charge of approximately 150 milliamperes to the battery.

6-Volt Battery Charger—This circuit can be used to charge 3-cell, 6-volt lead storage batteries at a maximum charging rate of 3.2 amperes. It is very similar to the 12-volt battery charger except for the rectifier configuration. In the 6-volt circuit, the four 1N5393 diodes are connected in a full-wave center-tapped rectifier circuit that provides the higher charging current of 3.2 amperes to the 6-volt battery. With the exception of the rectifier circuit, the indicator lamp, and the value used for R_1, the 6-volt charger is identical to the 12-volt charger and operates in the same way.

16-41 **AUTOMATIC SHUT-OFF AND ALARM**

Circuit Description

In this circuit, two T2700B or T2700D triacs and a CA3062 integrated circuit are interconnected to form an automatic shut-off and alarm.

The CA3062 integrated circuit consists of a photosensitive section, an amplifier, and a pair of high-current output transistors on a single monolithic chip. The photosensitive section consists of Darlington pairs and affords high sensitivity. The power amplifier has a differential configuration which provides complementing outputs in response to a light input—normally "ON" and normally "OFF". The separate photodetector, amplifier, and high-current switch provide flexibility of circuit arrangement. This feature plus the

16-41   **AUTOMATIC SHUT-OFF AND ALARM (cont'd)**

Circuit Description (cont'd)

high-current capability of the output section, can now provide the user with a complete system particularly useful in photoelectric control applications utilizing IR emitters and visible-light sources.

The CA3062 and resistors R_1, R_2, and R_3 are interconnected to form a latched memory system that can be used to stop clocks, record an intrusion, or activate light-actuated darkroom controls. The initial conditions are: terminal No. 2 at "high"-output-

voltage and terminal No. 6 at "low"-output-voltage. These conditions are reversed when a light pulse is received. Momentary interrupting of V^+ will reset the circuit.

In this system, ac is supplied to the load as long as the light source is "on". If the light path to the CA3062 is broken, then the ac to the load and light source is opened, thereby activating the alarm circuit. The system can be reset with the push-button S_1.

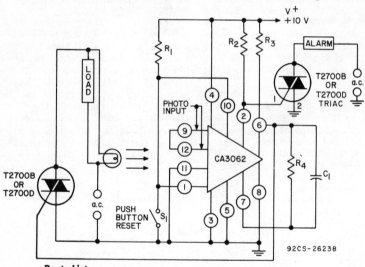

92CS-26238

Parts List

C_1 = 100 to 1000 pF
R_1, R_4 = 30000 ohms, 0.5 watt
R_2, R_3 = 100 ohms, 0.5 watt
S_1 = push-button reset switch

16-42   **PULSE GENERATOR**
With Provisions for Independent Control of "On" and "Off" Periods

Circuit Description

This pulse generator (astable multivibrator) includes provisions for independent control of the "on" and

"off" periods. The exceptionally high input resistance presented by the CA3130 integrated circuit is an at-

16-42   **PULSE-GENERATOR** (cont'd)

Circuit Description (cont'd)

tractive feature for multivibrator circuit design because it permits the use of timing circuits with high R/C ratios. Resistors R_1 and R_2 are used to bias the CA3130 to the mid-point of the supply voltage, and R_3 is the feedback resistor. The pulse repetition rate is selected by positioning S_1 to the desired position. The rate remains essentially constant when the resistors which determine "on-period" and "off-period" are adjusted.

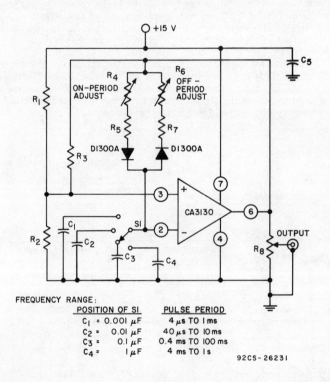

FREQUENCY RANGE:

POSITION OF SI		PULSE PERIOD
C_1 =	0.001 μF	4 μs TO 1 ms
C_2 =	0.01 μF	40 μs TO 10 ms
C_3 =	0.1 μF	0.4 ms TO 100 ms
C_4 =	1 μF	4 ms TO 1 s

92CS-26231

Parts List

$C_1 = 1$ μF
$C_2 = 0.1$ μF
$C_3, C_5 = 0.01$ μF
$C_4 = 0.001$ μF

$R_1, R_2, R_3 = 0.1$ megohm, 0.5 watt
$R_4, R_6 =$ potentiometer, 1 megohm, 0.5 watt

$R_5, R_7, R_8 = 2000$ ohms, 0.5 watt
$S_1 =$ range selector, four-position switch

16-43   **FUNCTION GENERATOR**

Circuit Description

  This function generator uses a CA3130 integrated circuit to provide integrator and threshold detector functions. The circuit generates a triangular or square-wave output that can be swept over a 1,000,000:1

16-43   **FUNCTION GENERATOR (cont'd)**

Circuit Description (cont'd)

range (0.1 Hz to 100 kHz) by means of a single control, R_7. A voltage-control input is also available for remote sweep-control.

The heart of the frequency-determining system is CA3080A integrated-circuit operational-transconductance-amplifier (OTA), IC_1, operated as a voltage-controlled current source. The output, I_0, is a current applied directly to the integrating capacitor, C_1, in the feedback loop of the CA3130 integrator IC_2 to provide the triangular-wave output. Potentiometer R_2 is used to adjust the circuit for slope symmetry of positive-going and negative-going signal excursions.

Another CA3130, IC_3 is used as a controlled switch to set the excursion limits of the triangular output from the integrator circuit. Capacitor C_3 is a "peaking adjustment" to optimize the high-frequency square-wave performance of the circuit.

Potentiometer R_{10} is adjustable to perfect the "amplitude symmetry" of the square-wave output signals. Output from the threshold detector is fed back via resistor R_8 to the input of IC_1 to toggle the current source from plus to minus in generating the linear triangular wave.

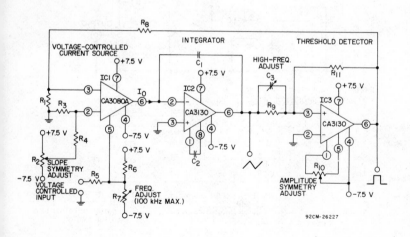

92CM-26227

Parts List

$C_1 = 100$ pF
$C_2 = 56$ pF
$C_3 =$ variable capacitor, 3 to 30 pF
$R_1, R_3 = 3000$ ohms, 0.5 watt

$R_2, R_{10} =$ potentiometer, 0.1 megohm, 0.5 watt
$R_4 = 10$ megohms, 0.5 watt
$R_5 = 10000$ ohms, 0.5 watt
$R_6 = 22000$ ohms, 0.5 watt

$R_7 =$ potentiometer, 10000 ohms, 0.5 watt
$R_8 = 270$ ohms, 0.5 watt
$R_9 = 39000$ ohms, 0.5 watt
$R_{11} = 0.15$ megohm, 0.5 watt

16-44   **VOLTAGE-PROGRAMMABLE TIMER**

Circuit Description

Designers frequently have a need for a "universal" timing circuit in which the timing cycle can be varied linearly over an extended range, preferably by simple linear variation of a control voltage as can be accomplished with this voltage-programmable timer. The timer employs two integrated circuits, a CA3099E programmable comparator and a CA3096E n-p-n/p-n-p transistor array. During the timing cycle, the output signal at terminal 3 of the CA3099E is a logic 0. (An internal transistor can sink a load current of 150 milliamperes into terminal 3.)

The timing capacitor C_1 is charged through a constant-current-source transistor Q_4. When capacitor C_1 is charged to a sufficiently high potential so that the voltage on terminal 14 of the CA3099E exceeds that on terminal 3, the comparator changes state, and terminal 3 goes "high" (logic 1) interrupting the load current. When switch S_1 is moved to the "reset" position, transistor Q_1 is driven into conduction, and thereby causes C_1 to discharge to reset the logic output of the CA3099E to 0. The timing cycle is re-initiated when switch S_1 is turned to the "time" position.

A constant current flow through transistor Q_4 (to charge the timing capacitor C_1) is established by the diode-connected transistors Q_2, Q_3, and Q_5, which are connected in a cur-

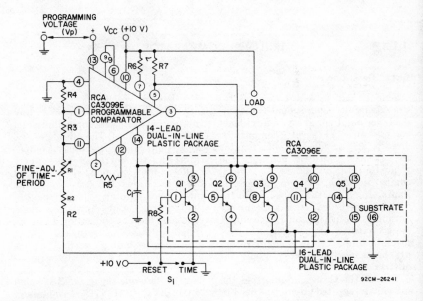

Parts List

$C_1 = 5$ μF, 10 V, Sprague 416P or equiv.
$R_1 =$ potentiometer, 2.5 megohms, 0.25 watt
$R_2 = 10$ megohms, 0.25 watt, $\pm5\%$

$R_3 = 1$ megohm, 0.25 watt, $\pm5\%$
$R_4 = 20000$ ohms, 0.25 watt, $\pm5\%$
R_5, $R_7 = 1000$ ohms, 0.25 watt, $\pm5\%$

$R_6 = 6200$ ohms, 0.25 watt, $\pm5\%$
$R_8 = 10000$ ohms, 0.25 watt, $\pm5\%$

16-44   VOLTAGE-PROGRAMMABLE TIMER (cont'd)

Circuit Description (cont'd)

rent-mirror configuration and are driven from terminal 5 (internal zener regulated) of the CA3099E to force current through resistors R_1 and R_2. Consequently, the magnitude of current flow in transistor Q_4 to charge capacitor C_1 is essentially determined by the total resistance of resistors R_1 and R_2. With a potential of +1 volt on terminal 3, the timing cycle T may be approximated as follows:

$$T = C_1(R_1 + R_2)/3.5 \text{ seconds}$$

The timing cycle, therefore, is 15 seconds when C_1 is 5 microfarads, R_1 is 1 megohm, and R_2 is 10 megohms. This time can be extended linearly by application of an increasingly positive programming voltage V_p to

terminal 13. In this way, the timing cycle can be programmed to vary linearly approximately as $1.15 TV_p$ when the programming voltage V_p is varied over the range from +1 to +7 volts. The current needed to supply terminal 13 is low; only 30 nanoamperes of current is required when V_p is 7 volts. Timing accuracy is relatively insensitive to changes in supply voltage. (Measured data indicate a typical change in time delay of only 1.5 per cent when the supply voltage is varied over the range from 9 to 12 volts.) Similarly, the time delay only varies by approximately 3.5 per cent when the temperature of the integrated circuit varies from 0°C to 70°C.

16-45   TRI-LEVEL COMPARATOR

Circuit Description

This circuit uses a CA3060 integrated-circuit operational-transconductance amplifier array in a tri-level comparator. Tri-level comparator circuits are an ideal application for the CA3060 since it contains the requisite three amplifiers. A tri-level comparator has three adjustable limits. If either the upper or lower limit is exceeded, the appropriate output is activated until the input signal returns to a selected intermediate limit. Tri-level comparators are particularly suited to many industrial control applications.

Two of the three amplifiers in the CA3060 integrated circuit are used to compare the input signal with the upper-limit and lower-limit reference voltages. The third amplifier is used to compare the input signal with a selected value of intermediate-limit reference voltage. By appropriate

selection or resistance ratios, this intermediate-limit may be set to any voltage between the upper-limit and lower-limit values. The output of the upper-limit and lower-limit comparator sets the corresponding upper- or lower-limit flip-flop. The activated flip-flop retains its state until the third comparator (intermediate-limit) in the CA3060 initiates a reset function, thereby indicating that the signal voltage has returned to the intermediate-limit selected. The flip-flops employ two CA3086 integrated-circuit transistor arrays with circuitry to provide separate "SET" and "POSITIVE OUTPUT" terminals.

Power is provided for the CA3060 via terminals 3 and 8 by ±6-volt supplies, and the built-in regulator provides amplifier-bias-current (I_{ABC}) to the three amplifiers via terminal 1. Lower-limit and upper-limit refer-

16-45   **TRI-LEVEL COMPARATOR (cont'd)**

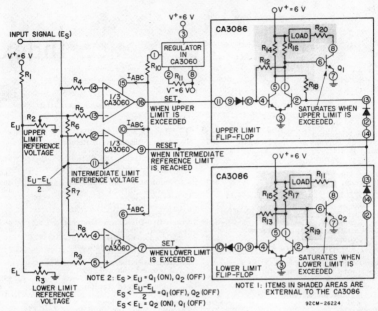

NOTE 2: $E_S > E_U = Q_1$ (ON), Q_2 (OFF)

$E_S < \dfrac{E_U - E_L}{2} = Q_1$ (OFF), Q_2 (OFF)

$E_S < E_L = Q_2$ (ON), Q_1 (OFF)

NOTE 1: ITEMS IN SHADED AREAS ARE EXTERNAL TO THE CA3086

92CM-26224

Parts List

R_1 = 13000 ohms, 0.5 watt
R_2, R_3 = potentiometer, 1000 ohms, 0.5 watt
R_4, R_5, R_8, R_9, R_{12}, R_{19} = 5100 ohms, 0.5 watt

R_6, $_7$, R_{14}, R_{17} = 10000 ohms, 0.5 watt
R_{10} = 20000 ohms, 0.5 watt
R_{11} = 25000 ohms, 0.5 watt
R_{13}, R_{18} = 150000 ohms, 0.5 watt

R_{15}, R_{16} = 4700 ohms, 0.5 watt
R_{20}, R_{21} = 100 ohms, 0.5 watt

Circuit Description (cont'd)

ence voltages are selected by appropriate adjustment of potentiometers R_3 and R_2, respectively. When resistors R_6 and R_7 are equal in value (as shown), the intermediate-limit reference voltage is automatically established at a value midway between the lower-limit and upper-limit values. Appropriate variation of resistors R_6 and R_7 permits selection of other values of intermediate-limit

voltages. Input signal (E_s) is applied to the three comparators via terminals 5, 12, and 14. The "SET" output lines trigger the appropriate flip-flop whenever the input signal reaches a limit value. When the input signal returns to an intermediate-value, the common flip-flop "RESET" line is energized. The loads in the circuits are 5-volt, 25-milliampere lamps.

Index